FISH
REPRODUCTION

FISH REPRODUCTION

Editors

Maria João Rocha

Department of Pharmaceutical Sciences
Superior Institute of Health Sciences–North
Interdisciplinary Centre for Marine and Environmental Research
Porto, Portugal

Augustine Arukwe

Department of Biology
Norwegian University of Science and Technology
Trondheim, Norway

B.G. Kapoor

Formerly Professor of Zoology
The University of Jodhpur
Jodhpur, India

Science Publishers

Enfield (NH)   Jersey   Plymouth

SCIENCE PUBLISHERS
An imprint of Edenbridge Ltd., British Isles.
Post Office Box 699
Enfield, New Hampshire 03748
United States of America

Website: *http://www.scipub.net*

sales@scipub.net (marketing department)
editor@scipub.net (editorial department)
info@scipub.net (for all other enquiries)

Library of Congress Cataloging-in-Publication Data

Fish reproduction/editors, Maria João Rocha, Augustine Arukwe, B.G. Kapoor.
  p. cm.
 Includes bibliographical references and index.
 ISBN-13: 978-1-57808-331-2
 ISBN-10: 1-57808-331-1
 1. Fishes--Reproduction. I. Rocha, Maria João. II. Arukwe, Augustine. III. Kapoor, B.
G.

QL639.2.F55 2007
573.6'17--dc22

2006053542

ISBN   978-1-57808-331-2

Published by Science Publishers, Enfield, NH, USA
An imprint of Edenbridge Ltd.
Printed in India

Preface

Reproduction is a continuous development process throughout ontogeny and therefore, it requires energetic, ecological, physiological, anatomical, biochemical and endocrinological adaptations. Reproduction has gained special significance because of the increasing importance of aquaculture which has become an important industry with the decline in world fish stocks. Fish spawning and nursery in aquaculture ponds raise a host of ecological issues for which knowledge of fish reproduction is fundamental. In another context fishes are often used as models in several research areas such as developmental biology, toxicology, etc. The understanding of reproductive systems is essential for such studies.

Fishes comprise over 28,000 species, with a remarkable variability in morphology, physiology and environmental adaptation. Knowledge on fish reproduction is scattered across numerous sources that shows a dynamic research field. The Editors believe it to be an opportune moment for an anthology of key topics collected under one knowledge umbrella that covers biological, cytological and physiological aspects of fish reproduction, with contributions from internationally recognized experts.

We envisaged that the first chapters would hook up some of the important issues affecting normal ways of reproductive development; details would focus on species living in contrasting environments, such as tropical and temperate fishes; distantly related, as are teleosts and cartilaginous fishes; and finally, fish having different reproductive strategies. Thereafter, since many fishes live in harsh environments, mainly induced by the continuous input of xenobiotic substances into waterways, we found it pertinent to include this topic. Here the authors focus their attention on the factors and mechanisms that may affect reproduction-related hormonal systems as also on known consequences for fish living in polluted environments. Finally, the interplay of modern concepts of fish reproduction in aquaculture is reviewed.

We are very pleased to see our ideas materialized into this compendium that, we hope, will be considered useful and valuable as a 'forum made of state-of-the-art review articles' in this rapidly growing research area which, in addition to the advances in molecular biology (as reflected in many chapters), incorporates the new genomic, proteomic and metabolomic approaches. This project was accomplished with the generous, enthusiastic and dedicated efforts of many colleagues who agreed to share their vision and knowledge. To them we express our most profound gratitude for their exceptional work.

Maria João Rocha
Augustine Arukwe
B.G. Kapoor

Contents

List of Contributors

Abe Shin-ichi
Department of Science, Kumamoto University, 2-39-1 Kurokami, Kumamoto 860-8555, Japan
E-mail: abeshin@gpo.kumamoto-u.ac.jp

Arukwe Augustine
Department of Biology, Norwegian University of Science and Technology (NTNU), Høgskoleringen 5, 7491 Trondheim, Norway.
E-mail: arukwe@bio.ntnu.no

Berg Ole Kristian
Department of Biology, Norwegian University of Science and Technology, N-7491 Trondheim, Norway.
E-mail: Ole.berg@ bio.ntnu.no

Bet-Sayad Will V.
Biology Department, The University of Mississippi, University, MS 38677, U.S.A.
E-mail: wbetsayad@hotmail.com

Bobe Julien
Institut National de la Recherche Agronomique (INRA), Campus de Beaulieu, F-35042 Rennes Cedex, France.
E-mail: Julien.Bobe@rennes.inra.fr

Croft Darren P.
School of Biology, University of Leeds, Leeds LS2 9JT, U.K.
E-mail: darren_croft@hotmail.com

De Silva Sena S.
School of Ecology & Environment, Deakin University, P.O. Box 423, Warrnambool, Victoria, Australia 3280.
E-mail: sena.desilva@deakin.edu.au

Falk-Petersen Inger-Britt
Norwegian College of Fishery Science, University of Tromsø, N-9037 Tromsø, Norway.
E-mail: ifa000@nfh.uit.no

Finstad Anders Gravbrat
Norwegian Institute for Nature Research, Tungasletta 2, N-7485 Trondheim, Norway.
E-mail: anders.finstad@nina.no

Fostier Alexis
Institut National de la Recherche Agronomique (INRA), Campus de Beaulieu, F-35042 Rennes Cedex, France.
E-mail: Alexis.Fostier@rennes.inra.fr

Hendon Jill M.
Biology Department, The University of Mississippi, University, MS 38677, U.S.A.
E-mail: jfrank@olemiss.edu

Griffiths Siân W.
Cardiff School of Biosciences, Main Building Cardiff University, PO Box 915, Cardiff, CF10 3TL, U.K.
E-mail: GriffithsSW@cardiff.ac.uk

Harboe Torstein
Institute of Marine Research, N-5392 Storebø, Norway.
E-mail: torstein.harboe@imr.no

Hoffmayer Eric R.
Center for Fisheries Research and Development, Gulf Coast Research Laboratory, Ocean Springs, MS 39564, U.S.A.
E-mail: eric.hoffmayer@usm.edu

Ingram Brett A.
PIRVic, Department of Primary Industries, Snobs Creek, Private Bag 20, Alexandra, Victoria, Australia 3714.
E-mail: Brett.Ingram@dpi.vic.gov.au

Jalabert Bernard
Institut National de la Recherche Agronomique (INRA), Campus de Beaulieu, F-35042 Rennes Cedex, France.
E-mail: Bernard.Jalabert@rennes.inra.fr

Karlsson Johnny
Department of Natural Sciences, Life Science Center, Orebo University, SE-70182 orebo, Sweden.
E-mail: johnny.karlsson@nat.oru.se

Kitano Takeshi
Department of Science, Faculty of Science, Kumamoto University, 2-39-1 Kurokami, Kumamoto 860-8555, Japan.
E-mail: tkitano@kumamoto-u.ac.jp

Knight Mairi E.
School of Biological Sciences, University of Plymouth, Drake Circus, Plymouth PL4 8AA, U.K.
E-mail: mairi.knight@plymouth.ac.uk

Korsgaard Bodil
Institute of Biology, University of Southern Denmark, Odense, Denmark.
E-mail: bodil@biology.sdu.dk

Koya Yasunori
Department of Biology, Faculty of Education, Gifu University, Yanagido, Gifu 501-1193, Japan.
E-mail: koya@gifu-u.ac.jp

Krause Jens
School of Biology, University of Leeds, Leeds LS2 9JT, U.K.
E-mail: j.krause@leeds.ac.uk

Larsson Anders
Department of Natural Sciences, Life Science Center, Örebro University, SE-701 82 Örebro, Sweden.
E-mail: anders.larsson@nat.oru.se

Magnhagen Carin
Dept. of Aquaculture, Swedish University of Agricultural Sciences, SE - 901 83 Umeå, Sweden.
E-mail: Carin.Magnhagen@vabr.slu.se

Magurran Anne E.
University of St. Andrews, Gatty Marine Laboratory, KY16 8BL, Scotland, U.K.
E-mail: aem1@st-andrews.ac.uk

Mangor-Jensen Anders
Institute of Marine Research, N-5392 Storebø, Norway.
E-mail: anders.mangor.jensen@imr.no

Matsuyama Michiya
Laboratory of Marine Biology, Faculty of Agriculture, Kyushu University, Hakozaki, Fukuoka 812-8581, Japan.
E-mail: rinya_m@agr.kyushu-u.ac.jp

Mayer Ian
Department of Biology, University of Bergen, N-5020 Bergen, Norway.
E-mail: ian.mayer@bio.uib.no

Meeren Terje van der
Institute of Marine Research, N-5392 Storebø, Norway.
E-mail: terje.van.der.meeren@imr.no

Modig Carina
Department of Natural Sciences, Life Science Center, Örebro University, SE-701 82 Örebro, Sweden.
E-mail: carina.modig@nat.oru.se

Mommsen Thomas P.
Department of Biology, University of Victoria, Victoria, B.C. Canada.
E-mail: tpmom@uvic.ca

Nguyen Thuy T.T.
Network of Aquaculture Centres in Asia-Pacific, P.O. Box 1040, Kasetsart Post Office, Bangkok 10903, Thailand.
E-mail: thuy.nguyen@enaca.org

Norberg Birgitta
Institute of Marine Research, N-5392 Storebø, Norway.
E-mail: birgittan@imr.no

Olsson Per-Erik
Department of Natural Sciences, Life Science Center, Örebro University, SE-701 82 Örebro, Sweden.
E-mail: per_erik.olsson@nat.oru.se

Pankhurst Ned W.
Fish Endocrinology Laboratory, Faculty of Science, Engineering and Information Technology, James Cook University, Townsville, Queensland 4811, Australia.
E-mail: n.pankhurst@griffith.edu.au

Parsons Glenn R.
Biology Department, The University of Mississippi, University, MS 38677, U.S.A.
E-mail: bygrp@olemiss.edu

Patzner Robert A.
Organismic Biology, University of Salzburg, Hellbrunnerstr. 34, A-5020 Salzburg, Austria.
E-mail: robert.patzner@sbg.ac.at

Planas Josep V.

Departament de Fisiologia, Facultad de Biologia, Universitat de Barcelona, Av. Diagonal 645, 08028 Barcelona, Spain.

E-mail: jplanas@ub.edu

Sorensen Peter W.

Department of Fisheries, Wildlife and Conservation Biology, University of Minnesota, St. Paul, Minnesota 55108, U.S.A.

E-mail: psorensen@umn.edu

Stacey Norm E.

Department of Biological Sciences, University of Alberta, Edmonton, Alberta, Canada T6G 2E9.

E-mail: norm.stacey@ualberta.ca

Swanson Penny

Northwest Fisheries Science Center, National Marine Fisheries Service 2725 Montlake Boulevard East, Seattle, Washington 98112, U.S.A.

E-mail: Penny.Swanson@noaa.gov

Taylor Martin I.

School of Biological Sciences, University of Wales Bangor, Bangor, Gwynedd, LL57 2DG, U.K.

E-mail: m.taylor@bangor.ac.uk

1

Oogenesis: Post-vitellogenic Events Leading to a Fertilizable Oocyte

Julien Bobe*, Bernard Jalabert and Alexis Fostier

INTRODUCTION

Fish oogenesis has already been thoroughly studied. The numerous contributions to this field have been reviewed by many authors. These reviews have specifically focused on the different steps of oogenesis including vitellogenesis, oocyte maturation and ovulation as well as their endocrine control (Wallace and Selman, 1981; Fostier and Jalabert, 1982; Goetz, 1983; Guraya, 1986; Goetz et al., 1991; Jalabert et al., 1991; Nagahama, 1994; Nagahama et al., 1994, 1995; Yaron, 1995; Goetz and Garczynski, 1997; Patino et al., 2001, 2003; Patino and Sullivan, 2002). In addition, the large number of fish species (over 28,000 identified so far), exhibiting a great variety of reproductive strategies, has incited many species-specific studies. On the other hand, the issue of fish egg quality has also generated vast interest from many investigators

Authors' address: Institut National de la Recherche Agronomique (INRA), Campus de Beaulieu, F-35042 Rennes Cedex, France.
E-mail: Bernard.Jalabert@rennes.inra.fr, Alexis.Fostier@rennes. inra.fr
*Corresponding author: E-mail: Julien.Bobe@rennes.inra.fr

(see Kjörsvik *et al.*, 1990; Brooks *et al.*, 1997 for review). Many studies on the determinism of fish egg quality have focused on the composition of yolk reserves accumulated during vitellogenesis, including the potential role of trophic elements such as fatty acids, free amino acids, vitamins, metabolic enzymes, etc. In contrast, the non-yolky content of egg cytoplasm, including RNAs, structure and regulatory proteins, and cortical alveoli content, has received far less attention. Yet, many of these constituents can be expected to play a role in yolk processing, fertilization, and control of early embryo development. On the other side, post-vitellogenic ovarian events, i.e., oocyte maturation and ovulation—that occur in a relatively short time at the end of yolk deposition and concern all egg compartments—are also known to be of great importance to ensure ovulation of a developmentally competent oocyte. However, most available analytical studies focus on the nature of endocrine, paracrine and autocrine signalling of meiosis resumption (oocyte maturation) and ovulation, but without paying too much attention to qualitative aspects of the results, i.e., the production of good quality eggs.

The purpose of the present chapter is to draw up a balance sheet of mechanisms operating throughout the pre-ovulatory follicular differentiation that participate in the production of a fertilizable and developmentally competent oocyte. In the first part, we will briefly present a global picture of the main mechanisms involved in the control of oocyte maturation, ovulation and in acquisition of follicular maturational competence. In the second part, we will review the existing literature showing the importance of the peri-ovulatory period in the determinism of fish egg quality. Finally, recently available gene expression data will be reviewed and discussed in order to tentatively characterize preovulatory follicular differentiation.

CONTROL OF MEIOSIS RESUMPTION AND OVULATION: THE GLOBAL PICTURE

The term 'oocyte maturation' designates a succession of complex cellular processes corresponding to meiosis resumption up to the second metaphase arrest (in most vertebrates) (Jalabert *et al.*, 1991). In fishes, morphological manifestations of this phenomenon are usually easy to observe and have already been described throughout their progress *in vivo* (Fulton, 1898; Bry, 1981). A methodology based on *in vitro* experimentation and observations was developed 30 years ago (Jalabert *et al.*, 1972; Jalabert, 1976, 1978), and was followed by numerous developments in the world.

Morphological features of maturation—more or less obvious, depending on the species—are mainly migration followed by disappearance (or 'breakdown') of germinal vesicle, yolk clarification, and an increase in oocyte volume. Germinal vesicle migration from the center of the oocyte to its cortex would be necessary, in large oocytes, for a normal expulsion of the first polar body (Lessman and Kavumpurath, 1984). This process should involve the oocyte cytoskeleton (Habibi and Lessman, 1986). Germinal vesicle breakdown (GVBD), a phenomenon common to all vertebrates, corresponds to the first meiotic division immediately followed by the preparation of the second division that will remain blocked at metaphase II until fertilization. Yolk clarification and oocyte volume increase are more particular to fishes, but with some species differences. These phenomena have been linked to a deep reorganization of lipoproteic yolk (Carnevali *et al.*, 1992; Iwamatsu *et al.*, 1992; Matsubara and Sawano, 1995) involving the action of proteolytic enzymes such as cathepsins (Carnevali *et al.*, 1999). In oocytes of marine fishes, that lay pelagic eggs, such as halibut (*Hippoglossus hippoglossus*), this results in an important rise of free amino acids and ions, leading to an osmotic disequilibrium. Besides, an active transport of potassium ion via gap junctions from granulosa cells to the maturing oocyte has been proposed in the salt marsh killifish, *Fundulus heteroclitus* (Cerda *et al.*, 1993). This osmotic gradient induces a water entry and a 4-fold increase of halibut oocyte volume (Finn *et al.*, 2002). This hydration was recently shown to involve the translocation to oocyte plasma membrane of a specific aquaporin (Fabra *et al.*, 2005). Water entry is less important in marine fishes laying demersal eggs such as *Fundulus heteroclitus*, in which a doubling of oocyte volume is observed (Wallace and Selman, 1978). In freshwater species such as rainbow trout (*Oncorhynchus mykiss*), a small but significant increase (25%) of oocyte wet mass is observed (Milla *et al.*, 2006).

Ovulation is the release of a mature oocyte from its follicle into the ovarian cavity (or the abdominal cavity in salmonids) (Fig. 1.1). This process requires the separation of the oocyte from the granulosa layer (Fig. 1.2), the rupture of follicle layers and the expulsion of the oocyte. This process needs a proper retraction of oocytes microvilli tightly connected to the granulosa cells during vitellogenesis (York *et al.*, 1993). Ever since the first data suggesting an implication of prostaglandin F2α (PGF2α) as a possible local ovulation regulator in trout (Jalabert and Szöllösi, 1975) and goldfish, *Carassius auratus* (Stacey and Pandey, 1975) a number of studies, recently reviewed (Goetz and Garczynski, 1997),

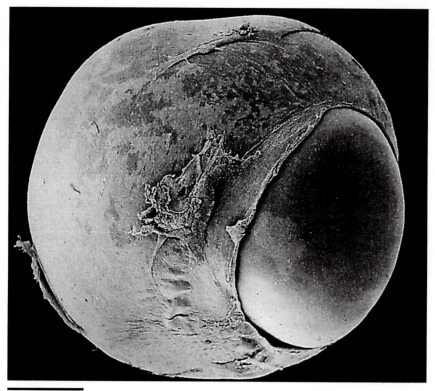

Fig. 1.1 Scanning electron microscopic view of a rainbow trout (*Oncorhynchus mykiss*) oocyte expelled from its follicular layers at the time of ovulation. Copyright INRA Cauty-Jalabert.

suggested that ovulation in fish probably involves the cooperation of various ovarian factors such as proteases and protease inhibitors, progestins, eicosanoids, catecholamines and vasoactive peptides. Ovulation in fish could, therefore, to some extent, be compared to what is observed in mammals, with an hypothesis drawing a parallel between ovulation and an inflammatory reaction (Espey, 1980, 1994).

As underlined in a recent review (Patiño *et al.*, 2003), maturation and ovulation are closely integrated and overlapping events. Dissociation between these events is obviously the result of experimental artefacts. This is the case *in vivo* after premature stimulation by exogenous hormones (de Montalembert *et al.*, 1978; Jalabert *et al.*, 1978a, b), before acquisition of what is called now 'follicular maturational competence', a concept that will be developed in the following paragraphs. This is also the case when ovarian follicles from various species are incubated *in vitro* in

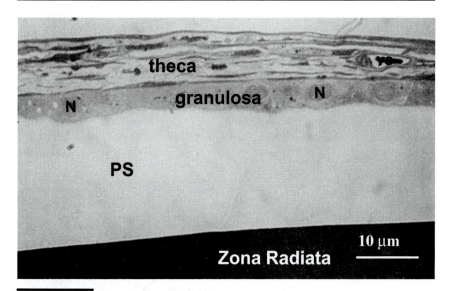

Fig. 1.2 Separation of the oocyte from the granulosa layer occurring prior to ovulation in rainbow trout (*Oncorhynchus mykiss*); N = Nucleus and PS = Peri-oocyte Space. Copyright INRA Cauty-Jalabert.

artificial mediums (Jalabert, 1976). Follicular tissues, isolated from vascular and possible nervous connexions (Wilmot *et al.*, 1993), may be lacking various factors essential for some of the complex mechanisms of ovulation, especially after already long-lasting maturation incubation. This technical difficulty is probably further increased in species such as salmonids in which oocyte hydration during maturation is limited, in contrast to marine species such as Atlantic croaker, *Micropogonias undulatus.*

In fishes as in other vertebrates, it is well admitted that both sequences of maturation and ovulation are triggered *in vivo* by the action of pituitary gonadotropins on the surrounding follicle layers. In fact, a typical photoperiod-regulated 'ovulatory discharge' of gonadotropin has been observed in some fish species such as the goldfish (Breton *et al.*, 1972; Stacey *et al.*, 1979; Hontela and Peter, 1983), but not in rainbow trout. Instead, in that particular species, plasma levels of both gonadotropins follicle-stimulating hormone (FSH) and luteinizing hormone (LH) exhibit a progressive and moderate rise, starting, about 6 and 4 days (at 12°C), respectively before the beginning of maturation (Breton *et al.*, 1998) (Fig. 1.3). Moreover, a second rise of plasma gonadotropins—in a much larger range than the first one—is observed after ovulation. However, differential profiles are observed, depending on evacuation or retention

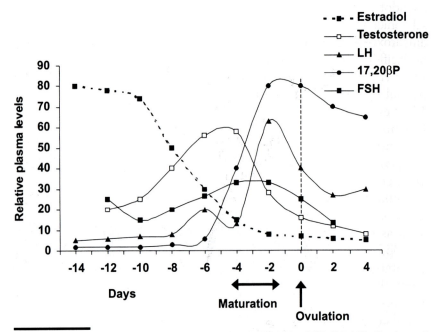

Fig. 1.3 Peri-ovulatory circulating levels of LH, FSH, the MIS (17,20β-dihydroxy-4-pregnen-3-one), estradiol and testosterone in rainbow trout (*Oncorhynchus mykiss*). Indicative plasma profiles redrawn (from Fostier and Jalabert, 1986; Breton *et al.*, 1998) are shown.

of ovulated eggs (Chyb *et al.*, 1999). When eggs are removed immediately after ovulation, a large increase of FSH levels with a decrease of LH are observed during two weeks following ovulation. In contrast, retention of the eggs in the body cavity induces a large increase of LH and only a weak increase of FSH during the same period. These differences in both gonadotropins post-ovulatory profiles might be related either to post-ovulatory ovarian reorganization, including gonial multiplications, a new wave of oocyte growth (de Mones *et al.*, 1989) and/or to the preservation of the quality of ovulated eggs when these are retained for an extended period in the body cavity. In many cases, it seems obvious that both gonadotropins are actually involved in the endocrine regulation of follicular events preceding, accompanying and following maturation and ovulation.

The gonadotropic action, which triggers oocyte maturation in fishes, is clearly relayed by a steroid hormone, designated by the term 'maturation-inducing steroid' or MIS. This hormone is synthesized by follicular layers (Jalabert *et al.*, 1991) through mediation of the cyclic AMP-protein kinase A (cAMP/PKA) pathway (Planas *et al.*, 1997) after activation of gonadotropin receptors (Breton *et al.*, 1986; Miwa *et al.*, 1994). In salmonids and some other freshwater fishes, the MIS is a progestin, 17,20β-dihydroxy-4-pregnen-3-one (17,20βP), which was first identified in plasma of Pacific salmon (*Oncorhynchus nerka*) during the peri-ovulatory period (Idler *et al.*, 1960). In rainbow trout, it appeared as the most efficient natural steroid able to trigger oocyte maturation *in vitro* (Fostier *et al.*, 1973) even in defolliculated oocytes (Table 1.1 and Fig. 1.4). It is secreted in *vitro* by ovarian follicles incubated with salmon gonadotropin in concentration-dependant amounts (Fostier *et al.*, 1981), and present in the plasma at an increasing level throughout oocyte maturation *in vivo* (Fostier and Jalabert, 1986). It was futher identified in the incubation medium of follicles from another salmonid, stimulated by a salmon gonadotropin (Nagahama and Adachi, 1985). In various marine species such as the spotted seatrout (*Cynoscion nebulosus*) another progestin, 17,20β,21-trihydroxy-4-pregnen-3-one (17,20β,21P), was identified as the maturation-inducing steroid MIS (Thomas and Trant, 1989). In rainbow trout, this steroid appears as efficient as 17,20βP to trigger maturation of oocytes incubated within intact follicles, but is much less efficient on defolliculated oocytes (Fig. 1.4). In spotted seatrout, two different MIS receptors have been identified: a membrane receptor (Patino and Thomas, 1990) and a nuclear receptor (Pinter and Thomas, 1995), exhibiting very different binding affinities for various C21 steroids (Thomas *et al.*, 2002). The membrane receptor would be involved in the non-genomic action of the MIS on oocyte maturation, whereas the nuclear receptor would be responsible for a genomic action on ovulation (Pinter and Thomas, 1999). Indeed, the MIS membrane receptor capacity of the spotted sea-trout ovary has been shown to be correlated with the oocyte responsiveness to 17,20β,21P for GVBD (Thomas *et al.*, 2001). The cDNA of the membrane receptor was recently cloned (Zhu *et al.*, 2003a). It seems to belong to a family, so far unknown, of putative membrane receptors genes that would be present in various vertebrates, including human, *Xenopus*, *Fugu*, and zebrafish, *Danio rerio* (Zhu *et al.*, 2003b). Details about the follicular mechanisms triggered by nuclear receptor activation are almost unknown. On the contrary, many studies have been devoted to mechanisms triggered in the oocyte of amphibians, mammals and fishes, by the non-genomic activation of a putative membrane receptor. Results tend to emerge into a general model involving the maturation promoting factor (MPF) which consists of two subunits: a serine/threonine protein kinase ($p34^{cdc2}$), acting as a catalytic component, and a cyclin B (Yamashita, 2000). Numerous studies have been developed on these mechanisms in

Table 1.1 Response of defolliculated rainbow trout oocytes to various steroids. Defolliculation of oocytes and *in vitro* incubation were performed according to the protocol described by Finet *et al.* (1988). Each steroid was assayed at the concentration of 0.5 μg/ml (i.e., 1.3 to 1.7 μM) in individual vials containing 15 defolliculated oocytes from the same female in 2 ml incubation medium. Maturational response observed after a 48 h-incubation at 15°C was either GVBD (+ +), partial maturation (+) (i.e., yolk clarification without GVBD), or no observable response (0).

Steroids (*trivial name*)	Maturational response
Control (ethanol)	
5-Cholesten-3β-ol (*cholesterol*)	-
5ß-Pregnane-3α,17,20β-triol	-
3α-Hydroxy-5β-pregnan-20-one	-
3α,17-Dihydroxy-5β-pregnan-20-one	-
17-Hydroxy-5β-pregnane-3,20-dione	-
20β-Hydroxy-4-pregnen-3-one (*20β-dihydro-progesterone*)	+
20α-Hydroxy-4-pregnen-3-one (*20α-dihydroprogesterone*)	-
17,20α-Dihydroxy-4-pregnen-3-one	-
17,20β-Dihydroxy-4-pregnen-3-one	+ +
17,20β,21-Trihydroxy-4-pregnen-3-one (*20β-dihydrocortexolone*)	+ +
11β,17,20β,21-Tetrahydroxy-4-pregnen-3-one (*20β-dihydrocortisol*)	-
17,20β,21-Trihydroxy-4-pregnene-3,11-dione (*20β-dihydrocortisone*)	-
4-Pregnene-3,20-dione (*progesterone*)	-
11α- Hydroxy-4-pregnene-3,20-dione (*11α-hydroxyprogesterone*)	-
17-Hydroxy-4-pregnene-3,20-dione (*17-hydroxyprogesterone*)	+
21-Hydroxy-4-pregnene-3,20-dione (*deoxycorticosterone*)	-
11β,21-Dihydroxy-4-pregnene-3,20-dione (*corticosterone*)	-
11β,17,21-Trihydroxy-4-pregnene-3,20-dione (*cortisol*)	-
4-Pregnene-3,11,20-trione (*11-keto-progesterone*)	-
17,21-Dihydroxy-4-pregnene-3,11,20-trione (*cortisone*)	-
5-Pregnene-3β,20β-diol	-
3β-Hydroxy-5-pregnen-20-one (*pregnenolone*)	+
3β,17-Dihydroxy-5-pregnen-20-one (*17-hydroxy-pregnenolone*)	-
5α-Androstane-3α,17β-diol (*17β-dihydroandrosterone*)	-
5α-Androstane-3β,17β-diol	-
17β-Hydroxy-5α-Androstan-3-one (*5α-dihydrotestosterone*)	-
3α-Hydroxy-5α-androstan-17-one (*Androsterone*)	-
3α-Hydroxy-5β-androstan-17-one (*etiocholanolone*)	-
3α-Hydroxy-5α-androstane-11,17-dione (*11-ketoandrosterone*)	-
17β-Hydroxy-4-androsten-3-one (*testosterone*)	-
11β,17β-Dihydroxy-4-androsten-3-one (*11β-hydroxytestosterone*)	-
17β-Hydroxy-4-androstene-3,11-dione (*11-ketotestosterone*)	-
4-Androstene-3,17-dione (*Androstenedione*)	-
11β-Hydroxy-4-androstene-3,17-dione (*11ß-hydroxyandrostenedione*)	-
4-Androstene-3,11,17-trione (*Adrenosterone*)	-
3β-Hydroxy-5-androsten-17-one (*dehydroepiandrosterone*)	-

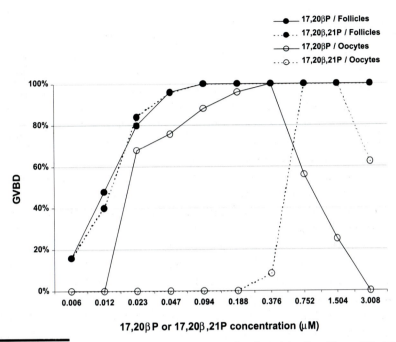

Fig. 1.4 Percent GVBD in rainbow trout oocytes incubated *in vitro* either within intact follicles or after defolliculation, in response to increasing concentrations of either the specific MIS 17,20β-dihydroxy-4-pregnen-3-one (17,20βP), or 17,20β,21-trihydroxy-4-pregnen-3-one (17,20β, 21P). The response of follicle-enclosed oocytes is almost identical for both steroids, whereas 17,20βP is much more efficient than 17,20β,21P on defolliculated oocytes. Moreover, both steroids exhibit a decreasing efficiency at higher concentrations (over 0.5 μM) on defolliculated oocytes, a paradoxical result that is never observed on intact follicles (Jalabert and Fostier, unpublished results). Defolliculation was performed according to the protocol described by Finet *et al.* (1988).

fish, following a very active research initiated by Y. Masui and C. Market in amphibians (Duesbery and Vande Woude, 1998). They will not be detailed here (see: Basu *et al.*, 2004 for a short state of the art).

Follicular and oocyte competence are defined as the ability to respond correctly to a hormonal stimulation by producing good quality oocytes, i.e., able to develop normally after fertilization. As already underlined (Jalabert *et al.*, 1991), this concept implies that 'each component is first able to receive and translate correctly an external signal (receptivity) within a normal physiological range (sensitivity), and then is capable of giving an appropriate response (responsiveness). Such a response might be the emission of another signal or the realization of the final biological events. Most of our knowledge concerns receptivity and sensitivity levels, which can be evaluated by measurable criteria capable of being stimulated by exogenous stimuli. The actual responsiveness of a biological system is more difficult to predict because the morphological or biochemical criteria usually available are only partial indicators in comparison to the complexity of the final response. Finally, correct evaluation of complete maturation is facing a last methodological difficulty: embryonic development, which is the best criterion for such an evaluation, requires other factors than those strictly dealing with oocyte maturation and ovulation, i.e., factors related to adequate fertilization and environmental conditions encountered by the embryo.

The acquisition of follicular and oocyte competence is a crucial practical problem, because fish farming constraints often require artificially controlling the end of reproductive cycles, and impose quality norms concerning the result of induction. Several factors have been involved in the control of competence acquisition, especially the state of homologous and heterologous gap-junctions, connexins, insulin-like growth factors (IGFs) and activins (Patiño and Sullivan, 2002). In rainbow trout ovary, analysis of expression profiles of various genes during pre-ovulatory period allows to associate the evolution of some of them with competence acquisition (Bobe *et al.*, 2003a, b, 2004). Most correspond to factors which have already been suggested—in other teleost or in mammals—to be involved in competence acquisition, modulation of follicular or oocyte sensitivity to hormones, yolk processing, or ovulation mechanisms. This is the case of IGF1 and IGF2, FSH receptors, members of activin and inhibin family, cathepsins, and enzymes of prostaglandins biosynthesis pathway. The multiplicity of factors putatively involved is not surprising, since competence acquisition necessarily corresponds in fact to the progressive coordinate differentiation of several ovarian compartments. The necessity of some physiological coordination between the acquisition of follicular competence as a whole and the emission of appropriate stimulating signals from the neuroendocrine system seems obvious. Furthermore, intraovarian regulations should be important in multispawner fish for the recruitment of successive groups of follicles for oocyte maturation during the whole spawning season. Besides, the complexity of such dialogues is increased by the necessary coordination between spawning by itself (strictly expulsion of oocytes outside the body) and fertilization, which involves signals exchange between females and males (Stacey, 2003). Whatever the level of regulation, insufficient attention has probably been paid, so far, to signals from the ovary, including non-steroid ones. Finally, the analysis of gonadotropic signals from the pituitary should not be limited to the mean absolute level of FSH and

LH, but should also take into account short-term profiles, thus renewing ancient works performed when available assays were probably giving only a mean value between both gonadotropins (Zohar *et al.*, 1986).

Factors other than gonadotropins and MIS have been shown to be involved in the control of intrafollicular oocyte maturation, depending on the species. However, most results were obtained from *in vitro* experiments performed on oocytes surrounded by follicular envelopes and ovarian tissues. Therefore, it is not always easy to know whether these factors play a role at the follicular level and/or at the oocyte level, and by which mechanisms. This is the case when putative MIS are assayed *in vitro* for their relative efficiency to trigger oocyte maturation. Indeed, results can be different whether follicle-enclosed or denuded oocytes are used as shown in Fig. 1.4.

Serotonin can induce oocyte maturation and ovulation *in vitro* in the medaka, *Oryzias latipes*, in the same manner as does LH by stimulating $17,20\beta P$ synthesis (Iwamatsu *et al.*, 1993). In contrast, serotonin inhibits intrafollicular oocyte maturation induced *in vitro* by the MIS in salt marsh killifish (Cerda *et al.*, 1995). This inhibition that appears to be mediated through an increase of follicular cAMP via a PKC pathway distinct from gonadotropin-induced PKA pathway (Cerda *et al.*, 1998a), is progressively released when oocyte competence (sensitivity to MIS) increases during the regular reproductive cycles of this fish (Cerda *et al.*, 1998b). It may be interesting to note that those species are characterized by a non-seasonal regular cyclic reproductive activity. Further studies would be required in such species as also others to explore the physiological relevance of these observations and the possible general occurrence of a serotonin regulation of oocyte maturation in fish.

Gonadotropin releasing hormones (GnRH) are known to play autocrine or paracrine roles in various organs—including gonads—of vertebrates. Several forms of these peptides, mainly salmon GnRH (sGnRH) and chicken II GnRH (cGnRH-II), have been detected in ovaries of catfish (Habibi *et al.*, 1994), goldfish (Pati and Habibi, 1998) and rainbow trout (Von Schalburg *et al.*, 1999). A thorough analysis of the variations of mRNA expression of these peptides and some splicing variants in rainbow trout ovary suggests that sGnRH peptide production may be tightly regulated, especially during the peri-ovulatory period (Uzbekova *et al.*, 2002). The actual primary site of action, i.e., oocyte *vs* follicular envelopes, and the respective signalling pathways of these two peptides are not fully clear, because most positive data have been coming, so far, from *in vitro* incubation of goldfish, *Carassius auratus*, and seabream, *Sparus aurata*, follicles. In these species, the proportion of maturing oocytes rarely exceeds 60% *in vitro*, suggesting that possible interactions between maturing and non-maturing follicles might complicate interpretations. In goldfish, however, either sGnRH or cGnRH-II appears to exert a stimulatory action on intrafollicular oocyte maturation when administered alone. The action of both forms is blocked by PKC inhibitors. However, sGnRH but not cGnRH-II exhibits an inhibitory action on gonadotropin-induced maturation (Pati and Habibi, 2002). Data in seabream showed a stimulatory action of sGnRH only, whereas cGnRH-II treatment was without effect (Nabissi *et al.*, 2000).

Insulin and insulin-like growth factors, their receptors and specific binding proteins have been identified in fish ovaries at various stages of development, especially at pre-ovulatory stages. They have been observed to induce intrafollicular oocyte maturation *in vitro*, probably by direct action at the oocyte level, as observed in seabream (Kagawa *et al.*, 1994a) and Indian carp, *Labeo rohita* (Dasgupta *et al.*, 2001), and in agreement with observations made in amphibians (El Etr *et al.*, 1979; Zhu *et al.*, 1998). However, these factors could also act through modulation of MIS synthesis as suggested by the effect of IGF1 on 17,20βP production by granulosa cells from coho salmon (Maestro *et al.*, 1995, 1997), *Oncorhynchus kisutch* or rainbow trout (Fostier *et al.*, 1994).

Finally, various other autocrine or paracrine factors, such as Transforming Growth Factor β (TGFβ) (Kohli *et al.*, 2003) activin, inhibin and follistatin (Pang and Ge, 1999; Wu *et al.*, 2000) have recently been observed to induce or modulate *in vitro* intrafollicular oocyte maturation in zebrafish. Although the same methodological criticism in goldfish and seabream might be expressed, it remains that all these factors probably interact in variable combinations, depending on the species, to modulate the main general mechanism of maturation induction and/or follicle competence acquisition (Pang and Ge, 2002).

THE IMPORTANCE OF POST-VITELLOGENIC EVENTS IN THE DETERMINISM OF EGG QUALITY

Egg quality can be defined as the ability of the egg to be fertilized and allow the development of a normal embryo. The peri-ovulatory period has a potential impact on the quality of ovulated eggs. More specifically, egg post-ovulatory ageing has a negative impact on the egg quality. In most species, ovulated eggs undergo a rapid (a few hours) decrease of their quality (Hirose *et al.*, 1979; Kjörsvik *et al.*, 1990). In contrast, salmonids can hold their eggs in the body cavity for at least a week without any significant loss of quality (Nomura *et al.*, 1974; Sakai *et al.*, 1975; Bry, 1981; Springate *et al.*, 1984; Aegerter and Jalabert, 2004). In addition, sub-optimal temperatures during the post-ovulatory period have a negative impact on egg survival after ovulation (Aegerter and Jalabert, 2004).

It is also clear that rearing conditions during the late stages of oogenesis can significantly impact egg quality. Rainbow trout females maintained at elevated temperature (15°, 18° or 21°C) for 2-3 months before ovulation exhibit lower egg survival than females maintained within a normal temperature range (9° or 12°C) (Pankhurst *et al.*, 1996). Similar observations were made using rainbow trout females held at 12° or 17°C for only a few weeks (2-4) before ovulation (Aegerter and Jalabert, 2004).

The level of maturational competence reached by the ovarian follicle (follicular maturational competence - FMC) at the time of gonadotropin stimulation is also critical for the production of good quality eggs. A precocious stimulation of oocyte maturation and ovulation can result in: (1) no oocyte maturation; (2) oocyte maturation not followed by ovulation; (3) partial ovulation; or (4) ovulation of low quality eggs. In Northern pike, *Esox lucius*, no response, maturation without ovulation, partial ovulation and full ovulation were observed after gonadotropic stimulation, depending on the dosage of partially purified salmon gondotropins (PPSG). Stimulation with the maturation-inducing steroid MIS, either alone or associated with a priming dose of PPSG did not induce full ovulation. In addition, eggs ovulated after MIS treatment (3 mg/kg, alone or after priming) were characterized by low and fluctuating fertilization rates (20-41%). In contrast, a single injection with PPSG (0.1 mg/kg) was characterized by a high fertilization rate (89%) (de Montalembert *et al.*, 1978). These results suggest that some mechanisms necessary for ovulation request a pituitary stimulation. In addition, they also state that these mechanisms are important not only for releasing the oocyte from its follicle but also for producing developmentally competent oocytes. Similar results were obtained in coho salmon (Jalabert *et al.*, 1978b). Female coho salmon injected with the maturation-inducing steroid 17,20βP (3 mg/kg) underwent oocyte maturation that was not followed by ovulation. In contrast, when a priming injection of (0.1 mg/kg SG-G100) was performed and followed by a second injection of either 17,20βP (3 mg/kg) or SG-G100 (1 mg/kg), ovulation was observed.

In female rainbow trout sampled before migration of the germinal vesicle and injected with 17,20βP (3 mg/kg on day 0 and day 2), oocyte maturation was observed for 94% of the females and ovulation for only 25% of the females. In contrast, all females primed with trout pituitary extract (TPE) and subsequently injected with 17,20βP (3 mg/kg), underwent oocyte maturation. In addition, 59% of those females ovulated (Jalabert *et al.*, 1978a). This observation further indicates that specific events, under gonadotropic control, are necessary for the completion of ovulation. In addition, 15 days after the first injection, ovaries were removed from females that had not ovulated and ovulation was performed *in vitro* using $PGF_{2\alpha}$. Alternatively, oocytes were removed from their follicle by manual dissection. Subsequently, oocytes obtained were fertilized and embryonic development was monitored. The authors observed that embryonic success was low and highly variable among females. They also observed that embryonic success after *in vitro* ovulation or manual dissection was extremely low compared to embryonic development observed after *in vivo* ovulation (Jalabert *et al.*, 1978a). Similarly, spawning induction of coho salmon females can induce a significantly lower embryonic success than in naturally ovulating females (control group) (Jalabert *et al.*, 1978b). In contrast, no difference was observed when such experiment was repeated using females close to natural ovulation (Jalabert *et al.*, 1978b).

Together, these observations and others (Kagawa *et al.*, 1994b; Patiño and Kagawa, 1999; Pang and Ge, 2002; Bobe *et al.*, 2003b) clearly indicate that the follicular maturational competence FMC reached at the time of gonadotropic stimulation is important not only for completion of oocyte maturation and ovulation, but also for the developmental competence of the ovulated egg. In addition, the above observations also indicate that some mechanisms involved in building egg fertility and/or quality are under the control of gonadotropin stimulation.

Interestingly enough, it was reported that the induction of oocyte maturation by gonadotropins requires RNA synthesis (Jalabert, 1976; Kagawa *et al.*, 1994b; King *et al.*, 1994; Sorbera *et al.*, 1999; Thomas *et al.*, 2001; Senthilkumaran *et al.*, 2002). RNA synthesis is probably also required for MIS-induced ovulation spotted seatrout, *Cynoscion nebulosus*, a marine species in which ovulation can be triggered *in vitro* by the MIS (Pinter and Thomas, 1999). Information on ovarian gene expression during FMC acquisition should therefore allow a better comprehension of the mechanisms involved in this process. For that reason, available gene expression data in the fish pre-ovulatory ovary have been reviewed and discussed next.

PRE-OVULATORY OVARIAN FOLLICULAR DIFFERENTIATION: NEW INSIGHTS FROM RECENT OVARIAN GENE EXPRESSION STUDIES

Expression data on genes involved in follicular maturational competence acquisition and regulation of oocyte maturation and ovulation originate from two different kinds of experiments.

Some species, such as several salmonids, spontaneously ovulate in captivity and exhibit a relatively slow kinetics of maturation. For those fish species, biological material can be sampled directly, without exogenous stimulation, at specific stages of maturational competence acquisition, oocyte maturation and ovulation. The large size of full-grown follicles also allows the use of deyolked ovarian tissue to perform gene expression studies (Garczynski and Goetz, 1997). Such technique allows the use of similar biological material (follicular layers of full-grown follicles and extra-follicular tissue) across pre-ovulatory and post-ovulatory ovarian stages.

In contrast, for other species such as Atlantic croaker, zebrafish and goldfish, it can be more convenient to use an hormonal stimulation to induce maturational competence acquisition, oocyte maturation and ovulation.

While both types of data are informative for the overall comprehension of involved mechanisms it is, however, possible that some gene expression differences occur between naturally occurring and hormonally induced events.

Steroid Receptivity and Steroidogenesis

In spotted seatrout, the action of the maturation-inducing steroid (MIS) is mediated by a receptor located on the oocyte plasma membrane (Patino and Thomas, 1990). Recently, a putative membrane progestin receptor was cloned and characterized (Zhu *et al.*, 2003b). The inhibition of progestin induction of oocyte maturation upon microinjection of antisense oligonucleotides in zebrafish was consistent with the identity of this receptor as an intermediary in oocyte maturation (Zhu *et al.*, 2003b). However, several forms (α, β and γ) were subsequently identified in several species. In zebrafish, both α and β forms are potentially involved in the control of oocyte maturation (Thomas *et al.*, 2004). In catfish, expression profiles of the three forms during reproductive cycle do not show a clear pattern associated with the timing of oocyte maturation (Kazeto *et al.*, 2004). Therefore, it remains unclear if an over-expression of MIS receptor occurs during FMC acquisition.

Cholesterol side-chain cleavage cytochrome P450 (P450scc), 3β-hydroxysteroid dehydrogenase (3βHSD) and 20β-hydroxysteroid dehydrogenase (20βHSD) are expressed in the fish pre-ovulatory ovary. In rainbow trout (*Oncorhynchus mykiss*), a gene expression profiling study suggested that P450scc and 3βHSD mRNA levels were higher in competent females than in poorly competent females (Bobe *et al.*, 2004). In the same study, 20βHSD expression appeared to be quite variable, depending on the female. This observation could be consistent with previous observations of strong daily fluctuations in MIS circulating levels observed in female rainbow trout undergoing oocyte maturation (Zohar *et al.*, 1986). In addition, no significant difference was observed between stages (Bobe *et al.*, 2004). Similar results were obtained in zebrafish (Wang and Ge, 2002). In this study, a constitutive expression of 20βHSD was observed in the ovarian follicle. Analysis of ovarian samples taken at different times before spawning showed no significant change of enzyme expression (Wang and Ge, 2002). In Nile tilapia, *Oreochromis niloticus*, 20βHSD expression is detected by Northern blot in ovarian follicles exhibiting migratory germinal vesicle and/or GVBD (Senthilkumaran *et al.*, 2002). Together, these results show that 20βHSD is expressed in the ovarian follicle prior to meiosis resumption, including follicles that do not undergo GVBD after gonadotropic stimulation. They also show that 20βHSD is expressed during meiosis resumption. However, there is no definitive quantitative observation that can lead to the conclusion that 20βHSD mRNA over-expression does occur during FMC acquisition and oocyte maturation. In fact, it is possible that 20βHSD over-expression occurs prior to FMC acquisition. However, it should be reminded that two different 20βHSD isoforms were reported in rainbow trout (Guan *et al.*, 1999). Based on bacterial expression assays, only type A form is believed to be active in rainbow trout and, to our knowledge, no information is available on the expression profiles of each form in the pre-ovulatory ovary of any fish.

In rainbow trout, it was shown that estradiol receptor α (ERα) mRNA levels were significantly lower in females undergoing oocyte maturation than in females exhibiting high FMC (Bobe *et al.*, 2004). In addition, high aromatase mRNA levels were observed in meiotically incompetent or poorly competent females while the lowest mRNA levels were observed in maturing or highly competent females (Bobe *et al.*, 2004). These observations are consistent with the drop in E2 circulating levels observed during FMC acquisition in rainbow trout (Fostier *et al.*, 1978; Scott *et al.*, 1980). However, this could not be a general picture for all fish species as suggested for the Japanese eel, *Anguilla japonica* (Matsubara *et al.*, 2003). Finally, in the view of recent structural studies a direct action of progestins on ER cannot be excluded (Mori *et al.*, 2000).

Both androgen receptors α and β expression appeared to decrease during rainbow trout oocyte maturation (Bobe *et al.*, 2004). These profiles are consistent with the drop in testosterone circulating levels observed in rainbow trout prior to ovulation (Scott *et al.*, 1983).

The steroidogenic acute regulatory protein (StAR) is responsible for the delivery of cholesterol to P450scc. In brook trout, *Salvelinus fontinalis*, it was shown that StAR mRNA was barely detectable in ovarian follicles before the resumption of meiosis and that its expression increased after germinal vesicle breakdown (Kusakabe *et al.*, 2002). In rainbow trout, a dramatic increase in StAR mRNA levels was observed in the last steps of FMC acquisition (Bobe *et al.*, 2004).

In rainbow trout, 11β-hydroxylase (11βH) mRNA was overexpressed during FMC acquisition. In vertebrates, this steroidogenic enzyme acts at the final steps of biosynthesis of glucocorticoids and mineralocorticoids. A mineralocorticoid receptor has been recently found to be expressed in immature rainbow trout ovaries (Sturm *et al.*, 2005). The observed increase in 11βH expression during FMC acquisition could be related to the hydration of the oocyte observed at that time in some fish species (Wallace *et al.*, 1992). However, a rise in cortisol circulating levels is not observed at the time of germinal vesicle breakdown (GVBD) but after ovulation (Bry, 1985) and cortisol has no detectable effect on MIS production *in vitro* when physiological levels are tested (Reddy *et al.*, 1999). On the other hand, the 11β-hydroxysteroid dehydrogenase (11βHSD) expression has been proposed to play a protective role of rainbow trout gonads against stress effects by cortisol deactivation (Kusakabe *et al.*, 2003). The rise in 11βH mRNA expression observed at the time of oocyte maturation could also be related to 11-oxo-androgens production. However, no ovarian synthesis of 11β-hydroxyandrostenedione from pregnenolone could be detected at a post-vitellogenic stage (Reddy *et al.*, 1999) and 11-ketotestosterone plasma levels were extremely low in female rainbow trout sampled during spawning season (Lokman *et al.*, 2002).

Growth and Differentiation Factor

IGF system

In vertebrates, the IGF systems is composed of two ligands IGF1 and IGF2, two receptors (type I and type II receptors), and six IGF-binding proteins (IGFBP-1 to 6). In fish, several studies reported germinal vesicle breakdown after *in vitro* stimulation by either recombinant human IGF1 or IGF2 depending on the species (Kagawa *et al.*, 1994a; Negatu *et al.*, 1998; Weber and Sullivan, 2000). In addition, human recombinant IGF1 significantly increased MIS receptor capacity in spotted seatrout (Thomas *et al.*, 2001), and the incidence of GAP junction in red seabream, *Pagrus major* (Patiño and Kagawa, 1999). In white bass, *Morone chrysops*, recombinant human IGF1 induced maturational competence but not oocyte maturation *in vitro* (Weber and Sullivan, 2005). In rainbow trout, the mRNA expression of IGF1 and 2 in deyolked ovarian tissue progressively increased during FMC acquisition and oocyte maturation (Bobe *et al.*, 2004). In contrast, very little information is available on IGFBP expression in the fish peri-ovulatory ovary. In rainbow trout deyolked ovary, IGFBP-2 expression profile is similar to those of IGF1 and 2, but no significant difference between groups was observed (Bobe *et al.*, 2004). Taken together, these results are consistent with the hypothesis of a role of the IGF system during FMC acquisition and oocyte maturation. It could be an autocrine regulation; however, the presence during the preovulatory period of a high level of receptor-type growth hormone (GH) binding in the rainbow trout ovary also suggests that IGFs could relay an endocrine GH effect (Gomez *et al.*, 1999). From recent expression profiling studies reporting a progressive over-expression of IGF, even in poorly competent females, we can infer that this role could be important quite early in the process of FMC acquisition.

Transforming growth factorβ (TGF-β) super family

The TGF-β super family comprises a large number of secreted growth factors. Three of the best characterized subfamilies are activins/inhibins, TGF-β and bone morphogenetic proteins (BMP).

Activin-inhibin system The activin-inhibin system has been suspected to play a role, in a paracrine manner, in the control of oocyte maturation in several fish species. In zebrafish, full-grown follicles—both activin βA and βB mRNAs—are expressed in the follicle layers (Wang and Ge, 2003). In rainbow trout, activin βA was detected by *in situ* hybridization in the theca cells of late vitellogenic follicles (Tada *et al.*, 2002) and its mRNA expression was observed to increase progressively during FMC acquisition whereas inhibin mRNA expression showed a sharp increase at the time of oocyte maturation (Bobe *et al.*, 2004). These observations are consistent with previous studies reporting an effect of human recombinant activin A and goldfish recombinant activin B on development of zebrafish oocyte competence (Pang and Ge, 2002), and GVBD induction in full-grown zebrafish ovarian follicles by recombinant human activin A and inhibin A (Wu *et al.*, 2000).

Bone morphogenetic proteins (BMPs) and their receptors are known to have important roles in mammalian ovarian folliculogenesis (Monget *et al.*, 2002; Knight and Glister, 2003). In rainbow trout, mRNA expression of BMP4 and BMP7 was recently reported in ovarian tissue during FMC acquisition or oocyte maturation (Bobe *et al.*, 2004). BMP4 and BMP7 are expressed by theca cells in rat (Shimasaki *et al.*, 1999) and cow (Knight and Glister, 2003). In chicken, both BMP4 and 7 were recently detected in both granulosa and theca cells (Onagbesan *et al.*, 2003). In rainbow trout deyolked pre-ovulatory ovary, BMP7 exhibits an increased expression in highly competent and maturing females, while BMP4 expression is increased later, at the time of oocyte maturation. During chicken follicular development, both BMP4 and BMP7 expression increase in granulosa cells while BMP7 expression is increased in the theca (Onagbesan *et al.*, 2003). Interestingly, BMP4 and 7 both stimulate IGF1-, and gonadotropin-stimulated progesterone production by cultured chicken granulosa cells (Onagbesan *et al.*, 2003). However, the specific roles of BMP4 and 7 in rainbow trout follicular differentiation require further investigations.

In addition, the expression of growth differentiation 9 (GDF9), another other TGFβ family member was recently reported in the trout preovulatory ovary (Bobe *et al.*, 2004). However, GDF9 did not exhibit an expression profile associated with the timing of FMC acquisition or oocyte maturation. As a matter of fact, GDF9 is an oocyte-derived growth factor involved in mammalian follicle development (Knight and Glister, 2003). The GDF9 mRNA expression detected here could, therefore, take place in previtellogenic follicles present in the trout periovulatory ovary.

TGF-β In mammals, 3 different forms of TGF-β (1-3) are expressed in the ovary and have been detected in both follicular cells and oocyte (Knight and Glister, 2003). In zebrafish, TGF-β1 and a type II receptor were cloned and their expression was detected in ovarian follicles through semi-quantitative RT PCR. Moreover, human recombinant TGF-β1 was able to inhibit both gonadotropin- and MIS-induced GVBD (Kohli *et al.*, 2003). In rainbow trout, expression of TGF-β monitored in deyolked ovarian tissue during FMC acquisition and oocyte maturation was not significantly correlated with the timing of follicular development (FMC acquisition or meiosis resumption). However, clustering analysis revealed that TGF-β belonged to a group of genes whose expression tended to decrease during FMC acquisition (Bobe *et al.*, 2004). Indeed, TGF-β expression profile was clustered with aromatase profile which was characterized by a strong expression in females unresponsive or poorly responsive to gonadotropic stimulation. The observations made in trout seem, therefore, to support the hypothesis of an inhibition of oocyte maturation by TGF-β.

TGF-β signaling pathway Two TGF-β receptors (type I and type II) are known and possess serine-threonine kinase activity. The Smad (identified as mediators of transcriptional activation by members of the TGF-β superfamily of cytokines) signalling proteins are involved in TGF-β signal transduction and act down-stream of the serine-threonine kinase receptors. In zebrafish full-grown follicles, type II receptor and Smad 2-4 mRNAs were detected using PCR (Kohli *et al.*, 2003).

Epidermal growth factor

Epidermal growth factor (EGF) is one of the many growth factors, produced in the ovary, that is suspected to play a role in regulating ovarian functions in vertebrates. In zebrafish full-grown follicles, EGF and EGF receptor (EGFR) are expressed in both the ovarian follicular cells and oocyte (Wang and Ge, 2004).

Fibroblast growth factor

In a recent gene expression survey performed during FMC acquisition, it was observed that fibroblast growth factor 2 (FGF2) mRNA expression increased, in deyolked ovarian tissue, during oocyte maturation (Bobe *et al.*, 2004). Interestingly, FGF2 is expressed in bovine follicles during final growth to pre-ovulatory follicle (Berisha *et al.*, 2000) where it could be involved in the proliferation of capillaries.

Cytokines

Several members of the tumor necrosis factor (TNF) and TNF receptor families are expressed in the peri-ovulatory ovary. In brook trout (*Salvelinus fontinalis*), a decoy TNF receptor is up-regulated within 24H after ovulation (Bobe and Goetz, 2000). In zebrafish, a TNF ligand called TRAIL-like and a TNF receptor, called ovarian TNF receptor (OTR), were expressed in the ovary at preovulatory stages (Bobe and Goetz,

2001b). In brook trout, another TNF ligand, called TNF-like, was detected in deyolked ovarian tissue during oocyte maturation (Bobe and Goetz, 2001b). These observations suggest that TNF ligands and receptors are involved in the ovarian physiological processes around the time of ovulation. These observations are consistent with the mammalian hypothesis of an inflammatory-like reaction during ovulation (Espey, 1980).

Gonadotropin, Gonadotropin Receptors and GnRH System

The expression of both gonadotropin receptors has been reported in the peri-ovulatory ovary of various fish species (Kumar *et al.*, 2001a, b; Laan *et al.*, 2002; Bobe *et al.*, 2003b; 2004 Vischer and Bogerd, 2003). Most investigators agree that LH receptor (LHr) is highly expressed during spawning season or in mature or peri-ovulatory ovary. In rainbow trout, LHr mRNA expression was downregulated during meiosis resumption when compared to the very last steps of FMC acquisition. However, the mRNA levels of LHr are not significantly lower in females at the time of spontaneous oocyte maturation than in females exhibiting no or little maturational competence (Bobe *et al.*, 2004). This is to put together with the little peak in LH circulating levels observed in rainbow trout 4 days before meiosis resumption (Fig. 1.3) (Breton *et al.*, 1998). In contrast, FSH receptor (FSHr) appeared to be over-expressed by the end of the FMC acquisition process (Bobe *et al.*, 2003b) and to remain elevated during oocyte maturation (Bobe *et al.*, 2004). This can be associated with the rise in FSH circulating levels observed in rainbow trout before oocyte maturation (Breton *et al.*, 1998) (Fig. 1.3). LH is the determining factor regulating the production of the maturation-inducing steroid and the induction of meiosis resumption in salmonids, while FSH has a significant but lower effect (Planas *et al.*, 2000). However, the physiological role of FSH in the pre-ovulatory ovary remains unclear and requires additional data, including cellular localization. FSHr mRNA expression could possibly occur in the previtellogenic follicles present in the ovary at that time. FSH and its receptor could, thus, be involved in recruiting the next generation of ovarian follicles. Interestingly, it was recently reported that gonadotropin subunits mRNAs were expressed in seabream oocyte during vitellogenesis. FSHβ subunit was also detected in ovulated oocytes (Wong and Zohar, 2004). The authors suggest that this expression could be related to intercellular communications between oocyte and follicular cells.

Connexins and GAP Junctions

GAP junctions are specialized regions of the plasma membrane of animal cells. They are characterized by intercellular cytoplasmic communication channels between adjacent cells. The homologous protein found in GAP junctions is called connexon, which is made of 6 connexin (Cx m, m=molecular mass) subunits (Beyer et al., 1990; Bruzzone et al., 1996). In Atlantic croaker, the expression of Cx32.2, but not of Cx32.7, is positively related to the oocyte responsiveness to 17,20β,21P for GVBD obtained by a pre-treatment with human chorionic gonadotropin (hCG) (Yoshizaki et al., 1994). In fact, in this species as also in red seabream, an increase in homologous and heterologous GAP junction contacts was observed after a gonadotropic stimulation of late vitellogenic follicles (York et al., 1993; Patino and Kagawa, 1999). More precisely, the hCG-dependent formation of GAP junctions observed in vitro was shown to be associated with increased levels of connexin 32.2mRNA levels in Atlantic croaker (Chang et al., 1999). In red seabream, Cx31.5 mRNA is over expressed by LH but not by human IGF1 (Choi and Takashima, 2000). In rainbow trout deyolked ovary, connexin 43 is strongly up-regulated at the time of oocyte maturation (Bobe et al., 2004). This suggests that specific connexins would be associated with specific steps of pre-ovulatory ovarian differentiation. Cx31.5 and 32.2 would increase during FMC acquisition in response to gonado-tropic stimulation, while Cx43 would be upregulated during meiosis resumption.

Proteases and Anti-proteases

Several proteases and proteases inhibitors are expressed in the fish peri-ovulatory ovary. A tissue kallikrein is highly expressed in brook trout ovarian tissue, including granulosa cells (Hajnik et al., 1998). A cysteine protease inhibitor is expressed in pre- and early-vitellogenic oocytes that are found in the brook trout ovary around ovulation (Bobe and Goetz, 2001a). Such anti-protease could, therefore, be involved in protecting some ovarian compartments (e.g., pre- and early vitellogenic oocytes) from ovarian proteases involved in the process of ovulation or in yolk protein processing related to oocyte hydration. Indeed, protease activity is observed in the pre-ovulatory ovary of several fishes (Berndtson and Goetz, 1988, 1990). Recent data showed that several cathepsins, some of them belonging to the cysteine protease family, are expressed in the pre-ovulatory ovary of salt marsh killifish (Fabra and Cerda, 2004). An ovarian expression of cathepsins B, D, K, L, S and Z in rainbow trout deyolked ovarian tissue was also repeated (Bobe *et al.*, 2004). Cathepsins D, L and Z exhibited increased mRNA expression at the time of oocyte maturation. These observations are in agreement with the report of an ovarian expression of cathepsins D, L and B in rainbow trout females undergoing oocyte maturation (Kwon *et al.*, 2001). In red seabream oocytes, cathepsin B, D and L enzymatic activities were reported during vitellogenesis or oocyte maturation (Carnevali *et al.*, 1999). In addition, these authors clearly showed that cathepsin D is responsible for processing vitellogenin into yolk proteins. However, it is noteworthy that in some species (e.g., rainbow trout), cathepsin expression is not limited to full-grown oocytes. Indeed, expression in deyolked ovarian tissue suggests that the action of these enzymes is not limited to yolk processing. Interestingly enough, cathepsin L, a member of the papain family, expression is selectively induced in granulosa cells of mice pre-ovulatory follicles by the LH surge (Robker *et al.*, 2000). It can, therefore, be hypothesized that cathepsins expressed in follicular layers or extra-follicular ovarian tissue are involved in the process of ovulation.

Other Non-steroidogenic Enzymes

Prostaglandin endoperoxide synthase-1 and 2 (PGS1 and PGS2)

PGS1 and 2, the two enzymes that catalyze the initial conversion of arachidonic acid in the biosynthetic pathway for prostaglandin synthesis are both expressed, although not differentially, during rainbow trout FMC acquisition and oocyte maturation (Bobe *et al.*, 2004). Using Northern blot analysis, a less-sensitive technique, it was previously reported that only PGS1 but not PGS2 was expressed in the brook trout pre-ovulatory ovary (Roberts *et al.*, 2000). Real-time PCR results in rainbow trout also suggested that PGS1 expression was much higher than PGS2 expression (Bobe *et al.*, 2004).

Conclusion

In summary, genes expressed in the pre-ovulatory ovary appear to be associated with several kinds of molecular events such as the direct control of meiosis resumption, the process of ovulation, ovarian functions outside the full-grown follicle, processing of yolk proteins, stockpiling in the oocyte of maternal RNA that will be involved in sustaining early embryonic development. Available expression data is a first step to a better comprehension of pre-ovulatory ovarian events.

Interestingly, some genes that are involved in the direct control of oocyte maturation, such as LH receptor and 20βHSD are not clearly upregulated during FMC acquisition. In fact, some data suggest that they could be already turned on in post-vitellogenic females, even when those are poorly responsive or unresponsive to gonadotropic stimulation. The most significant or limiting events of FMC acquisition could, therefore, be related to growth factor production (i.e., BMPs, IGFs), E2 synthesis (aromatase), intercellular communications (GAP junctions) or other limiting steps of steroid synthesis (i.e., Star).

It is also noteworthy that some genes are progressively over-expressed during FMC acquisition and follicular maturation. IGF2 for example is progressively over-expressed in deyolked pre-ovulatory ovary during FMC acquisition and oocyte maturation is probably important for both steps. In addition, the early expression of such genes (e.g., growth factors) in poorly competent females possibly occurs before the gonadotropic surge. This could suggest that some factors not regulated by LH would still be important for FMC acquisition.

Interestingly, gene expression studies suggest that some factors, putatively involved in the process of ovulation such as proteases (e.g., cathepsins), proliferation factors, enzymes involved in prostaglandin synthesis are already being upregulated during FMC acquisition or oocyte maturation.

Together, all these observations suggest that each functional compartment of the ovary may progressively differentiate at its own pace, and that a successful complete response to an endocrine stimulation can be obtained only when all these compartments are fully differentiated. This is in agreement with earlier studies reporting oocyte maturation not followed by ovulation when a precocious stimulation was performed.

Recently available genomic tools represent a great opportunity to decipher the mechanisms associated with pre-ovulatory ovarian follicular differentiation and draw a global view of the molecular events occurring in the various ovarian cellular compartments (oocyte, follicular layers, extrafollicular tissue, previtellogenic oocytes...). However, such approaches should be used on well-characterized samples (exact physiological stage, responsiveness to hormonal stimulation...) to be really informative. In addition, a specific attention should be devoted to the localisation of gene expression in ovarian tissue. While the use of *in situ* hybridization is valuable to identify the cells expressing each studied mRNA, its sensitivity can be somehow limiting. Alternatively, a quantitative expression analysis using only some compartments of ovarian tissue (e.g., deyolked ovarian tissue, granulosa, interstitial tissue...) could also be performed.

Future Directions/Perspectives

Some authors have proposed a two-stage model to describe FMC acquisition and oocyte maturation (Patiño *et al.*, 2001). In that particular model, oocyte maturational competence and the ability to produce MIS are acquired during the first step, while the second step corresponds to MIS-mediated meiosis resumption. This is probably the case in some species, especially those that get through a waiting period, that can fluctuate in duration, before encountering appropriate spawning conditions. This is not the case in species such as salmonids where *in vivo* an *in vitro* studies as well as gene expression data reviewed here strongly suggest a progressive differentiation of the pre-ovulatory ovarian follicle normally followed, when each ovarian compartment is ready, by its coordinate participation in endocrine, paracrine and cellular processes of maturation and ovulation. A recent review also suggested that oocyte maturation and ovulation are closely integrated and overlapping events (Patiño *et al.*, 2003). We suggest, in fact, to extend this concept of 'integrated and overlapping events' to include the acquisition of FMC.

Besides studies aiming to improve a successful spawning after hormonal stimulation, very few investigations have tried to understand the pre-ovulatory events able to influence the quality of ovulated eggs. It is, however, known in other vertebrates that oocyte maturation is associated with changes of the oocyte content in messenger RNAs. During this step, many transcripts are stockpiled or modified (adenylation/deadenylation) in relationship with oocyte developmental competence (Brevini-Gandolfi *et al.*, 1999). A specific attention should therefore be paid to pre-ovulatory modifications of oocyte transcriptome in relationship with the quality of subsequently ovulated eggs. A similar approach could also be performed at the proteomic level. If these events are indeed linked to egg quality, this could explain why poor egg quality can be observed when an early stimulation is performed. Events occurring in follicular layers should also be considered with great attention in order to better understand the oocyte-follicle dialogue and its importance for egg quality.

References

Aegerter, S. and B. Jalabert. 2004. Effects of post-ovulatory oocyte ageing and temperature on egg quality and on the occurrence of triploid fry in rainbow trout, *Oncorhynchus mykiss*. *Aquaculture* 231: 59-71.

Basu, D., A.K. Navneet, S. Dasgupta and S. Bhattacharya. 2004. Cdc2-cyclin B-induced G2 to M transition in perch oocyte is dependent on Cdc25. *Biology of Reproduction* 71: 894-900.

Berisha, B., D. Schams, M. Kosmann, W. Amselgruber and R. Einspanier. 2000. Expression and localisation of vascular endothelial growth factor and basic fibroblast growth factor during the final growth of bovine ovarian follicles. *Journal of Endocrinology* 167: 371-382.

Berndtson, A.K. and F.W. Goetz. 1988. Protease activity in brook trout (*Salvelinus fontinalis*) follicle walls demonstrated by substrate-polyacrylamide gel electrophoresis. *Biology of Reproduction* 38: 511-516.

Berndtson, A.K. and F.W. Goetz. 1990. Metallo-protease activity increases prior to ovulation in brook trout (*Salvelinus fontinalis*) and yellow perch (*Perca flavescens*) follicle walls. *Biology of Reproduction* 42: 391-398.

Beyer, E.C., D.L. Paul and D.A. Goodenough. 1990. Connexin family of gap junction proteins. *Journal of Membrane Biology* 116: 187-194.

Bobe, J. and F.W. Goetz. 2000. A tumor necrosis factor decoy receptor homologue is up-regulated in the brook trout (*Salvelinus fontinalis*) ovary at the completion of ovulation. *Biology of Reproduction* 62: 420-426.

Bobe, J. and F.W. Goetz. 2001a. Cysteine protease inhibitor is specifically expressed in pre- and early-vitellogenic oocytes from the brook trout periovulatory ovary. *Molecular Reproduction and Development* 60: 312-318.

Bobe, J. and F.W. Goetz. 2001b. Molecular cloning and expression of a TNF receptor and two TNF ligands in the fish ovary. *Comparative Biochemistry and Physiology* B129: 475-481.

Bobe, J., G. Maugars, T. Nguyen and B. Jalabert. 2003a. Specific gene expression profiles are associated with follicular maturational competence acquisition in rainbow trout (*Oncorhynchus mykiss*). *Fish Physiology and Biochemistry* 28: 309-311.

Bobe, J., G. Maugars, T. Nguyen, H. Rime and B. Jalabert. 2003b. Rainbow trout follicular maturational competence acquisition is associated with an increased expression of follicle stimulating hormone receptor and insulin-like growth factor 2 messenger RNAs. *Molecular Reproduction and Development* 66: 46-53.

Bobe, J., T. Nguyen and B. Jalabert. 2004. Targeted gene expression profiling in the Rainbow Trout (*Oncorhynchus mykiss*) ovary during maturational competence acquisition and oocyte maturation. *Biology of Reproduction* 71: 73-82.

Breton, B., R. Billard, B. Jalabert and G. Kann. 1972. Radioimmunologic assay of plasma gonadotropins in *Carassius auratus* during the nychthemeron and ovulation. *General and Comparative Endocrinology* 18: 463-468.

Breton, B., M. Govoroun and T. Mikolajczyk. 1998. GTH I and GTH II secretion profiles during the reproductive cycle in female rainbow trout: Relationship with pituitary responsiveness to GnRH-A stimulation. *General and Comparative Endocrinology* 111: 38-50.

Breton, B., F. Le Gac and E. Sambroni. 1986. Gonadotropin hormone (GtH) receptors in the ovary of the brown trout *Salmo trutta* L. *in vitro* studies. *General and Comparative Endocrinology* 64: 163-171.

Brevini-Gandolfi, T.A., L.A. Favetta, L. Mauri, A.M. Luciano, F. Cillo and F. Gandolfi. 1999. Changes in poly(A) tail length of maternal transcripts during *in vitro* maturation of bovine oocytes and their relation with developmental competence. *Molecular Reproduction and Development* 52: 427-433.

Brooks, S., C.R. Tyler and J.P. Sumpter. 1997. Quality in fish: What makes a good egg? *Reviews in Fish Biology and Fisheries* 7: 387-416.

Bruzzone, R., T.W. White and D.L. Paul. 1996. Connections with connexins: the molecular basis of direct intercellular signaling. *European Journal of Biochemistry* 238: 1-27.

Bry, C. 1981. Temporal aspects of macroscopic changes in rainbow trout (*Salmo gairdneri*) oocytes before ovulation and ova fertility during the post-ovulation period: Effect of treatment with 17 alpha-hydroxy-20 beta-dihydroprogesterone. *Aquaculture* 24: 153-160.

Bry, C. 1985. Plasma cortisol levels of female rainbow trout (*Salmo gairdneri*) at the end of the reproductive cycle: relationship with oocyte stages. *General and Comparative Endocrinology* 57: 47-52.

Carnevali, O., R. Carletta, A. Cambi, A. Vita and N. Bromage. 1999. Yolk formation and degradation during oocyte maturation in seabream *Sparus aurata*: involvement of two lysosomal proteinases. *Biology of Reproduction* 60: 140-146.

Carnevali, O., G. Mosconi, A. Roncarati, P. Belvedere, M. Romano and E. Limatola. 1992. Changes in the electrophoretic pattern of yolk proteins during vitellogenesis in the gilthead seabream, *Sparus aurata*l. *Comparative Biochemistry and Physiology* B103: 955-962.

Cerda, J.L., T.R. Petrino and R.A. Wallace. 1993. Functional heterologous gap-junctions in fundulus ovarian follicles maintain meiotic arrest and permit hydration during oocyte maturation. *Developmental Biology* 160: 228-235.

Cerda, J., T.R. Petrino, Y.W.P. Lin and R.A. Wallace. 1995. Inhibition of *fundulus heteroclitus* oocyte maturation *in-vitro* by serotonin (5-Hydroxytryptamine). *Journal of Experimental Zoology* 273: 224-233.

Cerda, J., G. Reich, R.A. Wallace and K. Selman. 1998a. Serotonin inhibition of steroid-induced meiotic maturation in the teleost *Fundulus heteroclitus*: Role of cyclic AMP and protein kinases. *Molecular Reproduction and Development* 49: 333-341.

Cerda, J., N. Subhedar, G. Reich, R.A. Wallace and K. Selman. 1998b. Oocyte sensitivity to serotonergic regulation during the follicular cycle of the teleost *Fundulus heteroclitus*. *Biology of Reproduction* 59: 53-61.

Chang, X., R. Patiño, P. Thomas and G. Yoshizaki. 1999. Developmental and protein kinase-dependent regulation of ovarian connexin mRNA and oocyte maturational competence in Atlantic croaker. *General and Comparative Endocrinology* 114: 330-339.

Choi, C.Y. and F. Takashima. 2000. Molecular cloning and hormonal control in the ovary of connexin 31.5 mRNA and correlation with the appearance of oocyte maturational competence in red seabream. *Journal of Experimental Biology* 203: 3299-3306.

Chyb, J., T. Mikolajczyk and B. Breton. 1999. Post-ovulatory secretion of pituitary gonadotropins GtH I and GtH II in the rainbow trout (*Oncorhynchus mykiss*): regulation by steroids and possible role of non-steroidal gonadal factors. *Journal of Endocrinology* 163: 87-97.

Dasgupta, S., D. Basu, L.R. Kumar and S. Bhattacharya. 2001. Insulin alone can lead to a withdrawal of meiotic arrest in the carp oocyte. *Journal of Bioscience* 26: 341-347.

de Mones, A., A. Fostier, C. Cauty and B. Jalabert. 1989. Ovarian early postovulatory development and oestrogen production in rainbow trout (*Salmo gairdneri* R.) from a spring-spawning strain. *General and Comparative Endocrinology* 74: 431-441.

de Montalembert, G., B. Jalabert and C. Bry. 1978. Precocious induction of maturation and ovulation in northern pike (*Esox lucius*). *Annales de Biologie Animale, Biochimie, Biophysique* 18: 969-975.

Duesbery, N.S. and G.F. Vande Woude. 1998. Cytoplasmic control of nuclear behavior during meiotic maturation of frog oocytes. *Biology of the Cell* 90: 461-466.

El Etr, M., S. Schorderet-Slatkine and E.E. Baulieu. 1979. Meiotic maturation in *Xenopus laevis* oocytes initiated by insulin. *Science* 205: 1397-1399.

Espey, L.L. 1980. Ovulation as an inflammatory reaction—A hypothesis. *Biology of Reproduction* 22: 73-106.

Espey, L.L. 1994. Current status of the hypothesis that mammalian ovulation is comparable to an inflammatory reaction. *Biology of Reproduction* 50: 233-238.

Fabra, M. and J. Cerda. 2004. Ovarian cysteine proteinases in the teleost *Fundulus heteroclitus*: molecular cloning and gene expression during vitellogenesis and oocyte maturation. *Molecular Reproduction and Development* 67: 282-294.

Fabra, M., D. Raldua, D.M. Power, P.M. Deen and J. Cerda. 2005. Marine fish egg hydration is aquaporin-mediated. *Science* 307: 345

Finet, B., B. Jalabert and S.K. Garg. 1988. Effect of defolliculation and 17 alpha-hydroxy, 20 beta-dihydroprogesterone on cyclic AMP level in full-grown oocytes of the rainbow trout, *Salmo gairdneri*. *Gamete Research* 19: 241-252.

Finn, R.N., G.C. Ostby, B. Norberg and H.J. Fyhn. 2002. *In vivo* oocyte hydration in Atlantic halibut (*Hippoglossus hippoglossus*); proteolytic liberation of free amino acids, and ion transport, are driving forces for osmotic water influx. *Journal of Experimental Biology* 205: 211-224.

Fostier, A. and B. Jalabert. 1982. Physiological basis of practical means to induce ovulation in fish. In: *Proceedings of the Second International Symposium on Reproductive Physiology of Fish*, C.C.J. Richter and H.J. Goos (eds.), Pudoc, Wageningen, pp. 164-173.

Fostier, A. and B. Jalabert. 1986. Steroidogenesis in rainbow trout (*Salmo gairdneri*) at various preovulatory stages: changes in plasma hormone levels and *in vivo* and *in vitro* responses of the ovary to salmon gonadotropin. *Fish Physiology and Biochemistry* 2: 87-99.

Fostier, A., B. Jalabert and M. Terqui. 1973. Predominant action of a hydroxylated derivative of progesterone on the *in vitro* maturation of oocytes of the rainbow trout (*Salmo gairdneri*). *Comptes Rendus de l'Académie des Sciences Paris* 277: 421-424.

Fostier, A., B. Jalabert, C.M. Campbell, M. Terqui and B. Breton. 1981. Cinétique de libération *in vitro* de la 17-hydroxy-20ß-dihydroprogestérone par des follicules de truite arc-en-ciel (*Salmo gairdnerii*). *Comptes Rendus de l'Académie des Sciences Paris* 292: 777-780.

Fostier, A., C. Weil, M. Terqui, B. Breton and B. Jalabert. 1978. Plasma estradiol 17 beta and gonadotropin during ovulation in rainbow trout *Salmo gairdneri*. *Annales de Biologie Animale, Biochimie, Biophysique* 18: 929-936.

Fostier, A., F. Le Gac and M. Loir. 1994. Insulin-like growth factors and gonadal regulation in fish. *Contraception, Fertilite, Sexualite* 22: 548-550.

Fulton, W. 1898. II-On the growth and maturation of the ovarian eggs of teleostean fishes. *Fish Board Scotland 16th annual report*: 88-124.

Garczynski, M.A. and F.W. Goetz. 1997. Molecular characterization of a ribonucleic acid transcript that is highly up-regulated at the time of ovulation in the brook trout (*Salvelinus fontinalis*) ovary. *Biology of Reproduction* 57: 856-864.

Goetz, F.W. 1983. Hormonal control of oocyte final maturation and ovulation in fishes. In: Fish Physiology, W.S. Hoar, D.J. Randall and E.M. Donaldson (eds.). Academic Press, New York, Vol. 9, pp. 117-170.

Goetz, F.W., A.K. Berndtson and M. Ranjan. 1991. Ovulation: Mediators at the ovarian level. In: Vertebrate Endocrinology: Fundamentals and Biomedical Implications, P.K.T. Pang and M.P. Schreibman (eds.), Academic Press, San Diego, pp. 127-203.

Goetz, F.W. and M. Garczynski. 1997. The ovarian regulation of ovulation in teleost fish. *Fish Physiology and Biochemistry* 17: 33-38.

Gomez, J.M., C. Weil, M. Ollitrault, P.Y. Le Bail, B. Breton and F. Le Gac. 1999. Growth hormone (GH) and gonadotropin subunit gene expression and pituitary and plasma changes during spermatogenesis and oogenesis in rainbow trout (*Oncorhynchus mykiss*). *General and Comparative Endocrinology* 113: 413-428.

Guan, G., M. Tanaka, T. Todo, G. Young, M. Yoshikuni and Y. Nagahama. 1999. Cloning and expression of two carbonyl reductase-like 20beta-hydroxysteroid dehydrogenase cDNAs in ovarian follicles of rainbow trout (*Oncorhynchus mykiss*). *Biochemical and Biophysical Research Communications* 255: 123-128.

Guraya, S.S. 1986. The cell and molecular biology of fish oogenesis. In: *Monographs in Developmental Biology*, H.W. Sauer (ed.). S. Karger, Basel, pp. 1-223.

Habibi, H.R. and C.A. Lessman. 1986. Effect of demecolcine (colcemid) on goldfish oocyte meiosis *in vitro. Gamete Research* 13: 103-114.

Habibi, H.R., D. Pati, M. Ouwens and H.J. Goos. 1994. Presence of gonadotropin-releasing hormone (GnRH) binding sites and compounds with GnRH-like activity in the ovary of African catfish, *Clarias gariepinus. Biology of Reproduction* 50: 643-652.

Hajnik, C.A., F.W. Goetz, S.Y. Hsu and N. Sokal. 1998. Characterization of a ribonucleic acid transcript from the brook trout (*Salvelinus fontinalis*) ovary with structural similarities to mammalian adipsin/complement factor D and tissue kallikrein, and the effects of kallikrein-like serine proteases on follicle contraction. *Biology of Reproduction* 58: 887-897.

Hirose, K., Y. Machida and E.M. Donaldson. 1979. Induced ovulation of Japanes flounder (*Limanda yokohoma*) with HCG and salmon gonadotropin, with special references to changes in the quality of eggs retained in the ovarian cavity after ovulation. *Bulletin of the Japanese Scientific Fisheries* 45: 31-36.

Hontela, A. and R.E. Peter. 1983. Characteristics and functional significance of daily cycles in serum gonadotropin hormone levels in the goldfish. *Journal of Experimental Zoology* 228: 543-550.

Idler, D.R., U.H.M. Fagerlund and A.P. Ronald. 1960. Isolation of pregn-4-ene-17á,20â-diol-3-one from the plasma of Pacific salmon (*Oncorhynchus nerka*). *Biochemical and Biophysical Research Communications* 2: 133-137.

Iwamatsu, T., S.Y. Takahashi, M. Ohishi, T. Yokochi and H. Maeda. 1992. Changes in electrophoretic patterns of oocyte proteins during oocyte maturation in *Oryzias latipes*. *Development Growth and Differentiation* 34: 173-179.

Iwamatsu, T., Y. Toya, N. Sakai, Y. Terada, R. Nagata and Y. Nagahama. 1993. Effect of 5-hydroxytryptamine on steroidogenesis and oocyte maturation in pre-ovulatory follicles of the medaka *Oryzias latipes*. *Development Growth and Differentiation* 35: 625-630.

Jalabert, B. 1976. *In vitro* oocyte maturation and ovulation in rainbow trout (*Salmo gairdneri*), northern pike (*Esox lucius*), and goldfish (*Carassius auratus*). *Journal of Fisheries Research Board of Canada* 33: 974-988.

Jalabert, B. 1978. Production of fertilizable oocytes from follicles of rainbow trout (*Salmo gairdneri*) following *in vitro* maturation and ovulation. *Annales de Biologie Animale Biochimie Biophysique* 18: 461-470.

Jalabert, B. and D. Szöllösi. 1975. *In vitro* ovulation of trout oocytes: Effect of prostaglandins on smooth muscle-like cells of the theca. *Prostaglandins* 9: 765-778.

Jalabert, B., B. Breton and C. Bry. 1972. Maturation and ovulation *in vitro* of rainbow trout *Salmo gairdnerii* ovocytes. *Comptes Rendus de l'Académie des Sciences Paris* 275: 1139-1142.

Jalabert, B., B. Breton and A. Fostier. 1978a. Precocious induction of oocyte maturation and ovulation in rainbow trout (*Salmo gairdneri*): Problems when using 17α-hydroxy-20β-dihydroprogesterone. *Annales de Biologie Animale, Biochimie, Biophysique* 18: 977-984.

Jalabert, B., F.W. Goetz, B. Breton, A. Fostier and E.M. Donaldson. 1978b. Precocious induction of oocyte maturation and ovulation in Coho salmon, *Oncorhynchus kisutch*. *Journal of Fisheries Research Board of Canada* 35: 1423-1429.

Jalabert, B., A. Fostier, B. Breton and C. Weil. 1991. Oocyte Maturation in Vertebrates. In: Vertebrate Endocrinology, Fundamentals and Biomedical Implications, P.K.T. Pang and M.P. Schreibman (eds.), Academic Press, San Diego, Vol. 1. pp. 23-90.

Kagawa, H., M. Kobayashi, Y. Hasegawa and K. Aida. 1994a. Insulin and insulin-like growth factors I and II induce final maturation of oocytes of red seabream, *Pagrus major, in vitro*. *General and Comparative Endocrinology* 95: 293-300.

Kagawa, H., H. Tanaka, K. Okuzawa and K. Hirose. 1994b. Development of maturational competence of oocytes of red seabream, *Pagrus major*, after human chorionic gonadotropin treatment in vitro requires RNA and protein synthesis. *General and Comparative Endocrinology* 94: 199-206.

Kazeto, Y., R. Goto-Kazeto, J.M. Trant. 2005. Membrane-bound progestin receptors in channel catfish and zebrafish ovary: changes in gene expression associated with the reproductive cycles and hormonal reagents. *General and Comparative Endocrinology* 142: 204-211.

King, W., P. Thomas and C.V. Sullivan. 1994. Hormonal regulation of final maturation of striped bass oocytes *in vitro*. *General and Comparative Endocrinology* 96: 223-233.

Kjörsvik, E., A. Mangor-Jensen and I. Homefjord. 1990. Egg quality in fishes. *Advances in Marine Biology* 26: 71-113.

Knight, P.G. and C. Glister. 2003. Local roles of TGF-beta superfamily members in the control of ovarian follicle development. *Animal Reproduction Science* 78: 165-183.

Kohli, G., S. Hu, E. Clelland, T. Di Muccio, J. Rothenstein and C. Peng. 2003. Cloning of transforming growth factor-beta 1 (TGF-beta 1) and its type II receptor from zebrafish ovary and role of TGF-beta 1 in oocyte maturation. *Endocrinology* 144: 1931-1941.

Kumar, R.S., S. Ijiri and J.M. Trant. 2001a. Molecular biology of channel catfish gonadotropin receptors: 1. Cloning of a functional luteinizing hormone receptor and preovulatory induction of gene expression. *Biology of Reproduction* 64: 1010-1018.

Kumar, R.S., S. Ijiri and J.M. Trant. 2001b. Molecular biology of the channel catfish gonadotropin receptors: 2. Complementary DNA cloning, functional expression, and seasonal gene expression of the follicle-stimulating hormone receptor. *Biology of Reproduction* 65: 710-717.

Kusakabe, M., I. Nakamura and G. Young. 2003. 11beta-hydroxysteroid dehydrogenase complementary deoxyribonucleic acid in rainbow trout: Cloning, sites of expression, and seasonal changes in gonads. *Endocrinology* 144: 2534-2545.

Kusakabe, M., T. Todo, H.J. McQuillan, F.W. Goetz and G. Young. 2002. Characterization and expression of steroidogenic acute regulatory protein and MLN64 cDNAs in trout. *Endocrinology* 143: 2062-2070.

Kwon, J.Y., F. Prat, C. Randall and C.R. Tyler. 2001. Molecular characterization of putative yolk processing enzymes and their expression during oogenesis and embryogenesis in rainbow trout (*Oncorhynchus mykiss*). *Biology of Reproduction* 65: 1701-1709.

Laan, M., H. Richmond, C. He and R.K. Campbell. 2002. Zebrafish as a model for vertebrate reproduction: characterization of the first functional zebrafish (*Danio rerio*) gonadotropin receptor. *General and Comparative Endocrinology* 125: 349-364.

Lessman, C.A. and S. Kavumpurath. 1984. Cytological analysis of nuclear migration and dissolution during steroid-induced meiotic maturation *in vitro* of follicle-enclosed oocytes of the goldfish (*Carassius auratus*). *Gamete Research* 10: 21-29.

Lokman, P.M., B. Harris, M. Kusakabe, D.E. Kime, R.W. Schulz, S. Adachi and G. Young. 2002. 11-Oxygenated androgens in female teleosts: Prevalence, abundance, and life history implications. *General and Comparative Endocrinology* 129: 1-12.

Maestro, M.A., J.V. Planas, J. Gutierrez, S. Moriyama and P. Swanson. 1995. Effects of insulin-like growth-factor-I (Igf-I) on steroid-production by isolated ovarian theca and granulosa layers of preovulatory Coho salmon. *Netherlands Journal of Zoology* 45: 143-146.

Maestro, M.A., J.V. Planas, S. Moriyama, J. Gutierrez, J. Planas and P. Swanson. 1997. Ovarian receptors for insulin and insulin-like growth factor I (IGF-I) and effects of IGF-I on steroid production by isolated follicular layers of the preovulatory Coho salmon ovarian follicle. *General and Comparative Endocrinology* 106: 189-201.

Matsubara, T. and K. Sawano. 1995. Proteolytic cleavage of vitellogenin and yolk proteins during vitellogenin uptake and oocyte maturation in barfin flounder (*Verasper moseri*). *Journal of Experimental Zoology* 272: 34-45.

Milla, S., B. Jalabert, H. Rime, P. Prunet and J. Bobe. 2006. Hydration of rainbow trout oocyte during meiotic maturation and in vitro regulation by 17,20β dihydroxy-4-pregnen-3-one and cortisol. *Journal of Experimental Biology* 209: 1147-1156.

Miwa, S., L. Yan and P. Swanson. 1994. Localization of two gonadotropin receptors in the salmon gonad by *in vitro* ligand autoradiography. *Biology of Reproduction* 50: 629-642.

Monget, P., S. Fabre, P. Mulsant, F. Lecerf, J.M. Elsen, S. Mazerbourg, C. Pisselet and D. Monniaux. 2002. Regulation of ovarian folliculogenesis by IGF and BMP system in domestic animals. *Domestic Animal Endocrinology* 23: 139-154.

Mori, T., S. Sumiya and H. Yokota. 2000. Electrostatic interactions of androgens and progesterone derivatives with rainbow trout estrogen receptor. *Journal of Steroid Biochemistry and Molecular Biology* 75: 129-137.

Nabissi, M., L. Soverchia, A.M. Polzonetti-Magni and H.R. Habibi. 2000. Differential splicing of three gonadotropin-releasing hormone transcripts in the ovary of seabream (*Sparus aurata*). *Biology of Reproduction* 62: 1329-1334.

Nagahama, Y. 1994. Molecular biology of oocyte maturation in fish. *Perspectives in Comparative Endocrinology*. pp. 193-198.

Nagahama, Y. and S. Adachi. 1985. Identification of maturation-inducing steroid in a teleost, the amago salmon (*Oncorhynchus rhodurus*). *Developmental Biology* 109: 428-435.

Nagahama, Y., M. Yoshikuni, M. Yamashita and M. Tanaka. 1994. Regulation of oocyte maturation in fish. In: *Fish Physiology, Molecular Endocrinology of Fish*, N.M. Sherwood and C.L. Hew (eds.), Academic Press, San Diego, Vol. 13, pp. 393-439.

Nagahama, Y., M. Yoshikuni, M. Yamashita, T. Tokumoto and Y. Katsu. 1995. Regulation of oocyte growth and maturation in fish. *Current Topics in Developmental Biology* 30: 103-145.

Negatu, Z., S.M. Hsiao and R.A. Wallace. 1998. Effects of insulin-like growth factor-I on final oocyte maturation and steroid production in *Fundulus heteroclitus*. *Fish Physiology and Biochemistry* 19: 13-21.

Nomura, M., K. Sakai and F. Takashima. 1974. The overripening phenomenon of rainbow trout, I. Temporal morphological changes of eggs retained in the body cavity after ovulation. *Bulletin of the Japanese Scientific Fisheries* 40: 977-984.

Onagbesan, O.M., V. Bruggeman, A.P. Vas, K. Tona, J. Williams and E. Decuypere. 2003. BMPs and BMPRs in chicken ovary and the effects of BMP-4 and -7 on granulosa cell proliferation and progesterone production *in vitro*. *American Journal of Endocrinology and Metabolism* 285: 973-983.

Pang, Y. and W. Ge. 1999. Activin stimulation of zebrafish oocyte maturation *in vitro* and its potential role in mediating gonadotropin-induced oocyte maturation. *Biology of Reproduction* 61: 987-992.

Pang, Y. and W. Ge. 2002. Gonadotropin and activin enhance maturational competence of oocytes in the zebrafish (*Danio rerio*). *Biology of Reproduction* 66: 259-265.

Pankhurst, N.W., G.J. Purser, G. Van der Kraak, P.M. Thomas and G.N.R. Forteath. 1996. Effect of holding temperature on ovulation, egg fertility, plasma levels of reproductive hormones and *in vitro* ovarian steroidogenesis in the rainbow trout *Oncorhynchus mykiss*. *Aquaculture* 146: 227-290.

Pati, D. and H.R. Habibi. 1998. Presence of salmon gonadotropin-releasing hormone (GnRH) and compounds with GnRH-like activity in the ovary of goldfish. *Endocrinology* 139: 2015-2024.

Pati, D. and H.R. Habibi. 2002. Involvement of protein kinase C and arachidonic acid pathways in the gonadotropin-releasing hormone regulation of oocyte meiosis and follicular steroidogenesis in the goldfish ovary. *Biology of Reproduction* 66: 813-822.

Patiño, R. and H. Kagawa. 1999. Regulation of gap junctions and oocyte maturational competence by gonadotropin and insulin-like growth factor-I in ovarian follicles of red seabream. *General and Comparative Endocrinology* 115: 454-462.

Patiño, R. and C.V. Sullivan. 2002. Ovarian follicle growth, maturation, and ovulation in teleost fish. *Fish Physiology and Biochemistry* 26: 57-70.

Patiño, R. and P. Thomas. 1990. Characterization of membrane receptor activity for 17 alpha, 20 beta, 21-trihydroxy-4-pregnen-3-one in ovaries of spotted seatrout (*Cynoscion nebulosus*). *General and Comparative Endocrinology* 78: 204-217.

Patiño, R., P. Thomas and G. Yoshizaki. 2003. Ovarian follicle maturation and ovulation: an integrated perspective. *Fish Physiology and Biochemistry* 28: 305-308.

Patiño, R., G. Yoshizaki, P. Thomas and H. Kagawa. 2001. Gonadotropic control of ovarian follicle maturation: The two-stage concept and its mechanisms. *Comparative Biochemistry and Physiology* B129: 427-439.

Pinter, J. and P. Thomas. 1995. Characterization of a progestogen receptor in the ovary of the spotted seatrout, *Cynoscion nebulosus*. *Biology of Reproduction* 52: 667-675.

Pinter, J. and P. Thomas. 1999. Induction of ovulation of mature oocytes by the maturation-inducing steroid 17,20beta, 21-trihydroxy-4-pregnen-3-one in the spotted seatrout. *General and Comparative Endocrinology* 115: 200-209.

Planas, J.V., F.W. Goetz and P. Swanson. 1997. Stimulation of brook trout ovarian steroidogenesis by gonadotropins I and II is mediated by the cyclic adenosine 3',5'-monophosphate/protein kinase A pathway. *Biology of Reproduction* 57: 647-654.

Planas, J.V., J. Athos, F.W. Goetz and P. Swanson. 2000. Regulation of ovarian steroidogenesis *in vitro* by follicle-stimulating hormone and luteinizing hormone during sexual maturation in salmonid fish. *Biology of Reproduction* 62: 1262-1269.

Reddy, P.K., R. Renaud and J.F. Leatherland. 1999. Effects of cortisol and triiodo-L-thyronine on the steroidogenic capacity of rainbow trout ovarian follicles at two stages of oocyte maturation. *Fish Physiology and Biochemistry* 21: 129-140.

Roberts, S.B., D.M. Langenau and F.W. Goetz. 2000. Cloning and characterization of prostaglandin endoperoxide synthase-1 and -2 from the brook trout ovary. *Molecular and Cellular Endocrinology* 160: 89-97.

Robker, R.L., D.L. Russell, L.L. Espey, J.P. Lydon, B.W. O'Malley and J.S. Richards. 2000. Progesterone-regulated genes in the ovulation process: ADAMTS-1 and cathepsin L proteases. *Proceedings of the National Academy of Sciences of the United States of America* 97: 4689-4694.

Sakai, K., M. Nomura and H. Oto. 1975. The over-ripening phenomenon of rainbow trout II, Changes in the percentage of eyed eggs, hatching rate and incidence of abnormal alevins during the process of over-ripening. *Bulletin of the Japanese Scientific Fisheries* 41: 855-860.

Scott, A.P., V.J. Bye and S.M. Baynes. 1980. Seasonal variations in sex steroids of female rainbow trout (*Salmo gairdneri* Richardson). *Journal of Fish Biology* 17: 587-592.

Scott, A.P., J.P. Sumpter and P.A. Hardiman. 1983. Hormone changes during ovulation in the rainbow trout (*Salmo gairdneri* Richardson). *General and Comparative Endocrinology* 49: 128-134.

Senthilkumaran, B., C.C. Sudhakumari, X.T. Chang, T. Kobayashi, Y. Oba, G. Guan, Y. Yoshiura, M. Yoshikuni and Y. Nagahama. 2002. Ovarian carbonyl reductase-like 20beta-hydroxysteroid dehydrogenase shows distinct surge in messenger RNA expression during natural and gonadotropin-induced meiotic maturation in Nile tilapia. *Biology of Reproduction* 67: 1080-1086.

Shimasaki, S., R.J. Zachow, D. Li, H. Kim, S. Iemura, N. Ueno, K. Sampath, R.J. Chang and G.F. Erickson. 1999. A functional bone morphogenetic protein system in the ovary.

Proceedings of the National Academy of Sciences of the United States of America 96: 7282-7287.

Sorbera, L.A., J.F. Asturiano, M. Carrillo, J. Cerda, D.E. Kime and S. Zanuy. 1999. In vitro oocyte maturation in the sea bass: Effects of hCG, pituitary extract and steroids. *Journal of Fish Biology* 55: 9-25.

Springate, J.R.C., N.R. Bromage, J.A.K. Elliott and D.L. Hudson. 1984. The timing of ovulation and stripping and their effects on the rates of fertilization and survival to eying, hatch and swim-up in the rainbow trout (*Salmo gairdneri* R.). *Aquaculture* 43: 313-322.

Stacey, N. 2003. Hormones, pheromones and reproductive behavior. *Fish Physiology and Biochemistry* 28: 229-235.

Stacey, N.E. and S. Pandey. 1975. Effects of indomethacin and prostaglandins on ovulation of goldfish. *Prostaglandins* 9: 597-607.

Stacey, N.E., A.F. Cook and R.E. Peter. 1979. Ovulatory surge of gonadotropin in the goldfish, *Carassius auratus*. *General and Comparative Endocrinology* 37: 246-249.

Sturm, A., N. Bury, L. Dengreville, J. Fagart, G. Flouriot, M.E. Rafestin-Oblin and P. Prunet. 2005. 11-deoxycorticosterone is a potent agonist of the rainbow trout (*Oncorhynchus mykiss*) mineralocorticoid receptor. *Endocrinology* 146: 47-55.

Tada, T., M. Endo, I. Hirono, F. Takashima and T. Aoki. 2002. Differential expression and cellular localization of activin and inhibin mRNA in the rainbow trout ovary and testis. *General and Comparative Endocrinology* 125: 142-149.

Thomas, P. and J. Trant. 1989. Evidence that 17 alpha, 20 beta, 21-trihydroxy-4-pregnen-3-one is a maturation-inducing steroid in spotted. *Fish Physiology and Biochemistry* 7: 185-192.

Thomas, P., J. Pinter and S. Das. 2001. Upregulation of the maturation-inducing steroid membrane receptor in spotted seatrout ovaries by gonadotropin during oocyte maturation and its physiological significance. *Biology of Reproduction* 64: 21-29.

Thomas, P., Y. Zhu and M. Pace. 2002. Progestin membrane receptors involved in the meiotic maturation of teleost oocytes: A review with some new findings. *Steroids* 67: 511-517.

Thomas, P., Y. Pang, Y. Zhu, C. Detweiler and K. Doughty. 2004. Multiple rapid progestin actions and progestin membrane receptor subtypes in fish. *Steroids* 69: 567-573.

Uzbekova, S., J.J. Lareyre, T. Madigou, B. Davail, B. Jalabert and B. Breton. 2002. Expression of prepro-GnRH and GnRH receptor messengers in rainbow trout ovary depends on the stage of ovarian follicular development. *Molecular Reproduction and Development* 62: 47-56.

Vischer, H.F. and J. Bogerd. 2003. Cloning and functional characterization of a gonadal luteinizing hormone receptor complementary DNA from the African catfish (*Clarias gariepinus*). *Biology of Reproduction* 68: 262-271.

Von Schalburg, K.R., C.M. Warby and N.M. Sherwood. 1999. Evidence for gonadotropin-releasing hormone peptides in the ovary and testis of rainbow trout. *Biology of Reproduction* 60: 1338-1344.

Wallace, R.A. and K. Selman. 1978. Oogenesis in *Fundulus heteroclitus*. I. Preliminary observations on oocyte maturation *in vivo* and *in vitro*. *Developmental Biology* 62: 354-369.

Wallace, R. and K. Selman. 1981. Cellular and dynamic aspects of oocyte growth in teleosts. *American Zoologist* 21: 325-343.

Wallace, R.A., M.S. Greeley, Jr. and R. McPherson. 1992. Analytical and experimental studies on the relationship between Na+, K+, and water uptake during volume increases associated with *Fundulus* oocyte maturation *in vitro*. *Journal of Comparative Physiology*, B 162: 241-248.

Wang, Y. and W. Ge. 2002. Cloning of zebrafish ovarian carbonyl reductase-like 20 beta-hydroxysteroid dehydrogenase and characterization of its spatial and temporal expression. *General and Comparative Endocrinology* 127: 209-216.

Wang, Y. and W. Ge. 2003. Spatial expression patterns of activin and its signaling system in the zebrafish ovarian follicle: Evidence for paracrine action of activin on the oocytes. *Biology of Reproduction* 69: 1998-2006.

Wang, Y. and W. Ge. 2004. Cloning of epidermal growth factor (EGF) and EGF receptor from the zebrafish ovary: Evidence for EGF as a potential paracrine factor from the oocyte to regulate activin/follistatin system in the follicle cells. *Biology of Reproduction* 71: 749-760.

Weber, G.M. and C.V. Sullivan. 2000. Effects of insulin-like growth factor-I on *in vitro* final oocyte maturation and ovarian steroidogenesis in striped bass, *Morone saxatilis*. *Biology of Reproduction* 63: 1049-1057.

Weber, G.M. and C.V. Sullivan. 2005. Insulin-like growth factor-I induces oocyte maturational competence but not meiotic resumption in white bass (*Morone chrysops*) follicles *in vitro*: evidence for rapid evolution of insulin-like growth factor action. *Biology of Reproduction*. (In Press).

Wilmot, G.R., P.A. Raymond and B.W. Agranoff. 1993. The expression of the protein p68/70 within the goldfish visual system suggests a role in both regeneration and neurogenesis. *Journal of Neuroscience* 13: 387-401.

Wong, T.T. and Y. Zohar. 2004. Novel expression of gonadotropin subunit genes in oocytes of the gilthead seabream (*Sparus aurata*). *Endocrinology* 145: 5210-5220.

Wu, T., H. Patel, S. Mukai, C. Melino, R. Garg, X. Ni, J. Chang and C. Peng. 2000. Activin, inhibin, and follistatin in zebrafish ovary: Expression and role in oocyte maturation. *Biology of Reproduction* 62: 1585-1592.

Yamashita, M. 2000. Toward modeling of a general mechanism of MPF formation during oocyte maturation in vertebrates. *Zoological Science* 17: 841-851.

Yaron, Z. 1995. Endocrine control of gametogenesis and spawning induction in the carp. *Aquaculture* 129: 49-73.

York, W.S., R. Patino and P. Thomas. 1993. Ultrastructural changes in follicle cell-oocyte associations during development and maturation of the ovarian follicle in Atlantic croaker. *General and Comparative Endocrinology* 92: 402-418.

Yoshizaki, G., R. Patino and P. Thomas. 1994. Connexin messenger ribonucleic acids in the ovary of Atlantic croaker: molecular cloning and characterization, hormonal control, and correlation with appearance of oocyte maturational competence. *Biology of Reproduction* 51: 493-503.

Zhu, L., N. Ohan, Y. Agazie, C. Cummings, S. Farah and X.J. Liu. 1998. Molecular cloning and characterization of *Xenopus* insulin-like growth factor-1 receptor: its role in mediating insulin-induced *Xenopus* oocyte maturation and expression during embryogenesis. *Endocrinology* 139: 949-954.

Zhu, Y., J. Bond and P. Thomas. 2003a. Identification, classification, and partial characterization of genes in humans and other vertebrates homologous to a fish membrane progestin receptor. *Proceedings of the National Academy of Sciences of the United States of America* 100: 2237-2242.

Zhu, Y., C.D. Rice, Y. Pang, M. Pace and P. Thomas. 2003b. Cloning, expression, and characterization of a membrane progestin receptor and evidence it as an intermediary in meiotic maturation of fish oocytes. *Proceedings of the National Academy of Sciences of the United States of America* 100: 2231-2236.

Zohar, Y., B. Breton and A. Fostier. 1986. Short-term profiles of plasma gonadotropin and 17 alpha-hydroxy, 20 beta-dihydroprogesterone levels in the female rainbow trout at the periovulatory period. *General and Comparative Endocrinology* 64: 189-198.

2

Physiological Function of Gonadotropins in Fish

Josep V. Planas[1,*] and Penny Swanson[2]

INTRODUCTION

In all vertebrate species, including fish, reproduction is controlled by pituitary gonadotropins (GTHs). In mammals, it is common knowledge that the two gonadotropic hormones—follicle stimulating hormone (FSH) and luteinizing hormone (LH)—stimulate gonadal function in both males and females. For a very long time, fish were believed to contain only one GTH that was the sole responsible for the stimulation of gametogenesis in both males and females (Fontaine *et al.*, 1987). The purification and identification of two distinct pituitary gonadotropins (GTHs) in fish (Suzuki *et al.*, 1988) forced a revision of the concept of one GTH regulating the entire process of sexual maturation (Swanson, 1991). It is now a well-established fact that fish pituitaries produce two distinct gonadotropic hormones—initially named GTH I and GTH II—that exhibit structural similarities with vertebrate FSH and LH. To date, two

Authors' addresses: [1]Departament de Fisiologia, Facultat de Biologia, Universitat de Barcelona, Av. Diagonal 645, 08028 Barcelona, Spain.
[2] Northwest Fisheries Science Centre, National Marine Fisheries Service, 2725 Montlake Boulevard East, Seattle, Washington 98112, U.S.A. E-mail: Penny.Swanson@noaa.gov
*Corresponding author: E-mail: jplanas@ub.edu

distinct GTHs have been identified and characterized from salmonid (chum and coho salmon: Suzuki *et al.*, 1988; Swanson *et al.*, 1991), perciform (red seabream, bonito, tuna and Mediterranean yellowfish: Koide *et al.*, 1993; Tanaka *et al.*, 1993; Okada *et al.*, 1994; García-Hernández *et al.*, 1997), cypriniform (Van der Kraak *et al.*, 1992) and pleuronectiform (halibut: Weltzien *et al.*, 2003) species and information on their biological activity has been obtained. In addition, the subunits for the two distinct GTHs have been identified by molecular cloning techniques for a number of other fish species. Therefore, the duality of fish GTHs appears to be well-established.

With the availability of purified fish GTHs, enough information has been obtained over the last few years on the biology of GTH I and GTH II to recognize the existence of a significant functional homology between GTH I and FSH and between GTH II and LH. In view of these findings, a change in terminology was recently proposed from GTH I to FSH and from GTH II to LH (Planas *et al.*, 2000). Consequently, fish GTHs are referred to in the existing literature as fish FSH and LH.

GONADOTROPIC REGULATION OF TESTICULAR FUNCTION

In male fish, the testis is the primary target organ for pituitary GTHs. In the testis, GTHs are known to regulate two processes that are the key for male reproduction: the production of sexual steroids (steroidogenesis) and the production of mature spermatozoa. These two processes, as will be examined in detail below, are deeply interdependent, for the generation of male gametes requires that testicular steroids are synthesized in response to pituitary GTHs. Again, it is a well-known fact that testicular steroids produced by somatic cells of the testis (i.e., Leydig cells) in response to GTHs act on neighboring germ cells in a paracrine fashion. Testicular steroids synthesized in response to GTHs are also secreted into the blood and exert typically endocrine effects such as the development of secondary sexual characteristics, induction of sexual behavior and feedback control of pituitary hormone secretion, including GTHs. In this section, we will only address the intratesticular actions of steroids produced in response to GTH stimulation.

Regulation of Testicular Steroidogenesis

In fish, the production of testicular steroids is primarily regulated by GTHs. From the initial descriptions of the plasma levels of sexual steroids in male fish (Baynes *et al.*, 1985), it was hypothesized that changes in sex steroids during sexual maturation were caused by the specific action of GTHs in the steroid-producing cells of the testis. Since then, a number of studies have shown that incubation of testicular tissue with GTH preparations containing predominantly LH (maturational GTH, SGA or SGG-100) or with LH-like activity (hCG) result in the stimulation of steroid production in several fish species (Ueda *et al.*, 1983, 1984; Schulz, 1986; Saad *et al.*, 1987; Miura *et al.*, 1991; Schulz *et al.*, 1994). The testicular production of 11-ketotestosterone (11-KT) and 17α,20β-dihydroxy-4-pregnen-3-one (17,20β-P), the two major steroids produced by male teleosts, can be stimulated by GTH both *in vivo* and *in vitro* and its effects change with the developmental stage of the testis (Sakai *et al.*, 1989; Schulz *et al.*, 1990).

The development of specific RIAs for FSH and LH has allowed us to observe that the plasma levels of FSH and LH display distinct patterns during male sexual maturation. The known correlation in salmonid fish between the plasma levels of testicular steroids (Scott *et al.*, 1982, 1989; Baynes *et al.*, 1985; Fitzpatrick *et al.*, 1986; Planas *et al.*, 1995) and the plasma levels of FSH and LH during sexual maturation (Swanson, 1991) supports the idea of a dual role for FSH and LH in regulating testicular steroidogenesis. During the mid and late stages of spermatogenesis, plasma FSH levels have been shown to increase in direct proportion to the plasma levels of 11-ketotestosterone (11-KT). But, as fish approach spermiation, plasma FSH levels decrease and plasma LH levels increase concomitantly with 17,20β-P.

In accordance with the fact that the fish testis is exposed to changing levels of FSH and LH during the course of sexual maturation, significant changes in the steroidogenic activity of pure preparations of FSH and LH *in vitro* have been demonstrated using testicular tissue at different stages during coho salmon spermatogenesis (Swanson *et al.*, 1989; Planas *et al.*, 1993, 1995). In testes from juvenile fish and during mid-spermatogenesis, coho salmon FSH and LH are equipotent in stimulating the production of 11-KT and 17,20β-P *in vitro*. However, as spermatogenesis proceeds and fish advance towards spermiation, LH increases its potency to stimulate 11-KT and 17,20β-P production, whereas FSH clearly loses its ability to stimulate 17,20β-P but not 11-KT production. The maturational increase in the LH-stimulated 17,20β-P production, previously pointed out by studies using partially purified GTH (Sakai *et al.*, 1989; Schulz *et al.*, 1990), is attributed to the appearance of the LH receptor (GTH-RII) in Leydig cells (Miwa *et al.*, 1994) and to the increase in the plasma levels of LH (Swanson, 1991) during spermiation. Therefore, one of the major actions of LH in the salmonid testis is to stimulate the production of 17,20β-P, which is known to be essential for spermiation and sperm maturation (Nagahama, 2000). On the other hand, in earlier stages of spermatogenesis, during which plasma LH levels are non-detectable and only the FSH receptor is found in the testis (Miwa *et al.*, 1994), FSH, and not LH, is considered to be the physiological stimulator of steroidogenesis in the salmonid testis. Therefore, FSH appears to be the major regulator of the production of testicular steroids (mostly 11-KT) in earlier stages of spermatogenesis in salmonids. Due to the known effects of 11-KT in promoting spermatogenesis in the teleost testis (see above), FSH can be said to be a major (although indirect) regulator of fish spermatogenesis.

The target cell for LH in the fish testis is the Leydig cell, as is indicated by the presence of LH receptors in salmon Leydig cells and also by the ability of salmon GTH (LH-like) to stimulate steroid production in a primary culture of trout Leydig cells (Loir, 1990). However, to date, it is not clear which cell type mediates the steroidogenic effects of FSH during spermatogenesis in teleosts. In mammals, FSH acts on Leydig cells and increases the number of LH receptors, allowing the Leydig cells to respond to LH. In the salmon testis, FSH receptors have only been detected during spermatogenesis in Sertoli cells, which are, in general, not considered to be steroidogenic. However, it is possible that FSH receptors exist in Leydig cells but they have not been detected due to limitations in the sensitivity of the *in vitro* ligand autoradiographical technique. In support of this possibility, *in vitro* studies with early spermatogenic testicular tissue from Japanese eel have shown that hCG (probably acting through a FSH receptor) stimulates both 11-KT production and the morphological activation of Leydig cells, further suggesting that hCG acts on Leydig cells (Miura *et al.*, 1991a, b). Molecular studies on the localization of transcripts for fish FSH and LH receptors in the testis will be necessary in order to resolve this issue.

In the steroidogenic target cells of the testis, GTHs increase the production of steroids by stimulating the activity of key enzymes in the steroidogenic pathway. In testicular tissue from mature rainbow trout, salmon GTH (LH) has been shown to stimulate the production of 11-KT and 17,20β-P—the two major steroids produced by the teleost testis—by stimulating the activity of several steroidogenic enzymes such as P450ssc, 3β-HSD, P45017c and 20β-HSD (Saad *et al.*, 1987). More recently, the expression of 11β-hydroxylase was shown to be stimulated *in vivo* by hCG treatment in Japanese eel testis (Jiang *et al.*, 1996) and localized in Leydig cells (Nagahama, 2000). However, despite the fact that most of all the key testicular steroidogenic enzymes have been cloned in salmonids, there is no information of the specific effects of FSH and LH on the expression of steroidogenic enzymes. Over the last few years, the localization and gonadotropic regulation of 20β-HSD in the salmonid testis has been a matter of controversy. Several studies have demonstrated that spermatozoa can convert 17α-hydroxyprogesterone produced in somatic cells into 17,20β-P due to the presence of 20β-HSD activity and that this conversion cannot be stimulated by GTH. However, testicular cells (most likely Leydig cells) also have 20β-HSD activity but there is some disagreement in the literature regarding whether this enzymatic activity can be stimulated by GTH. On one hand, Sakai *et al.* (1989) report that in spermiating amago salmon testicular fragments, a salmon GTH preparation does not stimulate 17,20β-P production by increasing 20β-HSD activity, which remains constant, but by stimulating the production of the precursor steroid: 17α-hydroxyprogesterone. On the other hand, pure coho salmon LH is more effective in stimulating 17,20β-P than 17α-hydroxy-progesterone production by the coho salmon testis *in vitro* (Planas *et al.*, 1995), indirectly suggesting that 20β-HSD activity is stimulated by LH. Finally, a salmon GTH preparation stimulates the enzymatic conversion of 17α-hydroxyprogesterone to 17,20β-P in the rainbow trout testis *in vitro* (Saad *et al.*, 1987). Therefore, in the mature salmonid testis, there are probably two different cellular sources for 17,20β-P: Leydig cells and mature spermatozoa. It appears that the former may respond to LH but not the latter. The functional significance of two different cellular sites of production of 17,20β-P and with divergent responsiveness to LH remains to be determined.

To date, there is scarce information on the biological activities of FSH and LH in non-salmonid species, which is primarily due to difficulties in the purification of FSH. The testicular steroidogenic activities of purified FSH and LH have been tested in carp (Van der Kraak *et al.*, 1992), seabream (Tanaka *et al.*, 1995) and turbot (Weltzien *et al.*, 2003). In these species, FSH and LH had very similar potency in stimulating the production of testicular androgens. Unfortunately, no studies on the possible changes in testicular sensitivity to the effects of FSH and LH during spermatogenesis have been conducted on non-salmonid species. In Japanese eel, cDNAs for both FSH and LHβ subunits have been identified and the fact that eel FSHβ subunit is expressed in the pituitary of immature animals (Yoshiura *et al.*, 1999) suggests that FSH may be produced and secreted to the blood. Consequently, this further supports the idea that FSH may stimulate spermatogenesis in this teleost. Recently, recombinant eel FSH has been produced and shown to be active in stimulating the production of 11-KT by the immature eel testis (Kamei et al., 2003).

Regulation of Spermatogenesis

Spermatogenesis is a developmental process by which testicular germ cells proliferate, differentiate and mature into haploid spermatozoa. The spermatogenic process consists of four distinct phases: (1) renewal of spermatogonial stem cells; (2) commitment of stem cells to spermatogenesis and proliferation of spermatogonia (mitosis); (3) meiosis; and (4) spermiogenesis (differentiation of spermatids into mature spermatozoa). As in most vertebrates, spermatogenesis in teleost fish is dependent on the action of pituitary GTHs. Classic studies using hypophysectomy and replacement therapy indicate that in most of the species examined, pituitary GTHs are required for the progression of spermatogenesis but that certain stages may be independent of GTH stimulation (reviewed by Billard et al., 1982; Billard, 1986). In addition, GTH treatment induces precocious spermatogenesis in several fish species.

As indicated above, the renewal of spermatogonial stem cells is considered the first as also a necessary stage in spermatogenesis. This mitotic process provides a population of spermatogonial stem cells from which cells can be recruited—upon stimulation by specific factors (see below)—to enter the spermatogenic process and initiate spermatogonial proliferation (mitosis). It is believed that mitotic proliferation of spermatogonial stem cells in the fish testis is not under the control of pituitary GTH (Billard, 1986) but under the control of estrogens (Miura et al., 1999). In immature Japanese eels, estradiol-17β (E_2) has been recently shown to stimulate the renewal of spermatogonial stem cells both in vivo and in vitro by acting on estrogen receptors localized in Sertoli cells (Miura et al., 1999). The cellular site of production of E_2 or its regulation is currently unknown.

During the past decade, knowledge about the stimulation of spermatogonial proliferation and the progression of spermatogenesis by GTHs in teleosts has increased vastly. Most of this information has been derived from studies on the Japanese eel testis, which has proven to be a good experimental model to study the endocrine regulation of spermatogenesis (reviewed by Nagahama, 2000). In the eel, the entire spermatogenic process can be induced both *in vivo* and *in vitro* by hCG (Miura *et al.*, 1991a, b). The major action of hCG in the eel testis is to stimulate Leydig cells to produce 11-KT, which triggers the initiation of spermatogenesis (Miura *et al.*, 1991c). In addition, hCG stimulates the appearance of androgen receptors on Sertoli cells (Ikeuchi *et al.*, 2001). Therefore, the model that exists today is that 11-KT, once produced in response to hCG, binds to the androgen receptors in Sertoli cells and stimulates the production of activin B, which stimulates the proliferation of spermatogonial cells. An important question that remains to be answered is the nature of the endogenous GTH produced by the eel pituitary that stimulates the production of 11-KT by Leydig cells. In Japanese eel, cDNAs for both FSH and LH β subunits have been identified. The very fact that eel FSHβ subunit is expressed in the pituitary of immature animals (Yoshiura *et al.*, 1999) suggests that FSH may be produced and secreted to the blood and, consequently, supports the idea that FSH may stimulate spermatogenesis in this teleost. Recently, recombinant eel FSH has been produced and shown to be active in stimulating the production of 11-KT by the immature eel testis (Yoshiura *et al.*, 1999).

In other teleost species, there is also information suggesting that one of the major functions of FSH in males is the stimulation of spermatogenesis. In rainbow trout, FSH has been shown to directly stimulate the proliferation of spermatogonia in a primary culture of testicular (somatic and germ) cells (Loir, 1999b). However, no effects of addition of 11-KT or human activin B on the proliferation of spermatogonia were found (Loir, 1999b), although the production of 11-KT in response to FSH was not measured in the study. For this reason, the mechanism by which FSH stimulates the proliferation of spermatogonia in trout testicular cells is not clear at this point. In other salmonid species such as Japanese huchen (Amer *et al.*, 2001) and Pacific salmon (Campbell *et al.*, 2003), circumstantial evidence suggests that FSH could be stimulating spermatogenesis. In salmonid fish, strikingly similar endocrine changes take place during early stages of spermatogenesis: the plasma levels of 11-KT in maturing fish increase concomitantly with the appearance of late-type B spermatogonia, which is an indication of the beginning of spermatogenesis (Loir, 1999a). In Pacific salmon, these changes are preceded by significant increases in the plasma levels of FSH (Campbell *et al.*, 2003). In addition, the spermatogonial proliferation *in vitro* in testicular tissue from the Japanese huchen has been shown to be stimulated by a crude GTH preparation as well as by 11-KT (Amer *et al.*,

2001). In view of the known ability of FSH to stimulate 11-KT production (Planas *et al.*, 1993, 1995) and of the presence of testicular FSH receptors in salmon during early spermatogenesis (Miwa *et al.*, 1994), it is reasonable to suggest that FSH plays an important role in the initiation of spermatogenesis in salmonid fish (Fig. 2.1). Similarly, treatment with 11-KT stimulates spermatogenesis *in vivo* in the African catfish (Cavaco *et al.*, 1998), suggests that in this species, 11-KT could be mediating the effects of putative catfish FSH.

In addition to 11-KT, other factors may also participate in the progression of spermatogenesis. For example, the insulin-like growth factor 1 (IGF-I) stimulates the proliferation of spermatogonia in a primary culture of testicular cells from rainbow trout (Loir, 1999b) and acts in conjunction with 11-KT in stimulating the progression of spermatogenesis in the Japanese eel testis (Nader *et al.*, 1999). Both IGF-I mRNA and IGF-I receptors have been detected in testicular cells from rainbow trout (Le Gac *et al.*, 1996), indicating that the stimulatory effects of IGF-I on spermatogenesis may take place via a paracrine fashion, although an endocrine action of IGF-I cannot be excluded. In this regard, it will be important to determine if FSH—either directly or indirectly through the

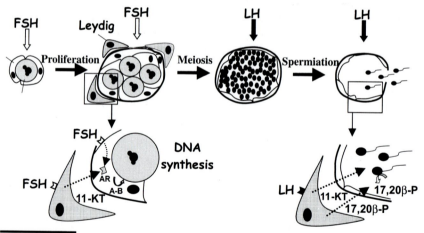

Fig. 2.1   Schematic representation of the major actions of FSH and LH in the fish testis. FSH stimulates spermatogonial proliferation indirectly by stimulating the production of 11-ketotestosterone (11-KT) in Leydig cells. FSH also acts on Sertoli cells inducing the expression of androgen receptors (AR) which will mediate the effects of 11-KT on activin B (A-B) production. Subsequently, A-B directly stimulates the proliferation of spermatogonia. During spermiation, LH acts on Leydig cells stimulating the production of 11-KT and $17\alpha,20\beta$-dihydroxy-4-pregnen-3-one ($17,20\beta$-P). Spermatozoa can also produce $17,20\beta$-P but it is not regulated by LH.

production of 11-KT—is involved in the regulation of the activation of the IGF system (peptides and receptors) in the teleost testis. Furthermore, the mechanism by which IGF-I stimulates the progression of spermatogenesis also needs to be investigated. Interestingly, 17,20β-P, a steroid produced in response to FSH by the salmon testis during spermatogenesis (Planas *et al.*, 1993, 1995), has been shown to stimulate the proliferation of spermatogonia in the Japanese huchen with even higher potency than 11-KT (Amer *et al.*, 2001). In support for a possible role in early spermatogenesis for FSH-induced 17,20β-P, the expression of the 17,20β-P receptor in the eel testis increases at days 3 and 6 after a single hCG injection (Todo *et al.*, 2000), coinciding with the time when spermatogonial proliferation has been observed (Miura *et al.*, 1991). In addition, the fibroblast growth factor 2 (FGF-2) also has weak proliferative effects in rainbow trout testicular cells (Loir, 1999b).

It is well known that in the lobular-type testis of most teleosts, the development of testicular germ cells is intimately associated with the presence of Sertoli cells. In the immature testis, Sertoli cells surround individual spermatogonia. Furthermore, as these proliferate within cysts, so do Sertoli cells proliferate to surround the entire cyst. It has been suggested that because of the constant developmental changes that a particular cyst or spermatocyst undergoes, the surrounding Sertoli cells must renew and proliferate at a high rate (Billard, 1986). The presence of FSH receptors in salmon Sertoli cells (Miwa *et al.*, 1994) is suggestive of the fact that FSH could stimulate the proliferation of Sertoli cells. Further research is clearly needed to describe the endocrine regulation of Sertoli cell function in teleosts.

At the completion of spermatogenesis, several important processes leading to the release of spermatozoa into the sperm duct (spermiation) and the acquisition of the capacity for sperm motility (maturation) appear to be regulated by pituitary GTHs. In a number of teleost species, *in vivo* treatment with GTH preparations or with GnRH analogs results in the stimulation of milt production (see Pankhurst, 1994). In addition, during natural spermiation, a significant increase in the plasma levels of LH has been observed in parallel to increases in 17,20β-P and 11-KT in a number of teleost species (Fitzpatrick *et al.*, 1986; Swanson *et al.*, 1991). It is now known that at this stage in testicular development, there is a significant increase in the sensitivity of the salmon testis to the *in vitro* effects of LH on 17,20β-P and 11-KT, coupled with the appearance of specific LH receptors (Miwa *et al.*, 1994; Planas *et al.*, 1995). Therefore, it has been suggested that 17,20β-P and 11-KT, produced in response to LH stimulation, may have an important role during spermiation. Miura *et al.* (1992) have demonstrated that 17,20β-P mediates the LH effects by causing increases in milt volume and the pH of the sperm duct, which in turn stimulates sperm motility by increasing the intracellular concentration of cAMP in mature spermatozoa. On the other hand, the role of 11-KT during spermiation in teleosts is not well established. Despite the lack of conclusive data, it has been suggested that 11-KT could be involved in regulating the rupture of cysts and release of spermatozoa into the sperm duct (Baynes *et al.*, 1985). In addition to steroid-mediated effects, LH may have direct effects in male reproductive tissues during spermiation. In trout, GTH (probably LH) acts directly on the sperm duct epithelium, modulating the ionic composition of the seminal plasma (Marshall *et al.*, 1993).

GONADOTROPIC REGULATION OF OVARIAN FUNCTION

In female fish, pituitary GTHs are the major regulators of ovarian function. GTHs are known to have a broad range of activities, including the regulation of steroid production by follicular cells, stimulation of growth and development of germ cells and ovulation. It has been classically thought that most of the ovarian actions of GTHs are mediated by steroids produced in response to GTH stimulation. However, now it is increasingly evident that there certain actions of GTHs in the ovary are independent of steroid production. Therefore, GTHs regulate all the aspects of ovarian development and function by a series of steroid-dependent as well as steroid-independent actions.

Regulation of Ovarian Steroidogenesis

The working model of the follicular production of steroids in the fish ovary is based on the two-cell type model proposed by Nagahama and collaborators 20 years ago, working on salmonid species. According to this model, which appears invalid for *Fundulus* and medaka, the coordinated steroidogenic activity of the two follicular layers—theca and granulosa—is orchestrated by pituitary GTH (Nagahama *et al.*, 1995). In theca layers, GTH stimulates the production of testosterone and 17α-hydroxyprogesterone (17-OHP), which are converted to E_2 and 17,20β-P in granulosa layers, also in response to GTH, during vitellogenesis and oocyte maturation, respectively (Nagahama *et al.*, 1995). As a result of the high correlation between the plasma levels of FSH and LH and of E_2

and 17,20β-P, respectively, it was initially proposed that FSH would be responsible for the follicular production of E_2 and LH for that of 17,20β-P (Swanson, 1991). With the availability of purified GTHs, the relative activities of FSH and LH on steroid production by follicular layers have been investigated. Initial *in vitro* studies reported similar steroidogenic activities for physiologically relevant concentrations of FSH and LH in ovarian follicles from juvenile coho salmon (Swanson *et al.*, 1989) and vitellogenic amago salmon (Suzuki *et al.*, 1988), tuna (Okada *et al.*, 1994), carp (Van der Kraak *et al.*, 1992), red seabream (Tanaka *et al.*, 1995) and brook trout (Planas *et al.*, 1997). In salmonids, similar steroidogenic activity of FSH and LH in stimulating the production of E_2 in vitellogenic follicles is attributed to the presence of the type I GTH receptor in both theca and granulosa layers, which bind FSH and LH with similar affinities (Yan *et al.*, 1992; Miwa *et al.*, 1994) despite FSH being the only physiologically relevant GTH during the ovarian growth phase. In contrast, in pre-ovulatory follicles, the steroidogenic activity of LH is significantly higher than FSH (Suzuki *et al.*, 1988; Van der Kraak *et al.*, 1992; Planas *et al.*, 1997) and is attributed to the appearance of the type II GTH receptor (specific for LH) during oocyte maturation (Yan *et al.*, 1992; Miwa *et al.*, 1994).

The differences in the steroidogenic activity of LH and FSH are also associated with the known changes in the ability of the ovary to synthesize E_2 and 17,20β-P during the transition from oocyte growth to final oocyte maturation (Nagahama *et al.*, 1995). During this transition, a shift from the production of E_2 to the production of 17,20β-P has been demonstrated in the ovary of a few salmonid (Nagahama *et al.*, 1995; Senthilkumaran *et al.*, 2004) and non-salmonid species such as Japanese yellowtail and red seabream (Ohta *et al.*, 2002). The steroidogenic shift in the ovary dictates the different patterns of E_2 and 17,20β-P plasma levels observed during female reproduction and, ultimately, controls the processes of oocyte growth and maturation (see below). However, because of the small number of teleost species examined to date, it is not known if the ovarian steroidogenic shift is a general mechanism in teleost fish. It has been hypothesized that the steroidogenic shift in the fish ovary is regulated by pituitary GTHs (Nagahama *et al.*, 1995). However, this hypothesis has only been tested in salmonid fish, where a comparison among plasma levels, biological activities and receptor levels of FSH and LH can be made. It is now known that the steroidogenic shift that takes place prior to oocyte maturation is associated with the increase in the plasma levels of LH, the appearance of receptors specific for LH in granulosa cells and the increase in the activity of LH in stimulating the production of 17,20β-P (Swanson, 1991; Yan et al., 1992; Miwa et al., 1994; Planas et al., 2000). The appearance of type II GTH receptor in granulosa cells is considered to be a critical regulatory point in the shift because it may mediate two LH-specific actions associated with oocyte maturation. First, LH inhibits the ovarian production of E_2 prior to oocyte maturation by directly inhibiting E_2 production by isolated granulosa layers (Suzuki et al., 1988; Maestro et al., 1997; Planas et al., 2000) and by inhibiting aromatase activity (Sire et al., 1981; Fostier, 1995). Second, LH, but not FSH, stimulates the production of 17,20β-P by isolated granulosa layers prior to oocyte maturation by stimulating the activity of 20β-HSD, as shown by the conversion of 17-OHP to 17,20β-P (Suzuki et al., 1988; Maestro et al., 1997; Planas et al., 2000). Interestingly, ovarian 20β-HSD also appears to be under gonadotropic control in non-salmonid species. In Japanese eel, administration of a salmon pituitary extract increased ovarian 20β-HSD activity (Kazeto et al., 2001) and in the Nile tilapia, hCG increased 20β-HSD mRNA levels in vitro (Senthilkumaran et al., 2002). Given the nature and the timing of the actions of LH during oocyte maturation, it is very important to identify the factor(s) that regulate the expression of the type II GTH receptor in the ovary.

One important question that has remained unanswered for a long time is whether the gonadotropic regulation of E_2 production in fish involves the stimulation of aromatase activity. Initial studies on isolated granulosa layers from amago salmon found no effect of a partially purified GTH preparation on the conversion of testosterone into E_2, used as an indirect measure of aromatase activity (Young et al., 1983). These results questioned whether aromatase activity is regulated by GTH in the salmonid ovary (Nagahama et al., 1995). Nevertheless, in other teleost species such as medaka and goldfish, it has been shown that partially purified GTH preparations or cyclic adenosine 3',5'-monophosphate (cAMP)-generating agents stimulate the conversion of testosterone into E_2 (Kagawa et al., 1984; Tan et al., 1986; Nagahama et al., 1991). A recent study evaluating the effects of salmon FSH on the direct activation of aromatase activity has shown that FSH stimulates E_2 production, aromatase activity and P450 aromatase mRNA levels in brown trout vitellogenic follicles in vitro (Montserrat et al., 2004). It is possible that these different results on the gonadotropic regulation of aromatase activity are the result of methodological or species-specific factors. However, one cannot rule out that the lack of stimulatory effects of GTH on aromatase activity in isolated salmon granulosa layers (Young et al., 1983) may be due to the absence of factors present in the intact follicle and required for GTH to stimulate aromatase activity. At this point, despite the small number of species examined, it is suggested that the stimulatory effects of FSH on ovarian aromatase activity may be a general mechanism in fish.

Regulation of Oogenesis

In fish, oocyte development proceeds through four major stages: (1) oogonial proliferation; (2) primary oocyte growth; (3) vitellogenesis; and (4) oocyte maturation (Fig. 2.2). Once the oocytes have undergone hormone-induced maturation, a series of events lead to the rupture of the follicular enclosure and the release of the fertilizable oocyte, a process

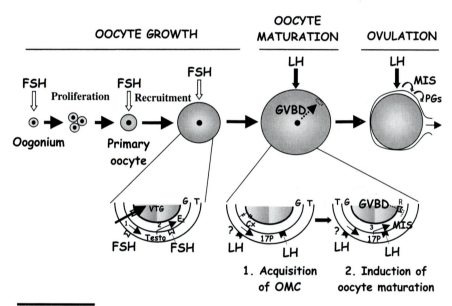

Fig. 2.2 Schematic representation of the major actions of FSH and LH in the fish ovary. During vitellogenesis, FSH increases the number of vitellogenin (VTG) receptors in the oocyte and stimulates the expression and activity of various steroidogenic enzymes (1) in theca layers (T) that lead to the production of testosterone (Testo), which is converted to estradiol-17β (E_2) in granulosa layers (G) by activation of aromatase (2). Later on, LH is believed to stimulate the acquisition of oocyte maturational competence (OMC) by stimulating the synthesis of 20β-HSD (3), the membrane maturation-inducing steroid (MIS) receptor (R) and connexins (Cx). Subsequently, LH stimulates the activity of 20β-HSD in granulosa cells which convert 17α-hydroxyprogesterone (17-OHP) into MIS, which, in turn, stimulates GVBD through its membrane receptor. After oocyte maturation, LH-induced MIS stimulates the production of prostaglandins (PGs), which trigger the release of the mature oocyte.

known as ovulation. In the ovary, oogenesis is initiated by the mitotic proliferation of oogonia and takes place throughout the life of the female in most iteroparous teleost species (Selman *et al.*, 1989). Oogonia differentiate into primary oocytes and a predetermined number (depending on the species) of primary oocytes are recruited into the primary growth phase. Simultaneously, with the initiation of the primary growth phase, primary oocytes start the first meiotic division and undergo the first stages of prophase but become arrested in the diplotene stage. This occurs precisely during meiotic arrest when most of the growth of the oocyte is accomplished by the accumulation of yolk proteins, primarily vitellogenin. Once vitellogenesis is completed the oocytes, upon appropriate hormonal stimulation, resume and complete the first meiotic division and initiate the second meiotic division, i.e., the oocytes undergo final maturation. In mature oocytes, the second meiosis is again arrested at metaphase until fertilization.

It has long been recognized that oocyte growth and development are both under gonadotropic control. However, most of the information available is on the gonadotropic control of oocyte maturation and very little data is available on the role of gonadotropic hormones during other developmental stages such as oogonial proliferation and oocyte growth. This lag in our knowledge on the regulation of ovarian function in fish is due to the slow progress in the purification of pituitary FSH that, in most teleosts, is the physiological GTH during the above-mentioned stages. Independently of the stage in which the oocyte is found, the gonadotropic control of oocyte growth and development does not appear to be the result of a direct action on the oocyte; rather, it is due to an indirect action through the surrounding follicular cells. For this reason, it is important to point out that oocyte growth and development take place in the context of developing somatic follicular cells (theca and granulosa) surrounding the oocytes. In fact, most of the known effects of GTHs on oogenesis are mediated through their effects on the production of steroids by the follicular cells (see above), which contain FSH and LH receptors (Yan *et al.*, 1992; Miwa *et al.*, 1994). In goldfish vitellogenic ovarian follicles, hCG has been shown to stimulate DNA synthesis presumably through the production of growth factors (Srivastava *et al.*, 1995). Unfortunately, the possible mediatory role of steroids in GTH-induced proliferation of follicular cells in fish has not been addressed to date.

Regulation of oocyte growth

To date, the precise role of pituitary GTHs on the regulation of oocyte growth in fish still remains largely unknown. From studies in salmonid species—in which FSH and LH plasma levels have been measured throughout the reproductive cycle—it is fairly clear that FSH is the only GTH that is positively correlated with the period of oocyte growth (Swanson, 1991; Oppen-Berntsen et al., 1994; Prat et al., 1996; Santos et al., 2001). Based on this information, it has been proposed that FSH is the GTH responsible for the regulation of oocyte growth. However, most of the current evidence for the role of FSH on oocyte growth in fish is indirect and derives from studies describing the effects of unilateral ovariectomy on the growth of the remaining ovary. In rainbow trout subjected to unilateral ovariectomy, Tyler et al. (1996) have shown that a second population of primary oocytes is recruited at the beginning of vitellogenesis to enter the growth phase. It has been suggested that FSH could be the hormone responsible for the recruitment of primary oocytes because plasma FSH levels increase after unilateral ovariectomy (Tyler et al., 1997). In naturally spawning rainbow trout, the increase in plasma FSH levels at ovulation has been linked to a possible role of FSH in the initiation of the following ovarian growth cycle (Prat et al., 1996). Similar is the case in Atlantic salmon, another iteroparous salmonid, where relatively high FSH levels have also been observed in several fish undergoing ovulation (Oppen-Berntsen et al., 1994). However, the plasma levels of FSH are low during ovulation in the semelparous coho salmon (Swanson, 1991). Furthermore, reduction of the plasma levels of FSH by passive immunoneutralization in vitellogenic rainbow trout has been reported to cause a decrease in oocyte volume, suggesting a role for FSH in oocyte growth (Santos et al., 2000). A possible hypothesis is that prior to the stimulation of recruitment of primary oocytes into the growth phase, FSH may have stimulated proliferation of the oogonias. It is known that, at least in iteroparous species, oogonial proliferation may take place during or right after ovulation, as shown by its induction by unilateral ovariectomy (Tyler et al., 1996). Clearly, further research is needed to identify the factors that regulate oogonial proliferation which will be essential to understand how fecundity is determined in fish.

It is well recognized that FSH, through its stimulation of the production of E_2 by follicular layers, is indirectly regulating two important processes of oocyte growth: the formation of the eggshell or vitelline envelope, and the accumulation of yolk proteins. It is known that E_2 produced in response to FSH, directly stimulates the synthesis of eggshell proteins in the liver in several fish species (reviewed in Hyllner et al., 1995). In Atlantic salmon, the plasma levels of FSH, E_2 and eggshell proteins show a high degree of correlation (Oppen-Berntsen et al., 1994). Interestingly, synthesis of eggshell proteins and their deposition in the oocyte are initiated during the primary oocyte growth phase and continue during vitellogenesis till such time as the oocyte matures. Similar to eggshell proteins, E_2 produced in response to FSH stimulates the hepatic production of vitellogenin (VTG), which is the major yolk protein in a number of teleost species. Furthermore, FSH specifically stimulates the uptake of VTG in rainbow trout ovarian follicles (Tyler et al., 1991). Presently, this constitutes the only direct demonstration of an effect of FSH in oocyte growth. In fish, VTG is sequestered by the oocyte by receptor mediated endocytosis and VTG receptors have been identified and characterized in several species (Tyler et al., 2000). Since the number and expression of VTG receptors increase during vitellogenesis (Tyler et al., 2000) parallel to the reported changes of FSH (Swanson, 1991; Oppen-Berntsen et al., 1994; Prat et al., 1996; Santos et al., 2001), it is possible that FSH stimulates the appearance of VTG receptors in the oocyte surface. However, there is no information to date on the specific effects of FSH on VTG receptor number or expression in fish ovaries.

During the process of oocyte growth in fish, another important aspect of ovarian function that is also possibly under gonadotropic control is follicular atresia. Unfortunately, very few studies have investigated the hormonal regulation of follicular atresia in fish. In rainbow trout, partially purified GTH and E_2 have been reported to inhibit apoptosis in vitellogenic ovarian follicles (Wood et al., 2002). Therefore, it is possible that FSH, either directly or indirectly through the production of E_2, may act as a survival factor during oogenesis.

Regulation of oocyte maturation

Oocyte maturation is characterized by a series of events that take place both in the nucleus and cytoplasm of the oocyte and which are necessary for the development of a fertilizable egg (Nagahama, 1987; Nagahama et al., 1995). These events begin with the migration of the oocyte nucleus or germinal vesicle (GV) to the animal pole and continue with the breakdown of the GV (GVBD), condensation of the chromosomes and extrusion of the first polar body. In fish, these processes are now known to be specifically stimulated in vivo by LH (Kagawa et al., 1998; Planas et al., 2000), which is found in high levels in the blood during oocyte maturation in all the teleosts examined. Not only is there a high correlation between the plasma levels of LH and the timing of initiation of oocyte maturation, but administration of GTH preparations has been shown to stimulate oocyte maturation in several species. Furthermore, a number of studies have shown that GTH preparations (partially purified or pituitary extracts) can stimulate oocyte maturation *in vitro* within intact ovarian follicles. More recently, with the availability of purified FSH and LH, the maturational activities of FSH and LH have been tested. In ovarian follicles from red seabream (Kagawa *et al.*, 1998) and coho salmon (Planas *et al.*, 2000), LH has been shown to induce GVBD *in vitro*, whereas FSH is completely ineffective. In contrast, carp FSH and LH were reported to have the same stimulatory effects on GVBD in goldfish ovarian follicles (Van der Kraak *et al.*, 1992), which may be suggestive of species-specific actions of fish GTHs on oocyte maturation. Therefore, LH appears to be the sole regulator of oocyte maturation in most teleost species.

It is well known that the stimulatory effects of LH on oocyte maturation are not direct but mediated by the maturation-inducing steroid (MIS), which is known to be produced by the follicular layers in response to LH (see above). An important aspect in the regulation of oocyte maturation is that the MIS (17,20β-P or 20β-S, depending on the species) acts directly on the oocyte to stimulate maturation by interacting with a membrane receptor (Nagahama *et al.*, 1995; Thomas *et al.*, 2004). Therefore, the working hypothesis regarding the regulatory action of LH on oocyte maturation has been that LH induces oocyte maturation because of its stimulation of MIS production. However, over the last few years, a substantial amount of information on several teleost species has accumulated, suggesting that the induction of oocyte maturation by LH may not depend exclusively on its ability to stimulate MIS production (Patiño *et al.*, 2001). From a number of *in vivo* and *in vitro* studies, it has been suggested that the effects of LH on the ovary may consist of an initial 'priming' effect and a subsequent 'maturational effect. The former would prepare the follicle for the maturational effect of MIS, i.e., for MIS to induce GVBD. It has been shown that the priming effect of LH is independent of steroid synthesis, in the sense that it requires protein and RNA synthesis and that it does not render the oocyte responsive to MIS. Furthermore, priming of the ovarian follicles with LH is necessary for MIS to induce GVBD. Therefore, it has been suggested that the priming effect of LH causes the acquisition of oocyte maturational competence (OMC). A two-stage model of the gonadotropic regulation of ovarian maturation has been formulated in which LH, during the first stage, would induce the acquisition of OMC and, during the second stage, would stimulate the production of MIS and its induction of GVBD (Patiño *et al.*, 2001). It is worth noting that the above-mentioned model has not yet been confirmed in salmonid fish, which could be related to the fact that full-grown ovarian follicles are already mature and competent. It is hoped that future studies will address the issue of acquisition of maturational competence in salmonid fish.

The protein and RNA synthesis requirement of the priming effects of LH suggests that LH may be stimulating the *de novo* synthesis of key proteins involved in acquisition of OMC. It is likely that one of these proteins is the enzyme 20β-HSD, since it is known that GTH (most likely LH) increases its *de novo* synthesis through a mechanism requiring protein and RNA synthesis (Nagahama, 1997). In post-vitellogenic immature tilapia follicles, hCG has been shown to increase 20β-HSD mRNA levels (Senthilkumaran *et al.*, 2002). Recent studies have pointed out three other possible candidate proteins responsible for OMC. One of these proteins is the oocyte receptor for MIS (MIS-R), which has been shown to be upregulated *in vivo* during oocyte maturation in several species (King *et al.*, 1997; Thomas *et al.*, 2001; Rahman *et al.*, 2002) and also in response to hCG *in vitro* in the spotted seatrout, in a process requiring new RNA and protein synthesis (Thomas *et al.*, 2001). A recent report on the cloning of a putative membrane receptor for MIS in spotted seatrout oocytes identifies the long-sought MIS receptor as a typical G protein-coupled receptor (Zhu *et al.*, 2003). This receptor is shown to be upregulated both at the protein and mRNA levels during oocyte maturation, coinciding with changes in receptor-binding activity, and its expression can be induced *in vitro* by hCG but also by 20β-S, the MIS for the spotted seatrout (Zhu *et al.*, 2003). These important findings support the hypothesis that LH stimulates acquisition of OMC by stimulating the expression of a putative MIS receptor. It will be important for future studies to confirm the presence of the MIS receptor in other species and demonstrate its regulation by endogenous LH. Another candidate protein responsible for OMC is connexin, which is the basic component of gap junctions (GJ). Homologous and heterologous GJ are intercellular channels that allow exchange of small substances between two granulosa cells and one granulosa cell and the oocyte, respectively. The number of homologous and heterologous GJ contacts in the ovary has been shown to increase during hCG-induced OMC acquisition (reviewed by Patiño *et al.*, 2001). Moreover, hCG has been shown to increase the mRNA levels of Cx32.2 in Atlantic croaker (Yoshizaki *et al.*, 1994) and of Cx31.5 in red seabream (Choi *et al.*, 2000) during OMC acquisition. Presently, the functional significance of increased levels of GJ and acquisition of OMC is not known. Finally, recent studies have suggested that activin could be involved in both the acquisition of OMC based on the ability of hCG to stimulate its expression (Pang *et al.*, 2002a) and on the ability of activin to induce GVBD in zebrafish ovarian follicles in a process dependent on new protein and RNA synthesis (Pang *et al.*, 1999, 2002b). Unfortunately, there is no information regarding the possible targets of activin; however, it is tempting to speculate that activin, in response to GTH stimulation, could act by increasing the expression of MIS-R.

Regulation of Ovulation

In the fish ovary, the release of a mature oocyte, a process known as ovulation, is known to be under gonadotropic control. It is believed that the physiological function of the pre-ovulatory elevation of plasma LH levels is to stimulate both oocyte maturation and ovulation. The known stimulatory effects of GTH preparations on ovulation appear to be indirect, primarily through their stimulation of 17,20β-P production by the follicular cells. Stimulation of *in vitro* ovulation by hCG in spotted sea-trout is dependent on steroid synthesis (Pinter *et al.*, 1999). A direct effect of 17,20β-P on ovulation *in vitro* has been demonstrated in yellow perch and spotted seatrout (Goetz *et al.*, 1979; Pinter *et al.*, 1999). Therefore, at least in yellow perch and spotted seatrout, LH induction of ovulation appears to be mediated by 17,20β-P. In contrast to the maturational effects of 17,20β-P mediated by a membrane MIS receptor, it has been suggested that 17,20β-P exerts its effects on ovulation through its binding to a nuclear MIS receptor (Pinter *et al.*, 1999). It is interesting to note that 17,20β-P exerts its maturational role as MIS by binding to a membrane receptor in the oocyte (Zhu *et al.*, 2003); whereas, it exerts its ovulation-inducing role by binding to a nuclear receptor most likely located in the follicular layers.

In yellow perch, the stimulatory effects of 17,20β-P on ovulation appear to be mediated by the follicular production of prostaglandin F (Goetz, 1997). In addition, 17,20β-P has also been shown to upregulate the expression of several proteins involved in ovulation (Langenau *et al.*, 1999). Therefore, several potential mediators of the effects of LH on ovulation have been identified in fish. However, it is important to point out the existence of differences in the involvement of key mediators of ovulation (e.g., prostaglandins) among various fish species (Goetz *et al.*, 1991). Whether these differences reflect different patterns of response to the pre-ovulatory effects of LH remains to be determined.

MECHANISM OF ACTION OF FISH GONADOTROPINS

As in other vertebrate groups, the effects of FSH and LH in fish gonadal cells are mediated by the second messenger cAMP. In a wide variety of fish species, increases in the intracellular concentration of cAMP caused either by the addition of exogenous cyclic nucleotides, phosphodiesterase inhibitors or activation of adenylyl cyclase by forskolin have been shown to mimic the effects of GTHs on steroid production (Bogomolnaya *et al.*, 1984; Salmon *et al.*, 1985; Tan *et al.*, 1986; Nagahama *et al.*, 1991) as well as other GTH-dependent process, such as expression of ovarian connexin Cx32.2 (Chang *et al.*, 1999). Furthermore, the production of cAMP in fish gonads is known to be stimulated by GTHs. Both FSH and LH stimulate the *in vitro* production of cAMP in the salmonid testis and ovary (Planas *et al.*, 1993, 1997), confirming early reports on the stimulation of cAMP production by partially purified GTH preparations in fish gonads (Fontaine *et al.*, 1972; Idler *et al.*, 1975). In the salmon ovary, the intracellular concentration of cAMP in theca and granulosa layers increases in response to a partially purified GTH preparation (Kanamori *et al.*, 1988). However, there is very little information on the modulation of cAMP levels by FSH and LH in the different cellular targets of the fish ovary. In isolated theca layers from pre-ovulatory brown trout, salmon LH stimulates the production of cAMP (Méndez *et al.*, 2003) and this effect is correlated with the stimulatory effects of LH and dibutyryl cAMP and forskolin on testosterone production. The gonadotropic stimulation of cAMP production in fish ovary suggests that GTH receptors are coupled to adenylyl cyclase, the enzyme responsible for the production of cAMP. In granulosa cells from amago salmon, the activity of adenylyl cyclase has been shown to be stimulated by a partially purified GTH preparation and by stimulatory and inhibitory G-proteins (Gs and Gi, respectively) (Mita *et al.*, 1994). In the testis, FSH and LH also stimulate the production of cAMP *in vitro* (Planas *et al.*, 1993) but the exact site of production of cAMP is not known.

One of the major functions of cAMP is to stimulate the activity of the cAMP-dependent protein kinase (PKA), which is known to mediate most of the effects of GTH. Several studies have indirectly shown the implication of PKA in mediating the actions of GTH in the fish gonads by using inhibitors of PKA or antagonistic analogues of cAMP. For example, inhibition of PKA has been shown to block the steroidogenic effects of FSH and LH in the coho salmon testis (Planas *et al.*, 1993) and ovary (Planas *et al.*, 1997) and the effects of a pituitary extract on *in vitro* steroid production by the *Fundulus* ovary (Cerda *et al.*, 1997). Inhibition of PKA has also been shown to block other GTH-dependent processes such as the expression of connexin Cx32.2 in the Atlantic croaker ovary (Chang *et al.*, 1999). Despite the well-known importance of PKA in mediating the action of GTHs in fish gonads, the characteristics and regulation of PKA have not been studied until very recently. In fish, the biochemical properties of PKA appear to be similar to those described in mammals. In pre-ovulatory trout theca cells, PKA has both kinase and cAMP-binding activities and its kinase activity is stimulated by salmon LH (Mendez *et al.*, 2003). Since LH stimulates the production of cAMP in fish gonadal cells, it has been suggested that cAMP binds the regulatory (R) subunits of PKA and causes the activation of the kinase activity probably by dissociating the catalytic (C) subunits from the holoenzyme. The latter is supported by the observation that LH causes a decrease in the half-life of the C subunits in trout theca layers, probably as a result of increased degradation of the active C subunits, released from the protective and inhibitory influence of the R subunits (Mendez *et al.*, 2003). Although the downstream targets of the activated C subunits in fish gonadal cells are not known, it is likely that LH-activated C subunits regulate the expression of genes that contain cAMP-response elements (CRE) in their promoter regions. To date, CREs have been identified in the regulatory regions of several GTH-regulated genes in fish: P450 aromatase (Yoshiura *et al.*, 2003), 20β-HSD (Senthilkumaran *et al.*, 2004) and connexin Cx32.2 (Yoshizaki *et al.*, 2000). Additional potential target genes for PKA are other steroidogenic genes known to be under GTH control, such as StAR, P450scc, 3β-HSD and P45017α (Nagahama, 1997). Of these genes, 3β-HSD is the only one that has been shown to increase its expression in response to cAMP in the fish gonads (Young *et al.*, 2000). In addition to regulating the expression of known GTH-responsive genes, LH can also regulate the level of PKA subunits. In trout theca cells, LH has been shown to increase the protein content of the RIIα, RIIβ and C subunits (Mendez *et al.*, 2003).

To date, there is no evidence for the involvement of other signaling pathways in mediating the actions of FSH and LH in fish gonads. Therefore, the current model of the mechanism of action of fish GTHs involves the coupling of adenylyl cyclase to G-protein coupled GTH receptors and the stimulation of the cAMP/PKA signaling pathway as the key mediator of GTH action.

Conclusions and Future Directions

Since the initial description of two distinct GTHs, FSH and LH, in fish a great deal has been learned about their biological activity. Unfortunately, this information is derived from studies using a reduced number of species due to the difficulties in obtaining pure preparations of FSH or LH from many other teleost species. It is foreseeable that in the near future, the generation of recombinant proteins will circumvent this problem and that biologically active gonadotropins will be available for a number of species.

Several areas regarding the biological function of gonadotropins will require specific attention in future studies. For example, the exact role of FSH in spermatogenesis and oogenesis needs to be determined, in particular those aspects related to the proliferation and recruitment of germ cells in early stages of development. In addition, there is a clear need for information on the role of GTHs in regulating the expression of GTH receptors in testicular and ovarian cells. Although it is clear that the appearance of the GTH type II receptor is crucial to gonadal maturation, the factor(s) that regulate either directly or indirectly the expression and/or activity of this receptor are completely unknown.

With the availability of either purified or recombinant fish FSH and LH, current genomic approaches such as genome sequencing and microarray analysis as well as proteomics, will be extremely useful in understanding the biological function of FSH and LH in fish.

References

Amer, M.A., T. Miura, C. Miura and K. Yamauchi. 2001. Involvement of sex steroid hormones in the early stages of spermatogenesis in Japanese huchen (*Hucho perryi*). *Biology of Reproduction* 65: 1057-1066.

Baynes, S. M. and A. P. Scott. 1985. Seasonal variations in parameters of milt production and in plasma concentration of sex steroids of male rainbow trout (*Salmo gairdneri*). *General and Comparative Endocrinology* 57: 150-160.

Billard, R. 1986. Spermatogenesis and spermatology of some teleost fish species. *Reproduction Nutrition and Development* 26: 877-920.

Billard, R., A. Fostier, C. Weil and B. Breton. 1982. Endocrine control of spermatogenesis in teleost fish. *Canadian Journal of Fish and Aquatic Sciences* 39: 65-79.

Bogomolnaya, A. and Z. Yaron. 1984. Stimulation *in vitro* of estradiol secretion by the ovary of a cichlid fish, *Sarotherodon aureus*. *General and Comparative Endocrinology* 53: 187-196.

Campbell, B., J.T. Dickey and P. Swanson. 2003. Endocrine changes during onset of puberty in male spring chinook salmon, *Oncorhynchus tshawytscha*. *Biology of Reproduction* 69: 2109-2117.

Cavaco, J.E., C. Vilrokx, V.L. Trudeau, R.W. Schulz and H.J. Goos. 1998. Sex steroids and the initiation of puberty in male African catfish (*Clarias gariepinus*). *American Journal of Physiology* 275: R1793-R1802.

Cerda, J., T.R. Petrino, A.M. Landin and W.P. Lin. 1997. Effects of isoquinolinesulfonamide H-8 on *Fundulus heteroclitus* ovarian follicles: role of cyclic nucleotide-dependent protein kinases on steroidogenesis and oocyte maturation *in vitro*. *Comparative Biochemistry and Physiology C Pharmacol Toxicol Endocrinol.* C117: 75-81.

Chang, X., R. Patiño, P. Thomas and G. Yoshizaki. 1999. Developmental and protein kinase-dependent regulation of ovarian connexin mRNA and oocyte maturational competence in Atlantic croaker. *General and Comparative Endocrinology*. 114: 330-339.

Choi, C.Y. and F. Takashima. 2000. Molecular cloning and hormonal control in the ovary of connexin 31.5 mRNA and correlation with the appearance of oocyte maturational competence in red seabream. *Journal of Experimental Biology* 203: 3299-3306.

Fitzpatrick, M.S., G. Van Der Kraak and C.B. Schreck. 1986. Profiles of plasma sex steroids and gonadotropin in coho salmon, *Oncorhynchus kisutch*, during final maturation. *General and Comparative Endocrinology* 62: 437-451.

Fontaine, Y.A. and S. Dufour. 1987. Current status of LH-FSH-like gonadotropin in fish. In: *Proceedings of the Third International Symposium on the Reproductive Physiology of Fish*, D.R. Idler, L.W. Crim and J.W. Walsh (eds.). Memorial University of Newfoundland, St. John's, Newfoundland, Canada, pp. 48-56.

Fontaine, Y.A., C. Salmon, E. Fontaine-Bertrand, E. Burzawa-Gerard and E.M. Donaldson. 1972. Comparison of the activities of two purified fish gonadotropins on adenyl cyclase activity in the goldfish ovary. *Canadian Journal of Zoology* 50: 1673-1676.

Fostier, A. 1995. Regulation of aromatase activity in the rainbow trout, *Oncorhynchus mykiss*, ovary. In: *Proceedings of the Fifth International Symposium on the Reproductive Physiology of Fish*, F.W. Goetz and P. Thomas (eds.). Fish Symposium 95, Austin, pp. 293-295.

García-Hernández, M.P., Y. Koide, M.V. Díaz and H. Kawauchi. 1997. Isolation and characterization of two distinct gonadotropins from the pituitary gland of Mediterranean yellowtail, *Seriola dumerilii*. *General and Comparative Endocrinology* 106: 389-399.

Goetz, F.W. 1997. Follicle and extrafollicular tissue interaction in $17\alpha,20\beta$-dihydroxy-4-pregnen-3-one-stimulated ovulation and prostaglandin synthesis in the yellow perch (*Perca flavescens*) ovary. *General and Comparative Endocrinology* 105: 121-126.

Goetz, F.W. and G. Theofan. 1979. *In vitro* stimulation of germinal vesicle breakdown and ovulation of yellow perch (*Perca flavescens*) oocytes. Effects of 17 alpha-hydroxy-20 beta-dihydroprogesterone and prostaglandins. *General and Comparative Endocrinology* 37: 273-285.

Goetz, F.W., A.K. Berndtson and M. Ranjan. 1991. Ovulation: mediators at the ovarian level. In: *Vertebrate Endocrinology, Fundamentals and Biomedical Implications*, P. K. T. Pang and M. Schreibman (eds.). Academic Press, New York, Vol. 4A, pp. 127-203.

Hyllner, S. J. and C. Haux. 1995. Vitelline envelope proteins in teleost fish. In: *Proceedings of the Fifth International Symposium on the Reproductive Physiology of Fish*, F.W. Goetz and P. Thomas (eds.). Fish Symposium 95, Austin, pp. 10-12.

Idler, D.R., S.J. Hwang and L.S. Bazar. 1975. Fish gonadotropin(s). I. Bioassay of salmon gonadotropin(s) *in vitro* with immature trout gonads. *Endocrine Research Communications* 2: 199-213.

Ikeuchi, T., T. Todo, T. Kobayashi and Y. Nagahama. 2001. Two subtypes of androgen and progestogen receptors in fish testes. *Comparative Biochemistry and Physiology* B 129: 449-455.

Jiang, J. Q., T. Kobayashi, W. Ge, H. Kobayashi, M. Tanaka, M. Okamoto, Y. Nonaka and Y. Nagahama. 1996. Fish testicular 11beta-hydroxylase: cDNA cloning and mRNA expression during spermatogenesis. *FEBS Letters* 397: 250-252.

Kagawa, H., G. Young and Y. Nagahama. 1984. *In vitro* estradiol-17 beta and testosterone production by ovarian follicles of the goldfish, *Carassius auratus*. *General and Comparative Endocrinology* 54: 139-143.

Kagawa, H., H. Tanaka, K. Okuzawa and M. Kobayashi. 1998. GTH II but not GTH I induces final maturation and the development of maturational competence of oocytes of red seabream *in vitro*. *General and Comparative Endocrinology* 112: 80-88.

Kamei, H., T. Ohira, Y. Yoshiura, N. Uchida, H. Nagasawa and K. Aida. 2003. Expression of a biologically active recombinant follicle stimulating hormone of Japanese eel *Anguilla japonica* using methylotropic yeast, *Pichia pastoris*. *General and Comparative Endocrinology* 134: 244-254.

Kanamori, A. and Y. Nagahama. 1988. Involvement of 3',5'-cyclic adenosine monophosphate in the control of follicular steroidogenesis of amago salmon (*Oncorhynchus rhodurus*). *General and Comparative Endocrinology* 72: 39-53.

Kazeto, Y., S. Adachi and K. Yamauchi. 2001. 20beta-Hydroxysteroid dehydrogenase of the Japanese eel ovary: Its cellular localization and changes in the enzymatic activity during sexual maturation. *General and Comparative Endocrinology* 122: 109-115.

King, W.T., S. Ghosh, P. Thomas and C.V. Sullivan. 1997. A receptor for the oocyte maturation-inducing hormone 17alpha,20beta,21-trihydroxy-4-pregnen-3-one on ovarian membranes of striped bass. *Biology of Reproduction* 56: 266-271.

Koide, Y., H. Itoh and H. Kawauchi. 1993. Isolation and characterization of two distinct gonadotropins, GTH I and GTH II, from bonito (*Katsuwonus plelamis*) pituitary glands. *International Journal of Peptide and Protein Research* 41: 52-65.

Langenau, D.M., F.W. Goetz and S.B. Roberts. 1999. The upregulation of messenger ribonucleic acids during 17alpha, 20beta-dihydroxy-4-pregnen-3-one-induced ovulation in the perch ovary. *Journal of Molecular Endocrinology* 23: 137-152.

Le Gac, F., M. Loir, P.Y. le Bail and M. Ollitrault. 1996. Insulin-like growth factor (IGF-I) mRNA and IGF-I receptor in trout testis and in isolated spermatogenic and Sertoli cells. *Molecular Reproduction and Development* 44: 23-35.

Loir, M. 1999a. Spermatogonia of rainbow trout: I. Morphological characterization, mitotic activity, and survival in primary cultures of testicular cells. *Molecular Reproduction and Development* 53: 422-433.

Loir, M. 1999b. Spermatogonia of rainbow trout: II. *in vitro* study of the influence of pituitary hormones, growth factors and steroids on mitotic activity. *Molecular Reproduction and Development* 53: 434-442.

Loir, M. 1990. Trout steroidogenic testicular cells in primary culture. I. Changes in free and conjugated androgen and progestagen secretions: effects of gonadotropin, serum, and lipoproteins. *General and Comparative Endocrinology* 78: 374-387.

Maestro, M.A., J.V. Planas, S. Moriyama, J. Gutierrez, J. Planas and P. Swanson. 1997. Ovarian receptors for insulin and insulin-like growth factor I (IGF-I) and effects of IGF-I on steroid production by isolated follicular layers of the preovulatory coho salmon ovarian follicle. *General and Comparative Endocrinology* 106: 189-201.

Marshall, W.S., S.E. Bryson and D.R. Idler. 1993. Gonadotropin action on brook trout sperm duct epithelium: Transport stimulation mediated by cAMP and Ca^{2+}. *General and Comparative Endocrinology* 90: 232-242.

Mendez, E., M. Maeland, B.S. Skalhegg and J.V. Planas. 2003. Activation of the cAMP-dependent protein kinase signaling pathway by luteinizing hormone in trout theca layers. *Molecular and Cellular Endocrinology* 205: 11-20.

Mita, M., M. Yoshikuni and Y. Nagahama. 1994. G-proteins and adenylyl cyclase in ovarian granulosa cells of amago salmon (*Oncorhynchus rhodurus*). *Molecular and Cellular Endocrinology* 105: 83-88.

Miura, T., K. Yamauchi, Y. Nagahama and H. Takahashi. 1991a. Induction of spermatogenesis in male Japanese eel, *Anguilla japonica*, by a single injection of human chorionic gonadotropin. *Zoological Science* 8: 63-73.

Miura, T., K. Yamauchi, H. Takahashi and Y. Nagahama. 1991b. Human chorionic gonadotropin induces all stages of spermatogenesis *in vitro* in the male Japanese eel (*Anguilla japonica*). *Developmental Biology* 146: 258-262.

Miura, T., K. Yamauchi, H. Takahashi and Y. Nagahama. 1991c. Hormonal induction of all stages of spermatogenesis in vitro in the male Japanese eel (*Anguilla japonica*). *Proceedings of the National Academy of Sciences of the United States of America* 88: 5774-5778.

Miura, T., K. Yamauchi, H. Takahashi and Y. Nagahama. 1992. The role of hormones in the acquisition of sperm motility in salmonid fish. *Journal of Experimental Zoology* 261: 359-363.

Miura, T., C. Miura, T. Ohta, M.R. Nader, T. Todo and K. Yamauchi. 1999. Estradiol-17beta stimulates the renewal of spermatogonial stem cells in males. *Biochemical Biophysical Research Communications* 264: 230-234.

Miwa, S., L. Yan and P. Swanson. 1994. Localization of two gonadotropin receptors in the salmon gonad by *in vitro* ligand autoradiography. *Biology of Reproduction* 50: 629-642.

Montserrat, N., A. Gonzalez, E. Mendez, F. Piferrer and J.V. Planas. 2004. Effects of follicle stimulating hormone on estradiol-17[beta] production and P-450 aromatase (CYP19) activity and mRNA expression in brown trout vitellogenic ovarian follicles *in vitro*. *General and Comparative Endocrinology* 137: 123-131.

Nader, M.R., T. Miura, N. Ando, C. Miura and K. Yamauchi. 1999. Recombinant human insulin-like growth factor I stimulates all stages of 11-ketotestosterone-induced spermatogenesis in the Japanese eel, *Anguilla japonica*, in vitro. *Biology of Reproduction* 61: 944-947.

Nagahama, Y. 1987. Gonadotropin action on gametogenesis and steroidogenesis in teleost gonads. *Zoological Science* 4: 209-222.

Nagahama, Y. 1997. 17 alpha,20 beta-dihydroxy-4-pregnen-3-one, a maturation-inducing hormone in fish oocytes: mechanisms of synthesis and action. *Steroids* 62: 190-196.

Nagahama, Y. 2000. Gonadal steroid hormones: major regulators of gonadal sex differentiation and gametogenesis in fish. In: *Proceedings of the Sixth International Symposium on the Reproductive Physiology of Fish*, B. Norberg, O.S. Kjesbu, G.L. Taranger, E. Auderson and S.O. Steffansson (eds.). John Grieg A/S, Bergen 2000, Bergen.

Nagahama, Y., A. Matsushita, T. Iwamatsu, N. Sakai and S. Fukada. 1991. A mechanism for the action of pregnant mare serum gonadotropin on aromatase activity in the ovarian follicle of the medaka, *Oryzias latipes*. *Journal of Experimental Zoology* 259: 53-58.

Nagahama, Y., M. Yoshikuni, M. Yamashita, T. Tokumoto and Y. Katsu. 1995. Regulation of oocyte growth and maturation in fish. *Current Topics in Developmental Biology* 30: 103-145.

Ohta, K., S. Yamaguchi, A. Yamaguchi, K. Gen, K. Okuzawa, H. Kagawa and M. Matsuyama. 2002. Biosynthesis of steroids in ovarian follicles of red seabream, *Pagrus major* (Sparidae, Teleostei) during final oocyte maturation and the relative effectiveness of steroid metabolites for germinal vesicle breakdown in vitro. *Comparative Biochemistry and Physiology Part B: Biochemistry and Molecular Biology* 133: 45-54.

Okada, T., I. Kawazoe, S. Kimura, Y. Sasamoto, K. Aida and H. Kawauchi. 1994. Purification and characterization of gonadotropin I and II from pituitary glands of tuna (*Thunnus obesus*). *International Journal of Peptide and Protein Research* 43: 69-80.

Oppen-Berntsen, D.O., S.O. Olsen, C.J. Rong, G.L. Taranger, P. Swanson and B.T. Walther. 1994. Plasma levels of eggshell zr-proteins, estradiol-17beta, and gonadotropins during an annual reproductive cycle of atlantic salmon (*Salmo salar*). *Journal of Experimental Zoology* 268: 59-70.

Pang, Y. and W. Ge. 1999. Activin stimulation of zebrafish oocyte maturation *in vitro* and its potential role in mediating gonadotropin-induced oocyte maturation. *Biology of Reproduction* 61: 987-992.

Pang, Y. and W. Ge. 2002a. Gonadotropin and activin enhance maturational competence of oocytes in the zebrafish (*Danio rerio*). *Biology of Reproduction* 66: 259-265.

Pang, Y. and W. Ge. 2002b. Gonadotropin regulation of activin betaA and activin type IIA receptor expression in the ovarian follicle cells of the zebrafish, *Danio rerio*. *Molecular and Cellular Endocrinology* 188: 195-205.

Pankhurst, N.W. 1994. Effects of gonadotropin releasing hormone analogue, human chorionic gonadotropin and gonadal steroids on milt volume in the New Zealand snapper, *Pagrus auratus* (Sparidae). *Aquaculture* 125: 185-197.

Patiño, R., G. Yoshizaki, P. Thomas and H. Kagawa. 2001. Gonadotropic control of ovarian follicle maturation: the two-stage concept and its mechanisms. *Comparative Biochemistry and Physiology Part B: Biochemistry and Molecular Biology* 129: 427-439.

Pinter, J. and P. Thomas. 1999. Induction of ovulation of mature oocytes by the maturation-inducing steroid 17,20beta,21-trihydroxy-4-pregnen-3-one in the spotted seatrout. *General and Comparative Endocrinology* 115: 200-209.

Planas, J.V. and P. Swanson. 1995. Maturation-associated changes in the response of the salmon testis to the steroidogenic actions of gonadotropins (GTH I and GTH II) *in vitro*. *Biology of Reproduction* 52: 697-704.

Planas, J.V., P. Swanson and W.W. Dickhoff. 1993. Regulation of testicular steroid production *in vitro* by gonadotropins (GTH I and GTH II) and cyclic AMP in coho salmon (*Oncorhynchus kisutch*). *General and Comparative Endocrinology* 91: 8-24.

Planas, J.V., F.W. Goetz and P. Swanson. 1997. Stimulation of brook trout ovarian steroidogenesis by gonadotropins I and II is mediated by the cyclic adenosine 3',5'-monophosphate/protein kinase A pathway. *Biology of Reproduction* 57: 647-654.

Planas, J.V., J. Athos, F.W. Goetz and P. Swanson. 2000. Regulation of ovarian steroidogenesis in vitro by follicle-stimulating hormone and luteinizing hormone during sexual maturation in salmonid fish. *Biology of Reproduction* 62: 1262-1269.

Prat, F., J.P. Sumpter and C.R. Tyler. 1996. Validation of radioimmunoassays for two salmon gonadotropins (GTH I and GTH II) and their plasma concentrations throughout the reproductive cycle in male and female rainbow trout (*Oncorhynchus mykiss*). *Biology of Reproduction* 54: 1375-1382.

Rahman, M., K. Ohta, M. Yoshikuni, Y. Nagahama, H. Chuda and M. Matsuyama. 2002. Characterization of ovarian membrane receptor for 17,20beta-dihydroxy-4-pregnen-3-one, a maturation-inducing hormone in yellowtail, *Seriola quinqueradiata*. *General and Comparative Endocrinology* 127: 71-79.

Saad, A. and J. Dépêche. 1987. *In vitro* effect of salmon gonadotropin on the testicular synthesis of androgens and of a progestin 17alpha-hydroxy-20beta-dihydroprogesterone, in the rainbow trout (*Salmo gairdneri*). *Reproduction Nutrition and Development* 27: 423-439.

Sakai, N., H. Ueda, N. Suzuki and Y. Nagahama. 1989. Steroid production by amago salmon (*Oncorhynchus rhodurus*) testes at different development stages. *General and Comparative Endocrinology* 75: 231-240.

Salmon, C., J. Marchelidon, E. Fontaine-Bertrand and Y.A. Fontaine. 1985. Human chorionic gonadotrophin and immature fish ovary: characterization and mechanism of the *in vitro* stimulation of cyclic adenosine monophosphate accumulation. *General and Comparative Endocrinology* 58: 101-108.

Santos, E.M., J.E. Harries and C.R. Tyler. 2000. Immunoneutralization as a technique to study the role of FSH in ovarian development in the rainbow trout (*Oncorhynchus mykiss*). In: *Proceedings of the Sixth International Symposium on the Reproductive Physiology of Fish*, B. Norberg, O.S. Kjesbu, G.L. Taranger, E. Andersson and S.O. Stefansson (eds.). Bergen 2000, Bergen.

Santos, E.M., M. Rand-Weaver and C.R. Tyler. 2001. Follicle-stimulating hormone and its alpha and beta subunits in rainbow trout (*Oncorhynchus mykiss*): Purification, characterization, development of specific radioimmunoassays, and their seasonal plasma and pituitary concentrations in females. *Biology of Reproduction* 65: 288-294.

Schulz, R. and V. Blüm. 1990. Steroid secretion of rainbow trout testis *in vitro*: variation during the reproductive cycle. *General and Comparative Endocrinology* 80: 189-198.

Schulz, R.W. 1986. *In vitro* secretion of testosterone and 11-oxotestosterone of testicular and spermiduct tissue from rainbow trout, *Salmo gairdneri* (Richardson). *Fish Physiology and Biochemistry* 1: 55-61.

Schulz, R.W., L. van der Corput, J. Janssen-Dommerholt and H.J.T. Goos. 1994. Sexual steroids during puberty in male African catfish (*Clarias gariepinus*): serum levels and gonadotropin-stimulated testicular secretion in vitro. *Journal of Comparative Physiology B* 164: 195-205.

Scott, A.P. and J.P. Sumpter. 1989. Seasonal variations in testicular germ cell stages and in plasma concentrations of sex steroids in male rainbow trout (*Salmo gairdneri*) maturing at two years old. *General and Comparative Endocrinology* 73: 46-58.

Scott, A.P., E.L. Sheldrick and A.P.F. Flint. 1982. Measurement of 17alpha,20beta-dihydroxy-4-pregnen-3-one in plasma of trout (*Salmo gairdneri*): seasonal changes and response to salmon pituitary extract. *General and Comparative Endocrinology* 46: 444-451.

Selman, K. and R.A. Wallace. 1989. Cellular aspects of oocyte growth in teleosts. *Zoological Science* 6: 211-231.

Senthilkumaran, B., C.C. Sudhakumari, X.T. Chang, T. Kobayashi, Y. Oba, G. Guan, Y. Yoshiura, M. Yoshikuni and Y. Nagahama. 2002. Ovarian carbonyl reductase-like 20beta-hydroxysteroid dehydrogenase shows distinct surge in messenger RNA expression during natural and gonadotropin-induced meiotic maturation in nile tilapia. *Biology of Reproduction* 67: 1080-1086.

Senthilkumaran, B., M. Yoshikuni and Y. Nagahama. 2004. A shift in steroidogenesis occurring in ovarian follicles prior to oocyte maturation. *Molecular and Cellular Endocrinology* 215: 11-18.

Sire, O. and J. Dépêche. 1981. *In vitro* effect of a fish gonadotropin on aromatase and 17beta-hydroxysteroid dehydrogenase activities in the ovary of the rainbow trout. *Reproduction Nutrition and Development* 21: 715-726.

Srivastava, R.K. and G. Van Der Kraak. 1995. Multifactorial regulation of DNA synthesis in goldfish ovarian follicles. *General and Comparative Endocrinology* 100: 397-403.

Suzuki, K., H. Kawauchi and Y. Nagahama. 1988. Isolation and characterization of subunits from two distinct salmon gonadotropins. *General and Comparative Endocrinology* 71: 302-306.

Suzuki, K., Y. Nagahama and H. Kawauchi. 1988. Steroidogenic activities of two distinct salmon gonadotropins. *General and Comparative Endocrinology* 71: 452-458.

Swanson, P. 1991. Salmon gonadotropins: reconciling old and new ideas. In: *Proceedings of the Fourth International Symposium on the Reproductive Physiology of Fish*, A.P. Scott, J.P. Sumpter, D.E. Kime and M.S. Rolfe (eds.). FishSymp, Sheffield, pp. 2-7.

Swanson, P., M. Bernard, M. Nozaki, K. Suzuki, H. Kawauchi and W.W. Dickhoff. 1989. Gonadotropins I and II in juvenile coho salmon. *Fish Physiology and Biochemistry* 7: 169-176.

Swanson, P., K. Suzuki, H. Kawauchi and W.W. Dickhoff. 1991. Isolation and characterization of two coho salmon gonadotropins, GTH I and GTH II. *Biology of Reproduction* 44: 29-38.

Tan, J.D., S. Adachi and Y. Nagahama. 1986. The *in vitro* effects of cyclic nucleotides, cyanoketone, and cycloheximide on the production of estradiol-17 beta by vitellogenic ovarian follicles of goldfish (*Carassius auratus*). *General and Comparative Endocrinology* 63: 110-116.

Tanaka, H., H. Kagawa, K. Okuzawa and K. Hirose. 1993. Purification of gonadotropins (Pm GTH I and II) from red seabream (*Pagrus major*) and development of a homologous radioimmunoassay for Pm GTH II. *Fish Physiology and Biochemistry* 10: 409-418.

Tanaka, H., H. Kagawa and K. Hirose. 1995. Steroidogenic activities of two distinct gonadotropins in red seabream, *Pagrus major*. In: *Proceedings of the Fifth International Symposium on the Reproductive Physiology of Fish*, F.W. Goetz and P. Thomas (eds.). Fish Symposium 95, Austin, pp. 10-12.

Thomas, P., J. Pinter and S. Das. 2001. Upregulation of the maturation-inducing steroid membrane receptor in spotted seatrout ovaries by gonadotropin during oocyte maturation and its physiological significance. *Biology of Reproduction* 64: 21-29.

Thomas, P., Y. Pang, Y. Zhu, C. Detweiler and K. Doughty. 2004. Multiple rapid progestin actions and progestin membrane receptor subtypes in fish. *Steroids* 69: 567-573.

Todo, T., T. Ikeuchi, T. Kobayashi, H. Kajiura-Kobayashi, K. Suzuki, M. Yoshikuni, K. Yamauchi and Y. Nagahama. 2000. Characterization of a testicular 17alpha, 20beta-dihydroxy-4-pregnen-3-one (a spermiation-inducing steroid in fish) receptor from a teleost, Japanese eel (*Anguilla japonica*). *FEBS Letters* 465: 12-17.

Tyler, C.R., J P. Sumpter, H. Kawauchi and P. Swanson. 1991. Involvement of gonadotropin in the uptake of vitellogenin into vitellogenic oocytes of the rainbow trout, *Oncorhynchus mykiss*. *General and Comparative Endocrinology* 84: 291-299.

Tyler, C.R., T.G. Pottinger, E. Santos, J.P. Sumpter, S.A. Price, S. Brooks and J.J. Nagler. 1996. Mechanisms controlling egg size and number in the rainbow trout, *Oncorhynchus mykiss*. *Biology of Reproduction* 54: 8-15.

Tyler, C.R., T.G. Pottinger, K. Coward, F. Prat, N. Beresford and S. Maddix. 1997. Salmonid follicle-stimulating hormone (GtH I) mediates vitellogenic development of oocytes in the rainbow trout, *Oncorhynchus mykiss*. *Biology of Reproduction* 57: 1238-1244.

Tyler, C.R., E.M. Santos and F. Prat. 2000. Unscrambling the egg-cellular, biochemical, molecular and endocrine advances in oogenesis. In: *Proceedings of the Sixth International Symposium on the Reproductive Physiology of Fish*, B. Norberg, O.S. Kjesbu, G.L. Taranger, E. Andersson, S.O. Stefansson (eds.), John Greig A/S, Bergen 2000, Bergen, pp. 273-280.

Ueda, H., G. Young, L.W. Crim, A. Kambegawa and Y. Nagahama. 1983. 17 alpha,20 beta-dihydroxy-4-pregnen-3-one: plasma levels during sexual maturation and *in vitro* production by the testes of amago salmon (*Oncorhynchus rhodurus*) and rainbow trout (*Salmo gairdneri*). *General and Comparative Endocrinology* 51: 106-112.

Ueda, H., O. Hiroi, A. Hara, K. Yamauchi and Y. Nagahama. 1984. Changes in serum concentrations of steroid hormones, thyroxine, and vitellogenin during spawning migration the chum salmon, *Oncorhynchus keta*. *General and Comparative Endocrinology* 53: 203-211.

Van der Kraak, G., K. Suzuki, R.E. Peter, H. Itoh and H. Kawauchi. 1992. Properties of common carp gonadotropin I and gonadotropin II. *General and Comparative Endocrinology* 85: 217-229.

Weltzien, F.-A., B. Norberg and P. Swanson. 2003. Isolation and characterization of FSH and LH from pituitary glands of Atlantic halibut (*Hippoglossus hippoglossus* L.). *General and Comparative Endocrinology* 131: 97-105.

Wood, A.W. and G. van der Kraak. 2002. Inhibition of apoptosis in vitellogenic ovarian follicles of rainbow trout (*Oncorhynchus mykiss*) by salmon gonadotropin, epidermal growth factor, and 17beta-estradiol. *Molecular Reproduction and Development* 61: 511-518.

Yan, L., P. Swanson and W.W. Dickhoff. 1992. A two-receptor model for salmon gonadotropins (GTH I and GTH II). *Biology of Reproduction* 47: 418-427.

Yoshizaki, G., R. Patino and P. Thomas. 1994. Connexin messenger ribonucleic acids in the ovary of Atlantic croaker: molecular cloning and characterization, hormonal control, and correlation with appearance of oocyte maturational competence. *Biology of Reproduction* 51: 493-503.

Yoshiura, Y., H. Suetake and K. Aida. 1999. Duality of gonadotropin in a primitive teleost, Japanese eel (*Anguilla japonica*). *General and Comparative Endocrinology* 114: 121-131.

Yoshizaki, G., W. Jin, R. Patino, P. Thomas and L. Janecek. 2000. Structural organization of the Atlantic Croaker Connexin 32.2 gene and its 5' flanking region. *Marine Biotechnology* 2: 154-160.

Yoshiura, Y., B. Senthilkumaran, M. Watanabe, Y. oba, T. Kobayashi and Y. Nagahama. 2003. Synergistic expression of Ad4BP/SF-1 and cytochrome P-450 aromatase (Ovarian type) in the ovary of Nile Tilapia, *Oreochromis niloticus*, during vitellogenesis suggests transcriptional interaction. *Biology of Reproduction* 68: 1545-1553.

Young, G., H. Kagawa and Y. Nagahama. 1983. Evidence for a decrease in aromatase activity in the ovarian granulosa cells of amago salmon (*Oncorhynchus rhodurus*) associated with final oocyte maturation. *Biology of Reproduction* 29: 310-315.

Young, G., T. Todo, M. Kusakabe, T. Kobayashi and Y. Nagahama. 2000. 3beta-hydroxysteroid dehydrogenase gene: sites of expression in trout gonads and cAMP-dependent regulation. In: *Proceedings of the Sixth International Symposium on the Reproductive Physiology of Fish*, B. Norberg, O.S. Kjesbu, G.L. Taranger, E. Andersson, S.O. Stefansson (eds.). John Greig A/S, Bergen 2000, Bergen p. 202.

Zhu, Y., C.D. Rice, Y. Pang, M. Pace and P. Thomas. 2003. Cloning, expression, and characterization of a membrane progestin receptor and evidence it is an intermediary in meiotic maturation of fish oocytes. *Proceedings of the National Academy of Sciences of the United States of America* 100: 2231-2236.

Gonadal Steroids: Functions and Patterns of Change

Ned W. Pankhurst

INTRODUCTION

Steroid hormones play a major role in the endocrine systems in all vertebrates, with actions that include regulation of mineral and water balance, modulation of immune response, and regulation of various facets of reproduction. Steroids are synthesized from the base molecule cholesterol via an enzymatically mediated cleavage pathway that involves sequential loss of carbon atoms, and the loss or addition of various active groups. There is considerable diversity of specific steroids produced by different vertebrate classes and taxa within classes, but some general considerations apply. Twenty-one-carbon (C_{21}) steroids occur as corticosteroids which are primarily concerned with metabolic and osmoregulatory function and only have indirect action in reproduction, and progestational steroids which have a primary role in reproduction in all groups. Effects of progestins variously include regulation of oocyte maturation, ovulation, and maintenance of pregnancy or fertilized eggs

Author's address: Science, Environment, Engineering and Technology Group, Griffith University, PMB 50 Gold Coast Mail Centre, Queensland, 9726, Australia. E-mail: n.pankhurst@griffith.edu.au

in internally fertilizing species. In aquatic vertebrates, progestins or their metabolites also appear to have widespread action as sexual pheromones, probably as a result of their capacity to unambiguously signal the occurrence of 'way points' in the reproductive process. C_{19} steroids (collectively termed androgens) are commonly considered to be 'male' steroids but, in fact, regulate a suite of reproductive processes in both sexes with expression of 'male' effects essentially being dependent on the metabolic fate of the androgens produced. Effects of androgens include regulation of gonadal growth and development of secondary sexual characteristics in males, and central effects on behaviour, hypothalamic and pituitary function in both sexes. C_{18} steroids (estrogens) can be synthesized in both sexes but occur at much higher levels in females where they regulate aspects of ovarian growth, development of female secondary sexual characteristics, and the hepatic synthesis of the egg-yolk precursor vitellogenin (Vtg) and egg shell proteins (collectively termed zona pellucida proteins).

Because of the ubiquity of their importance in vertebrate reproduction, the commonality of function and the relative ease with which plasma and tissue levels of steroids can be determined (in turn, a function of the structural commonality of steroids irrespective of the taxonomic group), gonadal steroids have been used more than any other class of hormones to describe endocrine states, make association between reproductive development and endocrine control, and assess environmental control of reproductive processes. For the same reasons, steroid hormone levels have also been commonly used as endocrine markers of response to hormone manipulation in controlled breeding, and disruption of endocrine function by environmental stress and pollutants, particularly the emerging concern over compounds that act as endocrine mimics. The result now is a large body of literature on steroid hormone profiles in most vertebrate groups but especially fish. There are over 28,000 extant species of fish (the great majority being teleostean or bony fish) and a correspondingly high diversity of reproductive styles. Owing to the specificity of peptide and protein hormones among different taxa, steroid hormone measurement has often been the only way to relatively easily assess the differences or commonalties in reproductive patterns and control processes. There is a caveat to this in the sense that the measurement of a small number of 'standard' hormones tends to show researchers what they are looking for with the result that current perceptions may be a simplification of the actual situation.

This chapter considers the above issues in relation to fish reproduction with view to summarizing current knowledge and directing attention to areas for further research. The chapter is not intended to be a comprehensive review of all literature on steroid hormone cycles in fish, and where possible readers will be directed to existing reviews in specialist topic areas.

STEROID HORMONE SYNTHESIS

Steroid hormones are synthesized from intracellular stores of cholesterol under the influence of the glycoprotein gonadotropins released into the circulation from the pituitary gland. As in other vertebrates, there are two gonadotropin forms. In common with other vertebrate terminology, these are now referred to as follicle-stimulating hormone (FSH) and luteinizing hormone (LH) (reviewed by Swanson *et al.*, 2003, Yaron *et al.*, 2003). Based on data from the small number of species for which assays for both FSH and LH are available, there appears to be a temporal separation of release of FSH and LH in species with *synchronous* ovarian development (i.e., only a single clutch matures at one time), with FSH being associated with the early stages of gonadal growth and development, and LH regulating maturational events (Swanson 1991; Swanson *et al.*, 2003). In contrast, in species with *asynchronous* ovarian development (multiple clutches at various developmental stages are present at any given time) both FSH and LH appear to change in concert (e.g., Elizur *et al.*, 1995; Jackson *et al.*, 1999). Both LH and FSH exert their action by binding to membrane receptors in the follicle cells of the ovary (Yan *et al.*, 1992: Miwa *et al.*, 1994), and the Leydig cells of the testis (Bogerd *et al.*, 2001; Vischer and Bogerd, 2003), resulting in elevation of intracellular cyclic AMP levels and subsequent activation of protein kinase A, and activation or synthesis of steroid synthesizing enzymes (Van Der Kraak and Wade, 1994; Nagahama, 2000).

A key regulatory step in the steroid synthesis process involves delivery of cholesterol molecules to the inner membranes of mitochondria via the action of a transfer protein called steroidogenic acute regulatory protein (StAR). Initial studies on mammals have shown the StAR is a 30kDa protein synthesized first in the cytosol as a 37kDa precursor in response to activation of cAMP-protein kinase A intracellular-signalling pathways. Binding to the outer membrane of the mitochondria and subsequent processing and phosphorylation to the 30kDa, mature protein generates a change in the membrane characteristics to allow the transfer of cholesterol to the inner mitochondrial membrane (inner and outer membranes are otherwise separated by an aqueous inter-membrane space which prevents transfer of the hydrophobic cholesterol) (reviewed by Stocco, 1997, 1999). More recently, it has been shown that only newly synthesized StAR is effective in mediating cholesterol transfer and this is often present at quite low levels, but the protein is highly active with an estimate of in excess of 400 molecules per minute of cholesterol being transferred by each StAR molecule (Artemenko *et al.*, 2001). Studies on fish indicate that the same mechanism operates in lower vertebrates with inter-renal tissue of rainbow trout *Oncorhynchus mykiss* showing PKA and StAR-dependent production of cortisol (Lacroix and Hontela, 2001; Kusakabe *et al.*, 2002; Geslin and Auperin, 2004). Ovarian StAR was also shown to peak in brook trout *Salvelinus fontinalis* in association with oocyte maturation and ovulation and the production of peak levels of maturational steroids (Kusakabe *et al.*, 2002).

Steroid nomenclature is formally based on structure through a hierarchy of hydrocarbon class, presence and level of unsaturation, and substituent atoms or groups (Kirk and Marples, 1995). However, the biological literature typically uses common or 'trivial' forms and unless there is scope for confusion or the absence of a common name, these will be used in the following discussion. Transfer of cholesterol to the inner mitochondrial membrane is followed by cleavage of a 6-carbon side chain to form the C_{21} steroid pregnenolone—the first steroid in the cleavage pathway—under the catalytic control of the cytochrome P450 side chain cleavage (P_{450} SCC) enzyme complex, with subsequent further enzymatic cleavage via two different possible routes. Pregnenolone may be converted to progesterone by 3β-hydroxysteroid dehydrogenase (3β-HSD) through what is referred to as the δ4 pathway, or to 17-hydroxypregnenolone by 17α-hydroxylase/C17,20 lyase cytochrome P450 ($P_{450}c17$) through the δ5 pathway (Norris, 1996; Nagahama, 2000). The subsequent metabolic fate of either progesterone or 17-hydroxypregnenolone depends to an extent on the tissue type and the species, but a general schema for fish is shown in Fig. 3.1. With some exceptions that will be discussed later, the active metabolic endpoint for synthesis during gametogenesis and gonad growth is 11-ketotestosterone (11KT) in males, and 17β-estradiol (E_2) in females. In both sexes, there is a shift to the production of C_{21} maturational steroids (typically 17,20β-dihydroxy-4-pregnen-3-one [17,20βP]) prior to gamete maturation, and ovulation and spermiation, in males and females, respectively (reviewed by Nagahama, 2000).

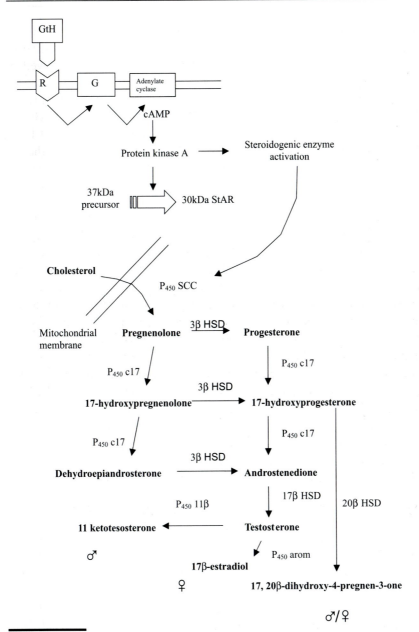

Fig. 3.1 General schema for steroid synthesis in teleosts fishes from Van der Kraak and Wade (1994), Stocco (1999) and Nagahama (2000). Abbreviations: GtH – gonadotropin; R – GtH receptor; StAR – steroidogenic acute regulatory protein; P_{450} SCC – side chain cleavage enzyme; 3β, 17β and 20βHSD – 3,17 or 20β-hydroxysteroid dehydrogenase; P_{450}c17 – 17α-hydroxylase/C17,20 lyase; P_{450}11β – 11β-hydroxylase; P_{450} arom – aromatase.

The chemical capacity for substitution of active groups in the steroid skeleton is substantial, and *in vitro* experiments using incubation of radioactive precursors with ovarian or testicular tissue do, in fact, show that a much larger number of compounds than shown in Fig. 3.1 can be synthesized (e.g., Schoonen *et al.*, 1987, 1989; Canario and Scott, 1989; Kime, 1990, 1992; Kime *et al.*, 1991; Borg *et al.*, 1992; Vermeulen *et al.*, 1993; Guigen *et al.*, 1995). This reflects the fact that intermediates will exist at all times in the steroid producing tissue, albeit often at low concentrations, and also that gonadal tissue is highly active in deactivation of steroids through further metabolism and conjugation (discussed in: Modes of steroid action and clearance). It also needs to be considered that *in vitro* systems have the capacity to generate patterns of steroid production that may not occur *in vivo*, as a result of the choice of incubation parameters. For example, in a series of studies, Kime and co-workers showed that the concentrations of substrate had a marked effect on the type and amounts of product produced. At low substrate concentrations, conjugated steroids and polar metabolites dominated whereas they were undetectable at high substrate concentrations (Kime, 1992; Kime and Abdullah, 1994; Ebrahimi *et al.*, 1995). This is interpreted as being a function of the relative capacity and activity of different enzymes in the cleavage pathway. An additional complication is provided by the fact that incubation temperatures also affect product mixes with higher amounts of conjugated steroids being produced at higher temperatures (Kime, 1979; Kime and Hyder, 1983). The outcome is that while *in vitro* studies can be a useful guide for possible synthesis pathways that may operate in gonadal tissues, results need to be validated by *in vivo* studies, and suitable demonstration of biological activity or effect.

MODES OF STEROID ACTION AND CLEARANCE

'Classical' or genomic steroid activity involves diffusion of hydrophobic steroid molecules across the lipid bi-layer of the cell membrane, and binding to cytosolic receptors. The steroid-receptor complex is then translocated to the nucleus where the binding of the complex to hormone response elements of specific genes regulates transcription and subsequent protein synthesis. The process typically has a timeframe of hours to days for the lag between the *de novo* synthesis of the steroid, and the appearance of its biological effect (reviewed in Norris, 1996). However, a number of steroid effects such as some of the acute behaviour-modifying effects of androgens and the maturational effects of progestins are too rapid to be explained through a mechanism of classical steroid action, leading to the view that steroids might also act via much more rapid membrane-receptor mediated intracellular signalling pathways. This has now been confirmed in several vertebrate groups with substantial evidence for the widespread non-genomic action of steroids in fish. Maturational effects of progestins have been shown to be dependent on binding to a G-protein like membrane receptor in the ovaries and testes of several species of fish, and membrane estrogen and androgen receptors have been identified in the testis and ovary of Atlantic croaker *Micropogonias undulatus* where they downregulate the production of androgens and estrogens, respectively (reviewed by Thomas, 2003). Steroids also exert effects on neural tissue and have the capacity to modulate behaviour, neural connectivity, and the presence or activity of ion channels in neural cell membranes (reviewed by Zakon, 1993). Mammalian studies indicate that this central action can be rapid (i.e., less than 30 minutes), suggesting the presence of non-genomic mechanism of steroid action here also (Pfaff et al., 2004).

Steroids have both *endocrine* action through transport from the hormone-producing tissue to the target organ via the circulation, and *paracrine* activity by direct diffusion into adjacent cells. The biological deactivation mechanisms vary in case of each type of action. Both androgens and estrogens can have paracrine effects but, typically, they also have extragonadal tissue targets, exposing them to metabolism in the plasma and non-target tissue. Steroids have a short plasma half life (e.g., $t_{1/2}$ is less than 30 minutes for E_2 in rainbow trout; Zohar, 1982) and are rapidly lost from the blood by conjugation with either sulphate or glucuronyl moieties, primarily in the liver but also in gonadal tissue, particularly at elevated temperatures (Scott and Vermeirssen, 1994; Haddy and Pankhurst, 1998a). Conjugated steroids are water soluble, allowing rapid clearance from the plasma and excretion in the bile or urine. Steroids are protected from conjugation by binding in the circulation to carrier proteins collectively known as sex steroid binding proteins (SBP). SBP have been characterized in a range of teleosts and tend to have high affinity for T and E_2, moderate affinity for some other androgens and low affinity for maturational C_{21} steroids and corticosteroids (Hobby et al., 2000a, b). The effect of this is for SBP to bind to steroids (mainly T and E_2) in the plasma and protect them from metabolism such that at typical plasma hormone values of T and E_2, all steroid is bound to SBP (Callard and Callard, 1987; Hobby et al., 2000a). An exception may be salmonids where peak plasma levels of T and E_2 can exceed the total binding capacity of SBPs (Hobby *et al.*, 2000a). The general effect of steroid binding to SBPs is to buffer the rate at which levels of T and E_2 change in the plasma, with the degree of buffering being a function of the level of saturation of SBP. Maturational steroids, in contrast, have primary tissue targets that are typically near to the tissue that produces them, and deactivation here appears to result both from *in situ* conjugation and further enzymatic conversion to highly polar (hydrophilic) metabolites that are excreted in the urine (Canario and Scott, 1989, 1990a; Vermeirssen *et al.*, 1998). A third route of excretion is the loss of free steroids to the water by diffusion across the gills. Exactly how significant this effect is in terms of total steroid budgets is unclear; however, it is known that appreciable amounts of free steroid do leak out the animal across the highly permeable epithelium of the gill filaments (Vermeirssen and Scott, 1996; Sorensen *et al.*, 2000). Fish can also absorb free steroids from the surrounding water. Vermeirssen and Scott (1996) reported the uptake of ^{3}H-17,20βP back across the gills after initial release to the water in rainbow trout, and Budworth and Senger (1993) described fish to fish transfer of injected T in rainbow trout held in a closed recirculation system.

ACTIONS OF STEROID HORMONES

Maturational Steroids

Maturational steroids (MS) have the primary role in females of stimulating the resumption of meiosis in oocytes that have completed vitellogenesis. As has been noted, MS bind to membrane receptors (Thomas, 2003), and initiate an intracellular signalling cascade that culminates in nuclear migration and breakdown. This is achieved through the action of a maturation-promoting factor (MPF) generated by the formation and subsequent phosphorylation of a cyclin-B cdc-2 kinase complex. The role of MS appears to be to stimulate cyclin-B synthesis from existing cyclin-B mRNA (Nagahama, 2000; review: Devlin and Nagahama, 2002). *In vitro* structure-activity studies in a number of species have shown that a wide range of C_{21} steroids possess potential maturational activity, but that the highest activity consistently resides with 17,20βP and 17,20β,21-trihydroxy-4-pregnen-3-one (20βS) (Jalabert, 1976; Pankhurst, 1985; Canario and Scott, 1988; Trant and Thomas, 1988; Ventling and Pankhurst, 1995). This is generally consistent with *in vivo* measurements showing that plasma levels of 17,20βP and, or 20βS typically peak during final oocyte maturation (reviewed in Thomas, 1994; Devlin and

Nagahama, 2002). In some species, notably flatfish, levels of other C_{21} steroids can be much higher. For example, in the dab *Limanda limanda*, 17,20α-dihydroxy-4-pregnen-3-one (17,20αP) is present at high levels in the plasma whereas free 17,20βP levels are low, although higher concentrations may be present as conjugates (Canário and Scott, 1990b, 1991). However, 17,20αP has low *in vitro* potency in stimulating oocyte maturation in a variety of species (Canário and Scott, 1988, 1990c; Ventling and Pankhurst, 1995) and the consensus is that 20β-hydroxylated forms are the active MS in most species (Canario and Scott, 1988; Trant and Thomas, 1988). The low levels of 17,20βP present in many asynchronous species are thought to reflect a process of short bursts of secretion, local paracrine action and rapid metabolism to biologically inactive forms (reviewed in Hobby and Pankhurst, 1997).

Final oocyte maturation is followed by ovulation and release of the mature ovum to the oviduct or in salmonids, to the body cavity. Ovulation is mediated in most species by LH stimulation of prostaglandin $F_{2\alpha}$ ($PGF_{2\alpha}$). However, in at least three species—the goldeye *Hiodon alosoides* (Pankhurst, 1985), the yellow perch *Perca flavescens* (Goetz, 1997) and the spotted seatrout *Cynoscion nebulosus* (Pinter and Thomas, 1999) 17,20βP is capable of inducing both final oocyte maturation and subsequent ovulation. In yellow perch at least, 17,20βP acts by stimulating production of $PGF_{2\alpha}$ through a mechanism that requires the presence of extrafollicular tissue (Goetz and Garczynski, 1997). The precise nature of the interaction between final oocyte maturation and ovulation, and the integrative role that MS appear to play in the process is the subject of ongoing research (Patiño *et al.*, 2003).

In males, MS regulate the processes of spermiation and milt hydration, and may also have a role in activation of sperm motility. In many species, plasma levels of 17,20βP or 20βS peak in association with the appearance of spermatozoa in the sperm duct (spermiation) and subsequent milt hydration (e.g., Scott and Baynes, 1982; Ueda *et al.*, 1983) and MS levels are also generally elevated in fish showing increased milt volumes after treatment with LH or LH-releasing hormone analogue (LHRH-A) (reviewed in Pankhurst and Poortenaar, 2000; King and Young, 2001; Lim *et al.*, 2004). In a number of other species, 17,20βP or 20βS levels are either very low or undetectable in spermiating fish (e.g., Scott *et al.*, 1991; Pankhurst and Kime, 1991; Dedual and Pankhurst, 1992; Barnett and Pankhurst, 1999; Pankhurst *et al.*, 1999). This appears, as in females, to be the result of short cycles of secretion and rapid metabolism (Canário and Scott, 1990a; Vermeirssen *et al.*, 1998). This is further supported by the finding that in both the greenback flounder (Pankhurst and Poortenaar, 2000) and the New Zealand snapper *Pagrus auratus* (Pankhurst, 1994), 17,20βP levels are typically quite low in naturally spermiating fish and increase only modestly in fish treated with LH or LHRH-A. However, injection of the immediate precursor 17-hydroxyprogesterone (17P) causes a marked elevation in plasma 17,20βP indicating that the normally low plasma levels result from the combination of limited substrate availability and rapid clearance after synthesis.

The cellular mechanisms involved in the stimulation of milt production are not known (Nagahama, 2000), but it is clear that MS stimulate two distinct processes—the release of spermatozoa to the sperm ducts and milt hydration (Pankhurst and Poortenaar, 2000). Treatment of mature fish with either acute or chronic doses of LHRH-A suggests that milt hydration is a rapid process, whereas spermiation requires longer exposure to elevated LH and MS levels (Mylonas *et al.*, 1997). There is also evidence that MS can regulate sperm motility through modulation of sperm duct pH and sperm intracellular cAMP concentrations (Miura *et al.*, 1991a; Morisawa *et al.*, 1993; Miura and Miura, 2003).

MS and their metabolites have widespread action as sexual pheromones producing both endocrine and behavioural responses (reviewed by Sorensen *et al.*, 2000; Stacey, 2003). An updated synthesis of the subject appears in this volume (see Chapter 6: N.E. Stacey and P.W. Sorensen). There is also evidence for a direct role for 17,20βP in modulating behaviour. Castrated rainbow trout implanted with 17,20βP showed partially restored spawning behaviour when exposed to nesting ovulated females (Mayer *et al.*, 1994). Additional effects of MS include central feedback action and short loop feedback effects on steroidogenesis and these effects are discussed together with the feedback effects of other steroids in the section on Steroid Feedback.

Androgens

Androgens occur in the plasma of both males and females. However, in females, androgens primarily serve as gonadal substrates for further metabolism by cytochrome P450 aromatase (P_{450} arom) to C_{18} steroids, principally E_2. Males typically have low levels of P_{450} arom, producing instead in most species, the active androgen 11-ketotestosterone (11KT) via several possible metabolic routes (reviewed by Borg, 1994; Nagahama, 2000).Other androgens do occur in the plasma with T, androstenedione (A), and 11β-hydroxytestosterone (OHT), all being reported as being elevated in the plasma of a range of sexually mature fish (reviewed in

Borg, 1994). Whether these androgens have direct effects themselves or are metabolic intermediates is less clear.

Males do produce low levels of estrogens, and E_2 appears to have a role in stimulating spermatogonial stem cell replacement in the testis via an estrogen receptor (ER) mediated process (Miura and Miura, 2003). 11KT appears to have a primary role in stimulating spermatogonial proliferation (Schulz et al., 2000). In the testis of the Japanese eel *Anguilla japonica*, at least, it is capable of stimulating all stages of spermatogenesis (Miura et al., 1991b). Androgen receptors have yet to be identified in germ cells but are present in Sertoli cells, suggesting that Sertoli cells mediate androgen effects on germ cell development through a molecular mechanism which is still not clearly understood (Nagahama et al., 2000; Schulz et al., 2000). Recent studies have shown that the onset of spermatogenesis is associated with the upregulation by 11KT of the growth factor, activin B, and the concomitant suppression of a spermatogenesis-inhibiting factor (Miura and Miura, 2003).

Androgens have also been reported to be associated with spermiation and milt hydration but the evidence for this is equivocal. Extended treatment of grey mullet *Mugil cephalus* with methyl testosterone was reported to result in spermiation (Weber and Lee, 1985), and similar results were obtained by long-term treatment of black porgy *Acanthopagrus schlegeli* (Lau et al., 1997) and Atlantic salmon *Salmo salar* parr (Berglund et al., 1995) with T. In contrast, androgens had no effect on spermiation or milt volume in rainbow trout (Billard et al., 1981), amago salmon *Oncorhynchus rhodurus*, and goldfish *Carassius auratus* (Ueda et al., 1985), NZ snapper (Pankhurst, 1994) or greenback flounder (Pankhurst and Poortenaar, 2000). It seems likely that the stimulatory effects of androgens reported by Weber and Lee (1985) and others have resulted from indirect effects (see Steroid Feedback, p. 80) rather than a direct gonadal effect of androgens on spermiation.

A significant effect of androgens in all vertebrates is the stimulation of the development of male secondary sex characteristics. The effects are varied but are generally associated with the assumption of external features that advertise reproductive readiness or fitness, or features which enhance the capacity to assume and retain a territory. For example, assumption of male coloration in stoplight parrotfish *Sparisoma viride* is correlated with increased plasma 11KT levels, and exogenous 11KT produces the same changes in females (Cardwell and Liley, 1991a). Breeding coloration in sticklebacks, guppies and Pacific salmon is also androgen-dependent, as is fin ray development and male fin coloration in cichlids, platyfish and medaka *Oryzias latipes* (reviewed in Borg, 1994), and assumption of spawning coloration in the demoiselle *Chromis dispilus* is associated with elevated plasma levels of T and 11KT (Pankhurst, 1990). Androgens cause epidermal thickening in salmonids (Pottinger and Pickering, 1985) and are also implicated in the increase in heart size seen in reproductively active male trout (Davie and Thorarensen, 1997). In blue-banded gobies, *Lythrypnus dalli*, treatment of females with 11KT results in the development of male-characteristic genital papilla morphology (Carlisle *et al.*, 2000). Other androgen-dependent secondary sexual characteristics include hypertrophy of the kidney of sticklebacks and production of a sticky glue that is used to secure nesting material (Borg, 1994), and the growth of sound-producing muscles in acoustically courting, territorial male plainfin midshipman *Porichthys notatus* under the influence of 11KT (Brantley *et al.*, 1993). Non-territorial sneak-spawning male midshipman have elevated plasma T levels and fully developed testes, but sonic muscles that resemble those of females and juveniles. A similar phenomenon occurs in the brown ghost knifefish *Apteronotus leptorhynchus*, where males produce a characteristic electric organ discharge or 'chirp'. Male-type chirps can be induced in females by treatment with T or dihydrotestosterone (Dulka, 1997).

Androgens are major modifiers of vertebrate behaviour and increases in plasma levels are usually accompanied by increased aggression and territoriality (Pfaff *et al.*, 1994). Elevated plasma levels of T and 11KT are closely associated with territorial behaviour in a wide range of fish species including bluegill *Lepomis macrochirus* (Kindler *et al.*, 1989), the demoiselle (Pankhurst, 1990), stoplight parrotfish (Cardwell and Liley, 1991b) garibaldi *Hypsypops rubicundus* (Sikkel, 1993), three-spined stickleback *Gasterosteus aculeatus* (Páll *et al.*, 2002), and the spiny damselfish *Acanthochromis polyacanthus* (Pankhurst *et al.*, 1999). Castration-replacement experiments show that these behaviours are often directly androgen-dependent (reviewed by Pankhurst, 1995). However, it is also clear that some aspects of androgen-mediated behaviour depend on the social context. For example, as noted above, nesting male demoiselles exhibit a marked increase in plasma T and 11KT levels in association with territorial display and spawning, but artificial elevation of androgens to the same levels in wild fish do not have any effect on behaviour if the fish are brooding eggs (Pankhurst and Carragher, 1995). Androgen levels can, in turn, be modulated by behaviour or social conditions along with factors such as density of conspecifics (Pankhurst and Barnett, 1993; Oliveira *et al.*, 2001a), the intensity of territorial challenge (Cardwell and Liley, 1991b) and the failure to acquire territory in sexually mature fish (Barnett and Pankhurst, 1994) all implicated in the modulation of plasma androgen levels. In general terms, the fish response confirms to the general vertebrate model of the 'challenge' hypothesis, whereby social conditions modulate androgen levels so as to make the levels of androgen-mediated behaviours appropriate to the prevailing social environment (reviewed by Oliveira *et al.*, 2002).

An additional facet of reproductive behaviour is the process of spawning itself. Chronic treatment of female goldfish with 11KT is capable of producing the male-specific behaviours of courtship and spawning, increased activity and reduced feeding behaviour in female fish (Stacey and Kobayashi, 1996). Implants of T produced secondary sexual characteristics but did not modify behaviour. In an extension of this research work, Kobayashi and Nakanishi (1999) demonstrated a similar effect of 11KT in producing male typical sexual behaviours of gynogenetic crucian carp *Carassius auratus langsdorfi*. These authors concluded that unlike mammals where behavioural or 'brain' sex is fixed early in life, teleosts retain bipotential brain sex and this can be modulated through exposure to androgens.

Estrogens

A major role for estrogens in fishes is the stimulation of hepatic synthesis of the phospholipoprotein yolk precursor, vitellogenin (Vtg). In a majority of species, the principal estrogen involved is 17β-estradiol (E_2). However, estrone (E_1) may have a minor vitellogenic role in some species (reviewed in Specker and Sullivan, 1994; Tyler *et al.*, 2000). E_2 secreted by the ovarian follicles is carried via the circulation to the liver where it binds to estrogen receptors (ER) in the hepatocyte cytoplasm. Three ER types have been identified in teleosts; α and β which are both hepatic receptors and γ, which have so far only been found in the ovary of Atlantic croaker (Hawkins *et al.*, 2000). E_2-ER binding stimulates the activity of Vtg gene(s), with various studies showing that teleosts may have one to several copies of the gene (reviewed in Watts *et al.*, 2003). The structure of Vtg varies among species but Vtg is typically a large dimeric molecule with molecular weights exceeding 500 kDa in some species (Sun *et al.*, 2003). Plasma titres of Vtg can be very high during vitellogenesis, with values being particularly high in salmonids where levels can exceed $50 \text{ mg} \cdot \text{ml}^{-1}$

(King *et al.*, 2003). In non-salmonids, plasma titres tend to be lower (e.g., ~ 2 mg·ml^{-1} in the greenback flounder *Rhombosolea tapirina*; Sun and Pankhurst, 1994). Vtg is carried in the bloodstream to the ovary, where it is taken up into the developing oocyte via Vtg-receptor mediated endocytosis, and cleaved into several yolk proteins (Tyler *et al.*, 2000). Males also possess genes for Vtg but expression is usually prevented by low plasma E_2 titres (Sun *et al.*, 2003).

The second important effect of estrogens is to stimulate hepatic synthesis of precursors of three structural proteins (ZP) that will form the zona pellucida of the developing oocyte and, subsequently, the chorion of the mature egg. These proteins are smaller than Vtg and typically have molecular weights in the 50-70 kDa range (Tyler *et al.*, 2000). Genes coding for ZP appear to show higher sensitivity to estrogens than Vtg genes, and shorter time frames for synthesis, and also appearance in the plasma after exposure to estrogens (Celius *et al.*, 2000; Berg *et al.*, 2004; Fujita *et al.*, 2004). Here also, hepatic sensitivity is greater to E_2 than to other estrogens (Celius and Walther, 2000). However, recent studies have shown that induction of ZP is not entirely estrogen-dependent, with cortisol having the capacity to potentiate the effects of E_2 on ZP expression (Westerlund *et al.*, 2001; Berg *et al.*, 2004).

Steroid Feedback

It is a well-established fact that gonadal steroids can modulate both the synthesis and release of gonadotropins by the pituitary. Studies on goldfish, rainbow trout, African catfish *Clarias gariepinus* and the European eel *Anguilla anguilla* have shown that gonadectomy increases plasma LH levels, and that T and, or E_2 attenuate this effect but increase LH accumulation by gonadotropes in the pituitary (reviewed in Kobayashi and Stacey, 1990). Similarly, studies on pubertal development in Atlantic salmon, rainbow trout, platyfish *Xiphophorus maculatus*, and European and Japanese eels show that estrogens and androgens increase LHβ subunit synthesis, but appear to have little effect on FSH synthesis (reviewed by Dufour *et al.*, 2000). Other recent studies have shown a differential effect of steroids on FSH, and LH expression, synthesis and release. Implantation of Atlantic salmon parr with T had no effect on FSH but increased the pituitary LH content (Antonopoulou *et al.*, 1999a). Treatment with an aromatase inhibitor decreased pituitary and plasma LH levels in a manner similar to castration, indicating the presence of aromatase-dependent positive feedback on LH synthesis and release. There was also evidence of aromatase-dependent negative feedback on FSH synthesis (Antonopoulou *et al.*, 1999b). Castration of three-spined stickleback reduced expression of LHβ subunit but had no effect on FSHβ, in concert with decreased plasma androgen levels (Hellqvist *et al.*, 2001). In goldfish, ovarectomy early in maturation resulted in increased FSHβ expression, and this could be suppressed by T, E_2 and 11KT. T and E_2 but not 11KT resulted in increased LHβ expression in intact fish leading to the conclusion that gonadal steroids inhibit FSHβ expression, and T and E_2 stimulate LHβ expression during early stages of maturation (Kobayashi *et al.*, 2000). E_2 treatment also increases pituitary LH mRNA levels and plasma LH in black porgy *Acanthopagrus schlegeli* (Du *et al.*, 2001), and high levels of T-induced LHβ expression and suppressed FSHβ expression in tilapia hybrids (Melamed *et al.*, 1997). The same study showed that at low concentrations, T also had a stimulatory effect on FSHβ expression, indicating the probability of a concentration component to the directional effect of steroid feedback on FSH expression.

Gonadal steroids also have regulatory effects at other levels in the endocrine cascade, with T and E_2 increasing pituitary responsiveness to GnRH in goldfish, E_2 having the same effect in rainbow trout, and 11KT reducing responsiveness in African catfish (reviewed in Trudeau and Peter, 1995). Steroids also have the capacity to modulate the inhibitory and stimulatory effects of dopamine and γ-aminobutyric acid, respectively, on the pituitary response to GnRH (Trudeau and Peter, 1995). Some of these effects are directly mediated by steroid modulation of the GnRH system. T and E_2 implants increase the GnRH content of hypothalamic neurons without changing the number of cells present in the African catfish. Similar effects are present in masu salmon *Oncorhynchus masou*, rainbow trout, platyfish (reviewed by Dubois *et al.*, 2001) and Indian catfish *Heteropneustes fossilis* (Tiwary *et al.*, 2002). Inhibitory effects have also been reported from Atlantic croaker where the MS 20βS reduces the pituitary responses of mature fish of both sexes, and regressed females to LHRH-A. This is accompanied by parallel decreases in brain pre-optic area and pituitary GnRH content (Mathews *et al.*, 2002).

An additional component of steroid feedback occurs at the gonadal level. As noted in (see p. 72: Modes of steroid action and clearance), membrane receptors for T and E_2 have been identified in the ovary and testis, respectively, of the Atlantic croaker, and these downregulate production of ovarian estrogen and testicular androgen synthesis via a rapid, presumably non-genomic mechanism (reviewed in Thomas, 2003). Govoroun *et al.*, (2001) reported reduced expression of P_{450} c17,

3β-HSD and 11β-hydroxylase in male rainbow trout given dietary E_2. Other steroidogenic enzymes and inter-renal steroid biosynthesis were unaffected, suggesting that this was a specific gonadal effect. These authors concluded that the time course for the effect implicated a classical genomic effect. *In vitro* studies showed a similar lack of effect of E_2 on inter-renal steroidogenesis in rainbow trout but suppressed capacity to convert pregnenolone to cortisol in chinook salmon *Oncorhynchus tschawytscha* (McQuillan et al., 2003). Barry *et al.* (1989) demonstrated a suppressive effect of 17,20βP on *in vitro* 11KT synthesis by testes of carp *Cyprinus carpio*, but the same effect was absent in Atlantic salmon (Antonopoulou, 1997). Collectively, these studies indicate that steroids themselves have the capacity to modulate gonadal steroidogenesis, probably through their effect on the activities of steroidogenic enzymes.

PATTERNS OF CHANGE

Reproductive cycles of fish are entrained by a combination of physical and biological variables and show high diversity across taxa but, typically, have a seasonal component (reviewed in Pankhurst and Porter, 2003). At high latitudes, reproduction tends to be of short duration and most species display synchronous gamete development. The timing of the reproductive phase is strongly regulated by photoperiod with temperature operating as a secondary modifying variable (Bromage *et al.*, 2001). At lower latitudes, fish are likely to display asynchronous ovarian and testicular development with multiple spawning cycles within a reproductive season. Photoperiod is still a significant regulatory variable but temperature change and proximate factors such as rainfall, food availability and social conditions are much more likely to impact on the timing of spawning (Pankhurst and Porter, 2003). Within this group, spawning periodicity tends to cluster into lunar or semi-lunar patterns, particularly among marine species (Robertson *et al.*, 1990), daily spawning (Zohar *et al.*, 1988; Hobby and Pankhurst, 1997), or species where social events such as mate attraction and brood presence determine the timing of subsequent egg clutches (Coward and Bromage, 2000). Mechanisms for transduction of environmental information to an endocrine signal are understood to the extent that it is clear the photoperiod signal is mediated by light phase suppression of secretion of melatonin from the pineal gland (Bromage *et al.*, 2001), and temperature effects appear to be regulated at least in part at the level of stimulation or suppression of steroidogenic enzyme activity or expression (Lim *et al.*, 2003; Pankhurst and Porter,

2003). However, the link between light-modulated melatonin secretion and the demonstrated light-induced changes that occur in LHRH gene expression (Amano et al., 1995; Davies et al., 1999) is yet to be understood. Neither is there any understanding of how lunar or semi-lunar information is transduced to produce corresponding cycles of gonadal activity (Pankhurst and Porter, 2003).

Species with Synchronous Gamete Development

The pattern of steroid secretion shown by fishes in this group is typified by salmonids where there is annual or bi-annual spawning and production of a single ovulatory clutch. The period from the recruitment of oocytes into vitellogenesis to ovulation extends for about 9 months, with the developing oocyte clutch initially showing quite high variability in individual oocyte size, but increasing synchrony as vitellogenesis proceeds (Tyler et al., 1990; King and Pankhurst, 2003). Prior to vitellogenesis, plasma levels of gonadal steroids are low or undetectable. During vitellogenesis, there is a gradual increase in plasma E_2 levels in females, with concomitant patterns of T, albeit generally at higher absolute concentrations. Elevated plasma E_2 levels are accompanied by high-plasma Vtg titres which roughly mirror the pattern of E_2 secretion. E_2 levels peak towards the end of vitellogenesis and then decline rapidly as P_{450} arom activity is suppressed. This results in a transient peak in plasma T levels. Plasma T levels decline as oocyte maturation proceeds, whereas plasma 17,20βP levels rise rapidly (Scott et al., 1980, 1982; Van Der Kraak et al., 1984; Fitzpatrick et al., 1986; Goetz et al., 1987; Mayer et al., 1992; Slater et al., 1994; Pankhurst and Thomas, 1998; Tveiten et al., 1998; King and Pankhurst, 2003). In the species where post-ovulatory measurements have been made, plasma levels of 17,20βP and LH tend to remain high after ovulation and for the period when eggs are held in the body cavity (Fitzpatrick et al., 1986; Liley and Rouger, 1990). This is thought to be related to the role of 17,20βP and LH in maintaining post-ovulatory egg viability prior to spawning (reviewed in Hobby and Pankhurst, 1997b). A synthesis of endocrine events and ovarian development in rainbow trout is shown in Fig. 3.2.

Males show a similar pattern of steroid secretion with an extended period of elevated plasma levels of T and 11KT during spermatogenesis followed by a decline in androgen levels as spawning approaches. Here also, there is a pre-spawning peak in plasma 17,20βP in association with spermiation and milt hydration (Scott et al., 1980; Ueda et al., 1983;

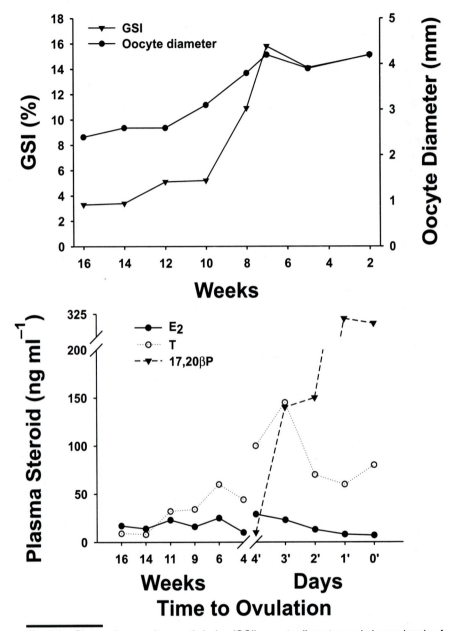

Fig. 3.2 Change in gonadosomatic index (GSI), oocyte diameter and plasma levels of *17β*-estradiol (E$_2$), testosterone (T) and 17,20β-dihydroxy-4-pregnen-3-one (17,20βP) in a Tasmanian population of rainbow trout *Oncorhynchus mykiss*. Composite data from Pankhurst *et al.* (1996), Pankhurst and Thomas (1998), and Pankhurst and Van Der Kraak (2000).

Baynes and Scott, 1985; Mayer *et al.*, 1992; Tveiten *et al.*, 1998). A consistent feature of salmonid reproductive cycles is the high levels of steroids that occur in the plasma. Among females, peak levels of E_2, T and 17,20βP can reach >50, 150-250 and >250 ng·ml^{-1}, respectively (Scott *et al.*, 1980, Van Der Kraak *et al.*, 1984; Pankhurst and Thomas, 1998), and males show peak levels of T, 11KT and 17,20βP of up to 150, 100 and 50 ng·ml^{-1}, respectively (Baynes and Scott, 1985; Mayer *et al.*, 1992). These levels are at the top end of the range of concentrations found in teleosts.

Similar patterns of development, albeit with lower absolute steroid levels are shown by a variety of cooler water species, including wolfish *Anarhichas lupus* (Tveiten *et al.*, 2000; Tveiten and Johnsen, 2001), yellow perch *Perca flavescens* (Dabrowski *et al.*, 1996; Ciereszko *et al.*, 1997; Sulistyo *et al.*, 2000), walleye *Stizostedion vitreum* (Malison *et al.*, 1994), goldeye *Hiodon alosoides* (Pankhurst *et al.*, 1986), and a range of cyprinids (Kagawa *et al.*, 1983; Kobayashi *et al.*, 1986; Aida, 1988; Barry *et al.*, 1990). Differences from the 'salmonid' pattern found among this group include low levels of 17,20βP found in some species along with often very high levels of C_{21} steroid conjugates (e.g., Tveiten *et al.*, 2000), and the behaviour- and temperature-dependent rapid transition to oocyte maturation and ovulation (and accompanying appearance of MS) in cyprinids (Aida, 1988). Goldeye show the additional curiosity of apparent absence of E_2 in the plasma during vitellogenesis (Pankhurst *et al.*, 1986).

The same patterns of reproductive development are shown by deeper water species at all latitudes where temperatures are generally low (<5°C) and stable, and periods of ovarian or testicular development are long (Pankhurst *et al.*, 1987). There is little information on endocrine function among this group of fishes. However, there exist limited data available for mid slope-dwelling species orange roughy *Hoplostethus atlanticus*. Orange roughy occur at depths of 800-1300 m and spawning occurs in mid winter, with the timing in New Zealand populations being extremely consistent from year to year. The precise date changes with latitude in such a way that the day length for particular stages of the spawning process is identical among different populations from different latitudes. This indicates the presence of a strong photoperiod cue used in seasonal phasing of reproduction (Pankhurst, 1988). Plasma steroid data from fish sampled live from trawl catches showed a steady increase in plasma T levels through gametogenesis in both sexes, to a peak at ~10 ng·ml^{-1} early in the spawning phase. Plasma E_2 levels remained low throughout

(< 1 ng·ml^{-1}), but were elevated in late stage vitellogenic females. Plasma 17P and 17,20βP levels were low at all times in females and non-detectable in males (Pankhurst and Conroy, 1988). There were, however, elevations in 11-deoxycortisol (11-DOC) in females undergoing final oocyte maturation, and 11-DOC does have maturational potency *in vitro* whereas 17,20βP does not (Pankhurst, 1987). However, as ovarian follicles do not appear to produce 11-DOC in response to gonadotropin (Pankhurst, 1987), the identity of the MS in orange roughy remains unconfirmed.

An additional group of fishes that has characteristics similar to high latitude annual spawners is the egg-laying or live-bearing elasmobranchs. There is much less information available on elasmobranch than teleost endocrine cycles but the pattern seems to be one of increased T and E$_2$ levels in association with vitellogenesis in females, a spike in progesterone (P) levels in association with ovulation and egg capsule deposition, and possibly elevated P levels in gestating live bearers (Sumpter and Dodd, 1979; Koob *et al.*, 1986; Heupel *et al.*, 1999; Koob and Callard, 1999; Rasmussen *et al.*, 1999). These patterns are seasonally synchronized in some species (e.g., Heupel *et al.*, 1999; Rasmussen *et al.*, 1999) but others appear to be aseasonal (Koob and Callard, 1999). This may reflect the reduced dependence for species producing large eggs or live young to synchronize hatching or birth with seasonal peaks in primary or secondary production. Males show elevated T in association with clasper elongation and calcification (which marks the transition to sexual maturity) and spermatogenesis (Heupel *et al.*, 1999). There is little information on the possible presence of other androgens. However, recent studies on the draughtboard shark *Cephaloscyllium laticeps* show that 11KT is never detectable in the plasma despite plasma T levels of up to 14 ng·ml^{-1} in mature fish. P levels reach 12 ng·ml^{-1} in females but are at low levels at all times in males (C. Awruch, N.W. Pankhurst and S.F. Thrusher, unpublished data). It is currently not clear whether there is an elasmobranch equivalent for the role of the MS in teleost males, or whether the whole process is androgen-mediated.

A comparable teleost group may be the live bearers. The limited evidence available suggests that endocrine patterns similar to those described for elasmobranchs also occur in this group. The guppy *Poecilia reticulata* is an ovo-viviparous species with fertilization and gestation occurring within the follicle. Plasma T and E$_2$ levels are elevated during vitellogenesis and also at parturition (when recruitment of the next oocyte batch occurs), but low through gestation. P, 17P and 17,20βP levels remain relatively constant, making it unclear whether C$_{21}$ steroids have a role in maintaining pregnancy (Venkatesh *et al.*, 1990). Sailfin molly *Poecilia latipinna* showed a similar pattern of T and E_2; 17P was variable and not apparently related to maturity or gestational stage, and 17,20βP was generally non-detectable (Kime and Groves, 1986). A more comprehensive examination of C_{21} steroids during gestation in the viviparous rockfish *Sebastes rastrelliger* showed elevated concentrations of 17,20β-dihydroxylated steroids and 17,20αP during gestation and low or undetectable levels at other times (Moore *et al.*, 2000). The authors of that study concluded that an array of C_{21} steroids might have progestational activity in the rockfish.

Species with Asynchronous Gamete Development

Patterns of ovarian and testicular development are more variable among this group but, as noted earlier, there is the common thread of extended seasonal spawning with multiple cycles of gamete maturation and spawning. Many species within this group are marine and the characteristics of endocrine patterns in marine fishes with asynchronous development have been previously reviewed by Pankhurst and Carragher (1991). Consistent features identified among this group were typically low plasma levels of androgens and estrogens (<1-2 ng·ml^{-1}). Where higher steroid levels occurred, they tended to be among species with longer ovulatory periodicity. MS levels were also generally low or undetectable, or showed little change across gonadal cycles. This was identified as probably resulting from the effects of further metabolism and clearance as noted (see p. 83: Species with Synchronous Gamete Development), and also sampling regimes that were too coarse to detect short-term changes in ovarian or testicular condition. The following discussion considers these themes in the light of more recent data, concentrating on two common patterns of spawning among asynchronous species; lunar or semi-lunar spawning, and daily spawning.

As discussed earlier, lunar and semi-lunar patterns are common among marine species and thought to serve a variety of functions including synchronization of adult reproductive cycles, and maximizing off-reef transport of eggs and larvae by matching spawning events to periods of maximum tidal flux (Robertson *et al.*, 1990). Species within this group tend to have clearly defined gonadal cycles with development of a discrete oocyte clutch for each spawning event and, in consequence, patterns of plasma steroids that are not dissimilar to those of group synchronous species. The main difference is that there is not necessarily a fall in plasma T and E_2 levels at ovulation, especially if there are further developing oocyte clutches in the ovary. In some species, highest T and E_2 levels occur in females undergoing oocyte maturation and ovulation (Pankhurst et al., 1999). In the killifish *Fundulus heteroclitus*, there is semi-lunar spawning with vitellogenesis and maturation of an oocyte clutch occurring over ~6 days prior to spawning. This is mirrored by an increase in plasma E_2 over the same period, with subsequently low E_2 levels for the 6 days following spawning (Bradford and Taylor, 1987; Cerdá et al., 1996). In the Gulf killifish *F. grandis*, there is a similar pattern of semi-lunar spawning but finer scale sampling showed that there were also daily changes in steroid levels with midnight peaks in plasma T in the few days prior to spawning (Emata et al., 1991). Tropical siganids (rabbitfishes) display strong lunar patterns of reproductive activity with spawning periods that extend over several months. Golden rabbitfish *Siganus guttatus* spawn at the first quarter of the moon and females show elevated plasma E_2, T, 17,20βP and 20βS levels leading up to the spawning period, with E_2 and T levels in excess of 10 ng·ml^{-1} (Rahman et al., 2000a). Males display a similar pattern but here, steroid levels reach ~20, 5 and 10 ng·ml^{-1} for T, 11KT and 17,20βP, respectively (Rahman et al., 2000b). This is consistent with an earlier view (Pankhurst and Carragher, 1991) that absolute steroid levels tend to be higher in species with longer intervals between spawning events. Other species of rabbitfish show similar trends but different timing for spawning. For example, forktail rabbitfish *S. argenteus* spawn in the last lunar quarter and have elevated plasma T, 11KT and 17,20βP levels in males (Rahman et al., 2003a), and E_2 and 17,20βP in females (Rahman et al., 2003b) in association with the spawning period.

At the shorter end of the scale is the daily spawning displayed by groups such as sparids (seabreams) and labrids (wrasses). The gilthead seabream *Sparus aurata* exhibits a peak in plasma E_2 6h before ovulation but no apparent change in 17,20βP levels (Zohar et al., 1988). Red seabream *Pagrus major* ovulate at 13:00–14:00h and spawn between 16:00 and 19:00h. There is an early morning peak (~0.4 ng·ml^{-1}) of 17,20βP in association with the onset of oocyte maturation, and a peak in E_2 (~1.2 ng·ml^{-1}) that appears to be associated with recruitment of the following vitellogenic clutch (Matsuyama et al., 1988; Kagawa et al., 1991). New Zealand snapper *Pagrus auratus* also spawn every evening with early afternoon ovulation (Scott et al., 1993). There is an extended spawning season of several months during which plasma E_2 and T levels remain elevated (Carragher and Pankhurst, 1993). Finer scale (3 hourly) sampling showed that there was no clear diel pattern in plasma levels of either T or E_2, but that ovarian levels of T and E_2 peaked from 21:00 to 24:00h in association with the final stages of vitellogenesis in the most mature clutch. In contrast, there was no change in ovarian levels of 17,20βP but a plasma peak at 09:00h as oocytes were undergoing hydration (Hobby and Pankhurst, 1997a). As with other sparids, peak hormone levels in New Zealand snapper were typically less than 2 ng·ml^{-1} (Carragher and Pankhurst, 1993; Hobby and Pankhurst, 1997a). A summary of annual and daily changes in plasma steroids in New Zealand snapper can be seen in Figs. 3.3 and 3.4.

Similar patterns have been described for the wrasse *Pseudolabrus japonicus* which spawns between 06:00 and 09:00h and has peak plasma levels of 17,20βP and 20βS at the time of initiation of oocyte maturation (Matsuyama *et al.*, 1998). In one of the few studies examining short-term changes in males, Matsuyama *et al.* (1995) described a daily cycle of spermatogenesis and spermiation in male *P. japonicus* with a midday peak in 11KT, and midday and midnight peaks in plasma 17,20βP. Sundaray *et al.* (2003) also found diel changes in plasma T, 11KT and 17,20βP in males of *P. sieboldi*. Patterns of this type are not restricted to labrids and sparids. Daily spawning Japanese whiting *Sillago japonica* complete oocyte maturation at ~16:00h and ovulate at 20:00h. Plasma E_2 and 17,20βP levels both peak at ~16:00h in association with vitellogenic recruitment of the following clutch, and maturation of the leading clutch, respectively (Matsuyama *et al.*, 1990). It also appears that diel changes in hormone levels can occur outside the period when such changes can be explained by corresponding diel shifts in ovarian or testicular condition. Lambda *et al.* (1983) described a circadian pattern of plasma T, E_2 and cortisol across a range of maturity states in catfish. More recently, Bayarri *et al.* (2004) found a daily cycle of brain sbGnRH content and plasma levels of LH and T in male sea bass *Dicentrarchus labrax* early in spermatogenesis.

From all of the above studies it is clear that ovarian and testicular changes occur over short timeframes in daily spawners, and there are correspondingly short episodes of elevations of plasma steroids. Hormone levels tend to be low and there may be no detectable changes in the plasma compartment despite steroidogenic activity at the gonadal level. This suggests that adequate description of endocrine events in species with short ovulatory cycles requires sampling strategies with short intervals between sampling, and measurement of hormone levels in both the plasma and ovarian or testicular compartment. There would also appear to be

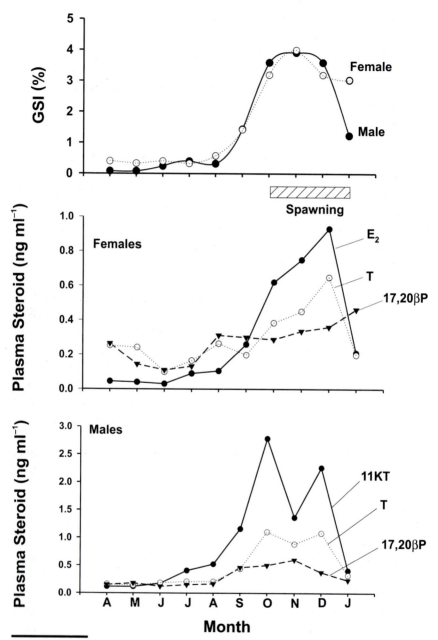

Fig. 3.3   Seasonal change in gonadosomatic index (GSI), and plasma levels of *17β*-estradiol (E$_2$), Testosterone (T) and 17,20β-dihydroxy-4-pregnen-3-one (17,20βP) in female, and T, 11-ketotestosterone (11KT) and 17,20βP in male new Zealand snapper *Pagrus auratus*. Data from Carragher and Pankhurst (1993).

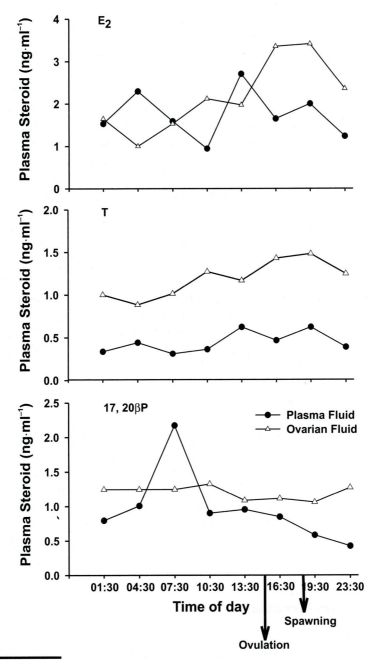

Fig. 3.4 Daily changes in plasma and ovarian fluid levels of 17β-estradiol (E_2), testosterone (T) and 17,20β-dihydroxy-4-pregnen-3-one (17,20βP) in daily spawning female New Zealand snapper *Pagrus auratus*. Data from Hobby and Pankhurst (1997a).

advantages in studies that utilize the measurement of steroidogenic enzyme expression as an alternative to measurement of steroid products, especially where periods of hormone synthesis are short. This approach is still relatively unexplored but there does appear to be a good correlation between plasma steroid levels and expression of some steroidogenic enzymes, further suggesting that steroidogenic enzyme gene transcription is the primary determinant of steroidogenesis (Kumar *et al.*, 2000).

There is a substantial number of additional studies from species with a known extended spawning season with multiple spawning events, but the ovulatory periodicity is unknown. Generally, these studies display extended periods of elevations of plasma T and E_2 in females across the period of vitellogenesis and spawning (e.g., Pankhurst and Conroy, 1987; Prat *et al.*, 1990; Berlinsky and Specker, 1991; Dedual and Pankhurst, 1992; Clearwater and Pankhurst, 1994; Murayama *et al.*, 1994; Haddy and Pankhurst, 1998; Johnston *et al.*, 1998; Rocha and Reis-Henriques, 1999; Asturiano *et al.*, 2000; Poortenaar *et al.*, 2001; Lee and Yang, 2002). Among the males, there are typically elevations in plasma T and 11KT throughout the period of reproductive development and spawning (Prat *et al.*, 1990; Dedual and Pankhurst, 1992; Berlinsky *et al.*, 1995; Haddy and Pankhurst, 1998b; Johnson *et al.*, 1998; Poortenaar *et al.*, 2001). The situation with MS is less clear with no detectable change being reported (e.g., Pankhurst and Conroy, 1987; Pankhurst and Kime, 1991; Johnson *et al.*, 1998; Lee and Yang, 2002). In other studies, there are elevated 17,20βP or 20βS levels most commonly in females in association with final oocyte maturation (e.g., Murayama *et al.*, 1994; Haddy and Pankhurst, 1998b; Rocha and Reis-Henriques, 1999; Asturiano *et al.*, 2000; Poortenaar *et al.*, 2001). The ambiguity of data on MS levels during maturation in asynchronous species presumably results from the difficulty of generating endocrine correlates from sampling intervals that are wider than the periods over which significant reproductive events occur. There is no question that the precision of the correlation of plasma steroid levels with reproductive processes improves in general proportion to the knowledge of the timing of those processes.

Other Considerations

As described earlier (p. refer: pages 74-79), both MS and androgens have the capacity to modulate behaviour and, in turn, be modulated by it. This means that social status and immediate behavioural history can affect the patterns of steroid hormones evolved during reproductive events. This effect appears to be most marked in males where dominance, territorial acquisition and defence, and size can all regulate absolute androgen levels (Pankhurst, 1995). For example, spermiated blue cod have elevated plasma T and 11KT levels but large males (which are loosely territorial) have absolutely higher levels of both hormones than smaller males (Pankhurst and Kime, 1991). Similarly, reproductively mature demoiselles have cyclically elevated plasma T and 11KT levels in association with cycles of spermiation, but these levels are always higher in territorial rather than non-territorial fish. Territory holders also show spawning-related increases in plasma 17,20βP which are absent from non-territorial fish (Barnett and Pankhurst, 1994).This type of behavioural disparity is most marked in species where males display alternative life history tactics, usually taking the form of large 'male coloration' territorial or 'Type I' males, and smaller 'female coloration' sneak spawning non-territorial 'Type II' males. Examples include the saddleback wrasse *Thalassoma duperrey* (Hourigan *et al.*, 1991), plainfin midshipman (Sisneros *et al.*, 2004) and the Azorean rock pool blenny *Parablennius parvicornis* (Oliveira *et al.*, 2001b). Type I males typically have much higher levels and longer periods of elevation of plasma 11KT levels, despite both Type I and Type II males having fully mature testes and elevated plasma T levels.

There is additional evidence of behavioural modulation of steroid hormone levels from studies conducted on salmonids. During spawning, both males and females show elevated plasma 17,20βP in association with courting and nest-digging behaviours (Liley *et al.*, 1986a,b; Liley and Roger, 1990). This is one of the few examples of behavioural modulation of steroid hormone levels in female fishes. Another interesting case of social or behavioural modulation of steroid cycles of females occurs in Tilapia *Oreochromis mossambicus*, where females mouthbrood spawned eggs for 2-4 weeks. Non-brooding fish display an ovulatory cycle of ~ 25 days with bimodal peaks in T and E_2 early in vitellogenesis and after spawning in association with recruitment of the next vitellogenic clutch, respectively. In brooding fish, the pattern of T is the same but with delay, whereas E_2 levels rise steadily through the brooding phase (Smith and Haley, 1988). In the closely related *Oreochromis niloticus*, brooders have a slower increase in both plasma T and E_2 during vitellogenesis than non-brooders, although here the absolute levels do appear to be directly related to ovarian stage (Tacon *et al.*, 2000). The message from this and other studies is that assessment of possible behavioural influences on hormone levels is a requisite for understanding of the significance of changes in hormone levels in association with reproductive events.

Another potentially confounding factor is that a considerable number of species undergo sex inversion at some stage in their reproductive lives and this, in turn, affects the patterns of steroid hormone production. Among protandrous species, male stages are dominated by production of 11KT, whereas transitional phases typically have low or reducing levels of 11KT and concomitant increases in E_2, often before vitellogenesis occurs. Among protogynous species, the pattern is reversed with increased 11KT levels, marking the transition to functional male status (reviewed by Frisch, 2004). A sub-set of sex-inverting species displays bi-directional sex change, further complicating the seasonal gonadal developmental pattern (Kroon et al., 2003). The effect here is that endocrine changes in potentially sex-inverting species need to be carefully benchmarked against gonadal condition of individual fish.

Finally, there are 'oddities' in certain species, suggesting that assignment of function to various steroids might not be as simple as originally thought. An example of this is provided by the New Zealand anguillid eels *Anguilla dieffenbachii* and *A. australis*. Sexually maturing migratory females have predictably elevated plasma T and E_2 levels, and males exhibit elevated T, 11KT and androstenedione (Lokman and Young, 1998; Lokman et al., 1998). However, females also show very high levels of 11KT, at variance with the notion of 11KT as the 'male' steroid. In eels, this may be associated with the possible androgen-mediation of the suite of somatic and sensory changes that occur in migrating anguillid eels (Pankhurst, 1984; Lokman et al., 1998). There is also evidence that 11KT potentiates the effect of E_2 in stimulating hepatic synthesis of Vtg (Asanuma et al., 2003). It does sound a warning against only measuring steroids that are expected to be present. In a similar vein, recrudescing goldeye show no evidence of measurable E_2 in the plasma, despite a synchronous ovarian developmental mode and high levels of T (Pankhurst et al., 1986). It is unclear whether this is due to another hormone being responsible for stimulating vitellogenesis, or a result of extremely restricted periods of elevated plasma E_2 levels. Red gurnard display the latter phenomenon, where despite a spawning period of over 6 months, plasma E_2 levels are only elevated at the beginning of vitellogenesis (Clearwater and Pankhurst, 1994). This indicates that elevated E_2 levels are not a requisite for maintaining vitellogenesis, although it also appears that an initial E_2 pulse is required to initiate Vtg gene expression (Vaillant et al., 1988). As Kime (1993) notes, there is high diversity of steroid products produced by the gonads of teleost fish but we tend to routinely measure only a few of these, with the result that there may well be other steroid products that play major roles in reproductive events. With that proviso, it is interesting to note that irrespective of function, the most commonly elevated steroid in the plasma of sexually mature fishes of both sexes is T (Pankhurst and Carragher, 1991). Fitzpatrick *et al.* (1987) noted that elevated plasma T was the best indicator of the potential responsiveness of coho salmon *Oncorhynchus kisutch* to LHRH-A. This suggests that even in species where T might not be the primary or active androgen, it remains perhaps the most useful and easily applied endocrine marker of reproductive events.

SUMMARY

Steroid hormones are ubiquitous features of vertebrate endocrine systems and have major roles in the regulation of gonadal growth and maturation. In fishes, gonadal steroids are synthesized in the follicle cells of the ovary, and the Leydig cells of the testis in females and males, respectively. As in other vertebrates, steroid synthesis occurs through the step-wise cleavage of the precursor molecule cholesterol, following the transfer of cholesterol to the inner membranes of mitochondria via the action of steroidogenic acute regulatory protein. Each cleavage product, in turn, has the capacity to act as an end-product hormone or the precursor for further metabolism. Suites of steroid-cleaving enzymes present in fishes exhibit considerable variation with corresponding variability in the steroids produced; however, some general patterns emerge. Twenty-one carbon (C_{21}) steroids (progestins) have maturational activity in both sexes. They also function as pheromones in many species. C_{19} steroids (androgens) regulate gonadal growth and secondary sexual characteristics in males, behaviour patterns in both sexes, and have central feedback action. Androgens also form the precursors for C_{18} (estrogen) steroid production. Estrogens stimulate the hepatic synthesis of yolk and eggshell precursors and also have central and local feedback action.

Patterns of change in plasma levels of steroid hormones are strongly influenced by specific characteristics of gonadal development. In species with synchronous ovarian development there is typically a slow increase in estrogens and androgens through gametogenesis, and a marked peak in C_{21} steroids in association with gamete maturation. Species displaying asynchronous gamete maturation often have more confused patterns of plasma steroids with seasonally elevated estrogen and androgen levels, but often no detectable peaks in C_{21} steroids. Short-term plasma and tissue measurements show that C_{21} peaks do occur but are often of small magnitude and short duration. Some species display eccentric patterns of steroid production with the apparent absence of estrogens during vitellogenesis, or the presence of high levels of 'male-specific' androgens in females. 'Oddities' of this type may be more widespread than currently thought, as most studies concentrate on the measurement of a small number of steroids for which assays are available, despite evidence that multiple steroidal metabolites are present in most species examined. With that proviso, measurement of plasma levels of key steroids remains an essential tool for determining reproductive endocrine status, and understanding the physiological control of reproductive processes.

Acknowledgements

Thanks are extended to Shilo Ludke for preparation of the figures and assistance with manuscript organization.

References

Aida, K. 1988. A review of plasma hormone changes during ovulation in cyprinid fishes. *Aquaculture* 74: 11-21.

Amano, M., S. Hyodo, S. Kitamura, K. Ikuta, Y. Suzuki, A. Urano and K. Aida. 1995. Short photoperiod accelerates preoptic and ventral telencephalic salmon GnRH synthesis and precocious maturation in underyearling male masu salmon. *General and Comparative Endocrinology* 99: 22-27.

Antonopoulou, E., S. Jakobsson, I. Mayer, P. Swanson and B. Borg. 1997. In vivo and in vitro effects of 17α,20β-dihydroxy-4-pregnen-3-one on testicular androgens in Atlantic salmon, *Salmo salar*, mature parr. *Journal of Experimental Zoology* 277: 66-71.

Antonopoulou, E., I. Mayer, B. Borg, P. Swanson, I. Murza and O. Christoforov. 1999a. Effects of testosterone on gonadotropins, testes and plasma 17α,20β-dihydroxy-4-pregnen-3-one levels in postbreeding mature Atlantic salmon, *Salmo salar*, male parr. *Journal of Experimental Zoology* 284: 425-436.

Antonopoulou, E., P. Swanson, I. Mayer and B. Borg. 1999b. Feedback control of gonadotropins in Atlantic salmon, *Salmo salar*, male parr. II. Aromatase inhibitor and androgen effects. *General and Comparative Endocrinology* 114: 142-150.

Artemenko, I.P., D. Zhao, D.B. Hales, K.H. Hales and C.R. Jefcoate. 2001. Mitochondrial processing of newly synthesized steroidogenic acute regulatory protein (StAR), but not total StAR mediates cholesterol transfer to cytochrome P450 side chain cleavage enzyme in adrenal cells. *Journal of Biological Chemistry* 276: 46583-46596.

Asanuma, H., H. Ohashi, H. Matsubara, S. Ijiri, T. Matsubara, T. Todo, S. Dachi and K. Yamauchi. 2003. 11-ketotestosterone potentiates estrogen-induced vitellogenin production in liver of Japanese eel (*Anguilla japonica*). *Fish Physiology and Biochemistry* 28: 383-384.

Asturiano, J.F., L.A. Sorbera, J. Ramos, D.E. Kime, M. Carrillo and S. Zanuy. 2000. Hormonal regulation of the European sea bass reproductive cycle: an individualized female approach. *Journal of Fish Biology* 56: 1155-1172.

Barnett, C.W. and N.W. Pankhurst. 1994. Changes in plasma levels of gonadal steroids and gonad morphology during the spawning cycle of male and female demoiselles *Chromis dispilus* (Pisces: Pomacentridae). *General and Comparative Endocrinology* 93: 260-274.

Barnett, C.W. and N.W. Pankhurst. 1999. Reproductive biology and endocrinology of greenback flounder *Rhombosolea tapirina* (Günther 1862). *Marine and Freshwater Research* 50: 35-42.

Barry, T.P., K. Aida and I. Hanyu. 1989. Effects of 17α,20β-dihydroxy-4-pregnen-3-one on the in vitro production of 11-ketotestosterone by testicular fragments of the common carp, *Cyprinus carpio. Journal of Experimental Zoology* 251: 117-120.

Barry, T.P., A.J.G. Santos, K. Furukawa, K. Aida and I. Hanyu. 1990. Steroid profiles during spawning in male common carp. *General and Comparative Endocrinology* 80: 223-231.

Bayarri, M.J., L. Rodríguez, S. Zanuy, J.A. Madrid, F.J. Sánchez-Vázquez, H. Kagawa, K. Okuzawa and M. Carrillo. 2004. Effect of photoperiod manipulation on the daily rhythms of melatonin and reproductive hormones in caged European sea bass *(Dicentrarchus labrax)*. *General and Comparative Endocrinology* 136: 72-81.

Baynes, S.M. and A.P.Scott. 1985. Seasonal variations in parameters of milt production and in plasma concentration of sex steroids of male rainbow trout (*Salmo gairdneri*). *General and Comparative Endocrinology* 57: 150-160.

Berg, A.H., L.Westerlund and P.E. Olsson. 2004. Regulation of Arctic char (*Salvelinus alpinus*) egg shell proteins and vitellogenin during reproduction and in response to 17β-estradiol and cortisol. *General and Comparative Endocrinology* 135: 276-285.

Berglund, I., E. Antonopoulou, I. Mayer and B. Borg. 1995. Stimulatory and inhibitory effects of testosterone on testes in Atlantic salmon male parr. *Journal of Fish Biology* 47: 586-598.

Berlinsky, D.L. and J.L. Specker. 1991. Changes in gonadal hormones during oocyte development in the striped bass, *Morone saxatilis*. *Fish Physiology and Biochemistry* 9: 51-62.

Berlinsky, D.L., L.F. Jackson, T.I.J. Smith and C.V. Sullivan. 1995. The annual reproductive cycle of the white bass *Morone chrysops. Journal of the World Aquaculture Society* 26: 252-260.

Billard, R., B. Breton and M. Richard. 1981. Inhibitory effect of some steroids on spermatogenesis in adult rainbow trout *Salmo gairdneri. Canadian Journal of Zoology* 59: 1479-1487.

Bogerd, J., M. Blomenrohr, E. Andersson, H.H.A.G.M. van der Putten, C.P.Tensen, H.F. Vischer, J.C.M. Granneman, C. Janssen-Dommerhold, H.J.Th. Goos and R.W. Schulz. 2001. Discrepancy between molecular structure and ligand selectivity of testicular follicle-stimulating hormone receptor of the African catfish (*Clarias gariepinus*). *Biology of Reproduction* 64: 1633-1643.

Borg, B. 1994. Androgens in teleost fishes. *Comparative Biochemistry and Physiology* C109: 219-245.

Borg, B., I. Mayer, J. Lambert, J. Granneman and R. Schulz. 1992. Metabolism of androstenedione and 11-ketotestosterone in the kidney of the three-spined stickleback, *Gasterosteus aculeatus. General and Comparative Endocrinology* 86: 248-256.

Bradford, C.S. and M.H. Taylor. 1987. Semilunar changes in estradiol and cortisol coincident with gonadal maturation and spawning in the killifish *Fundulus heteroclitus. General and Comparative Endocrinology* 66: 71-78.

Brantley, R.K., J.C. Wingfield and A.H. Bass. 1993. Sex steroid levels in *Porichthys notatus*, a fish with alternative reproductive tactics, and a review of the hormonal basis for male dimorphism among teleost fishes. *Hormones and Behaviour* 27: 332-347.

Bromage, N., M. Porter and C. Randall. 2001. The environmental regulation of maturation in farmed finfish with special reference to the role of photoperiod and melatonin. *Aquaculture* 197: 63-98.

Budworth, P.R. and P.L. Senger. 1993. Fish-to-fish transfer in a recirculating-water system. *Progressive Fish Culturist* 55: 250-254.

Callard, I.P. and G.V. Callard. 1987. Sex steroid receptors and non-receptor binding proteins. In: *Hormones and Reproduction in Fishes, Amphibians and Reptiles*, D.O. Norris and R.E. Jones (eds.). Plenum Press, New York, pp. 355-384.

Canário, A.V.M. and A.P. Scott. 1988. Structure-activity relationships of C_{21} steroids in an *in vitro* oocyte maturation bioassay in rainbow trout, *Salmo gairdneri. General and Comparative Endocrinology* 71: 338-348.

Canário, A.V.M. and A.P. Scott. 1989. Synthesis of 20α-hydroxylated steroids by ovaries of the dab (*Limanda limanda*). *General and Comparative Endocrinology* 76: 147-158.

Canário, A.V.M. and A.P. Scott. 1990a. Identification of, and development of radioimmunoassays for 17α,21-dihydroxy-4-pregnen-3-one and 3α,17α,21-trihydroxy-5β-pregnan-20-one in the ovaries of mature plaice *(Pleuronectes platessa)*. *General and Comparative Endocrinology* 78: 273-285.

Canário, A.V.M. and A.P. Scott. 1990b. Plasma levels of ovarian steroids, including 17α,20β-dihydroxy-4-pregnen-3-one and 3β,17α,20α-trihydroxy-5β-pregnane, in female dabs (*Limanda limanda*) marine flatfish-induced to mature and ovulate with human chorionic gonadotrophin. *General and Comparative Endocrinology* 77: 177-191.

Canário, A.V.M. and A.P. Scott. 1990c. Effects of steroids and human chorionic gonadotrophin on *in vitro* oocyte final maturation in two marine flatfish: The dab *Limanda limanda*, and the plaice, *Pleuronectes platessa. General and Comparative Endocrinology* 77: 161-176.

Canário, A.V.M. and A.P. Scott. 1991. Levels of 17α,20α-dihydroxy-4-pregnen-3-one, 3β,17α,20α-trihydroxy-5β-pregnane, and other sex steroids in blood plasma of male dab *Limanda limanda* (marine flatfish) injected with human chorionic gonadotrophin. *General and Comparative Endocrinology* 83: 258-264.

Carragher, J.F. and N.W. Pankhurst. 1993. Plasma levels of sex steroids during sexual maturation of snapper, *Pagrus auratus* (Sparidae), caught from the wild. *Aquaculture* 109: 375-388.

Cardwell, J.R. and N.R. Liley. 1991a. Hormonal control of sex and color change in the stoplight parrotfish *Sparisoma viride. General and Comparative Endocrinology* 81: 7-20.

Cardwell, J.R. and N.R. Liley. 1991b. Androgen control of social status in males of a wild population of stoplight parrotfish, *Sparisoma viride* (Scaridae). *Hormones and Behavior* 25: 1-18.

Carlisle, S.L., S.K. Marxer-Miller, A.V.M. Canario, R.F. Oliveira, L. Carneiro and M.S. Grober. 2000. Effects of 11-ketotestosterone on genital papilla morphology in the sex changing fish *Lythrypnus dalli. Journal of Fish Biology* 57: 445-456.

Celius, T. and B.T. Walther. 2000. Initiation of zonagenesis in Atlantic salmon (*S. salar* L.) reveals unusual potencies of estrogens. In: *Proceedings of the 6th International Symposium on the Reproductive Physiology of Fish*, B. Norberg, O.S. Kjesbu, G.L. Taranger, E. Andersson and S.O. Stefansson (eds.). John Grieg A/S, Bergen 2000, Bergen, p. 317.

Celius, T., J.B. Matthews, J.P. Giesy and T.R. Zacharewski. 2000. Quantification of rainbow trout (*Oncorhynchus mykiss*) zona radiata and vitellogenin mRNA levels using real-time

PCR after *in vivo* treatment with estradiol-*17β or α*-zearalenol. *Journal of Steroid Biochemistry* 75: 109-119.

Cerdá, J., B.G. Calman, G.J. LaFleur and S. Limesand. 1996. Pattern of vitellogenesis and follicle maturational competence during the ovarian follicular cycle of *Fundulus heteroclitus*. *General and Comparative Endocrinology* 103: 24-35.

Ciereszko, R.E., K. Dabrowski, A. Ciereszko, J. Ebeling and J.S. Ottobre. 1997. Effects of temperature and photoperiod on reproduction of female yellow perch *Perca flavescens*: Plasma concentrations of steroid hormones, spontaneous and induced ovulation, and quality of eggs. *Journal of the World Aquaculture Society* 28: 344-356.

Clearwater, S.J. and N.W. Pankhurst. 1994. Reproductive biology and endocrinology of female red gurnard, *Chelidonichthys kumu* (Lesson and Garnot) (Family Triglidae), from the Hauraki Gulf, New Zealand. *Australian Journal of Marine and Freshwater Research* 45: 131-139.

Coward, K. and N.R. Bromage. 2000. Reproductive physiology of female tilapia broodstock. *Reviews in Fish Biology and Fisheries* 10: 1-25.

Dabrowski, K., R.E. Ciereszko, A. Ciereszko, G.P. Toth, S.A. Christ, D. El-Saidy and J.S. Ottobre. 1996. Reproductive physiology of yellow perch (*Perca flavescens*): Environmental and endocrinological cues. *Journal of Applied Ichthyology* 12: 139-148.

Davie, P.S. and H. Thorarensen. 1997. Heart growth in rainbow trout in response to exogenous testosterone and 17-α methyltestosterone. *Comparative Biochemistry and Physiology* A111: 227-230.

Davies, B., N. Bromage and P. Swanson. 1999. The brain-pituitary-gonadal axis of female rainbow trout *Oncorhynchus mykiss*: Effects of photoperiod manipulation. *General and Comparative Endocrinology* 115: 155-166.

Dedual, M. and N.W. Pankhurst. 1992. Plasma steroid hormone concentrations in relation to the reproductive cycle of the sweep *Scorpis lineolatus* (Kyphosidae) caught from the wild. *Australian Journal of Marine and Freshwater Research* 43: 753-763.

Devlin, R.H. and Y. Nagahama. 2002. Sex determination and sex differentiation in fish: an overview of genetic, physiological and environmental influences. *Aquaculture* 208: 191-364.

Du, J-L., C-Y. Lee, P. Tacon, Y-H. Lee, F-P. Yen, H. Tanaka, S. Dufour and C-F. Chang. 2001. Estradiol-17β stimulates gonadotropin II expression and release in the protandrous male black porgy *Acanthopagrus schlegeli* Bleeker: A possible role in sex change. *General and Comparative Endocrinology* 121: 135-145.

Dubois, E.A., S. Slob, M.A. Zandbergen, J. Peute and H.J.Th. Goos. 2001. Gonadal steroids and the maturation of the species-specific gonadotropin-releasing hormone system in brain and pituitary of the male African catfish (*Clarias gariepinus*). *Comparative Biochemistry and Physiology* B129: 381-387.

Dufour, S., Y.S. Huang, K. Rousseau, M. Sbaihi, N. Le Belle, B. Vidal, J. Marchelidon, B. Quérat, E. Burzawa-Gérard, C.F. Chang and M. Schmitz. 2000. Puberty in teleosts: new insights into the role of peripheral signals in the stimulation of pituitary gonadotropins. In: *Proceedings of the 6th International Symposium on the Reproductive Physiology of Fish*, B. Norberg, O.S. Kjesbu, G.L. Taranger, E. Andersson and S.O. Stefansson (eds.). John Grieg A/S, Bergen 2000, Bergen, pp. 455-461.

Dulka, J.G. 1997. Androgen-induced neural plasticity and the regulation of electric-social behaviour in the brown ghost knifefish: Current status and future directions. *Fish Physiology and Biochemistry* 17: 195-202.

Ebrahimi, M., P.B. Singh and D.E. Kime. 1995. Biosynthesis of 17,20α-dihydroxy-4-pregnen-3-one, 17,20β-dihydroxy-4-pregnen-3-one, and 11-ketotestosterone by testicular fragments and sperm of the roach, *Rutilus rutilus. General and Comparative Endocrinology* 100: 375-384.

Elizur, A., I. Meiri, H. Rosenfeld, N. Zmora, W.R. Knibb and Y. Zohar. 1995. Seabream gonadotropins: Sexual dimorphism in gene expression. In: *Reproductive Physiology of Fish*, F.W. Goetz and P. Thomas (eds.). Fish Symposium 95, Austin, pp. 13-15.

Emata, A.C., A.H. Meier and S-M. Hsiao. 1991. Daily variations in plasma hormone concentrations during the semilunar spawning cycle of the Gulf killifish *Fundulus grandis. Journal of Experimental Zoology* 259: 343-354.

Fitzpatrick, M.S., G.Van Der Kraak and C.B. Schreck. 1986. Profiles of plasma sex steroids and gonadotropin in coho salmon *Oncorhynchus kisutch*, during final maturation. *General and Comparative Endocrinology* 62: 437-451.

Fitzpatrick, M.S., J.M. Redding, F.D. Ratti, and C.B. Schreck. 1987. Plasma testosterone concentration predicts the ovulatory response of coho salmon (*Oncorhynchus kisutch*) to gonadotropin-releasing hormone analog. *Canadian Journal of Fisheries and Aquatic Sciences* 44: 1351-1357.

Frisch, A. 2004. Sex-change and gonadal steroids in sequentially hermaphroditic teleost fish. *Reviews in Fish Biology and Fisheries* 14: 481-489.

Fujita, T., H. Fukada, M. Shimizu, N. Hiramatsu and A. Hara. 2004. Quantification of serum levels of precursors to vitelline envelope proteins (choriogenins) and vitellogenin in estrogen-treated masu salmon *Oncorhynchus masou. General and Comparative Endocrinology* 136: 49-57.

Geslin, M. and B. Auperin. 2004. Relationship between changes in mRNAs of the genes encoding steroidogenic acute regulatory protein and P450 cholesterol side chain cleavage in head kidney and plasma levels of cortisol in response to different kinds of acute stress in the rainbow trout *Oncorhynchus mykiss. General and Comparative Endocrinology* 135: 70-80.

Goetz, F.W. 1997. Follicle and extrafollicular tissue interaction in 17α,20β-dihydroxy-4-pregnen-3-one stimulated ovulation and prostaglandin synthesis in the yellow perch (*Perca flavescens*) ovary. *General and Comparative Endocrinology* 105: 121-126.

Goetz, F.W. and M. Garczynski. 1997. The ovarian regulation of ovulation in teleost fish. *Fish Physiology and Biochemistry* 17: 33-38.

Goetz, F.W., A.Y. Fostier, B. Breton and B. Jalabert. 1987. Hormonal changes during meiotic maturation and ovulation in the brook trout (*Salvelinus fontinalis*). *Fish Physiology and Biochemistry* 3: 203-211.

Govoroun, M., O.M. McMeel, H. Mecherouki, T.J. Smith and Y. Guigen. 2001. 17β-estradiol treatment decreases steroidogenic enzyme messenger ribonucleic acid levels in rainbow trout testis. *Endocrinology* 142: 1841-1848.

Guigen, Y., B. Jalabert, A. Benett and A. Fostier. 1995. Gonadal *in vitro* androstenedione metabolism and changes in some plasma and gonadal steroid hormones during sex inversion of the protandrous sea bass *Lates calcarifer. General and Comparative Endocrinology* 100: 106-118.

Haddy, J.A. and N.W. Pankhurst. 1998a. The dynamics of *in vitro* 17β-estradiol secretion by isolated ovarian follicles of the rainbow trout (*Oncorhynchus mykiss*). *Fish Physiology and Biochemistry* 18: 267-275.

Haddy, J.A. and N.W. Pankhurst. 1998b. Annual changes in reproductive condition and plasma concentrations of sex steroids in black bream, *Acanthopagrus butcheri* (Munro) (Sparidae). *Marine and Freshwater Research* 49: 389-397.

Hawkins, M.B., J.W. Thornton, D. Crewes, J.K. Skipper, A. Dotte and P. Thomas. 2000. Idenitification of a third distinct estrogen receptor and reclassification of estrogen receptors in teleosts. *Proceedings of the National Academy of Science of the United States of America* 97: 10751-10756.

Hellqvist, A., M. Schmitz, C. Lindberg, P-E. Olsson and B. Borg. 2001. LH-β and FSH-β mRNA expression in nesting and post-breeding three-spined stickleback, *Gasterosteus aculeatus*, and effects of castration on expression of LH-β, FSH-β and spriggin. *Fish Physiology and Biochemistry* 25: 311-317.

Heupel, M.R., J.M. Whittier and M.B. Bennett. 1999. Plasma steroid hormone profiles and reproductive biology of the epaulette shark, *Hemiscyllium ocellatum*. *Journal of Experimental Zoology* 284: 586-594.

Hobby, A.C. and N.W. Pankhurst. 1997a. The relationship between plasma and ovarian levels of gonadal steroids in the repeat spawning marine fish *Pagrus auratus* (Sparidae) and *Chromis dispilus* (Pomacentridae). *Fish Physiology and Biochemistry* 16: 65-75.

Hobby, A.C. and N.W. Pankhurst. 1997b. Post-ovulatory egg viability in the snapper *Pagrus auratus* (Sparidae). *Marine and Freshwater Research* 48: 385-389.

Hobby, A.C., D.P. Geraghty and N.W. Pankhurst. 2000a. Differences in binding characteristics of sex steroid binding protein in reproductive and nonreproductive female rainbow trout (*Oncorhynchus mykiss*), black bream (*Acanthopagrus butcheri*), and greenback flounder (*Rhombosolea tapirina*). *General and Comparative Endocrinology* 120: 249-259.

Hobby, A.C., N.W. Pankhurst and D.P. Geraghty. 2000b. A comparison of sex steroid binding protein (SBP) in four species of teleost fish. *Fish Physiology and Biochemistry* 23: 245-256.

Hourigan, T.F., M. Nakamura, Y. Nagahama, K. Yamauchi and G. Grau. 1991. Histology, ultrastructure and *in vitro* steroidogenesis of the testes of two male phenotypes of the protogynous fish, *Thalassoma duperrey* (Labridae). *General and Comparative Endocrinology* 83: 193-217.

Jackson, K., D. Goldberg, M. Ofir, M. Abraham and G. Degani. 1999. Blue gourami (*Trichogaster trichopterus*) gonadotropic β subunits (I and II) cDNA sequences and expression during oogenesis. *Journal of Molecular Endocrinology* 23: 177-187.

Jalabert, B. 1976. *In vitro* maturation and ovulation in rainbow trout (*Salmo gairdneri*), northern pike (*Esox lucius*) and goldfish (*Carassius auratus*). *Journal of the Fisheries Research Board of Canada* 33: 974-988.

Johnson, A.K., P. Thomas and R.R. Wilson. 1998. Seasonal cycles of gonadal development and plasma sex steroid levels in *Epinephelus morio*, a protogynous grouper in the eastern Gulf of Mexico. *Journal of Fish Biology* 52: 502-518.

Kagawa, H., G. Young and Y. Nagahama. 1983. Changes in plasma steroid hormone levels during gonadal maturation in female goldfish *Carassius auratus*. *Bulletin of the Japanese Society of Scientific Fisheries* 49: 1783-1787.

Kagawa, H., H. Tanaka, K. Okuzawa, M. Matsuyama and K. Hirose. 1991. Diurnal changes in plasma 17α,20β-dihydroxy-4-pregnen-3-one levels during spawning season in the red sea bream *Pagrus major*. *Nippon Suisan Gakkaishi* 57: 769.

Kindler, P.M., D.P. Philipp, M.R. Gross and J.M.Bahr. 1989. Serum 11-ketotestosterone and testosterone concentrations associated with reproduction in male bluegill (*Lepomis machrochirus*: Centrarchidae). *General and Comparative Endocrinology* 75: 446-453.

Kime, D.E. 1979. The effect of temperature on the steroidogenic enzymes of the rainbow trout. *General and Comparative Endocrinology* 39: 290-296.

Kime, D.E. 1990. *In vitro* metabolism of progesterone, 17-hydroxyprogesterone, and 17α, 20β-dihydroxy-4-pregnen-3-one by ovaries of the common carp *Cyprinus carpio*: Production rates of polar metabolites. *General and Comparative Endocrinology* 79: 406-414.

Kime, D.E. 1992. Progestogen metabolism by ovaries of the roach (*Rutilus rutilus* L.) and the rudd (*Scardinius erythophthalmus* L.). *Fish Physiology and Biochemistry* 9: 497-504.

Kime, D.E. 1993. 'Classical' and 'non-classical' reproductive steroids in fish. *Reviews in Fish Biology and Fisheries* 3: 160-180.

Kime, D.E. and M.A.S. Abdullah. 1994. The *in vitro* metabolism of 17-hydroxyprogesterone by ovaries of the goldfish, *Carassius auratus*, is affected by substrate concentration. *General and Comparative Endocrinology* 95: 109-116.

Kime, D.E. and D.J. Groves. 1986. Steroidogenesis by gonads of a viviparous teleost, the sailfin molly (*Poecilia latipinna*) *in vitro* and *in vivo*. *General and Comparative Endocrinology* 63: 125-133.

Kime, D.E. and M. Hyder. 1983. The effect of temperature and gonadotropin on testicular steroidogenesis in *Sarotherodon* (*Tilapia*) *mossambicus in vitro*. *General and Comparative Endocrinology* 50: 105-115.

Kime, D.E., B. Venkatesh and C.H. Tan. 1991. 5α-pregnane-3β,7α,17,20α- and 20β- tetrols as metabolites of progesterone and 17-hydroxyprogesterone in carp (*Cyprinus carpio*) ovarian incubations. *General and Comparative Endocrinology* 84: 401-404.

King, H.R. and N.W. Pankhurst. 2003. Ovarian growth and plasma sex steroid and vitellogenin profiles during vitellogenesis in Tasmanian female *Atlantic salmon* (*Salmo salar*). *Aquaculture* 219: 797-813.

King, H.R. and G. Young. 2001. Milt production by non-spermiating male Atlantic salmon (*Salmo salar*) after injection of a commercial gonadotropin releasing hormone analog preparation, 17α-hydroxyprogesterone or 17α,20β-dihydroxy-4-pregnen-3-one, alone or in combination. *Aquaculture* 193: 179-195.

King, H.R., N.W. Pankhurst, M. Watts and P.M. Pankhurst. 2003. Effect of elevated summer temperatures on gonadal steroid production, vitellogenesis and egg quality in female Atlantic salmon. *Journal of Fish Biology* 63: 153-167.

Kirk, D.N. and B.A. Marples. 1995. The structure and nomenclature of steroids. In: *Steroid Analysis*, H.L.J. Makin, D.B. Gower and D.N. Kirk (eds.). Blackie Academic and Professional, Glasgow, pp. 1-24.

Kobayashi, M. and T. Nakanishi. 1999. 11-ketotestosterone induces male-type sexual behaviour and gonadotropin secretion in gynogenetic crucian carp, *Carassius auratus langsdorfi*. *General and Comparative Endocrinology* 115: 178-187.

Kobayashi, M. and N.E. Stacey. 1990. Effects of ovariectomy and steroid hormone implantation on serum gonadotropin levels in female goldfish. *Zoological Science* 7: 715-721.

Kobayashi, M., K. Aida and I. Hanyu. 1986. Gonadotropin surge during spawning in male goldfish. *General and Comparative Endocrinology* 62: 70-79.

Kobayashi, M., Y.C. Sohn, Y. Yoshiura and K. Aida. 2000. Effects of sex steroids on the mRNA levels of gonadotropin subunits in juvenile and ovariectomized goldfish *Carassius auratus*. *Fisheries Science* 66: 223-231.

Koob, T.J. and I.P. Callard. 1999. Reproductive endocrinology of female elasmobranchs: Lessons from the little skate (*Raja erinacea*) and spiny dogfish (*Squalus acanthias*). *Journal of Experimental Zoology* 284: 557-574.

Koob, T.J., P. Tsang and I.P. Callard. 1986. Plasma estradiol, testosterone and progesterone levels during the ovulatory cycle of the skate *Raja erinacea*. *Biology of Reproduction* 35: 267-275.

Kroon, F.J., P.L. Munday and N.W. Pankhurst. 2003. Steroid hormone levels and bi-directional sex change in *Gobiodon histrio*. *Journal of Fish Biology* 62: 153-167.

Kumar, R.S., S. Ijiri and J.M. Trant. 2000. Changes in expression of genes encoding steroidogenic enzymes in the channel catfish (*Ictalurus punctatus*) throughout a reproductive cycle. *Biology of Reproduction* 63: 1676-1682.

Kusakabe, M., T. Todo, H.J. McQuillan, F.W. Goetz and G. Young. 2002. Characterization and expression of steroidogenic acute regulatory protein and MLN64 cDNAs in trout. *Endocrinology* 143: 2062-2070.

Lacroix, M. and A. Hontela. 2001. Regulation of acute cortisol synthesis by cAMP-dependent protein kinase and protein kinase C in a teleost species, the rainbow trout (*Oncorhynchus mykiss*). *Journal of Endocrinology* 169: 71-78.

Lamba, V.J., S.V. Goswami and B.I. Sundararaj. 1983. Circannual and circadian variations in plasma levels of steroids (cortisol, estradiol-17β, estrone and testosterone) correlated with annual gonadal cycle in the catfish, *Heteropneustes fossilis* (Bloch). *General and Comparative Endocrinology* 50: 205-225.

Lau, E-L., B-Y. Lin, F.Y. Lee, L-T. Sun, S. Dufour and C-F. Chang. 1997. Stimulation of testicular function by exogenous testosterone in male protandrous black porgy, *Acanthopagrus schlegeli*. *Journal of Fish Biology* 51: 327-333.

Lee, W-K. and S-W. Yang. 2002. Relationship between ovarian development and serum levels of gondal steroid hormones, and induction of oocyte maturation and ovulations in the cultured female Korean spotted sea bass *Lateolabrax maculatus* (Jeom-nong-eo). *Aquaculture* 207: 169-183.

Liley, N.R. and Y. Roger. 1990. Plasma levels of gonadotropin and 17α,20β-dihydroxy-4-pregnen-3-one in relation to spawning behaviour of rainbow trout, *Oncorhynchus mykiss* (Walbaum). *Journal of Fish Biology* 37: 699-711.

Liley, N.R., B. Breton, A. Fostier and E.S.P. Tan. 1986a. Endocrine changes associated with spawning behaviour and social stimuli in a wild population of rainbow trout (*Salmo gairdneri*). I. Males. *General and Comparative Endocrinology* 62: 145-156.

Liley, N.R., B. Breton, A. Fostier and E.S.P. Tan. 1986b. Endocrine changes associated with spawning behaviour and social stimuli in a wild population of rainbow trout (*Salmo gairdneri*). II. Females. *General and Comparative Endocrinology* 62: 157-167.

Lim, B-S., H. Kagawa, K. Gen and K. Okuzawa. 2003. Effects of water temperature on the gonadal development and expression of steroidogenic enzymes in the gonad of juvenile red seabream, *Pagrus major*. *Fish Physiology and Biochemistry* 28: 161-162.

Lim, H.K., N.W. Pankhurst and Q.P. Fitzgibbon. 2004. Effect of slow release gonadotropin releasing hormone analog on milt characteristics and plasma levels of gonadal steroids in greenback flounder, *Rhombosolea tapirina. Aquaculture* 240: 505-516.

Lokman, P.M. and G. Young. 1998. Gonad histology and plasma steroid profiles in wild New Zealand freshwater eels (*Anguilla dieffenbachii* and *A. australis*) before and at the onset of the natural spawning migration. II. Males. *Fish Physiology and Biochemistry* 19: 339-347.

Lokman, P.M., G.J. Vermeulen, J.G.D. Lambert and G. Young. 1998. Gonad histology and plasma steroid profiles in wild New Zealand freshwater eels (*Anguilla dieffenbachii* and *A. australis*) before and at the onset of the natural spawning migration. I. Females. *Fish Physiology and Biochemistry* 19: 325-338.

Malison, J.A., L.S. Procarione, T.P. Barry, A.R. Kapuscinski and T.B. Kayes. 1994. Endocrine and gonadal changes during the annual reproductive cycle of the freshwater teleost, *Stizostedion vitreum. Fish Physiology and Biochemistry* 13: 473-484.

Mathews, S., I.A. Khan and P. Thomas. 2002. Effects of the maturation-inducing steroid on LH secretion and the GnRH system at different stages of the gonadal cycle in Atlantic croaker. *General and Comparative Endocrinology* 126: 287-297.

Matsuyama, M., S. Adachi, Y. Nagahama and S. Matsuura. 1988. Diurnal rhythm of oocyte development and plasma steroid hormone levels in the female red sea bream, *Pagrus major*, during the spawning season. *Aquaculture* 73: 357-372.

Matsuyama, M., S. Adachi, Y. Nagahama, K. Maruyama and S. Matsuura. 1990. Diurnal rhythms of serum steroid hormone levels in the Japanese whiting *Sillago japonica*, a daily-spawning teleost. *Fish Physiology and Biochemistry* 8: 329-338.

Matsuyama, M., S. Morita, N. Hamaji, M. Kashiwagi and Y. Nagahama. 1995. Diurnal rhythm in testicular activity in the secondary male of a protogynous wrasse, *Pseudolabrus japonicus*. In: *Reproductive Physiology of Fish*, F.W. Goetz and P. Thomas (eds.). Fish Symposium 95, Austin, pp. 212-214.

Matsuyama, M., K. Ohta, S. Morita, M.M. Hoque, H. Kagawa and A. Kambegawa. 1998. Circulating levels and *in vitro* production of two maturation-inducing hormones in teleost: 17α,20β-dihydroxy-4-pregnen-3-one and 17α,20β, 21-trihydroxy-4-pregnen-3-one, in a daily spawning wrasse, *Pseudolabrus japonicus. Fish Physiology and Biochemistry* 19: 1-11.

Mayer, I., M. Schmitz, B. Borg and R. Schulz. 1992. Seasonal endocrine cycles in male and female Arctic charr (*Salvelinus alpinus*). 1. Plasma levels of three androgens, 17α,20β-dihydroprogesterone and 17β-estradiol. *Canadian Journal of Zoology* 70: 37-42.

Mayer, I., N.R. Liley and B. Borg. 1994. Stimulation of spawning behaviour in castrated rainbow trout (*Oncorhynchus mykiss*) by 17α,20β-dihydroxy-4-pregnen-3-one but not by 11-ketotestosterone. *Hormones and Behavior* 28: 181-190.

McQuillan, H.J., P.M. Lokman and G. Young. 2003. Effects of sex steroids, sex and sexual maturity on cortisol production: an *in vitro* comparison of chinook salmon and rainbow trout interrenals. *General and Comparative Endocrinology* 133: 154-163.

Melamed, P., G. Gur, H. Rosenfeld, A. Elizur and Z. Yaron. 1997. The mRNA levels of GtH Iβ, GtH IIβ and GH in relation to testicular development and testosterone treatment in pituitary cells of male tilapia. *Fish Physiology and Biochemistry* 17: 93-98.

Miura, T. and C.I. Miura. 2003. Molecular control mechanisms of fish spermatogenesis. *Fish Physiology and Biochemistry* 28: 181-186.

Miura, T., K. Yamauchi, H. Takahashi and Y. Nagahama. 1991a. The role of hormones in the acquisition of sperm motility in salmonid fish. *Journal of Experimental Zoology* 261: 359-363.

Miura, T., K. Yamauchi, H. Takahashi and Y. Nagahama. 1991b. Hormonal induction *in vitro* of all stages of spermatogenesis in the male Japanese eel (*Anguilla japonica*). *Proceedings of the National Academy of Science of the United States of America* 88: 5774-5778.

Miwa, S., L. Yan and P. Swanson.1994. Localisation of two gonadotropin receptors in the salmon gonad by *in vitro* ligand autoradiography. *Biology of Reproduction* 50: 629-642.

Moore, R.K., A.P. Scott. and P.M. Collins. 2000. Circulating C-21 steroids in relation to reproductive condition of a viviparous marine teleost, *Sebastes rastrelliger* (grass rockfish). *General and Comparative Endocrinology* 117: 268-280.

Morisawa, S., K. Ishida, M. Okuno and M. Morisawa. 1993. Role of pH and cyclic adenosine monophosphate in the acquisition of potential for sperm motility during migration from the sea to the river in chum salmon. *Molecular Reproduction and Development* 34: 420-426.

Murayama, T., M. Shiraishi and I. Aoki. 1994. Changes in ovarian development and plasma levels of sex steroid hormones in the wild female Japanese sardine (*Sardinops melanostictus*) during the spawning period. *Journal of Fish Biology* 45: 235-245.

Mylonas, C.C., A. Gissis, Y. Magnus and Y. Zohar. 1997. Hormonal changes in male white bass (*Morone chrysops*) and evaluation of milt quality after treatment with a sustained-release GnRHa delivery system. *Aquaculture* 153: 301-311.

Nagahama, Y. 2000. Gonadal steroid hormones: Major regulators of gonadal sex differentiation and gametogenesis in fish. In: *Proceedings of the 6th International Symposium on the Reproductive Physiology of Fish*, B. Norberg, O.S. Kjesbu, G.L. Taranger, E. Andersson and S.O. Stefansson (eds.). John Grieg A/S, Bergen, pp. 211-222.

Norris, D.O. 1996. *Vertebrate Endocrinology*. Academic Press, San Diego.

Oliveira, R.F., A.V.M. Canario, M.S. Grober and R. Serrão Santos. 2001b. Endocrine correlates of male polymorphism and alternative reproductive tactics in the Azorean rock pool blenny, *Parablennius sanguinolentus parvicornis*. *General and Comparative Endocrinology* 121: 278-288.

Oliveira, R.F., M. Lopes, L.A. Carneiro and A.V.M. Canario. 2001a. Watching fights raises fish hormone levels. *Nature* (*London*) 409: 475.

Oliveira, R.F., K. Hirschenhauser, L.A. Carneiro and A.V.M. Canario. 2002. Social modulation of androgen levels in male teleost fish. *Comparative Biochemistry and Physiology* B 132: 203-215.

Páll, M., I. Mayer and B. Borg. 2002. Androgens and behaviour in the male three-spined stickleback, *Gasterosteus aculeatus* I. Changes in 11-ketotestosterone levels during the nesting cycle. *Hormones and Behavior* 41: 377-383.

Pankhurst, N.W. 1984. Artificial maturation as a technique for investigating adaptations for migration in the European eel *Anguilla anguilla* (L.). In: *Mechanisms of Migration in Fishes*, J.D. McCleave, G.P. Arnold, J.J. Dodson and W.H. Neill (eds.). Plenum Publishing Corporation, New York, pp. 143-157.

Pankhurst, N.W. 1985. Final maturation and ovulation of oocytes of the goldeye *Hiodon alosoides* (Rafinesque) *in vitro*. *Canadian Journal of Zoology* 63: 1003-1009.

Pankhurst, N.W. 1987. *In vitro* steroid production by ovarian follicles of orange roughy *Hoplostethus atlanticus* Collett from the continental slope off New Zealand. In: *Reproductive*

Physiology of Fish, D.R. Idler, L.W. Crim and J.M. Walsh (eds.). Memorial University of Newfoundland, St John's pp. 266.

Pankhurst, N.W. 1988. Spawning dynamics of orange roughy, *Hoplostethus atlanticus*, in midslope waters of New Zealand. *Environmental Biology of Fishes* 21: 101-116.

Pankhurst, N.W. 1990. Changes in plasma levels of gonadal steroids during spawning behaviour in territorial male demoiselles *Chromis dispilus* (Pisces: Pomacentridae) sampled underwater. *General and Comparative Endocrinology* 79: 215-225.

Pankhurst, N.W. 1994. Effects of gonadotropin releasing hormone analogue, human chorionic gonadotropin and gonadal steroids on milt volume in the New Zealand snapper, *Pagrus auratus* (Sparidae). *Aquaculture* 125: 185-197.

Pankhurst, N.W. 1995. Hormones and reproductive behavior in male damselfish. *Bulletin of Marine Science* 57: 569-581.

Pankhurst, N.W. and C.W. Barnett. 1993. Relationship of population density, territorial interaction and plasma levels of gonadal steroids in spawning male demoiselles *Chromis dispilus* (Pisces: Pomacentridae). *General and Comparative Endocrinology* 90: 168-176.

Pankhurst, N.W. and J.F. Carragher. 1991. Seasonal endocrine cycles in marine teleosts. In: *Reproductive Physiology of Fish*, A.P. Scott, J.P. Sumpter, D.E. Kime and M.S. Rolfe (eds.). FishSymp 91, Sheffield, pp. 131-135.

Pankhurst, N.W., and J.F. Carragher. 1995. Effect of exogenous hormones on reproductive behaviour in territorial males of a natural population of demoiselles *Chromis dispilus* (Pisces: Pomacentridae). *Marine and Freshwater Research* 46: 1201-1209.

Pankhurst, N.W. and A.M. Conroy. 1987. Seasonal changes in reproductive condition and plasma levels of sex steroids in the blue cod, *Parapercis colias* (Bloch and Schneider) (Mugiloididae). *Fish Physiology and Biochemistry* 4: 15-26.

Pankhurst, N.W. and A.M. Conroy. 1988. Endocrine changes during gonadal maturation and spawning in the orange roughy (*Hoplostethus atlanticus* Collett), a teleost from the midslope waters off New Zealand. *General and Comparative Endocrinology* 70: 262-273.

Pankhurst, N.W. and D.E. Kime. 1991. Plasma sex steroid concentrations in male blue cod, *Parapercis colias* (Bloch and Schneider) (Pinguipedidae), sampled underwater during the spawning season. *Australian Journal of Marine and Freshwater Research* 42: 129-137.

Pankhurst, N.W. and C.W. Poortenaar. 2000. Milt characteristics and plasma levels of gonadal steroids in greenback flounder *Rhombosolea tapirina* following treatment with exogenous hormones. *Marine and Freshwater Behaviour and Physiology* 33: 141-159.

Pankhurst, N.W. and M.J.R. Porter. 2003. Cold and dark or warm and light: variations on the theme of environmental control of reproduction. *Fish Physiology and Biochemistry* 28: 385-389.

Pankhurst, N.W. and P.M. Thomas. 1998. Maintenance at elevated temperature retards the steroidogenic and ovulatory responsiveness of rainbow trout *Oncorhynchus mykiss* to luteinizing hormone releasing hormone analogue. *Aquaculture* 166: 163-177.

Pankhurst, N.W. and G. Van Der Kraak. 2000. Evidence that acute stress inhibits ovarian steroidogenesis in rainbow trout *in vivo*, through the action of cortisol. *General and Comparative Endocrinology* 117: 225-237.

Pankhurst, N.W., N.E. Stacey and G. Van Der Kraak. 1986. Reproductive development and plasma levels of reproductive hormones of goldeye *Hiodon alosoides* (Rafinesque) taken from the North Saskatchewan River during the open water season. *Canadian Journal of Zoology* 64: 2843-2849.

Pankhurst, N.W., P.J. McMillan and D.M. Tracey. 1987. Seasonal reproductive cycles in three commercially exploited fishes from slope waters off New Zealand. *Journal of Fish Biology* 30: 193-211.

Pankhurst, N.W., G.J. Purser, G. Van Der Kraak, P.M. Thomas and G.N.R. Forteath. 1996. Effect of holding temperature on ovulation, egg fertility, plasma levels of reproductive hormones, and *in vitro* ovarian steroidogenesis in the rainbow trout *Oncorhynchus mykiss*. *Aquaculture* 146: 277-290.

Pankhurst, N.W., P.I. Hilder and P.M. Pankhurst. 1999. Reproductive condition and behavior in relation to plasma levels of gonadal steroids in the spiny damselfish *Acanthochromis polyacanthus*. *General and Comparative Endocrinology* 115: 53-69.

Patiño, R., P. Thomas and G. Yoshizaki. 2003. Ovarian follicle maturation and ovulation: an integrated perspective. *Fish Physiology and Biochemistry* 28: 305-308.

Pfaff, D.W., M.I. Phillips and R.T. Rubin. 2004. *Principles of Hormone/Behaviour Relations*. Elsevier, Amsterdam.

Pinter, J. and P. Thomas. 1999. Induction of ovulation of mature oocytes by the maturation-inducing steroid 17α,20β,21-trihydroxy-4-pregnen-3-one in the spotted seatrout. *General and Comparative Endocrinology* 115: 200-209.

Poortenaar, C.W., S.H. Hooker and N. Sharp. 2001. Assessment of yellowtail kingfish (*Seriola lalandi lalandi*) reproductive physiology, as a basis for aquaculture development. *Aquaculture* 201: 271-286.

Pottinger, T.G. and A.D. Pickering. 1985. The effects of 11-ketotestosterone and testosterone on the skin structure of brown trout, *Salmo trutta* L. *General and Comparative Endocrinology* 59: 335-342.

Prat, F., S. Zanuy, M. Carrillo, A. de Mones and A. Fostier. 1990. Seasonal changes in plasma levels of gonadal steroids of sea bass *Dicentrarchus labrax* L. *General and Comparative Endocrinology* 78: 361-373.

Rahman, M.S., A. Takemura and K. Takano. 2000a. Correlation between plasma steroid hormones and vitellogenin profiles and lunar periodicity in the female golden rabbitfish, *Siganus guttatus* (Bloch). *Comparative Biochemistry and Physiology* B127: 113-122.

Rahman, M.S., A. Takemura and K. Takano. 2000b. Lunar synchronization of testicular development and plasma steroid hormone profiles in the golden rabbitfish. *Journal of Fish Biology* 57: 1065-1074.

Rahman, M.S., A. Takemura, S. Nakamura and K. Takano. 2003a. Rhythmic changes in testicular activity with lunar cycle in the forktail rabbitfish. *Journal of Fish Biology* 62: 495-499.

Rahman, M.S., A. Takemura, Y.J. Park and K. Takano. 2003b. Lunar cycle in the reproductive activity of the forktail rabbitfish. *Fish Physiology and Biochemistry* 28: 443-444.

Rasmussen, L.E.L., D.L. Hess and C.A. Luer. 1999. Alterations in serum steroid concentrations in the clearnose skate *Raja eglanteria*: Correlations with season and reproductive status. *Journal of Experimental Zoology* 284: 575-585.

Robertson, D.R., C.W. Petersen and J.D. Braun. 1990. Lunar reproductive cycles of benthic-brooding reef fishes: Reflections of larval biology or adult biology? *Ecological Monographs* 60: 311-329.

Rocha, M.J. and M.A. Reis-Henriques. 1999. Plasma levels of C_{18}-, C_{19}- and C_{21}-steroids in captive and feral female sea bass. *Journal of Fish Biology* 55: 26-34.

Schoonen, W.G.E.J., J.C.M. Granneman, J.G.D. Lambert and P.G.W.J. van Oordt. 1987. Steroidogenesis in the testes and seminal vesicles of spawning and non-spawning African catfish *Clarias gariepinus*. *Aquaculture* 63: 77-88.

Schoonen, W.G.E.J., M.T. Penders, G.H. Van Dam and J.G.D. Lambert. 1988. 5β-pregnane-3α,6α,17α,20β-tetrol and 5β-pregnane-3α,6α,17α-triol-20-one: Steroids of ovarian origin in the African catfish, *Clarias gariepinus* during oocyte maturation. *General and Comparative Endocrinology* 69: 181-187.

Schulz, R.W., J. Bogerd and H.J. Th. Goos. 2000. Spermatogenesis and its endocrine regulation. In: *Proceedings of the 6th International Symposium on the Reproductive Physiology of Fish*, B. Norberg, O.S. Kjesbu, G.L. Taranger, E. Andersson and S.O. Stefansson (eds.). John Grieg A/S, Bergen, pp. 225-232.

Scott, A.P. and A.M. Baynes. 1982. Plasma levels of sex steroids in relation to ovulation and spermiation in rainbow trout *(Salmo gairdneri)*. In: *Reproductive Physiology of Fish*, C.J.J. Richter and H.J. Th. Goos (eds.). Pudoc, Wageningen, pp. 103-106.

Scott, A.P. and E.L.M. Vermeirssen. 1994. Production of conjugated steroids by teleost gonads and their role as pheromones. In: *Perspectives in Comparative Endocrinology*, K.G Davey, R.E. Peter and S.S. Tobe (eds.). National Research Council of Canada, Ottawa, pp. 645-654.

Scott, A.P., V.J. Bye and S.M. Baynes. 1980a. Seasonal variation in sex steroids of female rainbow trout *(Salmo gairdneri* Richardson). *Journal of Fish Biology* 17: 587-592.

Scott, A.P., V.J. Bye, S.M. Baynes and J.R.C. Springate. 1980b. Seasonal variations in plasma concentrations of 11-ketotestosterone and testosterone in male rainbow trout, *Salmo gairdneri* Richardson. *Journal of Fish Biology* 17: 495-505.

Scott, A.P., E.L. Sheldrick and P.F. Flint. 1982. Measurement of 17α,20β-dihydroxy-4-pregnen-3-one in plasma of trout *(Salmo gairdneri* Richardson): Seasonal changes and response to salmon pituitary extract. *General and Comparative Endocrinology* 46: 444-451.

Scott, A.P., A.V.M. Canario, N.M. Sherwood and C.M. Warby. 1991. Levels of steroids including cortisol and 17α,20β-dihydroxy-4-pregnen-3-one, in plasma, seminal fluid, and urine of Pacific herring *(Clupea harengus pallasi)* and North Sea plaice *(Pleuronectes platessa* L.). *Canadian Journal of Zoology* 69: 111-116.

Scott, S.G., J.R. Zeldis and N.W. Pankhurst. 1993. Evidence of daily spawning in natural populations of the New Zealand snapper *Pagrus auratus* (Sparidae). *Environmental Biology of Fishes* 36: 149-159.

Sikkel, P.C. 1993. Changes in plasma androgen levels associated with changes in male reproductive behaviour in a brood cycling marine fish. *General and Comparative Endocrinology* 89: 229-237.

Sisneros, J.A., P.M. Forlano, R. Knapp and A.H. Bass. 2004. Seasonal variation of steroid hormone levels in an intertidal-nesting fish, the vocal plainfin midshipman. *General and Comparative Endocrinology* 136: 101-116.

Slater, C.H., C.B. Schreck and P. Swanson. 1994. Plasma profiles of the sex steroids and gonadotropins in maturing female spring chinook salmon *(Oncorhynchus tshawytscha)*. *Comparative Biochemistry and Physiology* A109: 167-175.

Smith, C.J. and S.R. Haley. 1988. Steroid profiles of the female Tilapia *Oreochromis mossambicus* and correlation with oocyte growth and mouth brooding behaviour. *General and Comparative Endocrinology* 69: 88-98.

Sorensen, P.W., A.P. Scott and R.L. Kitislinger. 2000. How common hormonal metabolites function as relatively specific pheromonal signals in the goldfish. In: *Proceedings of the 6th International Symposium on the Reproductive Physiology of Fish*, B. Norberg, O.S. Kjesbu, G.L. Taranger, E. Andersson and S.O. Stefansson (eds.). John Grieg A/S, Bergen 2000, Bergen, pp. 125-128.

Specker, J.L. and C.V. Sullivan. 1994. Vitellogenesis in fishes: status and perspectives. In: *Perspectives in Comparative Endocrinology*, K.G. Davey, R.E. Peter and S.S. Tobe (eds.). National Research Council of Canada, Ottawa, pp. 304-315.

Stacey, N.E. 2003. Hormones, pheromones and reproductive behaviour. *Fish Physiology and Biochemistry* 28: 229-235.

Stacey, N.E. and M. Kobayashi. 1996. Androgen induction of male sexual behaviours in female goldfish. *Hormones and Behaviour* 30: 434-445.

Stocco, D.M. 1997. A StAR search: implications in controlling steroidogenesis. *Biology of Reproduction* 56: 328-336.

Stocco, D.M. 1999. Steroidogenic acute regulatory protein. *Vitamins and Hormones* 55: 399-441.

Sulistyo, I., P. Fontaine, J. Rinchard, J-N. Gardeur, H. Migaud, B. Capdeville and P. Kestemont. 2000. Reproductive cycle and plasma levels of steroids in male Eurasian perch *Perca fluviatilis*. *Aquatic Living Resources* 13: 99-106.

Sumpter, J.P. and J.M. Dodd. 1979. The annual reproductive cycle of the female lesser spotted dogfish (*Scyliorhinus canicula*) and its endocrine control. *Journal of Fish Biology* 15: 687-695.

Sun, B. and N.W. Pankhurst. 2004. Patterns of growth, vitellogenin and gonadal steroid concentrations in greenback flounder. *Journal of Fish Biology* 64: 1399-1412.

Sun, B., N.W. Pankhurst and M. Watts. 2003. Development of an enzyme-linked immunosorbent assay (ELISA) for vitellogenin measurement in greenback flounder *Rhombosolea tapirina*. *Fish Physiology and Biochemistry* 29: 13-21.

Sundaray, J.K., K. Ohta, A. Yamaguchi, K. Suzuki and M. Matsuyama. 2003. Diurnal rhythm of steroid biosynthesis in the testis of terminal phase male of protogynous wrasse, *Pseudolabrus sieboldi*, a daily spawner. *Fish Physiology and Biochemistry* 28: 193-195.

Swanson, P. 1991. Salmon gonadotropins: reconciling old and new ideas. In: *Reproductive Physiology of Fish*, A.P. Scott, J.P. Sumpter, D.E. Kime and M.S. Rolfe (eds.). Fish Symposium 91, Sheffield, pp. 2-7.

Swanson, P., J.T. Dickey and B. Campbell. 2003. Biochemistry and physiology of fish gonadotropins. *Fish Physiology and Biochemistry* 28: 53-59.

Tacon, P., J.F. Baroiller, P.Y. Le Bail, P. Prunet and B. Jalabert. 2000. Effect of egg deprivation on sex steroids, gonadotropin, prolactin and growth hormone profiles during the reproductive cycle of the mouthbrooding cichlid fish *Oreochromis niloticus*. *General and Comparative Endocrinology* 117: 54-65.

Thomas, P. 1994. Hormonal control of final oocyte maturation in sciaenid fishes. In: *Perspectives in Comparative Endocrinology*, K.G. Davey, R.E. Peter and S.S. Tobe (eds.). National Research Council of Canada, Ottawa, pp. 619-625.

Thomas, P. 2003. Rapid, nongenomic steroid actions initiated at the cell surface: lessons from studies with fish. *Fish Physiology and Biochemistry* 28: 3-12.

Tiwary, B.K., R. Kirubagaran and A.K. Ray. 2002. Testosterone triggers the brain-pituitary-gonadal axis of juvenile female catfish *(Heteropneustes fossilis* Bloch) for precocious ovarian maturation. *General and Comparative Endocrinology* 126: 23-29.

Trant, J.M. and P. Thomas. 1988. Structure-activity relationships of steroids in inducing germinal vesicle breakdown of Atlantic croaker oocytes *in vitro. General and Comparative Endocrinology* 71: 307-317.

Trudeau, V.L. and R.E. Peter. 1995. Functional interactions between neuroendocrine systems regulating GtH-II release. In: *Reproductive Physiology of Fish*, F.W. Goetz and P. Thomas (eds.). Fish Symposium 95, Austin, pp. 44-48.

Tveiten, H. and H.K. Johnsen. 2001. Thermal influences on temporal changes in plasma testosterone and oestradiol-17β concentrations during gonadal recrudescence in female common wolffish. *Journal of Fish Biology* 59: 175-178.

Tveiten, H., I. Mayer, H.K. Johnsen and M. Jobling. 1998. Sex steroids, growth and condition of Arctic charr broodstock during an annual cycle. *Journal of Fish Biology* 53: 714-727.

Tveiten, H., A.P. Scott and H.K. Johnsen. 2000. Plasma-sulfated C_{21}-steroids increase during the periovulatory period in female common wolfish and are influenced by temperature during vitelleogenesis. *General and Comparative Endocrinology* 117: 464-473.

Tyler, C.R., J.P. Sumpter and P.R. Whitthames. 1990. The dynamics of oocyte growth during vitellogenesis in the rainbow trout *(Oncorhynchus mykiss). Biology of Reproduction* 43: 202-209.

Tyler, C.R., E.M. Santos and F. Prat. 2000. Unscrambling the egg—cellular, biochemical, molecular and endocrine advances in oogenesis. In: *Proceedings of the 6th International Symposium on the Reproductive Physiology of Fish*, B. Norberg, O.S. Kjesbu, G.L. Taranger, E. Andersson and S.O. Stefansson (eds.). John Grieg A/S, Bergen 2000, Bergen, pp. 273-280.

Ueda, H., G. Young, L.W. Crim, A. Kambegawa and Y. Nagahama. 1983. 17α,20β-dihydroxy-4-pregnen-3-one: Plasma levels during sexual maturation and *in vitro* production by testes of Amago salmon *(Oncorhynchus rhodurus)* and rainbow trout *(Salmo gairdneri). General and Comparative Endocrinology* 51: 106-112.

Ueda, H., A. Kambegawa and Y. Nagahama. 1985. Involvement of gonadotropin and steroid hormones in spermiation in the amago salmon, *Oncorhynchus rhodurus* and goldfish, *Carassius auratus. General and Comparative Endocrinology* 59: 24-30.

Vaillant, C., C. le Guellec, F. Pakdel and Y. Valotaire. 1988. Vitellogenin gene expression in primary culture of rainbow trout hepatocytes. *General and Comparative Endocrinology* 70: 284-290.

Van Der Kraak, G. and M.G. Wade. 1994. A comparison of signal transduction pathways mediating gonadotropin actions in vertebrates. In: *Perspectives in Comparative Endocrinology*, K.G. Davey, R.E. Peter and S.S.Tobe (eds.). National Research Council of Canada, Ottawa, pp. 56-93.

Van Der Kraak, G., H.M. Dye and E.M. Donaldson. 1984. Effects of LHRH and Des-Gly[10][D-Ala[6]]LH-RH-ethylamide on plasma sex steroid profiles in adult female coho salmon *(Oncorhynchus kisutch). General and Comparative Endocrinology* 55: 36-45.

Venkatesh, B., C.H. Tan and T.J. Lam. 1990. Steroid hormone profile during gestation and parturition of the guppy (*Poecilia reticulata*). *General and Comparative Endocrinology* 77: 476-483.

Ventling, A.R. and N.W. Pankhurst. 1995. Effects of gonadal steroids and human chorionic gonadotropin on final oocyte maturation *in vitro* in the New Zealand snapper *Pagrus auratus* (Sparidae). *Marine and Freshwater Research* 46: 467-473.

Vermeirssen, E.L.M. and A.P. Scott. 1996. Excretion of free and conjugated steroids in rainbow trout (*Oncorhynchus mykiss*): Evidence for branchial excretion of the maturation-inducing steroid, 17,20β-dihydroxy-4-pregnen-3-one. *General and Comparative Endocrinology* 101: 180-194.

Vermeirssen, E.L.M., A.P. Scott, C.C. Mylonas and Y. Zohar. 1998. Gonadotrophin releasing hormone agonist stimulates milt fluidity and plasma concentrations of 17,20β-dihydroxylated and 5β-reduced,3α-hydroxylated C_{21} steroids in male plaice (*Pleuronectes platessa*). *General and Comparative Endocrinology* 112: 163-177

Vermeulen, G.J., J.G.D. Lambert, M.J.P. Lenczowski and H.J. Th. Goos. 1993. Steroid hormone secretion by testicular tissue from African catfish, *Clarias gariepinus*, in primary cell culture—identification and quantification by gas chromatography-mass spectrometry. *Fish Physiology and Biochemistry* 12: 21-30.

Vischer, H.F. and J. Bogerd. 2003. Cloning and functional characterization of a gonadal luteinizing hormone receptor complementary DNA from the African catfish (*Clarias gariepinus*). *Biology of Reproduction* 68: 262-271.

Watts, M., N.W. Pankhurst, A.Pryce and B. Sun. 2003. Vitellogenin isolation, purification and antigenic cross-reactivity in three teleost species. *Comparative Biochemistry and Physiology* B134: 467-476.

Weber, G.M. and C.S. Lee. 1985. Effects of 17α-methyl-testosterone on spermatogenesis and spermiation in the grey mullet, *Mugil cephalus* L. *Journal of Fish Biology* 26: 77-84.

Westerlund, L., S.J. Hyllner, A. Schopen and P.E. Olsson. 2001. Expression of three vitelline envelope protein genes in Arctic charr. *General and Comparative Endocrinology* 122: 78-87.

Yan, L., P. Swanson and W.W. Dickhoff. 1992. A two receptor model for salmon gonadotropins (GTH I and GTH II). *Biology of Reproduction* 47: 418-427.

Yaron, Z., G. Gur, P. Melamed, H. Rosenfeld, A. Elizur and B. Levavi-Sivan. 2003. Regulation of fish gonadotropins. *International Review of Cytology* 225: 131-185.

Zakon, H.H. 1993. Weakly electric fish as model systems for studying long-term steroid action on neural circuits. *Brain, Behaviour and Evolution* 42: 242-251.

Zohar, Y. 1982. L'évolution de la pulsatilité et des cycles nycthéméraux de la sécrétion gonadotrope chez la truite arc-en-ciel femelle en relation avec le cycle sexuel annuel et par rapport á l'activité stéroidogène de l'ovaire. Ph.D. Thesis, Universite Pierre et Marie Curie, Paris.

Zohar, Y., G. Pagelson and M. Tosky. 1988. Daily changes in reproductive hormone levels in the female gilthead seabream *Sparus auratus* at the spawning period. In: *Reproduction in fish—Basic and applied aspects in endocrinology and genetics*, Y. Zohar and B. Breton (eds.). *Les Colloques de l'INRA* 44: 119-125.

Vitellogenesis

Thomas P. Mommsen[1],* and Bodil Korsgaard[2],*

INTRODUCTION

Vitellogenesis, the process of supplying maternal compounds to developing oocytes in oviparous animals, is a multifaceted process with finely tuned co-operation between a number of tissues and numerous metabolic pathways. It is also a seasonal physiological process involving multi-hormonal control. Regulation occurs not only at all levels of the hypothalamic-pituitary-gonad axis, but the overall process is also dependent on sufficiently functional metabolic processing of hepatic and extrahepatic tissues. The egg is a storage organ for the yolk, constituting the food supply for embryonic and limited post-embryonic development. Through the process of vitellogenesis, the mother fish supplies the yolk from her own energy sources through the precursor protein vitellogenin. Vitellogenin (Vtg) contains the main—albeit by no means exclusive—part of the nutrient material necessary for early growth and survival of the offspring. Vtg is a high molecular weight lipoglycophosphoprotein synthesized by maternal hepatocytes after induction by estradiol. Since Vtg

Authors' addresses: [1]Department of Biology, University of Victoria, Victoria, B.C. Canada.
E-mail: tpmom@uvic.ca
[2]Institute of Biology, University of Southern Denmark, Odense, Denmark.
E-mail: bodil@biology.sdu.dk
*Corresponding authors

also binds cations, especially calcium with high efficiency, it is not surprising that research has centred on Vtg, since it can provide protein, amino acids, various lipids and carbohydrates to the developing oocytes, as well as minerals and phosphates that achieve special prominence later on during skeletal development of the embryo. Subsequently, this large multifunctional compound with all its 'cling-ons' is released to the blood-circulation and sequestered by the ovary into the vitellogenic oocytes by receptor-mediated endocytosis and cleaved proteolytically into smaller units of egg yolk. Vitellogenesis and Vtg are specific to maturing females in all oviparous vertebrates, but the process of vitellogenesis is also part of the reproductive cycle preceding the maternal-embryonic trophic relationship observed in viviparous species of non-mammalian vertebrates.

While fish had been comfortable with this complex issue for millions of years, research has evolved at a different time scale. In a quarter of a century, it has gone from a seemingly simple to an amazingly intricate complex with a multitude of internal and external regulatory systems. Disregarding control mechanisms, the vitellogenic system itself has developed away from a single Vtg under the control of the straightforward estradiol, one type of nuclear estrogen receptor and a receiving oocyte. The current model involves multiple vitellogenin genes, each with unique promoter regions and varying sensitivity to induction by estradiol, and multiple Vtg proteins themselves, with variable degrees of post-translational modification. Further, the existence of three subtypes and variants of nuclear estrogen receptors, membrane actions of estradiol and at least two types of Vtg receptors fails to introduce any element of simplicity. And all this is before considering the potential of the effects of recent genome duplications found in quite a few of our finned friends or their ancient genome duplication. The only component of the system that may be as straightforward now as it was a quarter of a century ago is estradiol, the key steroid produced by the ovarian follicle cells in response to diverse inputs. However, even for estradiol, the picture has been complicated by the description of sex steroid binding globulins and other steroid-binding proteins that are likely to influence the availability of free estradiol and the uptake of the steroid into target organs and, not least, the dynamics of its removal from the circulation. To round this off, it is now obvious that estradiol is certainly not the only steroid inducing Vtg synthesis. Furthermore, for vitellogenesis, interesting interactions with other hormones, including other steroids, abound. A concise, by no means complete, picture of the various components is presented in Fig. 4.1.

While the importance of the hepatically produced Vtg as a precursor of components for the developing oocytes cannot be overstated, Vtg is clearly not the only transport molecule between the liver and oocytes, nor is the liver the only clearing house. Further, the actions of estradiol are not restricted to the liver or oocytes (ovary) and most other tissues are involved, ranging from brain, through gill, bone and intestine to muscle, adipose and skin, and reaching as far as ion transport, temperature adaptation, osmoregulation, coloration and behaviour.

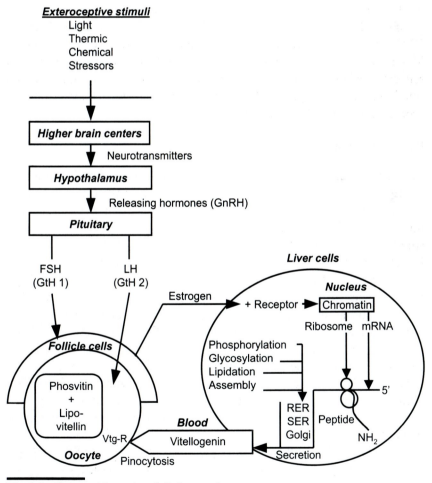

Fig. 4.1 The overall process of vitellogenesis.

In recent years, vitellogenesis, with special regard to the induced hepatic synthesis of Vtg, has received increasing attention due to the identification of several anthropogenic chemicals that have the ability to mimic or inhibit the effects of the estrogenic hormone and, thereby, are considered as endocrine disruptors. In the aquatic environment, teleosts displaying various ranges of reproductive strategies are ideal model systems with regard to endocrine disruption effects, thus accentuating the need for insight in (understanding) all parts of the process of vitellogenesis. Because of its importance to fish propagation, fish health, understanding of fish reproduction and endocrine regulation, as a model of steroid hormone action and not least as indicator of environmental health, vitellogenesis in fishes has been reviewed expertly in the last twenty years (Wallace, 1985; Mommsen and Walsh, 1988; Specker and Sullivan, 1995; Patiño and Sullivan, 2002; Arukwe and Goksøyr, 2003; Polzonetti-Magni et al., 2004; Hiramatsu et al., 2005). Therefore, rather than reiterating what has been comprehensively reviewed already, our approach is to briefly set the larger stage and then focus on a few seemingly ancillary processes and the players involved.

Environmental Impacts

Regulation of the reproductive cycle is under coordinated multi-hormonal control, the brain (hypothalamus) and pituitary constituting the upper level of the endocrine control. The brain/hypothalamus/pituitary axis integrates endogenous signals from the peripheral endocrine glands with stimuli from the environment. Hence, seasonality in teleost reproduction is under the influence not only from the endogenous positive and negative feedback exerted by the peripheral endocrine target organs, but also from changes in temperature, photoperiod, food availability, osmolarity, stressors and other natural impacts from the environment. The seasonal pattern in the onset and termination of the reproductive cycle in temperate regions necessitates species-dependent appropriate stimuli from the environment for the normal reproductive cycle to occur—also securing synchronicity in gamete production and spawning between sexes. Changing ambient temperatures and changes in daylight are two environmental factors known to strongly influence the reproductive cycle in nature.

In Atlantic salmon (*Salmo salar*) post-smolts, induction of Vtg synthesis by exogenous estradiol depended on the ambient temperature during hormone treatment but not on the previous short-term (1 month) acclimation temperature. Low ambient temperature (3°C) had a negative impact on the vitellogenic response to estradiol-treatment compared to controls held at 10°C, and this impact could not be overcome by higher doses of estradiol (Korsgaard *et al.*, 1986). Similar observations were made in mudskipper (*Periophthalmus modestus*), in which vitellogenesis was induced by high temperature treatment whereas gonadal development did not occur at low temperatures. In Atlantic salmon, however, exposure to elevated temperatures (22°C for one month) inhibited the vitellogenic response and the subsequent egg development (Watts *et al.*, 2004). The effect of maintaining female Atlantic salmon at an elevated temperature (16°C) on the responsiveness to luteinizing hormone releasing hormone analogue LHRHa indicated endocrine dysfunction by impaired pituitary responsiveness, leading to a lack of maturational competence in the female fish kept at the higher temperatures (King and Pankhurst, 2004 a, b). Observations on gonadal activity in killifish (*Fundulus heteroclitus*) exposed to various photoperiod and temperature conditions indicated a circannual rhythm of gonadal activity under constant temperature, and photoperiod conditions to be the basic trigger of gonadal development prior to spawning (Shimizu, 2003). Studies on the effects of photoperiod manipulation on the brain-pituitary-gonadal axis in female rainbow trout (*Oncorhynchus mykiss*) revealed a stimulatory long-short photoperiod advancing spawning by 3 to 4 months with correspondingly advancing peaks in GtH 2. However, earlier events in gonadal recrudescence appeared less affected by the photoperiod, but high levels of serum FSH were associated with rapid gonadal growth in the fish exposed to a stimulatory long-short photoperiod, indicating possible roles and interactions of sGNRH, gonadotropins and steroids with respect to normal and artificially stimulated ovarian maturation (Davies *et al.*, 1999). Similar observations were made in female rainbow trout exposed to accelerated photoperiod regimens, i.e., early spawning was achieved in accelerated groups, however with a reduction in egg size, which did not appear to be due to a decrease in plasma GtH 1 but rather to an alteration of ovarian follicle growth during the late stages in vitellogenesis. Furthermore, early and middle stages of ovarian growth appeared to be the photosensitive periods, whereas later stages appeared to be controlled more by an endogenous biological clock synchronized by the photoperiod (Bon *et al.*, 1999). Interactions of temperature and photoperiod on the annual reproductive cycle of the viviparous mosquitofish (*Gambusia affinis*) indicated ovarian recrudescence to be initiated by a rise in ambient temperature and ovarian regression to be caused by shorter day length (Koya and Kamiya, 2000).

The Hypothalamus-Pituitary Complex

The interactions between the hypothalamus and the pituitary represent the upper level of hormonal control of reproductive cycles in teleostean fish, the reproductive brain being the mediator of the aforementioned environmental cues affecting reproduction. The stimulatory neurohormone GnRH and the inhibitory dopamine are two important regulators. The main action of the former is the stimulation of gonadotropin secretion, the latter antagonizing the action of GnRH at the pituitary level. The ten amino acid residue GnRHs are expressed in the brain and gonads and, to date, two forms have been identified in salmonids (sGnRH and cGnRH-II), with sGnRH the form controlling gonadotropin release in rainbow trout. However, in most other teleosts, three forms of GnRH have been demonstrated, suggesting that the sGnRH may functionally compensate for the loss of the third form in salmonids (Adams *et al.*, 2002). In contrast, the antagonistic effect of dopamine on GnRH seems to be very much dependent on the fish species. The brain as a target tissue for an estrogenic response requires functional estrogen receptors (esrs), and is also known to express estrogen receptors (Shoham and Schachter, 1996). In the brain, estrogen receptors have been identified in dopaminergic neurons in the pre-optic area and expressed in GABA neurons in this area (Kah *et al.*, 1997; Anglade *et al.*, 1998). In nature, the removal of dopamine inhibition correlates with the decrease in circulating estradiol, suggesting that dopamine inhibitions may be mediated by high levels of estradiol (Anglade *et al.*, 1994). Turnover of dopamine appears to decrease in the advanced stage of exogenous vitellogenesis in the female rainbow trout, and pituitary dopaminergic activity to be significantly higher in immature than in the adult fish (Hernandez-Rauda *et al.*, 1999). Removal of dopamine inhibition is required to trigger the GnRH-stimulated LH synthesis and release, as well as ovarian development in the juvenile European eel (*Anguilla anguilla*) (Vidal *et al.*, 2004). Estradiol, however, may also play a major role in the regulation of GnRH secretion. A direct effect of estradiol on the regulation is mediated by estradiol receptors, which have been localized in the brain, but estradiol may also affect synthesis of GnRH indirectly mediated by dopaminergic neurons (Anglade *et al.*, 1994). The reproductive cycle may thus be disturbed not only by an imbalance in the feedback regulatory mechanisms due to changes in the steroid biosynthesis pathways but may also be due to agonistic or antagonistic actions on the brain (Jalabert *et al.*, 2000).

The enzyme aromatase is another factor implicated in the control of the reproductive cycle. The primary action of this enzyme is the conversion of androgens into estrogens by a mixed function oxidase reaction, aromatizing the A-ring of C19 androgens to C18 estrogens (Lephart, 1996). Two isoforms of a CYP19 gene have been identified so far—CYP19A1, expressed in the ovary and CYP19A2, expressed in brain and the pituitary. The promoter region of the brain variant contains two estrogen response elements similar to those of the Vtg gene (Kazeto et al., 2001). The enzyme is responsible for the synthesis of estrogen in the brain, although the physiological role of brain estrogen synthesis has not been fully resolved (Callard et al., 2001). Expression and activity of aromatase in the brain can be upregulated by estrogens and the activity of the enzyme is correlated with reproduction, as indicated by increasing aromatase activities during periods of reproductive activity (Gonzalez and Piferrer, 2003). Seasonal changes of neuroestrogens may also be implicated in the integration of signals from the environment in the opening of the reproductive cycle (Gelinas et al., 1998). However, the brain form is not entirely restricted to neural tissues. For instance, estradiol exerts direct feedback on aromatase expression in the ovary of adult fathead minnows (*Pimephales promelas*), by upregulating the expression of CYP19A2 (P450 aromB) in a dose-dependent manner (Halm et al., 2002). Finally, transcripts for mRNAs for both isoforms were present in unfertilized zebrafish (*Danio rerio*) eggs (Kishida and Callard, 2001); the fact that transcript abundance decreased following fertilization strongly indicates maternal transfer as a source of oocyte aromatase transcripts.

Pituitary Hormones

The pattern of the pituitary regulation of reproductive activity has been studied in several species of teleosts and, overall, appears to be fairly similar between species. Feedback regulation by gonadal steroids is also affected by directly altering the gonadotropin (GtH) release from the *pars distalis* of the pituitary. Two different GtHs have been identified in teleosts—GtH 1, corresponding to the follicle stimulating hormone (FSH) in mammals and GtH 2, corresponding to the luteinizing hormone (LH). GtH 1 is primarily involved in the process of vitellogenesis by inducing the follicular synthesis of estradiol, but is also observed to regulate the uptake of vitellogenin by vitellogenic oocytes (Santos et al., 2001). The primary action of GtH 2 is the regulation of the subsequent oocyte maturation, ovulation and spawning, by induction of the synthesis of the maturation inducing hormone (MIH) (Kawauchi *et al.*, 1989). Estradiol is the important regulator, exerting positive and negative feedback regulation on the two gonadotropins, downregulating secretion of GtH 1 and upregulating the secretion of GtH 2. The exact mechanism by which estradiol regulates the secretion of gonadotropin is not well understood, but steroid receptors are likely to be involved in this process. In fish, estrogen responsive elements have been identified in the promoter regions of genes for pituitary GtH 2, indicating that estradiol may act directly on the gonadotrophs (Xiong *et al.*, 1994; Kah *et al.*, 1997). In a study of the role of estradiol on the dopamine receptors and secretion of FSH and LH in the pituitary of female rainbow trout, estradiol (implanted) was observed to increase the circulating levels of LH, while decreasing the plasma levels of FSH in previtellogenic fish. Dopaminergic treatment exerted no effect on mRNA levels of the two forms of sGnRH 1 and 2, respectively, during previtellogenesis or vitellogenesis. It, therefore, appears that the development of the dopaminergic inhibition of gonadotropin release by onset of vitellogenesis requires factors other than estradiol and that estradiol should contribute partly to decrease the release of FSH (Vacher *et al.*, 2002). In the European sea bass (*Dicentrarchus labrax*), a cDNA encoding a GnRH receptor (GnRH-R) was obtained from the pituitary and a subsequent expression analysis during the reproductive cycle revealed the expression of the GnRH-R gene in all LH cells and some FSH cells of the proximal *pars distalis* of the pituitary. In the brain, the highest expression of the GnRH-R gene was observed at spawning. In contrast, expression of the gene in the pituitary was higher during late vitellogenesis when compared to spawning (Gonzalez-Martinez *et al.*, 2004), suggesting that the expression of the gene is regulated differentially in the brain and pituitary of the sea bass.

The seasonal variation of FSH cells and LH cells in the killifish (*Fundulus heteroclitus*) showed both cell types to be abundant during the spawning period, but decreasing in number during the post-spawn phase. During early gonadal development, the number of FSH cells were observed to increase, while the LH cells did not. Furthermore, FSH cells were the only abundant cells during active vitellogenesis. The results point towards different roles of the two GtHs also in female killifish reproduction (Shimizu *et al.*, 2003). In female Japanese eel (*Anguilla japonica*), the profiles of pituitary gonadotropin GtH 1-β and GtH 2-β mRNA transcript levels were investigated during different stages of ovarian development before vitellogenesis. It appeared that the differential transcription patterns of the two GtH subunit mRNAs were expressed already during silvering of female eels (Han *et al.*, 2003). In another experiment carried out on female Japanese eels, artificially induced to ovarian development by repeated injection with salmon GtH, the two GtH subunits also were expressed differentially, supplying evidence that the two GtHβ subunits are synthesized sequentially and have separate functional roles in reproduction of the Japanese eel (Suetake *et al.*, 2002). The fact that the two GtHs appear to have distinct functional roles in oogenesis in anguilliform fish is further indicated by the observation that mRNA levels of GtH 1-β peaked already at the primary yolk globule stage while GtH 2-β appeared to be first detectable at the oil droplet stage, showing a marked increase with the further progression of ovarian development (Kajimura *et al.*, 2001b).

The response of the GtH system to the removal of gonadal feedback in gonadectomized female striped bass (*Morone saxatilis*) was examined during mid vitellogenesis with subsequent estradiol replacement. The expression of two different GtHβ subunits in the pituitary was observed to be negatively controlled by gonadal feedback, as also indicated by restoration of subunit mRNA levels in the gonadectomized fish when compared to controls after estradiol replacement (Klenke and Zohar, 2003). In red seabream (*Pagrus major*), LHβ (GtH 2-β) mRNA appeared to be maintained at high levels from early gametogenesis through spawning, declining with gonadal regression (Gen *et al.*, 2003). In female seabream FSHβ (GtH 1-β) mRNA remained low throughout oocyte development, whereas LH seemed to be involved in regulation of both early and late oogenesis. A study of the regulation of ovarian steroidogenesis *in vitro* by FSH and LH during sexual maturation in coho salmon (*Oncorhynchus kisutch*) provided evidence for maturation-associated changes in the effects of FSH and LH in the salmonid ovary, further indicating that the two gonadotropins may have distinct functions in the teleostean ovary (Planas *et al.*, 2000). Similarly, two different types of gonadotropins (FSH and LH) were found in ayu (*Plecoglossus altivelis*) with FSHβ subunits appearing mainly during early vitellogenesis while LHβ subunit mRNA could be observed during the late phase of gonadal development. These observations reconfirm the idea that the primary function of FSH is to initiate oogenesis while LH regulates the subsequent development of the oocytes (Yoshida *et al.*, 2001). However, in immature females of Japanese flounder (*Paralichthys olivaceus*), the relative levels of GtH 1-β and GtH 2-β mRNA levels were low but showed a gradual increase concurrent with progressive ovarian development to reach the highest level at maturation. Expression of the two gonadotropin β subunits was highly correlated with estradiol concentrations in the circulation and with the gonadosomatic index (Kajimura *et al.*, 2001a). The similarity in the levels of the mRNA of the two subunits during oogenesis is considered characteristic of GtH synthesis of multiple spawners, as opposed to the differential changes observed in annual spawners such as salmonids.

Many other hormonal factors may affect the reproductive cycle at the pituitary level. The effect of recombinant human leptin on release and intracellular content of FSH and LH was examined using pituitary cells *in vitro* from female rainbow trout. Leptin was observed to have a direct effect at the pituitary level throughout the process of oogenesis stimulating the release of both gonadotropins, however with no effect on the intracellular content of gonadotropins (Weil *et al.*, 2003). The observation that leptin only affected FSH and LH release when oogenesis had already begun indicates that leptin is not a unique signal for the activation of the gonadotropic axis, but requires other promoting factors (Reidy and Weber, 2000).

In a study of potential interactions between somatotropin and gonadotropic axes in the rainbow trout, GtH 1 rose during the initial phases of vitellogenesis and then decreased during the major part of exogenous vitellogenesis to increase again during the final stages of gonadal maturation. GtH 2 showed a concomitant dramatic increase during final growth of the oocytes and maturation in the post-ovulated fish (Gomez *et al.*, 1999). The seasonal pattern of the two gonadotropins indicated a differential regulation at transcriptional and post-transcriptional levels. However, no simple relationship was observed between the somatotropic and gonadotropic axes at the pituitary level, as plasma GH levels could be related neither to sexual maturation nor to the described pattern of gonadotropic activity. This is in accordance with reports that expression of the GnRH-R gene was observed in LH and FSH cells, but not in the somatotrophs of the proximal *pars distalis* of the pituitary in the European sea bass (Gonzalez-Martinez *et al.*, 2004)

COMPONENTS OF VITELLOGENESIS

We continue our slightly broader approach of vitellogenesis by looking at associated components starting with the movement of estradiol, its binding the hepatocyte receptors, followed by a short focus on Vtg, its components and their provenance. Finally, we analyze the uptake, breakdown and reallocation of vitellogenin in the growing oocytes and also include maturation and hydration of oocytes. The coup de grace is a short synthesis of many of the components that make vitellogenesis 'work'.

Steroid-binding Proteins

With estradiol so central to the induction of vitellogenesis, a brief look at estrogen delivery to the liver and its local action seems appropriate. After synthesis in steroidogenic cells in the ovary (Young et al., 2005), and diffusion into the circulation, steroids tend to be bound to plasma-binding protein. Only 1 to 3% of total sex steroid circulate in the unbound form; the remainder is complexed with low affinity binding proteins such as albumin or corticosteroid-binding globulin (CBG) or the high affinity sex hormone-binding globulin (SHBG). The latter is predominantly produced by the fish liver and intestine and then secreted, but recently, transcripts for this protein have also been located in other tissues, implying a paracrine role for the protein (Miguel-Queralt et al., 2005). SHBG normally serve to protect steroid hormones from rapid degradation and also control the levels of steroid hormones available for binding to the hormone receptors in the target tissues (Hammond, 1995).

Steroid-binding proteins (SBPs) found in teleostean plasma preferentially bind C18 and C19 steroids such as estradiol or testosterone with affinities in the low nanomolar range, but relatively lower affinities for C21 steroids or the powerful xenoestrogen diethylstilbestrol (Laidley and Thomas, 1997). These binding characteristics set the steroid binding proteins apart from specific steroid receptors. The SHBGs also occur intracellularly, but the primary physiological roles have been attributed (to date) only to the circulating SHBGs that function as part of a novel steroid-signaling system initiated at the cell surface. This function is independent of the classical intracellular steroid receptors. The SHBGs are part of a signal transduction system for steroids at the cell membrane, even though initial binding of the SHBG to the receptor is only with the unliganded protein; subsequently, the receptor/binding protein complex binds the steroid and leads to activation of intracellular message transduction, often initiated by cAMP. Not surprisingly, the section of the binding globulin (BG) interacting with the receptor is highly conserved between mammals (TWDP/SEGVIFY) and teleosts (TxDPEGyIFY), where x can be F, Y or L and y can be A, L or V (Miguel-Queralt et al., 2005). With liganded steroid, SHBG does not bind to its receptor. Interestingly, fish SHBGs interact with potential endocrine-disrupting agents, such as the synthetic estrogen 17β-ethinylestradiol, in a species-specific manner. While in channel catfish (*Ictalurus punctatus*) and sea bass (*Dicentrarchus labrax*) binding affinity for estradiol and 17β-ethinylestradiol are similar, trout and salmon bind 17β-ethinylestradiol only rather weakly (Tollefsen *et al.*, 2002a, b).

The binding capacity for steroids due to the presence of SHBG in rainbow trout and black bream (*Spondyliosoma cantharus*) plasma was significantly greater in vitellogenic rather than in fish outside of the reproductive-stage (Hobby *et al.*, 2000; Tollefsen *et al.*, 2002b). Interference with the endocrine function of SBPs may thus introduce a novel mechanism for endocrine disruption, and provide additional answers to the question why some weakly acting xenoestrogens are causing 'estrogen-like' reproductive disturbances in developing males.

Further evidence suggests that levels of SHBG in the circulation change in the course of the reproductive cycle of fishes.

Estrogen Receptors

Fish possess two different subtypes of estrogen receptors, esr1 and esr2, previously known as ERα and ERβ. In addition, in some teleostean species, two variants of esr2 have been described, namely an a-variant (previously ERα2 or ERα) and a b-variant (previously ERβ1). The esr2a orthologue is thought to have evolved through gene duplication of esr2b after divergence of ray- and lobe-finned fishes. Interestingly, considerable expression and functional differences exist between these subtypes and variants. Considering the many estrogen-responsive tissues in fishes, the expression of estrogen receptors can be found in many tissues and cell types, including the obvious targets such as liver, brain, pituitary, gill, gonads and intestine. In fathead minnow, esr1 and esr2b are preferentially expressed in hepatic tissues, while both esr2 subtypes also occur in muscle, not normally considered a prime target for estradiol action in fishes. The highest expression of esr2a is noted for the intestine and the fewest number of transcripts for esr2a was located in the liver. Exposure of male fathead minnows to 0.4 nM estradiol in the surrounding water-induced estrogen receptor transcription in the liver. Surprisingly, induction (5-fold) was only noticed for esr1, while the rates of transcription for both esr2 variants remained unaltered by the estradiol treatment (Smith *et al.*, 2004). Some differences exist in ligand preference and binding properties (Menuet *et al.*, 2002; Hawkins and Thomas, 2004) that could turn out to be quite relevant at low physiological concentrations of estradiol, while more pronounced differences are noted for tissue distribution. For instance, the mRNAs for the β-orthologues are rather abundant in the zebrafish gonad, while the mRNA for esr1 is barely detectable (Menuet *et al.*, 2002). All three transcripts can be detected in liver, with a slight bias towards the β-orthologues.

Again, a similar picture has been reported for largemouth bass (*Micropterus salmoides*) (Sabo-Attwood *et al.*, 2004), with esr1 expression predominating in the liver and also being the most sensitive esr gene to induction by estradiol. In females of this species, mRNAs esr1 and esr2a are the most abundant in liver, while both esr2 variants are found to be non-hepatic. Times of increased plasma estradiol and Vtg during the annual reproductive cycle coincide with large peaks in esr1 mRNA abundance, with a minor increase in esr2b mRNA and absence of changes in esr2a expression. All three esrs incurred their highest level of expression in the ovary during early oocyte development, preceding peaks in plasma estradiol concentrations. Also, paralleling the changes in male sheepshead minnows (*Cyprinodon variegatus*), exposure of male bass to estradiol induced large increases in the mRNA for esr1 and lead to minor upregulation of expression of esr2ba, while esr2a mRNA remained unchanged.

Activation of target gene expression after exposure to estradiol is reported to include two mechanisms. The first, and standard model involves nuclear transcription factors, the so-called estrogen receptors (esr) introduced above. Binding of ligand, normally estradiol, to these receptors results in receptor dimerization followed by direct binding of the estradiol: 2-estrogen receptors complex to a specific DNA sequences called estrogen response elements (EREs) that are usually located in the promoter regions of susceptible genes. Esr1 is phosphorylated after binding of ligand resulting in increases in esr1-ERE binding. However, esrs can also be phosphorylated and activated without binding of ligand.

Subsequently, the complex reacts with a number of co-activator proteins and RNA polymerase II to increase the rate of transcription of genes containing the appropriate ERE. Ligand binding of an estrogen receptor dimer is a prerequisite of translocation to the ERE and is dependent on dissociation of HSP70 and HSP90 from the estrogen receptor. The second route of activation has been termed 'tethering', mediated by interaction of the estradiol-estrogen receptor complex with other DNA-bound transcription factors, without actually binding of the estradiol-estrogen receptor complex to the DNA. The result of the tethering is increased stability of the other transcription factors, again targeting additional coactivators and RNA polymerase II. Interestingly, such tethering is found, in mammals, for at least three genes that are relevant to our discussion on fishes, namely the estradiol-dependent activation of LDL-receptor genes, transcription of HSP70 and cross-talk with the cathepsin D promoter.

A third, non-genomic route of estradiol action involves membrane receptors on the cell surface and routes of intracellular message transduction such as protein kinases. This more direct route results in much shortened response times to the steroid hormones compared with routes involving activation of nuclear transcription machinery, although short reaction times can be supported with additional nuclear actions on different targets than those reached via their EREs. Although some attention has been given to non-genomic actions of other steroid hormones, especially cortisol (Borski, 2000), little information is currently available on tethering and non-genomic action of estradiol in fishes. Therefore, we largely restrict our discussion to the well-defined and more traditional actions mediated through EREs.

The generalized mechanisms of actions of estradiol in fish liver involving nuclear estrogen receptors Esr1 (ERα) are summarized in Fig. 4.2, together with affected genes and general observations on estrogen-dependent changes to liver.

However, estradiol is much more than a hormone activating transcription of susceptible genes. For instance, estradiol can also affect the binding of repressors to EREs, and stabilize Vtg or destabilize (fibrinogen, albumin) selective mRNAs. Curiously, estradiol exerts a two-pronged action on its own receptors. On the one hand, the steroid induces esr gene transcription; on the other hand, it enhances the degradation of esr protein by a proteasome-dependent mechanism (Nawaz et al., 1999). In fact, mechanisms concurrently activating protein breakdown through proteasomal or lysosomal pathways should not come as a surprise, considering that the massive output of Vtg (among many others) must come at a substantial metabolic cost, and energy for the biosynthesis for a protein as complex as Vtg and the diverse building blocks must be liberated and delivered to the liver prior to or concurrent with Vtg gene transcription.

Recently, the reversible regulation of protein function by S-nitrosylation, usually on cysteine residues, has been added to the important arsenal of post-translational modification such as phosphorylation or acetylation. Further, it has been reported at least for mammals that DNA binding and transcription of estrogen are strongly influenced by the redox status of target cells and also that nitric oxide brings about a structural change

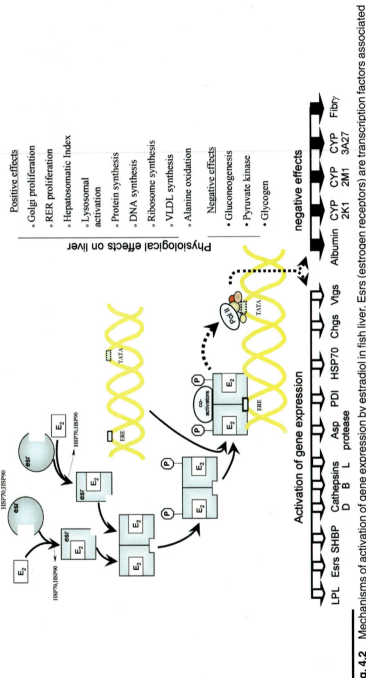

Fig. 4.2 Mechanisms of activation of gene expression by estradiol in fish liver. Esrs (estrogen receptors) are transcription factors associated with heat shock proteins inside the cell in the absence of ligand. Upon binding of estradiol, the HSPs are dissociated and the esrs dimerize, become phosphorylated and subsequently bind to an estrogen-responsive element on the nuclear DNA. In cooperation with cofactors and RNA polymerase II (pol II), the receptor-ligand complex initiates activation of gene expression of estradiol-sensitive genes.

Abbreviations—ERE : estrogen responsive element; HSP70/90: heat shock proteins 70 and 90; PolII: RNA polymerase II; TATA -box. mRNAs affected are LPL—lipoprotein lipase, SHBP—steroid hormones binding protein, Asp protease—aspartate protease (nothepsin); PDI—protein disulfide isomerase ; Chgs—choriogenins; Vtgs—vitellogenins; esrs—estrogen receptors; CYP—cytochrome P450; Fibrγ—fibrinogen γ.

of the estrogen receptor complex by S-nitrosylation, resulting in compromised DNA-binding and thus impeding estrogen-dependent gene transcription. Considering the highly conserved nature of estrogen receptors in vertebrates, it would be interesting to analyze the potential differences in S-nitrosylation and downstream function for the three esrs found in bony fishes. Of course, this particular behaviour of the estrogen receptor (and likely many others proteins) provides another potential access point for endocrine disruptors that impart oxidative stress.

Fish also express several estrogen-receptor related receptors (Bardet et al., 2004), that belong to the large family of orphan receptors and are still looking for a natural ligand and physiological role, but they may be important in adipocyte metabolism. These receptors have been shown to interact and potentially interfere with normal nuclear actions of estradiol (Giguere, 2002).

As indicated in the Fig. 4.2, the Vtg gene is just one of many hepatic genes under the control of estradiol. At times, Vtg may appear to be the most important, but it is the interplay of all these genes and their products that completes the picture. To provide a brief glimpse of the power of estradiol, we have included some of the physiological outcomes noticed for a vitellogenic fish liver associated with the production of Vtg, either temporally or functionally or both. Please note that the list is not meant to be complete and covers only the most obvious changes.

VITELLOGENIN

Vtg, the key gene controlled by estradiol, is a multifunctional protein transporting protein, lipid, phosphate and carbohydrate from the liver to the oocyte, delivering many of the basic building blocks for the growing oocyte. Vtg is a member of the large lipid transfer protein (LLTP) superfamily. As such, it is paralogous to vertebrate apolipoproteins and insect apolipophorins, signifying that the super family arose from an ancestral gene dedicated to the transport of lipids (Babin et al., 1999).

The protein backbone of a generalized piscine Vtg consists of six different domains, starting at the N-terminal with a signal peptide of about 16 residues. Originally, the remaining regions had been named partially after the functional groups carried on it or after the resulting yolk proteins. Adjacent to the signal peptide is a large (>350 residue) section that binds lipid, and has been termed lipovitellin H (heavy), followed by a variable-length domain containing polyserines that are highly phosphorylated when the mature Vtg leaves the liver—phosvitin. Next is a second lipidated section (lipovitellin L) followed by a fairly conserved cysteine rich domain, generally called the β'-component, and finally an also cysteine-rich, but more variable, C-terminal peptide. The idealized piscine Vtg gene follows the general exon-intron pattern established for amphibians and the chicken. However, in trout (Mouchel *et al.*, 1996), and easily confirmed for pufferfish (*Takifugu rubripes*) using the wealth of genomic information available for this species, exons 22 and 23 are fused to decrease the total number of exons from 35 to 34, and the number of introns to 33.

A general schematic of an idealized teleost Vtg and its derived functional groups is given in Fig. 4.3. However, over the years, the situation has been complicated by descriptions of multiple Vtgs that were obviously not closely related and other Vtgs that were lacking the phosvitin domain. As this point, three clear lineages of fish Vtgs can be distinguished, namely Vtg-A and Vtg-B, both following the pattern presented in the Fig. 4.3, and Vtg-C, a phosvitin-less Vtg, that additionally shows some 'shortcomings' in the C-terminal region. To confirm this idea of three orthologous Vtgs, we set out to conduct a quick phylogenetic analysis of some 30 Vtgs. We were surprised to find a different picture, one that clearly supported the idea first formulated by Sawaguchi and co-workers (Sawaguchi *et al.*, 2005), namely that not all Pv-containing teleostean Vtgs can be categorized as either Vtg-A or Vtg-B. In our analysis, we took the liberty to uniformly apply the nomenclature proposed by Hiramatsu *et al.* (2002), i.e., Vtg-A, -B and -C, to Vtgs previously labeled following different conventions, based on best amino acid alignments. As Fig. 4.4 shows quite convincingly, four distinct clusters can be distinguished, with a number of species falling into this as yet unassigned group, including zebrafish, fathead minnow, rainbow trout, Japanese eel, conger eel (*Conger myriaster*) and carp (*Cyprinus carpio*). These sequences do contain phosvitin, but their lipovitellin domain does not resemble Vtg-Cs. Perhaps, it is time to add a fourth (Vtg-D) sister group to the existing groups of teleostean Vtgs. To be able to do a proper alignment analysis, we only looked at the signal peptides and the large lipovitellin H domain and truncated all sequences before the phosvitin (if in existence) domain. The picture does not change if the full sequences are used.

A surprising variability seems to exist between the Vtgs present in different species or genera. Keeping in mind that genomic and expression data are still quite patchy, not a single species analyzed to date seems to have the capacity to produce all four groups of Vtgs. Some species possess

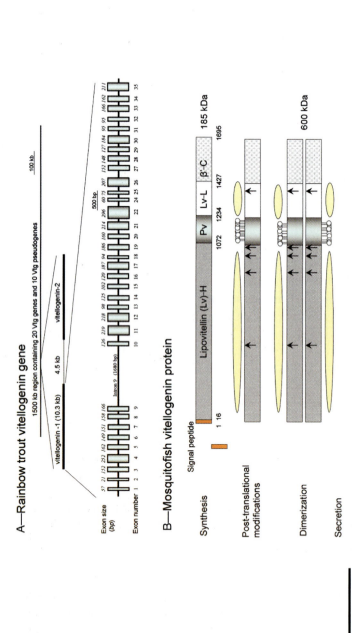

Fig. 4.3 Organization of teleostean vitellogenin gene and functional domains of fish vitellogenin. A. Organization of rainbow trout vitellogenin gene cluster and intron-exon arrangement. Note that exons corresponding to exons 22 and 23 in *Xenopus laevis* and *Gallus gallus* (part of the Pv region) appear to lack the intron in the trout. We have retained the amphibian/avian exon numbering, following the suggestion of Mouchel *et al.* (1997). B. Organization and intrahepatic processing of vitellogenin A (protein) from mosquitofish (*Gambusia affinis*). Abbreviations—SP—signal peptide; Lv-H—lipovitellin heavy chain; Pv—phosvitin; Lv-L—lipovitellin light chain; β-component (cysteine rich). Data redrawn from Sawaguchi *et al.* (2005). Note that the authors do not differentiate a cysteine-rich C-terminal domain as distinct from the α'-component (cf. Fig. 4.8).

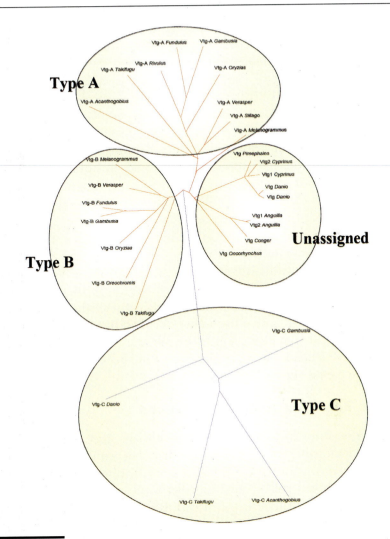

Fig. 4.4   Phylogenetic analysis of vitellogenin subtypes in teleostean fishes.

Vtg C *Acanthogobius flavimanus* (GenBank accession number **BAC06191**); Vtg C *T. rubripes* SINFRUP00000168025; Vtg C *Danio rerio* (**AAG30407**); Vtg C *Gambusia affinis* (**BAD93699**); Vtg A *A. flavimanus* (**BAC06190**); Vtg *Oncorhynchus mykiss* (**CAA63421**); Vtg *Pimephales promelas* (**AAD23878**); Vtg2 *Cyprinus carpio* (**BAD51933**); Vtg1 *C. carpio* (**AAL07472**); Vtg *Conger myriaster* (**BAD93275**); Vtg1 *Anguilla japonica* (**AAR8298**); Vtg2 *A. japonica* (**AAR8299**); Vtg B *Oreochromis aureus* (**T31095**); Vtg A *Rivulus marmoratus* (**AAQ16635**); Vtg A *Sillago japonica* (**BAC20186**); Vtg *D. rerio* (**AAK94945**); Vtg A *Verasper moseri* (**BAD93695**); Vtg B *V. moseri* (**BAD93696**); Vtg A *Oryzias latipes* (**BAB79696**); Vtg B *O. latipes* (**BAB79591**); Vtg B *Melanogrammus aeglefinus* (**AAK15157**); Vtg A *M. aeglefinus* (**AAK15158**); Vtg B *Fundulus heteroclitus* (**AAB17152**); Vtg A *F. heteroclitus* (**T43141**); Vtg A *Gambusia affinis* (**BAD93696**); Vtg B *G. affinis* (**BAD93697**); Vtg A *Takifugu rubripes* P00000176840; Vtg B *T. rubripes* P00000172281.

three Vtgs (Vtg-A, -B and -C) like the mosquitofish, a pufferfish and white perch (*Morone americana*). Others have only two, either VtgA and VtgB [killifish, haddock (*Melanogrammus aeglefinus*), Japanese medaka (*Oryzias latipes*), barfin flounder (*Verasper moseri*)] VtgA and VtgC (Zebrafish) or two yet to be classified Vtgs (genera *Salmo*, *Salvelinus*, *Coregonus* and *Thymallus*). The seemingly simplest situation is found in members of the genus *Oncorhynchus* that display only a single Vtg subtype—the as yet unnamed variety. Unique, to date, is a goby (*Acanthogobius flavimanus*) with a VtgC and an unknown Vtg. The 'unknown' contains Pv and has a large mass (530 kDa, native, 185 kDa calculated from cDNA) comparable to other Vtgs, but does not fall into any of the three Pv-positive clusters delineated in Figure 4.4.

C-type Vtgs, as found in the zebrafish, tend to be minor forms of Vtg. They either entirely lack a distinct Pv domain, or at best possess even shorter clusters of serines than rainbow trout Vtg. These Vtgs appear to be homologous to insect Vtgs or chicken VtgIII. Further, some of the VtgCs also lack the full β'-component and contain a shortened cysteine rich C-terminal region, resulting in less than half the amount of cysteines than Vtgs depicted in Figure 4.3. As a result, these Vtgs bind less calcium, tend to have much lower molecular masses and behave differently during anion-exchange chromatography.

Other Estrogen Responsive Genes

A study on estradiol actions in largemouth bass liver using differential display identified a few other proteins that changed after application of the hormone. One of these was fibrinogen γ, a secreted protein that was apparently down-regulated, and together with serum albumin, following the *Xenopus* model (Pastori *et al.*, 1991), is regulated by destabilization of the mRNA rather than direct transcriptional effects. Another equally fascinating target in bass liver was protein disulphide isomerase, whose induction tightly mirrored the pattern of estrogen receptor and Vtg mRNAs (Bowman *et al.*, 2002). The common abbreviation for this enzyme is, confusingly, ERp72, with ER signifying the endoplasmic reticulum. In eukaryotes, secreted (such as Vtg) and membrane proteins (such as esrs) fold in the endoplasmic reticulum and the enzyme responsible for disulphide formation in these proteins is the very protein disulphide isomerase. It is not much of a surprise, then, that both teleostean Vtgs and estrogen receptors contain numerous highly conserved cysteine residues that are accessible to serve as substrates for this enzyme. Focussing only on the highly conserved cysteine residues, we located between 20 and 24 conserved cysteines in Vtgs—not all of these are in the cysteine-rich C-terminal region—and either 14 or 15 in the three estrogen receptor variants.

Vitellogenin Genes

A common feature of vertebrate Vtgs appears to be a relatively recent, potentially multiple, Vtg gene duplication. For instance, rainbow trout possess tandem head-to-tail repeats of Vtg, leading to Vtgs that are more closely related to each other than to any Vtgs on the other clusters. While this was first described for Vtgs of rainbow trout, our phylogenetic tree includes such duplicate Vtgs, most likely representing tandem Vtgs, also for Japanese eel, zebrafish and carp. While it may at first appear that tandem repeats are restricted to the as yet unassigned group(s) of Vtgs, we also found what appeared to be numerous tandem repeats in the genomic DNA of the pufferfish (*Takifugu rubripes*), with seemingly tandem Vtgs (located to the same scaffold) of what we tentatively identified as VtgB. Altogether, the Vtg region of the rainbow trout genome contains 20 copies of the Vtg genes as well as 10 pseudogenes. All gene copies are a variation on the same theme, residing in a single 1500-kb region, occurring in tandems and are separated by a conserved 4.5-kb intronic sequence (see Fig. 4.2). In the pufferfish, the distance amounts to slightly less than 2 kb. The occurrence of such multiple genes, produced through gene amplification at a single locus, is not restricted to vitellogenin, but is also found for histones, some ribosomal RNAs and teleostean antifreeze proteins, possibly in the interest of the capacity to quickly ramp up production of closely related gene products.

Among the salmonids, the oncorhynchids occupy a unique position in that they only have a single type of vitellogenin, while the genera *Salmo*, *Salvelinus*, *Coregonus* and *Thymallus* possess two types of Vtg, but still have between 6 and 32 copies of the genes. Although partial salmonid Vtg sequences were useful in elucidating relationships between the different genera, unfortunately, exact identification of salmonid Vtg types have to wait until complete sequences—at least of the lipovitellin H domain—are available. A similar situation seems to exist for zebrafish. Using an EST scanning approach of a zebrafish cDNA library, Wang and co-workers (2000) identified at least seven different Vtg genes that included Vtgs A to C and, again, expression of phosvitinless Vtg-C was only about a tenth of the rate of expression of Vtg-A. Different relative sensitivities in response to exogenous estrogen and estrogen mimics exist for the production of mRNAs for estrogen receptor (ERα, esr1), VtgA and VtgB and in Japanese medaka (*Oryzias latipes*). Liver of the treated male fish responded with expression of esr1 at 10-fold lower concentrations than for the two Vtg genes. Further, in direct comparison, hepatic expression of VtgB gene was more sensitive to induction by estradiol than that of VtgA gene (Yamaguchi *et al.*, 2005), which could exert direct effects on the relative proportions of these two Vtgs in the circulation. If similar considerations apply to female medaka and other fish during natural vitellogenesis, some interesting effects of any hormone affecting estradiol availability can be expected, with potential repercussions on yolk composition.

Post-translational Processing

If we assume that teleostean and amphibian processes have some common themes, it appears that both phosphorylation of phosvitin and glycosaminylation of the lipovitellin region of Vtg happen inside of a smooth membrane compartment in the liver and fairly concurrently. Final phosphorylation of phosvitin and addition of galactose occur later on during the maturation of Vtg, namely during the packaging of Vtg into secretory vesicles in the Golgi apparatus. Interestingly, the mechanisms of acquisition of lipid components are still under debate. The assembly and secretion of triglyceride-rich lipoproteins in vertebrates requires apolipoprotein B (apoB) and an abundant cofactor, microsomal triglyceride transfer protein (MTP), that is located in the endoplasmic reticulum and is responsible for the transfer of TAGs into the lumen of the endoplasmic reticulum during the assembly of VLDL. The general motif of lipid-binding areas is common in MTP, apoB and vitellogenin and the hypothesis has been put forth that MTP should be considered the ancestral protein for such lipid transfer molecules. In this context, it should be pointed out that Vtg can also be looked at as a special case of a very low-density lipoprotein shuttling water-insoluble lipids from the liver to the ovary. As usual for lipoproteins, the focus has been on the protein domains, with much less emphasis on lipid portions or the mechanism and site of lipidation.

Lipidation of VLDL is known to change with diet, as does the composition of neutral lipids in fish eggs. The lipids carried on Vtg are predominantly phospholipids (Table 4.1), most likely destined for eventual incorporation into embryo membrane lipids. In multiply spawning fish that continue to feed during vitellogenesis and spawning seasons, the

Table 4.1   Lipid on vitellogenin A and yolk protein-400 (vitellogenin C) from *Gambusia affinis*.

Vitellogenin A			Yolk Protein-400
16.3		Percent Total Lipid	13.7
	74.3	Phospholipid (% of total)	75.7
	10.2	Triacylglycerols (% of total)	13.5
	15.5	Cholesterol (% of total)	10.8
0.44		Percent Total Phosphorus	0.05

Data from Sawaguchi *et al.* (2005). Yolk protein-400 is apparently not processed and hence represents Vtg-C found in plasma.

egg biomass may exceed the maternal body mass, setting up interesting lipid dynamics between diet, maternal stores and metabolic capabilities and egg composition. Some data hint that in such fish, dietary lipid exerts a larger influence on neutral lipids of eggs than on their phospholipids (Almansa *et al.*, 1999), implying that lipidation of Vtg may be controlled quite tightly by the maternal system.

The oil globules found in a number of teleosts eggs consist primarily of triacylglycerols, while lipids associated with the yolk platelets tend to be dominated by phospholipids, and smaller amounts of triacylglycerols, although in some species wax esters make up the largest proportion of egg lipids (Anderson *et al.*, 1990). A huge variability of lipid content exists in fish eggs, from a high of almost 50% in *Labeotrophus* sp. to a low of 0.1% (by mass) in plaice (*Pleuronectes platessa*) (Barry *et al.*, 1995). Following the August Krogh principle (Krebs, 1975) it is suggested that the plaice species would be a good model to look for a non-lipidated Vtg, or explore intricacies of the hepatic lipidation machinery. The limited capacity of Vtg to carry neutral lipids juxtaposed to the presence of considerable levels of triacylglycerols and/or wax esters in eggs—which must have reached the oocyte during vitellogenesis—leads to interesting questions about sources and mechanisms of transport to and uptake by the oocytes.

Even though the trout vitellogenin has a much shorter polyserine domain than other fish Vtgs, the overall phosphate content of trout Vtg is not substantially lower than in other teleosts, raising the question whether the degree of serine phosphorylation may be higher in trout than in other teleosts, at least partially compensating for the fewer serines in phosvitin.

In addition to lipids, phosphate, calcium and amino acids, fish vitellogenins also carry carbohydrates, varying between 2 and 20% of

Vtg mass. Unfortunately, any systematic analysis of the actual carbohydrates, covalent-binding sites to Vtg and processing remains to be done. For instance, even the question whether N-glycosylation of Vtg is a prerequisite for secretion from hepatocytes, as it is for many other secreted proteins, has not been addressed. However, some insights can be gained from a straightforward analysis of potential glycosylation sites, using common proteomics tools (http://www.cbs.dtu.dk/services/NetNGlyc/), allowing us to answer the latter question. Since four of the Vtgs listed in our evolutionary tree lack obvious glycosylation sites (rainbow trout, minnow, zebrafish Vtg-C, Japanese goby (*Tridentiger trigonocephalus*) Vtg-C), glycosylation apparently is not required for successful secretion of Vtg from the liver. Flow of carbohydrate from liver to ovary in zebrafish and goby will hardly be negatively affected since Vtg-Cs tend account for only a minor portion of the overall Vtg pool. However, the absence of an obvious glycosylation site from trout Vtg comes as a surprise. However, it is possible that the substantial microheterogeneity implicit in 10 tandem Vtgs in the trout may provide enough Vtgs containing accessible Asn sites for carbohydrate transfer to the ovary, especially since glycoproteins account for around 10 per cent of the total yolk proteins (Nagakawa, 1970).

In some of the other Vtgs, common themes concerning glycosylation can be noted. The VtgAs of *Rivulus*, *Fundulus* and *Gambusia* have a potential N-glycosylation site around position 448 (numbering for *Gambusia*), which would place this in the middle of Lv-H, making us wonder whether this will have any bearing on lipidation or whether lipidation may prevent access to this site by the glycosylation machinery. At any rate, most, but not all, other Vtgs from all four Vtg clusters depicted in Figure 4 have eligible asparagine residues N-terminally of and close to the phosvitin domain (if present), namely around positions 1050 and 1076 (numbering for *Verasper* VtgA), and thus well out of the way of the lipovitellins. One pattern, widespread among species and unique to VtgAs, is a potential glycosylation site at Asn[1396] (*Gambusia*) located some 30 residues before the C-terminal of lipovitellin L.

Sources of Amino Acids and Lipids

The role of peripheral, largely muscular, activation of proteolysis, involving activation of lysosomal enzyme, especially cathepsin D and other cathepsin has been reviewed elsewhere (Mommsen, 2004). It includes a discussion of the evolution of cathepsins and does not need to be repeated here.

Suffice to say that vitellogenic fish strike a fine balance between proteolytic enzymes and endogenous inhibitors to guide protein-derived amino acids towards hepatic vitellogenesis. The role of transporters of amino acids during uptake are in dire need of some attention and although specific funneling of amino acids during vitellogenesis has attracted some attention (Washburn *et al.*, 1992; Korsgaard and Mommsen, 1993), it seems too early to delineate specific patterns.

Although research on fish adipocyte has been slow to develop, the emerging picture seems to differ little from the situation in other vertebrate adipocytes. Because of the specific nature of fish eggs with their potentially high proportion of lipids, a discussion of adipocyte lipolysis seems warranted, first, as a proxy for similar processes in oocytes and, second, in the context of vitellogenesis, since lipid mobilization form an integral part of vitellogenesis and not least of estradiol action.

Generally, the major determinant of lipogenesis in adipocytes is the rate of re-esterification of fatty acids by lipoprotein lipase (LPL); the key role in lipolysis of triacylglycerols is played by hormone sensitive lipase (HSL). As expected, estradiol itself tends to reduce the activity of LPL while increasing enzyme markers of lipolysis, most likely at the transcriptional level. In addition, estradiol regulates the abundance of adrenoceptors on mammalian adipocytes. In fish adipocytes, as in mammals, glucagon, epinephrine (via $\alpha 1$-type receptors), growth hormone and tumour necrosis factor-α, all activate lipolysis, while insulin depresses lipolysis and counteracts the actions of glucagon. We have briefly summarized some of the actions and targets with emphasis on lipolysis in Fig. 4.5.

The main target in this scheme during vitellogenesis is HSL, a unique intracellular lipase that is regulated by hormones and is especially abundant in adipose and steroidogenic tissues. Protein kinase A (PKA) is activated through a cAMP dependent cascade, resulting in the phosphorylation of perilipin and HSL. Following phosphorylation, HSL is translocated to the surface of the lipid droplet. In fact, PKA-dependent translocation of the enzyme from the site of storage in the cytosol to its site of action on the lipid droplet appears to be the defining step of enzyme activation. Some interesting proteins are intricately associated with vertebrate lipid droplets, including the perilipins, a small family of chaperone-like proteins that regulate the binding of HSL to the surface of adipocyte lipid droplets, with a common N-terminal amino acid motif. This motif is shared with ADRP, another lipid-droplet associated protein and TIP47, often identified as a mannose 6-phosphate receptor binding protein that is also

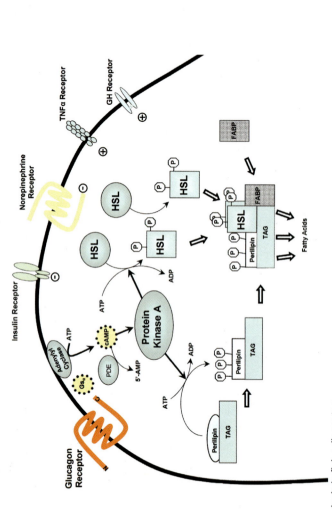

Fig. 4.5 Model of lipolysis in fish adipocytes.

Hormones such as catecholamines and glucagon activate the cAMP cascade. Protein kinase A activates perilipin and hormone-sensitive lipase through phosphorylation. After dimerization, and translocation to the surface of the lipid droplet, HSL can bind only in the presence of phosphorylated perilipin. In added presence of adipocyte fatty acid binding protein (FABP) hydrolyzes adipocyte triacylglycerols (TAG) into fatty acids. Among other things, the presence of FABP prevents product inhibition of HSL. GH and TNFα working through their own receptors also activates adipocyte lipolysis, while insulin and norepinephrine counteract the activation and decrease the rate of lipolysis, while at the same time upregulating the expression of lipoprotein lipase (LPL) and other components of synthetic routes. Mechanisms of action for the last four hormones in fishes are not known. Additional abbreviations—Gsα—stimulatory G-protein—α-subunit; PDE—phosphodiesterase.

allied with lipid droplets. Together, these lipid-droplet associated proteins have been termed the PAT-1 domain gene family, representing perilipin, APDR and Tip-47. In mammals, expression of perilipin is restricted to adipocytes and steroidogenic cells, two tissues likely to deliver lipid material to liver and oocytes.

It is clear by now that fish contain PAT proteins, and not only from our recent analysis of public databases that identified numerous proteins with PAT-1 domains in translated zebrafish and pufferfish databases. In addition, it has been reported that antibodies to mammalian perilipin bind to a lipid storage droplet (LSD) in striped bass adipocytes (Lu *et al.*, 2001).

Phosphorylation-dependent alterations in perilipin, that is attached to the lipid droplet, result in the binding of the lipase as a dimer and hydrolysis of adipocyte triacylglycerol (TAG). In addition to covalent modification by phosphorylating activating the enzyme, interaction with perilipin and translocation to the surface of lipid droplets, HSL is also regulated by oligomerization where the enzyme homodimer displays a much higher activity than the monomers. Further, HSL interacts with adipocyte-type fatty acid binding proteins (FABPs), increasing the activity of HSL and protecting the enzyme from product inhibition by removing the free fatty acids. FABPs belong to a family of proteins with over a dozen 15 paralogous members that bind fatty acids and other lipophilic ligands. These proteins are abundant in fish tissues, and even though none have yet been isolated from fish adipose tissue, some have been classified as 'adipocyte-type' (FABP H6) (Vayda *et al.*, 1998). We, therefore, postulate that the expression of FABP H6 is likely to change during vitellogenesis, especially in non-hepatic lipid storage tissues. Of course, as vitellogenesis proceeds and lipid substances are transferred to the growing oocytes, either on Vtg or on other lipoproteins, FABP presence will also be required in the ovary for lipolysis and in the oocytes for lipogenesis. Our idea is supported by expression of an alleged heart-type FABP in early and vitellogenic oocytes (Liu *et al.*, 2003) and abundance of this FABP changes in unison with accumulation of fatty acids in the oocyte. Binding to FABP within the oocyte may direct fatty acids towards lipid synthesis or storage.

While over half a dozen HSLs have been cloned and sequenced for fishes together with some physiological aspects, a proper structure/function analysis remains to be carried out. All vertebrate HSLs, including the teleostean enzymes as judged by a preliminary alignment, contain a catalytic triad consisting of Ser-Asp-His, the C-terminal half of hormone sensitive lipase contains an area of about 150 residues that have been termed the regulatory module. In this module reside key serine residues that are phosphorylated by protein kinase PKA.

At the same time as peripheral or hepatic lipid mobilization is activated, estradiol also stimulates hydroxymethylglutaryl CoA reductase activity—the committing enzyme in the biosynthesis of cholesterol—in frog liver without altering the concentrations of its mRNA and increases the abundance of low density lipoprotein (LDL) receptor protein and mRNA (Di Croce et al., 1997). Both compounds are essential during vitellogenesis, with its increased peripheral release of lipids to the processed by the liver and its increased turnover—mobilization and biosynthesis *via* mevalonate—of cholesterol.

The increased availability and turnover of free fatty acids during vitellogenesis can have additional secondary effects. The concentrations of polyunsaturated fatty acids such as arachidonate (in itself a potential source of important hormones, including prostaglandins) fluctuate in unison with plasma estradiol and peak during vitellogenesis (Booth et al., 1999; Cottrill et al., 2001). The fatty acids decrease the binding of estradiol to the SHBGs (van der Kraak and Biddiscombe, 1999), by decreasing the affinity without altering the overall binding capacity. As a result, the equilibrium between the bound and free steroid will be changed, leading to increased rates of steroid degradation and altered uptake of steroid by the target cells.

Sources of Minerals

As shown above for lipids, estrogen action and the delivery to the components of Vtg are highly coordinated between tissues. Another example is that both calcium and phosphate are mobilized from scales under the influence of estradiol and mediated by estrogen receptors, and while some portions are relocated to bone (Armour et al., 1997), others are available for hepatic syntheses. Estrogen action targets both osteoclasts and osteoblasts. In osteoclasts, mineral resorption is enhanced while osteoblasts downregulate the expression of osteonectin, a glycoprotein that, among other things, controls mineralization (Lehane et al., 1999). While freshly fertilized fish embryos do not express osteonectin, upregulation of osteonectin expression is especially important during embryonic development, when expression is enhanced by parathyroid hormone (PTH) -related peptide (Estevão et al., 2005).

The importance of phosphate and calcium to the synthesis of vitellogenin, under the control of estradiol and others, calls into question the role of two other hormones. The first is stanniocalcin, a glycoprotein produced by the corpuscles of Stannius that is usually a key endocrine regulator of phosphate and calcium homeostasis in fish. Obviously, during vitellogenesis, the normal anti-hypercalcemic effects of stanniocalcin in gill, gut and kidney and its role in renal phosphate uptake must be fine-tuned to the general mobilization of these minerals by estradiol. Similar interaction may exist with calcitonin produced by the ultimobranchial gland. Calcitonin duplicates the efforts of stanniocalcin as an antihypercalcemic factor, but in deference to the situation in mammals, also leads to increased mobilization of phosphate, resulting in hyperphosphatemia in fishes. In addition, these interactions are likely to be influenced by the availability of calcium in the surrounding water and possibly the amount of calcium sequestered by Vtg.

Although Vtg may transport iron to the oocytes, usually transferrin assumes this role. Little is known about the dynamics of transferrin during vitellogenesis, but the piscine transferrin contains an Estrogen response element (ERE) (Mikawa *et al.*, 1996).

Vitellogenin Uptake by the Oocyte

Vtg is taken up by the oocyte through receptor-mediated endocytosis (Tyler *et al.*, 1991). The Vtg receptor (Vtg-R) responsible recognizes Vtg-R, internalizes it and initiates degradation of the Vtg-receptor complex. The receptor belongs to the family of low-density lipoprotein receptors (LDL-R) that bind to different ligands and regulate lipid transport and tissue uptake (Davail *et al.*, 1998). LDL-Rs possess common structures: Ligand binding repeats of about 40 residues each (LBRs) that are rich in cysteine and contain an acidic Ser-Asp-Glu consensus sequence at their carboxyl ends. The LBR region, consisting of eight repeats in fishes—giving such receptors the acronym LR8—is followed by cysteine-rich repeats that bear similarity to epidermal growth factor (EGF) precursor. Following the first two of these EGF-precursor-like repeats (A and B) is a cysteine-poor spacer region containing five YWTD motifs (or closely related); so-called propeller repeats, involved in pH-dependent dissociation of receptor and ligands in acidic endosomal compartments (Fig. 4.6). Next is another EGF-precursor domain (C) adjacent to a short transmembrane domain. The cytoplasmic tail consists of around 50 amino acid residues and harbours NPXY motifs mediating clustering of ligand-receptor complexes and guiding them to coated pits and thus initiating endocytosis.

Vtg in the circulation is dimerized via binding sites located in its lipovitellin heavy chain (Lv-H) region. Some debate seems to exist about

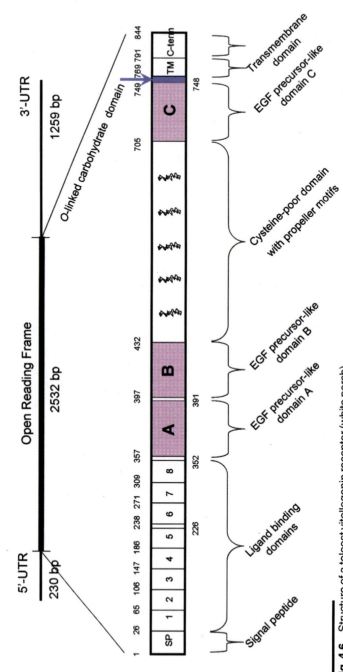

Fig. 4.6   Structure of a teleost vitellogenin receptor (white perch). Top portion—mRNA arrangement (base pairs). Bottom portions—amino acid numbering and identification of the various sections of the receptor. C-term—indicates the C-terminal of the peptide facing the cytosol. Not all fish vitellogenin receptors contain the O-linked carbohydrate domain (dark arrow), located in the plasma portion adjacent to the transmembrane domain (see text and Table 4.3). Drawn from data published by Hiramatsu *et al.* (2004). GenBank accession number AAO92396.

the stoichiometry of the Vtg-R/Vtg complex, with the most convincing model for fish suggesting that a single Vtg-R can transport a Vtg dimer and that the dimer is captured preferentially by the first three LBRs (Ding, 2005) on the receptor.

Vtg and other ligands such as apoA and apoE bind to receptors of the LDL-R super family (LDL-R, very-low-density-lipoprotein receptor-VLDLR, Vtg-R) through lysine or arginine residues that target clusters of acidic amino acids in the LBR domains of their respective receptors. The Vtg dimers bind to their receptors through key areas in the lipovitellin-1 portion, especially the sequence between amino acids 162 and 248 (in tilapia). Using site-directed mutagenesis on the Vtg of tilapia (*Oreochromis aureus*), Li *et al.* (2003) identified Lys[181], a basic residue that forms part of a conserved octapeptide (HLTKTKDL), as a key amino acid in this interaction. Not surprisingly, this Lys is completely conserved in most teleostean Vtgs—and other ligands of this super family of receptors—as is the other basic residue (all VtgAs have R instead of K) in the octapeptide. In a major deviation from this model, the four Vtg-Cs identified to date (see Fig. 4.4) contain numerous substitutions in the positions corresponding to these two lysines, even though the surrounding amino acids are either identical or represent single base changes, usually conserving the chemical character of the residue. Table 4.2 presents the 11-mer of the double Lys region for a representative Vtg-A and for the four Vtg-Cs from Fig. 4.4. We have added the 11-mer from a tentatively identified Vtg-C from spotted green pufferfish (*Tetraodon nigroviridis*) to the table (Table 4.2) to strengthen our point. This systematic difference makes us wonder about the importance of this particular area to receptor

Table 4.2 Receptor binding region of teleostean vitellogenins.

Species	Vitellogenin	11-mer around Lys[181]
Oreochromis aureus	B-type	HLT**K**TKDLNHC
Acanthogobius flavimanus	C-type	DLTQVVDVTNC
Danio rerio	C-type	IVTRIVDITNC
Gambusia affinis	C-type	IVTQVVDVTNC
Takifugu rubripes[a]	C-type	SVTQVVDIDNC
Tetraodon nigroviridis[a]	C-type	SITQVVDVSNC
		-:*-:**:-**

Comparison of C-type Vgs: * conserved residue; : chemical nature conserved; - not conserved. The boxed amino acid in the tilapia sequence indicates Lys[181]. Positions corresponding to the two lysines in Vtg-A and Vtg-B are bolded. [a] Data retrieved from the corresponding cDNAs (*cf.* Fig. 4.4 for GenBank entries) or genomic databases (Ensembl or TIGR) for the two pufferfish.

binding or about the specificity of the receptor itself. It is quite possible that Vtgs of the C-type fail to bind to receptors of the 'tilapia' type and either specific Vtg receptors exist in the five species listed in the table or other receptors of the superfamily accept and transport the less abundant C-type Vtgs.

Two forms of Vtg-R cDNAs have been identified in fishes, differing only by the presence or absence of a short Ser/Thr-rich area, the so-called O-linked sugar domain, located between epidermal growth factor (EGF) precursor C and the transmembrane domain. The two transcripts are splice-variants of the same mRNA. Interestingly, in the white perch (and possibly the Japanese eel), no evidence has been found for transcripts containing the O-linked sugar domain and in rainbow trout the shorter mRNA has been identified to be ovary-specific (Table 4.3). Apart from these minor variations on the theme, it is now established that production of Vtg-R-mRNA precedes the vitellogenic growth phase and peaks in early vitellogenesis, and the number of Vtg receptors per follicle increases with increasing follicle size. Expression gradually trails off during the vitellogenic growth phase and receptor protein shows a marked decrease by the time of ovulation (Lancaster and Tyler, 1994; Hieble et al., 1995; Prat et al., 1998; Li et al., 2003). The observed timing of the transcripts for Vtg-R may indicate that the receptor is recycled to the oocyte surface during the vitellogenic growth phase and that the longevity of the receptor changes with different phases of vitellogenesis.

When we queried public databases for torafugu (*Takifugu rubripes*), green spotted pufferfish (*Tetraodon nigroviridis*) and zebrafish (*Danio rerio*) with the protein sequence for white perch Vtg-R, we located Vtg receptors

Table 4.3 The O-linked carbohydrate domain in teleostean vitellogenin receptors.

Species	Vtg-R with O-linked carbohydrate domain (domain length)	Vtg-R lacking O-linked carbohydrate domain (tissue)
Oncorhynchus mykiss	yes (35 aa)	yes (ovary)
Oreochromis aureus	yes (20 aa)	yes (ovary)
Morone americana	no	yes (ovary and others)
Anguilla japonica	?	yes (ovary)
Takifugu rubripes	yes (20 aa)	no expression data

The Vtg-R cDNA available for *Conger myriaster* (GenBank **BAB64338**) is incomplete and truncated before the region potentially containing the O-linked carbohydrate domain. Other entries are: *Oncorhynchus mykiss* (**CAD10640**), *Oreochromis aureus* (**AAO27569**), *Morone americana* (**AAO92396**), *Anguilla japonica* (**BAB64337**). *Takifugu rubripes*—deduced from the genomic database (http://fugu.hgmp.mrc.ac.uk/blast/blast.html)

for the two pufferfish (including one containing an O-glycosylation domain, (cf. Table 4.3), and also sequences with large homologies to LDL-R (both pufferfish), low-density lipoprotein receptor-related protein-1 (α2-macroglobulin receptor) for *Torafugu* and LPR2 (megalin) for zebrafish and *Torafugu* (T. Mommsen and B. Korsgaard, unpubl.). Since all three receptor types could be relevant for the uptake of oocyte building blocks, this type of query provides an interesting starting point for future research.

At least in trout, the Vtg-R seems to be fairly specific to Vtg (Tyler and Lubberink, 1996), while other vertebrate Vtg-Rs including white perch Vtg-R may be not quite as unifunctional, and accept and transport other lipids, including low-density-lipoproteins (LDLs). Of course, the trout is one of those fish species where Vtg seems to be responsible for the overwhelming majority of oocyte growth. In this context, it should also be reiterated that the oncorhynchids only possess a single type of Vtg (Trichet *et al.*, 2000). Many teleosts drastically deviate from this simple picture, and hence may require a less single-minded Vtg-R than the trout. As an alternate strategy to the trout, which spawn demersal eggs without a large oil globule, most teleosts spawn pelagic eggs where neutral lipids—consisting of triglycerides, wax and stearyl esters—account for the bulk of the total lipids in the egg. Thus, their composition differs considerably from lipids carried by Vtg that tend to be dominated by phospholipids such as phosphatidylcholine and phosphatidylethanolamine (cf. Table 4.1). It is, therefore, conceivable that the bulk of neutral lipids reach the oocyte via LDLs or very-low-density-lipoproteins (VLDLs), requiring either the presence of additional receptors or less stringent Vtg-Rs.

Lipoprotein Lipase

An interesting relationship between LPL and VLDL-R has been noted for mammals, and considering the similarities between these players in mammals and other vertebrates, comparable mechanisms can be expected for the fishes. First, VLDL-R and LPL are usually expressed in the same tissues; second, LPL binds directly to the receptors and thus the receptor controls LPL internalization and degradation; finally, LPL reduces circulating VLDLs to smaller remnants before they can be endocytosed via the receptor. Considering the specific dynamics of ovarian LPL around vitellogenesis, perhaps a new angle of approach should be devoted to VLDL-R and Vtg-R, since it is more than likely that ovarian LPL facilitates the hydrolysis of lipoprotein other than Vtg, delivering fatty acids for uptake and re-esterification to the growing oocyte.

Given the importance of lipid to the growing oocyte either via vitellogenin or other lipoproteins, it does not come as a surprise that fish ovaries display high titers of LPL during vitellogenesis. LPL is usually anchored to the vascular endothelium, where it binds plasma lipoproteins, hydrolyzing some of their lipids, resulting in the uptake of the products into the adjacent tissues. Clearly, the vitellogenic oocyte is one such target and expresses LPL mRNA during all phases of vitellogenesis, but especially during the later stages. Interestingly, expression of enzyme is either non-existent or very low during embryogenesis (Fig. 4.7). However, keeping in mind that expression and translation into functional protein are not necessarily correlated, it is well possible that the enzyme is still active during embryogenesis. Clearly, some sort of lipolytic machinery must be activated to account for the ensuing lipolysis during embryogenesis, although it is unclear as to how lipidated the Vtg remnants are at this point or whether LPL-type enzymes are still required or whether other lipases suffice.

Oocyte Growth

Large species variation exists in the pattern of oocyte growth in teleostean fish specifically relating to the final size of the oocyte by the end of vitellogenesis and also to egg type and ambient salinity. However, all oocytes undergo the same basic stages of growth, i.e., oogenesis followed by a previtellogenic stage of arrested oocytes slowly acquiring vitelline envelope proteins, cortical alveolar stage. Next is the largely estradiol-driven vitellogenic stage, characterized by massive uptake of vitellogenin and other maternal materials. Postvitellogenic oocytes then enter into the maturation phase. Until this point, the cells are arrested at the G2/M border of prophase 1 and their first meiotic division is incomplete. During maturation, meiosis is resumed and eventually completed, initiated by a surge in LH, followed by ovulation. Indirectly, gonadotropin can also induce the final phase of oocyte maturation by stimulating the production of some C21 progestins known as maturation-inducing steroids (MIS). MIS, acting very quickly through a membrane-bound receptor, are prime examples of steroids mediating their actions through non-genomic routes. In this MIS are not alone. It is now well established that estradiol (and xenoestrogens) will exert rapid actions on intracellular message transduction systems, including the classical adenylyl cyclase cascade, in fish ovarian membranes (Thomas, 2003). Similar data for other vertebrate systems lead to new insights into interactions between nuclear

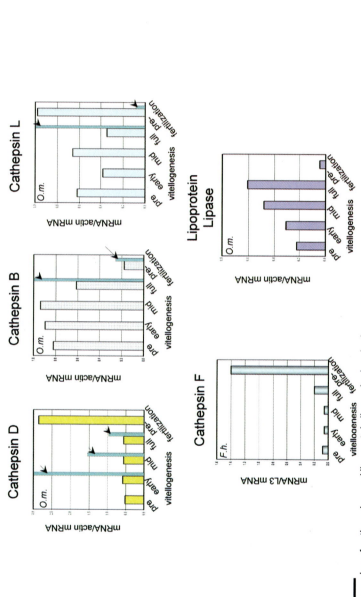

Fig. 4.7 Dynamics of cathepsins and lipoprotein lipase in developing oocytes. Data on mRNA expression for rainbow trout (*Oncorhynchus mykiss*) and killifish (*Fundulus heteroclitus*) are expressed as mRNA of enzyme in question over the mRNA of reference genes (actin for trout and ribosomal protein L3 for killifish). The figure (arrows) also includes enzyme activity measurements for cathepsin D (*Sparus aurata*), and cathepsins L and B (*F. heteroclitus*); these are expressed as a percentage of the maximum, set to 100 per cent. Data recalculated from (Carnevali *et al.*, 1999), (Kwon *et al.*, 2001), (Fabra and Cerda, 2004) and (LaFleur, Jr. *et al.*, 2005).

and membrane actions of numerous steroid hormones (Watson *et al.*, 2005). Further, non-genomic actions are also involved in the androgen-dependent inhibition of estradiol synthesis in fish ovarian cells (Braun and Thomas, 2003).

The largest increase in oocyte size normally occurs during the process of vitellogenesis. Whereas the role of gonadotropins in the uptake of vitellogenin into the oocytes is well elucidated—that GtH I has a primary function in stimulating the uptake of Vtg (Prat *et al.*, 1998)—less is known about the processing pathways of vitellogenin in the oocytes. In the ovary, the circulating vitellogenin enters the capillary network of the follicle, penetrates the basal lamina to reach the oolemma, passes the channels of the zona radiata to be finally internalized and stored in the ooplasm in yolk granules, globules or platelets as phosvitin and lipovitellin.

In marine and brackish water fish that produce pelagic-type eggs, a second dramatic increase in egg mass is observed. Especially during final maturation, the water content rapidly increases to 85 to 95 per cent, together with a 3- to 5-fold rise in volume, a phase termed 'hydration' (Reith *et al.*, 2001). This increase in oocyte volume is driven by large amounts of free amino acids, liberated by complete or partial hydrolysis from oocyte proteins derived from Vtg (Thorsen and Fyhn, 1996; Matsubara andand Koya, 1997; Polzonetti-Magni *et al.*, 2004) (see Fig. 4.7) and made possible through the active expression and presence of aquaporin in the oocyte (Fabra *et al.*, 2005). In addition to facilitating the uptake of water, aided in this by concurrent changes in acid-base status and other osmolytes, these free amino acids will assist in early osmoregulation, and in providing buoyancy for floating eggs, and subsequently serve as important substrates driving aerobic metabolism and protein synthesis during early development.

To supply the oocytes with the required components, ovaries support tissue and oocytes must be set up to handle any of these compounds. For instance, LPL mentioned the above works in the capillary epithelium surrounding the oocytes and, thus, is ideally placed to deliver protein and lipid to the oocytes from lipoproteins. While in the oocytes, specific enzymes are involved mediating the processing of yolk proteins in the oocyte after Vtg has been sequestered through receptor-mediated endocytosis. In fish, the lysosomal system of the oocyte holds all enzymes necessary for the degradation of Vtg and formation and rearrangement of compounds into yolk. The proteolytic processing of the precursor protein vitellogenin and the yolk proteins is likely to be mediated by lysosomal enzymes that attack peptide bonds, glycosidic bonds and phosphate bonds.

One of the enzymes involved in fragmentation of Vtg inside the oocyte is cathepsin D, an acidic aspartate protease produced by the lysosomes. Teleosts seem to express two paralogues of the enzymes, likely to be remnants of the initial genome duplication of fishes (Taylor *et al.*, 2003). Cathepsin Ds are expressed in most fish tissues, including muscle, liver and oocyte. Variable roles have been described to these enzymes, ranging from control of muscle proteolysis to vitellogenin breakdown in developing oocytes, both essential processes during vitellogenesis. The enzyme is synthesized with an 18-amino acid signal peptide that is cleaved off in the endoplasmic reticulum prior to translocation to the Golgi apparatus. Following glycosylation, the enzyme is shuttled to the lysosomes and enzyme activity is developed through attack by a cysteine protease that cleaves off an N-terminal 46 residue prosequence (Nielsen and Nielsen, 2001). Cathepsin D expression is responsive to estradiol as well as other steroids and clearly accounts for some, possibly the bulk, of the early phases in the breakdown of Vtg in developing fish oocytes. Its mRNA abundance (together with cathepsin L, below) peaks around the time of fertilization (Kwon *et al.*, 2001), even though high activity and abundance of mRNA do not always coincide and one cannot tacitly extrapolate from mRNA to active protein. Cathepsin D actions are not restricted to dealing with Vtg, but the enzyme is also actively involved—at least in birds—in the fragmentation of apoB which can reach the oocyte via endocytosis through the Vtg-Rs. Involvement of the enzyme in degradation of any proteolipid carrier transported by the Vtg-Rs is therefore likely. Cathepsin D and Vtg can be co-localized in the multivesicular bodies during vitellogenesis in the rainbow trout (Sire *et al.*, 1994), suggesting that cathepsin D is the key lysosomal enzyme responsible for the first degradative step of Vtg. During later phases, a more generalized activation of the lysosomes occurs, with increased transcription and activation of other hydrolases, including acidic phosphatase, likely required for the hydrolysis of phosvitin-derived yolk proteins.

Cathepsin D is not alone in assuming a role in the breakdown of vitellogenin and its components. Far from it. At least five other proteases are likely to be involved. One is a female specific aspartate protease, variously termed nothepsin or liver-specific aspartic protease (LAP) which was originally described for fish liver, but recently found to be also expressed in ovary (Kurokawa *et al.*, 2005). The enzyme is closely related to cathepsin D (Riggio *et al.*, 2000) and also inducible by estrogen and, as such, has been proposed to be included as a potential indicator of fish exposure to estrogen mimics (Mommsen, 2004). Cathepsins L and B, cysteine proteases are abundant in many fish tissues, including the ovary and have long been associated with oocyte and egg proteolysis (Kestemont et al., 1999); especially cathepsin L was implicated in the secondary processing of lipovitellin (Carnevali et al., 1999). In trout, abundance of mRNAs for both enzymes peak in early and mid-vitellogenesis (Fig. 4.7) (Kwon et al., 2001). These and other cysteine proteases have most recently been assigned prominent roles in oocyte development and maturation in a killifish (F. heteroclitus). For instance, the mRNA for cathepsin F in F. heteroclitus undergoes a drastic increase in abundance during oocyte maturation (Fig. 4.7), while mRNAs for cathepsins K and H decrease during this period. Interestingly, the dynamics of mRNA abundance and enzyme activity do not seem to correlate very tightly (Fig. 4.7), although it should be pointed out that species difference may account for the differences in peak abundance/activity, since the data in the figure are derived from different species. In post-vitellogenic follicles undergoing $17\alpha,20\beta$-dihydroxy-4-pregnen-3-one ($17,20\beta$P) induced maturation, mRNAs for cathepsins F, L, S and H all peak 20-30 h after hormone application, coinciding with or slightly preceding the phase of germinal vesicle breakdown (Fig. 4.7).

Just like during fish muscle proteolysis under the control of estradiol, endogenous protease inhibitors may play a role in directing and regulating proteolytic activity. Protease inhibitors with multiple specificities are abundant in fish tissues, including muscle, skin, liver and ovary (Ohkubo et al., 2004a). Two protease inhibitors were directly associated with the uptake of yolk proteins in the vitellogenic oocytes in the rainbow trout. The proteins encoded by these transcripts were found to be localized in the cortical ooplasm of the vitellogenic oocytes and directly associated with sites of vitellogenin processing (Wood et al., 2004). Two oocyte protease inhibitor cDNAs were detected throughout the vitellogenic growth phase of the oocytes, and expression declined coinciding with oocyte degeneration. A second, apparently quite targeted, wave of proteolysis occurs during the hydration phase, producing the free amino acids required to help drive the massive influx of water.

Interestingly, similar mechanisms as in the oocyte disassembling and rearranging vitellogenin and other components of yolk appear to be acting in the periphery of the maternal system. As in the oocyte, cathepsin D plays an overwhelming role in the mobilization of amino acids from muscle, with contributions of cathepsin L, calpains and cathepsin L-like enzymes (Mommsen, 2004).

The enzymatic machinery involved in processing of Vtg in the growing oocytes can clearly distinguish between the different forms of Vtg, leading to a plethora of different products during oocyte growth and further processing in the course of oocyte maturation—well recognized as the different yolk proteins that originally lent their names to the different sections of Vtg. A stylized lineage of oocyte proteins is depicted for the marine teleost Barfin flounder (*Verasper moseri*) that produces pelagophil eggs (Fig. 4.8). These eggs incur a massive influx of water before ovulation, driven by increasing concentrations of free amino acids (over ten-fold), derived from Vtg, and osmolytes.

The amount of Vtg degradation differs substantially in species with benthophil eggs, that experience only modest increases in free amino acids (3-fold) and with it limited uptake of water. Using the killifish as a model teleost, VtgA appears to be the main source of free amino acids through degradation of LvH, although major portions of LvH also remain as a 103 kDa component in the oocytes after maturation. In contrast, VtgB is processed in the growing oocytes to smaller components, but no further degradation is noticed during final maturation. As such, VtgB does not seem to contribute to the increased amino acid pool during final maturation. The fate of any phosvitin-less VtgCs has been analyzed in two species. In the Japanese common goby, a 320 kDa dimeric VtgC remains unchanged during yolk accumulation, while in the mosquitofish *Gambusia affinis*, a 400 kDa dimeric VtgC is proteolytically processed into distinct lipovitellin H and lipovitellin L subunits, but apparently not any further (Ohkubo *et al.*, 2004b; Sawaguchi *et al.*, 2005).

CHORIOGENINS

A fascinating group of estrogen-sensitive gene products relevant to growing oocytes are the vitelline envelope proteins (VEPs), a.k.a. choriogenins. The chorion (also called vitelline envelope, or *zona radiata*, or by analogy to mammalian structures *zona pellucida* [ZP] proteins) of teleostean eggs is assembled from two to four different VEPs and consists of a thin outer layer and a thicker inner layer, forming an extracellular envelope that is produced early during ovarian growth. The inner layer of the chorion, characterized by choriogenins and ZP proteins, is laid down largely during the previtellogenic phase and the proteins are eventually processed into the hardened chorion. The much thinner outer layer tends to be dissolved after fertilization. Once the VEPs are crosslinked, they provide mechanical and chemical protection for the

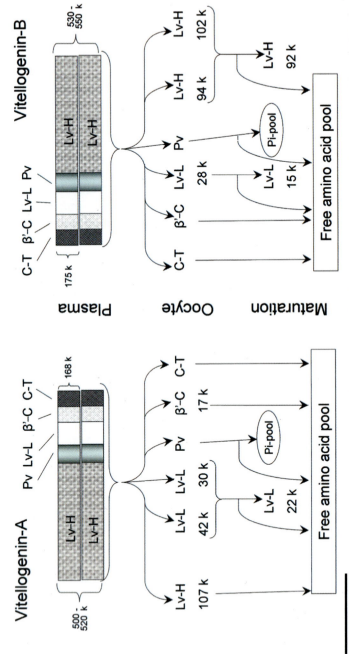

Fig. 4.8 Processing of vitellogenins in oocytes and during oocyte maturation. Vitellogenin monomers (before post-translational modification) have masses of 168 kDa (Vtg-A) and 175 kDa (Vtg-B), respectively. Vtg-A dimers taken up by the oocytes have masses between 500 and 520 kDa, while Vtg-B dimers range from 530 to 550 kDa. Modified from Matsubara *et al.* (1999).

Abbreviations: Lv-H—lipovitellin heavy chain; Pv—phosvitin; Lv-L—lipovitellin light chain; β'-component (cysteine rich); C-T—cysteine-rich C-terminal domain; k = kDa.

developing embryo, protection from microbial infection and they are involved in fertilization, and play various roles in the course of embryonic development. Eventually, the heavily cross-linked sialoglycoproteins are hydrolyzed by a dedicated, aptly named, hatching enzyme (Hiroi et al., 2004).

Unfortunately, the nomenclature of these egg envelope proteins is a bit of a mess. Although the term 'choriogenin' has been suggested for peptides synthesized by the fish liver, this recommendation has not yet found general acceptance. Three choriogenins (VEP, α, β, γ) present in salmonid fishes and the Japanese medaka have attracted the most attention. As in the case of most other teleosts, salmonids and medaka synthesize these proteins in the liver in response to estradiol. Subsequently, these proteins are transported in the plasma to be taken up by growing ovary. In medaka, the homology of three deduced, female-specific proteins to mammalian ZP proteins has been pointed out. Hence, these three peptides have also been called ZPA, ZPB and ZPC. However, some additional proteins of the vitelline envelope may be contributed by the ovary. In contrast to many other teleosts, cDNAs for zebrafish and carp (Cyprinus carpio), vitelline envelope proteins have been localized to the ovary, and in carp, genes homologous to ZPB and ZPC are expressed only in the ovary (Wang and Gong, 1999). While expression of these genes is female specific, it remains to be seen whether zebrafish and carp genes are as estrogen sensitive as the corresponding genes in liver.

One characteristic of choriogenins is their relatively high content of glutamate. Following fertilization, the hardening of the chorion is brought about by cross-linking between glutamate and lysine catalyzed by an oocyte/embryo-specific transglutaminase in rainbow trout (Ha and Iuchi, 1998). However, in a similar study on carp oocytes that convincingly shows the importance of the oocyte transglutaminase to cross-linking of choriogenins H and L, the carp transglutaminase responsible was not distinguishable from other tissue transglutaminases (Chang et al., 2002). Either way, transglutaminase transcripts are already present in previtellogenic oocytes, in matured oocytes and after ovulation and, therefore, of maternal origin.

The initiation of choriogenins and Vtg synthesis and their respective dynamics of inducibilities are species-dependent, but a few generalizations can be made. (1) Choriogenin induction and production precedes Vtg induction and biosynthesis; (2) chg induction is more sensitive to estradiol than Vtg, but the mechanism may differ; and (3) species differences exist in the types of chgs induced preferentially.

Higher sensitivity of choriogenins to estradiol is in agreement with the annual cycles of plasma VEPs and Vtg in cherry salmon (*Oncorhynchus masou*), where VEPs dominate the picture of estrogen-sensitive plasma proteins during the previtellogenic growth phase when circulating estradiol levels are below 0.4 nM. When salmon enter the vitellogenic phase driven by increasing levels of estradiol and are reflected in noticeable increases in the gonadosomatic index, Vtg levels increase and ratios Vtg to choriogenin H and choriogenin L change to 18:1.6:1, respectively (Fujita *et al.*, 2004, 2005). This observation agrees with the idea that chorion formation precedes vitellogenesis and that low levels of estradiol are sufficient to induce their synthesis. Similar expression patterns are noted for sheepshead minnows, where estrogenic compounds, including estradiol, lead to 10-fold stronger induction of VtgA than for VtgB, while VEPγ is preferentially induced over VEPβ (Knoebl *et al.*, 2004). In Sakhalin taimen (*Hucho perryi*) serum choriogenin L exceeded choriogenin H levels through oocyte growth, while the expression of medaka choriogenin L mRNA was more sensitive to induction by xenoestrogens than choriogenin H (Lee *et al.*, 2002).

In medaka, the L- (or VEP γ = ZPC=ZP3) form of choriogenin is more sensitive to induction by estradiol than the H-form, while in other species induction of the H-form (=ZPB=ZP2) seems to take precedence. In Atlantic salmon, it appears that preferentially the VEPα and β forms are induced. These two forms of vitelline envelopes are closely related and show the closest relationship to the minor form of chg H found in medaka and some other species. Obviously, a thorough systematic analysis of choriogenin types remains to be done before we can shed light on specific relationships and hormonal sensitivity.

One of the first observations on these estrogen-sensitive hepatic genes was that their transcription was somewhat more sensitive to estradiol than the Vtg gene and that it occurred earlier following exposure. This led to the suggestion of supplanting Vtgs with choriogenins as even more sensitive biomarkers for exposure of fish to estrogenic compounds (Arukwe and Goksøyr, 2003). Unfortunately, the observation of VEPβ expression in naïve Arctic char (*Salvelinus alpinus*) has somewhat dampened the enthusiasm for such wholesale usage of choriogenin mRNAs (Westerlund *et al.*, 2001) for this purpose. It also implies that regulation of choriogenin genes is more complex than initially envisaged, and their expression does not simply precede expression of vitellogenin (Berg *et al.*, 2004). Further, research from our own laboratory has shown that rainbow trout embryos respond to xenoestrogens exposure by dramatically augmenting the stationary concentrations of all three choriogenin mRNAs at very high sensitivity, but without ever synthesizing the actual proteins, making us wonder about the metabolic implications of such seemingly frivolous transcriptional activity.

Differences exist in estrogen sensitivity in the expression of two VEP precursor genes in medaka, with choriogenin L mRNA responding at lower estradiol doses than mRNA of the larger choriogenin H (Lee *et al.*, 2002). Similarly, in Atlantic salmon, expression of VEPs and Vtg was, as expected, estradiol sensitive, but induction for the zona radiata proteins preceded Vtg synthesis and occurred more rapidly and at lower doses of estradiol than the induction and synthesis of Vtg (Celius and Walther, 1998). Overall, it appears that priming of the ER/estradiol system is not necessarily required for successful upregulation of the VEP system and that basal concentrations of ERs may suffice for the estradiol-dependent synthesis of VEPs. Alternatively, other mechanisms have to be invoked to explain selective choriogenin induction without affecting other estradiol-dependent genes such as ER or Vtg. One option would be that the 'hepatic memory' that is so important to secondary induction of estrogen-responsive genes, may be at work in the liver due to previous exposure to some hormones specifically sensitizing the VEP genes, possibly through changes in chromatic structure. A thorough analysis of the promoter regions of the piscine VEPs may shed light on this hypothesis. However, it already appears that non-genomic actions can be excluded, since rates of choriogenin induction are not quite rapid enough although non-genomic actions of estradiol, as they apply to oogenesis, should be reassessed.

Phosvitin Breakdown

Complete degradation of phosvitin, described for the maturing oocytes of the barfin flounder (Matsubara *et al.*, 1999) can be expected to lead specifically to increases in both phosphate and serine during final maturation. Indeed, such increases have been noted in Atlantic halibut (*Hippoglossus hippoglossus*) during oocyte maturation and at early stages of hydration, with increases in free inorganic phosphate preceding those for serine (Fig. 4.9), implying that during this phase, phosphatases are more active than proteases hydrolyzing the polyserine backbone of phosvitin (Finn *et al.*, 2002). Both hydrolytic processes most likely involve activation of the lysosomal machinery, with acid phosphatase attacking the phosphoserines and the usual cocktail of cathepsins taking care of the proteins.

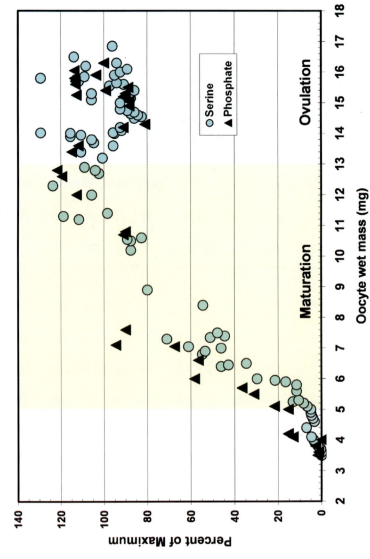

Fig. 4.9 Changes in phosphate and serine concentrations in halibut (*Hippoglossus hippoglossus*) oocytes during maturation. Data are expressed in per cent of average maximum concentration reached. Data recalculated from Finn *et al.* (2002).

HYOSOPHORIN

Fish eggs also contain hyosophorin, a common >100 kDa glycoprotein exclusively transcribed and synthesized in early and growing oocytes, where the carbohydrate moieties account for up to 90 per cent of the total mass. This unique glycoprotein, that can be localized to the cortical granules of oocytes, is thought to play a role in water accumulation in the perivitelline space. Furthermore, since some fish hyosophorins can bind calcium, they have been proposed to serve as a potential calcium reservoir. While their specific functions are still a matter debate, the hyosophorins undergo drastic changes immediately following fertilization in carp (*Cyprinus carpio*). In the course of the cortical reaction, these glycoproteins are launched into the perivitelline space, where they are quickly degraded to 20 to 30 kDa fragments (Tsao *et al.*, 1999). The interesting question here concerns the source of all the carbohydrate moieties—in this case plenty of sialic acid. Although a 'carbohydrate balance sheet' has yet to be done for vitellogenesis, it seems unlikely that Vtg could deliver the sialic acid units required for the synthesis of hyosophorins and the many other abundant glycoproteins found in the oocytes, since the overall carbohydrate content of Vtg is relatively small, yet sialoglycoproteins alone are more than 10 times as abundant as Pv-derived components in growing oocytes (Tsao *et al.*, 1999). From the abundance of hyosophorin mRNA early in vitellogenesis, it appears that protein backbone synthesis and carbohydrate side-chain assembly is controlled by the oocytes and the flux of carbohydrate from extra-ovarian tissues to the oocyte must be substantial, and can only partially involve Vtg as a carrier. It has also been shown that oocytes in mid-vitellogenesis already contain the enzymatic machinery, especially two key sialyltransferases (Kitazume *et al.*, 1994), in order to successfully synthesize numerous complex sialylglycoconjugates.

Vasa-like Genes

Vasa-like genes, key regulators of primordial germ cell development and useful markers for germ cells, are expressed in oogonia to diplotene oocytes of tilapia and can be localized in the cytoplasm of previtellogenic oocytes. Subsequently, expression decreases from vitellogenic to postvitellogenic oocytes (Kobayashi *et al.*, 2002). Interestingly, vasa protein does not always co-localize with its mRNA (Knaut *et al.*, 2000). During a later stage of maturation, fish ovaries increase the expression of vasa genes. Similar increases can be evoked by injection of fish with estradiol or gonadotropin-releasing hormone (GnRH), while estradiol in fish co-injected with growth hormone exhibit lowered concentrations of vasa gene mRNA (Cardinali *et al.*, 2004)

Riboflavin-binding Protein

One interesting compound yet to be looked at in the fishes is riboflavin (Vitamin B2) binding protein, a protein produced by the chicken liver in response to estradiol that regulates the transport of riboflavin to the growing oocytes. Riboflavin is found in salmon oocytes at similar concentrations as in chicken eggs (H.B. White and M.A. Letavic, unpubl.) and the corresponding binding protein has been sequenced for Nile tilapia (*Oreochromis niloticus*) (GenBank AAP42246), but its induction by estradiol and role in vitamin transport in teleosts remain to analyzed. However, it should be mentioned that the riboflavin-binding protein binds to Vtg-R in chicken and a similar role of the binding protein chicken can be envisaged for fish.

Spindlin-like gene

The product of the spindlin-like gene, a protein thought to be important during the transition from mature oocyte to early embryo, was detected in mature carp (*Carassius auratus gibelio*) eggs. The content decreases after fertilization, suggesting that the carp-spindlin represents a maternal protein that is expressed during oocyte maturation and assumes a similar function as in mammals (Wang *et al.*, 2005).

SUMMARY

An integrated picture of vitellogenesis in fishes is presented in Fig. 4.10. Here, emphasis has been put on sources of vitellogenin-building units, treating the vitellogenic liver as the most important clearing house for intermediates. Peripheral mobilization and hepatic biosynthetic routes are all under the control of estradiol. In the previtellogenic stage, the liver supplies parts for the production of the vitelline envelope. During vitellogenesis, vitellogenin, lipoproteins and other carriers are required to deliver the building blocks for the growing oocytes. Some of these compounds—not vitellogenin—may be funnelled through other ovarian cells. During final maturation, vitellogenins are partially or completed degraded to produce the various yolk protein, the lipid droplets and free amino acids to drive hydration (if required).

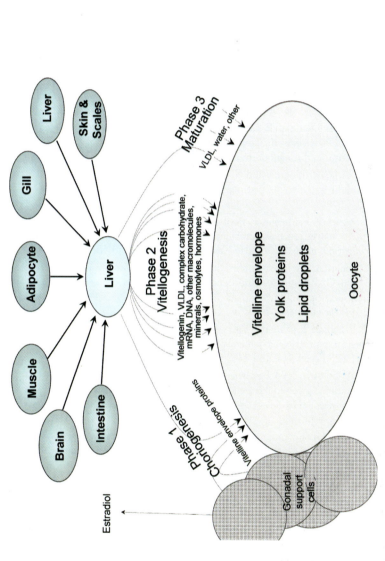

Fig. 4.10   Simplified summary of metabolite flux from tissues *via* the liver to the oocyte. A substantial portion of this flux is driven by estradiol. For roles of other hormones, see Fig. 4.1. Every tissue listed here is capable of producing hormones that can influence hepatic and/or oocyte metabolism. Also, it is not inconceivable that other tissues deliver compounds directly to the oocyte or to the oocyte *via* the gonadal support cells.

Acknowledgements

We thank Tom Moon and Mathilakath Vijayan for useful discussions and Gord Brown for help with the phylogenetic analyses.

References

Adams, B.A., E.D. Vickers, C. Warby, M. Park, W.H. Fischer, C.A. Grey, J.E. Rivier and N.M. Sherwood. 2002. Three forms of gonadotropin-releasing hormone, including a novel form, in a basal salmonid, *Coregonus clupeaformis*. *Biology of Reproduction* 67: 232-239.

Almansa, E., M.J. Pérez, J.R. Cejas, P. Badia, J.E. Villamandos and A. Lorenzo. 1999. Influence of broodstock gilthead seabream (*Sparus aurata* L.) dietary fatty acids on egg quality and egg fatty acid composition throughout the spawning season. *Aquaculture* 170: 323-336.

Anderson, A.J., A.H. Arthington and S. Anderson. 1990. Lipid classes and fatty acid composition of the eggs of some Australian fish. *Comparative Biochemistry and Physiology B* 96: 267-270.

Anglade, I., F. Pakdel, T. Bailhache, F. Petit, G. Salbert, P. Jego, Y. Valotaire and O. Kah. 1994. Distribution of estrogen receptor-immunoreactive cells in the brain of the rainbow trout (*Oncorhynchus mykiss*). *Journal of Neuroendocrinology* 6: 573-583.

Anglade, I., V. Bouard, C. Le Jossic-Corcos, E.L. Mananos, D. Mazurais, D. Michel and O. Kah. 1998. The GABAergic system: a possible component of estrogenic feedback on gonadotropin secretion in rainbow trout (*Oncorhynchus mykiss*). *Bulletin Francais de la Peche et de la Pisciculture* 350: 647-654.

Armour, K.J., D.B. Lehane, F. Pakdel, Y. Valotaire, R. Graham, G. Russell and I.W. Henderson. 1997. Estrogen receptor mRNA in mineralized tissues of rainbow trout: Calcium mobilization by estrogen. *FEBS Letters* 411: 145-148.

Arukwe, A. and A. Goksøyr. 2003. Eggshell and egg yolk proteins in fish: Hepatic proteins for the next generation: Oogenetic, population, and evolutionary implications of endocrine disruption. *Comparative Hepatology* 2: 4.

Babin, P.J., J. Bogerd, F.P. Kooiman, W.J. Van Marrewijk and D.J. Van der Horst. 1999. Apolipophorin II/I, apolipoprotein B, vitellogenin, and microsomal triglyceride transfer protein genes are derived from a common ancestor. *Journal of Molecular Evolution* 49: 150-160.

Bardet, P.L., S. Obrecht-Pflumio, C. Thisse, V. Laudet, B. Thisse and J.M. Vanacker. 2004. Cloning and developmental expression of five estrogen-receptor related genes in the zebrafish. *Development Genes and Evolution* 214: 240-249.

Barry, T.P., J.A. Malison, J.A. Held and J.J. Parrish. 1995. Ontogeny of the cortisol stress response in larval rainbow trout. *General and Comparative Endocrinology* 97: 57-65.

Berg, A.H., L. Westerlund and P.E. Olsson. 2004. Regulation of Arctic char (*Salvelinus alpinus*) egg shell proteins and vitellogenin during reproduction and in response to 17beta-estradiol and cortisol. *General and Comparative Endocrinology* 135: 276-285.

Bon, E., B. Breton, M.S. Govoroun and F. Le Menn. 1999. Effects of accelerated photoperiod regimes on the reproductive cycle of the female rainbow trout: II Seasonal variations of plasma gonadotropins (GTH I and GTH II) levels correlated with ovarian follicle growth and egg size. *Fish Physiology and Biochemistry* 20: 143-154.

Booth, R.K., R.S. McKinley and J.S. Ballantyne. 1999. Plasma non-esterified fatty acid profiles in wild Atlantic salmon during their freshwater migration and spawning. *Journal of Fish Biology* 55: 260-273.

Borski, R.J. 2000. Nongenomic membrane actions of glucocorticoids in vertebrates. *Trends in Endocrinology and Metabolism* 11: 427-436.

Bowman, C.J., K.J. Kroll, T.G. Gross and N.D. Denslow. 2002. Estradiol-induced gene expression in largemouth bass (*Micropterus salmoides*). *Molecular and Cellular Endocrinology* 196: 67-77.

Braun, A.M. and P. Thomas. 2003. Androgens inhibit estradiol-17beta synthesis in Atlantic croaker (*Micropogonias undulatus*) ovaries by a nongenomic mechanism initiated at the cell surface. *Biology of Reproduction* 69: 1642-1650.

Callard, G.V., A.V. Tchoudakova, M. Kishida and E. Wood. 2001. Differential tissue distribution, developmental programming, estrogen regulation and promoter characteristics of cyp19 genes in teleost fish. *Journal of Steroid Biochemistry and Molecular Biology* 79: 305-314.

Cardinali, M., G. Gioacchini, S. Candiani, M. Pestarino, G. Yoshizaki and O. Carnevali. 2004. Hormonal regulation of vasa-like messenger RNA expression in the ovary of the marine teleost *Sparus aurata*. *Biology of Reproduction* 70: 737-743.

Carnevali, O., R. Carletta, A. Cambi, A. Vita and N. Bromage. 1999. Yolk formation and degradation during oocyte maturation in seabream *Sparus aurata*: involvement of two lysosomal proteinases. *Biology of Reproduction* 60: 140-146.

CCottriil, R.A., R.S. McKinley, G. van der Kraak, J.-D. Dutil, K.B. Reid and K.J. McGrath. 2001. Plasma non-esterified fatty acid profiles and 17β-oestradiol levels of juvenile immature and maturing adult American eels in the St Lawrence River. *Journal of Fish Biology* 59: 364-379.

Celius, T. and B.T. Walther. 1998. Oogenesis in Atlantic salmon (*Salmo salar* L.) occurs by zonagenesis preceding vitellogenesis in vivo and in vitro. *Journal of Endocrinology* 158: 259-266.

Chang, Y.S., Y.W. Wang and F.L. Huang. 2002. Cross-linking of ZP2 and ZP3 by transglutaminase is required for the formation of the outer layer of fertilization envelope of carp egg. *Molecular Reproduction and Development* 63: 237-244.

Davail, B., F. Pakdel, H. Bujo, L.M. Perazzolo, M. Waclawek, W.J. Schneider and F. Le Menn. 1998. Evolution of oogenesis: The receptor for vitellogenin from the rainbow trout. *Journal of Lipid Research* 39: 1929-1937.

Davies, B., N. Bromage and P. Swanson. 1999. The brain-pituitary-gonadal axis of female rainbow trout *Oncorhynchus mykiss*: Effects of photoperiod manipulation. *General and Comparative Endocrinology* 115: 155-166.

Di Croce, L., G. Bruscalupi and A. Trentalance. 1997. Independent responsiveness of frog liver low-density lipoprotein receptor and HMGCoA reductase to estrogen treatment. *European Journal of Physiology* 435: 107-111.

Ding, J. L. 2005. Vitellogenesis and vitellogenin uptake into oocytes. In: *Molecular Aspects of Fish and Marine Biology*, P. Melamed and N.M. Sherwood (eds.), World Scientific, London, Vol. 4: *Hormones and Their Receptors in Fish Reproduction*. pp. 254-276.

Estevão, M.D., B. Redruello, A.V. Canário and D.M. Power. 2005. Ontogeny of osteonectin expression in embryos and larvae of seabream (*Sparus auratus*). *General and Comparative Endocrinology* 142: 155-162.

Fabra, M. and J. Cerda. 2004. Ovarian cysteine proteinases in the teleost *Fundulus heteroclitus*: molecular cloning and gene expression during vitellogenesis and oocyte maturation. *Molecular Reproduction and Development* 67: 282-294.

Fabra, M., D. Raldua, D.M. Power, P.M.T. Deen and J. Cerda. 2005. Marine fish egg hydration is aquaporin-mediated. *Science* 307: 545.

Finn, R.N., G.C. Ostby, B. Norberg and H.J. Fyhn. 2002. *In vivo* oocyte hydration in Atlantic halibut (*Hippoglossus hippoglossus*); proteolytic liberation of free amino acids, and ion transport, are driving forces for osmotic water influx. *Journal of Experimental Biology* 205: 211-224.

Fujita, T., H. Fukada, M. Shimizu, N. Hiramatsu and A. Hara. 2004. Quantification of serum levels of precursors to vitelline envelope proteins (choriogenins) and vitellogenin in estrogen treated masu salmon, *Oncorhynchus masou*. *General and Comparative Endocrinology* 136: 49-57.

Fujita, T., H. Fukada, M. Shimizu, N. Hiramatsu and A. Hara. 2005. Annual changes in serum levels of two choriogenins and vitellogenin in masu salmon, *Oncorhynchus masou*. *Comparative Biochemistry and Physiology* B 141: 211-217.

Gelinas, D., G.A. Pitoc and G.V. Callard. 1998. Isolation of a goldfish brain cytochrome P450 aromatase cDNA: mRNA expression during the seasonal cycle and after steroid treatment. *Molecular and Cellular Endocrinology* 138: 81-93.

Gen, K., S. Yamaguchi, K. Okuzawa, N. Kamakura, H. Tanaka and H. Kagawa. 2003. Physiological roles of FSH and LH in red seabrean, *Pagrus major*. *Fish Physiology and Biochemistry* 28: 77-80.

Giguere, V. 2002. To ERR in the estrogen pathway. *Trends in Endocrinology and Metabolism* 13: 220-225.

Gomez, J.M., C. Weil, M. Ollitrault, P.Y. Le Bail, B. Breton and F. Le Gac. 1999. Growth hormone (GH) and gonadotropin subunit gene expression and pituitary and plasma changes during spermatogenesis and oogenesis in rainbow trout (*Oncorhynchus mykiss*). *General and Comparative Endocrinology* 113: 413-428.

Gonzalez, A. and F. Piferrer. 2003. Aromatase activity in the European sea bass (*Dicentrarchus labrax* L.) brain. Distribution and changes in relation to age, sex, and the annual reproductive cycle. *General and Comparative Endocrinology* 132: 223-230.

Gonzalez-Martinez, D., T. Madigou, E. Mananos, J.M. Cerda-Reverter, S. Zanuy, O. Kah and J.A. Munoz-Cueto. 2004. Cloning and expression of gonadotropin-releasing hormone receptor in the brain and pituitary of the European sea bass: An *in situ* hybridization study. *Biology of Reproduction* 70: 1380-1391.

Ha, C.R. and I. Iuchi. 1998. Enzyme responsible for egg envelope (chorion) hardening in fish: Purification and partial characterization of two transglutaminases associated with their substrate, unfertilized egg chorion, of the rainbow trout, *Oncorhynchus mykiss*. *Journal of Biochemistry (Tokyo)* 124: 917-926.

Halm, S., N. Pounds, S. Maddix, M. Rand-Weaver, J.P. Sumpter, T.H. Hutchinson and C.R. Tyler. 2002. Exposure to exogenous 17β-oestradiol disrupts P450aromB mRNA expression in the brain and gonad of adult fathead minnows (*Pimephales promelas*). *Aquatic Toxicology* 60: 285-299.

Hammond, G.L. 1995. Potential functions of plasma steroid-binding proteins. *TEM* 6: 298-304.

Han, Y.S., I.C. Liao, Y.S. Huang, W.N. Tzeng and J.Y. Yu. 2003. Profiles of PGH-alpha, GTH I-beta, and GTH II-beta mRNA transcript levels at different ovarian stages in the wild female Japanese eel *Anguilla japonica*. *General and Comparative Endocrinology* 133: 8-16.

Hawkins, M.B. and P. Thomas. 2004. The unusual binding properties of the third distinct teleost estrogen receptor subtype ERbetaa are accompanied by highly conserved amino acid changes in the ligand binding domain. *Endocrinology* 145: 2968-2977.

Hernandez-Rauda, R., G. Rozas, G.P. Rey, J. Otero and M. Aldegunde. 1999. Changes in the pituitary metabolism of monoamines (dopamine, norephinephrine, and serotonin) in female and male rainbow trout (*Oncorhynchus mykiss*) during gonadal recrudescence. *Physiological and Biochemical Zoology* 72: 352-359.

Hieble, J.P., D.B. Bylund, D.E. Clarke, D.C. Eikenburg, S.Z. Langer, R.J. Lefkowitz, K.P. Minneman and R.R. Jr. Ruffolo. 1995. International Union of Pharmacology. X. Recommendation for nomenclature of α_1-adrenoceptors: Consensus update. *Pharmacological Reviews* 47: 267-270.

Hiramatsu, N., R.W. Chapman, J.K. Lindzey, M.R. Haynes and C.V. Sullivan. 2004. Molecular characterization and expression of vitellogenin receptor from white perch (*Morone americana*). *Biology of Reproduction* 70: 1720-1730.

Hiramatsu, N., A.O. Cheek, C.V. Sullivan, T. Matsubara and A. Hara. 2005. Vitellogenesis and endocrine disruption. In: *Biochemistry and Molecular Biology of Fishes*, T.P. Mommsen and T.W. Moon (eds.), Elsevier, Amsterdam, *Vol. 6. Environmental Toxicology*, pp. 431-471.

Hiramatsu, N., T. Matsubara, A. Hara, D.M. Donato, K. Hiramatsu, N.D. Denslow and C.V. Sullivan. 2002. Identification, purification and classification of multiple forms of vitellogenin from white perch (*Morone americana*). *Fish Physiology and Biochemistry* 26: 355-370.

Hiroi, J., K. Maruyama, T. Kawazu, T. Kaneko, R. Ohtani-Kaneko and S. Yasumasu. 2004. Structure and developmental expression of hatching enzyme genes of the Japanese eel *Anguilla japonica*: An aspect of the evolution of fish hatching enzyme gene. *Development Genes and Evolution* 214: 176-184.

Hobby, A.C., D.P. Geraghty and N.W. Pankhurst. 2000. Differences in binding characteristics of sex steroid binding protein in reproductive and nonreproductive female rainbow trout (*Oncorhynchus mykiss*), black bream (*Acanthopagrus butcheri*), and greenback flounder (*Rhombosolea tapirina*). *General and Comparative Endocrinology* 120: 249-259.

Jalabert, B., J.F. Baroiller, B. Breton, A. Fostier, F. Le Gac, Y. Guiguen and G. Monod. 2000. Main neuro-endocrine, endocrine and paracrine regulations of fish reproduction, and vulnerability to xenobiotics. *Ecotoxicology* 9: 25-40.

Kah, O., I. Anglade, B. Linard, F. Pakdel, G. Salbert, T. Bailhache, B. Ducouret, C. Saligaut, P. Le Goff, Y. Valotaire and P. Jego. 1997. Estrogen receptors in the brain-pituitary complex and the neuroendocrine regulation of gonadotropin release in rainbow trout. *Fish Physiology and Biochemistry* 17: 53-62.

Kajimura, S., Y. Yoshiura, M. Suzuki and K. Aida. 2001a. cDNA cloning of two gonadotropin beta subunits (GTH-Ibeta and -IIbeta) and their expression profiles during gametogenesis in the Japanese flounder (*Paralichthys olivaceus*). *General and Comparative Endocrinology* 122: 117-129.

Kajimura, S., Y. Yoshiura, M. Suzuki, T. Utoh, N. Horie, H. Oka and K. Aida. 2001b. Changes in the levels of mRNA coding for gonadotropin Ibeta and IIbeta subunits during vitellogenesis in the common Japanese conger *Conger myriaster*. *Fisheries Science* 67: 1053-1062.

Kawauchi, H., K. Suzuki, L. Hiromichi, P. Swanson, N. Naito, Y. Nagahama, M. Nozaki, Y. Nakai and S. Itoh. 1989. The duality of teleost gonadotropin. *Fish Physiology and Biochemistry* 7: 29-38.

Kazeto, Y., S. Ijiri, A.R. Place, Y. Zohar and J.M. Trant. 2001. The 5'-flanking regions of CYP19A1 and CYP19A2 in zebrafish. *Biochemical and Biophysical Research Communications* 288: 503-508.

Kestemont, P., J. Cooremans, A. Abi-Ayad and C. Mélard. 1999. Cathepsin L in eggs and larvae of perch *Perca fluviatilis*: Variations with developmental stage and spawning period. *Fish Physiology and Biochemistry* 21: 59-64.

King, H.R. and N.W. Pankhurst. 2004a. Effect of maintainance at elevated temperatures on ovulation and luteinizing hormone releasing hormone analogue responsiveness of female Atlantic salmon (*Salmo salar*) in Tasmania. *Aquaculture* 233: 583-597.

King, H.R. and N.W. Pankhurst. 2004b. Effect of short-term temperature reduction on ovulation and LHRHa responsiveness in female Atlantic salmon (*Salmo salar*) maintained at elevated water temperatures. *Aquaculture* 238: 421-436.

Kishida, M. and G.V. Callard. 2001. Distinct cytochrome P450 aromatase isoforms in zebrafish (*Danio rerio*) brain and ovary are differentially programmed and estrogen regulated during early development. *Endocrinology* 142: 740-750.

Kitazume, S., K. Kitajima, S. Inoue, Y. Inoue and F.A. Troy. 1994. Developmental expression of trout egg polysialoglycoproteins and the prerequisite alpha 2,6-, and alpha 2,8-sialyl and alpha 2,8-polysialyltransferase activities required for their synthesis during oogenesis. *Journal of Biological Chemistry* 269: 10330-10340.

Klenke, U. and Y. Zohar. 2003. Gonadal regulation of gonadotropin subunit expression and pituitary LH protein content in female hybrid striped bass. *Fish Physiology and Biochemistry* 28: 25-27.

Knaut, H., F. Pelegri, K. Bohmann, H. Schwarz and C. Nüsslein-Volhard. 2000. Zebrafish vasa RNA but not its protein is a component of the germ plasm and segregates asymmetrically before germline specification. *Journal of Cell Biology* 149: 875-888.

Knoebl, I., M.J. Hemmer and N.D. Denslow. 2004. Induction of zona radiata and vitellogenin genes in estradiol and nonylphenol exposed male sheepshead minnows (*Cyprinodon variegatus*). *Marine Environmental Research* 58: 547-551.

Kobayashi, T., H. Kajiura-Kobayashi and Y. Nagahama. 2002. Two isoforms of vasa homologs in a teleost fish: their differential expression during germ cell differentiation. *Mechanisms of Development* 111: 167-171.

Korsgaard, B. and T.P. Mommsen. 1993. Gluconeogenesis in hepatocytes of immature rainbow trout (*Oncorhynchus mykiss*): Control by estradiol. *General and Comparative Endocrinology* 89: 17-27.

Korsgaard, B., T.P. Mommsen and R.L. Saunders. 1986. The effect of temperature on the vitellogenic response in Atlantic salmon post-smolts (*Salmo salar*). *General and Comparative Endocrinology* 62: 193-201.

Koya, Y. and E. Kamiya. 2000. Environmental regulation of annual reproductive cycle in the mosquitofish, *Gambusia affinis*. *Journal of Experimental Zoology* 286: 204-211.

Krebs, H.A. 1975. The August Krogh Principle: 'For many problems there is an animal on which it can be most conveniently studied'. *Journal of Experimental Zoology* 194: 221-226.

Kurokawa, T., S. Uji and T. Suzuki. 2005. Identification of pepsinogen gene in the genome of stomachless fish, *Takifugu rubripes. Comparative Biochemistry and Physiology B* 140: 133-140.

Kwon, J.Y., F. Prat, C. Randall and C.R. Tyler. 2001. Molecular characterization of putative yolk processing enzymes and their expression during oogenesis and embryogenesis in rainbow trout (*Oncorhynchus mykiss*). *Biology of Reproduction* 65: 1701-1709.

LaFleur, G.J., Jr., D. Raldua, M. Fabra, O. Carnevali, N. Denslow, R.A. Wallace and J. Cerda. 2005. Derivation of major yolk proteins from parental vitellogenins and alternative processing during oocyte maturation in *Fundulus heteroclitus. Biology of Reproduction* 73: 815-824.

Laidley, C.W. and P. Thomas. 1997. Changes in plasma sex steroid-binding protein levels associated with ovarian recrudescence in the spotted seatrout (*Cynoscion nebulosus*). *Biology of Reproduction* 56: 931-937.

Lancaster, P.M. and C.R. Tyler. 1994. Developmental expression and modulation of the vitellogenin receptor in ovarian follicles of the rainbow trout, *Oncorhynchus mykiss. Journal of Experimental Zoology* 269: 458-466.

Lee, C., J. Na, K. Lee and K. Park. 2002. Choriogenin mRNA induction in male medaka, *Oryzias latipes* as a biomarker of endocrine disruption. *Aquatic Toxicology* 61: 233.

Lehane, D.B., N. McKie, R.G. Russell and I.W. Henderson. 1999. Cloning of a fragment of the osteonectin gene from goldfish, *Carassius auratus*: its expression and potential regulation by estrogen. *General and Comparative Endocrinology* 114: 80-87.

Lephart, E.D. 1996. A review of brain aromatase cytochrome P450. *Brain Research Reviews* 22: 1-26.

Li, A., M. Sadasivam and J.L. Ding. 2003. Receptor-ligand interaction between vitellogenin receptor (VtgR) and vitellogenin (Vtg), implications on low density lipoprotein receptor and apolipoprotein B/E. The first three ligand-binding repeats of VtgR interact with the amino-terminal region of Vtg. *Journal of Biological Chemistry* 278: 2799-2806.

Liu, R.Z., E.M. Denovan-Wright and J.M. Wright. 2003. Structure, linkage mapping and expression of the heart-type fatty acid-binding protein gene (fabp3) from zebrafish (*Danio rerio*). *European Journal of Biochemistry* 270: 3223-3234.

Lu, X., J. Gruia-Gray, N.G. Copeland, D.J. Gilbert, N.A. Jenkins, C. Londos and A.R. Kimmel. 2001. The murine perilipin gene: The lipid droplet-associated perilipins derive from tissue-specific, mRNA splice variants and define a gene family of ancient origin. *Mammalian Genome* 12: 741-749.

Matsubara, T. and Y. Koya. 1997. Course of proteolytic cleavage in three classes of yolk proteins during oocyte maturation in barfin flounder *Verasper moseri*, a marine teleost spawning pelagic eggs. *Journal of Experimental Zoology* 278: 189-200.

Matsubara, T., N. Ohkubo, T. Andoh, C.V. Sullivan and A. Hara. 1999. Two forms of vitellogenin, yielding two distinct lipovitellins, play different roles during oocyte maturation and early development of barfin flounder, *Verasper moseri*, a marine teleost that spawns pelagic eggs. *Developmental Biology* 213: 18-32.

Menuet, A., E. Pellegrini, I. Anglade, O. Blaise, V. Laudet, O. Kah and F. Pakdel. 2002. Molecular characterization of three estrogen receptor forms in zebrafish: binding characteristics, transactivation properties, and tissue distributions. *Biology of Reproduction* 66: 1881-1892.

Miguel-Queralt, S., G.V. Avvakumov, M. Blazquez, F. Piferrer and G.L. Hammond. 2005. Sea bass (*Dicentrarchus labrax*) sex hormone binding globulin: molecular and biochemical properties and phylogenetic comparison of its orthologues in multiple fish species. *Molecular and Cellular Endocrinology* 229: 21-29.

Mikawa, N., I. Hirono and T. Aoki. 1996. Structure of medaka transferrin gene and its 5'-flanking region. *Molecular Marine Biology and Biotechnology* 5: 225-229.

Mommsen, T.P. 2004. Salmon spawning migration and muscle protein metabolism: the August Krogh principle at work. *Comparative Biochemistry and Physiology* B 139: 383-400.

Mommsen, T.P. and P.J. Walsh. 1988. Vitellogenesis and oocyte assembly. In: *Fish Physiology*, W.S. Hoar, and D.J. Randall (eds.), Academic Press, New York, Vol. 11A, pp. 347-406.

Mouchel, N., V. Trichet, A. Betz, J.P. Le Pennec and J. Wolff. 1996. Characterization of vitellogenin from rainbow trout (*Oncorhynchus mykiss*). *Gene* 174: 59-64.

Nagakawa, H. 1970. Studies on rainbow trout egg (*Salmo gairdneri irideus*). II. Carbohydrate in the egg protein. *Journal of the Faculty of Fish and Animal Husbandry of Hiroshima University* 8: 77-84.

Nawaz, Z., D.M. Lonard, A.P. Dennis, C.L. Smith and B.W. O'Malley. 1999. Proteasome-dependent degradation of the human estrogen receptor. *Proceedings of the National Academy of Sciences of the United States of America* 96: 1858-1862.

Nielsen, L.B. and H.H. Nielsen. 2001. Purification and characterization of cathepsin D from herring muscle (*Clupea harengus*). *Comparative Biochemistry and Physiology* B 128: 351-363.

Ohkubo, M., K. Miyagawa, K. Osatomi, K. Hara, Y. Nozaki and T. Ishihara. 2004a. Purification and characterization of myofibril-bound serine protease from lizard fish (*Saurida undosquamis*) muscle. *Comparative Biochemistry and Physiology* B 137: 139-150.

Ohkubo, N., T. Andoh, K. Mochida, S. Adachi, A. Hara and T. Matsubara. 2004b. Deduced primary structure of two forms of vitellogenin in Japanese common goby (*Acanthogobius flavimanus*). *General and Comparative Endocrinology* 137: 19-28.

Pastori, R.L., J.E. Moskaitis, S.W. Buzek and D.R. Schoenberg. 1991. Coordinate estrogen-regulated instability of serum protein-coding messenger RNAs in *Xenopus laevis*. *Molecular Endocrinology* 5: 461-468.

Patiño, R. and C.V. Sullivan. 2002. Ovarian follicular growth, maturation, and ovulation in teleost fish. *Fish Physiology and Biochemistry* 26: 57-70.

Planas, J.V., J. Athos, F.W. Goetz and P. Swanson. 2000. Regulation of ovarian steroidogenesis in vitro by follicle-stimulating hormone and luteinizing hormone during sexual maturation in salmonid fish. *Biology of Reproduction* 62: 1262-1269.

Polzonetti-Magni, A.M., G. Mosconi, L. Soverchia, S. Kikuyama and O. Carnevali. 2004. Multihormonal control of vitellogenesis in lower vertebrates. *International Review of Cytology* 239: 1-46.

Prat, F., K. Coward, J.P. Sumpter and C.R. Tyler. 1998. Molecular characterization and expression of two ovarian lipoprotein receptors in the rainbow trout, *Oncorhynchus mykiss*. *Biology of Reproduction* 58: 1146-1153.

Reidy, S.P. and J. Weber. 2000. Leptin: An essential regulator of lipid metabolism. *Comparative Biochemistry and Physiology* A 125: 285-298.

Reith, M., J. Munholland, J. Kelly, R.N. Finn and H.J. Fyhn. 2001. Lipovitellins derived from two forms of vitellogenin are differentially processed during oocyte maturation in haddock (*Melanogrammus aeglefinus*). *Journal of Experimental Zoology* 291: 58-67.

Riggio, M., R. Scudiero, S. Filosa and E. Parisi. 2000. Sex- and tissue-specific expression of aspartic proteinases in *Danio rerio* (zebrafish). *Gene* 260: 67-75.

Sabo-Attwood, T., K.J. Kroll and N.D. Denslow. 2004. Differential expression of largemouth bass (*Micropterus salmoides*) estrogen receptor isotypes alpha, beta, and gamma by estradiol. *Molecular and Cellular Endocrinology* 218: 107-118.

Santos, E.M., M. Rand-Weaver and C.R. Tyler. 2001. Follicle-stimulating hormone and its alpha and beta subunits in rainbow trout (*Oncorhynchus mykiss*): Purification, characterization, development of specific radioimmunoassays, and their seasonal plasma and pituitary concentrations in females. *Biology of Reproduction* 65: 288-294.

Sawaguchi, S., Y. Koya, N. Yoshizaki, N. Ohkubo, T. Andoh, N. Hiramatsu, C.V. Sullivan, A. Hara and T. Matsubara. 2005. Multiple vitellogenins (Vgs) in mosquitofish (*Gambusia affinis*): Identification and characterization of three functional Vg genes and their circulating and yolk protein products. *Biology of Reproduction* 72: 1045-1060.

Shimizu, A. 2003. Effect of photoperiod and temperature on gonadal activity and plasma steroid levels in a reared strain of the mummichog (*Fundulus heteroclitus*) during different phases of its annual reproductive cycle. *General and Comparative Endocrinology* 131: 310-324.

Shimizu, A., H. Tanaka and H. Kagawa. 2003. Immunocytochemical applications of specific antisera raised against synthetic fragment peptides of mummichog GtH subunits: Examining seasonal variations of gonadotrophs (FSH cells and LH cells) in the mummichog and applications to other acanthopterygian fishes. *General and Comparative Endocrinology* 132: 35-45.

Shoham, Z. and M. Schachter. 1996. Estrogen biosynthesis—regulation, action, remote effects, and value of monitoring in ovarian stimulation cycles. *Fertility and Sterility* 65: 687-701.

Sire, M.-F., P.J. Babin and J.-M. Vernier. 1994. Involvement of the lysosomal system in yolk protein deposit and degradation during vitellogenesis and embryonic development in trout. *Journal of Experimental Zoology* 269: 69-83.

Smith, A.J., M.A. Sanders, B.R. Thompson, C. Londos, F.B. Kraemer and D.A. Bernlohr. 2004. Physical association between the adipocyte fatty acid-binding protein and hormone-sensitive lipase: A fluorescence resonance energy transfer analysis. *Journal of Biological Chemistry* 279: 52399-52405.

Specker, J.L. and C.V. Sullivan. 1995. Vitellogenesis in fishes: status and perspectives. In: *Perspectives in Comparative Endocrinology*, K.G. Davey, R.E. Peter and S.S. Tobe (eds.). National Research Council of Canada. Ottawa, pp. 304-315.

Suetake, H., K. Okubo, N. Sato, Y. Yoshiura, Y. Suzuki and K. Aida. 2002. Differential expression of two gonadotropin (GTH) beta subunit genes during ovarian maturation induced by repeated injection of salmon GTH in the Japanese eel *Anguilla japonica*. *Fisheries Science* 68: 290-298.

Taylor, J.S., I. Braasch, T. Frickey, A. Meyer and P.Y. Van de. 2003. Genome duplication, a trait shared by 22000 species of ray-finned fish. *Genome Res.* 13: 382-390.

Thomas, P. 2003. Rapid, nongenomic steroid actions initiated at the cell surface: lessons from studies with fish. *Fish Physiology and Biochemistry* 28: 3-12.

Thorsen, A. and H.J. Fyhn. 1996. Final oocyte maturation *in vivo* and *in vitro* in marine fishes with pelagic eggs: Yolk protein hydrolysis and free amino acid content. *Journal of Fish Biology* 48: 1195-1209.

Tollefsen, K.E., R. Mathisen and J. Stenersen. 2002a. Estrogen mimics bind with similar affinity and specificity to the hepatic estrogen receptor in Atlantic salmon (*Salmo salar*) and rainbow trout (*Oncorhynchus mykiss*). *General and Comparative Endocrinology* 126: 14-22.

Tollefsen, K.E., J.F.A. Meys, J. Frydenlund and J. Stenersen. 2002b. Environmental estrogens interact with and modulate the properties of plasma sex steroid-binding proteins in juvenile Atlantic salmon (*Salmo salar*). *Marine Environmental Research* 54: 697-701.

Trichet, V., N. Buisine, N. Mouchel, P. Moran, A.M. Pendas, J.P. Le Pennec and J. Wolff. 2000. Genomic analysis of the vitellogenin locus in rainbow trout (*Oncorhynchus mykiss*) reveals a complex history of gene amplification and retroposon activity. *Molecular and General Genetics* 263: 828-837.

Tsao, C.C., F.L. Huang and Y.S. Chang. 1999. Purification, characterization, and molecular cloning of carp hyosophorin. *Molecular Reproduction and Development* 52: 288-296.

Tyler, C.R. and K. Lubberink. 1996. Identification of four ovarian receptor proteins that bind vitellogenin but not other homologous plasma lipoproteins in the rainbow trout, *Oncorhynchus mykiss*. *Journal of Comparative Physiology* 166: 11-20.

Tyler, C.R., J.P. Sumpter, H. Kawauchi and P. Swanson. 1991. Involvement of gonadotropin in the uptake of vitellogenin into vitellogenic oocytes of the rainbow trout, *Oncorhynchus mykiss*. *General and Comparative Endocrinology* 84: 291-299.

Vacher, C., F. Ferriere, M.H. Marmignon, E. Pellegrini and C. Saligaut. 2002. Dopamine D2 receptors and secretion of FSH and LH: role of sexual steroids on the pituitary of the female rainbow trout. *General and Comparative Endocrinology* 127: 198-206.

van der Kraak, G.J. and S. Biddiscombe. 1999. Polyunsaturated fatty acids modulate the properties of the sex steroid binding protein in the goldfish. *Fish Physiology and Biochemistry* 20: 115-123.

Vayda, M.E., R.L. Londraville, R.E. Cashon, L. Costello and B.D. Sidell. 1998. Two distinct types of fatty acid-binding protein are expressed in heart ventricle of Antarctic teleost fishes. *Biochemical Journal* 330: 375-382.

Vidal, B., C. Pasqualini, N. Le Belle, M.C. Holland, M. Sbaihi, P. Vernier, Y. Zohar and S. Dufour. 2004. Dopamine inhibits luteinizing hormone synthesis and release in the juvenile European eel: A neuroendocrine lock for the onset of puberty. *Biology of Reproduction* 71: 1491-1500.

Wallace, R.A. 1985. Vitellogenesis and oocyte growth in non-mammalian vertebrates. In: *Developmental Biology*, L. Browder (ed.). Pergamon Press, New York, Vol. 1. pp. 127-177.

Wang, H. and Z. Gong. 1999. Characterization of two zebrafish cDNA clones encoding egg envelope proteins ZP2 and ZP3. *Biochimica et Biophysica Acta* 1446: 156-160.

Wang, H., T. Yan, J.T. Tan and Z. Gong. 2000. A zebrafish vitellogenin gene (vg3) encodes a novel vitellogenin without a phosvitin domain and may represent a primitive vertebrate vitellogenin gene. *Gene* 256: 303-310.

Wang, X.L., M. Sun, J. Mei and J.F. Gui. 2005. Identification of a spindlin homolog in gibel carp (*Carassius auratus gibelio*). *Comparative Biochemistry and Physiology* B 141: 159-167.

Washburn, B.S., M.L. Bruss, E.H. Avery and R.A. Freedland. 1992. Effects of estrogen on whole animal and tissue glucose use in female and male rainbow trout. *American Journal of Physiology* 263: R1241-R1247.

Watson, C.S., N.N. Bulayeva, A.L. Wozniak and C.C. Finnerty. 2005. Signaling from the membrane via membrane estrogen receptor-alpha: estrogens, xenoestrogens, and phytoestrogens. *Steroids* 70: 364-371.

Watts, M., N.W. Pankhurst and H.R. King. 2004. Maintenance of Atlantic salmon (*Salmo salar*) at elevated temperature inhibits cytochrome P450 aromatase activity in isolated ovarian follicles. *General and Comparative Endocrinology* 135: 381-390.

Weil, C., P.Y. Le Bail, N. Sabin and F. Le Gac. 2003. In vitro action of leptin on FSH and LH production in rainbow trout (*Onchorynchus mykiss*) at different stages of the sexual cycle. *General and Comparative Endocrinology* 130: 2-12.

Westerlund, L., S.J. Hyllner, A. Schopen and P.E. Olsson. 2001. Expression of three vitelline envelope protein genes in arctic char. *General and Comparative Endocrinology* 122: 78-87.

Wood, A.W., J. Matsumoto and G. Van Der Kraak. 2004. Thyroglobulin type-1 domain protease inhibitors exhibit specific expression in the cortical ooplasm of vitellogenic rainbow trout oocytes. *Molecular Reproduction and Development* 69: 205-214.

Xiong, F., D. Liu, Y. Le Drean, H.P. Elsholtz and C.L. Hew. 1994. Differential recruitment of steroid hormone response elements may dictate the expression of the pituitary gonadotropin II beta subunit gene during salmon maturation. *Molecular Endocrinology* 8: 782-793.

Yamaguchi, A., H. Ishibashi, S. Kohra, K. Arizono and N. Tominaga. 2005. Short-term effects of endocrine-disrupting chemicals on the expression of estrogen-responsive genes in male medaka (*Oryzias latipes*). *Aquatic Toxicology* 72: 239-249.

Yoshida, D., M. Nagae, F. Ito and K. Soyano. 2001. Molecular cloning of cDNAs encoding pituitary glycoprotein hormone alpha, FSH beta and LH beta subunits in ayu, *Pecoglossus altivelis*. *Zoological Science* 18: 929-939.

Young, G., M. Kusakabe, I. Nakamura, P.M. Lokman and F.W. Goetz. 2005. Gonadal steroidogenesis in teleost fish. In: *Molecular Aspects of Fish and Marine Biology*, P. Melamed and N.M. Sherwood (eds.), World Scientific, London, Vol. 4: *Hormones and Their Receptors in Fish Reproduction*, pp. 155-223.

Follicular Steroidogenesis in Fish

Michiya Matsuyama

INTRODUCTION

In teleosts as in other vertebrates, pituitary gonadotropins (GtHs) are the major hormones that stimulate various ovarian activities. In most cases, however, GtH action on ovarian development is not direct, but through the biosynthesis of steroid hormones by the ovarian follicles that, in turn, mediates various stages of oogenesis including oocyte growth (vitellogenesis) and oocyte maturation (Nagahama, 1993).

During vitellogenesis, oocytes are arrested in the first meiotic prophase, estradiol-17β (E2) synthesized in the follicular cells is introduced into the vascular system and stimulates the hepatic synthesis and secretion of vitellogenin (reviewed by Specker and Sullivan, 1994; Tyler and Sumpter, 1996). E2 was also shown to induce the synthesis of egg membrane (vitellin envelope) proteins, choriogenins, in the liver (LaFleur *et al.*, 1995; Tyler *et al.*, 2000). Over the last two decades, a number of studies on the steroidogenesis during vitellogenesis in various teleosts have incubated intact ovarian follicles or fragments or measured blood steroid levels. *In vitro* incubation of ovarian follicles (amago salmon *Oncorhynchus*

Author's address: Laboratory of Marine Biology, Faculty of Agriculture, Kyushu University, Hakozaki, Fukuoka 812-8581, Japan. E-mail: rinya_m@agr.kyushu-u.ac.jp

rhodurus, Kagawa *et al.*, 1982; goldfish *Carassius auratus*, Kagawa *et al.*, 1984; guppy *Poecilia reticulata*, Venkatesh *et al.*, 1992; medaka *Oryzias latipes*, Kobayashi *et al.*, 1996) or ovarian fragment (e.g., Atlantic croaker *Micropogonias undulatus*, Trant and Thomas, 1989) demonstrated that E2 was synthesized via testosterone (T). In numerous species, it has been shown that blood E2 levels increase during vitellogenesis, along with T levels, e.g., plaice *Pleuronectes platessa* (Wingfield and Grimm, 1977), white sucker *Catostomus commersoni* (Scott *et al.*, 1984), spotted sea-trout *Cynoscion nebulosus* (Thomas *et al.*, 1987), red seabream *Pagrus major* (Mastuyama *et al.*, 1988a), gulf killifish *Fundulus grandis* (Greeley *et al.*, 1988), sea bass *Dicentrarchus labrax* (Prat *et al.*, 1990), demoiselles *Chromis dispilus* (Barnett and Pankhurst, 1994), striped bass *Morone saxatilis* (King *et al.*, 1994), and spiny damselfish *Acanthochromis polyacanthus* (Pankhurst *et al.*, 1999). These studies have led us to the accepted view that T is a precursor for E2 production in the teleost ovary. In addition, among teleost females, serum E2 and T exert negative or positive feedback effects on GtH synthesis and secretion (reviewed in Devlin and Nagahama, 2002; Yaron *et al.*, 2003). Thus, T seems to play important roles in endocrine control of oogenesis in many teleost. However, complete synthetic pathway of E2 within ovarian follicles has been demonstrated in only a few species.

After the oocyte completes vitellogenesis, it becomes ready for the next process, i.e., the resumption of meiosis, called oocyte maturation. It consists of the germinal vesicle breakdown (GVBD), chromosome condensation, and formation of the first polar body. It has now become accepted that oocyte maturation in teleosts is induced by maturation-inducing hormone (MIH) produced in the follicle cells by the stimulation of maturational GtH (now referred to as LH, luteinizing hormone). In a number of teleost species, C21-steroids, such as progesterone and its derivatives, have been shown to be potent initiators of GVBD *in vitro* and to be present at high levels in plasma of fish undergoing oocyte maturation (Scott and Canario, 1987). Among the C21-steroids, however, only two progestins have been identified as the naturally occurring MIH in fish: 17,20β-dihydroxy-4-pregnen-3-one (17,20β-P) and 17,20β,21-trihydroxy-4-pregnen-3-one (20β-S).

Thus, under the regulation by GtH, a distinct shift in steroidogenesis from E2 to MIH production occurs in the follicular cells immediately prior to oocyte maturation. The mechanism of synthesis and action of fish steroid hormones, particularly in MIH, has been studied and reviewed in mainly salmonids (Nagahama, 1993, 1997) and sciaenids (Thomas and Trant, 1989; Trant and Thomas, 1989). However, detailed synthetic pathways of steroid hormones produced in the fish ovarian follicle cells and their functions throughout vitellogenesis and oocyte maturation are largely unknown. This chapter attempts to summarize research into follicular steroidogenesis in fish, with a focus on our recent studies of five marine teleosts including gonochoristic and hermaphroditic species.

REPRODUCTIVE CHARACTERISTICS OF FISH

The reproductive characteristics of five marine species referred in this chapter are described briefly as follows.

1. Chub mackerel, *Scomber japonicus*: The chub mackerel is a pelagic fish that belongs to Scombridae. Chub mackerel has asynchronous ovaries, i.e., oocytes at different stages of development are present simultaneously, and field data suggest that single female can spawn every two days in the prime of spawning season (Yamada *et al.*, 1998). Like hatchery-reared broodstock of many commercial fish species (Zohar, 1989), female chub mackerel are unable to complete their reproductive cycle in captivity. In sea cages or indoor tanks under a natural photoperiod and water temperature, female fish retain yolked oocytes in their ovaries throughout the spawning season from April to June (Shiraishi *et al.*, 2005). These oocytes never undergo oocyte maturation and ovulation. The major endocrinological dysfunction in captive chub mackerel seems therefore to involve a lack of LH secretion from the pituitary, which is responsible for inducing oocyte maturation after the completion of vitellogenesis. We can obtain maturaing ovarian follicles of chub mackerel by using of human chorionic gonadotropin (hCG) which has LH-like activity (Shiraishi *et al.*, 2005).

2. Yellowtail, *Seriola quinqueradiata*: Japanese yellowtail is a carangid fish and has been regarded as one of the most important consumable marine fish in Japan. Yellowtail is a serial spawner, but oocyte maturation does not occur under captivity like chub mackerel. Maturing follicles of yellowtail can be also obtained by hCG injection (Matsuyama *et al.*, 1996).

3. Pufferfish, *Takifugu rubripes*: Japanese pufferfish, belongs to a Family Tetraodontidae, has been used as a new model for genome studies because of their compact genome (Brenner *et al.*, 1993; Venkatesh *et al.*, 2000). Ovary of the pufferfish during spawning season contains two clutches of oocyte; the majority of oocytes are vitellogenic and

there are a few residual pre-vitellogenic oocytes. Wild female spawns adhesive eggs one time per spawning season. Oocyte maturation in captive pufferfish is usually induced by using continuously releasing of gonadotropin-releasing hormone analogue (GnRHa) (Matsuyama *et al.*, 2001), which induces the discharge of innate LH from pituitary gonadotrophs.

4. Bambooleaf wrasse, *Pseudolabrus sieboldi*: The bambooleaf wrasse is a diandric protogynous labrid fish. Functional females change their sex to secondary males with terminal-phase appearance (Nakazono, 1979). Each female spawns almost everyday during the two-month spawning season, and has a diurnal rhythm in oocyte growth, maturation, ovulation, and spawning (Matsuyama *et al.*, 1998a). Recently, *Pseudolabrus japonicus* was divided into two species, *P. sieboldi* and *P. eoethinus*, based on the morphological characteristics and mitochondrial DNA sequences (Mabuchi *et al.*, 2000). The bambooleaf wrasse population that we studied previously clearly exhibits the morphological characteristics of *P. sieboldi*, as do fish examined in the current study. Therefore, we use the term bambooleaf wrasse to indicate *P. sieboldi* in this chapter, consistent with our previous studies (Matsuyama *et al.*, 1997, 1998a, b)

5. Red seabream, *Pagrus major*: The red seabream, a sparid fish, is one of the most important object species as well as yellowtail in the aquaculture industry of Japan. The sexual pattern of red seabream is gonochorism with a bisexual juvenile stage; testes originate from the undeveloped ovary via a bisexual gonad in the juvenile stage (Matsuyama *et al.*, 1988b). Adult female spawns daily over two months in captivity during the spawning season (Matsuyama *et al.*, 1988a).

STEROIDOGENESIS IN THE VITELLOGENIC FOLLICLES

The nomenclature of the steroids used in the present study is shown in Table 5.1. Removed ovaries were placed in ice-cold Leibovitz's L-15 culture medium, and then cut into small fragments and pipetted to disperse follicles-enclosed oocytes. The follicles-enclosed oocytes at different developmental stages were gathered and stored at –80°C until incubation. There was little difference in the steroid metabolic patterns during the incubation with frozen and intact follicles. Two hundred follicles at the same stage were incubated with different radiolabeled steroid precursors

($[^3H]$ or $[^{14}C]$) and the metabolites were separated by the thin layer chromatography (TLC). Radiolabeled steroid metabolites were identified by their chromatographic mobility in TLC and by recrystallization as described by Axelrod *et al.* (1965).

Chub mackerel

When vitellogenic follicles were incubated with $[^3H]$P5 and $[^3H]$17-P, a total of six radioactive metabolites were identified as P5, 17-P5, 17-P, DHEA, AD, and T (see Table 5.1 for steroid abbreviations). $[^{14}C]$AD was metabolized into T and E2 was produced after incubation with $[^3H]$T. The major metabolites produced from $[^3H]$P5 were 17-P5, 17-P, and DHEA. However, the radioactivity of 17-P from $[^3H]$P5 was 8.6 times that of DHEA from $[^3H]$P5, indicating that P5 is largely metabolized via 17-P5 to 17-P but not to DHEA. Along with the results for another three different radioactive precursors, this led to the conclusion that E2 was synthesized through a major pathway from P5 via 17-P5, 17-P, AD, and T (Matsuyama *et al.*, 2005) (Fig. 5.1).

Table 5.1   Nomenclature of steroids.

Systematic name	Trivial name	Abbreviation
3-hydroxy-1,3,5(10)-estratrien-17-one	estrone	E1
1,3,5(10)-estratriene-3,17β-diol	estradiol-17β	E2
4-androstene-3,17-dione	androstenedione	AD
5-androstene-3β, 17β-diol	androstenediol	A5
3β-hydroxy-5-androsten-17-one	dehydroepiandrosterone	DHEA
17β-hydroxy-4-androsten-3-one	testosterone	T
11β,17β-dihydroxy-4-androsten-3-one	11β-hydroxytestosterone	11β-T
17β-hydroxy-4-androstene-3,17-dione	11-ketotestosterne	11-KT
3β-hydroxy-5-pregnen-20-one	pregnenolone	P5
3β,17-dihydroxy-5-pregnen-20-one	17-hydroxypregnenolone	17-P5
5-pregnene-dihydroxy-3β,17,20β-triol	-	17,20β-P5
4-pregnene-3,20-dione	progesterone	P
17-hydroxy-4-pregnene-3,20-dione	17-hydroxyprogesterone	17-P
17,20β-dihydroxy-4-pregnen-3-one	-	17,20β-P
17,20α-dihydroxy-4-pregnen-3-one	-	17,20α-P
17,21-dihydroxy-4-pregnene-3,20-dione	11-deoxycortisol	17,21-P
17,20β,21-trihydroxy-4-pregnen-3-one	-	20β-S
17-hydroxy-5β-pregnane-3,20-dione	-	17-P-5β
17,20β-dihydroxy-5β-pregnan-3-one	-	17,20β-P-5β
17,21-dihydroxy-5β-pregnane-3,20-dione	-	17,21-P-5β

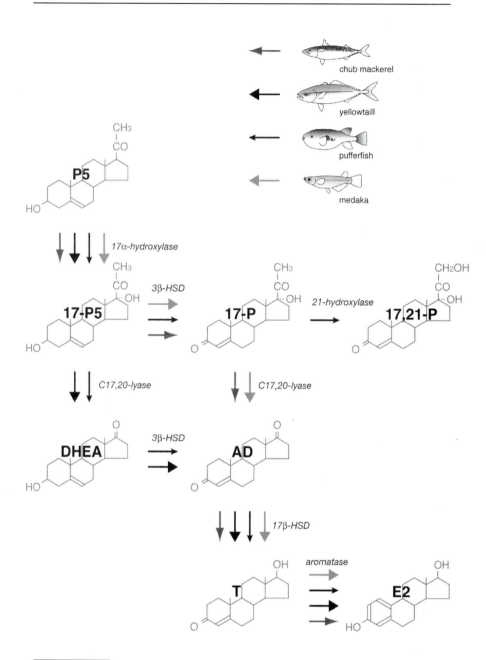

Fig. 5.1 Main steroidogenic pathways in the vitellogenic ovarian follicles of four gonochoristic teleosts, chub mackerel, yellowtail, pufferfish and medaka.

Yellowtail

When vitellogenic follicles were incubated with [^{3}H]P5, six metabolites could be detected. The steroid metabolites were P5 (precursor), 17-P5, 17-P, DHEA, AD, and 17,20β-P5 (see Table 5.1 for steroid abbreviations). The biosynthetic intensity of DHEA, however, was four times greater than that of 17-P, it is likely that pathway mainly followed through DHEA to AD, that is different from the case of chub mackerel. When [^{14}C]AD was used as precursor, E2, T, A5, two 5-reduced (5α- and 5β-) forms of androstane and two unidentified highly polar bands were produced. This result clearly indicates that AD was converted into E2 via T. Small production of A5 may be due to the reversible reaction by 3β-hydroxysteroid dehydrogenase (3β-HSD) from T to A5. The E2 synthetic pathway in the vitellogenic follicles of yellowtail is also shown in Fig. 5.1 in which E2 was produced through a major pathway from P5 via 17-P5, DHEA, AD and T (Rahman *et al.*, 2002a).

Pufferfish

The major pathway from P5 to E2 in the vitellogenic follicles of pufferfish is same as in yellowtail, i.e., E2 was synthesized from P5 via 17-P5, DHEA, AD and T (Matsuyama *et al.*, 2001, along with unpublished data). In the vitellogenic follicles of pufferfish, coincident with the E2 synthesis, 17,21-P was actively converted from 17-P by 21-hydroxylase. As a result, two final products, E2 and 17,21-P, were synthesized in the vitellogenic follicles of pufferfish (Fig. 5.1).

Bambooleaf wrasse

When vitellogenic follicles were incubated with [^{3}H]P5, eight major fractions appeared on TLC after development in benzen:acetone (4:1) (Fig. 5.2). Fraction 4 was divided into DHEA and 17-P in chloroform:ethyl acetate (2:1), and the biosynthetic intensity of DHEA was 7.5-times greater than that of 17-P. Fraction 7 was also divided into 17,20β-P and 17,21-P in benzene:chloroform:diethyl ether:methanol (2:2:2:1). Fractions 1, 2, 3, 5, 6, and 8 were identified as E1, AD, P5, E2, 17-P5, and 17,20β-P5, respectively (see Table 5.1 for steroid abbreviations). To clarify the synthetic pathway to E2, additional five radiolabeled steroids, 17-P, DHEA, AD, E1 and T, were used as precursors. Of these five precursors, E2 was synthesized from DHEA, AD and E1. In all incubations, T production could not be detected, and T was not entirely converted to

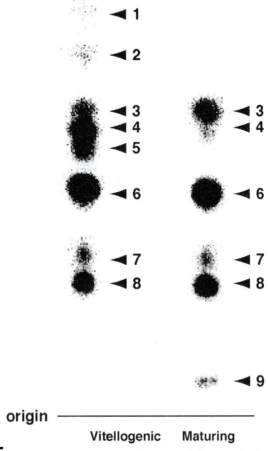

Fig. 5.2 Autoradiograms of steroid metabolites from vitellogenic and maturing ovarian follicles of the bambooleaf wrasse incubated with [7-^{3}H]pregnenolone. Nine metabolites were separated by thin layer chromatography and developed with benzene:acetone (B:A=4:1) mixture. 1, E1; 2, AD; 3, P5; 4, DHEA and 17-P; 5, E2; 6, 17-P5; 7, 17,20β-P and 17,21-P; 8, 17,20β-P5; and 9, 20β-S. See Table 5.1 for steroid abbreviations.

E2 (Ohta *et al.*, 2001), clearly indicating that in the vitellogenic follicles of bambooleaf wrasse T is not the precursor for E2 production. In contrast, E1 was produced instead of T from P5, DHEA, and AD, and exogenous E1 was converted directly into E2. These results demonstrate that E2 is synthesized from E1 in the vitellogenic follicles of bambooleaf wrasse. Like in the case of pufferfish, 17,21-P was also produced as a final product as well as E2 in the vitellogenic follicles of the bamboo- leaf wrasse owing to continuously high 21-hydroxylase activity which converts 17-P to 17,21-P.

The steroidogenic pathways in the vitellogenic follicles of bambooleaf wrasse are shown in Fig. 5.3 (Ohta *et al.*, 2001).

Daily changes in two clutches of oocytes, the largest and the second-largest oocytes, and serum levels of three steroids are summarized in Fig. 5.4. Serum E2 levels exhibited a diurnal rhythm peaked at 03:00 hr, when the developmental stages of largest and the second-largest follicles were late germinal vesicle migration (GVM) and mid yolk (MY) stages, respectively. Bambooleaf wrasse MY follicles have active vitellogenic capacity but late GVM cannot produce E2, as described later. The high

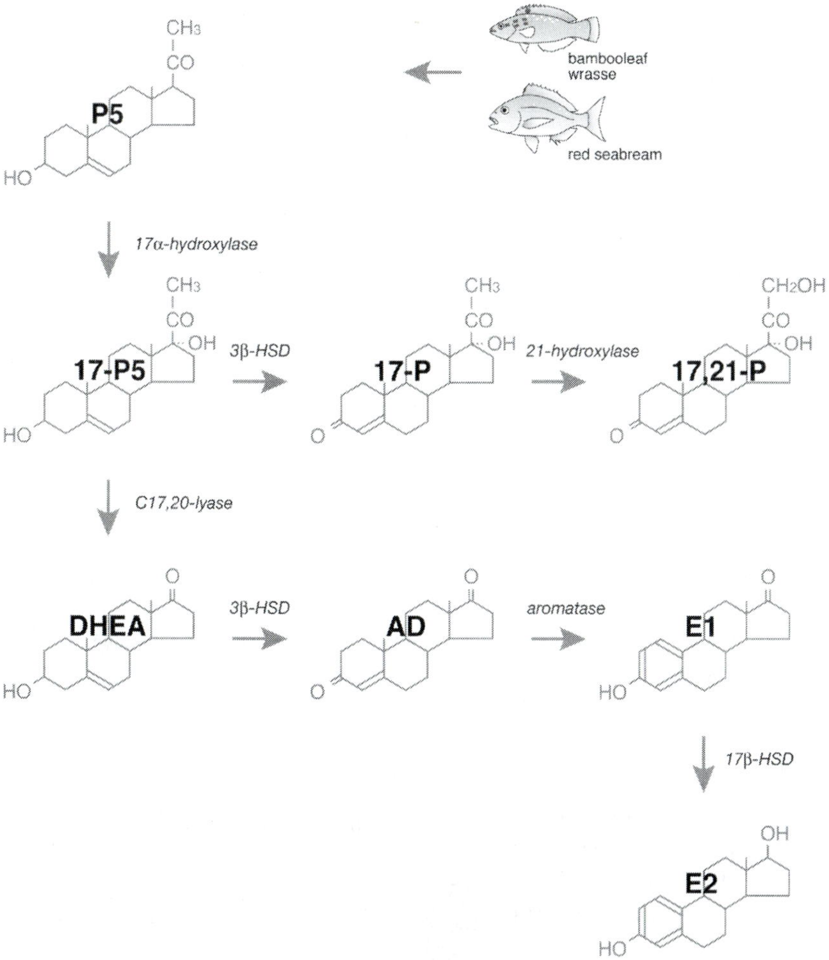

Fig. 5.3 Main steroidogenic pathways in the vitellogenic ovarian follicles of two hermaphroditic teleosts, bambooleaf wrasse and red seabream.

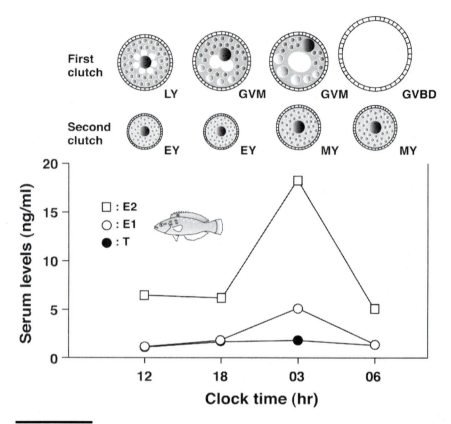

Fig. 5.4 Diurnal changes of serum levels of estrone (E1), estradiol-17β (E2), and testosterone (T) in female bambooleaf wrasse. EY, early yolk; MY, mid yolk; LY, late yolk; GVM, germinal vesicle migration; and GVBD, germinal vesicle breakdown.

serum E2 levels are likely originated from the second-largest follicles. The serum levels of E1 showed a diurnal rhythm that paralleled E2 levels but not in T levels, supporting the *in vitro* data that E2 is converted from E1 but not T.

Recently, the steroidogenic pathways in the testis of the sex-changed secondary male of bambooleaf wrasse has also been determined. In the secondary testis 11-KT, a major androgen in male teleosts (Borg, 1994), was produced as one of end products, in which AD is synthesized from P5 in a similar pathway to that in ovarian follicles, then AD is converted to 11-KT via T and 11β-T (Ohta *et al.*, 2003).

Red seabream

When radiolabeled 17-P, DHEA, or AD were used as precursors, both T and E1 were synthesized by follicles, leading to E2 production. *In vitro* conversion of E1 into E2, however, was 16-fold greater than T conversion into E2, suggesting that E2 is synthesized mainly via E1 rather than through T. The metabolites from different radiolabeled precursors and main steroidogenic pathways in the vitellogenic follicles were almost the same as those of bambooleaf wrasse. Namely, E2 was synthesized through a major pathway from P5 via 17-P5, DHEA, AD, and E1 (Ohta *et al.*, 2002a) (Fig. 5.3). In parallel with the E2 production, 17,21-P syntyhesis was also advanced through the pathway from P5 via 17-P5 and 17-P. Both T and E1 appeared with E2 in the circulation throughout the day, where serum T levels were similar to those of E1. The relatively high serum T levels in the red seabream may be a reflection of extragonadal production of T.

The follicular steroidogenesis of medaka during vitellogenesis (Kobayashi *et al.*, 1996), is also shown in Fig. 5.1, where E2 is synthesized in a same pathway of chub mackerel; from P5 via 17-P5, 17-P, AD and T. Chub mackerel, yellowtail, pufferfish, and medaka are gonochoric, while bambooleaf wrasse and red seabream have a bisexual phase in their life history. Interestingly, T is the substrate precursor of E2 in vitellogenic ovarian follicles of chub mackerel, yellowtail, pufferfish and medaka. In contrast, E2 is synthesized via E1 rather than T in bambooleaf wrasse and red seabream. Gonadal steroid hormones are likely to play an important role in sex changes of hermaphroditic fish (Baroiller *et al.*, 1999; Devlin and Nagahama, 2002). In protogynous species, androgen treatment has generally been shown to be effective in inducing female-to-male inversions (Reinboth, 1975; Kramer *et al.*, 1988; Grober *et al.*, 1991). We have also succeeded in inducing a gonadal sex change from female to male in bambooleaf wrasse by the implantation of T and 11-KT (Sakai *et al.*, 2004). Thus, despite limited information, we hypothesize that low levels or lack of production of T in ovarian follicles of red seabream and bambooleaf wrasse may be related to sex reversals in their life cycles. In other words, ovarian T production (over a certain threshold) may affect the maintenance of ovarian structure and function in the female phase of hermaphroditic species. In the ovary of gonochoristic species, E2 is converted from AD via T.

E2 is synthesized in both the ovary and sex-changed (secondary) testis throughout the life of a single bambooleaf wrasse (Ohta *et al.*, 2003). Serum E2 levels of the secondary male showed an extreme increase at around 15:00 h (Sundaray *et al.*, 2003). This time-specific production of E2 has allowed us to hypothesize that E2 may involved in spermatogonial proliferation since the number of B-type spermatogonia and spermatocytes was maximal (Matsuyama *et al.*, 1997). Interestingly, the synthetic pathway of E2 in the testis was different from that in the ovary, i.e., in the ovarian follicles E2 is synthesized from AD via E1, whereas in the secondary testis E2 is produced from AD via T. The shift in E2 production pathways between ovarian follicles and secondary testis may be closely related with the gonadal sex change of bambooleaf wrasse. To analyze the mechanism of shift in the steroidogenic pathways, cloning of cDNAs encoding some key-enzymes and their mRNA expressions were investigated by RT-PCR. Complementary DNAs encoding two P450 aromatase isoforms (CYP19a and CYP19b) and four types of 17β-hydroxysteroid dehydrogenase (17β-HSD-1, -5, -7, and -12) were isolated and cloned. RT-PCR showed that 17β-HSD-1 expression was limited to the ovary, while the other enzymes were detected in both ovary and testis. Moreover, 17β-HSD-1 expressed in human embryonic kidney (HEK) 293 cells selectively converted E1 to E2 (Ohta *et al.*, unpublished data). These results, combined with our previous studies, suggest that in the ovarian follicles of bambooleaf wrasse E2 is synthesized from AD via E1 by CYP19a and 17β-HSD-1, while in the secondary testis AD is converted to T by another isoform of 17β-HSD. Determination of CYP19 isoform that converts T to E2 in the secondary testis is underway. The steroidogenic enzymes regulating the steroidogenic shift that involves in the protogynous sex change in the bambooleaf wrasse will be clarified in the near future.

E2 production pathways differ between even gonochoristic species. Chub mackerel ovarian follicles synthesize E2 using the same pathways as medaka, namely P5, 17-P5, 17-P, AD, T, and E2. In yellowtail and pufferfish, AD is synthesized from 17-P5 via DHEA but not 17-P. The reasons for these species-specific differences in E2 synthetic pathways is presently unknown. Gonadal steroids have been shown to exert negative and/or positive effects on GtH secretion in a variety of teleost species (Devlin and Nagahama, 2002; Yaron *et al.*, 2003). Besides the numerous studies on regulation of GtH synthesis and secretion by E2 and T, few studies have investigated the effects of intermediate products in E2 synthesis to clarify their functions. Recently, it has been reported that AD produced in the vitellogenic follicles of the goldfish acts as a primary pheromone inducing agonistic behavior among males (Poling *et al.*, 2001). Future studies should look at potential physiological roles for intermediate products other than T in the E2 synthetic pathway, such as steroid feedback or pheromonal activity.

STEROIDOGENESIS IN THE MATURING FOLLICLES

Chub mackerel

The incubation of maturing follicles at the GVM stage with [^{3}H]P5 resulted in a total of ten synthesized steroids including precursor P5. They are AD, P5, 17-P, 17-P5, 17,20β-P-5β, 20β-S, 17,20β-P, 17,21-P, and two unknown polar metabolites (see Table 5.1 for steroid abbreviations). When [^{3}H]17-P was used as a precursor, 17-P, 17-P-5β, AD, 17,20β-P, 17,21-P, 20β-S were identified (see Table 5.1 for steroid abbreviations). When [^{3}H]17,20β-P was used as the precursor, 17,20β-P-5β was largely produced, but 20β-S was not synthesized. When [^{3}H]17,21-P was used, 20β-S was synthesized along with 17,21-P-5β. Thus, 20β-S is synthesized from 17-P via 17,21-P but not via 17,20β-P. The steroidogenic pathways in chub mackerel maturing follicles are shown in Fig. 5.5. Of the steroids produced in the follicles undergoing oocyte maturation, only 17,20β-P was highly effective for induction of GVBD *in vitro*. The ability of 20β-S to induce GVBD *in vitro* was much lower that that of 17,20β-P. These results strongly suggest that 17,20β-P acts as MIH in chub mackerel (Matsuyama *et al.*, 2005).

Yellowtail

Incubations at the GVM stage with [^{3}H]P5 yielded nine steroid metabolites, namely P5 (precursor), 17-P5, 17,20β-P-5, 17-P, 17-P-5β, AD, 17,21-P, 17,20β-P, and 17,20β-P-5β. During incubation with [^{3}H]17-P, only the six 4-pregnene steroid metabolites identified above were produced, but 5-pregnene steroids were not produced. From just prior to GVM to GVBD, there was a five- to seven-fold increase in the production of 17,20β-P. In addition to 17,20β-P production, its 5-reduced metabolite, 17,20β-P-5β was produced in significant amounts. In contrast, no AD was produced during GVBD. 20β-S and its metabolites were not detected. The steroidogenic pathways clarified in the maturing follicles of yellowtail are shown in Fig. 5.5. Among the metabolites, 17,20β-P was the most effective inducer of GVBD *in vitro*, and the secondmost effective inducer was 17,20-P-5β. 17-P, 17-P-5β, and 17,21-P were always less effective (Rahman *et al.*, 2001).

Pufferfish

At GVBD stage, there were increasing productions of 17,21-P, 20β-S, 17,20β-P, and 17,20β-P from 17-P, in particular, a massive scale of 20β-S production was observed (Matsuyama *et al.*, 2001). 20β-S was mainly

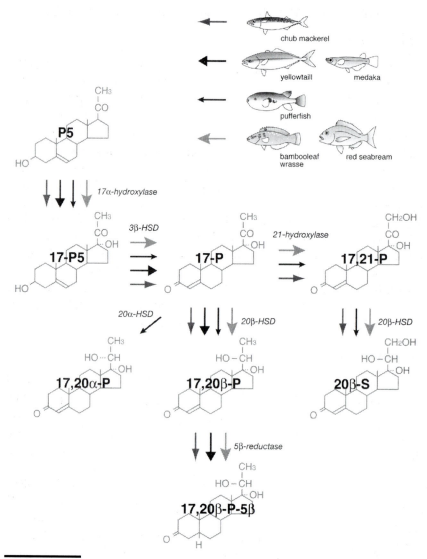

Fig. 5.5 Main steroidogenic pathways in the maturing ovarian follicles of six teleost species.

synthesized through the pathway from 17-P via 17,21-P. The steroidogenic pathways in the maturing follicles of pufferfish are shown in Fig. 5.5. Of the steroids produced during final oocyte maturation (FOM), 17,20β-P and 20β-S exhibited maturation-inducing activity *in vitro*, however, 20β-S was overwhelmingly more effective at inducing GVBD than 17,20β-P. These results provide strong evidence that 20β-S has a physiological role as an MIH in the pufferfish. Continuously increasing 21-hydroxylase activity throughout vitellogenic and maturing stages was distinct in pufferfish follicles. During vitellogenesis, active 21-hydroxylase likely enables ovarian follicles to accumulate enough 17,21-P, which leads to 20β-S production on a massive scale by 20β-hydroxysteroid dehydrogenase (20β-HSD) during GVBD.

The production of 17,20α-P, an isomer of 17,20β-P, (converted from 17-P by 20α-hydroxysteroid dehydrogenase, 20α-HSD) was dramatically increased in the GVBD follicles, and its peak was higher than that of 17,20β-P, but lower than that of 20β-S. A large amount of 17,20α-P has been found in the blood and ovarian incubates of mature flatfish species (Canário and Scott, 1990; Scott and Canário, 1987, 1990), cyprinids (Kime, 1992), and catfish (Zairin et al., 1992). The role of this steroid is still unknown. It is not a potent inducer of oocyte maturation, but it may be a behavioral steroid or pheromone (Scott and Canário, 1987).

Bambooleaf wrasse

The autoradiograms of steroid metabolites from maturing follicles incubated with [³H]P5 was shown in Figure 5.2. In the maturing follicles of wrasse, the production of E2, E1, AD, and DHEA decreased or disappeared. In contrast, 17,20β-P5, 17,20β-P, and 20β-S production increased. When follicles were incubated with [³H]17-P, three additional metabolites appeared, 17-P-5β, 17,21-P-5β and 17,20β-P-5β. Although both [³H]17,21-P and [³H]17,20β-P were converted to 20β-S, the production of 20β-S from 17,21-P was 28 times greater than that of 17,20β-P. On the other hand, [³H]17,20β-P5 was not converted into 17,20β-P. These results indicate that 20β-S was mainly synthesized from 17-P via 17,21-P, and 17,20β-P was directly converted from 17-P but not from 17,20β-P5. The steroidogenic pathways in the bambooleaf wrasse maturing follicles are shown in Figure 5.5 (Ohta and Matsuyama, 2002). Of the steroids produced in the wrasse maturing follicles, 17,20β-P and 20β-S showed an equally highest effectiveness on GVBD in vitro. Since 20β-S is not synthesized from 17,20β-P, high effectiveness of 17,20β-P at inducing GVBD in vitro is due to its own potency. Thus, in this species, two 20β-hydroxylated progestins, 17,20β-P and 20β-S, likely act as MIH. In succession from the vitellogenic stage, continuous high 21-hydroxylase activity throughout the FOM was observed, and in all likelihood, this 21-hydroxylase activity enables ovarian follicles to synthesize enough 17,21-P. Consequently, activated 20β-HSD converted 17-P and 17,21-P to 17,20β-P and 20β-S, respectively, during FOM.

Red seabream

The metabolites from different radiolabeled precursors and main steroidogenic pathways in the maturing follicles of red seabream were almost the same as those of bambooleaf wrasse (Ohta *et al.*, 2002b) as well as in vitellogenic stage (Fig. 5.5). *In vitro* GVBD assay showed that 17,20β-P and 20β-S exhibited the greatest effect for inducing GVBD, suggesting that both 17,20β-P and 20β-S act as MIH in this species like in the bambooleaf wrasse.

The pattern of steroidogenesis in the medaka maturing follicles (Kobayashi *et al.*, 1996) is similar to that in yellowtail (Fig. 5.5). Among the 13 metabolites synthesized from 17-P, 17,20β-P was the most potent inducer of FOM *in vitro*, indicating that 17,20β-P is the MIH in the medaka (Fukada *et al.*, 1994). After the formation of 17,20β-P from P5 via 17-P5 and 17-P, 17,20β-P-5β appeared in the follicles (Fukada *et al.*, 1994). As well as in medaka in the maturing follicles of chub mackerel, yellowtail, bambooleaf wrasse, and red seabream the production of 17,20β-P-5β, which is the 5β-reduced form of 17,20β-P, increased with 17,20β-P production. The 17,20β-P-5β was much less effective than 17,20β-P at GVBD induction in these species. The quick conversion of 17,20β-P probably represents the inactivation process of MIH. Similarly, all of the metabolites of progesterone (P) in amphibian ovarian follicles were found to be less effective than P, a proposed MIH in amphibians (Schuetz and Glad, 1985), in induction of oocyte maturation, and it has been suggested that the metabolism of P by amphibian ovaries, particularly 5α- and 5β-reduced derivatives, represents an inactivation process (Ozon *et al.*, 1975; Schatz and Morrill, 1975; Thibier-Fouchet *et al.*, 1976).

In the case of bambooleaf wrasse and red seabream, 17,20β-P is not the only active steroid to induce oocyte maturation; 20β-S is as potent as 17,20β-P. However, the 5β-reduced form of 20β-S could not be detected, suggesting that the 5β-reductase of these fish selectively catalyzes 17,20β-P into its inactivated from, but not 20β-S. It is a well-known fact that various C21-steroids and their glucuronide and sulfate conjugates play important roles as pheromones (Scott and Vermeirssen, 1994). The gonads are the major sites of glucuronation and sulfation for teleost reproductive steroids. In goldfish, pre-ovulatory females release 17,20β-P, the assumed MIH in this species, into the water. This stimulates the release of GtH, milt production, and spawning behavior in males (Stacey *et al.*, 1994). Sulfated 17,20β-P, 17,20β-dihydroxy-4-pregnen-3-one-20-sulfate (17,20β-P-S), is also released and has a similar effect (Sorensen *et al.*, 1995). In addition, greater amounts of 20β-S and its sulfated form were also produced and released into the water compared with 17,20β-P (Scott and Sorensen, 1994). Both forms of 20β-S were relatively less effective than 17,20β-P and 17,20β-P-S, as determined by electro-olfactogram recording, suggesting that the effect is likely to be in addition to those of free and sulfated 17,20β-P. These studies suggest that the metabolites of 17,20β-P and 20β-S may have any roles in fish reproduction. A future study on the breakdown of 17,20β-P and 20β-S and the physiological roles of their metabolites in each fish species is necessary.

Recent studies have demonstrated that 20β-hydroxylated progestins not only induce FOM, but also ovulation. The induction of ovulation is under genomic control, probably via the nuclear progestin receptor (Pinter and Thomas, 1995, 1999), while the induction of maturation is non-genomic, via the oocyte plasma membrane receptor, as described later. The membrane and nuclear progestin receptors in spotted sea trout ovaries have binding affinities to certain steroids, consistent with their abilities to induce maturation and ovulation, respectively (Pinter and Thomas, 1999). Of the different steroids tested, 17,20β-P had a higher affinity for nuclear receptors and the ability to induce ovulation, whereas 20β-S had a much higher affinity for membrane receptors and the ability to induce maturation. However, 17,20β-P was not produced by spotted sea-trout ovarian follicles and the plasma concentrations of immunoreactive 17,20β-P remained low during oocyte maturation (Thomas et al., 1987; Thomas and Trant, 1989). Therefore, it has been suggested that both oocyte maturation and ovulation of the spotted seatrout are regulated by 20β-S (Pinter and Thomas, 1999). In the maturing follicles of bamboo-leaf wrasse and red seabream, 17,20β-P and 20β-S were synthesized simultaneously. Accordingly, in these species, both 17,20β-P and 20β-S may play roles not only in maturation but also in ovulation, although the ability of 17,20β-P and 20β-S to induce ovulation is unknown. On the other hands, both 17,20β-P and 20β-S were also produced in the maturing follicles of chub mackerel and pufferfish, one of two progestins likely act as MIH in respective species; 17,20β-P in chub mackerel and 20β-S in pufferfish, but the function of other was unclear. Further studies on the ability of 20β-hydroxyprogestins to induce ovulation and the characterization of both membrane and nuclear progestin receptors are needed to resolve the different functions of 17,20β-P and 20β-S in teleosts.

MIH RECEPTOR

Under the regulation by LH, during the first stage of oocyte maturation the follicle (somatic) cells acquire the ability to produce MIH and the oocyte to respond to MIH (i.e., oocyte maturational competence, OMC), whereas in the second stage the follicle cells produce MIH and, consequently, the oocyte is released from meiotic arrest (Patiño et al., 2001). The increased oocyte MIH receptor activity is presumably associated with OMC acquisition. MIH induces oocyte matuartion by acting through a plasma membrane receptor on the oocyte surface (Nagahama, 1997; Thomas et al., 2002). This receptor, unrelated to the classical steroid nuclear receptors, mediates rapid non-genomic steroid activity (Thomas et al., 2002). Recently, we characterized the receptor binding to the ovarian plasma membrane in yellowtail undergoing GVM (Rahman, 2002b).

Scatchard analysis of 17,20β-P binding to the ovarian membrane fractions revealed the presence of a single class of high affinity (K_D = 22.9 nM) and limited capacity (B_{max} = 2.1 pmol/g tissue) binding site for 17,20β-P (Fig. 5.6). Competitions among various steroids for the displacement of 17,20β-P binding sites in the ovarian membrane receptor revealed that, of the steroids tested, only 17,20β-P was an effective competitor at lower concentration. The relative binding affinities (RBAs) of various steroids for yellowtail ovarian membrane receptor are shown in Table 5.2. Except for 17,20β-P, all steroids tested showed the competition with an order of magnitude less affinity for the receptor with RBAs. An increase in specific 17,20β-P binding in the ovaries was observed that was associated with FOM, and which rapidly disappeared just after ovulation (Fig. 5.7). Thus, the receptor with high affinity, limited capacity and highly specific to the 17,20β-P was characterized in the plasma membrane of yellowtail maturing follicles. In addition to the previous findings of follicular steroidogenesis and in vitro GVBD assay, these results indicate that 17,20β-P is the MIH in yellowtail.

An oocyte 17,20β-P receptor has been characterized on ovarian membranes of rainbow trout Oncorhynchus mykiss in which 17,20β-P acts as MIH (Yoshikuni et al., 1993). Similarly, ovarian 20β-S receptors have been characterized for the spotted seatrout (Patiño and Thomas, 1990b) and striped bass (King et al., 1997) in which 20β-S has been identified as MIH.

Interestingly, in bambooleaf wrasse and red seabream maturing follicles, 17,20β-P and 20β-S were produced and showed an equally high effectiveness on GVBD in vitro, suggesting both two progestins act as MIHs in these species. Binding assays for the oocyte MIH receptor have been performed on the bambooleaf wrasse and red seabream; however, the receptor has not yet been characterized in both species because the

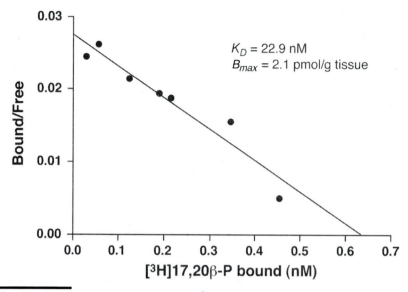

Fig. 5.6 Scatcharad analysis of specific [³H]17,20β-P binding to ovarian membranes prepared from the yellowtail oocytes under germinal vesicle migration.

Table 5.2 Relative binding affinities (RBAs) of various steroids for yellowtail ovarian membrane receptor.

Competitor	Concentration[a] (nM)	RBA[b] (%)
17,20β-P	19.1	100.0
17-P5	500.0	3.8
17,21-P	602.6	3.2
17-P	701.5	2.7
17-P-5β	1088.9	1.8
17,20β-P-5β	2360.5	0.8
20β-S	2460.4	0.8
17,20β-P5	5794.3	0.3
P5	ND[c]	0.3

[a] Values were calculated from the competitor concentration required to displace 50% of 100 nM [³H]17,20β-P specifically bound to the membrane.

[b] Calculated from the ratio of 17,20β-P concentration to competitor concentration x 100.

[c] ND, no-displacement.

amount of specific binding of 17,20β-P and 20β-S for the oocyte membrane fraction is very low. This low binding activity of the oocyte membrane causes us to speculate that the MIH receptors of these daily spawning species breaks easily during preparation of the membrane fraction, or

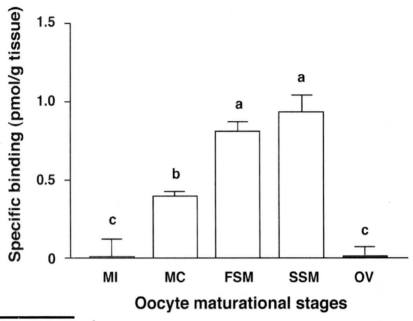

Fig. 5.7 Specific [^{3}H]17,20β-P binding to the membrane preparations from different maturational stages of yellowtail oocyte. Vertical bar represents the mean and SEM for triplicate determinations from three fish. Error bars with different letter superscripts are significantly different ($p<0.05$). Cytoplasmic and nuclear events of each maturation stage of oocytes are as follows (Rahman et al., 2002b). MI (maturationally incompetent), central germinal vesicle surrounded by scattered oil droplets in the cytoplasm; MC (maturationally competent), central germinal vesicle surrounded by large united oil droplets; FSM (first stage of maturation), germinal vesicle already migrated; SSM (second stage of maturation), completion of germinal vesicle breakdown; OV (ovulation), oocytes free from follicular tissue.

appears only for a very short period so that we could not collect ovaries that have a highly active receptor. Other technique is required to clarify the involvement of 17,20β-P and 20β-S in oocyte maturation in the bambooleaf wrasse and red seabream because the traditional binding assay method could not be applied to these species.

STEROIDOGENIC SHIFT

In teleosts, two 20β-hydroxylated progestins are known to be MIHs, although a naturally occurring MIH has been conclusively identified in a limited number of species. 17,20β-P has been identified as an MIH in amago salmon (Nagahama and Adachi, 1985), the first MIH identified in teleosts, and in Indian catfish *Clarias batrachus* (Haider and Rao, 1992), a killifish *Fundulus heteroclitus* (Petrino et al., 1993), medaka (Fukada *et al.*, 1994), and yellowtail (Rahman *et al.*, 2001, 2002b). 20β-S has also been identified as an MIH in Atlantic croaker and spotted seatrout (Trant *et al.*, 1986; Thomas and Trant, 1989; Trant and Thomas, 1989; Patiño and Thomas, 1990a), striped bass (King *et al.*, 1994, 1997), and a pufferfish (Matsuyama *et al.*, 2001). Both steroids, 17,20β-P and 20β-S, are 20β-hydroxylated progestins, and the only structural difference between the two is a hydroxyl group at the 21-position. As shown above, 17,20β-P and 20β-S are mainly synthesized from 17-P and 17,21-P, respectively by 20β-HSD.

Data presented in this chapter indicate a distinct shift in the steroidogenic pathway from E2 to MIH by ovarian follicles, although the steroidogenic pathways in E2 and MIH are different among species. Regardless of the difference in E2 synthetic pathways and whether the MIH is 17,20β-P or 20β-S, there is a common enzymatic kinetics in MIH production; the decrease in C17,20-lyase activity involved in the cessation of E2 production and increase in 20β-HSD leads to a dramatic elevation in MIH.

In the ovarian follicles of teleost, the steroidogenic enzyme which serves as a branch point for E2 and MIH is cytochrome P-450 17α-hydroxylase/C17,20-lyase (P-450c17). P-450c17 is a single cytochrome P-450 enzyme mediating both 17α-hydroxylase and C17,20-lyase activities (Chung *et al.*, 1987; Hanukoglu, 1992) in which only C17,20-lyase activity is required for the synthesis of C19-steroids such as DHEA and AD. Therefore, depression of one of two functions of P450c17 is involved in the steroidogenic shift from the pathway leading to E2 to MIH synthesis. To date, however, there have been no studies in fish to explain the mechanism regulating this phenomenon. Several studies in mammals have shown that C17,20-lyase activity is regulated by P450 oxidoreductase, cytochrome *b*5, or Ser/Thr phosphorylation (Miller, 1999). Further studies have demonstrated that the selective elevation of 17,20-lyase activity is brought about by a cAMP-dependent protein kinase through Ser/Thr phosphorylation while dephosphorylation leaves alone the 17α-hydroxylase activity for P450c17 (Zhang *et al.*, 1995; Pandey *et al.*, 2003). In fish, although cDNAs encoding P450c17 gene of rainbow trout (Sakai *et al.*, 1992) and Japanese eel *Anguilla japonica* (Kazeto *et al.*, 2000), little is known about the mechanism regulating the P450c17 gene expression.

The selective productions of MIH, 17,20β-P and/or 20β-S, are caused by elevation of 20β-HSD activity. The dramatic increase of 20β-HSD enzyme activity induced by GtH/hCG in isolated ovarian follicles of teleosts (Nagahama, 1997; Kazeto *et al.*, 2001) is inhibited by actinomycin D, suggesting that 20β-HSD is transcriptionally regulated (Nagahama, 1997). Recently, cDNA encoding a functional carbonyl reductase-like 20β-HSD was cloned from the ovary of rainbow trout (Guan *et al.*, 1999), ayu *Plecoglossus altivelis* (Tanaka *et al.*, 2002), and Nile tilapia *Oreochromis niloticus* (Senthilkumaran *et al.*, 2002) in which 17,20β-P is produced as MIH. It has been demonstrated that in Nile tilapia the stage-specific activation of 20β-HSD is achieved by enhancing the transcription of 20β-HSD (Senthilkumaran *et al.*, 2002), and suggested that CRE binding protein cDNAs (CREBs) are the probable candidate transcriptional factors for 20β-HSD (Senthilkumaran *et al.*, 2004). More molecular studies are needed to explain the mechanism underlying the steroidogenic shift, which should determine how GtH and other hormonal factors act on ovarian follicles to turn expression of these specific genes on and off at specific times during oocyte maturation.

Conclusions

In this chapter, I briefly review our studies on the follicular steroidogenesis in five marine teleosts including gonochoristic and hermaphroditic species with available studies of another fish. The basic concept of follicular steroidogenesis in teleost was common between species, where E2 is synthesized during vitellogenesis while MIH is produced during oocyte maturation. The synthetic pathways in E2 and MIH, however, are different among species. One of the interesting findings is that T is not synthesized in the vitellogenic follicles of protogynous bambooleaf wrasse in which E2 is converted from E1 but not T. This E2 synthetic pathway via E1 was also found in the vitellogenic follicles of red seabream showing juvenile hermaphroditism. E2 production pathways differ between even gonochoristic species in which E2 is synthesized from T. The reason for the species-specific differences in E2 synthetic pathways is unknown. To date, only two 20β-hydroxylated progestins, 17,20β-P and 20β-S, have been identified as the naturally occurring MIH in fish. Besides the maturation-inducing function, feedback effects of them on LH secretion have been reported in rainbow trout (Weil and Marcuzzi, 1990) and Atlantic croaker (Mathews *et al.*, 2002). The biology and ecology of fish is sufficiently diverse and reproductive physiology, including the strategy of follicular steroidogenesis, may be also diverse in response to them. In the viviparous rockfishes of the genus *Sebastes*, some functional roles of 17,20β-dihydroxylated steroids in regulating viviparity in addition to inducing oocyte maturation have been suggested (Nagahama *et al.*, 1991;

Moore *et al.*, 2000). Physiological actions of steroid hormones can be understood intensively in the regulation of transcription of the target genes. Accordingly, to understand the mechanism of multifarious functions of sex steroid hormones, structure and function of the target genes should be evolved in parallel to the research on steroidogenic enzyme and steroid hormone receptor genes.

Acknowledgements

The author appreciates the very useful scientific and editorial comments provided by two anonymous reviewers who spent a great deal of time reviewing the manuscript. The studies from our laboratory described in this chapter have been supported in part by Grant-in-Aid for Scientific Research (10660182, 16380136) and Exploratory Research (15658059) from the Japan Society for the Promotion of Science. Research support by a grant from the Tohwa Shokuhin Bounty is also acknowledged.

References

Axelrod, L.R., C. Matthijssen, J.M. Goldzieher and J.E. Pulliam. 1965. Definitive identification of microquantities of radioactive steroids by recrystallization to constant specific activity. *Acta Endocrinologica* S49: 7-8 Suppl. 99.

Barnett, C.W. and N.W. Pankhurst. 1994. Changes in plasma levels of gonadal steroids and gonad morphology during the spawning cycle of male and female demoiselles *Chromis dispilus* (Pisces: Pomacentridae). *General and Comparative Endocrinology* 93: 260-274.

Baroiller, J.F., Y. Guiguen and A. Fostier. 1999. Endocrine and environmental aspects of sex differentiation in fish. *Cellular and Molecular Life Sciences* 55: 910-931.

Borg, B. 1994. Androgens in teleost fishes. *Comparative Biochemistry and Physiology* 109C, 219-245.

Brenner, S., G. Elgar, R. Sanford, A. Macrae, B. Venkatesh and S. Aparicio. 1993. Characterization of the pufferfish (Fugu) genome as a compact model vertebrate genome. *Nature (London)* 336: 265-268.

Canário, A.V.M. and A.P. Scott. 1990. Plasma levels of ovarian steroids, including $17\alpha,20\beta$-dihydroxy-4-pregnen-3-one and $3\beta,17\alpha,20\beta$-trihydroxy-5β-pregnane, in female dabs (*Limanda limanda*)—marine flatfish—induced to mature and ovulate with human chorionic gonadotropin. *General and Comparative Endocrinology* 77: 177-191.

Chung, B.C., J. Picado-Leonard, M. Haniu, M. Bienkowski, P.F. Hall, J.E. Shively and W.L. Miller. 1987. Cytochrome P450c17 (steroid 17α-hydroxylase/17,20 lyase): cloning of human adrenal and testis cDNAs indicates the same gene is expressed in both tissues. *Proceedings of the National Academy of Sciences of the United States of America* 84: 407-411.

Devlin, R.H. and Y. Nagahama. 2002. Sex determination and sex differentiation in fish: An overview of genetic, physiological, and environmental influences. *Aquaculture* 208: 191-364.

Fukada, S., N. Sakai, S. Adachi and Y. Nagahama. 1994. Steroidogenesis in the ovarian follicle of medaka (*Oryzias latipes*, a daily spawner) during oocyte maturation. *Development, Growth and Differentiation* 36: 81-88.

Greeley, M.S., R. MacGregor III and K.R. Marion. 1988. Variation in plasma oestrogens and androgens during the seasonal and semilunar spawning cycles of female gulf killifish, *Fundulus grandis* (Baird and Girard). *Journal of Fish Biology* 33: 419-429.

Grober, M.S., I.M.D. Jackson and A.H. Bass. 1991. Gonadal steroids affect LHRH preoptic cell number in a sex/role changing fish. *Journal of Neurobiology* 22: 734-741.

Guan, G., M. Tanaka, T. Todo, G. Young, M. Yoshikuni and Y. Nagahama. 1999. Cloning and expression of two carbonyl reductase-like 20β-hydroxysteroid dehydrogenase cDNAs in ovarian follicles of rainbow trout (*Oncorhynchus mykiss*). *Biochemical and Biophysical Research Communications* 255: 123-128.

Haider, S. and N.V. Rao. 1992. Oocyte maturation in *Clarias batrachus*. III. Purification and characterization of maturation-inducing steroid. *Fish Physiology and Biochemistry* 9: 505-512.

Hanukoglu, I. 1992. Steroidogenic enzymes: structure function, and role in regulation of steroid hormone biosynthesis. *The Journal of Steroid Biochemistry and Molecular Biology* 43: 779-804.

Kagawa, H., G. Young, S. Adachi and Y. Nagahama. 1982. Estradiol-17β production in amago salmon (*Oncorhynchus rhodurus*) ovarian follicles: role of the thecal and granulosa cells. *General and Comparative Endocrinology* 47: 440-448.

Kagawa, H., G. Young and Y. Nagahama. 1984. *In vitro* estradiol-17β and testosterone production by ovarian follicles of the goldfish, *Carassius auratus*. *General and Comparative Endocrinology* 54: 139-143.

Kazeto, Y., S. Ijiri, T. Todo, S. Adachi and K. Yamauchi. 2000. Molecular cloning and characterization of Japanese eel ovarian P450c17 (CYP17) cDNA. *General and Comparative Endocrinology* 118: 123-133.

Kazeto, Y., S. Adachi and K. Yamauchi. 2001. 20β-Hydroxysteroid dehydrogenase of the Japanese eel ovary: Its cellular localization and changes in the enzymatic activity during sexual maturation. *General and Comparative Endocrinology* 122: 109-115.

Kime, D.E., A.P. Scott and A.V.M. Canário. 1992. *In vitro* biosynthesis of steroids, including 11-deoxycortisol and 5α-pregnane-3β, 7α,17,20β-tetrol, by ovaries of the goldfish *Carassius auratus* during the stage of oocyte final maturation. *General and Comparative Endocrinology* 87: 375-384.

King, V.W., P. Thomas, R.M. Harrell, R.G. Hodson and C.V. Sullivan. 1994. Plasma levels of gonadal steroids during final oocyte maturation of striped bass, *Morone saxatilis* L. *General and Comparative Endocrinology* 95: 178-191.

King, V.W., S. Ghosh, P. Thomas and C.V. Sullivan. 1997. A receptor for the oocyte maturation-inducing hormone 17α,20β,21-trihydroxy-4-pregnen-3-one on ovarian membranes of striped bass. *Biology of Reproduction* 56: 266-271.

Kobayashi, D., M. Tanaka, S. Fukada and Y. Nagahama. 1996. Steroidogenesis in the ovarian follicles of the medaka (*Oryzias latipes*) during vitellogenesis and oocyte maturation. *Zoological Science* 13: 921-927.

Kramer, C.R., S. Koulish and P.L. Bertacchi. 1988. The effects of testosterone implants on ovarian morphlogy in bluehaed wrasse, *Thalassoma bifasciatum* (Bloch) (Teleostei: Labridae). *Journal of Fish Biology* 32: 397-407.

LaFleur, G. Jr., B.M. Byrne, C. Haux, R.M. Greenberg and R.A. Wallace. 1995. Liver-derived cDNAs: Vitellogenin and vitelline envelope protein precursors (choriogenins). In: *Proceedings of the Fifth International Symposium on the Reproductive Physiology of Fish*, F.W. Goetz and P. Thomas (eds.). Fish Symposium 95, Austin, pp. 336-338.

Mabuchi, K., T. Kobayashi and T. Nakabo. 2000. Genetic differences between two *Pseudolabrus* species (Osteichthyes: Labridae) from the southern coasts of Japan. *Species Diversity* 5: 163-172.

Mathews, S., I.A. Khan and P. Thomas. 2002. Effects of the maturation-inducing steroid on LH secretion and the GnRH system at different stages of the gonadal cycle in Atlantic croaker. *General and Comparative Endocrinology* 126: 287-297.

Matsuyama, M., S. Adachi, Y. Nagahama and S. Matsuura. 1988a. Diurnal rhythm of oocyte development and plasma steroid hormone levels in the female red seabream, *Pagrus major*, during the spawning season. *Aquaculture* 73: 357-372.

Matsuyama, M., R. Torres and S. Matsuura. 1988b. Juvenile bisexuality in the red sea bream *Pagurus major*. *Environmental Biology of Fishes* 21: 27-36.

Matsuyama, M., H. Kagawa, M. Arimoto, K. Maruyama, H. Hirotsuji, M. Kashiwagi and R. Liu. 1996. Changes of ovarian maturity in the yellowtail *Seriola quinqueradiata* induced by a human chorionic gonadotropin. *Suisanzoshoku* 44: 189-195.

Matsuyama, M., S. Morita, N. Hamaji, M. Kashiwagi, K. Ohta and Y. Nagahama. 1997. Diurnal spermatogenesis and spawning in the secondary male of a protogynous wrasse (*Pseudolabrus japonicus*). *Zoological Science* 14: 1001-1008.

Matsuyama, M., S. Morita, T. Nasu and M. Kashiwagi. 1998a. Daily spawning and development of sensitivity to gonadotropin and maturation-inducing steroid in the oocytes of the bamboo-leaf wrasse *Pseudolabrus japonicus*. *Environmental Biology of Fishes* 52: 281-290.

Matsuyama, M., K. Ohta, S. Morita, M.M. Hoque, H. Kagawa and A. Kambegawa. 1998b Circulating levels and *in vitro* production of two maturation-inducing hormones in teleost: $17\alpha,20\beta$-dihydroxy-4-pregnen-3-one and $17\alpha,20\beta,21$-trihydroxy-4-pregnen-3-one, in a daily spawning wrasse, *Pseudolabrus japonicus*. *Fish physiology and Biochemistry* 19: 1-11.

Matsuyama, M., A. Sasaki, K. Nakagawa, T. Kobayashi, Y. Nagahama and H. Chuda. 2001. Maturation-inducing hormone of the tiger puffer, *Takifugu rubripes* (Tetraodontidae, Teleostei): biosynthesis of steroids by the ovaries and the relative effectiveness of steroid metabolites for germinal vesicle breakdown *in vitro*. *Zoological Science* 18: 225-234.

Matsuyama, M., T. Shiraishi, J.K. Sundaray, M.A. Rahman, K. Ohta and A. Yamaguchi. 2005. Steroidogenesis in ovarian follicles of chub mackerel, *Scomber japonicus*. *Zoological Science* 22: 101-110.

Miller, W.L. 1999. P450c17 – The qualitative regulator of steroidogenesis. In: *Molecular and Cellular Pediatric Endocrinology*, S. Handwerger (ed.). Humana Press, Totowa, pp. 139-152.

Moore, R.K., A.P. Scott and P.M. Collins. 2000. Circulation C-21 steroids in relation to reproductive condition of a viviparous marine teleost, *Sebastes rastrelliger* (grass rockfish). *General and Comparative Endocrinology* 117: 268-280.

Nagahama, Y. 1993. Molecular endocrinology of oocyte growth and maturation in fish. *Fish Physiology and Biochemistry* 11: 1-6.

Nagahama, Y. 1997. $17\alpha,20\beta$-Dihydroxy-4-pregnen-3-one, a maturation-inducing hormone in fish oocytes: Mechanisms of synthesis and action. *Steroids* 62: 190-196.

Nagahama, Y. and S. Adachi. 1985. Identification of maturation-inducing steroid in teleost, the amago salmon (*Oncorhynchus rhodurus*). *Developmental Biology* 109: 425-435.

Nagahama, Y., A. Takemura, K. Takano, S. Adachi and M. Kusakari. 1991. Serum steroid hormone levels in relation to the reproductive cycle of *Sebastes taczanowskii* and *S. schlegeli*. *Environmental Biology of Fishes* 30: 31-38.

Nakazono, A. 1979. Studies on the sexual reversal and spawning behavior of five species of Japanese labrid fishes. *Report of Fishery Research Laboratory, Kyushu University* 4: 1-64.

Ohta, K. and M. Matsuyama. 2002. Steroidogenic pathways to 17,20β-dihydroxy-4-pregnen-3-one and 17,20β,21-trihydroxy-4-pregnen-3-one in the ovarian follicles of the bambooleaf wrasse, *Pseudolabrus sieboldi*. *Fishery Science* 68: 41-50.

Ohta, K., T. Mine, A. Yamaguchi and M. Matsuyama. 2001. Steroidogenic pathway to estradiol-17β synthesis in the ovarian follicles of the protogynous wrasse, *Pseudolabrus sieboldi*. *Zoological Science* 18: 937-945.

Ohta, K., S. Yamaguchi, A. Yamaguchi, K. Gen, K. Okuzawa, H. Kagawa and M. Matsuyama. 2002a. Biosynthesis of estradiol-17β in the ovarian follicles of the red seabream, *Pagrus major* during vitellogenesis. *Fisheries Science* 68: 680-687.

Ohta, K., S. Yamaguchi, A. Yamaguchi, K. Gen, K. Okuzawa, H. Kagawa and M. Matsuyama. 2002b. Biosynthesis of steroids in ovarian follicles of red seabream, *Pagrus major* (Sparidae, Teleostei) during final oocyte maturation and the relative effectiveness of steroid metabolites for germinal vesicle breakdown *in vitro*. *Comparative Biochemistry and Physiology* B 133: 45-54.

Ohta, K., J.K. Sundaray, T. Okita, M. Sakai, T. Kitano, A. Yamaguchi, T. Takeda and M. Matsuyama. 2003. Bi-directional sex change and its steroidogenesis in the wrasse, *Pseudolabrus sieboldi*. *Fish Physiology and Biochemistry* 28: 173-174.

Ozon, R., R. Belle, C. Serres and C. Fouchet. 1975. Mechanism of action of progesterone on amphibian oocytes. A possible biological role for progesterone metabolism. *Molecular and Cellular Endocrinology* 3: 221-231.

Pandey, A.V., S.H. Mellon and W.L. Miller. 2003. Protein phosphatase 2A and phosphoprotein SET regulate androgen production by P450c17. *The Journal of Biological Chemistry* 278: 2837-2844.

Pankhurst, N.W., P.I. Hilder and P.M. Pankhurst. 1999. Reproductive condition and behavior in relation to plasma levels of gonadal steroids in the spiny damselfish *Acanthochromis polyacanthus*. *General and Comparative Endocrinology* 115: 53-69.

Patino, R. and P. Thomas. 1990a. Gonadotropin stimulates 17α,20β,21-trihydroxy-4-pregnen-3-one production from endogenous substrates in Atlantic croaker ovarian follicles undergoing final oocyte maturation *in vitro*. *General and Comparative Endocrinology* 78: 474-478.

Patino, R. and P. Thomas. 1990b. Characterization of membrane receptor activity for 17α,20β,21-trihydroxy-4-pregnen-3-one in ovaries of spotted seatrout (*Cynoscion nebulosus*). *General and Comparative Endocrinology* 78: 204-217.

Patino, R., G. Yoshizaki, P. Thomas and H. Kagawa. 2001. Gonadotropic control of ovarian follicle maturation: the two-stage concept and its mechanism. *Comparative Biochemistry and Physiology* B 129: 427-439.

Petrino, T.R., Y.W. Lin, J.C. Netherton, D.H. Powell and R.A. Wallace. 1993. Steroidogenesis in *Fundulus heteroclitus* V.: Purification, characterization, and metabolism of 17α,20β-dihydroxy-4-pregnen-3-one by intact follicles and its role in oocyte maturation. *General and Comparative Endocrinology* 92: 1-15.

Pinter, J. and P. Thomas. 1995. Characterization of a progestogen receptor in the ovary of the spotted seatrout, *Cynoscion nebulosus. Biology of Reproduction* 52: 667-675.

Pinter, J. and P. Thomas. 1999. Induction of ovulation of mature oocytes by the maturation inducing steroid 17α,20β,21-trihydroxy-4-pregnen-3-one in the spotted seatrout. *General and Comparative Endocrinology* 115: 200-209.

Poling, K.R., E.J. Fraser and P.W. Sorensen. 2001. The three steroidal components of the goldfish preovulatory pheromone signal evoke different behaviors in males. *Comparative Biochemistry and Physiology* B 129: 645-651.

Prat, F., Z.M. Carrillo, A. de Mones and A. Fostier. 1990. Seasonal changes in plasma levels of gonadal steroids of sea bass, *Dicentrarchus labrax* L. *General and Comparative Endocrinology* 78: 361-373.

Rahman, M.A., K. Ohta, H. Chuda, S. Nakano, K. Maruyama and M. Matsuyama. 2001. Gonadotropin-induced steroidogenic shift towards maturation-inducing hormone in Japanese yellowtail during final oocyte maturation. *Journal of Fish Biology* 58: 462-474.

Rahman, M.A., K. Ohta, H. Chuda, A. Yamaguchi and M. Matsuyama. 2002a. Steroid hormones and their synthetic pathways in the vitellogenic ovarian follicles of yellowtail, *Seriola quinqueradiata. Journal of the Faculty of Agriculture, Kyushu University* 46: 311-319.

Rahman, M.A., K. Ohta, M. Yoshikuni, Y. Nagahama, H. Chuda and M. Matsuyama. 2002b. Characterization of ovarian membrane receptor for 17,20β-dihydroxy-4-pregnen-3-one, a maturation-inducing hormone in yellowtail, *Seriola quinqueradiata. General and Comparative Endocrinology* 127: 71-79.

Reinboth, R. 1975. Spontaneous and hormone-induced sex-inversion in wrasses (Labridae). *Publications Stazione Zoological Napoli* 39 (Supplement): 550-573.

Sakai, N., M. Tanaka, S. Adachi, W.L. Miller and Y. Nagahama. 1992. Rainbow trout cytochrome P450c17 (17α-hydroxylase/17,20-lyase)-cDNA cloning, enzymatic properties and temporal pattern of ovarian P450c17 mRNA expression during oogenesis. *FEBS Letters* 301: 60-64.

Sakai, M., K. Ohta, A. Yamaguchi and M. Matauyama. 2004. Induction of the bi-directional sex change by steroid hormone implantation in the protogynous wrasse, *Pseudolabrus sieboldi.* In: *Trend in Comparative Endocrinology*, T. Oishi, K. Tsutsui, S. Tanaka and S. Kikuyama (eds.). AOSCE, Nara, Japan, pp. 402-404.

Schatz, F. and G.A. Morrill. 1975. 5β-Reductive pathway of progesterone metabolism in the amphibian ovarian cytosol. *Biology of Reproduction* 13: 408-414.

Schuetz, A.W. and R. Glad. 1985. *In vitro* production of meiosis inducing substance. *Development, Growth and Differentiation* 27: 201-211.

Scott, A.P. and A.V.M. Canario. 1987. Status of oocyte maturation-inducing steroids in teleosts. In: *Proc. 3rd Int. Symp. Reproductive Physiology of Fish*, D.R. Idler, L.W. Crim and J.M. Walsh (eds.). Memorial University of Newfoundland, St. John's, Canada, pp. 224-234.

Scott, A.P. and A.V.M. Canario. 1990. Plasma levels of ovarian steroids, including 17α,21-dihydroxy-4-pregnene-3,20-dione (11-deoxycortisol) and 3α,17α,21-trihydroxy-5β-pregnan-20-one, in female plaice (*Pleuronectes platessa*) induced to mature with human chorionic gonadotropin. *General and Comparative Endocrinology* 78: 286-298.

Scott, A.P. and P.W. Sorensen. 1994. Time course of release of pheromonally active gonadal steroids and their conjugates by ovulatory goldfish. *General and Comparative Endocrinology* 96: 309-323.

Scott, A.P. and E.L.M. Vermeirssen. 1994. Production of conjugated steroids by teleost gonads and their role as pheromones. In: *Perspective in Comparative Endocrinology*, K.G. Davey, R.E. Peter and S.S. Tobe (eds.). National Research Council of Canada, Ottawa. pp. 645-654.

Scott, A.P., D.S. MacKenzie and N.E. Stacey. 1984. Endocrine changes during natural spawning in the white sucker, *Catostomus commersoni*. II. Steroid hormones. *General and Comparative Endocrinology* 56: 349-359.

Senthilkumaran, B., C.C. Sudhakumari, X.T. Chang, T. Kobayashi, Y. Oba, G. Guan, Y. Yoshiura, M. Yoshikuni and Y. Nagahama. 2002. Ovarian carbonyl reductase-like 20β-hydroxysteroid dehydrogenase shows distinct surge in messenger RNA expression during natural and gonadtropin-induced meiotic maturation in Nile tilapia. *Biology of Reproduction* 67: 1080-1086.

Senthilkumaran, B., M. Yoshikuni and Y. Nagahama. 2004. A shift in steroidogenesis occurring in ovarian follicles prior to oocyte maturation. *Molecular and Cellular Endocrinology* 215: 11-18.

Shiraishi, T., K. Ohta, A. Yamaguchi, M. Yoda, H. Chuda and M. Matsuyama. 2005. Reproductive parameters of the chub mackerel *Scomber japonicus* estimated from human chorionic gonadotropin-induced final oocyte maturation and ovulation in captivity. *Fisheries Science* 71: 531-542.

Sorensen, P.W., A.P. Scott, N.E. Stacy and L. Bowdin. 1995. Sulfated 17α,20β-dihydroxy-4-pregnen-3-one functions as a potent and specific olfactory stimulant with pheromonal actions in the goldfish. *General and Comparative Endocrinology* 100: 128-142.

Specker, J.L. and C.V. Sullivan. 1994. Vitellogenesis in fishes: status and perspectives. In: *Perspectives in Comparative Endocrinology*, K.G. Davey, R.E. Peter and S.S. Tobe (eds.). National Research Council of Canada, Ottawa, pp. 304-315.

Stacey, N.E., J.R. Cardwell, N.R. Liley, A.P. Scott and P.W. Sorensen. 1994. Hormones as sex pheromones in fish. In: *Perspectives in Comparative Endocrinology*, K.G. Davey, R.E. Peter and S.S. Tobe (eds.). National Research Council of Canada, Ottawa, pp. 438-448.

Sundaray, J.K., K. Ohta, A. Yamaguchi, K. Suzuki and M. Matsuyama. 2003. Diurnal rhythm of steroid biosynthesis in the testis of terminal phase male of protogynous wrasse, *Pseudolabrus sieboldi*, a daily spawner. *Fish Physiology and Biochemistry* 28: 193-195.

Tanaka, M., S. Nakajin, D. Kobayashi, S. Fukada, G. Guan, T. Todo, B. Senthilkumaran and Y. Nagahama. 2002. Teleost ovarian carbonyl reductase-like 20β-hydroxysteroid dehydrogenase: potential role in production of maturation-inducing hormone during final oocyte maturation. *Biology of Reproduction* 66: 1498-1504.

Thibier-Fauchet, C., O. Mulner and R. Ozon. 1976. Progesterone biosynthesis and metabolism by ovarian follicles and isolated oocytes *Xenopus laevis*. *Biology of Reproduction* 14: 317-326.

Thomas, P. and J.M. Trant. 1989. Evidence that 17α,20β,21-trihydroxy-4-pregnen-3-one is a maturation-inducing steroid in spotted seatrout. *Fish Physiology and Biochemistry* 7: 185-191.

Thomas, P., Y. Zhu and M. Pace. 2002. Progestin membrane receptors involved in the meiotic maturation of teleost oocytes: a review with some new findings. *Steroids* 67: 511-517.

Thomas, P., N.J. Brown and J.M. Trant. 1987. Plasma levels of gonadal steroids during the reproductive cycle of female spotted seatrout *Cynoscion nebulosus*. In: *Reproductive*

Physiology of Fish, D.R. Idler, L.W. Crim and J.M. Walsh (eds.). Memorial University of Newfoundland, St. Johns, Canada, p. 219.

Trant, J.M. and P. Thomas. 1989. Isolation of a novel maturation-inducing steroid produced *in vitro* by ovaries of Atlantic croaker. *General and Comparative Endocrinology* 75: 397-404.

Trant, J.M., P. Thomas and C.H.L. Shackeleton. 1986. Identification of 17α,20β,21-trihydroxy-4-pregnen-3-one as the major ovarian steroid produced by the teleost *Micropogonias undulatus* during final oocyte maturation. *Steroids* 47: 89-99.

Tyler, C.R. and J.P. Sumpter. 1996. Oocyte growth and development in teleosts. *Reviews in Fish Biology and Fisheries* 6: 287-318.

Tyler, C.R., E.M. Santos and F. Prat. 2000. Unscrambling the egg—cellular, biochemical, molecular and endocrine advances in oogenesis. In: *Proceedings of the Sixth International Symposium on the Reproductive Physiology of Fish.* pp. 273-280. B. Norberg, O.S. Kjesbu, G.L. Taranger, E. Andersson and S.O. Stefansson (eds.), John Greig A/S, Bergen 2000, Bergen, pp. 273-280.

Venkatesh, B., C.H. Tan, D.E. Kime, G.L. Loy and T.J. Lam. 1992. Steroid metabolism by ovarian follicles and extrafollicular tissue of the guppy (*Poecilia reticulata*) during oocyte growth and gestation. *General and Comparative Endocrinology* 86: 378-394.

Venkatesh, B., P. Gilligan and S. Brenner. 2000. Fugu: A compact vertebrate reference genome. *FEBS Letters* 476: 3-7.

Weil, C. and O. Marcuzzi. 1990. Cultured pituitary cell GtH response to GnRH at different stages of rainbow trout oogenesis and influence of steroid hormones. *General and Comparative Endocrinology* 79: 483-491.

Wingfield, J.C. and A.S. Grimm. 1977. Seasonal changes in plasma cortisol, testosterone and oestradiol-17β in the plaice, *Pleuronectes platessa* L. *General and Comparative Endocrinology* 31: 1-11.

Yamada, T., I. Aoki and I. Mitani. 1998. Spawning time, spawning frequency and fecundity of Japanese chub mackerel, *Scomber japonicus* in the waters around the Izu islands, Japan. *Fishery Research* 38: 83-89.

Yaron, Z., G. Gur, P. Melamed, H. Rosenfeld, A. Elizur and B. LevaviSivan. 2003. Regulation of fish gonadotropins. *International Review of Cytology* 225: 131-185.

Yoshikuni, M., N. Shibata and Y. Nagahama. 1993. Specific binding of [^{3}H] 17α,20β-dihydroxy-4-pregnen-3-one to oocyte cortices of rainbow trout (*Oncorhynchus mykiss*). *Fish Physiology and Biochemistry* 11: 15-24.

Zairin, M. Jr., K. Asahina, K. Furukawa and K. Aida. 1992. Plasma steroid hormone profiles during HCG induced ovulation in female walking catfish *Clarias batrachus*. *Zoological Science* 9: 607-617.

Zhang, L.H., H. Rodriguez, S. Ohno and W.L. Miller. 1995. Serine phosphorylation of human P450c17 increases 17,20-lyase activity: Implications for adrenarche and the polycystic ovary syndrome. *Proceedings of the National Academy of Sciences of the United States of America* 92: 10619-10623.

Zohar, Y. 1989. Fish reproduction: its physiology and artificial manipulation. In: *Fish Culture in Warm Water System: Problems and Trends,* M. Shilo and S. Sarig (eds.). CRC Press, Boca Raton, pp. 65-119.

Hormonally Derived Sex Pheromones in Fish

Norm E. Stacey[1,*] and Peter W. Sorensen[2]

INTRODUCTION

Throughout the vertebrates, reproductive hormones are well known to perform two critical functions that synchronize sexual interactions between conspecifics. They act on the brain to induce synchrony between an individual's reproductive behavior and the maturation of its gametes, and they also act on effectors to generate behavioral and other signals that induce reproductive synchrony between conspecifics (Pfaff *et al.*, 2002). In fish, however, hormones can have a third set of actions: after release to the environment, they can serve as sex pheromones. Many sex steroids, prostaglandins and their metabolites are detected with great sensitivity and specificity by a diverse array of fishes, and exert key effects on the reproductive behavior and physiology of such ecologically and economically important taxa as the carps, salmonids, and gobies (Stacey and Sorensen, 2002, 2006).

Authors' addresses: [1]Department of Biological Sciences, University of Alberta, Edmonton, Alberta, Canada T6G 2E9.

[2]Department of Fisheries, Wildlife and Conservation Biology, University of Minnesota, St. Paul, Minnesota 55108, U.S.A. E-mail: psorensen@umn.edu

Corresponding author: E-mail: norm.stacey@ualberta.ca

Although sex pheromones have long been known to mediate diverse reproductive functions in fish (Liley, 1982), evidence that many of these pheromones might be steroids and prostaglandins is comparatively recent. Soon after Døving (1976) first hypothesized that fish are predisposed to evolve pheromonal functions for released sex hormones, Colombo *et al.* (1980) reported pheromonal effects of a steroid glucuronide in the black goby (*Gobius niger* = *G. jozo*). Since then, pheromonal actions of sex steroids and prostaglandins have been reported in a diversity of fishes including cypriniforms (zebrafish, *Danio rerio*; Van Den Hurk and Resink, 1992: goldfish, *Carassius auratus*; Kobayashi *et al.*, 2002: crucian carp, *C. carassius*; Bjerselius *et al.*, 1995: common carp, *Cyprinus carpio*; Irvine *et al.*, 1993; Stacey *et al.*, 1994: tinfoil barbs, *Puntius* spp.; Cardwell *et al.*, 1995; oriental weatherfish loach, *Misgurnus anguillicaudatus*; Kitamura *et al.*, 1994a), salmoniforms (Atlantic salmon, *Salmo salar*; Moore and Waring, 1996: brown trout, *Salmo trutta*: Moore *et al.*, 2002: lake whitefish, *Coregonus clupeaformis*; Laberge and Hara, 2003: arctic char, *Salvelinus alpinus*; Sveinsson and Hara, 2000), and perciforms (round goby, *Neogobius melanostomus*; Murphy *et al.*, 2001).

Furthermore, underwater electro-olfactogram (EOG) screening studies using a large number of sex steroids and prostaglandins have demonstrated sensitive and specific olfactory responsiveness in diverse euteleost species. Best studied are the ostariophysins in which virtually all of the more than 100 cypriniform, characiform and siluriform species tested detected at least one prostaglandin or steroid at the low (typically pM) concentrations expected of a pheromone. However, because such EOG responsiveness has also been observed in other diverse taxa such as the primitive Elopiformes (*Megalops cyprinoides*: tarpons; see Stacey and Sorensen, 2002), the Osmeriformes (*Plecoglossus altivelis*; Kitamura *et al.*, 1994b), and the Cyprinodontiformes (Family Aplocheilidae; see Stacey and Sorensen, 2002), it does not seem unreasonable to expect that the use of steroids and prostaglandins as sex pheromones will be found to occur in a far wider array of fishes than has been currently documented.

The discovery that fish have commonly evolved pheromonal functions for released sex steroids and prostaglandins has at least three significant implications for further studies of fish reproduction. First, studies of endocrine functions in fishes should be expanded to include exogenous as well as endogenous actions, as these can be intimately linked and are equally important. Second, because they serve as a direct link between an individual's endocrine system and the nervous system(s) of conspecifics, steroid and prostaglandin pheromones function as a special class of sexual signals. Third, because these pheromones are derived from a homologous and evolutionarily conserved vertebrate endocrine system, they raise fascinating questions about species specificity of chemical cues and signals and the co-evolution of internal and external signaling systems. In this chapter we shall first briefly review the current evidence that released hormones and related compounds have pheromonal functions in fishes, and then address key issues that might focus future studies.

Throughout this chapter, we will use the term 'hormonal pheromone' to describe chemicals that function as hormones and are also known to be detected by and/or to induce reproductive responses in conspecifics. However, we will also apply this term to situations in which the odorant is likely a hormone precursor or metabolite, or simply a component of a hormone biosynthetic pathway. Implicit in this terminology is the unproven assumption that changes in production and release of these non-hormonal odorants transmit information about similar changes in hormone synthesis.

ORDER CYPRINIFORMES

The order Cypriniformes is found within the superorder Ostariophysi, which contains nearly 5,700 largely freshwater fishes that also comprise the orders Characiformes (tetras, piranhas) and Siluriformes (catfishes). The cypriniforms include two superfamilies, the Cyprinoidea, containing family Cyprinidae (over 2,000 species of carps and minnows), and the Cobitoidea (approximately 350 species), containing the families Catostomidae (suckers), Cobitidae (loaches), Gyrinochelidae (algae eaters), and Balitoridae (river loaches) (Nelson, 1994). Much of our current understanding of hormonal pheromones is based on studies of cypriniform and other ostariophysin fishes, many of which are ecologically and economically important.

Goldfish and Related Carps

Due to the ready availability of goldfish and the ease with which they reproduce in the laboratory, their reproductive pheromones are currently the best characterized of any fish. Although there is evidence for pheromone release by male goldfish (Stacey et al., 2001; Fraser and Stacey, 2002, Sorensen et al., 2005), far more is known about females, which sequentially release at least three distinct pheromones, two of which are mixtures of hormones and related compounds (Fig. 6.1) EOG recordings and behavioral and endocrine bioassays (Irvine and Sorensen, 1993; Stacey et al., 1994; Bjerselius et al., 1995) strongly suggest that the reproductive pheromones of the goldfish are remarkably similar to those of the closely related crucian and common carps.

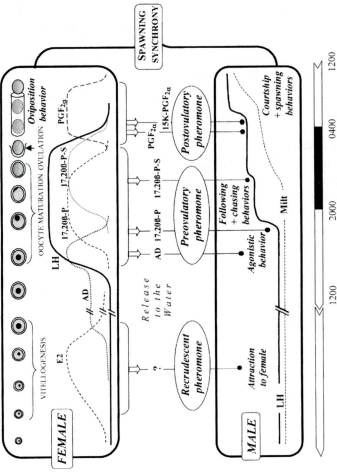

Fig. 6.1 Schematic model of female goldfish pheromones and their primer and releaser effects on males. 17β-Estradiol (E$_2$) in vitellogenic females stimulates urinary release of an unidentified recrudescent pheromone that attracts males. In postvitellogenic females, exogenous cues induce a luteinizing hormone surge stimulating the release of a dynamic pre-ovulatory pheromone containing androstenedione (AD), the maturation inducing steroid 17,20βP, and its sulfated metabolite, 17,20βP-S. Early in the luteinizing hormone (LH) surge, androstenedione induces agonistic behaviors among males. As the 17,20βP:androstenedione ratio increases, males increase luteinizing hormone and begin to follow and chase conspecifics. Late in the luteinizing hormone surge, 17,20βP-S dominates the pre-ovulatory pheromone mixture, enhancing its behavioral and endocrine effectiveness. Males exposed to the pre-ovulatory pheromone increase both the quantity and quality of sperm in the sperm ducts prior to ovulation. At ovulation, eggs in the oviduct induce synthesis of prostaglandin F$_{2\alpha}$ (PGF$_{2\alpha}$) which acts in the brain to stimulate female sex behavior and is released with its major metabolite (15K-PGF$_{2\alpha}$) as a post-ovulatory pheromone stimulating both male courtship and spawning behaviors and additional luteinizing hormone increase. Reprinted from Kobayashi *et al.* (2002), Sorensen and Stacey (2004), and Stacey and Sorensen (2005) with kind permission of the Royal Society of New Zealand and Elsevier Ltd.

The goldfish mating system is typical of many non-territorial, non-parental cyprinids, and appears identical to that of the common carp, with which it will hybridize (Taylor and Mahon, 1977). Goldfish, crucian, and common carp are temperate species that undergo vitellogenesis in winter and ovulate large numbers of oocytes several times over a protracted spring-summer spawning season. Ovulation is triggered when a combination of increasing temperature and submerged vegetation (a preferred substrate for the adhesive eggs) induces a photoperiodically synchronized pre-ovulatory surge of luteinizing hormone (LH) in postvitellogenic females (e.g., Stacey et al., 1989). Ovulation occurs several hours before dawn, at which time small groups of males compete for access to ovulated females, as they repeatedly enter aquatic vegetation to oviposit unguarded eggs over a period of several hours (Fig. 6.1).

In the evolution of this promiscuous mating system, where spawning typically occurs in dimly lit and turbid waters, and where a male's reproductive success is determined only by the number of eggs he fertilizes, the origin and evolution of hormonal pheromones likely has been shaped through selection for males best able to detect and respond to chemicals that indicate the presence and reproductive condition of females. Indeed, from vitellogenesis until the completion of spawning, males respond to three distinct female pheromones (Kobayashi et al., 2002). First to be released is an uncharacterized recrudescence pheromone, which attracts males, is induced by estradiol (E_2), and appears to be released in the urine throughout vitellogenesis (Yamazaki, 1990). In contrast, the pre-ovulatory steroid pheromone and the post-ovulatory prostaglandin pheromone are released during the brief peri-ovulatory period of final oocyte maturation and spawning.

Female pre-ovulatory steroid pheromone

For simplicity, we use only one term (pre-ovulatory steroid pheromone) to describe the pheromonal odor of females undergoing final oocyte maturation. However, both the behavioral and endocrine responses of males indicate that they perceive multiple pre-ovulatory pheromone odors that result from the pheromone varying both temporally, due to shifts in steroid synthesis during final oocyte maturation, and spatially, due to differential release of the pheromone's components across the gills and in urine.

Females release the pre-ovulatory steroid pheromone during a dramatic LH surge that commences in photophase and terminates during the following scotophase (Stacey *et al.* 1979, 1989; Fig. 6.1). If exposed to females or their odor during this period, males rapidly increase their LH and steroid concentrations and within several hours have increased volumes of milt (sperm and seminal fluid) in the sperm ducts (Kobayashi *et al.*, 1986a; Stacey *et al.*, 1989), effects that are blocked by cutting the olfactory tracts (Kobayashi *et al.*, 1986b). These primer endocrine effects, together with releaser effects on male behavior, dramatically enhance male fertility (DeFraipont and Sorensen, 1993; Zheng *et al.*, 1997). The odor of ovulatory females also appears to induce ovulation in other post-vitellogenic females (Kobayashi *et al.*, 2002), as discussed at the end of this section.

EOG and steroid release studies indicate that male responses to the pre-ovulatory pheromone result from three key steroids released by females: androstenedione (AD); the maturation-inducing steroid 4-pregnen-17,20β-dihydroxy-3-one (17,20βP); and its 20β-sulfated metabolite 17,20βP-S. Each of these steroids is effective when tested alone and acts through specific and sensitive (picomolar threshold) olfactory receptors (Sorensen *et al.*, 1990, 1995a). However, the female also releases additional but less potent steroid odorants (e.g., 17,20βP-glucuronide; 17,20β,21-trihydroxy-4-pregnen-3-one, 17,20β,21P) that are likely to augment the action of 17,20βP, 17,20βP-S and AD. Furthermore, it is important to appreciate that, because synthesis and release rates of the steroid odorants change individually during final maturation, the pre-ovulatory pheromone is a shifting mixture in which AD, 17,20βP, and then 17,20βP-S reach peak release rates during the early, mid and late portions of the LH surge, respectively (Stacey *et al.*, 1989; Scott and Sorensen, 1994; Sorensen and Scott, 1994).

If tested in all-male groups, males display agonistic behavior when exposed to the low 17,20βP:AD ratios that females release early in their LH surge, but exhibit prolonged (up to 12 hours) inspection and following behavior in response to the higher 17,20βP:AD ratios released nearer to ovulation (Poling *et al.*, 2001). These higher 17,20βP:AD ratios also increase male LH, leading to plasma steroid changes that are likely to be responsible for the increased milt production and enhanced paternity observed at spawning (Dulka *et al.*, 1987a; DeFraipont and Sorensen, 1993; Zheng *et al.*, 1997). In response to the significant quantities of 17,20βP-S that females release late in the LH surge, males display brief but intense bouts of chasing and following and increased plasma LH that, as with 17,20βP, increase milt with several hours (Sorensen *et al.*, 1995a; Poling *et al.*, 2001).

Although the endocrine and behavioral effects of the pre-ovulatory steroid pheromone are dramatic and well documented, much less is known of the mechanisms mediating these responses. The unconjugated components of the pheromone (AD and 17,20βP) are released almost exclusively across the gills, and estimated on the basis of EOG studies to be detectable only in the immediate vicinity of the female. In contrast, 17,20βP-S is released in controlled pulses of urine, and should be detectable at greater distances (Sorensen *et al.*, 2000). Interestingly, however, males evidently can integrate these gill and urine odors that normally would be spatially separated, because AD suppresses behavioral responses to 17,20βP-S, and 17,20βP and 17,20βP-S will synergize each other's actions (P. W. Sorensen, in preparation). 17,20βP appears to increase LH by reducing tonic dopaminergic inhibition (Zheng and Stacey, 1997) but nothing is known of the mechanisms underlying either 17,20βP-S induced LH increase, nor has it been investigated. Water-borne AD inhibits the endocrine-gonadal response to 17,20βP (Stacey, 1991), but neither the mechanism nor the biological function of this effect is known. Finally, although microsatellite paternity studies show that 17,20βP exposure dramatically increases paternity during competitive spawning, it is unclear whether this effect is due to increases in competitive behaviors, sperm motility, or sperm release, all of which occur in response to 17,20βP exposure (DeFraipont and Sorensen, 1993; Zheng *et al.*, 1997; D. Hoysak and N. E. Stacey, unpublished results). However, the fact that 17,20βP also induces similar effects on paternity in competitive *in vitro* fertilization (Zheng *et al.*, 1997) suggests that increased sperm quality is a major component of the pheromone's effect.

As noted above, there is some evidence that the odor of pre-ovulatory females affects females as well as males: exposure to 17,20βP increases the incidence of ovulation (Kobayashi *et al.*, 2002). Since male goldfish release only small quantities of 17,20βP and related C21 odorants (Sorensen and Scott, 1994; Sorensen *et al.*, 2005), these results suggest that 17,20βP release by females undergoing final oocyte maturation might mediate the synchronous ovulations observed in goldfish held in the laboratory (Kobayashi *et al.*, 1988), in cultured common carp (see Stacey *et al.*, 1994) and in crucian carp under natural conditions (N. E. Stacey, personal observations). Such ovulatory synchrony might serve as a predator swamping strategy, or might reduce the potentially disruptive effects on spawning of a high male:female sex ratio.

Female post-ovulatory prostaglandin pheromone

When ovulation occurs in the latter half of scotophase, movement of eggs into the oviduct stimulates the synthesis of prostaglandin $F_{2\alpha}$ ($PGF_{2\alpha}$) that rapidly exerts two critical and simultaneous effects which link male and female sexual behaviors to the presence of ovulated eggs (Stacey and Liley, 1974; Sorensen *et al.*, 1988, 1995b). First, plasma $PGF_{2\alpha}$ functions as a hormone that acts within minutes in the brain to stimulate female spawning behaviors; spawning of ovulated fish is blocked by the prostaglandin (PG) synthetase inhibitor indomethacin, and restored in indomethacin-treated fish, or induced in non-ovulated fish, by $PGF_{2\alpha}$ injection (Stacey, 1976, 1981). At the same time, $PGF_{2\alpha}$ and a metabolite, 15-keto-prostaglandin $F_{2\alpha}$ ($15K\text{-}PGF_{2\alpha}$), are released in the urine to function as a post-ovulatory pheromone that induces both releaser effects on male courtship and primer effects that further increase milt production (Sorensen *et al.*, 1988, 1989; Zheng and Stacey 1996, 1997; Fig. 6.1).

EOG and aquarium studies show that although $PGF_{2\alpha}$ and $15K\text{-}PGF_{2\alpha}$ are detected by separate olfactory receptor mechanisms at nM and pM thresholds, respectively, they appear to induce similar effects on male behavior and physiology. Crucian and common carp also may use PGFs as female post-ovulatory pheromones, because EOG studies show that these species detect PGFs with approximately the same sensitivities and specificities as do goldfish (Bjerselius and Olsén, 1993; Irvine and Sorensen, 1993). In goldfish, $PGF_{2\alpha}$ is released both in urine and across the gills, whereas $15K\text{-}PGF_{2\alpha}$ appears to be released almost exclusively in urine pulses, whose frequency changes during spawning (Appelt *et al.*, 1995; Appelt and Sorensen, 1999). Based on studies of PGF release from ovulated goldfish, EOG detection thresholds and urinary pulse frequency, it is estimated that pheromonal $PGF_{2\alpha}$ would be detectable only in the immediate vicinity of the female, whereas pheromonal $15K\text{-}PGF_{2\alpha}$, which is released at rates as high as $100 \text{ ng} \cdot \text{h}^{-1}$, should create 15 active spaces more than 50 liters in volume during each hour of spawning (Sorensen *et al.*, 2000).

Although $PGF_{2\alpha}$ and 15-keto-$PGF_{2\alpha}$ induce courtship and LH increase in males that appear qualitatively similar to the responses induced by the pre-ovulatory steroid pheromone, the modes of action of the pre-ovulatory and post-ovulatory pheromones are very different. For example, 17,20βP increases LH and milt in isolated fish (Sorensen *et al.*, 1989; Fraser and Stacey, 2002). However, PGFs do not increase LH in isolated males (Sorensen *et al.*, 1989), but do increase LH and/or milt when a male or males spawn with a $PGF_{2\alpha}$-injected female that would be releasing pheromonal PGFs (Kyle *et al.*, 1985; Zheng and Stacey, 1996), or when the males are exposed to PGFs in the presence of other males (Sorensen *et al.*, 1989). Although this is a complex issue that requires further study (particularly because the pre-ovulatory pheromonal steroids AD and 17,20βP-S have not been examined in this context), the simplest interpretation appears to be that pheromonal 17,20βP acts directly on individual males to trigger a neuroendocrine reflex, whereas PGFs induce their endocrine effects indirectly through the socio-sexual behaviors they trigger among conspecifics. Consequently, studies of the effects induced by the pre-ovulatory steroid pheromone have typically involved the simple addition of steroids to aquarium water, whereas the effects of the post-ovulatory prostaglandin pheromone have been assessed either by placing males with non-ovulated females that have been injected with $PGF_{2\alpha}$ or by exposing groups of males to PGFs.

The pre-ovulatory and post-ovulatory hormonal pheromones of goldfish differ not only in exerting their effects directly or indirectly, but also in the physiological mechanisms through which they induce endocrine and testicular responses in males. Specifically, the two pheromones act through distinct mechanisms to increase LH (Zheng and Stacey, 1997), and 17,20βP appears to increase milt only by stimulating LH secretion, whereas pheromonal PGFs can also increase milt through a more rapid, extra-pituitary pathway (Zheng and Stacey, 1996).

LH release in goldfish is controlled by hypothalamic neurons that innervate the gonadotropes to stimulate LH release through gonadotropin-releasing hormone or inhibit LH release through dopamine (DA), which acts through DA type-2 (D2) receptors (see Trudeau, 1997). The pre-ovulatory pheromone appears to increase LH by reducing tonic DA inhibition on gonadotropes, whereas the post-ovulatory pheromone does not. For example, D2 receptor agonists such as bromocryptine and LY171555 block 17,20βP, induced LH and milt increase, but do not affect these responses to the post-ovulatory pheromone (Zheng and Stacey, 1997). Consistent with these results, the ratio of dihydroxyphenylacetic acid (DOPAC; the primary DA metabolite in goldfish)/DA in the pituitary decreases within 20 min. of pheromone exposure (Dulka *et al.*, 1992). These findings of distinct pheromonal mechanisms are in accord with the observations that, even though supra-threshold concentrations of 17,20βP induce no further increase in LH (Dulka *et al.*, 1987a), further LH increase does occur if males are also exposed to the post-ovulatory pheromone (Sorensen *et al.*, 1989). As well, 17,20βP exposure at any time of the day induces equivalent LH increase, whereas exposure to the post-ovulatory pheromone through spawning with a $PGF_{2\alpha}$-injected female PG pheromone is most effective at night (Dulka et al., 1987b).

The clearest evidence that only the post-ovulatory pheromone increases milt through an extra-pituitary mechanism is that hypophysectomy blocks the milt increase normally induced by $17,20\beta P$ exposure, but does not block milt increase induced by spawning with a $PGF_{2\alpha}$-injected female (Dulka et al., 1987a; Zheng and Stacey, 1996). Furthermore, although the two pheromones can increase LH within 15-20 min (Kyle et al., 1985; Dulka et al., 1987a), the latency of the spawning-induced milt increase ($<$1h; Kyle et al., 1985) is markedly shorter than that of the $17,20\beta P$-induced milt release (Dulka et al., 1987a). The spawning (PG)-induced milt is also not delayed at lower water temperatures as the case with the milt increase mediated by exposure to the $17,20\beta P$ pheromone.

In summary, the peri-ovulatory hormonal pheromones of female goldfish induce in males a complex suite of responses that presumably have evolved as critical components of sperm competition. Behavioral responses first alert males to the presence of impending ovulation and then enable them to locate the ovulated female(s) within a spawning group. At the same time, physiological responses first activate an endocrine mechanism that increases both the quantity and quality of releasable sperm in anticipation of imminent spawning opportunities, and then activate both endocrine and apparently non-endocrine mechanisms that likely function to replenish sperm stores during spawning.

The male pheromone

Male goldfish also release a hormonal sex pheromone that is comprised at least partially of AD and that evidently functions to reduce milt (Stacey, 1991; Fraser and Stacey, 2002) and to induce agonistic behaviors (Poling et al., 2001) in other males. For example, it now is clear that mature males normally release large quantities of AD (50 ng $\cdot$ h^{-1}), and that this release can increase tenfold during spawning (Sorensen et al., 2005). Males are likely to use mixture composition to identify the male AD cue that, unlike the AD released by females (Scott and Sorensen, 1994), is released in the virtual absence of C21 steroids (Sorensen et al., 2005). Exposure to AD induces agonistic behavior among males (Poling et al., 2001) that presumably would normally mediate male-male recognition and consequent behavioral interactions. Evidence that a male AD cue inhibits milt comes from studies comparing grouped and isolated males. Thus, in the presence of males with low LH, male goldfish appear to inhibit their potential milt production, because they dramatically increase milt volume if isolated (Fraser and Stacey, 2002). These milt responses to withdrawal of male cues are distinct from those induced by female pre-ovulatory and post-ovulatory pheromones because they are not accompanied by LH increase and have longer latencies (Stacey et al., 2001; Fraser and Stacey, 2002).

Despite this evidence that males release AD and increase and decrease milt production in response to isolation and AD exposure, respectively, we have been unable to reduce milt by exposing isolated males to male odor (Fraser and Stacey, 2002). These results seem to suggest that either the male AD pheromone might be effective only at an extremely short range, or that it must be perceived in conjunction with other male cues.

In addition to reducing milt production in the presence of mature males with low LH, male goldfish also increase milt in the presence of a gonadotropin-injected or $17,20\beta$P-exposed male (Stacey et al., 2001; Fraser and Stacey, 2002). Although there is no information as to what chemical cue(s) might be released by stimulated males with elevated LH, these results suggest that an ovulatory female not only induces a male's milt increase directly by exposing him to her pre-ovulatory steroid pheromone, but also exerts an indirect effect through the males in which she induces LH increase. If true, then selective pressures of sperm competition appear to have led to mechanisms enabling a male goldfish to increase both his absolute fertility, by responding to cues from ovulatory females, and his relative fertility, by responding to cues from male competitors.

To summarize, we feel it appropriate to view mature goldfish as members of a complex network that uses hormonal pheromones to co-ordinate multiple functions within and between genders. Prior to ovulation and spawning, males are attracted to females by an unknown recrudescent pheromone and are both the source and receivers of inhibitory cues that suppress milt production. However, when environmental cues trigger a female ovulatory LH surge, the tonic reciprocal inhibition among males is briefly transformed into a positive feedback system in which the release of the pre-ovulatory pheromone induces LH increase in both males and other females, amplifying and disseminating the original stimulus and potentially promoting synchronous final maturation among local conspecifics. Although we assume that the evolution of this chemical network has been driven primarily by sperm competition, it is premature to exclude the possibility that interactions among females have also played a role.

Other Cypriniforms

Evidence from EOG and behavioral and endocrine bioassay studies suggest that, because the components and functions of the goldfish hormonal pheromone system appear to be remarkably similar to the pheromones of closely related crucian and common carps (Bjerselius and Olsén, 1993; Irvine and Sorensen, 1993; Stacey *et al.*, 1994; Bjerselius *et al.*, 1995), they also are likely to be similar to pheromones of other species within the sub-tribe Cyprini (see Rainboth, 1991) which have not been examined. As well, because all of more than 80 cypriniform species that have been examined in EOG screening detect some form of F prostaglandin at nM concentrations, it is possible that a post-ovulatory PGF pheromone similar to that in the goldfish and related carps might be ubiquitous among the cypriniforms. Although a hormonal role for $PGF_{2\alpha}$ in inducing female spawning behavior has been demonstrated in only a few cypriniform species (e.g., Liley and Tan, 1985; Cardwell *et al.*, 1995), such a role for PGs might be expected to be widespread given that it has also been demonstrated in such distantly related perciform taxa as a cichlid (*Cichlasoma bimaculatum*, Family Cichlidae; Cole and Stacey, 1984) and the paradise fish (*Macropodus opercularis*, Family Belontiidae; Villars *et al.*, 1985). However, the only cypriniform other than goldfish and related carps (Superfamily Cyprinoidea) in which a pheromonal role for PGFs has been demonstrated is a loach (*Misgurnus anguillicaudatus*, Family Cobitidae, Superfamily Cobitoidea).

In *Misgurnus*, as in goldfish (Kobayashi *et al.*, 2002), PGFs function as a potent post-ovulatory pheromone that stimulates male sexual behaviors. In the presence of either ovulated females, or non-ovulated females injected with $PGF_{2\alpha}$, 15-keto-$PGF_{2\alpha}$, or 13,14-dihydro-15-keto-$PGF_{2\alpha}$, males display a similar suite of courtship and spawning behaviors that are virtually abolished if the nares are occluded (Kitamura *et al.*, 1994a).

When injected with hCG, female *Misgurnus* release negligible amounts of PGFs for up to 8 h, but soon after ovulation they release very large quantities of 13,14-dihydro-15-keto-$PGF_{2\alpha}$ (>2 $\mu g \cdot h^{-1}$), smaller quantities of $PGF_{2\alpha}$ (~ 10 $ng \cdot h^{-1}$), and virtually no 15-keto-$PGF_{2\alpha}$ (Ogata *et al.*, 1994). These release data are consistent with EOG studies (Kitamura *et al.*, 1994a), showing that male loach are more sensitive to 13,14-dihydro-15-keto-$PGF_{2\alpha}$ (1 pM detection threshold) than to $PGF_{2\alpha}$ or 15-keto-$PGF_{2\alpha}$ (1 nM detection threshold); it is not known if these three PGFs are detected by the same olfactory receptor mechanism.

In contrast to the apparently ubiquitous detection of PGFs among cypriniforms, EOG screening studies show that detection of steroids is restricted to certain higher taxa, and that the specific compounds detected appear to be related to phylogeny. Most striking is the fact that of the 11 species tested from the Superfamily Cobitoidea (from all four cobitoid families: Gyrinocheilidae, Balitoidae, Homalopteridae and Catostomidae; Nelson, 1994), none detected any steroids. Steroids also were not detected by any of the five bitterling species (from the cyprinid subfamily Acheilognathinae, Superfamily Cyprinoidea), but were detected by at least one species in six other cyprinid subfamilies that have been examined. In subfamily Cyprininae, where data are sufficient to discern patterns of detection, Tribe Labeonini, which appears to detect primarily 17,20βP-S and related C21 sulfates, is distinctly different from Tribes Systomini, Cyprinini and Squaliobarbini, which detect a variety of free and conjugated C21s, as well as C19s. It is important to understand that these EOG screening studies are based on small numbers of individuals from only a fraction of extant cypriniforms, and that they have been unable to test all PGs and steroids that cypriniforms release and thus might detect. Nonetheless, these studies clearly demonstrate that detection of hormones is widespread among cypriniforms and that the specific compounds detected vary considerably from those detected in the goldfish and related carps. Moreover, the striking correlation between phylogeny and the pattern of detected compounds suggests that intensive study of a few well-chosen model species might provide a broad insight into the hormonal pheromone function of cypriniforms.

ORDER SALMONIFORMES

The salmoniform fishes, currently comprised only of family Salmonidae (Nelson, 1994), are divided into three subfamilies: Thymallinae (grayling), Coregoninae (whitefish), and Salmoninae (salmon and trout). Hormonal pheromones appear to have been studied only in the latter two taxa, with the great majority of the work being done on the salmonins. Indeed, the only information on hormonal pheromones in coregonids comes from the lake whitefish (*Coregonus clupeaformis*), which exhibits sensitive (10 nM threshold) EOG responses to 15K-PGF$_{2\alpha}$ and 13,14-dihydro-PGF$_{2\alpha}$, which act through a common receptor mechanism and induce larger EOG responses in males than in females (Hara and Zhang, 1997; LaBerge and Hara, 2003). PGFs also increase locomotory behavior in response to water-borne PGFs, although LaBerge and Hara (2003) do not report whether males and females exhibit equivalent behavioral responses, and the source of pheromonal PGF in whitefish is unknown.

Genus *Salmo*: Atlantic Salmon and Brown Trout

Most hormonal pheromone studies in Subfamily Salmoninae have been conducted on the genus *Salmo*, although *Oncorhynchus* and *Salvelinus* species also have been investigated. Hormonal pheromones of Atlantic salmon and the brown trout appear to be very similar which has been suggested as one reason for their ability to hybridize (Youngson *et al.*, 1992; Olsén *et al.*, 2000). In Atlantic salmon, hormonal pheromone studies have focused on the precociously mature male parr as a convenient laboratory model. These studies have shown that male parr detect both PGFs and reproductive steroids including testosterone (T) and 17,20βP-S, although the possible pheromonal roles of these steroids are unclear. For example, EOG recordings (Moore and Scott, 1991) show that precocious males can be extremely sensitive to T, but that this occurs for only a brief period prior to spawning, a situation not reported in any other species, including brown trout. T also stimulates positive rheotaxis in male parr (Moore, 1991), but does not affect hormone concentrations or milt volume (Waring *et al.*, 1996). EOG studies (Moore and Scott, 1992) also indicate that although male parr do not normally detect 17,20βP-S, even brief (5 sec.) exposure to urine from ovulated females immediately induces extreme sensitivity to this steroid, a rapid sensitization of the olfactory epithelium not reported in any other vertebrate. Although attempts to demonstrate a reproductive effect of water-borne 17,20βP-S in male Atlantic parr have so far been unsuccessful (Waring and Moore, 1995; Waring *et al.*, 1996), it will be important to determine if this sensitization phenomenon can be replicated and, if so, the underlying mechanisms responsible.

Unlike the situation with steroidal pheromones, there is clear evidence that mature male parr are acutely sensitive to PGFs ($PGF_{1\alpha}$ and $PGF_{2\alpha}$), which, as in goldfish (Kobayashi *et al.*, 2002), induce primer endocrine effects that include increases in plasma LH, steroids, and/or milt (Moore and Waring, 1995, 1996; Olsén *et al.*, 2001). Also, a carbamate pesticide that reduces EOG responses to $PGF_{2\alpha}$ blocks $PGF_{2\alpha}$-induced endocrine responses (Waring and Moore, 1997). Despite abundant evidence for reproductive responses to water-borne PGFs, however, the origin(s) and function(s) of the putative pheromone remain unclear. For example, urine from ovulated Atlantic salmon has been proposed as the source of pheromonal PGFs because urine and PGFs can exert similar priming effects, and because ovulated urine contains significant immunoreactive $PGF_{2\alpha}$ (Moore and Waring, 1996). However, Olsén *et al.* (2001, 2002) report that although ovulated salmon urine induces strong releaser effects (which they attribute to unidentified pheromones), it has relatively weak primer effects; instead, they propose that ovarian fluid, which has much higher $PGF_{2\alpha}$ than does ovulated urine, is the major source of the priming pheromone. These apparent inconsistencies should be clarified if definitive biochemical techniques are used in future studies to determine what specific PG compounds are released and by which route(s).

Studies of the hormonal pheromones of brown trout suggest similarities to what has been reported for Atlantic salmon, as might be expected from congeners. For example, as in Atlantic salmon, the odor of ovulated brown trout induces endocrine-priming effects in males, and there is evidence that these effects are due to PGFs. Thus, EOG studies not only show that brown trout are acutely sensitive to PGFs but also are in general agreement both as to the relative olfactory potencies of PGFs in male trout (e.g., $PGF_{2\alpha} \geq PGF_{1\alpha} > 15K\text{-}PGF_{2\alpha} > 13,14\text{-}dihydro\text{-}15K\text{-}PGF_{2\alpha}$), and as to the likelihood that all detected PGFs act through a single olfactory receptor mechanism (Essington and Sorensen, 1996; Moore and Waring, 1996; Hara and Zhang, 1997; Moore *et al.*, 2002; Laberge and Hara, 2003). As in Atlantic salmon (Olsén *et al.*, 2001, 2002), immunoreactive $PGF_{2\alpha}$ is in much greater quantities in ovulatory fluid (>175 ng $\cdot$ ml^{-1}) than in urine (<5 ng $\cdot$ ml^{-1}), both of which are able to increase plasma hormones and/or milt when added to aquarium water (Olsén *et al.*, 2000; Moore *et al.*, 2002). Laberge and Hara (2003) report that PGFs induce similar EOG responses not only in mature males and females but also in undifferentiated juveniles; these findings are consistent with those of Essington and Sorensen (1996) in brown trout and of Sveinsson and Hara (2000) in arctic charr. Thus, it is surprising that Moore *et al.* (2002) report that PGF does not induce EOG response in immature male parr, even at extreme concentrations (100 nM), results that are consistent with a report of increased sensitivity to PGFs in Atlantic salmon (Moore and Waring, 1995). However, it seems clear that pheromonal PGFs in brown trout induce both releaser effects (increased locomotory activity in mature males and females and induced digging and nest probing in females; Laberge and Hara, 2003) and primer endocrine effects (e.g., Moore *et al.*, 2002), and that it will be important in future studies to determine the specific PGFs involved, their origins and routes of release, and their specific pheromonal functions. It will also be important for future studies to confirm reports (Essington and Sorensen, 1996; Hara and Zhang, 1997; Laberge and Hara, 2003) that brown trout do not detect steroids (T and 17,20βP) that have been proposed to be pheromones in Atlantic salmon; such apparently striking differences between congeners is markedly different than what has been observed in cypriniforms, as discussed above.

Genus *Salvelinus* (charrs)

EOG studies, using a smaller number of PGs and steroids than have been used in cypriniforms, indicate that charrs commonly detect PGFs but that, except for T glucuronide, which is detected in brook charr (Essington and Sorensen, 1996) and etiocholanolone glucuronide, which is detected by all salmonids that have been studied, they do not detect steroids (Essington and Sorensen 1996; Hara and Zhang, 1997; Laberge and Hara, 2003). Neither the stage of maturity nor gender dramatically affects EOG responsiveness to $PGF_{2\alpha}$ in Arctic charr (Sveinsson and Hara, 2000) or brook charr, *S. fontinalis* (Essington and Sorensen, 1996). This is consistent with one study of brown trout (Laberge and Hara, 2003), but is in marked contrast to other studies of brown trout (Moore *et al.*, 2002) and Atlantic salmon parr (Moore and Waring, 1995)

Based on their observations that adult male Arctic charr release immunoreactive PGFs, and that ovulated Arctic charr are attracted to $PGF_{2\alpha}$, Sveinsson and Hara (2000) propose that PGFs in this species function as a male hormone affecting females. If true, this pheromonal function is surprisingly different than that proposed for *Salmo*, where post-ovulatory PGFs are proposed to affect males (e.g., Moore *et al.*, 2002; Olsén *et al.*, 2002), particularly considering that brook charr and brown trout hybridize (Sorensen *et al.*, 1995c).

Pheromonal studies in charrs also provide evidence that stream-resident individuals release bile acids that serve as pheromonal attractants for maturing adults, a pheromonal function that could clearly be adaptive in species with anadromous life cycles (Døving *et al.*, 1980; Selset and Døving, 1980; Jones and Hara, 1985; Zhang *et al.*, 2001). However, currently, there is no information on the specific bile acids that charr release or the specific functions these putative reproductive pheromones perform.

Genus *Oncorhynchus* (Pacific salmon)

Ovulatory fluid has long been known to attract males of several *Oncorhynchus* species (Emmanuel and Dodson, 1979; Honda, 1980,

1982a). As well, male rainbow trout (*Oncorhynchus mykiss*), exhibit primer endocrine responses to bile fluid (containing both bile and sex steroid conjugates; Vermeirssen and Scott, 2001) and urine from ovulated females (Scott *et al.*, 1994; Vermeirssen *et al.*, 1997). Behavioral responses to urinary pheromone(s) are androgen-dependent not only in male rainbow trout (Yambe and Yamazaki, 2001) but also in masu salmon (*O. masou*) (Yambe *et al.*, 1999, 2003; Yambe and Yamazaki, 2000). Although female *Oncorhynchus* are known to release sex pheromones with well-documented effects on males, the chemical identities of these pheromones are unknown. However, *Oncorhynchus* pheromones appear to be rather different than those of *Salmo* and *Salvelinus* insofar as there is little evidence that *Oncorhynchus* detect either PGs or steroids other than etiocholanolone glucuronide.

Despite one EOG study (Pottinger and Moore, 1997) reporting that immature rainbow trout detect T, and a behavioral study (Dittman and Quinn, 1994) in which chinook salmon (*O. tshawytscha*) avoided 17,20βP, all other studies of *Oncorhynchus* species indicate that rainbow trout, chinook and amago salmon (*O. rhodurus*) do not detect a range of PGs and steroids including T and 17,20βP (Kitamura *et al.*, 1994b; Hara and Zhang, 1997; Laberge and Hara, 2003; A. H. Dittman and P. W. Sorensen, unpublished results). Furthermore, although urine from ovulated masu salmon contains a modest concentration of immunoreactive $PGF_{2\alpha}$, the urinary pheromone in this species does not have chemical characteristics of known PGs or sex steroids (Yambe *et al.*, 1999). In summary, the available information on reproductive pheromones in *Oncorhynchus* suggests either that this group employs novel prostaglandins or other compounds (Yambe *et al.*, 1999), or that *Oncorhynchus*, which is proposed to be more derived than either *Salvelinus* or *Salmo* (Stearley and Smith, 1993), may have lost the use of PGFs (Laberge and Hara, 2003).

To summarize, current information on salmonid hormonal pheromones indicates the widespread occurrence of PGF pheromones that induce both releaser and primer effects, and may operate in more than one gender combination. In contrast, evidence for steroidal pheromones is limited to the Atlantic salmon. Given that salmonid life histories often include diandry (typically involving small and precocious parasitic males competing for females that are monopolized by large, territorial *bourgeois* males; see Taborsky, 2001), it is hoped that future studies will determine whether hormonal pheromone functions differ in parasitic and bourgeois salmonids, as will be discussed below in the black goby (Locatello *et al.*, 2002).

ORDER PERCIFORMES

The ecologically and economically important perciform fishes (more than 9,000 species comprising 40 per cent of living teleosts) are the dominant vertebrate group in oceans and many tropical and subtropical freshwaters (Nelson, 1994). Although the first example of a fish hormonal pheromone came from a perciform, the black goby (Colombo et al., 1980), there has since been relatively little additional information on hormonal pheromones in this important taxon. Most of the available information comes from studies of the black and round gobies (family Gobiidae) that describe olfactory sensitivity and behavioral responses to water-borne sex steroids. Additionally, EOG studies in an African cichlid (Haplochromis burtoni; family Cichlidae) show that this fish detects and discriminates among a number of conjugated sex steroids (Robison et al., 1998; Cole and Stacey, 2003, 2006), while EOG and behavioral studies of the ruffe (Gymnocephalus cernuus; family Percidae: Sorensen et al., 2004) suggest that it employs novel and unidentified steroids as a female pheromone.

The Black and Round Gobies

The mating systems of gobies typically involve male territoriality and paternal behaviors, as well as alternative male tactics in which parasitic males attempt to fertilize eggs in the nest of bourgeois males (e.g., Miller, 1984; Rasotto and Mazzoldi, 2002; Immler et al., 2004). Although it has been known for many years that ovulated gobies release an unidentified pheromone that triggers male behavior (Tavolga, 1956), hormonal pheromone studies in gobies have focused on steroid pheromones that seem to attract ovulated females to the male's nest (Colombo et al., 1980, 1982; Murphy et al., 2001; Zielinski et al., 2003; Bélanger et al., 2004). In laboratory studies of the Mediterranean black goby (Gobius niger), ovulated females are attracted to, and will oviposit in, benthic shelters releasing etiocholanolone glucuronide, suggesting that either this or similar steroids also attract females to the male's nest under natural situations (Colombo et al., 1980). Etiocholanolone glucuronide is a major steroid product of the mesorchial gland, a specialized, non-spermatogenic and Leydig cell-rich portion of the testis commonly found among the gobies (Locatello et al., 2002; Arbuckle et al., 2005). Although there is no direct evidence that the olfactory system of the black goby detects etiocholanolone glucuronide, EOG studies (Murphy et al., 2001) of the round goby show that it detects this and other steroid odorants.

The round goby, a native of the Ponto-Caspian area of eastern Europe, appears to have been recently introduced in ship ballast water to the Laurentian Great Lakes (Charlebois *et al.*, 2001), from which it also has the potential to invade the Mississippi drainage (Corkum *et al.*, 2004). Because the round goby eats the eggs and young of Great Lakes fishes, and has been linked with declines in some native species (see Corkum *et al.*, 2004), there is current interest in controlling or eliminating this invasive species. Indeed, the fact that female round gobies use male odor to locate nest sites appears to provide the opportunity to employ pheromonal manipulation as a means of biological control, an approach that is also being explored to control invasive sea lamprey (*Petromyzon marinus*) and carps (Twohey *et al.*, 2003; Sorensen and Stacey, 2004).

Evidence that nesting male round gobies release a pheromone attracting females comes from recent studies showing that vitellogenic females, but not nonvitellogenic females, are attracted to the odor of mature males, and exhibit larger EOG responses to the odor (either whole odor or C-18 extracts) of reproductively active males than to the odor of reproductively inactive males (Bélanger *et al.*, 2004). Although the nature of the putative pheromonal component(s) of male odor has not been determined, EOG recordings using a large number of sex steroids and prostaglandins suggest that round gobies do not detect prostaglandins, but that mature males and females exhibit equivalent olfactory responses to more than a dozen free and conjugated C18-, C19- and C21-steroids including etiocholanolone glucuronide. Furthermore, EOG cross-adaptation tests indicate the detected steroids act through four olfactory receptor mechanisms for which the most potent known odorants are estrone, estradiol-3β-glucuronide, etiocholanolone, and dehydroepiandrosterone-3-sulphate (Murphy *et al.*, 2001). Although it is not yet known whether any of these putative pheromonal steroids are released by round gobies, recent *in vitro* studies (Arbuckle *et al.*, 2005) show that the round goby testis synthesizes etiocholanolone as well as a variety of conjugated and unconjugated 5β-reduced-11-oxo-androgens that are yet to be tested for olfactory activity.

Although round gobies are not known to exhibit overt reproductive behaviors in response to steroid odorants, they do increase the frequency of ventilation (opercular and buccal pumping), which likely facilitates odor detection by increasing water flow through the olfactory organ, as has been proposed in other benthic fish (Nevitt, 1991; Belanger *et al.*, 2003). Male round gobies increase their ventilation rate when exposed to etiocholanolone, estrone, and estradiol-3β-glucuronide, whereas females increase ventilation only in response to etiocholanolone (Murphy *et al.*, 2001). This behavioral dimorphism appears to be induced by plasma androgen in males, because females display male-typical responsiveness to estrone and estradiol-3β-glucuronide within two weeks of an androgen implant (Murphy and Stacey, 2002). Also, because ventilation responses adapt rapidly (typically within 15 min.), they provide a simple behavioral assay to show that the steroids indicated by EOG cross-adaptation studies to be discriminated at the level of the sensory epithelium also are discriminated behaviorally (Murphy *et al.*, 2001).

Although it is not known what specific reproductive effects might be induced by the putative steroidal pheromones of round gobies, the nature of the detected compounds suggests more than one function. Given that females both detect and respond behaviorally to etiocholanolone, it seems reasonable to assume that, as has been proposed in the black goby (Colombo *et al.*, 1980), this and related 5β-reduced androgens acting through the same olfactory receptor mechanism (Murphy *et al.*, 2001) are released by nest-holding males to attract ovulated females. However, because etiocholanolone induces behavioral response in males (Murphy *et al.*, 2001), this steroid might also play some role in male-male interactions such as those discussed below in the black goby. Similarly, we suspect that detection of C18 steroids by round gobies is indicative of a male-typical response to the odor of females (e.g., Tavolga, 1956), because behavioral response to these compounds normally is seen only in males (Murphy *et al.*, 2001) but can be induced in females by androgen implant (Murphy and Stacey, 2002), and because goby testes appear not to synthesize C18 steroids (Colombo *et al.*, 1977; Asahina *et al.*, 1989; Arbuckle *et al.*, 2005). Furthermore, given the evidence for separate olfactory receptor mechanisms for free (estrone and E2) and conjugated C18 steroids (estradiol-3β-glucuronide) in round gobies, and the evidence that free and conjugated steroids of fish are released across the gill and in the urine/bile, respectively (Scott and Vermeirssen, 1994; Vermeirssen and Scott, 1996; Sorensen *et al.*, 2000), it is interesting to speculate that, as discussed above for free and sulfated 17,20βP in goldfish, female round gobies might release two distinct estrogenic pheromones.

In the black goby, where males employ alternate reproductive tactics (Mazzoldi and Rasotto, 2002; Rasotto and Mazzoldi, 2002), there is evidence of distinct odors for bourgeois and parasitic males. Thus, bourgeois male odor induces aggression in other bourgeois males, whereas parasitic male odor does not (Locatello *et al.*, 2002). Although it is not known what chemical differences in the two odors are responsible for their differential effects, the much larger mesorchial glands of bourgeois males might suggest differential release of etiocholanolone or other steroids (Rasotto and Mazzoldi, 2002). From the progress made in recent years, the gobies appear likely to become increasingly valuable models for hormonal pheromone research given the fact that they are speciose, widely available, and well suited to laboratory studies.

The Eurasian Ruffe

As with the round goby, the recent introduction of the ruffe (Family Percidae) to the Laurentian Great Lakes, and the consequent detrimental effects on native fauna (Gunderson *et al.*, 1998), have prompted investigations to determine the feasibility of utilizing pheromones to control it. The resulting information on alarm (Maniak *et al.*, 2000) and female reproductive pheromones (Sorensen *et al.*, 2004) already have made the pheromones of ruffe better understood than those of any other percid. As with many percids, ruffe are a non-territorial and sexually isomorphic species that engage in a single, nocturnal spring spawning and, as with the goldfish, employ a promiscuous mating system in which males engage in scramble competition.

Bioassays using ruffe holding water show that females undergoing final oocyte maturation release a pheromone that stimulates male reproductive activities (chasing and nudging; Sorensen *et al.*, 2004) and is temporally correlated with circulating 17,20β,21P (also known as '20β-S'), the proposed maturation-inducing steroid for this and other perciforms (Thomas, 2003). Although EOG studies (Sorensen *et al.*, 2004) have shown that ruffe do not detect 17,20β,21P, or any other tested steroids and prostaglandins, injecting female ruffe with 17,20β,21P causes rapid pheromone release in the urine, an effect not seen following 17,20βP injection of females or 17,20β,21P injection of males.

These results suggest that the pheromone(s) released by female ruffe undergoing final maturation is an unknown metabolite of 17,20β,21P that might play a role in formation of the aggregations that immediately precede spawning in this species (Sorensen *et al.*, 2004). If so, an important lesson from this study is that we should be prepared to find that the steroid, prostaglandin, and bile acid pheromones of many fish are novel metabolites that cannot be elucidated by simple EOG or immunoassay studies alone. Although it is difficult to assess the degree to which this possibility already has impacted research on hormonal pheromones, one likely example comes from a study of the Mozambique tilapia (*Oreochromis mossambicus*), in which the presence of pheromonal steroid sulfates was suggested by EOG responses to fractionated body odors, but in which all tested steroids lacked olfactory activity (Frade *et al.*, 2002). Here, as in many aspects of fish pheromone research, progress undoubtedly is being impeded by our poor understanding of hormone metabolism.

HORMONAL REGULATION OF HORMONAL PHEROMONE FUNCTIONS

The reproductive endocrine system functions not only as the source of hormonal odorants released by the pheromone sender but also as a regulator of pheromonal responsiveness in conspecifics. In the pheromone sender, research has focused on the direct effect of endocrine activity on pheromone production by correlating changes in plasma hormones with changes in hormonal pheromone release (e.g., Stacey *et al.*, 1989; Scott and Sorensen, 1994; Sorensen and Scott, 1994). However, it should also be expected that the sender's hormones exert indirect effects on pheromone function; for example, androgens that regulate a male's visual (Borg, 1994) or acoustic signals (Remage-Healey and Bass, 2004) could enhance his pheromonal functions by attracting females to the active space of his pheromones. Similarly, hormones of pheromone receivers might exert indirect effects by inducing behaviors that increase the likelihood of pheromone encounter. Although this possibility has not been explored, there is evidence that the reproductive hormones of receiving individuals regulate their ability to respond to hormonal pheromones. In the Indian barb (*Barilius bendelisis*), for example, androgen treatment enhances behavioral response to 15K-PGF$_{2\alpha}$ and an ovarian extract believed to contain steroid sulfates (Bhatt and Sajwan, 2001; Bhatt *et al.*, 2002). Also, in masu salmon, male parr develop behavioral attraction to ovulated urine as they undergo precocious maturity (Yambe *et al.*, 1999); androgen appears to mediate not only this behavioral change (Yambe and Yamazaki, 2001) but also a steroidal primer response to urine (Yambe *et al.*, 2003). Although it is not known where or how androgen treatments mediate behavioral and endocrine responses to pheromones in *Barilius* and masu salmon, evidence from other species suggests that androgens of receiving individuals regulate the response to hormonal pheromones through actions on the olfactory epithelium (Cardwell *et al.*, 1995) and at central neural sites (Murphy and Stacey, 2002).

EOG recording studies provide the clearest evidence that hormones can act at the peripheral sensory level to modulate pheromone detection. In particular, these studies show that olfactory response to PGs is greater in males than in females in both cypriniforms (goldfish; Sorensen and Goetz, 1993: tinfoil barbs; Cardwell et al., 1995: red fin sharks, *Epalzeorhynchus frenatus*; Stacey et al., 2003) and salmoniforms (lake whitefish and brown trout; Laberge and Hara, 2003). In contrast, EOG response to steroids is typically equivalent in males and females, although Lower et al. (2004) report that male roach, *Rutilus rutilus*, are markedly more sensitive to steroids than are females.

It seems likely that gender differences in EOG responses to PGs are androgen-dependent, because androgen treatment of juvenile tinfoil barbs (Cardwell et al., 1995) and red fin sharks (Stacey et al., 2003), which does not affect the response to steroids, greatly enhances response to 15-keto-$PGF_{2\alpha}$. These androgen effects on PG-induced EOG responses are consistent with reports of nuclear androgen receptors in fish olfactory epithelium (Pottinger and Moore, 1997). However, it is not known whether they result from actions on existing receptor neurons or recruitment of new receptor neurons, which are rapidly replaced in fish (Zippel et al., 1997).

Evidence that responsiveness to hormonal pheromones can be modulated by central actions of androgens is provided by situations in which steroid pheromones induce equivalent EOG responses in males and females, but induce biological response only in males. Studies in this area have exploited the bisexual potential of the adult teleost brain (Kobayashi et al., 2000) to explore androgen effects on pheromone-induced male-typical responses in adult female fish. Perhaps the clearest example is seen in the round goby, in which EOG recordings show that the olfactory epithelia of males and females are equally responsive to three steroids (etiocholanolone, estrone, and estradiol-3β-glucuronide) which have been shown in EOG cross-adaptation studies to act through distinct olfactory receptor mechanisms (Murphy et al., 2001). Here, the fact that males transiently increase ventilation in response to all three steroid odorants, whereas female respond only to etiocholanolone (Murphy et al., 2001), suggests that the mechanism responsible for gender-typical behavioral response resides not at the peripheral sensory level but at more central sites of olfactory processing. Androgen evidently plays a role in the normal male-typical pattern of behavioral responsiveness because, within several weeks of androgen implant, females not only increase ventilation in response to all three steroid odorants but also, as with males (Murphy *et al.*, 2001), can be shown through behavioral cross-adaptation experiments to discriminate among these odors (Murphy and Stacey, 2002).

There also is evidence for central androgen actions in mediating endocrine response to hormonal pheromones. In goldfish, for example, 17,20βP exposure induces gender-typical endocrine responses: exposed males increase LH at all times of day (Dulka *et al.*, 1987b), whereas females evidently respond only at night (Kobayashi *et al.*, 2002). Thus, because males and females exhibit equivalent EOG response to 17,20βP (Sorensen *et al.*, 1987), the fact that androgen-implanted females exhibit a male-typical LH response (Kobayashi *et al.*, 1997) suggests a central androgen effect.

Given the clear evidence that androgens play important roles in modulating male responses to hormonal pheromones in many species, it is unfortunate that the underlying mechanisms mediating these androgenic effects are completely unknown and that the potential for hormones to modulate female responses to pheromones has not been explored. Nonetheless, it is clear that physiological mechanisms have evolved to synchronize pheromonal responsiveness with those life history stages in which reproductive responses to conspecifics are adaptive. In addition to optimizing pheromone functions among conspecifics, such mechanisms also might function to reduce heterospecific interactions, as discussed in the following section.

HORMONAL PHEROMONES AND THE ISSUE OF SPECIES SPECIFICITY

The likelihood that many fish species have evolved to employ hormones and structurally-related compounds as sex pheromones raises the important question of species specificity (Sorensen and Stacey, 1999; Sorensen *et al.*, 2000). In particular, given the fact that the chemical structures of sex steroid and prostaglandin hormones appear to have been conserved throughout the evolutionary history of the fishes, and that their metabolic pathways also appear to be relatively conserved, it is reasonable to question whether and how hormonal compounds might function as species-specific pheromones. This question is particularly significant because hormonal pheromones appear to be commonly employed by many of the more speciose groups that live in proximity to each other (e.g., the cypriniforms, characiforms, salmonins and siluriform groups; Stacey and Cardwell, 1995; Stacey and Sorensen, 2002; Laberge and Hara, 2003; Narayanan and

Stacey, 2003). Indeed, it could be argued that studies addressing species-specificity are those most needed to advance our understanding of sex pheromone function.

The basic question of species specificity appears straightforward insofar as it asks whether and how a species' pheromones might enable it to recognize and respond to conspecifics and not heterospecifics. Answering this question is far from straightforward, however, because it actually encompasses at least four corollary questions. First, what is the full complement of chemicals detected by the study species? Second, which and how much of these detected chemicals also are released by the sympatric species, and when are they released? Third, to what degree are responses to pheromones influenced by other sensory modalities (visual, auditory) that also might transmit conspecific information (e.g., McLennan and Ryan, 1997). Fourth, what is the frequency and fitness cost of responses to heterospecifics? The need for this information, which is virtually non-existent for any fish pheromone system, will be implicit in the following brief discussion, which focuses on how research on hormonal pheromones might further our understanding of species specificity. For additional discussion of this complex topic, the reader is referred to Sorensen and Stacey (1999).

Selective Pressure for Specificity

Reduction in fitness that might result from inappropriate (maladaptive) response to heterospecific odor is likely to represent a considerable pressure for the evolution of species-specific pheromonal information. From this perspective, it is important to assess not only the likelihood that such heterospecific responses could occur and the fitness costs that might be associated with them, but also the mechanisms through which selection might alter pheromone specificity, and costs associated with these mechanisms. Although a great variety of scenarios could theoretically result in pheromone specificity (Sorensen and Stacey, 1999, and below), no study has yet addressed this question in a comprehensive manner.

Likelihood and costs of heterospecific responses

Evolutionary selection for species-specificity is expected to exist only for species that exist in *functional reproductive sympatry*, i.e., species that, under natural spawning conditions, might detect heterospecific odors. Although a key factor determining the number of fish species expected to experience such functional reproductive sympatry is the active space of hormonal products released by sympatric heterospecifics, very little is known about the active spaces of hormonal products in species other than the goldfish. However, because these products are derived from an endocrine system employing low concentrations of endogenous hormones, active spaces in many cases might be relatively small. For example, laboratory studies of hormonal release rates and EOG detection thresholds in the goldfish suggest that 17,20βP released during the LH surge by a single 25 g female creates an active space of $\sim 30 \ 1 \cdot h^{-1}$ (Scott and Sorensen, 1994), whereas post-ovulatory 15-keto-PGF$_{2\alpha}$ creates an active space of $\sim 300 \ 1 \cdot h^{-1}$ (Sorensen *et al.*, 2000). However, AD released by 25 g sexually active males creates an active space greater than $500 \ 1 \cdot h^{-1}$ (Sorensen *et al.*, 2005). In natural waters, active spaces may be even smaller, given the fact that degradation could occur (Sorensen *et al.*, 2000) and that humic acids can attenuate EOG responses to steroids and prostaglandin in goldfish (Hubbard *et al.*, 2002; Mesquite *et al.*, 2003). Furthermore, even when a species enters the active space of a heterospecific, it will experience functional reproductive sympatry only if in a reproductive state in which it is physiologically capable of detecting or responding to the odor (section on **Hormonal Regulation of Hormonal Pheromone Functions**). From this perspective, it would appear that for many hormonal pheromones, the likelihood of functional reproductive sympatry may be insufficient to exert a strong selective pressure for specificity. Notably, this possibility has not been tested.

Selective pressures to evolve specificity should be proportional to the fitness cost of responding to heterospecific odor which, in turn, will be determined by aspects of individual life history and the nature of the heterospecific response. To cite extreme examples, a semelparous female that ovulates in response to a single exposure to heterospecific odor could experience complete reproductive failure, whereas in an iteroparous male such as the goldfish, an equivalent neuroendocrine response might result in negligible reduction in fitness. In the former case, selection would be expected to lead to specificity or to other changes that reduce the likelihood of heterospecific exposure; in the latter case, non-specificity might persist, particularly if the frequency of functional reproductive sympatry is low.

Mechanisms by which specificity might evolve

Both the extent and nature of adaptive responses to 'costly' non-specific hormonal pheromones are expected to vary depending on the nature of the pheromone function. In particular, we have proposed (Stacey and

Sorensen, 2002; Wisenden and Stacey, 2005) that, in terms of the effects that an individual's released hormonal products exert on conspecifics, it is important to distinguish three functionally distinct conditions: *ancestral*, *spying*, and *communication* (Fig. 6.2). In the ancestral phase, individuals (*originators*) release hormones and related product(s) that are not detected by conspecifics, and thus do not function as pheromones. This pre-pheromonal condition can then progress to *spying* if the receivers evolve the ability to detect and respond adaptively to the originator's released hormonal product(s), which we term a cue(s). Amongst species that spy, cue originators may or may not benefit from the receiver's response but, importantly, remain in an unspecialized state with respect to production and release of pheromonal cues. Finally, spying can progress to *communication* if there is a mechanism whereby responses of receivers can exert selective pressure for specialization in production and/or release of the detected cue(s), which we now term a pheromonal *signal(s)*, released by a *signaler* (Fig. 6.2). We consider the use by goldfish of seemingly unmodified hormones as pheromones (e.g., 17,20βP and AD) to be an example of spying, whereas the use by the black goby of etiocholanolone glucuronide (Etio-g) to be produced by the specialized mesorchial gland (Rasotto and Mazzoldi, 2002) may be an example of communication.

There are at least two reasons why the distinction between spying and communication is highly relevant to the issue of species specificity. First, in spying, the onus for evolving specificity likely falls entirely on the receiving individual. In goldfish, for example, it is possible that males exhibit an endocrine-gonadal response to 17,20βP regardless of whether it is encountered in the presence of conspecifics or heterospecifics. Such heterospecific responses are unlikely to carry high costs for males, as only a small fraction of annual sperm production might potentially be lost. Heterospecific responses are also unlikely to exert any selective pressure on females, given the fact that male response to conspecific 17,20βP is a component of sperm competition, and that females appear to achieve full fertility when spawning with males that have not been exposed to 17,20βP (Zheng *et al.*, 1997). The situation in which hormonal pheromones function as communicatory signals, however, may be very different. In these instances, heterospecific responses could reduce fitness of both signaler and receiver and thus lead to more complex adaptational responses (see Sorensen and Stacey, 1999).

The spying-communication dichotomy also is relevant to species-specificity because, in communication, evolutionary selection by receivers

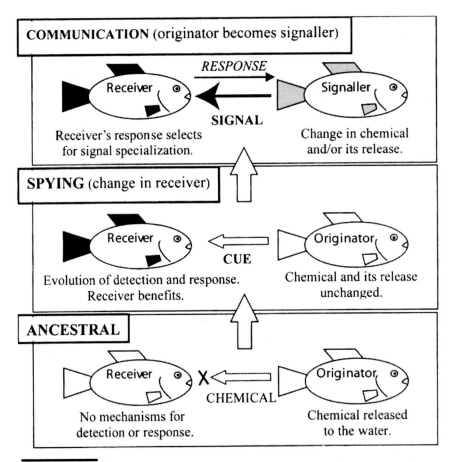

Fig. 6.2 Proposed stages in the evolution of hormonal pheromones. Redrawn from Stacey and Sorensen (1991, 1999, 2002, 2005) with kind permission from Elsevier Ltd., and from Sorensen and Stacey (1999) with kind permission from Springer-Verlag GmbH.

on signalers might result in signal amplification, thereby increasing pheromonal active space and increasing the number of species in functional reproductive sympatry. For example, in species such as gobies, where females evidently use olfactory cues to locate nesting males, it seems likely that female mate selection based on differential steroid release by males has been responsible for the evolution of the specialized testicular tissue (mesorchial gland) synthesizing pheromonal steroids (e.g., Locatello *et al.*, 2002; Arbuckle *et al.*, 2005). Such female selection is likely to have increased the active space of male pheromone, although this appears not to have been studied.

Is the Diversity of Hormonal Pheromones Sufficient to Yield Specificity?

For any particular species, the potential for released hormones and related products to generate species-specific pheromones is assumed to depend on three key factors: the number and nature of species with which it experiences functional reproductive sympatry; the diversity of hormonal products released by these species; and whether the pheromones consist of single compounds, or are mixtures that might include compounds unrelated to hormones. No study has yet examined the hormonal pheromones of a suite of species that lives in functional reproductive sympatry. However, given that the typical active space of hormonal pheromones is expected to be small, that opportunities for heterospecific interaction should be further reduced by restricting pheromonal responsiveness to brief periods of reproductive maturity (Moore and Scott, 1991; Cardwell *et al.*, 1995; section on **Hormonal Regulation of Hormonal Pheromone Functions**), and that the fish olfactory system exhibits a remarkable ability to discriminate minor differences in steroid and prostaglandin structure (Sorensen *et al.*, 1988, 1990, 1995b; Murphy *et al.*, 2001), the potential chemical diversity of released hormonal products does not appear to be particularly problematic for the evolution of species specific pheromones.

Diversity of hormonal products released and detected by fish

Fish appear to produce a small and common suite of hormonal sex steroids: E_2, T, and 11-ketotestosterone (11-KT) during gonadal development and $17,20\beta P$ or $17,20\beta,21P$ during ovarian and testicular maturation (Borg, 1994; Thomas, 2003). However, they release not only these hormonal steroids but also their glucuronidated and sulfated conjugates as well as unconjugated and conjugated forms of their precursors and metabolites (e.g., Scott *et al.*, 1991; Van Weerd *et al.*, 1991; Sorensen and Scott, 1994; Parks and Leblanc, 1998). Moreover, these studies are likely to reflect only a portion of the diversity of steroids released by fish because only a few species have been studied and then using a few immunoassays with restricted specificities.

EOG studies also reflect the great diversity of steroidal compounds that fish detect, and therefore presumably release (Stacey and Sorensen 2002; Fig. 6.3). In addition to detecting unconjugated hormones such as E_2 (round goby; Murphy *et al.*, 2001), T (Atlantic salmon; Moore and

Scott, 1991), 11-KT (tinfoil barbs; Cardwell *et al.*, 1995) and 17,20βP (goldfish; Sorensen *et al.*, 1990), fish detect glucuronidated E_2 (round goby, Murphy *et al.*, 2001; Fig. 6.3A) and 17,20βP (northern redbelly dace, *Phoxinus eos*, order Cypriniformes; Fig. 6.3B) and sulfated E_2 (*H. burtoni*; Cole and Stacey, 2003; Fig. 6.3C), T (*H. burtoni*; Cole and Stacey, 2003; Fig. 6.3D), and 17,20βP (red fin shark, *Epalzeorhynchus frenatus*, order Cypriniformes; Fig. 6.3E). Fish also detect diverse unconjugated and conjugated metabolites of androgens (large spot catfish, *Synodontis ocellifer*, order Siluriformes; Narayanan and Stacey, 2003; Fig. 6.3F) and C21 steroids (*S. ocellifer*, Narayanan and Stacey, 2003; Fig. 6.3G). Although there often is good correspondence between EOG and biological response (e.g., Sorensen *et al.*, 1995a; Murphy *et al.*, 2001), it is emphasized that EOG is an extracellular measure of olfactory receptor neuron function that need not directly reflect receptor specificity. Cloning and expression of fish pheromone receptors should enable a more direct approach to determining how the fish olfactory system discriminates between odors. The C21 steroids might be particularly important in generating species-specific odours because of the large number of metabolites they have the potential to generate. For example, with the exception of C17-glucuronidation, all five carbons in the C21 structure (C3, 11, 17, 20, and 21) that commonly bear oxygen groups can be conjugated with glucuronide and/or sulfate. When combined with metabolic reduction of the A-ring to yield 5β, 3α-, 5β, 3β-, 5α, 3α- and 5α,3β-configurations, fish could potentially release an enormous diversity of C21 metabolites. Indeed, because there are few studies of C21 metabolism and release in fish (e.g., Schoonen *et al.*, 1988; Asahina *et al.*, 1989; Van Weerd *et al.*, 1991; Scott and Sorensen, 1994), it seems reasonable to expect that important pheromonal C21 metabolites await discovery.

Fig. 6.3 Electro-olfactogram studies show that diverse steroids and prostaglandins are detected by the olfactory epithelia of fish. EOG data are presented either as a percentage (mean ± S. E. M.) of response to an amino acid standard (B, C, D, F, G, H, I, J) or as mean (± S. E. M.) base to peak mV change (A, E). A: Responses of mature male and female round gobies (*N* = 6) to 17β-estradiol-3β-glucuronide (redrawn from Murphy *et al.* (2001) with kind permission of Springer); B: Responses of adult male and female *Phoxinus eos* (*N* = 5) to 4-pregnen-17,20β-diol-3-one-20β-glucuronide (redrawn from Stacey *et al.* (2003) with permission from NRC Research Press); C and D: Responses of mature male cichlids (*N* = 9-10) to 17β-estradiol-3,17-disulfate and testosterone-sulfate (redrawn from Stacey *et al.* (2003) with kind permission of NRC Research Press); E: Responses of male rainbow sharks (N = 4) to 4-pregnen-17,20β-diol-3-one-20β-sulfate (redrawn from Stacey *et al.* (2003) with kind permission from NRC Research Press); F and G: Responses of *Synodontis* catfish (*N* = 4) of unknown sex and maturity to etocholanolone and 5β-pregnan-3α,17,20β-triol

Fig. 6.3 Contd.

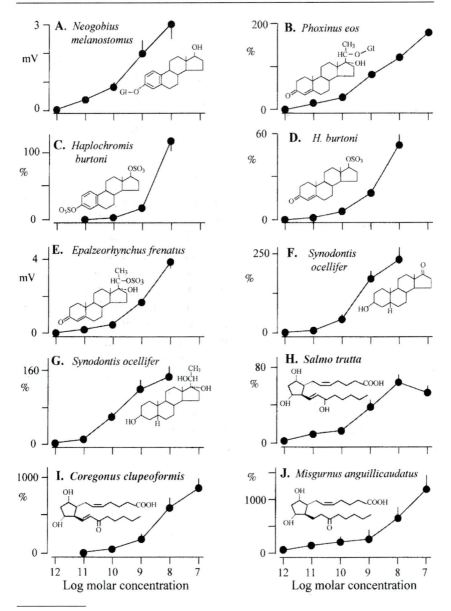

A. *Neogobius melanostomus*

B. *Phoxinus eos*

C. *Haplochromis burtoni*

D. *H. burtoni*

E. *Epalzeorhynchus frenatus*

F. *Synodontis ocellifer*

G. *Synodontis ocellifer*

H. *Salmo trutta*

I. *Coregonus clupeoformis*

J. *Misgurnus anguillicaudatus*

Log molar concentration

Fig. 6.3 Contd.

(redrawn from Stacey *et al.* (2003) with kind permission of NRC Research Press); H and I: Responses of male and female brown trout (N = 11-12) and lake whitefish (N = 10-11) to prostaglandin $F_{2\alpha}$ and 15-keto-prostaglandin $F_{2\alpha}$ (redrawn from Laberge and Hara, 2003) with kind permission of the authors and Blackwell Publishing); J: Responses of male oriental weatherfish to 13,14-dihydro-15-keto-prostaglandin $F_{2\alpha}$ (redrawn from Kitamura *et al.*, 1994a with kind permission of the authors and Elsevier).

The potential diversity of prostaglandin pheromones is particularly unclear, because prostaglandin metabolism in fish is very poorly understood (Sorensen *et al.*, 1995b). EOG studies indicate that olfactory sensitivity is restricted to F-series prostaglandins, although even here there is diversity in the most potent PGF detected: e.g., $PGF_{2\alpha}$ and 15-keto-$PGF_{2\alpha}$ in brown trout and lake whitefish (Laberge and Hara, 2003; Fig. 6.3H,I); 13,14-dihydro-15-keto-$PGF_{2\alpha}$ in Oriental weatherfish loach (Kitamura *et al.*, 1994a; Fig. 6.3J).

Hormonal pheromones consisting of mixtures

The potential for hormonal pheromones to be species-specific increases dramatically if these odors normally function not as single compounds but as mixtures that can be discriminated by the nature and ratios of their components. Thus, it is significant that EOG and pheromone bioassay studies clearly show that some fish species not only detect but also discriminate a variety of hormonal products (goldfish; Kobayashi *et al.*, 2002: round goby; Murphy *et al.*, 2001: *H. burtoni*; Cole and Stacey, 2003, 2006), suggesting they may also be capable of discriminating between conspecific and heterospecific mixtures. In insects, species specificity of pheromones typically results from the use of mixtures that can include inhibitory compounds released by heterospecifics (Sorensen *et al.*, 1998). Although this phenomenon has not been definitively demonstrated in fish, pheromonal response can be inhibited by heterospecific odor. Thus, whereas male goldfish normally increase socio-sexual activity when a 15-keto-$PGF_{2\alpha}$ solution is added to aquaria (Sorensen *et al.*, 1988), this response is blocked if the prostaglandin solution also contains blue gourami (*Trichogaster trichopterus*) odor (Sorensen *et al.*, 2000). These results are highly significant because they clearly show that even though single hormonal pheromone compounds can elicit biological response in many fishes, this does not necessarily mean that heterospecifics releasing these compounds will induce the same response. Moreover, they raise the distinct possibility that pheromones containing hormonal products might also contain—or be influenced by—non-hormonal compounds, a scenario which greatly increases the mechanisms whereby such pheromones can confer species specificity.

Evidence for hormonal pheromone specificity

Although it is intuitive that selective pressures to recognize conspecifics and/or avoid heterospecifics should lead to species-specific sex pheromones, there is surprisingly little information for such specificity in fish. In salmonids (*Oncorhynchus masou* and *rhodurus*; Honda, 1982), cyprinid bitterlings (*Rhodeus ocellatus* and *Acheilognathus lanceolatus*; Honda, 1982b), and poeciliid swordtails (*Xiphophorus cortezi, nigrensis* and *montezumae*; McLennan and Ryan, 1997), choice-test studies have shown that conspecific odor is more attractive than heterospecific odor (see also Liley, 1982; Stacey *et al.*, 1986; Sorensen and Stacey, 1999). Such studies provide clear evidence that some fish have evolved the ability to discriminate between conspecific and heterospecific odor. However, it is not clear that the results of choice tests are directly relevant to natural situations, in which specificity would be displayed not by a preference for conspecific over heterospecific odors, but rather by a response to conspecific odor and a lack of response to heterospecific odor. This issue is particularly important where it has been shown not only that conspecific odor is preferred over heterospecific odor, but also that heterospecific odor is preferred over a blank control (e.g., McLennan and Ryan, 1997). From this perspective, a study (McKinnon and Liley, 1986) of male response to female odor in two *Trichogaster* species (*pectoralis* and *trichopterus*) makes a particularly clear case for specificity, both because male response (air gulping at the surface) was evaluated in the presence of only conspecific or heterospecific odor, and was shown to be influenced by female reproductive state. McKinnon and Liley (1986) found that, whereas male *trichopterus* increased air gulping in response to both conspecific and heterospecific female odor, male *pectoralis* responded only to conspecific odor, an asymmetry in responsiveness they attributed to the fact that the male *pectoralis* used in the study had been sympatric with congeners, whereas the male *trichopterus* had not.

There are two reasons why hormonal pheromones appear to have great potential to further our understanding of species-specific sex pheromones in fish: these hormonal products are currently the only identified odorants in teleost fish that appear capable of transmitting reproductive information (Sorensen and Caprio, 1998), and they are detected by many species that are diverse in phylogeny and life history characteristics. Although both steroid and prostaglandin pheromones would be expected to contribute to specificity, only for steroids is information on metabolism and release sufficient to speculate on what these contributions might be.

Although there clearly exist great differences in the hormonal steroids and related compounds that fishes are known to detect, there is virtually no evidence that such differences have been driven by selection for conspecific recognition (or heterospecific avoidance) in situations of reproductive sympatry. Indeed, although EOG studies clearly show that entirely different suites of hormonal compounds are detected by distantly related fish such as gobies (Murphy *et al.*, 2001) and goldfish (Kobayashi *et al.*, 2002), it is perhaps more important that these studies also show very similar hormonal compounds being detected by species within lower taxa (tribe, genera), where the potential for interspecific interaction should be greatest (Stacey and Cardwell, 1995; Stacey and Sorensen, 2002). Such patterns of steroid detection suggest that once the ability to detect a specific hormonal product has evolved, there is a strong tendency for that particular ability to be retained by descendent species, and that recent speciation events are not typically driven or facilitated by changes in specificity of olfactory receptor neurons. To cite several of the clearest examples, goldfish and common carp (subtribe Cyprini), which can hybridize in the Laurentian Great Lakes (Taylor and Mahon, 1977), not only detect very similar PGs and steroids (Irvine and Sorensen, 1993), but also exhibit similar LH and milt responses to 17,20βP and 17,20βP-S (Stacey *et al.*, 1994; Sorensen *et al.*, 1995b). In the African *Synodontis* catfishes (Family Mochokidae; Fig. 6.3F,G), EOG studies indicate that although southern species detect a slightly different suite of C19 and C21 steroids than do species from the northern and western parts of the continent, there is little if any difference in the steroids detected by species within each of these broad geographical areas (Narayanan and Stacey, 2003). Similarly, more preliminary EOG studies indicate that a number of broadly sympatric genera of Malaysian cyprinids in tribe Labeonini (e.g., *Epalzeorhynchus*, Fig. 6.3E) appear to detect only 17,20βP-S (Stacey and Sorensen, 2002), whereas numerous characiform 'tetras' in tribe Tetragonopterinae appear to detect only estradiol-3-sulfate (Cardwell and Stacey, 1995). It is important to stress that even if EOG studies of carp, catfish, tetras, and labeonines have identified all steroidal compounds that these fish detect, the findings that the same or similar steroid odors are detected in no way eliminates the possibility of species specificity, which could be achieved by species-specific blends of a common mixture, or by the use of non-hormonal components of a pheromonal mixture. What these EOG data do suggest, however, is that if specificity occurs within these groups, it likely does not involve the use of distinctly different steroidal compounds.

In summary, the strongest selection pressures to evolve species-specific sex pheromones are expected for those related species in functional reproductive sympatry which rely heavily upon olfaction to mediate intraspecific reproductive activities, and for which response to heterospecifics bears significant cost. Both the diversity of steroids released by fish, and the specificity with which many of these steroids are detected, suggest clear potential for species-specific hormonal pheromone systems employing released products of reproductive steroidogenesis. However, patterns of steroid detection suggest that species differences have arisen through passive evolutionary divergence and that, if pheromones containing hormonal products are species specific, such specificity likely has evolved through the use of mixtures that may contain non-hormonal compounds. We hope that further research will explore this fundamental problem that is of such broad relevance to fish reproductive biology.

Acknowledgements

Norm E. Stacey gratefully acknowledges many years of support from the Natural Sciences and Engineering Research Council of Canada (N.S.E.R.C.). Peter W. Sorensen thanks the Alberta Heritage Foundation for Medical Research, Minnesota Agricultural Experiment Station, Minnesota Sea Grant, Great Lakes Fishery Commission, The National Institutes of Health (NIH/DC03792), and the National Science Foundation (NSF/IBN9723798), all of which have generously supported research on fish hormonal pheromones over the past two decades.

References

Appelt, C.W. and P.W. Sorensen. 1999. Freshwater fish release urinary pheromones in a pulsatile manner. In: *Advances in Chemical Signals in Vertebrates*, R.E. Johnston, D. Müller-Schwarze and P.W. Sorensen (eds.). Kluwer Academic/Plenum Publishers, New York, pp. 247-256.

Appelt, C.W., P.W. Sorensen and R.G. Kellner. 1995. Female goldfish appear to release pheromonally active F-prostaglandins in urinary pulses. In: *Proceedings of the Fifth International Symposium on the Reproductive Physiology of Fish*, F.W. Goetz and P. Thomas (eds.). Fish Symposium 95, Austin Texas, p. 270.

Arbuckle, W.J., A.J. Belanger, L.D. Corkum, B.S. Zielinski, W. Li, S.-S. Yun, S. Bachynski and A.P. Scott. 2005. *In vitro* biosynthesis of novel 5ß-reduced steroids by the testis of the round goby, *Neogobius melanostomus*. *General and Comparative Endocrinology* 140: 1-13.

Asahina, K., K. Suzuki, T. Hibiya and B. Tamaoki. 1989. Structure and steroidogenic enzymes of the seminal vesicles of the urohaze-goby (*Glossogobius olivaceus*). *General and Comparative Endocrinology* 74: 385-391.

Bélanger, A.J., W.J. Arbuckle, L.D. Corkum, D.B. Gammon, W. Li, A.P. Scott and B.S. Zielinski. 2004. Behavioral and reproductive responses by reproductive female *Neogobius melanostomus* to odours released by conspecific males. *Journal of Fish Biology* 65: 933-946.

Belanger, R.M., C.M. Smith, L.D. Corkum and B.S. Zielinski. 2003. Morphology and histochemistry of the peripheral olfactory organ in the round goby, *Neogobius melanostomus* (Teleostei: Gobiidae). *Journal of Morphology* 257: 62-71.

Bhatt, J.P. and M.S. Sajwan. 2001. Ovarian steroid sulphate functions as priming pheromone in male *Barilius bendelisis* (Ham.). *Journal of Bioscience* 26: 253-263.

Bhatt, J.P., J.S. Kandwal and R. Nautiyal. 2002. Water temperature and pH influence olfactory sensitivity to pre-ovulatory and post-ovulatory ovarian pheromones in male *Barilius bendelisis*. *Journal of Bioscience* 27: 273-281.

Bjerselius, R. and K.H. Olsen. 1993. A study of the olfactory sensitivity of crucian carp (*Carassius carassius*) and goldfish (*Carassius auratus*) to $17\alpha,20\beta$ dihydroxyprogesterone and prostaglandin $F_{2\alpha}$. *Chemical Senses* 18: 427-436.

Bjerselius, R., K.H. Olsén and W. Zheng. 1995. Endocrine, gonadal and behavioral responses of male crucian carp (*Carassius carassius*) to the hormonal pheromone $17\alpha,20\beta$-dihydroxy-4-pregnen-3-one. *Chemical Senses* 20: 221-230.

Borg, B. 1994. Androgens in teleost fishes. *Comparative Biochemistry and Physiology* C 109: 219-245.

Cardwell, J.R. and N.E. Stacey. 1995. Hormonal sex pheromones in characiform fishes: An evolutionary case study. In: *Fish Pheromones: Origins and Modes of Action*, A.V.M. Canário and D.M. Power (eds.). University of Algarve Press, Faro, Portugal, pp. 47-55.

Cardwell, J.R., N.E. Stacey, E.S.P. Tan, D.S.O. McAdam and S.L.C. Lang. 1995. Androgen increases olfactory receptor response to a vertebrate sex pheromone. *Journal of Comparative Physiology* A 176: 55-61.

Charlebois, P.M., L.D. Corkum, D.J. Jude and C. Knight. 2001. The round goby (*Neogobius melanostomus*) invasion: Current research and future needs. *Journal of Great Lakes Research* 27: 263-266.

Cole, K.S. and N.E. Stacey. 1984. Prostaglandin induction of female spawning behavior in *Cichlosoma bimaculatum* (Pisces, Cichlidae). *Hormones and Behavior* 18: 235-248.

Cole, T.B. and N.E. Stacey. 2003. Olfactory and endocrine response to steroids in an African cichlid fish, *Haplochromis burtoni*. *Fish Physiology and Biochemistry* 28: 265-266.

Cole, T.B. and N.E. Stacey. 2006. Olfactory response to steroids in an African mouth-brooding cichlid, *Haplochromis burtoni* (Günther). *Journal of Fish Biology* 68: 661-680.

Colombo, L., P.C. Belvedere and A. Pilati. 1977. Biosynthesis of free and conjugated 5β-reduced androgens by the testis of the black goby, *Gobius jozo* L. *Bolletino di Zoologia* 44: 131-144.

Colombo, L., A. Marconato, P.C. Belvedere and C. Frisco. 1980. Endocrinology of teleost reproduction: A testicular steroid pheromone in the black goby, *Gobius jozo* L. *Bolletino di Zoologia* 47: 355-364.

Colombo, L., P.C. Belvedere, A. Marconato and F. Bentivegna. 1982. Endocrinology of teleost reproduction. In: *Proceedings of the Second International Symposium on the Reproductive Physiology of Fish*, C.J.J. Richter and H.J. Th. Goos (eds.). Pudoc, Wageningen, pp. 84-94.

Corkum, L.D., M.R. Sapota and K.E. Skora. 2004. The round goby, *Neogobius melanostomus*, a fish invader on both sides of the Atlantic Ocean. *Biological Invasions* 6: 173-181.

DeFraipont, M. and P.W. Sorensen. 1993. Exposure to the pheromone $17\alpha,20\beta$-dihydroxy-4-pregnen-3-one enhances the behavioural spawning success, sperm production, and sperm motility of male goldfish. *Animal Behavior* 46: 245-256.

Dittman, A.H. and T.P. Quinn. 1994. Avoidance of a putative pheromone, 17α,20β-dihydroxy-4-pregnen-3-one, by precociously mature chinook salmon (*Oncorhynchus tshawytscha*). *Canadian Journal of Zoology* 72: 215-219.

Døving, K. 1976. Evolutionary trends in olfaction. In: *The Structure-Activity Relationships in Olfaction*, G. Benz (ed.). IRL Press, London, pp. 149-159.

Døving, K.J.B., R. Selset and G. Thommesen. 1980. Olfactory sensitivity to bile acids in salmonid fishes. *Acta Physiologica Scandinavica* 108: 123-131.

Dulka, J.G., N.E. Stacey, P.W. Sorensen and G.J. Van Der Kraak. 1987a. A sex steroid pheromone synchronizes male-female spawning readiness in goldfish. *Nature (London)* 325: 251-253.

Dulka, J.G., P.W. Sorensen and N.E. Stacey. 1987b. Socially-stimulated gonadotropin release in male goldfish: Differential circadian sensitivities to a steroid pheromone and spawning stimuli. In: *Proceedings of the Third International Symposium on the Reproductive Physiology of Fish*, D.R. Idler, L.W. Crim and J.M. Walsh (eds.). Memorial University Press, St. John's, Newfoundland, Canada, p. 160.

Dulka, J.G., B.D. Sloley, N.E. Stacey and R.E. Peter. 1992. A reduction in pituitary dopamine turnover is associated with sex pheromone-induced gonadotropin increase in goldfish. *General and Comparative Endocrinology* 86: 496-505.

Emanuel, M.E. and J.J. Dodson. 1979. Modification of rheotropic behavior of male rainbow trout (*Salmo gairdneri*) by ovarian fluid. *Journal of the Fisheries Research Board of Canada* 36: 63-68.

Essington, T.E. and P.W. Sorensen. 1996. Overlapping sensitivities of brook trout and brown trout to putative hormonal pheromones. *Journal of Fish Biology* 48: 1027-1029.

Frade, P., P.C. Hubbard, E.N. Barata, and A.V.M. Canário. 2002. Olfactory sensitivity of the Mozambique tilapia to conspecific odours. *Journal of Fish Biology* 61: 1239-1254.

Fraser, E.J. and N.E. Stacey. 2002. Isolation increases milt production in goldfish. *Journal of Experimental Zoology* 293: 511-524.

Gunderson, J.L., M.R. Klepinger, C.R. Bronte and J.E. Marsden. 1998. Overview of the international symposium on Eurasian ruffe (*Gymnocephalus cernuus*) biology, impacts and control. *Journal of Great Lakes Research* 24: 165-169.

Hara, T.J. and C. Zhang. 1997. Topographic bulbar projections and dual neural pathways of the primary olfactory neurons in salmonid fishes. *Neuroscience* 82: 301-313.

Honda, H. 1980. Female sex pheromone of rainbow trout, *Salmo gairdneri*, involved in courtship behavior. *Bulletin of the Japanese Society of Scientific Fisheries* 46: 1109-1112.

Honda, H. 1982a. On the female sex pheromones and courtship behavior in the salmonids, *Oncorhynchus masou* and *O. rhodurus*. *Bulletin of the Japanese Society of Scientific Fisheries* 48: 47-49.

Honda, H. 1982b. On the female pheromones and courtship behavior in the bitterlings *Rhodeus ocellatus ocellatus* and *Acheilognathus lanceolatus*. *Bulletin of the Japanese Society of Scientific Fisheries* 48: 43-45.

Hubbard, P.C., E.N. Barata and A.V.M. Canário. 2002. Possible disruption of pheromonal communication by humic acid in the goldfish, *Carassius auratus*. *Aquatic Toxicology* 60: 169-183.

Immler, S., C. Mazzoldi and M.B. Rasotto. 2004. From sneaker to parental male: change of reproductive traits in the black goby, *Gobius niger* (Teleostei, Gobiidae). *Journal of Experimental Zoology* A 301: 177-185.

Irvine, I.A.S. and P.W. Sorensen. 1993. Acute olfactory sensitivity of wild common carp, *Cyprinus carpio*, to goldfish sex pheromones is influenced by gonadal maturity. *Canadian Journal of Zoology* 71: 2199-2210.

Jones, K.A. and T.J. Hara. 1985. Behavioural responses of fishes to chemical cues: Results from a new bioassay. *Journal of Fish Biology* 27: 495-504.

Kitamura, S., H. Ogata and F. Takashima. 1994a. Activities of F-type prostaglandins as releaser sex pheromones in cobitide loach, *Misgurnus anguillicaudatus*. *Comparative Biochemistry and Physiology* A 107: 161-169.

Kitamura, S., H. Ogata and F. Takashima. 1994b. Olfactory responses of several species of teleost to F-prostaglandins. *Comparative Biochemistry and Physiology* A 107: 463-467.

Kobayashi, M., K. Aida and I. Hanyu. 1986a. Gonadotropin surge during spawning in male goldfish. *General and Comparative Endocrinology* 62: 70-79.

Kobayashi, M., K. Aida and I. Hanyu. 1986b. Pheromone from ovulatory female goldfish induces gonadotropin surge in males. *General and Comparative Endocrinology* 63: 451-455.

Kobayashi, M., K. Aida and I. Hanyu. 1988. Hormone changes during the ovulatory cycle in goldfish. *General and Comparative Endocrinology* 69: 301-307.

Kobayashi, M., K. Furukawa, M.-H. Kim and K. Aida. 1997. Induction of male-type gonadotropin secretion by implantation of 11-ketotestosterone in female goldfish. *General and Comparative Endocrinology* 108: 434-445.

Kobayashi, M., N.E. Stacey, K. Aida and S. Watabe. 2000. Sexual plasticity of behavior and gonadotropin secretion in goldfish and gynogenetic crucian carp. In: *Proceedings of the Sixth International Symposium on the Reproductive Physiology of Fish*, B. Norberg, O.S. Kjesbu, G.L. Taranger, E. Andersson and S.O. Stefansson (eds.). John Grieg A/S, Bergen 2000, Bergen, pp. 117-124.

Kobayashi, M., P.W. Sorensen and N.E. Stacey. 2002. Hormonal and pheromonal control of spawning behavior in the goldfish. *Fish Physiology and Biochemistry* 26: 71-84.

Kyle, A.L., N.E. Stacey, R.E. Peter and R. Billard. 1985. Elevations in gonadotrophin concentrations and milt volumes as a result of spawning behavior in the goldfish. *General and Comparative Endocrinology* 57: 10-22.

Laberge, F. and T.J. Hara. 2003. Behavioral and electrophysiological responses to F-prostaglandins, putative spawning pheromones, in three salmonid fishes. *Journal of Fish Biology* 62: 206-221.

Liley, N.R. 1982. Chemical communication in fish. *Canadian Journal of Fisheries and Aquatic Sciences* 39: 22-35.

Liley, N.R. and E.S.P. Tan. 1985. The induction of spawning behavior in *Puntius gonionotus* (Bleeker) by treatment with prostaglandin $F_{2\alpha}$. *Journal of Fish Biology* 26: 491-502.

Locatello, L., C. Mazzoldi and M.B. Rasotto. 2002. Ejaculate of sneaker males is pheromonally inconspicuous in the black goby, *Gobius niger* (Teleostei, Gobiidae). *Journal of Experimental Zoology* 293: 601-605.

Lower, N., A.P. Scott and A. Moore. 2004. Release of sex steroids into the water by roach. *Journal of Fish Biology* 64: 16-33.

Maniak, P.J., R. Lossing and P.W. Sorensen. 2000. Injured Eurasian ruffe, *Gymnocephalus cernuus*, release an alarm pheromone which may prove useful in their control. *Journal of Great Lakes Research* 26: 183-195.

Mazzoldi, C. and M.B. Rasotto. 2002. Alternative mating tactics in *Gobius niger*. *Journal of Fish Biology* 61: 157-172.

McKinnon, J.F. and N.R. Liley. 1986. Asymmetric species-specificity in responses to female sexual pheromone by males of two species of *Trichogaster* (Pisces: Belontiidae). *Canadian Journal of Zoology* 65: 1129-1134.

McLennan, D.A. and M.J. Ryan. 1997. Responses to conspecific and heterospecific olfactory cues in the swordtail *Xyphophorus cortezi*. *Animal Behaviour* 54: 1077-1088.

Mesquite, R.M.R.S., A.V.M. Canário and E. Melos. 2003. Partition of fish pheromones between water and aggregates of humic acid. Consequences for sexual signaling. *Environmental Science and Technology* 37: 742-746.

Miller, P.J. 1984. The tokology of Gobiid fishes. In: *Fish Reproduction: Strategies and Tactics*, G.W. Potts and R.W. Wootton (eds.). Academic Press, New York, pp. 119-153.

Moore, A. 1991. Behavioral and physiological responses of precocious male Atlantic salmon (*Salmo salar* L.) parr to testosterone. In: *Proceedings of the Fourth International Symposium on the Reproductive Physiology of Fish*, A.P. Scott, J.P. Sumpter, D.E. Kime and M.S. Rolfe (eds.). FishSymp 91, Sheffield, pp. 194-196.

Moore, A. and A.P. Scott. 1991. Testosterone is a potent odorant in precocious male Atlantic salmon (*Salmo salar* L.) parr. *Philosophical Transactions of the Royal Society of London*, B 332: 241-244.

Moore, A. and A.P. Scott. 1992. 17α,20β-dihydroxy-4-pregnen-3-one-20-sulphate is a potent odorant in precocious male Atlantic salmon parr which have been pre-exposed to the urine of ovulated females. *Proceedings of the Royal Society of London*, B 249: 205-209.

Moore, A. and C.P. Waring. 1995. Seasonal changes in olfactory sensitivity of mature male Atlantic salmon (*Salmo salar* L.) parr to prostaglandins. In: *Proceedings of the Fifth International Symposium on the Reproductive Physiology of Fish*, F.W. Goetz, and P. Thomas (eds.). Fish Symposium 95, Austin Texas, p. 273.

Moore, A., and C.P. Waring. 1996. Electrophysiological and endocrinological evidence that F-series prostaglandins function as priming pheromones in mature male Atlantic salmon (*Salmo salar*) parr. *Journal of Experimental Biology* 199: 2307-2316.

Moore, A., K.H. Olsén, N. Lower and H. Kindahl. 2002. The role of F-series prostaglandins as reproductive priming pheromones in the brown trout. *Journal of Fish Biology* 60: 613-624.

Murphy, C.A. and N.E. Stacey. 2002. Methyl-testosterone induces male-typical behavioral responses to putative steroidal pheromones in female round gobies (*Neogobius melanostomus*). *Hormones and Behavior* 42: 109-115.

Murphy, C.A., N. Stacey and L.D. Corkum. 2001. Putative steroidal pheromones in the round goby, *Neogobius melanostomus*: olfactory and behavioral responses. *Journal of Chemical Ecology* 27: 443-470.

Narayanan, A. and N.E. Stacey. 2003. Olfactory responses to putative steroidal pheromones in allopatric and sympatric species of mochokid catfish. *Fish Physiology and Biochemistry* 28: 275-276.

Nelson, J.S. 1994. *Fishes of the World*, Third Edition. John Wiley and Sons, New York.

Nevitt, G.A. 1991. Do fish sniff? A new mechanism of olfactory sampling in pleuronectid flounders. *Journal of Experimental Biology* 157: 1-18.

Ogata, H., S. Kitamura and F. Takashima. 1994. Release of 13,14-dihydro-15-keto-prostaglandin $F_{2\alpha}$, a sex pheromone, to water by cobitid loach following ovulatory stimulation. *Fisheries Science* 60: 143-148.

Olsén, K.H., R. Bjerselius, E. Petersson, E. Jarva, I. Mayer and M. Hedenskog. 2000. Lack of species-specific primer effects of odours from female Atlantic salmon, *Salmo salar*, and brown trout, *Salmo trutta*. *Oikos* 88: 213-220.

Olsén, K.H., R. Bjerselius, I. Mayer and H. Kindahl. 2001. Both ovarian fluid and female urine increase sex steroid levels in mature Atlantic salmon (*Salmo salar* L.) male parr. *Journal of Chemical Ecology* 27: 2337-2349.

Olsén, K.H., A.-K. Johanssen, R. Bjerselius, I. Mayer and H. Kindahl, H. 2002. Mature Atlantic salmon (*Salmo salar* L.) male parr are attracted to ovulated female urine but not ovarian fluid. *Journal of Chemical Ecology* 28: 29-40.

Parks, L.G. and G.A. Leblanc. 1998. Involvement of multiple biotransformation processes in the metabolic elimination of testosterone by juvenile and adult fathead minnows (*Pimephales promelas*). *General and Comparative Endocrinology* 112: 69-79.

Pfaff, D.W., A.P. Arnold, A.M. Etgen, S.E. Fahrbach and R.T. Rubin (eds.). 2002. *Hormones, Brain and Behavior*. Academic Press, New York.

Poling, K.R., E.J. Fraser and P.W. Sorensen. 2001. The three steroidal components of the goldfish preovulatory pheromone signal evoke different behaviours in males. *Comparative Biochemistry and Physiology* B 129: 645-651.

Pottinger, T.G. and A. Moore. 1997. Characterization of putative steroid receptors in the membrane, cytosol and nuclear fractions from the olfactory tissue of brown and rainbow trout. *Fish Physiology and Biochemistry* 16: 45-63.

Rainboth, W.J. 1991. Cyprinids of South East Asia. In: *Cyprinid Fishes: Systematics, Biology and Exploitation*, I.J. Winfield and J.S. Nelson (eds.). Chapman and Hall, London, pp. 156-210.

Rasotto, M.B. and C. Mazzoldi. 2002. Male traits associated with alternative reproductive tactics in *Gobius niger*. *Journal of Fish Biology* 61: 173-184.

Remage-Healey, L. and A.H. Bass. 2004. Rapid, hierarchical modulation of vocal patterning by steroid hormones. *Journal of Neuroscience* 24: 5892-5900.

Robison, R.R., R.D. Fernald and N.E. Stacey. 1998. The olfactory system of a cichlid fish responds to steroidal compounds. *Journal of Fish Biology* 53: 226-229.

Schoonen, W.J.E.J., J.G.D. Lambert and P.G.W.J. Van Oordt. 1988. Quantitative analysis of steroids and steroid glucuronides in the seminal vesicle fluid of feral spawning and feral and cultivated non-spawning African catfish, *Clarias gariepinus*. *General and Comparative Endocrinology* 70: 91-100.

Scott, A.P. and P.W. Sorensen. 1994. Time course of release of pheromonally active steroids and their conjugates by ovulatory goldfish. *General and Comparative Endocrinology* 96: 309-323.

Scott, A.P. and E.L.M. Vermeirssen. 1994. Production of conjugated steroids by teleost gonads and their role as pheromones. In: *Perspectives in Comparative Endocrinology*, K.G. Davey, R.E. Peter and S.S. Tobe (eds.). National Research Council, Ottawa, pp. 645-654.

Scott, A.P., A.V.M. Canário, N.M. Sherwood and C.M. Warby. 1991. Levels of steroids, including cortisol and 17α,20ß-dihydroxy-4-pregnen-3-one, in plasma, seminal fluid and urine of Pacific herring (*Clupea harengus pallasi*) and North Sea plaice (*Pleuronectes platessa* L.). *Canadian Journal of Zoology* 69: 111-116.

Scott, A.P., N.R. Liley and E.L.M. Vermeirssen. 1994. Urine of reproductively mature female rainbow trout, *Oncorhynchus mykiss* (Walbaum), contains a priming pheromone which enhances plasma levels of sex steroids and gonadotrophin II in males. *Journal of Fish Biology* 44: 131-147.

Selset, R. and K.B. Døving. 1980. Behaviour of mature anadromous char (*Salvelinus alpinus* L.) towards odorants produced by smolts of their own populations. *Acta Physiologica Scandinavica* 108: 113-122.

Sorensen, P.W. and J. Caprio. 1998. Chemoreception. In: *The Physiology of Fishes*, D.H. Evans (ed.), Second Edition, CRC Press, Boca Raton, pp. 375-405.

Sorensen, P.W. and A.P. Scott. 1994. The evolution of hormonal sex pheromones in teleost fish: poor correlation between the pattern of steroid release by goldfish and olfactory sensitivity suggests that these cues evolved as a result of chemical spying rather than signal specialization. *Acta Physiologica Scandinavica* 152: 191-205.

Sorensen, P.W and N.E. Stacey. 1999. Evolution and specialization of fish hormonal pheromones. In: *Advances in Chemical Signals in Vertebrates*, R.E. Johnston, D. Muller-Schwarze and P.W. Sorensen (eds.). Kluwer Academic/Plenum Publishers, New York, pp. 15-47.

Sorensen, P.W. and N.E. Stacey. 2004. Brief review of fish pheromones and discussion of their possible uses in the control of non-indigenous teleost fishes. *New Zealand Journal of Marine and Freshwater Research* 38: 399-417.

Sorensen, P.W., T.J. Hara and N.E. Stacey. 1987. Extreme olfactory sensitivity of mature and gonadally-regressed goldfish to a potent steroidal pheromone, 17α,20β-dihydroxy-4-pregnen-3-one. *Journal of Comparative Physiology* A 160: 305-313.

Sorensen, P.W., T.J. Hara, N.E. Stacey and F.W. Goetz. 1988. F prostaglandin function as potent olfactory stimulants that comprise the postovulatory female sex pheromone in goldfish. *Biology of Reproduction* 39: 1039-1050.

Sorensen, P.W., K.J. Chamberlain and N.E. Stacey. 1989. Differing behavioral and endocrinological effects of two female sex pheromones on male goldfish. *Hormones and Behavior* 23: 317-332.

Sorensen, P.W., T.J. Hara, N.E. Stacey and J.G. Dulka, 1990. Extreme olfactory specificity of the male goldfish to the preovulatory steroidal pheromone 17α,20β-dihydroxy-4-pregnen-3-one. *Journal of Comparative Physiology* A 166: 373-383.

Sorensen, P.W., A.P. Scott, N.E. Stacey and L. Bowdin. 1995a. Sulfated 17α,20β-dihydroxy-4-pregnen-3-one functions as a potent and specific olfactory stimulant with pheromonal actions in the goldfish. *General and Comparative Endocrinology* 100: 128-142.

Sorensen, P.W., A.R. Brash, F.W. Goetz, R.G. Kellner, L. Bowdin and L.A. Vrieze. 1995b. Origins and functions of F prostaglandins as hormones and pheromones in the goldfish. In: *Proceedings of the Fourth International Symposium on the Reproductive Physiology of Fish*, F. W. Goetz and P. Thomas (eds.). FishSymp 95, Austin, Texas, pp. 252-254.

Sorensen, P.W., J.R. Cardwell, T. Essington and D.E. Weigel. 1995c. Reproductive interactions between brook and brown trout in a small Minnesota stream. *Canadian Journal of Fisheries and Aquatic Sciences* 52: 1958-1965.

Sorensen, P.W., T.A. Christensen and N.E. Stacey. 1998. Discrimination of pheromonal cues in fish: Emerging parallels with insects. *Current Opinion in Neurobiology* 8: 458-467.

Sorensen, P.W., A.P. Scott and R.L. Kihslinger. 2000. How common hormonal metabolites function as relatively specific pheromonal signals in goldfish. In: *Proceedings of the Sixth International Symposium on the Reproductive Physiology of Fish*, B. Norberg, O.S. Kjesbu, G.L. Taranger, E. Andersson and S.O. Stefansson (eds.). John Grieg A/S, Bergen 2000, Bergen, pp. 125-128.

Sorensen, P.W., C.A. Murphy, K. Loomis, P. Maniak and P. Thomas. 2004. Evidence that 4-pregnen-17,20β,21-triol-3-one functions as a maturation-inducing hormone and pheromone precursor in the percid fish, *Gymnocephalus cernuus*. *General and Comparative Endocrinology* 139: 1-11.

Sorensen, P.W., M. Pinillos and A.P. Scott. 2005. Sexually mature male goldfish release large quantities of androstenedione into the water where it functions as a pheromone. *General and Comparative Endocrinology* 140: 164-175.

Stacey, N.E. 1976. Effects of indomethacin and prostaglandins on spawning behavior of female goldfish. *Prostaglandins* 12: 113-128.

Stacey, N.E. 1981. Hormonal regulation of female reproductive behavior in fish. *American Zoologist* 21: 305-316.

Stacey, N.E. 1991. Hormonal pheromones in fish: Status and prospects. In: *Proceedings of the Fourth International Symposium on the Reproductive Physiology of Fish*, A.P. Scott, J.P. Sumpter, D.S. Kime and M.S. Rolfe (eds.). FishSymp 91, Sheffield, pp. 177-181.

Stacey, N.E. and J.R. Cardwell. 1995. Hormones as sex pheromones in fish: widespread distribution among freshwater species. In: *Proceedings of the Fifth International Symposium on the Reproductive Physiology of Fish*, F.W. Goetz and P. Thomas (eds.). Symposium 95, Austin, Texas, pp. 244-248.

Stacey, N.E. and N.R. Liley. 1974. Regulation of spawning behavior in the female goldfish. *Nature (London)* 247: 71-72.

Stacey, N.E. and P.W. Sorensen. 2002. Fish hormonal pheromones. In: *Hormones, Brain, and Behavior*, D.W. Pfaff, A.P. Arnold, A.M. Etgen, S.E. Fahrbach and R.T. Rubin (eds.). Volume 2. Academic Press, New York, pp. 375-435.

Stacey, N.E. and P.W. Sorensen. 2006. Reproductive pheromones. In: *Fish Physiology, Vol. 24: Behaviour and Physiology of Fish*, K.A. Sloman, R.W. Wilson and S. Balshine (eds.). Elsevier Press, Amsterdam, pp. 359-412.

Stacey, N.E., A.F. Cook and R.E. Peter. 1979. Spontaneous and gonadotropin induced ovulation in the goldfish, *Carassius auratus*. L.: Effects of external factors. *Journal of Fish Biology* 15: 305-316.

Stacey, N.E., A.L. Kyle and N.R. Liley. 1986. Fish reproductive pheromones. In: *Chemical Signals in Vertebrates*, D. Duvall, D. Müller-Schwarze and R.M. Silverstein (eds.). Plenum Press, New York, Vol. 4, pp. 117-133.

Stacey, N.E., P.W. Sorensen, G.J. Van Der Kraak and J.G. Dulka. 1989. Direct evidence that 17α,20ß-dihydroxy-4-pregnen-3-one functions as a goldfish primer pheromone: Preovulatory release is closely associated with male endocrine responses. *General and Comparative Endocrinology* 75: 62-70.

Stacey, N.E., W.B. Zheng and J.R. Cardwell. 1994. Milt production in common carp (*Cyprinus carpio*): Stimulation by a goldfish steroid pheromone. *Aquaculture* 127: 265-276.

Stacey, N.E., E.J. Fraser, P.W. Sorensen and G.J. Van Der Kraak. 2001. Milt production in goldfish: Regulation by multiple social stimuli. *Comparative Biochemistry and Physiology* C 130: 467-476.

Stacey, N.E., A. Chojnacki, A. Narayanan, T.B. Cole and C.A. Murphy. 2003. Hormonally-derived sex pheromones in fish: exogenous cues and signals from gonad to brain. *Canadian Journal of Physiology and Pharmacology* 81: 329-341.

Stearley, R.F. and G.R. Smith. 1993. Phylogeny of the Pacific trouts and salmons (*Oncorhynchus*) and the genera of the family Salmonidae. *Transactions of the American Fisheries Society* 122: 1-33.

Sveinsson, T. and T.J. Hara. 2000. Olfactory sensitivity and specificity of arctic char, *Salvelinus alpinus*, to a putative male pheromone, prostaglandin $F_{2\alpha}$. *Physiology and Behavior* 69: 301-307.

Taborsky, M. 2001. The evolution of bourgeois, parasitic, and cooperative reproductive behaviors in fishes. *Journal of Heredity* 92: 100-110.

Tavolga, W.N. 1956. Visual, chemical and sound stimuli as cues in the sex discriminatory behavior of the gobiid fish, *Bathygobius soporator*. *Zoologica* 41: 49-64.

Taylor, J. and R. Mahon. 1977. Hybridization of *Cyprinus carpio* and *Carassius auratus*, the first two exotic species in the lower Laurentian Great Lakes. *Environmental Biology of Fishes* 1: 205-208.

Thomas, P. 2003. Rapid, nongenomic steroid actions initiated at the cell surface: lessons from studies in fish. *Fish Physiology and Biochemistry* 28: 3-12.

Trudeau, V.L. 1997. Neuroendocrine regulation of gonadotrophin II release and gonadal growth in goldfish. *Reviews in Reproduction* 2: 55-68.

Twohey, M.B., P.W. Sorensen and W. Li. 2003. Possible applications of pheromones in an integrated sea lamprey management program. *Journal of Great Lakes Research* 29 (Supplement 1): 794-800.

Van Den Hurk, R. and J.W. Resink. 1992. Male reproductive system as sex pheromone producer in teleost fish. *Journal of Experimental Zoology* 261: 204-213.

Van Weerd, J.H., M. Sukkel, J.G.D. Lambert and C.J.J. Richter. 1991. GCMS-identified steroids and steroid glucuronides in ovarian growth-stimulating holding water from adult African catfish, *Clarias gariepinus*. *Comparative Biochemistry and Physiology* 98: 303-311.

Vermeirssen, E.L.M. and A.P. Scott. 1996. Excretion of free and conjugated steroids in rainbow trout (*Oncorhynchus mykiss*): Evidence for branchial excretion of the maturation-inducing steroid 17α,20β-dihydroxy-4-pregnen-3-one. *General and Comparative Endocrinology* 101: 180-194.

Vermeirssen, E.L.M. and A.P. Scott. 2001. Male priming pheromone is present in bile, as well as urine, of female rainbow trout. *Journal of Fish Biology* 58: 1039-1045.

Vermeirssen, E.L.M., A.P. Scott and N.R. Liley. 1997. Female rainbow trout urine contains a pheromone which causes a rapid rise in plasma 17,20β-dihydroxy-4-pregnen-3-one levels and milt amounts in males. *Journal of Fish Biology* 50: 107-119.

Villars, T.A., N. Hale and D. Chapnick. 1985. Prostaglandin $F_{2\alpha}$ stimulates reproductive behavior of female paradise fish (*Macropodus opercularis*). *Hormones and Behavior* 19: 21-35.

Waring, C. and A. Moore. 1995. F-series prostaglandins have a pheromonal priming effect on mature male Atlantic salmon (*Salmo salar*) parr. In: *Proceedings of the Fifth International Symposium on the Reproductive Physiology of Fish*, F.W. Goetz and P. Thomas (eds.). Fish Symposium 95, Austin, Texas, pp. 255-257.

Waring, C.P., A. Moore and A.P. Scott. 1996. Milt and endocrine responses of mature male Atlantic salmon (*Salmo salar* L.) parr to water-borne testosterone, $17\alpha,20\beta$-dihydroxy-4-pregnen-3-one 20-sulfate, and the urines from adult female and male salmon. *General and Comparative Endocrinology* 103: 142-149.

Wisenden, B.D. and N.E. Stacey. 2005. Fish semiochemicals and the evolution of communication networks. In: *Animal Communication Networks*, P. McGregor (ed.). Cambridge University Press, London, pp. 540-567.

Yamazaki, F. 1990. The role of urine in sex discrimination in the goldfish *Carassius auratus*. *Bulletin of the Faculty of Fisheries*, Hokkaido University 41: 155-161.

Yambe, H. and F. Yamazaki. 2000. Urine of ovulated female masu salmon attracts immature male parr treated with methyltestosterone. *Journal of Fish Biology* 57: 1058-1064.

Yambe, H. and F. Yamazaki. 2001. A releaser pheromone that attracts methyltestosterone-treated immature fish in the urine of ovulated female trout. *Fisheries Science* 67: 214-220.

Yambe, H., M. Shindo and F. Yamazaki. 1999. A releaser pheromone that attracts males in the urine of mature female masu salmon. *Journal of Fish Biology* 55: 158-171.

Yambe, H., A. Munakata, S. Kitamura, K. Aida and N. Fusetani. 2003. Methyltestosterone induces male sensitivity to both primer and releaser pheromones in the urine of ovulated female masu salmon. *Fish Physiology and Biochemistry* 28: 279-280.

Youngson, A.F., D. Knox and R. Johnstone. 1992. Wild adult hybrids of *Salmo salar* L. and *Salmo trutta* L. *Journal of Fish Biology* 40: 817-820.

Zhang, C., S.B. Brown and T.J. Hara. 2001. Biochemical and physiological evidence that bile acids produced and released by lake char (*Salvelinus namaycush*) function as chemical signals. *Journal of Comparative Physiology* B 171: 161-171.

Zheng, W. and N.E. Stacey. 1996. Two mechanisms for increasing milt volume in male goldfish. *Journal of Experimental Zoology* 276: 287-295.

Zheng, W. and N.E. Stacey. 1997. A steroidal pheromone and spawning stimuli act *via* different neuroendocrine mechanisms to increase gonadotropin and milt volume in male goldfish (*Carassius auratus*). *General and Comparative Endocrinology* 105: 228-235.

Zheng, W., C. Strobeck and N.E. Stacey. 1997. The steroid pheromone 4-pregnen-$17\alpha,20\beta$-diol-3-one increases fertility and paternity in goldfish. *Journal of Experimental Biology* 200: 2833-2840.

Zielinski, B., W. Arbuckle, A. Belanger, L.D. Corkum, W. Li and A.P. Scott. 2003. Evidence for the release of sex pheromones by male round gobies (*Neogobius melanostomus*). *Fish Physiology and Biochemistry* 28: 237-239.

Zippel, H.P., P.W. Sorensen and A. Hansen. 1997. High correlation between microvillous olfactory receptor cell abundance and sensitivity to pheromones in olfactory nerve-sectioned goldfish. *Journal of Comparative Physiology* A 180: 39-52.

Reproductive Physiology in Viviparous Teleosts

Yasunori Koya

INTRODUCTION

Recently, studies on reproductive physiology in fishes have progressed rapidly through research using model species. From these studies, all the mechanisms of endocrine control of sex differentiation, puberty, oogenesis and oocyte maturation are being elucidated in female oviparous fishes. In viviparous teleosts, numerous morphological studies had been made on the maternal-embryonic trophic relationship by the 1980s, and several excellent reviews have been published (Wourms, 1981; Wourms *et al.*, 1988; Schindler and Hamlett, 1993). However, as Wourms *et al.* (1988) pointed out, there have been few studies on reproductive endocrinology of viviparous teleosts; consequently, there was not enough information to formulate a review. Some easy-to-breed poeciliid fish, including the guppy (*Poecilia reticulata*) and mosquitofish (*Gambusia affinis*), have been used as experimental animals in endocrinological research for many years, and the number of references related to hypophysial hormones is quite

Author's address: Department of Biology, Faculty of Education, Gifu University, Yanagido, Gifu 501-1193, Japan. E-mail: koya@gifu-u.ac.jp

large. Based on the progress of studies on reproductive endocrinology in oviparous fish for the past 20 years, and the use of Poeciliidae and *Zoarces* (Zoarcidae) as experimental animals for endocrine disruptors in recent years, there is now a considerable body of cumulative knowledge on reproductive endocrinology in viviparous teleosts. Moreover, as an object of aquaculture, research on the reproductive physiology of *Sebastes* (Scorpaenidae) has progressed quickly since the 1980s, and still continues to grow.

Against this background, this chapter aims to introduce current knowledge concerning the reproductive endocrinology in female viviparous Poeciliidae, Zoarcidae, and Scorpaenidae, and trace the research progress comparatively as widely as possible. Many investigations to date have been summarized for each family concerning the reproductive cycle, features of viviparity, secretion cycle and function of hypophysial hormones including gonadotropin (GTH), and ovarian sex steroids and prostaglandins, as endocrine factors regulating reproduction. Particularly in the case of poeciliids, the number of references introduced has become quite large. Here, the author aims to clarify the present state of research on the reproductive endocrinology in viviparous teleosts, elucidating those subject areas that have been received little attention to aid further progress of this research field.

POECILIIDAE

Gonadal Physiology and Reproductive Cycle

All of approximately 300 viviparous species of Poeciliidae are distributed in mild climate zones, and the females repeat the vitellogenesis, oocyte maturation, pregnancy, and delivery cycle at approximately one-month intervals. In this group, there are some species in which embryos of multiple stages are conceived at once by superfetation, and which give birth repeatedly in a short cycle, like a *Heterandria* (Turner, 1937). The parturition intervals are affected by not only temperature but also photoperiod, and it is suggested that temperature primarily influences the rate of embryonic development, while the photoperiod primarily affects the progress of vitellogenesis in the mosquitofish (Koya *et al.*, 2004a). In mosquitofish, sex differentiation of the gonad has occurred at the time of birth, and paired ovaries fuse a few days after birth to become a single cystovary (Koya *et al.*, 2003a). The fish mature sexually 90 days after birth in the male, and 110 days in the female. During the process of first reproduction of the female, a single group of vitellogenic oocytes first appear at 90 days, and develop to the pre-maturation stage by 110 days after birth (Koya et al., 2003a).

Mature male poeciliids transfer their sperm to the female in the form of sperm balls (or spermatozeugmata) by copulation. It is considered that the sperm balls break down just after they are introduced into the female genital tract, and that each spermatozoon reaches the ovary by way of the oviduct. The breakdown of the sperm ball needs an environment of isotonic osmolality containing ions, including sodium and potassium (Morisawa and Suzuki, 1981). There are two main sites for sperm preservation in the ovary. One is the forefront end of the oviduct, called the 'seminal receptacle' (Jalabart and Billard, 1963). The other is a synaptic knob-shaped space on the follicle surface which is an expanded blind alley of a small tract branching off from the oviduct, called the 'delle' (Purser, 1938), or 'sperm pocket' (Kobayashi and Iwamatsu, 2002). It is believed that the epithelium of the main oviduct has a function for sperm storage, based on electron microscopic observations of sperm in which the head is deeply mounted in the cytoplasm of the epithelium of the oviduct (Potter and Kramer, 2000). It is further thought that the sperms directly participating in fertilization are the ones stored in the sperm pocket. The sperm stored in the sperm pocket is observed with electron microscopy (Kobayashi and Iwamatsu, 2002). However, it is not known how a spermatozoon is introduced into an egg in the surrounding follicle cells. There is one histological photograph that caught the moment of fertilization by Fraser and Renton (1940). The follicle layer is divided ultrastructurally into two layers separated by basement membrane (Jollie and Jollie, 1964). The first layer consists of a very thin thecal cell layer, and the second layer consists of granulosa cell layer. A very thin chorion (about 0.2 μm)—in which the micropyle is absent—is underneath the follicle cells (Kobayashi and Iwamatsu, 2002).

In Poeciliidae, the mature oocyte is not ovulated—it is fertilized within the follicle—and the fertilized egg develops within the follicle and is ovulated just before delivery. There are various types of maternal dependency of nutrition for embryonic development, from types mainly dependent on the egg yolk to types dependent on the maternal nutrition supply. However, it seems that embryos of all species of poeciliids obtain nutrition from the mother to a certain extent. In *Poecilia* and *Gambusia*, the yolk accumulation in the oocytes for the next gestation starts during the late gestation period (Stolk, 1951a; Sokol, 1955; Ishii, 1961; Young and Ball, 1983a; Koya et al., 2000). Vitellogenins (VTGs), precursors of yolk proteins, are identified and purified in the mosquitofish (Tolar et al., 2001; Sawaguchi et al., 2003, 2005). Sawaguchi et al. (2005) identified three distinct forms of VTGs from the plasma of mosquitofish, and suggested that two of these VTGs (m.w. 600 kDa in both VTGs) are processed as two different types of lipovitellin, phosvitin, and β'-component, and one VTG (m.w. 400 kDa) lacking the phosvitin domain is taken up intact into the oocyte. Under constant rearing conditions (25°C, 16L8D), mosquitofish repeat reproduction (vitellogenesis, pregnancy and delivery) at approximately 22-day intervals (Koya et al., 2000). Vitellogenesis of the oocytes for the next gestation starts one week after fertilization (10 days after the last delivery), and yolk accumulation rapidly progresses from just before delivery (20 days after the last delivery) to immediately after delivery, and then the yolky oocytes reach maturation and fertilization 3 days after delivery (Koya et al., 2000). Thus, the overlapping of pregnancy and vitellogenesis stands in the way of discussing the significance of endocrine changes during pregnancy. Mosquitofish adapted to the temperate zone have an annual reproductive cycle (Koya et al., 1998). The first vitellogenesis of the season begins with the rise of temperature in the spring. The above-stated ovarian cycle is repeated during spring and summer. Vitellogenesis of the next clutch of oocytes ceases with the shortening of the day-length during late summer, and finally reproduction concludes with the last delivery of the season (Koya and Kamiya, 2000). It is possible to analyze the process of vitellogenesis without an overlap of pregnancy through the artificial induction of vitellogenesis by raising the temperature during early spring (Koya et al., 2003b).

Gonadotropins

Intense research on the role of the pituitary hormones in reproduction, along with morphological studies of light microscopic levels in parallel with experimental studies, e.g., hypophysectomy and injection of pituitary extract, were conducted in the 1960s to the first half of the 1970s. From the second half of the 1970s onwards, the ultrastructures of each type of hormone producing cell in the pituitary were observed. Immunohistochemical studies using antibodies to hormones were done in the late 1980s. The relation of GTH with reproduction will be introduced first. Stolk (1951b) reported that the size of the anterior lobe of the pituitary changes during pregnancy reflected the activity of gonadotropic cells (GTH cells) in the guppy. In further research, Sokol (1961) showed that six kinds of functional cell types were discriminated in the adenohypophysis of the guppy, among which the GTH cells occupied the ventral half of meso-adenohypophysis (proximal pars distalis; PPD). The same distributions of the GTH cells in adenohypophysis are shown in platyfish, *Xiphophorus maclatus* (Schreibman, 1964), molly or sailfin molly, *Poecilia latipinna* (Batten *et al.*, 1975; Peute *et al.*, 1976), and guppy (Chambolle, 1977). It is suggested that the GTH cells are distributed in the PPD in the platyfish, because injected radiolabeled estradiol concentrated in this area, which was immunostainable with antiserum to ovine luteinizing hormone (LH) (Kim *et al.*, 1979). Schreibman and Margolis-Kazan (1979) observed the pituitary of the platyfish immunohistochemically using antiserum against carp (*Cyprinus carpio*) GTH-β, and they found that the specific cells not only in the PPD but also in the pars intermedia (PI) show a positive immunoreaction. However, it is now time to specify whether or not the immuno-positive cells in the PI are the GTH cells. Then, since the specific cells of the PI having shown a positive immunoreaction by electron microscopy of the adult (Margolis-Kazan *et al.*, 1981) and by immunohistochemistry of the neonate (Schreibman *et al.*, 1982a) in the platyfish using anti-carp GTH-β and rabbit luteinizing hormone-releasing hormone (LH-RH), there is the possibility that the GTH cells exist not only in the PPD but also in the PI. Schreibman *et al.* (1982a) considered the possibility of the existence of two kinds of GTH cells, i.e., follicle-stimulating hormone (FSH) producing cells (FSH cell) and LH-producing cells (LH cell), in these two immuno-positive areas. Then, it was shown that cells made immuno-positive by antiserum against mammalian LH are distributed in both the PPD and PI in the pituitary of the guppy, and that these LH cells are seen at all stages from neonate to adult (Zentel *et al.*, 1987).

The reports concerning the relationship between the activity of the GTH cells in the PPD and the ovarian cycle in poeciliid indicate that the GTH cells are activated at the time of active yolk formation immediately after delivery. The activity decreases temporarily during early pregnancy, and cells are again activated from the middle of pregnancy to just before delivery in relation to resumption of yolk formation (Sokol, 1961; Ball and Baker, 1969; Sage and Bromage 1970a; Young and Ball, 1983a). It seems that GTH participates neither in pregnancy nor delivery, although it has a role in vitellogenesis. This is supported by the facts that oocyte development is prevented by hypophysectomy, although pregnancy and delivery are normally performed in the molly (Ball, 1962). Furthermore, not pregnancy but vitellogenesis is inhibited by methallibure (gonadotropin-blocking agent) treatment from the second week (middle of pregnancy). However, pregnancy is prevented by its treatment from immediately after delivery (vitellogenic period) in the guppy (Lam et al., 1985). And finally, prevention of oocyte growth in the resulting estrogen treatment is weakened by the injection of homogenate of the anterior pituitary of frog in the mosquitofish (Ishii, 1961).

From research using oviparous teleosts in the 1980s, it became clear that two kinds of GTHs, i.e., FSH and LH exist in teleosts as well as tetrapods (Suzuki et al., 1988). Magliulo-Cepriano et al. (1994) demonstrated for the first time in poeciliid fish that the two kinds of GTH cells distributed in different areas in the pituitary, and appeared at different developmental stages, by immunostaining of the pituitary in platyfish using antiserum against FSH and LH of coho salmon (Oncorhynchus kisutch). Thus, FSH-immunoreactive cells are distributed over the PI from the ventral side of the PPD, and are seen in fish at all developmental stages, whereas LH-immunoreactive cells are distributed in the ventral PPD and in scattered clusters in the PI, and are seen in pubertal to mature fish (Magliulo-Cepriano et al., 1994). The author and others recently investigated the distribution of both FSH and LH cells in the pituitary of mosquitofish using the specific antibody to FSH-β and LH-β of mummichog (Fundulus heteroclitus) (Shimizu and Yamashita, 2002), which is the closest related species to poeciliids in previous reports. The results demonstrated that FSH-immunoreactive cells are distributed in the dorsal side of the PPD, and LH-immunoreactive cells are in the ventral side of PPD and all parts of PI (Koya unpublished data; Fig. 7.1). Moreover, when the immunohistochemical staining against anti-FSH and anti-LH antibody was investigated during the ovarian cycle of the mosquitofish, the activity of both cells showed a different change, respectively (Koya unpublished data; Table 7.1). The activity of FSH cells showed a maximum level immediately after delivery, decreased gradually from 5 days to 10 days (fertilization and early pregnancy), and increased from 15 days to 20 days (mid- to late pregnancy). This change was in agreement with the activity change of the GTH cells of the ventral side of the PPD reported previously. On the other hand, activity of LH cells showed a maximum level immediately after delivery, fell to a minimum on day 5, and recovered gradually from day 10 to 20 after that. Thus, the FSH cells maintained high activity during the vitellogenic period from before to after delivery, whereas the LH cells showed a remarkable change in activity before and after fertilization (oocyte maturation). Furthermore, although both FSH- and LH-immunoreactive cells were not observed before the breeding season in early spring,

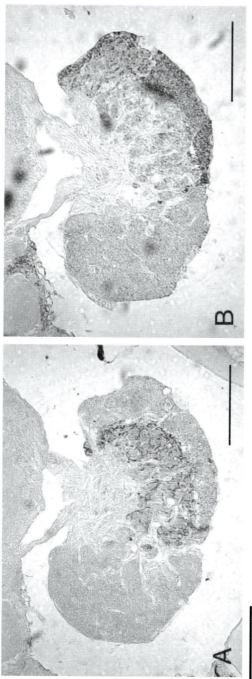

Fig. 7.1 Lateral sagittal sections of a pituitary gland of an adult female mosquitofish stained with anti-FSHβ (A) and anti-LHβ (B) of mummichog. Each section is a serial of the same sample, respectively. The anterior end is to the left. Note the numerous immunoreactive cells on the dorsal side of PPD in A, and on the ventral side of PPD and all parts of PI in B. Scale bar=100 µm.

Table 7.1   Intension of immunohistochemical staining against *mummichog* anti-FSHβ and anti-LHβ antibody during the ovarian cycle of mosquitofish.

Days after delivery	antibody	
(ovarian cycle)	FSHβ	LHfβ
0 day	+++	+++
(vitellogenesis)	+++	+++
	ND	+++
5 days	++	+
(maturation and	+++	+
fertilization)	+	+
10 days	+	++
(early pregnancy)	++	+
	+	++
15 days	++	++
(mid-pregnancy)	+++	++
	+	+++
20 days	+++	++
(late pregnancy and	+++	++
vitellogenesis)	++	+++

Observation is carried out on three females in each day. Staining intensity is divided into three grades indicated by number of +. ND indicates a fish which could not observe the section.

FSH-immunoreactive cells appeared with yolk formation initiated by the rise of water temperature, whereas LH-immunoreactive cells did not (Koya *et al.*, 2003b). In summary, it is considered that FSH participates in vitellogenesis and LH participates in final oocyte maturation in viviparous poeciliids as well as in several oviparous teleosts reported previously, including salmonids (Swanson, 1991). Moreover, it is probable that GTHs do not have an important role in pregnancy.

Thyrotropin, Thyroid Hormone and their Interaction with Pituitary-gonadal Axis

Thyroid-stimulating hormone (TSH) and thyroid hormones, with which secretion is adjusted by TSH, have also been researched in relation to reproduction in viviparous poeciliids. Stolk (1951c) reported for the first time that the thyroid gland changes periodically in relation to the ovarian cycle in the guppy. Grosso (1961) showed that as a result of thiourea treatment of the adult female guppy vitellogenesis was inhibited, miscarriage occurred, the number of deliveries decreased, and the delivery interval grew longer. He suggested that thyroxin has a certain effect on gametogenesis and pregnancy of viviparous poeciliids through the direct action of thyroxin or the interaction of TSH and GTH. The thyroid gland activity of the guppy was investigated in detail by interferometry of the colloid and histological methods, and it was shown definitely that the thyroid gland changes in relation to the ovarian cycle (Bromage and Sage, 1968). Thus, the thyroid gland activity relates to the exogenous yolk accumulation in the oocytes, and the peak of activity is seen immediately after delivery (Bromage and Sage, 1968). It is shown by light microscopy (Schreibman, 1964), electron microscopy (Batten et al., 1975), and immunohistochemical methods using the anti-human TSH-β antibody (Margolis-Kazan and Schreibman, 1981) that TSH producing cells (TSH cells) are distributed on the dorsal side of the PPD. It was revealed in detail by light microscopy (Sage and Bromage, 1970a) and electron microscopy (Young and Ball, 1983b) that the TSH cells show parallel changes reflecting periodic changes of the thyroid gland. Since the activity of the TSH cells indicates a parallel change with not only thyroid activity but also the GTH cells, Young and Ball (1983b) have suggested the possibility that thyroid hormones have an active role in the regulation of vitellogenesis.

Neurosecretory substances from the hypothalamus and/or the feedback action of steroid hormones adjust regulation of the secretion of hypophysial hormones including GTH. Secretion of GTH was reportedly stimulated by mammalian luteinizing hormone-releasing hormone (LH-RH), and is inhibited by dopamine in an organ culture experiment of the pituitary in the molly (*Poecilia latipinna*) (Groves and Batten, 1986a). Moreover, it is also indicated that estradiol-17β (E_2) and testosterone promote GTH secretion in the activated GTH cells, and they inhibit GTH secretion in the inactive GTH cells in the same culture experiment (Groves and Batten, 1986b). Furthermore, it seems that progesterone generally works to restrain GTH secretion, and that 17,20β-dihydroxy-4-pregnen-3-one (17,20β-DP) has no feedback action effect (Groves and Batten, 1986b). Some experiments have investigated the interaction of GTH and TSH, using cultures of the pituitary or administration of live fish with steroid and thyroid hormones. It is suggested that thyroxin inhibits the GTH cells *in vitro* or *in vivo*, and that estrogen activates the TSH cells, while androgen directly activates the thyroid gland, inhibiting the activity of the TSH cells *in vivo*, although both steroids inhibit the TSH cells *in vitro* (Sage and Bromage, 1970b). From these results, Sage and Bromage

(1970b) advocated a model of interaction of the hypothalamus-pituitary-gonadal axis and hypothalamus-pituitary-thyroid gland axis (Fig. 7.2). When the female guppy is immersed in thyroxin immediately after delivery, the delivery interval is shortened (Lam and Loy, 1985), suggesting the possibility that thyroxin stimulates the gonad directly, and accelerates yolk formation and early embryonic development. In this way, thyroid hormones may influence not only vitellogenesis but also the growth of embryos. In oviparous teleosts, it is suggested that thyroid hormones are taken up into the oocyte with the yolk (Monteverdi and Di Giulio, 2000), and have an important role in early embryonic development (Tagawa et al., 1990). It may be possible that such a phenomenon occurs also in viviparous teleosts.

Adrenocorticotropin

The relation between reproduction and adrenocorticotropin (ACTH) among adenohypophysial hormones will be described last. Sage and Bromage (1970a) reported that ACTH-producing cells (ACTH cells) remain inactive during pregnancy, as shown by light microscopic observations of the pituitary in the guppy. However, based on the results of electron microscopic observations of the pituitary in the molly, clear

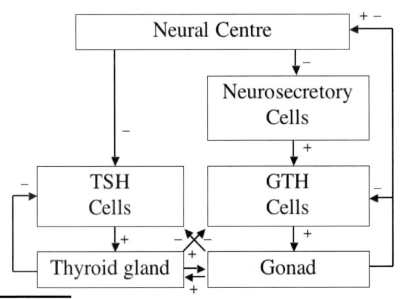

Fig. 7.2 Interactions in the endocrine system of *Poecilia* (altered Sage and Bromage, 1970b). +: stimulation; –: inhibition.

changes of activity in the ACTH cells during pregnancy are now accepted (Young and Ball, 1983c). That is, the ACTH cells remain relatively inactive during vitellogenesis, and become moderately active during late pregnancy, irrespective of whether vitellogenic oocyte growth has commenced or not. From these results, the possibility has been discussed that ACTH has a certain role through the secretion of corticosteroids at the time of delivery (Young and Ball, 1983c). From unpublished data of measured plasma levels of cortisol during the ovarian cycle of the guppy (see Venkatesh *et al.*, 1992a), cortisol concentration in the blood plasma—which remains high during the late gestation—decreases rapidly just before delivery. The fall of the plasma cortisol level during late gestation may possibly trigger the delivery as described below. Since hypophysectomy does not prevent delivery (Ball, 1962), it seems that the cyclic change of the ACTH cells during the ovarian cycle is not indispensable to delivery.

Steroid Hormones

Lambert (1970a) investigated steroidogenesis in the poeciliid ovary in relation to the distribution of enzymes for steroid biosynthesis within the ovary of the guppy. This research showed that enzymes indispensable for steroid biosynthesis, such as 3β-hydroxy-steroid dehydrogenase (3β-HSD), 3α-HSD, and 17β-HSD, are located in the granulosa cells, and suggested that the granulosa cells are the main sites for steroidogenesis. Moreover, it is also suggested that the atretic follicles, which are observed frequently in the ovary of the guppy, do not show steroidogenesis activity since the existence of enzymes required for steroid biosynthesis has not been proved (Lambert, 1970b). Schreibman *et al.* (1982b) investigated the distribution of 3β-HSD and glucose-6-phosphate dehydrogenase using platyfish, and showed that both enzymes are localized not in the granulosa cells but in the stromal cells of the ovary. Thus, in poeciliids, the two-cell type model of the steroidogenesis accepted by some oviparous teleosts (Nagahama, 1987) has not yet been proven.

Lambert and Pot (1975) suggested that the delta-5 pathway, i.e., from pregnenolone via 17α-hydroxypregnenolone, dehydroepiandrostendione, and androstendione to testosterone, is the predominant steroidogenic pathway in the ovary of the guppy, from the results of the $^3H/^{14}C$ ratios of metabolites incubated homogenates of the ovaries with two kinds of radio-labeled precursor steroids, pregnenolone-7α-^3H and progesterone-4-^{14}C. However, it has been suggested that steroidogenesis of the ovary of the guppy proceeds on the delta-4 pathway as shown from the study results of the steroid metabolism of intact follicles (Venkatesh *et al.*, 1992b). It is thought that the difference in the results depends on incubation methods, i.e., whether using homogenate of the whole ovary or isolated follicles. Venkatesh *et al.* (1992b) discovered that not only the ovarian follicle but also the extrafollicular tissue (EF tissue), which is the remaining tissue after isolating the follicles from the ovary, had the ability of steroidogenesis in the guppy. It is possible that the steroidogenic pathway differs in the follicle from the EF tissue.

The ovary of the molly transformed radiolabeled testosterone into 5β-reduced androgens and glucuronides, and since the yields of the metabolites were highest in vitellogenic follicles, it is suggested that these conjugates have a certain physiological activity such as pheromones (Kime and Groves, 1986). It is similarly shown by the ovarian follicle of the guppy that some polar 7-hydroxylated steroids are synthesized besides 5β-reduced metabolites and glucuronides (Venkatesh *et al.*, 1992b). However, classical steroids are also detected in the guppy, and the amount of synthesis of E_2 is high in the vitellogenic follicle (Venkatesh *et al.*, 1992c). In accordance with this result, a high plasma concentration of E_2 is maintained during vitellogenesis from immediately after delivery to fertilization (Venkatesh *et al.*, 1990). Moreover, the same results are also obtained in the molly (Kime and Groves, 1986), and it is further indicated that in viviparous poeciliids vitellogenesis is controlled by E_2. It is shown in the molly that the high plasma concentration of E_2 during vitellogenesis decreases rapidly by ovariectomy or hypophysectomy (Kime and Groves, 1986). From the results of this experiment, it is thought that active secretion of GTH occurs from the GTH cells in the pituitary, which show high activity during the vitellogenic period as aforementioned, and that synthesis of E_2 in the ovary is stimulated by GTH. When yolk accumulation of oocytes is induced by raising the temperature in mosquitofish, the FSH cells become active the following day after the rise in water temperature, the amount of E_2 production by the ovarian follicle increases on the second day, plasma concentration of VTG also increases rapidly, and yolk accumulation into oocytes is observed on the third day (Koya *et al.*, 2003b). Thus, the mechanisms of endocrine control of vitellogenesis in viviparous poeciliids and oviparous teleosts are considerably clarified.

In the guppy, production of E_2 decreases rapidly after vitellogenesis, and in turn, $17,20\beta$-DP is produced (Venkatesh *et al.*, 1992c). Although its plasma concentration does not show any change during the growth of oocytes or pregnancy, higher values are shown at the time of fertilization than at late gestation (Venkatesh *et al.*, 1990). From these results, the possibility is suggested that 17,20β-DP function as a maturation-inducing hormone (MIH) in viviparous poeciliid as well as in many oviparous teleosts. However, in the ovarian follicle after completed vitellogenesis, 5β-reduced progestins and androgens come to be synthesized in addition to 17,20β-DP, and the amounts of these metabolites are higher than that of 17,20β-DP (Venkatesh *et al.*, 1992b). It is suggested that the production of the 5β-reduced androgens have a role in the inhibition of aromatase activity during this period. However, the function of 5β-reduced progestins detected in large quantities is presently unknown. It is difficult to think that these compounds have functions specific to viviparity, since some reports have indicated that such compounds are produced also in oviparous teleosts (see Venkatesh *et al.*, 1992b). When it enters the gestation period, the steroid metabolism changes dramatically, precursors are metabolized into higher polar and water-soluble compounds (see Venkatesh *et al.*, 1992b), and the amounts of 17,20β-DP, 17α-hydroxyprogesterone, and progesterone produced become low (Venkatesh *et al.*, 1992c). Moreover, the plasma concentrations of these three kinds of progestins do not show any significant changes during the gestation period (Venkatesh *et al.*, 1990). This indicates that in viviparous poeciliids 17,20β-DP does not participate in the maintenance of pregnancy like *Sebastes* held, as mentioned later. It is indeed possible that highly polar compounds are chemical signals, indicating that the female is pregnant (Venkatesh *et al.*, 1992b).

Hormonal Control of Parturition

The experimental investigations on endocrine control of parturition have included hormone injection done since 1960s, and most research questions regarding poeciliids were solved by the research of Venkatesh and his collaborators from the second half of the 1980s to the first half of the 1990s. Ishii (1961, 1963) has conducted some experiments on hormone treatment and temperature control using mosquitofish. His research revealed that in estrogen treatment during late pregnancy (Ishii, 1961), a sudden fall of water temperature and injection of homogenate of neurohypophysis or oxytocin (Ishii, 1963) are factors which cause premature delivery. Kujala (1978) has reported that in the guppy, injection of deoxycortisol, oxytocin, vasotocin, and neurohypophysis of carp induced premature delivery, and deoxycortisol induced follicular rupture *in vitro*. Although various environmental and endocrinological factors are related to delivery of poeciliids, it is possible that neurohypophysial hormones have a direct influence on delivery. Then, Venkatesh *et al.* (1991) showed that in the guppy, E_2 and antiprogestin (RU 486) cause premature delivery, whereas progesterone, 17,20β-DP, cortisol, and aromatase inhibitor (4-hydroxyandrost-4-ene-3,17-dione) delay delivery.

Since mature embryos are discharged from the follicle (equivalent to ovulation in oviparous teleosts) in advance of delivery in poeciliid fish, it is necessary to clarify the mechanisms of ovulation in order to clarify the mechanisms of delivery. It has been suggested that prostaglandins (PGs) have a role in ovulation in some oviparous teleosts including yellow perch (*Perca flavescens*) (Goetz and Theofan, 1979) and brook trout (*Salvelinus fontinalis*) (Goetz *et al.*, 1982). Tan *et al.* (1987) investigated the synthesis of PGs using the post-partum follicle (PPF) after delivery in the guppy, and showed that PPF had the ability to synthesize both prostaglandin E (PGE) and prostaglandin F (PGF) from arachidonic acid, the precursor of PGs. The same investigators suggested the possibility that PGs participate in ovulation as well as oviparous teleosts. That PGs induce ovulation and successive delivery was confirmed by the injection of prostaglandin E_2 (PGE_2) or prostaglandin F_2 ($PGF_{2\alpha}$) during late pregnancy, which induced premature delivery (Venkatesh *et al.*, 1992d), and both PGs induced the discharge of the embryo from the follicle *in vivo* and *in vitro* (Venkatesh *et al.*, 1992e). Venkatesh *et al.* (1992a) further investigated the synthesizing ability of PGs in the follicle in the ovarian cycle in the guppy, and showed the ability for PG synthesis was low during vitellogenesis, oocyte maturation, and early pregnancy, increased during middle to late pregnancy, but then declined after delivery. Moreover, they discovered that in the guppy, EF tissue has the ability to synthesize a large amount of PGs, and pointed out the possibility that PGs which control ovulation and delivery may be mainly synthesized and secreted by the EF tissue. Production of PGs by the PPF is inhibited by cAMP and forskolin in a dose-dependent manner (Tan *et al.*, 1987), while PG production by the ovarian follicle and EF tissue is inhibited by cAMP and cortisol (Venkatesh *et al.*, 1992a). These results have suggested that cAMP and cortisol participate in inhibition of ovulation immediately after oocyte maturation. Moreover, since the plasma concentration of cortisol remains high during pregnancy and decreases rapidly just before delivery (see Venkatesh *et al.*, 1992a), it is suggested that cortisol cancels the block of PG production and increases the same. This is supported by the report of Venkatesh *et al.* (1991) that cortisol delays delivery. In delivery of poeciliids, it is considered that PGs synthesized in ovarian tissue may be the first trigger, after which the discharge of the embryo from the follicle is induced, and then subsequent contraction of the smooth muscle of the ovary by secretion of neurohypophysial hormones induces discharge of the embryo to outside the ovary.

In such a scenario, how is the action of E_2, which induces premature delivery, explained? It is thought that production of E_2 by the follicle becomes active just before delivery since vitellogenesis for the next pregnancy has begun during late pregnancy. In fact, in the guppy, the E_2 concentration in the blood plasma begins to increase during late pregnancy (Venkatesh et al., 1990), and production of E_2 by the follicle also increases during this period (Venkatesh et al., 1992c). However, since E_2 and other sex steroids do not influence PG production (Venkatesh et al., 1992a), it is conceivable that the delivery-inducing action by E_2 depends on an antagonism with cortisol, or mediates a route other than PG synthesis. This is a subject that requires further research.

ZOARCIDAE

Reproductive Cycle

There are two viviparous species of Zoarcidae among about 65 existing species (Wourms, 1981). Both species belong to *Zoarces*, Northeast Atlantic species (*Z. viviparus*), and Northwest Pacific species (*Z. elongatus*). Both species are the bottom fish, which inhabit the blackish water of the coast to the offshore waters of the temperate to sub-Arctic zones. *Z. viviparus* has long been known to be viviparous, and since this fish also inhabits a comparatively polluted environment it is used as a marker species, indicating the influence of endocrine disrupting substances in recent years (see Rasmussen et al., 2002). The annual reproductive cycle of *Zoarces* will be outlined using the example of *Z. elongatus* whose annual changes of ovarian histology have been observed in detail (Koya et al., 1993a). In *Z. elongatus*, accumulation of endogenous yolk (yolk vesicle or cortical alveoli) occurs during the gestation period between December to April; after that, exogenous yolk formation occurs from May to August. The oocyte, which matures in September, ovulates and is fertilized in the ovarian cavity. The fertilized egg hatches within the ovary in October, the embryo continues to grow within the ovary after hatching, and is delivered between March to April. Copulation by the male is presumed to take place around August, just before ovulation based on testicular development (Koya et al., 1993b). No specialized structure for sperm storage is found in the ovary (Koya *et al.*, 1993a). It seems that the reproductive cycle of *Z. viviparus* is almost the same as that of *Z. elongatus*; vitellogenesis generally starts in May or June, the mature oocyte is ovulated and fertilized in August, the fertilized egg develops within the ovarian cavity until December, and birth occurs from January to February (see Schmidt, 1920; Korsgaard, 1983; Kosior and Kuczyński, 1997; Larsson *et al.*, 2002). However, in the North Sea in the northern Netherlands, it is reported that there are two gestation periods per year, i.e., from September to January and from April to July (Bretschneider and DeWit, 1947).

The oocyte of *Z. elongatus* grows remarkably during the vitellogenic period, and the egg (about 4.4 mm in diameter) is comparatively large (Koya *et al.*, 1993a). Identified VTG include the 540 kDa protein in *Z. elongatus* (gel filtration, Koya *et al.*, 1997), and 500 kDa protein (Native-PAGE, Korsgaard and Pedersen, 1998) and 170 kDa protein in *Z. viviparus* (SDS-PAGE, Larsson *et al.*, 2002). In *Z. elongatus*, serum levels of VTG begin to increase in May when vitellogenesis begins, they peak in September, decrease rapidly in October, and are maintained around the lower detection limit thereafter (Koya *et al.*, 1997). Moreover, in *Z. viviparus* VTG is detected at high concentrations in June and September in accordance with the advance of yolk formation (Larsson *et al.*, 2002).

The embryo hatched within the ovary during early pregnancy grows remarkably after absorbing its own egg yolk during mid- to late pregnancy (Koya *et al.*, 1994). The wet weight of the embryo increases 12 times, from 20 mg of the fertilized egg to 240 mg in *Z. viviparus* (Bretschneider and DeWit, 1947), and from 45 mg to 540 mg (12 times) and dry weight from 7 mg to 65 mg (9.3 times) in *Z. elongatus* (Koya *et al.*, 1994). The increase in dry weight provides evidence that the embryo is dependent on the nutrition supply from the mother. The hindgut of the embryo is well developed (Kristoffersson *et al.*, 1973), and has been shown experimentally to have the ability to absorb protein from outside fluid (Koya *et al.*, 1994). Moreover, it is indicated that embryo has the ability to take up amino acid though the ovarian fluid and to metabolize it as proven by research using a tracer (Korsgaard, 1992).

Pituitary Hormones

Histological investigations of the pituitary of *Zoarces* are few, but Öztan (1966) identified six kinds of hormone-producing cells in the pars distalis and two in the PI using light and electron microscopes. It has been suggested that the two kinds of acidophilic cells among them are GTH cells since their activity corresponds with that of gonad (Öztan, 1966). In order to show whether the hypophysial hormones are required for maintenance of pregnancy, the author and others tried an experiment using hypophysectomies from *Z. elongatus* in mid-pregnancy (late November) (Koya *et al.*, unpublished data). We prepared four experimental groups: of hypophysectomized fish (hypox group); fish injected with male pituitary homogenate once per week after hypophysectomy (injection group); sham-operated fish (sham group); and intact control fish (intact group). As a result, the embryo did not die or was not released even after 60 days in the hypox group. However, premature delivery occurred within 20 days in the sham group. Moreover, the dry weight of embryos increased significantly during the experimental period in the intact and injection groups, whereas there was no significant increase in the hypox group. These results suggest that hypophysial hormones participate in the nutrition supply to the embryo from the mother, and that the hormone which triggers delivery is possibly secreted from the pituitary since the premature delivery, which was considered to be the result of stress in the sham group, did not occur in the hypox group.

Steroid Hormones

Reports about the steroid metabolism in the ovary of *Zoarces* suggested that 11-hydroxytestosterone is synthesized via the delta-5 pathway (Kristoffersson *et al.*, 1976), but there is no research on the changes of steroid synthetic activity accompanying the reproductive cycle. Research on the blood concentration of sex steroids indicated that E_2 is high during vitellogenesis and declines during pregnancy in both *Z. elongatus* and *Z. viviparus* (Koya *et al.*, 1997; Larsson *et al.*, 2002). It is clear that E_2 regulates vitellogenesis from the fact that VTG in the blood (Korsgaard and Petersen, 1979; Korsgaard, 1994; Koya *et al.*, 1997; Korsgaard and Pedersen, 1998) and estrogen receptor in the liver (Andreassen and Korsgaard, 2000) are induced by E_2 injection. Moreover, it is shown that the vitelline envelope protein is also induced in the blood by E_2 in *Z. viviparus* (Larsson *et al.*, 2002). In *Z. viviparus*, testosterone is contained in female blood at the same or higher concentrations as male blood (Larsson *et al.*, 2002). However, the physiological significance is unknown. Moreover, 11-ketotestosterone, a functional androgen in teleosts, is also detected at very low levels in female blood (Larsson *et al.*, 2002).

In order to investigate the influence of steroid hormones on pregnancy, some experiments have been attempted using hormone treatment on pregnant females. Although VTG and total calcium concentration in the serum of pregnant females was increased by the injection of E_2 (Korsgaard, 1994), the calcium of ovarian fluid, which is considered to be taken up directly by the embryo, decreased by the injection of ethynyl-estradiol (Korsgaard et al., 2002). The VTG induced by E_2 injection in the blood is not detected in the ovarian fluid during pregnancy (Korsgaard, 1983). These results indicate that E_2 is able to induce VTG. Rasmussen et al. (2002) showed that when a pregnant female was immersed in a solution containing E_2 or octhylphenol, which is an estrogenic substance, the calcium of the ovarian fluid decreased prominently, free amino acid in the blood of the mother fish decreased, and the growth of the embryo was disturbed remarkably. As mentioned above, it seems that the injection of E_2 into the pregnant female blocks the nutrition transfer from the mother to the embryos. It is possible that E_2 has an inhibiting effect on action or secretion (by negative feedback) of a hormone that participates in the nutrition supply to the embryo, which is considered to secrete from pituitary. Korsgaard (1994) reported that E_2 injection of pregnant females inhibits calcium absorption by embryos; thus, it may be possible for maternal E_2 to directly affect the embryo. Andreassen et al. (2003) cloned the estrogen receptor α of Z. viviparus, and showed that the receptor exists not only in the brain and pituitary of the mother but also in the embryo by RT-PCR. This suggests that exogenous E_2 may indeed affect the early embryonic development.

There is little research on the delivery of Zoarces. Only a few studies reported that the injection of progesterone alone. or progesterone and E_2 during late pregnancy stimulate delivery, whereas E_2 alone does not (Korsgaard and Petersen, 1979). This is contrary to the action of E_2 and progesterone in poeciliid. In order to clarify the endocrine mechanisms of the maintenance of pregnancy and delivery, it is necessary to investigate endocrine changes with the reproductive cycle and to conduct planned hormone treatment experiments.

SCORPAENIDAE

Reproductive Cycle

Sebastes and Sebastiscus are known in Scorpaenidae as a viviparous genus. Among them there are approximately 100 species of Sebastes, which equals one third of all Scorpaenidae. Sebastes has an annual reproductive cycle, which will be outlined using examples of the white-edged rockfish (Sebastes taczanowskii) (Takemura et al., 1987), and black rockfish (Sebastes schlegeli) (Mori et al., 2003). These fish have been extensively investigated including the histological changes of the gonads. The male fish usually matures functionally and copulates in November, and from this time onwards the female fish starts vitellogenesis. In March and April, vitellogenesis is completed, and the developed oocytes mature and are ovulated. It is thought that fertilization occurs following oocyte maturation and ovulation within the ovarian cavity. The fertilized egg develops within the ovarian cavity thereafter, and is then delivered in June. It is clear that the stimulation of copulation by the male is not necessary for vitellogenesis and oocyte maturation, since vitellogenesis and oocyte maturation occur in females isolated from males (Koya *et al.*, 2004b). The embryo grows by obtaining nutrition from its own egg yolk during pregnancy, and there is no special organ for nutrition transfer in either the ovary or embryo (Wourms, 1991). However, it has been clearly shown by subsequent experimental research using tracers, that the embryos absorb nutrients originating from the mother (MacFarlane and Bowers, 1995; Takemura *et al.*, 1995). The timing of delivery differs somewhat with the habitat region or among species (Takemura *et al.*, 1987; Takahashi *et al.*, 1991; Moore *et al.*, 2000; Mori *et al.*, 2003). Hatching occurs within the ovary just before delivery and larvae, which are free from the fertilization membrane, are brought forth all at once (Kusakari, 1995). The next clutch of oocytes does not start vitellogenesis during pregnancy (Takemura *et al.*, 1987; Mori *et al.*, 2003).

Gonadotropin-releasing Hormones

Although there is almost no research on GTH of *Sebastes*, there has been some investigation of the gonadotropin-releasing hormone (GnRH), which regulates GTH secretion. Collins *et al.* (2001), who analyzed the molecular forms of GnRH of grass rockfish (*Sebastes rastrelliger*), by radioimmunoassay (RIA) for identified GnRH molecules and high performance liquid chromatography (HPLC), detected four forms of GnRH: seabream GnRH (sbGnRH), chicken GnRH-II, salmon GnRH, and pejerrey GnRH. Among these GnRHs, sbGnRH is considered to participate in the regulation of GTH secretion since considerable amounts of GnRH have been detected in the pituitary. The brain concentration of the sbGnRH increases rapidly after delivery in female grass rockfish (Collins *et al.*, 2001). This has suggested that the amount of secretion of the sbGnRH exceeded the amount of synthesis, and it is possible that the secretion of GTH is promoted by GnRH secretion during pregnancy. The role of hypophysial hormones in reproduction, especially pregnancy in *Sebastes*, requires further investigation.

Steroid Hormones

Some research has investigated the changes of steroid hormone concentrations in the blood during the reproductive cycle of *Sebastes*. The blood concentrations of E_2 increase gradually during vitellogenesis, peak just before oocyte maturation, and decline rapidly during pregnancy (Nagahama *et al.*, 1991; Kwon *et al.*, 1999; Mori *et al.*, 2003). While a single group of oocytes is accumulated in the egg yolk, the serum E_2 concentration is high in *Sebastiscus* also, which reproduce more than twice in one breeding season (Takano *et al.*, 1991). Moreover, serum levels of VTG change almost in parallel to that of E_2 (Takemura *et al.*, 1991; Mori *et al.*, 2003). It is certain that vitellogenesis is promoted by E_2 in viviparous scorpaenids as well as oviparous teleosts.

The blood concentration of testosterone shows parallel changes to E_2 one month later, and peaks during oocyte maturation or early pregnancy, and decreases rapidly thereafter (Nagahama *et al.*, 1991; Takano *et al.*, 1991; Kwon *et al.*, 1999; Mori *et al.*, 2003). It can be imagined that testosterone is the precursor of E_2 from such a serum profile pattern. However, it is shown that maternal thyroxin is transferred from mother to oocytes during vitellogenesis to early pregnancy in *Sebastes inermis*, and that ovarian thyroxin levels are higher than reported in oviparous teleosts (Kwon *et al.*, 1999). This would suggest that testosterone might participate in the transportation of thyroxin to oocyte (Kwon *et al.*, 1999). It is thought that the high concentration of thyroxine in the ovary contributes to the maintenance of pregnancy or embryonic development. Kang and Chang (2004) reported the presence of a significant increase of whole body levels of triiodothyronine (T3) in new-born larvae by maternal T3 injection, and that the rates of growth and survival of offspring spawn from T3 injected females were higher that those of offspring spawn from control females in black rockfish.

In white-edged and black rockfish, serum 17,20β-DP is hardly detected during vitellogenesis, but increases rapidly in place of E_2 during oocyte maturation. It then maintains a high concentration during pregnancy, and then decreases rapidly around delivery (Nagahama *et al.*, 1991; Mori *et al.*, 2003). Moore *et al.* (2000) showed that in grass rockfish the plasma concentration of C21 steroids increases during pregnancy and decreases rapidly after delivery. However, in this species, it seems that the rate of increase is not high for 17,20β-DP, but is for 17,20β-dihydroxy-5β-pregnen-3-one and 17,20α-dihydroxy-4-pregnen-3-one. At any rate, it seems that C21 steroid has a certain function during oocyte maturation to pregnancy in *Sebastes*. On the other hand, it is thought that progesterone does not have any specific function for pregnancy, since its serum level changes little during pregnancy (Nagahama *et al.*, 1991). Takemura *et al.* (1989) indicated that in white-edged rockfish, 17,20β-DP has the highest maturation-inducing effect *in vitro* of all the other steroids tested, that human chorionic gonadotropin (HCG) also has the same effect, and that further HCG stimulates production of 17,20β-DP in oocyte follicle. From these results, it is considered that 17,20β-DP is synthesized by the follicle by stimulus of GTH, and has a role in MIH in *Sebastes* as well as several oviparous teleosts.

Where is 17,20β-DP, which is detected at high concentrations in the blood during pregnancy, synthesized? In black rockfish, the post-ovulatory follicle (POF) formed by the ovulation of mature oocyte maintains the cell structure on the ovarian hilus throughout pregnancy (Koya *et al.*, 2004b). This is in contrast to the POF of oviparous teleosts, which usually degenerates and is absorbed immediately after spawning (Kagawa and Takano, 1979; Matsuyama *et al.*, 1988). The POF in black rockfish has a low ability for E_2 synthesis, but a high ability for 17,20β-DP synthesis *in vitro* (Koya *et al.*, 2004b). This shows that POF of *Sebastes* is functionally homologous to the mammalian corpus luteum. What is the function of 17,20β-DP synthesized by POF? We do not yet have the data to answer this question clearly. However, the decline of 17,20β-DP is possibly a certain signal on the occasion of delivery of the embryo or unfertilized egg, since serum 17,20β-DP levels abruptly decrease to basal levels prior to the release of the embryo or egg in pregnant or non-pregnant black rockfish (Koya *et al.*, 2004b). Furthermore, pregnant females delivered embryos within one week after an abrupt decrease in serum 17,20β-DP levels, whereas non-pregnant females released unfertilized egg a little later, within two weeks, after the decline of 17,20β-DP levels (Koya *et al.*, 2004b). This implies that both a decline in 17,20β-DP levels in the mother and some response from the embryos is necessary for smooth parturition.

CONCLUSION

It has become clear that in viviparous poeciliids and scorpaenids, there is no fundamental difference from oviparous teleosts in the endocrine mechanism of gametogenesis, especially the role of steroid hormones. In zoarcids, the endocrine mechanism of final oocyte maturation has not been fully investigated. Although there are still a few reports on the role

---

of GTH in the process of gametogenesis—which strictly divided FSH and LH—it could be that in poeciliids FSH participates in vitellogenesis and LH participates in the final oocyte maturation. Elucidation of the functions of the two kinds of GTHs in other families of viviparous teleosts is needed. There are differences in the mechanisms of endocrine control of maintenance of pregnancy, at least in the role of steroid hormones, among each family. These seem to depend on whether pregnancy occurs within the follicles before ovulation or within the ovarian cavity after ovulation (Fig. 7.3). In poeciliids, pregnancy occurs within follicles with inhibition of ovulation, and the production and blood levels of progestins including 17,20β-DP, which are considered to participate final oocyte maturation, decrease immediately during pregnancy (Fig. 7.3A). In scorpaenids, however, pregnancy occurs within the ovarian cavity after ovulation with the inhibition of egg release, and the production and blood levels of 17,20β-DP are maintained at a high level during pregnancy (Fig. 7.3B). It is thought that the mechanism of pregnancy maintenance is probably different and that the role of 17,20β-DP in the process from oocyte maturation to pregnancy differs among families. Still, 17,20β-DP may have a positive function in the maintenance of pregnancy in scorpaenids.

Although hypophysial hormones participate in neither maintenance of pregnancy nor growth of the embryo during pregnancy in poeciliids, it is likely to be related to the growth of the embryo in zoarcids. Moreover, in scorpaenids, GTH is thought to be secreted during pregnancy, judging from the changes in the detected amount of GnRH. Thus, the participation of hypophysial hormones to pregnancy is different in families. The level of nutrition supply from mother to embryos during pregnancy differs with the family, and the dependence of embryos on the maternal nutrition is shown as scorpaenids<poeciliids<zoarcids. Not only the embryo but also the ovarian structure or metabolism of trophic substances of the mother are needed for the large alteration, as the degree of nutrition dependence of the embryo on the mother increases, and various hormones must become involved as factors regulating the change. Moreover, it is suggested that 17,20β-DP has a certain role in pregnancy as in scorpaenids; thus it may be possible that GTH is involved as a stimulus factor in steroid production. This point also calls for future research.

Detailed experiments focused on delivery have been conducted in poeciliids, and it is clear that PGs produced in the ovary induce ovulation prior to delivery. In oviparous teleosts, PG production in the ovary may be regulated by MIH (17,20β-DP, etc.) following the oocyte maturation,

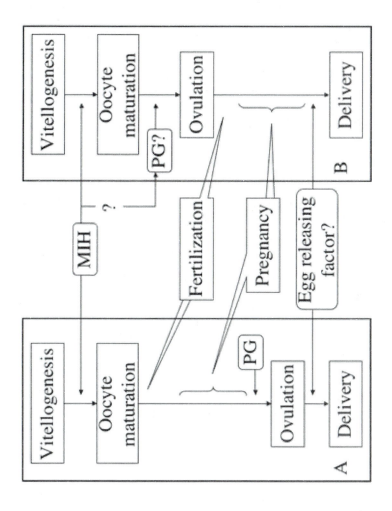

Fig. 7.3 Differences in timing of fertilization in the ovarian cycle and expected hormonal involvement in reproductive events among poeciliids (A), zoarcids (B), and scorpaenids (B).

or by another endocrine system completely different from oocyte maturation. In poeciliids at least, there is a different mechanism from oocyte maturation involved in PG production, since intrafollicular gestation is realized due to the fertilization and pregnancy occurs between oocyte maturation and ovulation (Fig. 7.3A). However, the production of PGs immediately after oocyte maturation is inhibited, and the release of the inhibition triggers pre-delivery ovulation. In zoarcids and scorpaenids, intraluminal gestation is realized due to the ovulation immediately after oocyte maturation, and it is expected that delivery occur by the same mechanism as egg release in oviparous fishes (Fig. 7.3B). It is possible to maintain the embryo over a long period within the ovarian cavity by inhibiting the stimulus for egg release. In scorpaenids, a high concentration of 17,20β-DP may have inhibited the stimulus for egg release. Further research on this point needs to be carried out.

Acknowledgments

I would like to thank Drs K. Yamauchi and S. Adachi (Hokkaido University), Dr. T. Matsubara (Hokkaido National Fisheries Research Institute, Fisheries Research Agency of Japan), Dr T. Ikeuchi (Nagahama Institute of Bio-science and Technology), Dr A. Shimizu (National Research Institute of Fisheries Science, Fisheries Research Agency of Japan), and Ms. K. Shimizu and Ms. M. Kawada (Gifu University), for their helpful advice and assistance with the unpublished research on *Zoarces elongatus* and *Gambusia affinis*. Special thanks to Drs A. Shimizu and T. Ikeuchi for their critical reading of the manuscript, and valuable counsel. I am grateful to Dr M. Nakagawa (National Center for Stock Enhancement, Fisheries Research Agency of Japan, Goto Station), and to the many graduates and alumni of the Department of Biology, Faculty of Education, Gifu University, for their cooperation in the research on viviparous fish.

References

Andreassen, R.K. and B. Korsgaard. 2000. Characterization of a cytosolic estrogen receptor and its up-regulation by 17β-estradiol and the xenoestrogen 4-tert-octylphenol in the liver of eelpout (*Zoarces viviparus*). *Comparative Biochemistry and Physiology* C 125: 299-313.

Andreassen, T.K., K. Skjoedt, I. Anglade, O. Kah and B. Korsgaard. 2003. Molecular cloning, characterisation, and tissue distribution of oestrogen receptor alpha in eelpout (*Zoarces viviparus*). *General and Comparative Endocrinology* 132: 356-368.

Ball, J.N. 1962. Brood-production after hypophysectomy in the viviparous teleost *Mollienisia latipinna* Le Sueur. *Nature (London)* 194: 787.

Ball, J.N. and B.I. Baker. 1969. The pituitary gland: anatomy and histophysiology. In: *Fish Physiology*, W.S. Hoar and D.J. Randall (eds.). Academic Press, New York, Vol. 2, pp. 1–110.

Batten, T., J.N. Ball and M. Benjamin. 1975. Ultrastructure of the adenohypophysis in the teleost *Poecilia latipinna*. *Cell and Tissue Research* 161: 239-261.

Bretschneider, L.H. and J.J.D. DeWit. 1947. *Sexual Endocrinology of Non-mammalian Vertebrates*. Elsevier, Amsterdam.

Bromage, N.R. and M. Sage. 1968. The activity of the thyroid gland of *Poecilia* during the gestation cycle. *Journal of Endocrinology* 41: 303-311.

Chambolle, P. 1977. Hypophyse et reproduction chez *Gambusia* sp. (Poisson Téléostéen, Poeciliidae vivipare). *Investigación Pesquera* 41: 1-13.

Collins, P.M., D.F. O'Neill, B.R. Barron, R.K. Moore and N.M. Sherwood. 2001. Gonadotropin-releasing hormone content in the brain and pituitary of male and female grass rockfish (*Sebastes rastrelliger*) in relation to seasonal changes in reproductive status. *Biology of Reproduction* 65: 173-179.

Fraser, E.A. and R.M. Renton. 1940. Observation on the breeding and development of the viviparous fish, *Heterandria formosa*. *Quarterly Journal of Microscopical Science* 81: 479-502.

Goetz, F.W., D.C. Smith and S.P. Krickle. 1982. The effects of prostaglandins, phosphodiestrase inhibitors and cyclic AMP on ovulation of brook trout (*Salvelinus fontinalis*) oocytes. *General and Comparative Endocrinology* 48: 154-160.

Goetz, F.W. and G. Theofan. 1979. *In vitro* stimulation of germinal vesicle breakdown and ovulation of yellow perch (*Perca flavescens*) oocytes. Effects of 17α-hydroxy-20β-dihydroprogesterone and prostaglandins. *General and Comparative Endocrinology* 37: 273-285.

Grosso, L.L. 1961. The effect of thiourea, administrated by immersion of the maternal organism, on the embryos of *Lebistes reticulatus*, with notes on the adult gonadal changes. *Biological Bulletin* 121: 481-496.

Groves, D.J. and T.F.C. Batten. 1986a. Direct control of the gonadotroph in a teleost, *Poecilia latipinna*. II. Neurohormones and neurotransmitters. *General and Comparative Endocrinology* 62: 315-326.

Groves, D.J. and T.F.C. Batten. 1986b. Direct control of the gonadotroph in a teleost, *Poecilia latipinna*: gonadal steroids. *General and Comparative Endocrinology* 61: 402-416.

Ishii, S. 1961. Effects of some hormones on the gestation of the top minnow. Journal of Faculty of Science, University of Tokyo, Section 4, 9: 279-290.

Ishii, S. 1963. Some factors involved in the delivery of the young in the top-minnow, *Gambusia affinis*. Journal of Faculty of Science, University of Tokyo, Section 4, 10: 181-187.

Jalabert, B. and R. Billard. 1969. Étude ultrastructurale du site de conservation des spermatozoïdes dans l'ovaire de *Poecilia reticulata* (Poisson Téléostéen). *Annales de Biologie Animale Biochimie et Biophysique* 9: 273-280.

Jollie, W.P. and L.G. Jollie. 1964. The fine structure of the ovarian follicle of the ovoviviparous poeciliid fish, *Lebistes reticulates*. II. Formation of follicular pseudoplacenta. *Journal of Morphology* 114: 503-526.

Kagawa, H. and K. Takano. 1979. Ultrastructure and histochemistry of granulosa cells of pre- and post-ovulatory follicles in the ovary of the medaka, *Oryzias latipes*. *Bulletin of Faculty of Fisheries*, Hokkaido University 30: 191-204. (In Japanese with English Abstract).

Kang, D.-Y. and Y.J. Chang. 2004. Effects of maternal injection of 3,5,3'-triiodo-L-thyronine (T₃) on growth of newborn offspring of rockfish, *Sebastes schlegeli*. *Aquaculture* 234: 641-655.

Kim, Y.S., M. Sar and W.E. Stumpf. 1979. Estrogen target cells in the pituitary of platyfish, *Xiphophorus maculates. Cell and Tissue Research* 198: 435-440.

Kime, D.V. and D.J. Groves. 1986. Steroidogenesis by gonads of a viviparous teleost, the sailfin molly (*Poecilia latipinna*), *in vitro* and *in vivo*. *General and Comparative Endocrinology* 63: 125-133.

Kobayashi, K. and T. Iwamatsu. 2002. Fine structure of the storage micropocket of spermatozoa in the ovary of the guppy *Poecilia reticulata*. *Zoological Science* 19: 545-555.

Korsgaard, B. 1983. The chemical composition of follicular and ovarian fluids of the pregnant blenny (*Zoarces viviparus* (L.)). *Canadian Journal of Zoology* 61: 1101-1108.

Korsgaard, B. 1992. Amino acid uptake and metabolism by embryos of the blenny *Zoarces viviparus. The Journal of Experimental Biology* 171: 315-328.

Korsgaard, B. 1994. Calcium metabolism in relation to ovarian functions during early and late pregnancy in the viviparus blenny *Zoarces viviparus. Journal of Fish Biology* 44: 661-672.

Korsgaard, B. and I. Petersen. 1979. Vitellogenin, lipid and carbohydrate metabolism during vitellogenesis and pregnancy, and after hormonal induction in the blenny *Zoarces viviparus* (L.). *Comparative Biochemistry and Physiology* 63B: 245-251.

Korsgaard, B. and K.L. Pedersen. 1998. Vitellogenin in *Zoarces viviparus*: Purification, quantification by ELISA and induction by estradiol-17β and 4-nonylphenol. *Comparative Biochemistry and Physiology* C 120: 159-166.

Korsgaard, B., T.K. Andreassen and T.H. Rasmussen. 2002. Effects of an environmental estrogen, 17α-ethynyl-estradiol, on the maternal-fetal trophic relationship in the eelpout *Zoarces viviparus* (L). *Marine Environmental Research* 54: 735-739.

Kosior, M. and J. Kuczyński. 1997. Microscopic observations of a maturation process in eelpout (*Zoarces viviparus* L.) from the southern Baltic. A proposed gonadal maturation scale. *Fisheries Research* 30: 151-155.

Koya, Y. and E. Kamiya. 2000. Environmental regulation of annual reproductive cycle in the mosquitofish, *Gambusia affinis. Journal of Experimental Zoology* 286: 204-211.

Koya, Y., S. Ohara, T. Ikeuchi, S. Adachi, T. Matsubara and K. Yamauchi. 1993a. The reproductive cycle of female *Zoarces elongatus*, a viviparous teleost. *Bulletin of the Hokkaido National Fisheries Research Institute* 57: 9-20. (In Japanese with English Abstract).

Koya, Y., S. Ohara, T. Ikeuchi, S. Adachi, T. Matsubara and K. Yamauchi. 1993b. Testicular development and sperm morphology in the viviparous teleost, *Zoarces elongatus. Bulletin of the Hokkaido National Fisheries Research Institute* 57: 21-31. (In Japanese with English Abstract).

Koya, Y., T. Ikeuchi, T. Matsubara, S. Adachi and K. Yamauchi. 1994. Embryonic growth during gestation of the viviparous eelpout, *Zoarces elongatus. Japanese Journal of Ichthyology* 41: 338-342.

Koya, Y., T. Matsubara, T. Ikeuchi, S. Adachi and K. Yamauchi. 1997. Annual changes in serum vitellogenin concentrations in viviparous eelpout, *Zoarces elongatus. Comparative Biochemistry and Physiology* A118: 1217-1223.

Koya, Y., T. Itazu and M. Inoue. 1998. Annual reproductive cycle based on histological changes in the ovary of the female mosquitofish, *Gambusia affinis*, in central Japan. *Ichthyological Research* 45: 241-248.

Koya, Y., M. Inoue, T. Naruse and S. Sawaguchi. 2000. Dynamics of oocyte and embryonic development during ovarian cycle of the viviparous mosquitofish *Gambusia affinis*. *Fisheries Science* 66: 63-70.

Koya, Y., A. Fujita, F. Niki, E. Ishihara and H. Miyama. 2003a. Sex differentiation and pubertal development of gonads in the viviparous mosquitofish, *Gambusia affinis*. *Zoological Science* 20: 1231-1242.

Koya, Y., S. Sawaguchi, K. Shimizu and A. Shimizu. 2003b. Endocrine changes during the onset of vitellogenesis in spring in the mosquitofish. *Fish Physiology and Biochemistry* 28: 349-350.

Koya, Y., S. Ishikawa and S. Sawaguchi. 2004a. Effects of temperature and photoperiod on ovarian cycle in the mosquitofish, *Gambusia affinis*. *Japanese Journal of Ichthyology* 51: 43-50. (In Japanese with English Abstract).

Koya, Y., H. Mori and M. Nakagawa. 2004b. Serum 17,20β-dihydroxy-4-pregnen-3-one levels in pregnant and non-pregnant female rockfish, *Sebastes schlegeli*, viviparous teleost, and its production by post-ovulatory follicles. *Zoological Science* 21: 565-573.

Kristoffersson, R., S. Broberg and M. Pekkarinen. 1973. Histology and physiology of embryotrophe formation, embryonic nutrition and growth in the eel-pout, *Zoarces viviparus* (L.). *Annales Zoologici Fennici* 10: 467-477.

Kristoffersson, R., S. Pesonen and T. Teräväinen. 1976. Patterns of steroid metabolism in the ovarian tissue of a viviparous teleost fish, *Zoarces viviparus* (L.). *Annales Zoologici Fennici* 13: 189-194.

Kujala, G.A. 1978. Corticosteroid and neurohypophyseal hormone control of parturition in the guppy, *Poecilia reticulata*. *General and Comparative Endocrinology* 36: 286-296.

Kusakari, M. 1995. Studies on the reproductive biology and artificial juvenile production of kurosoi *Sebastes schlegeli*. *Scientific Reports of Hokkaido Fisheries Experimental Station* 47: 41-124. (In Japanese with English Abstract).

Kwon, J.K., Y.J. Chang, Y.C. Sohn and K. Aida. 1999. Plasma and ovarian thyroxine levels in relation to sexual maturation and gestation in female *Sebastes inermis*. *Journal of Fish Biology* 54: 370-379.

Lam, T.L. and G.L. Loy. 1985. Effects of L-thyroxine on ovarian development and gestation in the viviparous guppy, *Poecilia reticulata*. *General and Comparative Endocrinology* 60: 324-330.

Lam, T.J., S.C. Pavri and K.L. Ng. 1985. Effects of methallibure on gestation in the guppy, *Poecilia reticulata*. *General and Comparative Endocrinology* 60: 315-323.

Lambert, J.G.D. 1970a. The ovary of the guppy *Poecilia reticulata*. The granulosa cells as site of steroid biosynthesis. *General and Comparative Endocrinology* 15: 464-476.

Lambert, J.G.D. 1970b. The ovary of the guppy, *Poecilia reticulata*. The atretic follicle, a *corpus atreticum* or *corpus luteum praeovulationis*. *Zeitschrift für Zellforschung und mikroskopische Anatomie* 107: 54-67.

Lambert, J.D.G. and M.G.E. Pot. 1975. Steroidogenesis in ovarian tissue of a viviparous teleost, the guppy *Poecilia reticulata*. *Comparative Biochemistry and Physiology* B50: 585-589.

Larsson, D.G.J., I. Mayer, S.J. Hyllner and L. Förlin. 2002. Seasonal variations of vitelline envelope proteins, vitellogenin, and sex steroids in male and female eelpout (*Zoarces viviparus*). *General and Comparative Endocrinology* 125: 184-196.

MacFarlane, R. and M.J. Bowers. 1995. Matrotrophic viviparity in the yellowtail rockfish *Sebastes flavidus*. *Journal of Experimental Biology* 198: 1197-1206.

Margolis-Kazan, H., J. Peute, M.P. Schreibman and L.R. Halpern. 1981. Ultrastructural localization of gonadotropin and luteinizing hormone releasing hormone in the pituitary gland of a teleost fish (the platyfish). *Journal of Experimental Zoology* 215: 99-102.

Margolis-Kazan, H. and M.P. Schreibman. 1981. Cross-reactivity between human and fish pituitary hormones as demonstrated by immunocytochemistry. *Cell and Tissue Research* 221: 257-267.

Magliulo-Cepriano, L., M.P. Schreibman and V. Blüm. 1994. Distribution of variant forms of immunoreactive gonadotropin-releasing hormone and β-gonadotropin I and II in the platyfish, *Xiphophorus maculatus*, from birth to sexual maturity. *General and Comparative Endocrinology* 94: 135-150.

Matsuyama, M., S. Adachi, Y. Nagahama and S. Matsuura. 1988. Diurnal rhythm of oocyte development and plasma steroid hormone levels in female red seabream, *Pagrus major*, during the spawning season. *Aquaculture* 73: 357-372.

Monteverdi, G.H. and R.T. Di Giulio. 2000. Vitellogenin association and oocytic accumulation of thyroxine and 3,5,3'-triiodothyronine in gravid *Fundulus heteroclitus*. *General and Comparative Endocrinology* 120: 198-211.

Moore, R.K., A.P. Scott and P.M. Collins. 2000. Circulating C-21 steroids in relation to reproductive condition of a viviparous marine teleost, *Sebastes rastrelliger* (grass rockfish). *General and Comparative Endocrinology* 117: 268-280.

Mori, H., M. Nakagawa, K. Soyano and Y. Koya. 2003. Annual reproductive cycle of black rockfish, *Sebastes schlegeli*, in captivity. *Fisheries Science* 69: 910-932.

Morisawa, M. and K. Suzuki. 1981. Osmolality and potassium ion: Their roles in initiation of sperm motility in teleosts. *Science* 210: 1145-1146.

Nagahama, Y. 1987. Gonadotropin action on gametogenesis and steroidogenesis in teleost gonads. *Zoological Science* 4: 209-222.

Nagahama, Y., A. Takemura, K. Takano, S. Adachi and M. Kusakari. 1991. Serum steroid hormone levels in relation to the reproductive cycle of *Sebastes taczanowskii* and *S. schlegeli*. *Environmental Biology of Fishes* 30: 31-38.

Öztan, N. 1966. The fine structure of the adenohypophysis of *Zoarces viviparus* L. *Zeitschrift für Zellforschung und mikroskopische Anatomie* 69: 699-718.

Peute, J., M.G.A. de Bruyn, R. Seldenrijk and P.G.W.J. van Oordt. 1976. Cytolophysiology and innervation of gonadotropic cells in the pituitary of the black molly (*Poecilia latipinna*). *Cell and Tissue Research* 174: 35-54.

Potter, H. and C.R. Kramer. 2000. Ultrastructural observations on sperm storage in the ovary of the platyfish, *Xiphophorus maculatus* (Teleostei: Poeciliidae): The role of the duct epithelium. *Journal of Morphology* 245: 110-129.

Purser, G.L. 1938. Reproduction in *Lebistes reticulatus*. *Quarterly Journal of Microscopical Science* 81: 150-158.

Rasmussen, T.H., T.K. Andreassen, S.N. Pedersen, L.T.M. Van der Ven, P. Bjerregaard and B. Korsgaard. 2002. Effects of waterborne exposure of octylphenol and oestrogen on pregnant viviparous eelpout (*Zoarces viviparus*) and her embryos in ovario. *Journal of Experimental Biology* 205: 3857-3876.

Sage, M. and N.R. Bromage. 1970a. The activity of the pituitary cells of the teleost *Poecilia* during the gestation cycle and the control of the gonadotropic cells. *General and Comparative Endocrinology* 14: 127-136.

Sage, M. and N.R. Bromage. 1970b. Interactions of the TSH and thyroid cells with the gonadotropic cells and gonads in Poecilid fishes. *General and Comparative Endocrinology* 14: 137-140.

Sawaguchi, S., Y. Koya and T. Matsubara. 2003. Deduced primary structures of three types of vitellogenin in mosquitofish (*Gambusia affinis*), a viviparous fish. *Fish Physiology and Biochemistry* 28: 363-364.

Sawaguchi, S., Y. Koya, N. Yoshizaki, N. Ohkubo, T. Andoh, N. Hiramatsu, C. V. Sullivan, A. Hara and T. Matsubara. 2005. Multiple vitellogenins (Vgs) in mosquitofish (*Gambusia affinis*): Identification and characterization of three functional Vg genes and their circulating and yolk protein products. *Biology of Reproduction* 72: 1045-1060.

Schreibman, M.P. 1964. Studies on the pituitary gland of *Xiphophorus maculates* (the platyfish). *Zoologica* 49: 217-243.

Schreibman, M.P. and H. Margolis-Kazan. 1979. The immunocytochemical localization of gonadotropin, its subunits, and thyrotropin in the teleost, *Xiphophorus maculates*. *General and Comparative Endocrinology* 39: 467-474.

Schreibman, M.P., H. Margolis-Kazan and L. Halpern-Sebold. 1982a. Immunoreactive gonadotropin and luteinizing hormone releasing hormone in the pituitary gland of neonatal platyfish. *General and Comparative Endocrinology* 47: 385-391.

Schreibman, M.P., E.J. Berkowitz and R. vandenHurk. 1982b. Histology and histochemistry of the testis and ovary of the platyfish, *Xiphophorus maculates*, from birth to sexual maturity. *Cell and Tissue Research* 224: 81-87.

Schindler, J.F. and W.C. Hamlett. 1993. Maternal-embryonic relations in viviparous teleosts. *Journal of Experimental Zoology* 266: 378-393.

Schmidt, J. 1920. Racial investigations V. Experimental investigations with *Zoarces viviparous* L. *Comptes Rendus des Travaux du Laboratoire Carlsberg* 14: 1-14.

Shimizu, A. and M. Yamashita. 2002. Purification of mummichog (*Fundulus heteroclitus*) gonadotropins and their subunits, using an immunochemical assay with antisera raised against synthetic peptides. *General and Comparative Endocrinology* 125: 79-91.

Sokol, H.W. 1955. Experimental demonstration of thyrotropic and gonadotropic activity in the adenohypophysis of the guppy, *Lebistes reticulates* (Peters). *Anatomical Record* 122: 451

Sokol, H.W. 1961. Cytological changes in the teleost pituitary gland associated with the reproductive cycle. *Journal of Morphology* 109: 219-235.

Stolk, A. 1951a. Histo-endocrinological analysis of gestation phenomena in the cyprinodont *Lebistes reticulates* Peters. IV. The oocyte-cycle during pregnancy. *Proceedings of the Koninklijke Nederlandse Akademie van Wetenschappen, C* 54: 574-578.

Stolk, A. 1951b. Histo-endocrinological analysis of gestation phenomena in the cyprinodont *Lebistes reticulates* Peters. III. Changes in the pituitary gland during pregnancy. *Proceedings of the Koninklijke Nederlandse Akademie van Wetenschappen, Series C* 54: 566-573.

Stolk, A. 1951c. Histo-endocrinological analysis of gestation phenomena in the cyprinodont *Lebistes reticulates* Peters. I. Thyroid activity during pregnancy. *Proceedings of the Koninklijke Nederlandse Akademie van Wetenschappen, Series C* 54: 550-557.

Suzuki, K., H. Kawauchi and Y. Nagahama. 1988. Isolation and characterization of two distinct gonadotropins form chum salmon pituitary glands. *General and Comparative Endocrinology* 71: 292-301.

Swanson, P. 1991. Salmon gonadotropins: Reconciling old and new ideas. In: *Reproductive Physiology of Fish*, A.P. Scott, J.P. Sumpter, D.M. Kime and M.S. Rolfe (eds.), FishSymp91, University of East Anglia, UK, pp. 2-7.

Takahashi, H., K. Takano and A. Takemura. 1991. Reproductive cycles of *Sebastes taczanowskii*, compared with those of other rockfishes of the genus *Sebastes. Environmental Biology of Fishes* 30: 23-29.

Takano, K., A. Takemura, M. Furihata, T. Nakanishi and A. Hara. 1991. Annual reproductive and spawning cycles of female *Sebastiscus marmoratus. Environmental Biology of Fishes* 30: 39-48.

Takemura, A., K. Takano and H. Takahashi. 1987. Reproductive cycle of a viviparous fish, the white-edged rockfish, *Sebastes taczanowskii. Bulletin of the Faculty of Fisheries*, Hokkaido University 38: 111-125.

Takemura, A., K. Takano and K. Yamauchi. 1989. The *in vitro* effects of various steroid hormones and gonadotropin on oocyte maturation of the viviparous rockfish, *Sebastes taczanowskii. Bulletin of the Faculty of Fisheries*, Hokkaido University 40: 1-7.

Takemura, A., K. Takano and H. Takahashi. 1995. The uptake of macromolecular materials in the hindgut of viviparous rockfish embryos. *Journal of Fish Biology* 46: 485-493.

Takemura, A., A. Hara and K. Takano. 1991. Immunochemical identification and partial characterization of female-specific serum proteins in white-edged rockfish, *Sebastes taczanowskii. Environmental Biology of Fishes* 30: 49-56.

Tagawa, M., M. Tanaka, S. Matsumoto and T. Hirano. 1990. Thyroid hormones in eggs of various freshwater, marine and diadromous teleosts and their changes during egg development. *Fish Physiology and Biochemistry* 8: 515-520.

Tan, C.H., T.J. Lam, L.Y. Wong and M.K. Pang. 1987. Prostaglandin synthesis and its inhibition by cyclic AMP and forskolin in postpartum follicles of the guppy (*Poecilia reticulata*). *Prostaglandins* 34: 697-715.

Tolar, J.F., A.R. Mehollin, R.D. Watson and R.A. Angus. 2001. Mosquitofish (*Gambusia affinis*) vitellogenin: Identification, purification, and immunoassay. *Comparative Biochemistry and Physiology* C 128: 237-245.

Turner, C.L. 1937. Reproductive cycles and superfetation in Poeciliid fishes. *Biological Bulletin* 72: 145-164.

Venkatesh, B., C.H. Tan and T.J. Lam. 1990. Steroid hormone profile during gestation and parturition of the guppy (*Poecilia reticulata*). *General and Comparative Endocrinology* 77: 476-483.

Venkatesh, B., C.H. Tan and T.J. Lam. 1991. Progestins and cortisol delay while estradiol-17β induces early parturition in the guppy, *Poecilia reticulata. General and Comparative Endocrinology* 83: 297-305.

Venkatesh, B., C.H. Tan and T.J. Lam. 1992a. Prostaglandin synthesis *in vitro* by ovarian follicles and extrafollicular tissue of the viviparous guppy (*Poecilia reticulata*) and its regulation. *Journal of Experimental Zoology* 262: 405-413.

Venkatesh, B., C.H. Tan, D.E. Kime, G.L. Loy and T.J. Lam. 1992b. Steroid metabolism by ovarian follicles and extrafollicular tissue of the guppy (*Poecilia reticulata*) during oocyte growth and gestation. *General and Comparative Endocrinology* 86: 378-394.

Venkatesh, B., C.H. Tan and T.J. Lam. 1992c. Steroid production by ovarian follicles of the viviparous guppy (*Poecilia reticulata*) and its regulation by precursor substrates, dibutyryl cAMP and forskolin. *General and Comparative Endocrinology* 85: 450-461.

Venkatesh, B., C.H. Tan. and T.J. Lam. 1992d. Prostaglandins and teleost neurohypophyseal hormones induce premature parturition in the guppy, *Poecilia reticulata*. *General and Comparative Endocrinology* 87: 28-32.

Venkatesh, B., C.H. Tan and T.J. Lam. 1992e. Effects of neurohypophyseal and adrenohypophyseal hormones, steroids, eicosanoids, and extrafollicular tissue on ovulation in vitro of guppy (*Poecilia reticulata*) embryos. *General and Comparative Endocrinology* 87: 20-27.

Wourms, J.P. 1981. Viviparity: The maternal–fetal relationship in fishes. *American Zoologist* 21: 473-515.

Wourms, J.P. 1991. Reproduction and development of *Sebastes* in the context of the evolution of piscine viviparity. *Environmental Biology of Fishes* 30: 111-126.

Wourms, J.P., B.D. Grove and J. Lombardi. 1988. The maternal-embryonic relationship in viviparous fishes. In: *Fish Physiology*, W.S. Hoar and D.J. Randall (eds.). Academic Press, San Diego Vol. 11B, pp. 1-134.

Young, G. and J.N. Ball. 1983a. Ultrastructural changes in the adenohypophysis during the ovarian cycle of the viviparous teleost *Poecilia latipinna*. I. The gonadotrophic cells. *General and Comparative Endocrinology* 48: 39-59.

Young, G. and J.N. Ball. 1983b. Ultrastructural changes in the adenohypophysis during the ovarian cycle of the viviparous teleost *Poecilia latipinna*. II. The thyrotrophic cells and the thyroid gland. *General and Comparative Endocrinology* 51: 24-38.

Young, G. and J.N. Ball. 1983c. Ultrastructural changes in the adenohypophysis during the ovarian cycle of the viviparous teleost *Poecilia latipinna*. III. The growth hormone, adrenocorticotoropic, and prolactin cells and the pars intermedia. *General and Comparative Endocrinology* 52: 86-101.

Zentel, H.J., L. Jennes, R. Reinboth and W.E. Stumpf. 1987. Ontogeny of gonadotropin releasing hormone and gonadotropin immunoreactivity in brain and pituitary of normal and estrogen-treated guppies, *Poecilia reticulata* Peters. *Cell and Tissue Research* 249: 227-234.

Mating Systems in Fishes

Martin I. Taylor[1],* and Mairi E. Knight[2]

INTRODUCTION

With more than 28,800 described species (FISHBASE, 1996), fishes are the most diverse group of vertebrates, occupying habitats ranging from ephemeral puddles to the depths of the ocean abyss. Selection pressures operating on species within these contrasting habitats are likely to be very different and, not unexpectedly, fishes display a huge diversity of mating and parental care systems from genetic monogamy, to polygynandry and from no parental care through extended biparental mouthbrooding to male pregnancy. This chapter will describe the different types of mating systems observed among sexually reproducing non-hermaphroditic fish species, review the patterns and distributions of these mating strategies and go on to discuss the potential factors shaping and maintaining them. The primary focus will be on bony fishes (Osteichthys), although Elasmobranchs will be discussed where their inclusion is informative.

Authors' addresses: [1]School of Biological Sciences, University of Wales Bangor, Bangor, Gwynedd, LL57 2UW, U.K.
[2]School of Biological Sciences, University of Plymouth, Drake Circus, Plymouth PL4 8AA, U.K.
E-mail: mairi.knight@plymouth.ac.uk
*Corresponding author: E-mail: m.taylor@bangor.ac.uk

Research into fish mating systems has lagged behind those of birds and insects, principally due to the inherent difficulties in studying subjects which live both underwater and frequently at depths or in environments that are difficult to access. Despite the physical challenges, there have been regular and significant advances in our understanding of fish mating systems over the last 40 years. Breder and Rosen's (1966) *Modes of Reproduction in Fishes* and Thresher's (1984) *Reproduction in Coral Reef Fishes* have proved to be invaluable resources. Since then, there have been several key reviews summarizing recent advances, notably by Barlow (1984, 1986), Berglund (1997), Avise *et al.* (2002) and Whiteman and Cote (2004). We direct interested readers to these publications.

Many advances in the field over the past two decades have come through the application of genetic markers. Microsatellites, also known as variable number tandem repeats (VNTRs), and simple sequence repeats (SSRs), have in particular revolutionized numerous aspects of evolutionary biology, the study of fish mating systems being no exception. Most significant has been the ability to assign parentage to offspring with certainty. A considerable number of studies have now taken such an approach, and have often yielded revelatory results. Such studies have highlighted the need to describe mating systems both in social and genetic terms since they have exposed the fact that the two often do not concur. In particular, behaviours such as cuckoldry and sneaking are now known to be extremely common. Molecular markers have been intensively used in the study of a number of fish families, most notably the cichlids (Kellogg *et al.*, 1995, 1998; Parker and Kornfield, 1996; Knight *et al.*, 1998; Taylor *et al.*, 2003), syngnathids (Jones and Avise, 1997, 2001; Kvarnemo *et al.*, 2000), sticklebacks (Jones *et al.*, 1998b; Blais *et al.*, 2004) and guppies (Becher and Magurran, 2004), among others. While molecular techniques have revolutionized the field, it is important to be cautious in interpreting results. For example, using allozymes, Chesser *et al.* (1984) reported 56% of female mosquitofish (*Gambusia affinis*) mating with multiple males, whereas a later study using microsatellites reported that 100% of females mate multiply (Zane *et al.*, 1999). While there are several potential explanations for this discrepancy, the most likely cause is that allozymes lack sufficient variability to detect multiple mating in a large percentage of cases. Thankfully, there is now an extensive literature addressing the resolving powers of genetic markers (e.g., Neff and Pitcher, 2002).

Mating systems are strongly influenced by the form of parental care. A full review of patterns and evolutionary transitions of parental care is beyond the scope of this chapter (see e.g., Blumer, 1982 for an overview), but we provide a brief summary as a primer. In teleost fishes, a lack of parental care by either parent appears to be the ancestral state (Gittleman, 1981; Gross and Sargent, 1985; Reynolds *et al.*, 2002). Approximately, 80% of teleost fishes and all elasmobranchs display no parental care at all (Gross, 2005). Uniparental care, biparental care and mouthbrooding appear to have evolved multiple times from the no-care state (Reynolds *et al.*, 2002). Where parental care does occur in the fishes, approximately 50% of these species exhibit paternal care, with approximately 30% displaying maternal care and 20% biparental care (Gross, 2005). Livebearing in fishes is relatively common and appears to have evolved from egg laying at least twelve times in teleost fishes (Goodwin *et al.*, 2002) and nine to ten times in Elasmobranchs (Dulvy and Reynolds, 1997). Mouthbrooding appears to have evolved many times from biparental substrate guarding in the cichlid fishes (Goodwin *et al.*, 1998, see also Koblmuller *et al.*, 2004) but its origins in the eight other families it is found in are currently largely uncharacterized (see Rüber *et al.*, 2004).

DEFINING MATING SYSTEMS

Monogamy

Monogamy appears to be the most clear-cut of mating systems: a single male mates with a single female and together they produce offspring (Fig. 8.1). In reality, situations are rarely this simple and many variations on that basic theme exist that are frequently specific to particular taxa. At its most extreme, the term monogamy describes long-term pair-bonding with strict partner fidelity, but in fishes such systems are extremely rare, and appear to be restricted to the butterfly fishes (Chaetodontidae; Fricke, 1973). Pairing need not be permanent for a mating system to qualify as monogamous. Thus, monogamy also refers to situations where individuals pair up and mate exclusively with each other for one mating bout, be that with or without parental care. Such cases are often described as 'serially' or 'sequentially' monogamous. Species may also be classified as 'socially' or 'genetically' monogamous. Many species may establish apparently monogamous pair-bonds but indulge in extra-pair copulations and are, thus, only socially monogamous. Genetic monogamy is usually restricted to cases where an exclusive mating relationship between a single male and a single female exists. This may be for a single breeding season or longer. Genetic monogamy is the rarest form of monogamy and

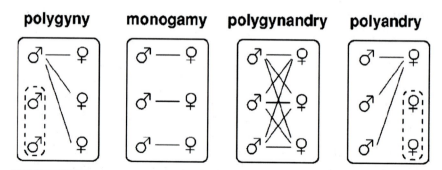

Fig. 8.1 Schematic representation of mating systems, revised from Avise *et al.* (2002).

to date has only been documented in Syngnathidae (Jones *et al.*, 1998a; Kvarnemo *et al.*, 2000), Cichlidae (Taylor *et al.*, 2003), and Centrarchidae (DeWoody *et al.*, 2000).

Polygamy

There is some confusion within the existing literature regarding the terminology associated with polygamous mating systems. Polygamy itself is a descriptive term covering all situations where at least one sex mates with more than one partner. Within polygamy, there are three possible states (Fig. 8.1):

> *polygyny*, where a single male mates with several females, but each female only mates with a single male
> *polyandry*, where a single female mates with several males, but each male only mates with a single female
> *polygynandry*, where both males and females mate with more than one member of the opposite sex

As for monogamy, both polygyny and polyandry can be either simultaneous or serial/sequential, although at what point sequential polyandry and sequential monogamy fuse is to a large extent arbitrary. As such, we will generally confine our discussion of mating systems to the behaviour of individuals within any one reproductive cycle, unless otherwise stated. Social and genetic differences are likewise also frequently found, so, for example, social polygyny in fishes is common but parentage studies using molecular markers often reveal systems to be genetically polygynandrous (see, e.g., Avise *et al.*, 2002).

Problems with Defining Mating Systems

In real life situations, the above descriptive definitions are frequently found to be inappropriate due to the complexity of the mating systems, which rather than falling into neat categories, exist as a continuum. In this context, there are a few points worth highlighting.

Mating systems are not absolute

Within a species, there may be variation in individual behaviour. In the Gulf pipefish (*Syngnathus scovelli*), males brood eggs, and females mate with multiple males. At first glance, it would appear to be a straightforward case of polyandry, but males occasionally brood eggs from more than one female (Jones and Avise, 1997). While technically this system is actually polygynandrous, the tendency for polygamy is so female-biased that it is more usefully termed polyandrous. In the apparently monogamous Western Australian seahorse (*Hippocampus subelongatus*), some males form long-term pair-bonds with females and appear monogamous across spawning bouts. However, other males switch mates between spawning bouts and can, therefore, be classified as polygynous (Kvarnemo *et al.*, 2000). In the bonnethead shark (*Sphyrna tiburo*), 80% of females appear to be genetically monogamous, but some 20% of larger individuals mate multiply (Chapman *et al.*, 2004). The authors conclude this system to be monogamous. In the apparently monogamous filefish (*Oxymonacanthus longirostris*), some females that do not secure a mate (and, therefore, a territory) solicit males that already have partners. The already partnered females are extremely aggressive towards other females. Kokita (2002) refers to this as bigamy, and 5-29% of males actually mate polygynously.

Mating systems are often facultative

Mating systems will often vary, depending on local conditions. For example, when resources are abundant individuals may mate multiply, whereas when resources are limited individuals may switch to monogamy. Thus, mating systems can differ on a spatial scale, with significant variation in reproductive strategy between populations within species, such as has been documented in sailfin mollies (*Poecilia latipinna*, Trexler *et al.*, 1997) and guppies (*Poecilia reticulata*, Kelly *et al.*, 1999). They may also vary significantly on a temporal scale with, for example, changes in resource availabillty or predation pressure. In two-spotted gobies (*Gobiusculus flavescens*), changes in the operational sex ratio throughout the season due to heavily biased predation on males has been reported to lead to complete sex role reversal throughout a single breeding season (Forsgren et al., 2004).

Alternative mating strategies

Several different reproductive strategies may exist simultaneously in one population as distinct evolutionary stable strategies (see Henson and Warner, 1997 for a comprehensive review) as opposed to 'second best options' (where systems are often not absolute, see above discussion). In ocellated wrasse (*Symphodus ocellatus*), there are four distinct male reproductive strategies within a single population, each of which appear to have similar reproductive success (Taborsky et al., 1987). In St Peters fish (*Sarotherodon galilaeus*), a mouthbrooding tilapia, coexisting strategies span uni- and bi-parental care and monogamy vs polygamy (Ros et al., 2003). In some cases, strategies may be 'hard wired', with individuals either restricted to one state over their lifetimes, or more commonly progressing to different strategies depending on age and/or size. Strategies can also, however, be plastic. Leiser and Itzkowitz (2004), for example, experimentally manipulated population density in pupfish (*Cyprinodon variegatus*) and found male strategy (whether to be a sneaker or a primary male) to vary with such manipulation. It is becoming increasingly clear that species displaying complete uniformity in this respect may be the exception rather than the rule.

There is an inherent bias in the literature

There are two main aspects to this bias. First, considerably more is known about mating systems in the more accessible groups such as reef fishes and shallow-dwelling freshwater species than about pelagic or deep-sea fish. Moreover, only a small minority of systems have actually been rigorously studied. Second, with particular, but not exclusive reference to genetic techniques, the mating system itself may affect the power of the research. For example, in lemon sharks (*Negaprion brevirostris*), Feldheim et al. (2004) could reliably establish that philopatric females were polyandrous as they stayed in one place and sampling of both females and their offspring was relatively straightforward. In contrast, males were not found to sire more than one brood, but since they ranged further, additional copulation may well have gone undetected and thus these animals may well not be truly polyandrous. Thus a system may, for example, be termed polygandrous when in fact we know very little about female mating patterns.

Despite these caveats, we will continue to use these categories throughout as they provide a useful framework for discussion.

MONOGAMY

Monagamous marine and freshwater species have very different characteristics. In marine reef fishes, monogamy typically consists of long-term pair bonds where spawning occurs on a daily basis, or more often, over the course of the spawning season. Parental care is not provided, and eggs have a planktonic dispersal phase (Barlow, 1984). In freshwater fishes, monogamy typically involves biparental care with the pair remaining together until the fry are independent, when parents may or may not part company (Barlow, 1984). Monogamous freshwater and marine fish may be fundamentally different in their reproductive behaviours. However, there are several characteristics that are common between the two environments. Monogamous species in both environments are usually territorial and strongly associated with the substrate, although there are differences in the way that the two groups are territorial. Monogamous marine species tend to defend feeding territories, whereas monogamous freshwater species usually defend breeding territories only. Monogamous freshwater species can be further divided into three main categories, substrate spawners, mouthbrooders and bubblenesters. Monogamous mouthbrooding species are found in many different lineages of cichlid fishes on different continents. South American monogamous mouthbrooding genera include *Aequidens*, *Geophagus*, *Gymnogeophagus* and *Satanoperca*. Examples from Africa include *Eretmodus*, *Tanganicodus* and *Xenotipalia* from Lake Tanganyika. Biparental care is thought to evolve in mouthbrooders when the brood is exceptionally large and cannot be cared for by a single parent, or when both parents are required to guard the free-swimming young (Perrone and Zaret, 1979; Clutton-Brock, 1991), although other suggestions have been advanced (Morley and Balshine, 2003). Monogamous substrate spawning species are perhaps the most common and taxonomically widely spread group and it is found in five of the six monogamous freshwater fish families. Bubblenest builders are the rarest and appear to be confined to the Channidae.

Monogamous Freshwater Fish Families

Barlow (1984, 1986) identified five families of freshwater fishes that definitely exhibit social monogamy: Cichlidae, Osteoglossidae,

Heteropneustidae, Channidae, and Bagridae, with a further three families having uncertain claims to monogamous members—Charicidae, Aphredoderidae (at least some members of which have since been shown to be polygamous, Fletcher et al., 2004) and Ictaluridae. Since his review, we can add Centrarchidae (DeWoody et al., 2000).

Cichlidae is a species-rich and behaviourally and morphologically diverse family of teleost fishes which are distributed naturally in sub-Saharan Africa, neotropical America, and parts of the Indian subcontinent (Keenleyside, 1991). They display a wide variety of breeding systems and strategies including substrate guarding, maternal mouthbrooding and biparental mouthbrooding. Monogamy appears to be the ancestral character state of the family (Barlow 1963, 1964; Gittleman, 1981; Goodwin et al., 1998), and social monogamy remains particularly common within the family (Barlow, 1984, 2000). Cichlids inhabiting sub-Saharan Africa can be roughly divided into the tilapiine lineage, the haplochromine lineage, and a number of lineages inhabiting Lake Tanganyika. Cichlids inhabiting Lakes Malawi and Victoria are almost entirely haplochromine species, all known species of which are polygynandrous maternal mouthbrooders, and thus monogamous cichlids are very rare in these lakes. Social monogamy is common in Lake Tanganyika, however, with several lineages utilizing monogamous breeding systems (the Eretmodinae, Lamprologinae, Ectodinae and *Boulengerochromis*). The largest known cichlid *Boulengerochromis microlepis* is endemic to Lake Tanganyika, and is a socially monogamous biparental guarder (Matthes, 1961; Kuwamura, 1986b). Males and females vigorously defend eggs and offspring which are laid on a large flat rock surface before being transferred by the female to a large sand pit where the young are guarded until they reach 10 cm in length (Kuwamura, 1997). Most other monogamous species in Lake Tanganyika are members of the Lamprologine lineage, which are all substrate spawners and guarders. Pairs jointly defend territories in among the rocky substrate where the eggs are laid. At least twenty such species are thought to be socially monogamous (Kuwamura, 1997) and there are several species of biparental mouthbrooding cichlids in Lake Tanganyika, which are socially and probably genetically monogamous. *Eretmodus cyanostictus* is a biparental mouthbrooding cichlid endemic to Lake Tanganyika. The pairs jointly defend breeding territories in the shallow rocky shore 'surge zone'. After spawning, females incubate the eggs and then the young for eight to twelve days before passing the young to the male who continues the incubation for another 10 to 16 days (Neat and Balshine-Earn, 1999; Morley and Balshine, 2002). No further care is given to the offspring after they are released by the male. *E. cyanostictus* has been shown to be genetically as well as socially monogamous at least within a single breeding season (Taylor *et al.*, 2003). There are likely to be many further cichlid species within Lake Tanganyika where social monogamy also translates to genetic monogamy. Many cichlid species from Central and South America are also monogamous, for example, the well-studied Midas cichlid complex (*Amphilophus* sp.) (McKaye, 1977; McKaye and McKaye, 1977), and the convict cichlid (*Archocentrus nigrofasciatum*) (Makareth and Keenleyside, 1993; Wisenden, 1994, 1995).

Osteoglossidae is a family of large freshwater fishes found in Africa and South America. There are at least two monogamous species in the family. The arapaima (*Arapaima gigas*) is the largest freshwater fish in the world, and is socially and possibly genetically monogamous. A spawning nest is dug by both parents in sand. The female deposits up to 50,000 eggs that the male then fertilizes. The parents protect the young for about three months (Luling, 1964), the male protecting the eggs while the female guards the territory (Lowe-McConnell, 1987). The African arowana (*Heterotis niloticus*) is also monogamous. Large, circular nests are built from plants and mud in shallow pools by both parents. Both parents participate in brood care, guarding both eggs and fry (Bard, 1973).

Bagridae, a family of catfish, is widely distributed in Asia and Africa. Kampango, *Bagrus meridionalis* grows to about 1.5 m in length and is endemic to Lake Malawi (Snoeks, 2004). Both parents defend a breeding territory consisting of a large sand crater—sometimes excavated underneath an overhanging rock—and provide complex care for the free swimming offspring (McKaye and Oliver, 1980). Social monogamy is also known in the Lake Tanganyikan catfish (*Phyllonemus filinemus*). Within a breeding pair, the male or female mouthbrood the offspring until they are approximately 1 cm in length, after which joint mouthbrooding and guarding by both parents takes place (Ochi *et al.*, 2001). It is likely that other species in the Bagrid family are also monogamous (see, e.g., Ochi *et al.*, 2000), although only paternal egg-brooding is documented (Ochi *et al.*, 2002).

Channidae (snakeheads) is a family of large African and Asian fishes that either build bubblenests or mouthbrood eggs. Bubblenest builders form monogamous pairs that remain together throughout the spawning season. Pairs build elaborate nests by clearing an area of surrounding vegetation and weave remaining vegetation into a column (Soin, 1960). During spawning, the pair move up the column during which time the female releases eggs for the male to fertilize. One or both the parents fiercely guard the eggs (Lowe McConnell, 1987; Talwar and Jhingran, 1992). *Channa gachua* and *C. orentalis* are the only two species of snakehead that are mouthbrooders. In these species, the males mouthbrood eggs and fry, but both sexes appear to defend the free-swimming offspring (Ettrich and Schmidt, 1989).

Heteropneustidae (Air sac catfish) consists of two species, one of which the Indian catfish (*Heteropneustes fossilis*) is known to be monogamous. *H. fossilis* is found in Thailand, Sri Lanka and Myanmar, and grows to some 50 cm. Yellow-green eggs are laid in a hole in the substrate prepared by both parents who defend both eggs and fry (Holly, 1928, 1931).

Centrarchidae contains the second species of freshwater fish where genetic monogamy has been demonstrated, the externally fertilizing large-mouth bass (*Micropterus salmoides*). This species appears to be predominately monogamous; 88% of broods have been determined to be the product of a single adult pair (DeWoody *et al.*, 2000). In this species, the males defend crude nests in which females deposit their eggs. The males then defend the eggs and free swimming larvae for up to 30 days. There are reports of biparental care in this species (DeWoody *et al.*, 2000), although male care appears to be more common. The closely related small-mouth bass (*Micropterus dolomieui*) appears to be at least socially monogamous and may also be genetically monogamous (Weigman *et al.*, 1992; Weigman and Bayliss, 1995).

Monogamous Marine Families

Whiteman and Cote (2004) identified 18 marine fish families where monogamy occurs, to which we can add the hammerhead sharks (Sphyrnidae) (Chapman *et al.*, 2004). The families are: Acanthuridae, Apogonidae, Balistidae, Canthigastrinae, Cirrhitidae Chaetodontidae, Gobiidae, Labridae, Malacanthidae, Microdesmidae, Monacanthidae, Opistognathidae, Pegasidae, Pomacanthidae, Pomacentridae, Serranidae Solenostimidae and Syngnathidae. Barlow (1984) concluded that monogamous marine species are usually small bodied, strongly site attached, non- or weakly seasonal and have a planktonic dispersal phase and no parental care. Whiteman and Cote (2004) supported Barlow's conclusions, although did not find evidence that monogamous species were small bodied. Since their sample size was very small, we await further analyses before discarding Barlow's (1984) conclusion of an association between body size and reproductive strategy.

While social monogamy and planktonic spawning are the norm in monogamous marine reef fishes, genetic monogamy with parental care has been identified in some seahorses (Syngnathidae). The male-pregnant syngnathids are one of a select group of lineages that display sex-role reversal (see following discussion). Females transfer unfertilized eggs to the male's ventral surface where they are fertilized. The eggs and larvae are then incubated in a brood pouch in the case of the seahorses, or are attached to the ventral surface of the male in the case of pipefish. Two species of seahorse appear to be genetically monogamous: *H. subelongatus* (Kvarnemo *et al.*, 2000) and *H. angustus* (Jones *et al.*, 1998a). Several other species of seahorses are likely to be genetically monogamous as they show long-term pair bonds in the wild (*H. comes*, Perante *et al.*, 2002; *H. whitei*, Vincent and Sadler, 1995) or have demonstrated monogamy in the laboratory (*H. fuscus*, Vincent, 1994; *H. zoterae*, Masonjones and Lewis, 1996). It is also possible that some pipefish species are genetically monogamous, as these appear to be socially monogamous (Vincent *et al.*, 1992 but see also the following discussion on polyandrous systems). Genetic monogamy is not confined to the teleost fishes. While many elasmobranchs are considered to be polyandrous due to their ability to store sperm such as in the lemon shark (*Negaprion brevirostris*, Feldheim *et al.*, 2001, 2002) and nurse shark (*Ginglymostoma cirratum*, Saville *et al.*, 2002), the bonnethead shark (*Sphyrna tiburo*), appears to be predominately genetically monogamous. Females give birth to between three and eighteen embryos, and 81% of the females have been found to have embryos sired by a single male (Chapman *et al.*, 2004).

POLYGAMY

Among fishes, polygamous mating systems of one kind or another predominate. As such, it is not possible to present a comprehensive review of all known examples of polygamy in the fishes, and instead we will use illustrative examples and, where relevant, outline general patterns. Unlike monogamous mating systems, distinguishing between species inhabiting freshwater and marine systems appears unnecessary in polygamous species.

Polygyny

Strict polygynous mating systems, where one male mates with several females but each female only mates with one male, are often subdivided into two sub-categories:

(a) *Resource-defence polygyny*, where males control access to a good and/or safe resource. Often, in fishes, this takes the form of males defending territories and/or nests. In lingcod (*Ophiodon elongatus*), for example, females lay a single mass of eggs and males guard these egg masses, sometimes guarding several at a time laid by different females (King and Withler, 2005). Three-spined sticklebacks (*Gasterosteus aculeatus*) are another commonly cited example where the males defend nests which may contain the eggs of six or more females (Wootton, 1984).

(b) *Female-defence polygyny*, where one male directly defends groups of females. Often these are referred to as harems and one male will mate with all of the fertile females he guards as is the case, for example, in chevron butterflyfishes (*Chaetodon trifascialis*) (Yabuta and Kawashima, 1997).

The wide-eyed flounder (*Bothus podas*) provides an interesting combination of these two systems. In this species, both sexes are territorial throughout the year, but male territories encompass several females and females apparently show mating fidelity to that male (Carvalho *et al.*, 2003).

Lekking systems, found among a diverse array of fish species (e.g., some Lake Malawi cichlids (*Copadichromis eucinostomus*) (McKaye, 1983, 1984; Taylor *et al.*, 1998) are also often classified as a third category of polygynous system. Most such systems, however, are probably better described as polygynandrous (see below). Generally, among fish taxa, resource-defence polygyny is extremely common, more so than female-defence polygyny.

Polyandry and Polygynandry: Females Mating with More Than One Male

Before proceeding to discuss these two other categories of polygamous mating system, we first consider the question of why females might mate with several different males. For comprehensive reviews on female multiple mating see Hosken and Blackenthorn (1998), Jennions and Petrie (2000) and Zeh and Zeh (2003).

Anisogamy and a related typically male-skewed operational sex ratio (Emlin and Oring, 1977) dictate that in a great majority of cases male, but not female, reproductive output increases with the number of mates (Bateman's classic gradient, 1948). This, taken with the associated costs to most matings (see Orsetti and Rutowski, 2003; Maklakov *et al.*, 2005), has underpinned the convention that while it may be in a male's interest to mate with more than one individual, it is not in a females.

Burke and Bruford (1987) and Wetton *et al.* (1987) have experimented with simultaneous exposure of high rates of cuckoldry in two separate bird species; they have challenged this view and triggered a flood of studies focused on re-evaluating animal mating systems using genetic markers. We now know that a female's mating with multiple males is not only common across all major animal groups, including the fishes (reviewed in Birkhead, 2000), but may turn out to be the norm. This has, in turn, driven a dramatic reassessment of both male and female reproductive strategies.

We refer to a female mating with multiple males as a scenario where more than one male is mated with per reproductive cycle, after Jennions and Petrie (2000). There are cases where females will mate repeatedly with the same male (so under a monogamous system) but since this is exceedingly rare among fishes we confine our discussion to the former scenario. For clarity, female multiple mating behaviour is divided into several different categories with respect to the likely underlying reasons for its evolution, although boundaries are not absolute and categories not mutually exclusive (Fedorka and Mousseau, 2002). Examples given are not restricted to fishes although the scenarios likely to be of particular importance among fish species are outlined later.

Sex role reversal

Here the conventional male and female roles are simply reversed. Sometimes referred to as 'classical' polyandry, females will mate with several males and invest nothing in parental care, but males mate only once and invest heavily. There are a few well-known examples among fishes notably among Syngnathids (Berglund and Rosenqvist, 2003). Such systems present few conceptual difficulties since they sit with conventional interpretations of, for example, biased operational sex ratios. They are discussed at greater length in the following section on polyandry.

Passive multiple mating

A female may be to a large extent passive in the paternity of her offspring such as in cases of broadcast spawning. This is extremely common among fishes, where neither male nor female have direct control over fertilization and many males are likely to fertilize the clutch of one female. With urogenetic openings in their throats to enable spawning into root masses, female pirate perch (*Aphredus sayanus*), for example, lay their eggs and males then rush to cast their sperm. Here, single clutches typically have multiple fathers (Fletcher *et al.*, 2004).

Females may alternatively be the victims of males 'sneaking' in and casting their sperm when females are mating with 'preferred', often dominant, males. Such behaviour is extremely common in many animal taxa, including fish and while it results in multiply sired offspring, it does not usually involve active solicitation of different males by females. Both broadcast spawning and sneaking are discussed further in the following section on polygynandry.

A final scenario that can be categorized as passive multiple mating is when females mate with more than one male simply to avoid or minimize harassment, 'convenience polyandry', as seems to be the case in some turtle species (Lee and Hays, 2004). Clearly, in such cases, the costs of harassment exceed the benefits of the female of mating more than once.

Active multiple mating

In contrast, females of many species are active in their solicitation of several mating partners. Generally, the benefits to a female displaying such behaviour can be split into two main categories: direct and indirect benefits.

(a) Direct benefits
  (i) *material gains*—Females either receive nutritional benefit from the male sperm or spermatophore itself (some butterflies, for example, Torres-Vila and Jennions, 2005) and bushcrickets (Hockham *et al.*, 2004), or males may present females with food items, 'nuptial gifts' such as in dance flies (LeBas *et al.*, 2004). In such cases, the advantages to females of mating with more than one male may be simply energetic. Generally common among insect systems, other direct material gains may include access to male territories, or even transfer of anti-predator defence chemicals as has been reported from moth species (Gonzalez *et al.*, 1999).
  (ii) *fertility insurance*—Mating with more than one male may minimize the percentage of a female's gametes that remain unfertilized either due to sperm depletion from recent mating with other females, completely sterile males, or where the sperm of one male is not enough to reliably fertilize an entire clutch of eggs. The mating systems of some newt species appear to illustrate this (Osikowski

and Rafinski, 2001). Male bluehead wrasse (*Thalasomma bifasciatum*) provide an interesting twist to such a scenario. Males show significant variation in the amount of sperm that they produce per spawning event and the males most desired by females actually produce less sperm (Warner *et al.*, 1995).

(b) Indirect/genetic benefits   In the above scenarios, advantages to females are direct and easy to comprehend. Often, however, there are no obvious benefits to the female. It is most likely in such cases that females are acting to enhance the genetic quality of their offspring, thus indirectly benefiting themselves. There is growing empirical support for this being the case (Evans and Magurran, 2000). The suggested ways that such genetic enhancement may be brought about are now numerous, broadly summarized as:

  (i) *Trading up.* After mating with one male, a female encounters subsequent potentially genetically superior males that she then also mates with. There is some evidence supporting this being an element of female multiple mating strategies in guppies (*Poecilia reticulata* (Pitcher *et al.*, 2003).

 (ii) *Facilitating sperm competition.* Females mate with several males and the genetic quality is assessed through sperm efficacy as found in dung flies (Hosken *et al.*. 2003). Post-copulatory choice may also involve direct action by the female, 'cryptic female choice'. Female rock-shrimp, for example, have been observed to differentially manipulate male spermatophores according to the perceived quality of the male (Thiel and Hinojosa, 2003).

(iii) *Bet-hedging.* Unable to assess male quality directly, the females 'hedge their bets' and mate with several males to ensure at least some high genetic quality offspring. This has been suggested to be driving multiple mating patterns in female pupfish (*Cyprinidon variegatus*) (Draud and Itzkowitz, 2004).

(iv) *Avoiding inbreeding.* If females have a high chance of mating with relatives but relatives are not recognizable, multiple mating minimizes the chances of the majority of any brood being sired by a close relative. Such avoidance has been suggested to contribute to the mating behaviour of bank voles (Ratkiewicz and Borkowska, 2000) as well as field crickets (Tregenza and Wedell, 2002).

 (v) *Genetic complementarity.* Females mate with multiple males to minimize the costs of their eggs being fertilized by genetically incompatible sperm. Again there are studies supporting such a hypothesis: from pseudoscorpions (Newcomer *et al.*, 1999) and fruit flies (Clark, 2002).

(vi) *Maximizing heterozygosity.* Females mate with multiple males to maximize heterozygosity in their offspring at many or a few key loci, as a direct correlate of vigour and/or secondary sexually selected traits (Brown, 1998).

Although the majority of research on female multiple mating behaviour has to date been carried out on bird species, clearly, it is also widespread among fish. As the research effort continues to diversify, patterns may start to emerge. In the meantime, only a few generalizations can be made. Sex role reversal is restricted to only a few fish taxa, notably Syngnathidae. In contrast, passive multiple mating in the form of broadcast spawning and sneaker males is extremely common among a diverse array of fish taxa. Convenience polyandry is probably less common since in fishes males are less able to control females directly. Notable exceptions to this are livebearing (viviparous) species where sexual coercion and convenience polyandry may be significantly contributing to the evolution of female mating patterns (Magurran, 2001; Plath *et al.*, 2003). Active multiple mating is also known to be widespread. Direct food provisioning is unusual among fishes and so material gains are unlikely to have had a major influence. Fertility insurance is the more likely direct benefit underpinning female multiple mating behaviour in fishes. Indirect mechanisms where females ensure genetic quality of their offspring are difficult to assess on an individual basis. Nevertheless, they are likely to be common and particularly strongly selected for in species where the male quality is difficult to assess and/or where offspring suffer a severe fitness loss from genetic incompatibility or inbreeding.

Polyandry

A female's offspring being sired by more than one male may arise from a suit of scenarios that can have quite different forces driving them as outlined above. Here, as in the section on polygyny, we refer only to strict polyandry where females mate with several males in one reproductive cycle, but the males only mate with one female.

In a few fish species, complete role reversal has occurred, with males providing all the care for offspring and sexual selection being stronger on females than males, so-called 'classic' polyandry. Some species of pipefish appear to be classically polyandrous, and a substantial amount of recent research has focused on this group for this reason (see Berglund *et al.*, 2005). Andersson (2005) proposes three steps for classic polyandry to evolve. Firstly, the male care must arise, a situation not difficult to envisage among fishes where paternal care is the most common state of parental care. Secondly, the females must then become able to lay more eggs in a single clutch than a male can accommodate in a nest or territory. Finally, the females must then compete among each other for access to males. In the Gulf pipefish (*Syngnathus scovelli*) females do indeed produce more eggs than a male can house (Berglund *et al.*, 1989; Berglund and Rosenqvist, 2003) and molecular markers have revealed that up to four males may carry eggs from one female (Jones and Avise, 1997).

There are some cases where only some of these criteria are met, and although complete role reversal has not occurred, polyandry may still be observed. The polyandrous cardinal fish (*Apogon notatus*) provides a good example. In this species, sex role reversal appears not to have occurred because the operational sex ratio (OSR) remains male-biased due to high female mortality (Okuda, 1999). In *Julidochromis marlieri*, a cichlid fish from Lake Tanganyika, larger females may produce more eggs than a single male can defend, and in such cases these females will deposit eggs in the nests of more than one male (Yamagishi and Kohda, 1996). Another potential example comes from the Isaza goby (*Gymnogobius isaza*), a freshwater species endemic to Lake Biwa in Japan. In any one breeding cycle, the males never accept eggs from more than one female probably to minimize fungal infections and there appears to be an optimum number of eggs that males prefer to defend (Takahashi *et al.*, 2004).

Anemone fishes (*Amphirion* spp.) illustrate a rare system of resource-defence polyandry, where one female dominates several subordinate males per anemone. Occasionally the dominant female may be able to dominate more than one anemone, thereby mating polyandrously (Moyer and Sawyers, 1973). Berglund (1997) also cites an unusual case of polyandry in the deep-sea angler fish (sub-order Ceratioidei) where the males actually physically attach to females to mate and more than one male can attach to a single female, thus technically this is a polyandrous mating system.

Polygynandry

Sometimes referred to as promiscuity, we here prefer the less loaded term 'polygynandry', referring to situations where both males and females mate with more than one partner during one reproductive cycle.

We have discussed that polygynandry covers many different types of the mating scenario. Individuals may play an active role in soliciting several mates, such as in Lake Malawi rock-dwelling cichlids, where males attract multiple females to spawn in territories (sequentially), but females visit several males per clutch of eggs (Kellogg *et al.*, 1995; Parker and Kornfield, 1996). Alternatively the mating scenario may be largely passive, as is the case in broadcast spawning, where 'explosive breeding assemblages' form commonly and no individual is making an actual choice of an individual mate (Thresher, 1984). Sperm competition and genetic compatibility may be important in such systems, although it is likely to be difficult to measure.

A case of genetic polygynandry that lies somewhere in-between the two highlighted above is where some males 'sneak' matings, an apparently common tactic among fishes, which can have very significant effects on apparently polygynous mating dynamics (Jones *et al.*, 2001). Often, both sexes of a mating pair may be unaware that their eggs are being sired by extra males, in which sense they are passive, but sneaking males clearly are not. Sneaking as an alternative mating strategy may be fixed, or it can be plastic, often as a stage of a life cycle. Frequently, younger or smaller males that are unable to secure and defend territories adopt sneaking strategies. Alternatively, in some species, all males will opportunistically sneak, for example, in sandperch (*Parapercis snyderi*) sneaking males were often neighbouring harem masters (Ohnishi *et al.*, 1997). As a rule, sneakers are male, although there is a single documented case where females appear to sneak, cited in Henson and Warner (1997, citing Johnston, 1996). In the brown chromis damselfish (*Chromis multilineata*), males deny parasitized females access to their nests, but such females 'dart in' and lay their eggs before males are able to stop them. In some species, such as the bluegill sunfish (*Lepomis macrochirus*) males may go to great lengths to avoid sneaking by other males, indicating that it is very costly (Neff *et al.*, 2004). In other species, such as the rusty blenny (*Parablennius sanguinolentus*), sneaking is tolerated by males, suggesting that either the males may reap some additional benefit, or that sneakers are not significantly successful (Oliviera *et al.*, 2002).

MATING SYSTEM DETERMINANTS

Having discussed the incidence of various breeding systems in fishes, we will address the possible reasons why different breeding systems are found in different species and environments. The basic assumption is that the systems that we observe are those that on average maximize individual lifelong reproductive success. In fishes, the vast majority of species display some kind of polygamy, and polygynandry is likely to be broadly the most common genetic, if not social, mating system. As outlined above, within these categories there is wide variation both among and within species as to the mating behaviour that individuals display. This is due to a complex combination of factors exerting varying (and fluctuating) degrees of influence on reproductive behaviour (see Reynolds, 1996 for review). These are difficult to separate into discrete entities as often they are not independent and causality is impossible to assign. With this in mind, however, a broad summary of the potential influences on mating systems is discussed below.

Phylogenetic Constraints

While many fish families display a variety of mating strategies, they are also to a greater or lesser extent constrained by phylogeny (Thresher, 1984; Godwin, 1995; Gross, 2005). Thus, the ancestral state of a group is likely to have considerable influence on extant group members since some strategies are more likely to evolve from some ancestral states than from others (see Reynolds *et al.*, 2002; Ah-King *et al.*, 2005). Monogamy, for example, is less likely to evolve from an ancestral polygynandrous state than it is from, for example, a polygynous state. The increasing availability of robust molecular phylogenies of fish genera and families will help to control for phylogenetic influences, and the development of this area is advancing (Goodwin *et al.*, 1998, 2001; Reynolds *et al.*, 2002).

Biological Factors

Barlow (1984) observed that in freshwater fishes at least, monogamous taxa appear to be large. Small fishes have higher metabolic rates than larger fishes and, consequently, small females can theoretically increase their fitness by deserting a male after spawning, and spawn with another male. There may be insufficient time for large females to produce a second clutch, and thus the best strategy may be to remain with the male and assist with the brood defence.

Monogamy may also be favoured over polygamy if the reproductive output is higher due to reduced inter-clutch intervals as a result of greater familiarity between the partners. This appears to be the case in *Eretmodus cyanostictus*, a cichlid fish from Lake Tanganyika. Females unassisted by males release smaller and less developed young, which may have reduced survival, as well as increasing the inter-spawning interval (Gruter and Taborsky, 2004). This hypothesis may also help to explain monogamy in the seahorse (*Hippocampus fuscus*), where the number of offspring produced increases with the time since pair formation (Vincent, 1994).

Intersexual Conflict and Sex Ratio Biases

The potentially potent influence of sexual conflict on mating systems is a relatively new, but lively, field of research (see, e.g., Henson and Warner, 1997; Cameron *et al.*, 2003; Chapman *et al.*, 2003; Zeh and Zeh, 2003; Pizzari and Snook, 2004; Härdling and Kaitala, 2005). The evolutionary interests of males and females may be quite different and can bring about 'arms races' of adaptation and counter adaptation. In a Lake Tanganyikan cichlid (*Perissodus microlepis*) both males and females guard the young but the males have been observed 'farming-out' offspring to other pairs, possibly as an alternative more lucrative strategy to brood desertion (Ochi and Yanagisawa, 2005). Whether this is an evolutionary stable strategy or not remains to be seen. Similar sexual conflicts have been observed occurring in other biparental cichlid species (Gruter and Taborsky, 2005). Generally, however, sexual conflict would be expected to have particularly profound effects in polygamous and especially polygynandrous systems where male and females may behave contrary to the other sex's evolutionary interests, thus leaving such systems particularly prone to the evolution of antagonistic mating tactics (Chapman *et al.*, 2003; Morrow and Arnqvist, 2003).

The density and relative abundances of males and females in a population will affect mate availability both temporally and spatially (Vincent and Sadler, 1995). This has been suggested to lead to the evolution and maintenance of monogamy in systems where mate availability is low due to either synchronous breeding or low densities. In these situations, once a partner is located, the best strategy may be to stick with them (Knowlton, 1979; Wittenberger and Tilson, 1980; Wickler and Seibt, 1981; Nakai *et al.*, 1990). Alternatively, high densities of one sex can also lead to different mating systems predominating. If one sex is more conspicuous either in a sensory or behavioural context, for example, a severely biased sex ratio may result due to differential predation pressures and this may in turn have a significant effect on the mating system itself (Forsgren *et al.*, 2004). If the population is extremely male skewed high levels of polygyny and sneaking behaviour might be predicted.

Ecological and Environmental Factors

Mating systems will also depend heavily on the distribution and availability of essential resources, including food and spawning or nesting sites (Emlin and Oring, 1977). If resources cannot be easily defended, as is the case for many pelagic and deep water species, then polygynandrous mating systems are expected to predominate (Berglund, 1997). In contrast, monogamy may be promoted if pairs of individuals are necessary to defend a feeding territory against conspecific or heterospecific competitors. For territory defence to be important in maintaining monogamy, pairs should be able to defend larger territories than single individuals, as is the case in the powder-blue tang (*Acanthurus leucosternon*) (Robertson *et al.*, 1979). Removal of one of the pairs did not appear to result in a reduction of territory size, but females did appear to incur a cost through a reduced feeding rate. Males did not appear to suffer such feeding-related costs (Fricke, 1986; Hourigan, 1989). Resources may affect mating systems in less direct ways. For example, freedom from mate guarding in resource-rich habitats have been suggested to govern monogamy in gobies (Reavis and Barlow, 1998). Effects are often complex. Monogamy in the marine cleaner goby (*Gobiosoma evelynae*) is likely to be maintained by intrasexual aggression which is in turn associated with both mate availability and other resources (Harding *et al.*, 2003).

In low resource environments, females are widely dispersed, and thus males are unable to mate with more than one female. In marine fishes, this situation appears to result in facultative monogamy: where resources are abundant, individuals are polygamous, whereas in resource poor areas, individuals are constrained to monogamy. In some hawkfishes (Cirrhitidae), individuals are monogamous when the coral heads they are associated with are scarce or small in size, while harem formation is found in areas with high densities or large coral heads (Donaldson, 1989). A similar situation occurs in the halfmoon triggerfish (*Sufflamen chrysopterus*) which has both monogamous pairs and polygamous spawning depending on female density (Fricke, 1980; Kawase and Nakazono, 1994).

Predation pressures may also have a significant effect on the mating strategy adopted, as highlighted above. High predator pressure may require eggs and offspring to be guarded by both parents, favouring monogamy. In marine fishes, monogamy is unlikely to be maintained by the need for biparental care, as almost all marine species display either no parental care, or frequently uniparental care (usually male). The spiny chromis damsel fish (*Acanthochromis polyacanthus*) is an exception to this.

It is also unusual among the reef fish in that it does not have a pelagic larval dispersal phase, and parents defend offspring for several months (Robertson, 1973).

Biparental care is far more common in freshwater species although it is still relatively rare. This may be associated with the low incidence of planktonic dispersal. There appears to be a tight link between monogamy and the biparental care of offspring in freshwater species, particularly when fry as well as eggs are guarded (Barlow, 1974). If the need for biparental care is ultimately responsible for the evolution and mainentance of monogamy, lower offspring survival is expected following the removal of one parent. There has been extensive investigation of this hypothesis in cichlid fishes. In the Lake Tanganyikan substrate-spawning species toae cichlid (*Neolamprologus toae*), the removal of males decreased offspring survival (Nagoshi, 1987). However, the effect was age specific, removal of males guarding fry shorter than 6 mm resulted in the loss of the entire brood, whereas fry longer than 18 mm were usually not affected by the males removal. Similar results were found in the scale eating cichlid (*Perissodus microlepis*) that engages in a primitive form of mouthbrooding in which eggs are laid on the substrate before being incubated in the mouth by both parents (Yagisawa and Nshombo, 1983). In the biparentally mouthbrooding (*Eretmodus cyanostictus*), low offspring survival was found when the male parent was removed, suggesting there is reduced offspring survival when only one of the parents is present (Morley and Balshine, 2002). In the midas cichlid (*Amphilophus citrinellus*), both parents appear to be necessary to raise offspring successfully (McKaye, 1977; McKaye and McKaye, 1977; Barlow, 1984). Biparental care may be necessitated in mouthbrooders as a result of large brood sizes that do not fit into a single parent's oral cavity, or a necessity for post-release guarding of fry (Kuwamura, 1986; Balshine-Earn, 1995).

Shifts in mating strategies may also occur along physical environmental gradients indicating that these also may exert a strong influence on the evolution of mating systems. Thresher (1984), for example, observed a frequent latitudinal shift from pelagic to demersal spawners in families of reef fishes. The constraints placed upon individuals as a result of extreme external environmental factors may lead to the evolution and/or maintenance of monogamy. In high latitudes, seasonally cold temperatures and short day lengths may prevent a female producing more than a single clutch of eggs in a spawning season, as may the pronounced wet and dry seasons found in low latitudes (Lowe-McConnell, 1975). In systems where male parental care has already evolved, females are predicted to remain with the male and assist in caring for the eggs and offspring when there is insufficient time during the breeding season for a second clutch. Thus, selection for biparental care will also select for monogamy.

Summary and Future Directions

The fishes illustrate a wide diversity of mating strategies but the overwhelming majority display some form of polygamous mating system, with polygynandry being the most common genetic, if not social, strategy. Monogamy is rare, and appears to be most commonly found in large bodied, substrate spawning species. Several strategies may be employed simultaneously, varying between individuals within both populations and species, depending on a host of factors such as the distribution and availability of resources, the operational sex ratio (itself influenced by factors such as predation), the extent of sexual conflict, and physically driven local environmental conditions. Many fishes are also facultative in their mating decisions and may be, for example, monogamous when resources are at a premium, and polygamous when resources are abundant. Consequently, mating strategies as a whole exist on a continuum, and attempting to rigidly categorize species into such classifications as 'monogamous' or 'polygamous' may be unrealistic: although these terms are useful in a locally descriptive sense they will always require qualification on a case by case basis.

As more detailed information is being gathered on the details of mating systems, our understanding of the reasons why some species remain monogamous, and why females in some species mate multiply will become more complete. The use of molecular markers such as microsatellites in addition to experimental approaches will accelerate the accumulation of knowledge in species which are particularly difficult to observe. The development of robust molecular phylogenies for many currently poorly understood groups will also allow us to rigorously test evolutionary hypotheses using comparative methods. Further, the adoption of several fish species (Zebrafish, Takifugu and Medaka) as model organisms in genome-sequencing projects will allow us to study, in unprecedented detail, the effects of mate choice strategies at the molecular level. Advances in underwater technologies will also increasingly allow us access to species that have previously been beyond our grasp due to the extremes of the environments in which they live.

Acknowledgments

We would like to thank Marjorie Maillard for proofreading and editorial assistance and Martin Genner, Jonathan Ellis, and Derek Dunn for their helpful comments.

References

Ah-King, M., C. Kvarnemo and B.S. Tullberg. 2005. The influence of territoriality and mating system on the evolution of male care: a phylogenetic study on fish. *Journal of Evolutionary Biology* 18: 371-382.

Andersson, M. 2005. Evolution of classical polyandry: Three steps to female emancipation. *Ethology* 111: 1-23.

Avise, J.C., A.G. Jones, D. Walker and J.A. DeWoody. 2002. Genetic mating systems and reproductive natural histories of fishes: Lessons for ecology and evolution. *Annual Review of Genetics* 36: 19-45.

Balshine-Earn, S. 1995. The costs of parental care in Galilee St Peter's fish, *Sarotherodon galilaeus*. *Animal Behaviour* 50: 1-7.

Bard, J. 1973. Le poisson de la famille des Osteoglossidae et la pisciculture. *Bois et Forets des Tropiques* 147: 63-70.

Barlow, G.W. 1963. Evolution of behaviour. *Science* 139: 851-852.

Barlow, G.W. 1964. Evolution of the Asian teleost *Badis badis* V. Dynamics of fanning and other parental activities with comments on the behaviour of the larvae and post larvae. *Zeitschrift fur Tierpsychologie* 21: 99-123.

Barlow, G.W. 1974. Contrasts in social behavior between Central American cichlid fishes and coral-reef surgeon fishes. *American Zoologist* 14: 9-34.

Barlow, G.W. 1984. Patterns of monogamy among teleost fishes. *Archiv-fiir FischerecWissenschaft. Beih* 1: 75-123.

Barlow, G.W. 1986. A comparison of monogamy among freshwater and coral reef fishes. In: *Indo-Pacific Fish Biology: Proceedings of the Second International Conference on Indo-Pacific Fishes*, T. Uyeno, R. Arai, T. Taniuchi and K. Matsuura (eds.). Ichthyological Society of Japan, pp. 767-775.

Barlow, G.W. 2000. *Cichlid Fishes. Nature's Grand Experiment in Evolution*. Perseus Books, Cambridge.

Bateman, A.J. 1948. Intra-sexual selection in *Drosophila*. *Heredity* 2: 349-368.

Becher, S.A. and A.E. Magurran. 2004. Multiple mating and reproductive skew in Trinidadian guppies. *Proceedings of the Royal Society of London* B 271: 1009-1014.

Berglund, A. 1997. Mating systems and sex allocation. In: *Behavioral Ecology of Teleost Fishes*, J-G.J. Godin (ed.). Oxford University Press, Oxford, pp. 237-265.

Berglund, A. and G. Rosenqvist. 2003. Sex role reversal in pipefish. In: *Advances in the Study of Behaviour*, P.J.B. Slater, J.S. Rosenblatt, C.T. Snowden and T.J. Rober (eds.). Elsevier, Amsterdam, Vol. 32, pp. 131-167.

Berglund, A., G. Rosenqvist and I. Svensson. 1989. Reproductive success of females limited by males in 2 pipefish species. *American Naturalist* 133: 506-516.

Berglund, A., M.S. Widemo and G. Rosenqvist. 2005. Sex-role reversal revisited: choosy females and ornamented, competitive males in a pipefish. *Behavioral Ecology* 16: 649-655.

Birkhead, T.R. 2000. *Promiscuity: An Evolutionary History of Sperm Competition and Sexual Conflict*. Faber and Faber, London.

Blais, J., C. Rico and L. Bernatchez. 2004. Nonlinear effects of mate choice in wild three-spined sticklebacks. *Evolution* 58: 2498-2510.

Blumer, L.S. 1982. A bibliography and categorization of bony fishes exhibiting parental care. *Zoological Journal of the Linnaen Society* 76: 1-22.

Breder, C.M. and D.E. Rosen. 1966. *Modes of Reproduction in Fishes*. Natural History Press, Garden City, New York.

Brown, J.A. 1998. The new heterozygosity theory of mate choice and the MHC. *Genetica* 104: 215-221.

Burke, T. and M.W. Bruford. 1987. DNA fingerprinting in birds. *Nature (London)* 327: 149-152.

Cameron, E., T. Day and L. Rowe. 2003. Sexual conflict and indirect benefits. *Journal of Evolutionary Biology* 16: 1055-1060.

Carvalho, N., P. Afonso and R.S.A. Santos. 2003. The haremic mating system and mate choice in the wide-eyed flounder, *Bothus podas. Environmental Biology of Fishes* 66: 249-258.

Chapman, D.D., P.A. Prodohl, J. Gelsleichter, C.A. Manire and M.S. Shivji. 2004. Predominance of genetic monogamy by females in a hammerhead shark, *Sphyrna tiburo*: Implications for shark conservation. *Molecular Ecology* 13: 1965–1974.

Chapman, T., G. Arnqvist, J. Bangham and L. Rowe. 2003. Sexual conflict. *Trends in Ecology and Evolution* 18: 41-47.

Chesser, R.K., M.W. Smith and M.H. Smith. 1984. Biochemical genetics of mosquitofish. III. Incidence and significance of multiple paternity. *Genetica* 74: 77-81.

Clark, A.G. 2002. Sperm competition and the maintenance of polymorphism. *Heredity* 88: 148-153.

Clutton-Brock, T.H. 1991. *The Evolution of Parental Care*. Princeton University Press, Princeton.

DeWoody, J.A., D. Fletcher, S.D. Wilkins, W.S. Nelson and J.C. Avise. 2000. Genetic monogamy and biparental care in an externally fertilizing fish, the largemouth bass (*Micropterus salmoides*). *Proceedings of the Royal Society of London* B 267: 2431–2437.

Donaldson, T.J. 1989. Facultative monogamy in obligate coral-dwelling hawkfishes (Cirrhitidae). *Environmental Biology of Fishes* 26: 295-302.

Draud, M.J. and M. Itzkowitz. 2004. Mate numbers or mate quality: female mate choice in the polygynandrous variegated pupfish (*Cyprinodon variegatus*). *Ethology, Ecology and Evolution* 16: 1-13.

Dulvy, N.K. and J.D. Reynolds. 1997. Evolutionary transitions among egg-laying, live-bearing and maternal inputs in sharks and rays. *Proceedings of the Royal Society of London* B 264: 1309-1315.

Emlin, S.T. and L.W. Oring. 1977. Ecology, sexual selection and the evolution of mating systems. *Science* 197: 215-223.

Ettrich, G. and J. Schmidt. 1989. *Channa gachua* aus Südostasien und *Channa orientalis* von Sri Lanka - zwei gute Arten. *Die Aquarien und Terrarien-Zeitschrift* 42: 465-467.

Evans, J.P. and A.E. Magurran. 2000. Multiple benefits of multiple mating in guppies. *Proceedings of the National Academy of Sciences of the United States of America* 97: 10074-10076.

Fedorka, K.M. and T.A. Mosseau. 2002. Material and genetic benefits of multiple mating and polyandry. *Animal Behaviour* 64: 361-367.

Feldheim, K.A., S.H. Gruber and M.V. Ashley. 2001. Multiple paternity of a lemon shark litter (Chondrichthyes: Carcharhinidae). *Copeia* 2001: 781-786.

Feldheim, K A., S.H. Gruber and M.V. Ashley. 2002. The breeding biology of lemon sharks at a tropical nursery lagoon. *Proceedings of the Royal Society of London* B 269: 1655-1661.

Feldheim, K.A., S.H. Gruber and M.V. Ashley. 2004. Reconstruction of parental microsatellite genotypes reveals female polyandry and philopatry in the lemon shark, *Negaprion brevirostris*. *Evolution* 58: 2332-2342.

FishBase. 1996. FishBase 96. CD-ROM. ICLARM, Manila.

Fletcher, D.E., E.F. Dakin, B.A. Porter and J.C. Avise. 2004. Spawning behavior and genetic parentage in the pirate perch (*Aphredoderus sayanus*), a fish with an enigmatic reproductive morphology. *Copeia* 2004: 1-10.

Forsgren, E., T. Amundsen, Å.A. Borg and J. Bjelvenmark. 2004. Unusually dynamic sex roles in a fish. *Nature (London)* 429: 551-554.

Fricke, H.W. 1973. Individual partner recognition in fish: field studies on *Amphiprion bicinctus*. *Naturwissenschaften* 60: 204-205.

Fricke, H.W. 1980. Control of differing mating systems in a coral reef fish by one environmental factor. *Animal Behaviour* 28: 561-569.

Fricke, H.W. 1986. Partner swimming and mutual partner guarding in monogamous butterflyfish (Pisces: Chaetodontidae): A joint advertisement for territory. *Ethology* 73: 307-333.

Gittleman, J.L. 1981. The phylogeny of parental care in fishes. *Animal Behaviour* 29: 936-941.

Godwin, J. 1995. Phylogenetic and habitat influences on mating system structure in the humbug damselfishes (Dascyllus: Pomacentridae). *Bulletin of Marine Science* 57: 637-652.

Gonzalez, A., C. Rossini, M. Eisner and T. Eisner. 1999. Sexually transmitted chemical defense in a moth (*Utetheisa ornatrix*). *Proceedings of the National Academy of Sciences of the United States of America* 96: 5570-5574.

Goodwin, N.B., S. Balshine-Earn and J.D. Reynolds. 1998. Evolutionary transitions in parental care in cichlid fish. *Proceedings of the Royal Society of London* B 265: 2265-2272.

Goodwin, N.B., S. Balshine and J.D. Reynolds. 2001. Using phylogenies to test evolutionary hypotheses about cichlid fishes. *Journal of Aquaculture and Aquatic Sciences* 9: 256-268.

Goodwin, N.B., N.K. Dulvy and J.D. Reynolds. 2002. Life-history correlates of the evolution of live-bearing in fishes. *Philosophical Transactions of the Royal Society of London* B 357: 259-267.

Gross, M.R. 2005. The evolution of parental care. *Quarterly Review of Biology* 80: 35-45.

Gross, M.R. and R.C. Sargent. 1985. The evolution of male and female parental care in fishes. *American Zoologist* 25: 807-822.

Gruter, C. and B. Taborsky. 2004. Mouthbrooding and biparental care: An unexpected combination, but male brood care pays. *Animal Behaviour* 68: 1283-1289.

Gruter, C. and B. Taborsky. 2005. Sex ratio and the sexual conflict about brood care in a biparental mouthbrooder. *Behavioural Ecology and Sociobiology* 58: 44-52.

Härdling, R. and A. Kaitala. 2005. The evolution of repeated mating under sexual conflict. *Journal of Evolutionary Biology* 18: 106-115.

Harding, J.A., G.R. Almany, L.D. Houck and M.A. Hixon. 2003. Experimental analysis of monogamy in the Caribbean cleaner goby, *Gobiosoma evelynae*. *Animal Behaviour* 65: 865-874.

Henson, S.A. and R.R. Warner. 1997. Male and female alternative reproductive behaviours in fishes: a new approach using intersexual dynamics. *Annual Review of Ecology and Systematics* 28: 571-592.

Hockham, L.R., J.A. Graves and M.G. Ritchie. 2004. Sperm competition and the level of polyandry in a bushcricket with large nuptial gifts. *Behavioral Ecology and Sociobiology* 57: 149-154.

Holly, M. 1928. Die importierten Welse. *Blatterfür Aquarier and Terrarienkande* 39: 329-334.

Holly, M. 1931. Die importierten Welse. *Blatterfür Aquarier and Terrarienkande*. 42: 361-363.

Hosken, D.J. and W.U. Blackenthorn. 1998. Female multiple mating, inbreeding avoidance, and fitness: It is not only the magnitude of costs and benefits that counts. *Behavioral Ecology* 10: 462-464.

Hosken, D.J., T.W.J. Garner, T. Tregenza, N. Wedell and P.I. Ward. 2003. Superior sperm competitors sire higher-quality young. *Proceedings of the Royal Society of London* B 270: 1933-1938.

Hourigan, T.F. 1989. Environmental determinants of butterflyfish social systems. *Environmental Biology of Fishes* 25: 61-78.

Jennions, M.D. and M. Petrie. 2000. Why do females mate multiply? A review of the genetic benefits. *Biological Reviews* 75: 21-64.

Jones, A.G. and Avise, J.C. 1997. Microsatellite analysis of maternity and the mating system in the Gulf pipefish *Syngnathus scovelli*, a species with male pregnancy and sex-role reversal. *Molecular Ecology* 6: 203-213.

Jones, A.G. and J.C. Avise. 2001. Mating systems and sexual selection in male pregnant pipefishes and seahorses: insights from microsatellite-based studies of maternity. *Journal of Heredity* 92: 150-158.

Jones, A.G., C. Kvarnemo, G.I. Moore, L.W. Simmons and J.C. Avise. 1998a. Microsatellite evidence for monogamy and sex-biased recombination in the Western Australian seahorse *Hippocampus angustus*. *Molecular Ecology* 7: 1497-1506.

Jones, A.G., S. Ostlund-Nilsson and J.C. Avise. 1998b. A microsatellite assessment of sneaked fertilization and egg thievery in the fifteenspine stickleback. *Evolution* 5: 848-858.

Jones, A.G., D. Walker, C. Kvarnemo, K. Lindstrom and J.C. Avise. 2001. How cuckoldry can decrease the opportunity for sexual selection: Data and theory from a genetic parentage analysis of the sand goby, *Pomatoschistus minutus*. *Proceedings of the National Academy of Sciences of the United States of America* 98: 9151-9156.

Kawase, H. and A. Nakazono. 1994. Reproductive behavior of the flagtail triggerfish, *Sufflamen chrysopterus*. *Proceedings 7th International Coral Reef Symposium, Guam*. 905-907.

Keenleyside, M.H.A. 1991. *Cichlid Fishes—Behaviour, Ecology and Evolution*. Chapman and Hall, London.

Kellogg, K.A., J.A. Markert, J.R. Stauffer and T.D. Kocher. 1995. Microsatellite variation demonstrates multiple paternity in lekking cichlid fishes from Lake Malawi, Africa. *Proceedings of the Royal Society of London* B 260: 79-84.

Kellogg, K.A., J.A. Markert, J.R. Stauffer and T.D. Kocher. 1998. Intraspecific brood mixing and reduced polyandry in a maternal mouth-brooding cichlid. *Behavioral Ecology* 9: 309-312.

Kelly, C.D., J.G.J. Godin and J.M. Wright. 1999. Geographical variation in multiple paternity within natural populations of the guppy (*Poecilia reticulata*). *Proceedings of the Royal Society of London* B 266: 2403-2408.

King, J.R. and R.E. Withler. 2005. Male nest site fidelity and female serial polyandry in lingcod (*Ophiodon elongatus*, Hexagrammidae). *Molecular Ecology* 14: 653-660.

Knight, M.E., G.F. Turner, C. Rico, M.J.H. van Oppen and G.M. Hewitt. 1998. Microsatellite paternity analysis on captive Lake Malawi cichlids supports reproductive isolation by direct mate choice. *Molecular Ecology* 7: 1605-1610.

Knowlton, N. 1979. Reproductive synchrony, parental investment, and the evolutionary dynamics of sexual selection. *Animal Behaviour* 27: 1022-1033.

Koblmuller, S., W. Salzburger and C. Sturmbauer. 2004. Evolutionary relationships in the sand-dwelling cichlid lineage of Lake Tanganyika suggest multiple colonization of rocky habitats and convergent origin of biparental mouthbrooding. *Journal of Molecular Evolution* 58: 79-96.

Kokita, T. 2002. The role of female behaviour in maintaining monogamy of a coral-reef filefish. *Ethology* 108: 157-168.

Kuwamura, T. 1986a. Parental care and mating systems of cichlid fishes in lake Tanganyika: A preliminary field survey. *Journal of Ethology* 4: 129-146.

Kuwamura, T. 1986b. Substratum spawning and biparental guarding of the Tanganyikan cichlid fish *Boulengerochromis microlepis*, with notes on its life history. *Physiological Ecology of Japan* 23: 31-43.

Kuwamura, T. 1997. The evolution of parental care and mating systems among Tanganyikan cichlids. In: *Fish Communities in Lake Tanganyika*. H. Kawanabe, M. Hori and N. Nagoshi (Eds.) Kyoto University Press, Kyoto, pp. 57-86.

Kvarnemo, C., G.I. Moore, A.G. Jones, W.S. Nelson and J.C. Avise. 2000. Monogamous pair-bonds and mate switching in the Western Australian seahorse *Hippocampus subelongatus*. *Journal of Evolutionary Biology* 13: 882-888.

LeBas, N.R., L.R. Hockham and M.G. Ritchie. 2004. Sexual selection in the gift-giving dance fly, *Rhamphomyia sulcata*, favors small males carrying small gifts. *Evolution* 58: 1763-1772.

Lee, P.L.M. and G.C. Hays. 2004. Polyandry in a marine turtle: females make the best of a bad job. *Proceedings of the National Academy of Sciences of the United States of America* 101: 6530-6535.

Leiser, J.K. and M. Itzkowitz. 2004. To defend or not to defend? Size, residence, and conditional mating in male variegated pupfish *Cyprinodon variegatus*. *Ethology* 16: 299-313.

Lowe-McConnell, R.H. 1975. *Fish Communities in Tropical Freshwaters: Their Distribution, Ecology and Evolution*. Longman, New York.

Lowe-McConnell, R. 1987. *Ecological Studies in Tropical Fish Communities*. Cambridge University Press, London.

Luling, V. 1964. Sur Biologie und okoligie von *Arapaima gigas* (Pisces, Osteoglossidae). *Zeitschrift fur Morphologie und Okologie der Tiere* 54: 436-530.

Magurran, A.E. 2001. Sexual conflict and evolution in Trinidadian guppies. *Genetica* 112: 463-474.

Maklakov, A.A., T. Bilde and Y. Lubin. 2005. Sexual conflict in the wild: Elevated mating rate reduces female lifetime reproductive success. *American Naturalist* 165: S38-S45.

Mackereth, R.W. and M.H.A. Keenleyside. 1993. Breeding territoriality and pair formation in the Convict cichlid (*Cichlasoma nigrofasciatum*; Pisces, Cichlidae). *Canadian Journal of Zoology* 71: 960-967.

Masonjones, H.D. and S.M. Lewis. 1996. Courtship behavior in the dwarf seahorse, *Hippocampus zosterae*. *Copeia* 1996: 634-640.

Matthes, H. 1961. *Boulengerochromis microlepis*, a Lake Tanganyikan fish of economic importance. *Bulletin of Aquatic Biology* 3: 1-15.

McKaye, K.R. 1977. Competition for breeding sites between the cichlid fishes of Lake Jiloa, Nicaragua. *Ecology* 58: 291-302.

McKaye, K.R. 1983. Ecology and breeding behaviour of a cichlid fish, *Cyrtocara eucinostomus* on a large lek in Lake Malawi, Africa. *Environmental Biology of Fishes* 8: 81-96.

McKaye, K. 1984. Behavioral aspects of cichlid reproductive strategies: patterns of territoriality and brood defence in Central American substratum spawners and African mouth brooders. In: *Fish Reproduction: Strategies and Tactics*, G.W. Potts and R.J. Wootton (eds.). Academic Press, London, pp. 245-273.

McKaye, K.R. and N.M. McKaye. 1977. Communal care and kidnapping of young by parental cichlids. *Evolution* 31: 674-681.

McKaye, K.R. and M.K. Oliver. 1980. Geometry of a selfish school: defence of cichlid young by bagrid catfish in Lake Malawi, Africa. *Animal Behaviour* 28 : 1287-1290.

Morley, J.I. and S. Balshine. 2002. Faithful fish: territory and mate defence favour monogamy in an African cichlid fish. *Behavioral Ecology and Sociobiology* 52: 326-331.

Morley, J.I. and S. Balshine. 2003. Reproductive biology of *Eretmodus cyanostictus*, a cichlid fish from Lake Tanganyika. *Environmental Biology of Fishes* 66: 169-179.

Morrow, E.H. and G. Arnqvist. 2003. Costly traumatic insemination and a female counter-adaptation in bed bugs. *Proceedings of the Royal Society of London* B 270: 2377-2381.

Moyer, J.T. and C.E. Sawyers. 1973. Territorial behavior of the anemonefish *Amphiprion xanthurus* with notes on the life history. *Japanese Journal of Ichthyology* 20: 85-93.

Nagoshi, M. 1987. Survival of broods under parental care and parental roles of cichlid fish, *Lamprologus toae*, in Lake Tanganyika. *Japanese Journal of Ichthyology* 35: 392-395.

Nakai, K., Y. Yanagisawa, T. Sato, Y. Niimura and M.M. Gashagaza. 1990. Lunar synchronization of spawning in cichlid fishes of the tribe Lamprologini in Lake Tanganyika. *Journal of Fish Biology* 37: 589-598.

Neat, T.B. and S. Balshine-Earn. 1999. A field survey of the breeding habits of *Eretmodus cyanostictus*, a biparental mouthbrooding cichlid in Lake Tanganyika. *Environmental Biology of Fishes* 55: 333-338.

Neff, B.D. and T.E. Pitcher. 2002. Assessing the statistical power of genetic analyses to detect multiple mating in fishes. *Journal of Fish Biology* 61: 739-750.

Neff, B.D., L.M. Cargnelli and I.M. Côté. 2004. Solitary nesting as an alternative breeding tactic in colonial nesting bluegill sunfish (*Lepomis macrochirus*). *Behavioral Ecology and Sociobiology* 56: 381-387.

Newcomer, S.D., J.A. Zeh and D.W. Zeh. 1999. Genetic benefits enhance the reproductive success of polyandrous females. *Proceedings of the National Academy of Sciences of the United States of America* 96: 10236-10241.

Ochi, H., A. Rossiter and Y. Yanagisawa. 2000. The first record of a biparental mouthbrooding catfish. *Journal of Fish Biology* 57: 1601-1604.

Ochi, H. and Y. Yanagisawa. 2005. Farming-out of offspring is a predominantly male tactic in a biparental mouthbrooding cichlid *Parissodus microlepis*. *Environmental Biology of Fishes* 73: 335-340.

Ochi, H., A. Rossiter and Y. Yanagisawa. 2001. Biparental mouthbrooding of the catfish *Phyllonemus filinemus* in Lake Tanganyika. *Ichthyological Research* 48: 225-229.

Ochi, H., A. Rossiter and Y. Yanagisawa. 2002. Paternal mouthbrooding in bagrid catfishes in Lake Tanganyika. *Ichthyological Research* 49: 270-273.

Ohnishi, N., Y. Yanagisawa and M. Kohda. 1997. Sneaking by harem masters of the sandperch, *Parapercis snyderi*. *Environmental Biology of Fishes* 50: 217-223.

Okuda, N. 1999. Sex roles are not always reversed when the potential reproductive rate is higher in females. *American Naturalist* 153: 540-548.

Oliveira, R.F., N. Carvalho, J. Miranda, E.J. Goncalves, M. Grober and R.S. Santos. 2002. The relationship between the presence of satellite males and nest-holders' mating success in the Azorean rock-pool blenny *Parablennius sanguinolentus parvicornis*. *Ethology* 108: 223-235.

Orsetti, D.M. and R.L. Rutowski. 2003. No material benefits, and a fertilization cost, for multiple mating by female leaf beetles. *Animal Behaviour* 66: 477-484.

Osikowski, A. and J. Rafinski. 2001. Multiple insemination increases reproductive success of female Montandon's newt (*Triturus montandoni*, Caudata, Salamandridae). *Behavioral Ecology and Sociobiology* 49: 145-149.

Parker, A. and I. Kornfield. 1996. Polygynandry in *Pseudotropheus zebra*, a cichlid fish from Lake Malawi. *Environmental Biology of Fishes* 47: 345-352.

Perante, N.C., M.G. Pajaro, J.J. Meeuwig and A.C.J. Vincent. 2002. Biology of a seahorse species, *Hippocampus comes* in the central Philippines. *Journal of Fish Biology* 60: 821-837.

Perrone, M. and T.M. Zaret. 1979. Parental care patterns of fishes. *American Naturalist* 113: 351-361.

Pitcher, T.E., B.D. Neff, H. Rodd and L. Rowe. 2003. Multiple mating and sequential mate choice in guppies: females trade up. *Proceedings of the Royal Society of London* B 270: 1623-1629.

Plath, M., J. Parzefall and I. Schlupp. 2003. The role of sexual harassment in cave and surface dwelling populations of the Atlantic molly, *Poecilia mexicana* (Poeciliidae, Teleostei). *Behavioral Ecology and Sociobiology* 54: 303-309.

Pizzari, T. and R.R. Snook. 2004. Sexual conflict and sexual selection: Measuring antagonistic coevolution. *Evolution* 58: 1389-1393.

Ratkiewitcz, M. and A. Borkowska. 2000. Multiple maternity in the bank vole (*Clethrionomys glareolus*): Field and experimental data. Z. *Säugetierkunde (International Journal of Mammalian Biology)* 65: 6-14.

Reavis, R.H. and G.W. Barlow. 1998. Why is the coral-reef fish *Valenciennea strigata* (Gobiidae) monogamous? *Behavioral Ecology and Sociobiology* 43: 229-237.

Reynolds, J.D. 1996. Animal breeding systems. *Trends in Ecology and Evolution* 11: 68-72.

Reynolds, J.D., N.B. Goodwin and R.P. Freckleton. 2002. Evolutionary transitions in parental care and live bearing in vertebrates. *Philosophical Transactions of the Royal Society of London* B 357: 269-281.

Robertson, D.R. 1973. Field observation on the reproductive behaviour of a pomacentrid fish, *Acanthochromis polyacanthus*. *Zeitschrift fur Tierpsychologie* 32: 319-324.

Robertson, D.R., N.V.C. Polunin and K. Leighton. 1979. The behavioral ecology of three Indian Ocean Surgeon fishes (*Acanthus lineatus, A. leucosternon* and *Zebrasoma scopes*): Their feeding strategies, and social and mating systems. *Environmental Biology of Fishes* 4: 125-170.

Ros, A.F.H., I. Zeilstra and R.F. Oliviera. 2003. Mate choice in the Galilee St. Peter's fish, *Sarotherodon galilaeus*. *Behaviour* 140: 1173-1188.

Rüber, L., R. Britz, H.H. Tan, P.K.L. Ng and R. Zardoya. 2004. Evolution of mouthbrooding and life-history traits in the fighting fish genus *Betta*. *Evolution* 58: 799-813.

Saville, K.J, A.M. Lindley, E.G. Maries, J.C. Carrier and H.L. Pratt. 2002. Multiple paternity in the nurse shark, *Ginglymostoma cirratum*. *Environmental Biology of Fishes* 63: 347-351.

Snoeks, J. 2004. The non-cichlid fishes of the Lake Malawi system: a compilation. In: *The Cichlid Diversity of Lake Malawi/Nyasa/Niassa: Identification, Distribution and Taxonomy*, J. Snoeks (ed.). Cichlid Press, El Paso, Texas, pp. 20-26.

Soin, S.G. 1960. Reproduction and development of the snakehead *Ophiocephalus argus warpachowskii* (Berg). *Issues in Ichthyology* 15: USSR Academy of Sciences, pp. 127-137. (In Russian).

Taborsky, M., B. Hudde and P. Wirtz. 1987. Reproductive behaviour and ecology of *Symphodus* (*Crenilabrus*) *ocellatus*, a European wrasse with four types of male behaviour. *Behaviour* 102: 82-118.

Takahashi, D., H. Asada, T. Takeyama, M. Takahata, R. Katoh, S. Awata and M. Kohda. 2004. Why egg-caring males of Isaza (*Gymnogobius isaza*, Gobiidae) refuse additional females: preliminary field observations. *Journal of Ethology* 22: 153-159.

Talwar, P.K. and A.G. Jhingran. 1992. *Inland Fishes of India and Adjacent Countries*. Balkema Publishers, Rotterdam, Vol. 2. pp. 543-1158.

Taylor, M.I., G.F. Turner, R. Robinson and J.R. Stauffer, Jr. 1998. Parasites, sexual selection and bower height skew in a bower-building cichlid. *Animal Behaviour* 56: 377-382.

Taylor, M.I., J.I. Morley, C. Rico and S. Balshine. 2003. Evidence for genetic monogamy and female-biased dispersal in the biparental mouthbrooding cichlid *Eretmodus cyanostictus* from Lake Tanganyika. *Molecular Ecology* 12: 3173-3177.

Thresher, R.E. 1984. *Reproduction in Reef Fishes*. T.F.H. Publications, Inc., Neptune City.

Thiel, M. and I.A. Hinojosa. 2003. Mating behaviour of female rock shrimp *Rhynchocinetes typus* (Decapoda: Caridea)—indication for convenience polyandry and cryptic female choice. *Behavioral Ecology and Sociobiology* 55: 113-121.

Torres-Vila, L.M. and M.D. Jennions. 2005. Male mating history and female fecundity in the Lepidoptera: Do male virgins make better partners? *Behavioral Ecology and Sociobiology* 57: 318-326.

Tregenza, T. and N. Wedell. 2002. Polyandrous females avoid costs of inbreeding. *Nature (London)* 415: 71-73.

Trexler, J.C., J. Travis and A. Dinep. 1997. Variation among populations of the sailfin molly in the rate of concurrent multiple paternity and its implications for mating-system evolution. *Behavioral Ecology and Sociobiology* 40: 297-305.

Vincent, A.C.J. 1994. Operational sex ratios in seahorses. *Behaviour* 128: 153-167.

Vincent, A., I. Ahsnejo, A. Berglund and G. Rosenqvist. 1992. Pipefish and seahorses: Are they all sex role reversed? *Trends in Ecology and Evolution* 7: 237-241.

Vincent, A.C.J. and L.M. Sadler. 1995. Faithful pair bonds in wild seahorses, *Hippocampus whitei. Animal Behaviour* 50: 1557-1569.

Warner, R.R., D.Y. Shapiro, A. Marcanato and C.W. Petersen. 1995. Sexual conflict: males with highest mating success convey the lowest fertilization benefits to females. *Proceedings of the Royal Society of London* B 262: 135-139.

Weigmann, D.D. and J.R. Baylis. 1995. Male body size and paternal behaviour in small mouth bass, *Micropterus dolomieui* (Pisces: Centrarchidae). *Animal Behaviour* 50: 1543-1555.

Weigmann, D.D., J.R. Baylis and M.H. Hoff. 1992. Sexual selection and fitness variation in a population of smallmouth bass, *Micropterus dolomieui* (Pisces: Centrarchidae). *Evolution* 46: 1740-1753.

Wetton, J.H., R.E. Carter, D.T. Parkin and D. Walters. 1987. Demographic study of a wild house-sparrow population by DNA fingerprinting. *Nature (London)* 327: 147-149.

Whiteman, E.A. and I.M. Cote. 2004. Monogamy in reef fishes. *Biological Reviews* 79: 351-375.

Wickler, W. and U. Seibt. 1981. Monogamy in crustacea and man. *Zeitschrift fur Tierpsychologie* 57: 215-234.

Wisenden, B.D. 1994. Factors affecting mate desertion by males in free-ranging convict cichlids (*Cichlasoma nigrofasciatum*). *Behavioral Ecology* 5: 439-447.

Wisenden, B.D. 1995. Reproductive behavior of free-ranging convict cichlids, *Cichlasoma nigrofasciatum. Environmental Biology of Fishes* 43: 121-134.

Wittenberger, J.F. and R.L. Tilson. 1980. The evolution of monogamy: Hypotheses and evidence. *Annual Review of Ecology and Systematics* 11: 197-232.

Wootton, R.J. 1984. Introduction: tactics and strategies in fish reproduction. In: *Fish Reproduction: Strategies and Tactics*, G.W. Potts and R.J. Wootton (eds.). Academic Press, London, pp. 1-12.

Yabuta, S. and M. Kawashima. 1997. Spawning behavior and haremic mating system in the corallivorous butterflyfish, *Chaetodon trifascialis*, at Kuroshima island, Okinawa. *Ichthyological Research* 44: 183-188.

Yagisawa, Y. and M. Nshombo. 1983. Reproduction and parental care of the scale-eating cichlid fish *Perissodus microlepis* in Lake Tanganyika. *Physiology and Ecology Japan* 20: 23-31.

Yamagishi, S. and M. Kohda. 1996. Is the cichlid fish *Julidochromis marlieri* polyandrous? *Ichthyological Research* 43: 469-471.

Zane, L., W.S. Nelson, A.G. Jones and J.C. Avise. 1999. Microsatellite assessment of multiple paternity in natural populations of a live-bearing fish, *Gambusia holbrooki*. *Journal of Evolutionary Biology* 12: 61-69.

Zeh, J.A. and D.W. Zeh. 2003. Toward a new sexual selection paradigm: Polyandry, conflict and incompatibility. *Ethology* 109: 929-950.

Reproductive Strategies of Fish

Robert A. Patzner

Comprising around 30,000 species, fish have developed a large variety not only of morphology and mode of life but also of reproductive strategies. Many different reproductive styles have been created in the course of evolution. Reviews on modes of reproduction in fish are given, e.g., by Breder and Rosen (1966), Balon (1975, 1981) and Thresher (1984). This chapter provides an overview over the variety of life history phenomena associated with different reproductive styles. The different reproductive strategies present in fish are summarized in a table.

RELEASE OF OFFSPRING

Depending on whether fertilization is internal or external, fish may produce eggs or free-swimming young, respectively larvae. In 94% of all fish species, unfertilized eggs are released by the female. Six per cent of fish species are 'bearers', i.e. 'viviparous' (Fig. 9.1, Table 9.1). In some rare cases, fertilized eggs are laid after copulation. In Table 9.1, this reproductive mode is not listed separately but included under the heading **oviparity**.

Author's address: Organismic Biology, University of Salzburg, Hellbrunnerstr. 34, A-5020 Salzburg, Austria. E-mail: robert.patzner@sbg.ac.at

Table 9.1 Different reproductive strategies in fishes. Orders, families and species are ordered alphabetically within classes (after Froese and Pauly, 2004). ● = all or nearly all species; ◗ = some species; ? = not known.

Class and Order	Family	No. of species	Oviparity	Viviparity	Nest spawners	Demersal eggs	Pelagic eggs	Special breeders	Gonochorism	Sex change (m to f)	Sex change (f to m)	simult. hermaprodit
Myxini												
Myxiniformes	Myxinidae	69	●			●			●	◗?		
Cephalaspidomorphi												
Petromyzontiformes	Geotriidae	4	●			●			●			
Petromyzontiformes	Petromyzontidae	41	●			●			●			
Chondrichthyes												
Carcharhiniformes	Carcharhinidae	50		●					●			
Carcharhiniformes	Hemigaleidae	7		●					●			
Carcharhiniformes	Proscylliidae	6	●	◗		●			●			
Carcharhiniformes	Pseudotriakidae	1		●					●			
Carcharhiniformes	Scyliorhinidae	89	●	◗		●			●			
Carcharhiniformes	Sphyrnidae	8		●					●			
Carcharhiniformes	Triakidae	34		●					●			
Chimaeriformes	Callorhynchidae	4	●			●			●			
Chimaeriformes	Chimaeridae	20	●			●			●			
Heterodontiformes	Heterodontidae	8	●			●			●			
Hexanchiformes	All families (2)	5		●					●			
Lamniformes	All families (7)	16		●					●			
Orectolobiformes	Brachaeluridae	2		●					●			
Orectolobiformes	Ginglymonostomatidae	3		●					●			
Orectolobiformes	Hemiscyllidae	11	●			●			●			
Orectolobiformes	Orectolobidae	6		●					●			
Orectolobiformes	Parascylliidae	7	●			●			●			
Orectolobiformes	Rhincodontidae	1		●					●			
Orectolobiformes	Stegostomatidae	1	●			●			●			
Pristiophoriformes	Pristiophoridae	5		●					●			
Rajiformes	Anacanthobatidae	7	●			●			●			
Rajiformes	Dasyatidae	70		●					●			
Rajiformes	Gymnuridae	12		●					●			
Rajiformes	Hexatrygonidae	4	?	?		?			●			
Rajiformes	Myliobatidae	42		●					●			
Rajiformes	Plesiobatidae	1	?	?		?			●			
Rajiformes	Potamotrygonidae	20		●					●			

(Table 9.1 Contd.)

(*Table 9.1 Contd.*)

Class *and* Order	Family	No. of species	Oviparity	Viviparity	Nest spawners	Demersal eggs	Pelagic eggs	Special breeders	Gonochorism	Sex change (m to f)	Sex change (f to m)	simult. hermaprodit
Rajiformes	Rajidae	200	●			●			●			
Rajiformes	Rhinobatidae	45		●					●			
Rajiformes	Urolophidae	35		●					●			
Torpediniformes	Narcinidae	24		●					●			
Torpediniformes	Torpedinidae	14		●					●			
Squaliformes	All families (4)	74		●					●			
Squatiniformes	Squatinidae	13		●					●			
Actinopterygii												
Acipenseriformes	All families (2)	25	●			●			●			
Amiiformes	Amiidae	1	●		●				●			
Albuliformes	Albulidae	5	●				●		●			
Anguilliformes	Anguillidae	15	●				●		●			
Anguilliformes	Muraenidae	200	●				●		▸	▸	▸	▸
Anguilliformes	Other families (13)	522	●				●		●			
Atheriniformes	Atherinidae	165	●			●			●			
Atheriniformes	Melanotaeniidae	53	●			▸	▸		●			
Atheriniformes	Atherinopsidae	104	●			●			●			
Atheriniformes	Other families (5)	8	●				●		●			
Aulopiformes	All families (13)	56	●				●					●
Batrachoidiformes	Batrachoididae	69	●		●				●			
Beloniformes	Adrianichthyidae	11	●			●			●			
Beloniformes	Belonidae	34	●			●			●			
Beloniformes	Exocoetidae	52	●			▸	▸		●			
Beloniformes	Hemiramphidae	85	▸	▸		▸	▸		●			
Beloniformes	Scomberesocidae	4	●				●		●			
Beryciformes	All families (7)	121	●				●		●			
Characiformes	Anostomidae	110	●		●				●			
Characiformes	Characidae	776	●		▸	▸			●			
Characiformes	Erythrinidae	10	●		●				●			
Characiformes	Hemiodontidae	50	●		●				●			
Characiformes	Hepsetidae	1	●		●				●			
Characiformes	Other families (7)	615	●				●		●			
Clupeiformes	Clupeidae	216	●			▸	▸		●	▸		▸
Clupeiformes	Other families (4)	181	●					●	●			
Cypriniformes	Catostomidae	68	●		▸	●			▸	▸		
Cypriniformes	Cobitidae	110	●			●			▸	▸		

(*Table 9.1 Contd.*)

(Table 9.1 Contd.)

Class and Order	Family	No. of species	Oviparity	Viviparity	Nest spawners	Demersal eggs	Pelagic eggs	Special breeders	Gonochorism	Sex change (m to f)	Sex change (f to m)	simult. hermaprodit
Cypriniformes	Cyprinidae	2,010	●		◗	●		◗	●		◗	
Cypriniformes	Other families (3)	577	●			●			●			
Cyprinodontiformes	Anablepidae	9		●					●			
Cyprinodontiformes	Cyprinodontidae	100	●			●			●			
Cyprinodontiformes	Fundulidae	48	●			●			●			
Cyprinodontiformes	Goodeidae	40		●					●			
Cyprinodontiformes	Poeciliidae	293		●					◗	◗		
Cyprinodontiformes	Profundulidae	5	●			●			●			
Cyprinodontiformes	Rivulidae	255	●			●			●			●
Cyprinodontiformes	Valenciidae	2	●			●			●			
Elopiformes	All families (2)	7	●				●		●			
Esociformes	Umbridae	5	●		●				●			
Esociformes	Esocidae	5	●			●			●			
Gadiformes	Gadidae	22	●			◗	●		●			
Gadiformes	Other families (9)	453	●				●		●			
Gasterosteiformes	Aulorhynchidae	2	●		◗			◗	●			
Gasterosteiformes	Gasterosteidae	7	●		●				●			
Gasterosteiformes	Other families (3)	7	●		●				●			
Gobiesociformes	Gobiesocidae	120	●		●				●			
Gonorynchiformes	Kneriidae	24	●		●				●			
Gonorynchiformes	Other families (3)	3	●				●		●			
Gymnotiformes	Gymnotidae	9	●		◗			◗	●			
Gymnotiformes	Rhamphichthyidae	6	●		●				●			
Gymnotiformes	Sternopygidae	41	●		◗	◗			●			
Gymnotiformes	Other families (3)	100	●			●			●			
Lampriformes	All families (7)	18	●				●		●			
Lophiiformes	Antennariidae	41	●			◗	◗		●			
Lophiiformes	Other families (17)	259	●				●		●			
Myctophiformes	All families (2)	241	●				●		●			
Notacanthiformes	Halosauridae	15	●				●		●			
Notacanthiformes	Notacanthidae	10	●				●		●			
Ophidiiformes	Aphyonidae	21	◗	◗				◗	●			
Ophidiiformes	Bythitidae	90		●					●			
Ophidiiformes	Carapidae	31	●				●		●			
Ophidiiformes	Ophidiidae	209	◗	◗				◗	●			
Ophidiiformes	Parabrotulidae	3	●				●		●			

(Table 9.1 Contd.)

(*Table 9.1 Contd.*)

Class and Order	Family	No. of species	Oviparity	Viviparity	Nest spawners	Demersal eggs	Pelagic eggs	Special breeders	Gonochorism	Sex change (m to f)	Sex change (f to m)	simult. hermaprodit
Osmeriformes	Alepocephalidae	63	●				●		●			
Osmeriformes	Argentinidae	19	●				●		●			
Osmeriformes	Galaxiidae	40	●			●			●			
Osmeriformes	Lepidogalaxiidae	1	●			●			●			
Osmeriformes	Microstomatidae	8	●				●		●			
Osmeriformes	Opisthoproctidae	10	●				●		●			
Osmeriformes	Osmeridae	13	●			●			●			
Osmeriformes	Plecoglossidae	1	●			●			●			
Osmeriformes	Salangidae	11	●			●			●			
Osmeriformes	Other families (4)	61	●			?	?		●			
Osteoglossiformes	Gymnarchidae	1	●		●				●			
Osteoglossiformes	Hiodontidae	2	●				●		●			
Osteoglossiformes	Mormyridae	198	●		▶	▶			●			
Osteoglossiformes	Notopteridae	5	●		●				●			
Osteoglossiformes	Osteoglossidae	5	●		▶			▶	●			
Osteoglossiformes	Pantodontidae	1	●		●				●			
Perciformes	Acanthuridae	72	●			▶	●		●			
Perciformes	Acropomatidae	40	●			?	?		?			
Perciformes	Amarsipidae	1	●			?	?		?			
Perciformes	Ambassidae	41	●			●			●			
Perciformes	Ammodytidae	18	●			●			●			
Perciformes	Anabantidae	30	●		▶	▶	▶	▶	●			
Perciformes	Anarhichadidae	5	●		●				●			
Perciformes	Aplodactylidae	5	●			?	?		?			
Perciformes	Apogonidae	207	●					●	●			
Perciformes	Ariommatidae	6	●				●		●			
Perciformes	Arripidae	2	●				●?		●			
Perciformes	Artedidraconidae	24				●?						
Perciformes	Badidae	15	●			●			●			
Perciformes	Banjosidae	1	●			?	?		?			
Perciformes	Bathyclupeidae	4	●			?	?		?			
Perciformes	Bathydraconidae	15	●					●?	●			
Perciformes	Bathymasteridae	7	●			●			●			
Perciformes	Blenniidae	345	●		●				●			
Perciformes	Bovichtidae	11	●			?	?		?			
Perciformes	Bramidae	18	●			?	?		?			

(*Table 9.1 Contd.*)

(*Table 9.1 Contd.*)

Class and Order	Family	No. of species	Oviparity	Viviparity	Nest spawners	Demersal eggs	Pelagic eggs	Special breeders	Gonochorism	Sex change (m to f)	Sex change (f to m)	simult. hermaprodit
Perciformes	Caesionidae	20	●				●		●			
Perciformes	Callanthiidae	9	●				●				●	
Perciformes	Callionomidae	130	●				●		●			
Perciformes	Carangidae	140	●				●		●			
Perciformes	Caristiidae	4	●				●		●			
Perciformes	Centracanthidae	7	●		●						●	
Perciformes	Centrarchidae	27	●		●				●			
Perciformes	Centrogeniidae	1	●			?	?		?			
Perciformes	Centrolophidae	27	●				●		●			
Perciformes	Centropomidae	22	●			▶	▶			●		
Perciformes	Cepolidae	19	●				●		●		▶	
Perciformes	Chaenopsidae	56	●		●				●			
Perciformes	Chaetodontidae	114	●				●		●			
Perciformes	Champsodontidae	5	●				●		●			
Perciformes	Channichthyidae	17	●			?	?		?			
Perciformes	Channidae	21	▶	▶	▶			▶	●			
Perciformes	Cheilodactylidae	18	●				●		●			
Perciformes	Cheimarrhichthyidae	1	●			?	?		?			
Perciformes	Chiasmodontidae	15	●				●		●			
Perciformes	Chironemidae	4	●			?	?		?			
Perciformes	Cichlidae	1,300	●		▶			▶	●		▶	
Perciformes	Cirrhitidae	32	●			●			●		▶	
Perciformes	Clinidae	73	▶	▶	▶				●			
Perciformes	Coiidae	5	●			?	?		?			
Perciformes	Coryphaenidae	2	●				●		●			
Perciformes	Creediidae	16	●				●			●		
Perciformes	Cryptacanthodidae	4	●		●				●			
Perciformes	Dactyloscopidae	41	●		●				●			
Perciformes	Dichistiidae	3	●			?	?		?			
Perciformes	Dinolestidae	1	●				●		●			
Perciformes	Dinopercidae	2	●			?	?		?			
Perciformes	Draconettidae	7	●				●		●			
Perciformes	Drepaneidae	3	●				●		●			
Perciformes	Echeneidae	8	●				●		●			
Perciformes	Elassomatidae	5	●			●			●			
Perciformes	Eleginopidae	1	●			?	?		?			

(*Table 9.1 Contd.*)

(*Table 9.1 Contd.*)

Class *and* Order	*Family*	*No. of species*	*Oviparity*	*Viviparity*	*Nest spawners*	*Demersal eggs*	*Pelagic eggs*	*Special breeders*	*Gonochorism*	*Sex change (m to f)*	*Sex change (f to m)*	*simult. hermaprodit*
Perciformes	Eleotridae	150	●		●				●			
Perciformes	Embiotocidae	24		●					●			
Perciformes	Emmelichthyidae	14	●				●					●
Perciformes	Enoplosidae	1	●				●		●			
Perciformes	Ephippidae	20	●				●		●			
Perciformes	Epigonidae	15	●				●		●			
Perciformes	Gempylidae	23	●				●		●			
Perciformes	Gerreidae	40	●				●		●			
Perciformes	Glaucosomatidae	4	●				●		●			
Perciformes	Gobiidae	1,875	●		●				▶		▶	
Perciformes	Grammatidae	9	●		●			▶	●			
Perciformes	Haemulidae	150	●				●		●			
Perciformes	Harpagiferidae	5	●		●				●			
Perciformes	Helostomatidae	1	●				●		●			
Perciformes	Icosteidae	1	●				●		●			
Perciformes	Inermiidae	2	●				●		●			
Perciformes	Istiophoridae	11	●				●		●			
Perciformes	Kraemeriidae	8	●		●				●			
Perciformes	Kurtidae	2	●					●	●			
Perciformes	Kyphosidae	42	●				●		●			
Perciformes	Labridae	500	●		▶	▶			▶		●	
Perciformes	Labrisomidae	102	▶	▶	▶				●			
Perciformes	Lactariidae	2	●				●		●			
Perciformes	Lateolabracidae	2	●				●		●			
Perciformes	Latridae	9	●				●		●			
Perciformes	Leiognathidae	24	●				●		●			
Perciformes	Leptobramidae	1	●			?	?		?			
Perciformes	Leptoscopidae	4	●						●			
Perciformes	Lethrinidae	39	●				●				●	
Perciformes	Lobotidae	1	●				●		●			
Perciformes	Lutjanidae	103	●				●		●			
Perciformes	Luvaridae	1	●				●		●			
Perciformes	Malacanthidae	40	●				●		●			
Perciformes	Menidae	1	●				●		●			
Perciformes	Microdesmidae	60	●		●				●			
Perciformes	Monodactylidae	5	●				●		●			

(*Table 9.1 Contd.*)

(*Table 9.1 Contd.*)

Class *and* Order	Family	No. of species	Oviparity	Viviparity	Nest spawners	Demersal eggs	Pelagic eggs	Special breeders	Gonochorism	Sex change (m to f)	Sex change (f to m)	simult. hermaprodit
Perciformes	Moronidae	6	●				●		●			
Perciformes	Mugilidae	80	●				●		●			
Perciformes	Mullidae	55	●				●		●			
Perciformes	Nandidae	10	●		◐	◐		◐	●			
Perciformes	Nematistiidae	1	●			?	?		?			
Perciformes	Nemipteridae	62	●				●				●	
Perciformes	Nomeidae	15	●				●		●			
Perciformes	Notograptidae	2	●		●?				●			
Perciformes	Nototheniidae	50	●			●			●			
Perciformes	Odacidae	12	●				●		●			
Perciformes	Odontobutidae	5	●		●				●			
Perciformes	Opistognathidae	60	●					●	●			
Perciformes	Oplegnathidae	6	●				●		●			
Perciformes	Osphronemidae	49	●		◐			◐	●			
Perciformes	Ostracoberycidae	3	●			?	?		?			
Perciformes	Parascorpididae	1	●				●		●			
Perciformes	Pempheridae	25	●			●	●		●			
Perciformes	Pentacerotidae	12	●			●	●		●			
Perciformes	Percichthyidae	22	●			●			●			
Perciformes	Percidae	159	●			●			●			
Perciformes	Perciliidae	2	●			●			●			
Perciformes	Percophidae	40	●				●		●			
Perciformes	Pholidae	15	●		●				●			
Perciformes	Pholidichthyidae	2	●		●				●			
Perciformes	Pinguipedidae	50	●				●				●	
Perciformes	Plesiopidae	38	●		●			◐	●			
Perciformes	Polycentridae	2	●		◐	◐			●			
Perciformes	Polynemidae	33	●				●			●		
Perciformes	Polyprionidae	9	●				●				●	
Perciformes	Pomacanthidae	74	●				●		●			
Perciformes	Pomacentridae	321	●		●				◐	◐	◐	
Perciformes	Pomatomidae	3	●				●		●			
Perciformes	Priacanthidae	18	●				●		●			
Perciformes	Pseudaphritidae	3	●			?	?	◐	?			
Perciformes	Pseudochromidae	98	●		◐			◐	●		◐	
Perciformes	Ptilichthyidae	1	●		●?				●			

(*Table 9.1 Contd.*)

(*Table 9.1 Contd.*)

Class *and Order*	*Family*	*No. of species*	*Oviparity*	*Viviparity*	*Nest spawners*	*Demersal eggs*	*Pelagic eggs*	*Special breeders*	*Gonochorism*	*Sex change (m to f)*	*Sex change (f to m)*	*simult. hermaprodit*
Perciformes	Rachycentridae	1	●				●		●			
Perciformes	Rhyacichthyidae	1	●		●				●			
Perciformes	Scaridae	83	●				●		▶	●		
Perciformes	Scatophagidae	4	●		●				●			
Perciformes	Schindleriidae	2	●		●				●			
Perciformes	Sciaenidae	270	●				●		●			
Perciformes	Scombridae	51	●				●		●			
Perciformes	Scombrolabracidae	1	●				●		●			
Perciformes	Scombropidae	2	●				●?		●			
Perciformes	Scytalinidae	1	●		●?				●			
Perciformes	Serranidae	449	●			▶	●			▶	▶	
Perciformes	Siganidae	25	●				●		●			
Perciformes	Sillaginidae	31	●				●		●			
Perciformes	Sparidae	112	●		▶	▶	▶		▶	▶	▶	▶
Perciformes	Sphyraenidae	18	●				●		●			
Perciformes	Stichaeidae	65	●		●				●			
Perciformes	Stromateidae	13	●				●		●			
Perciformes	Symphysanodontidae	5	●			?	?		?			
Perciformes	Terapontidae	45	●				●		▶		▶	
Perciformes	Tetragonuridae	3	●				●		●			
Perciformes	Toxotidae	6	●				●?		●			
Perciformes	Trachinidae	8	●				●		●			
Perciformes	Trichiuridae	32	●				●		●			
Perciformes	Trichodontidae	2	●			●			●			
Perciformes	Trichonotidae	6	●				●		●			
Perciformes	Tripterygiidae	115	●		●				●			
Perciformes	Uranoscopidae	50	●				●		●			
Perciformes	Xenisthmidae	19	●		●?				●			
Perciformes	Xiphiidae	1	●				●		●			
Perciformes	Zanclidae	1	●				●		●			
Perciformes	Zaproridae	1	●			?	?		?			
Perciformes	Zoarcidae	220	▶	▶	▶				●			
Pleuronectiformes	All families (11)	572	●				●		●			
Polymixiiformes	Polymixiidae	1	●				●		●			
Saccopharyngiformes	All families (4)	26	●				●		●			
Salmoniformes	Salmonidae	66	●			●			●			

(*Table 9.1 Contd.*)

(*Table 9.1 Contd.*)

Class *and* Order	Family	No. of species	Oviparity	Viviparity	Nest spawners	Demersal eggs	Pelagic eggs	Special breeders	Gonochorism	Sex change (m to f)	Sex change (f to m)	simult. hermaprodit
Scorpaeniformes	Abyssocottidae	20	●			?			●			
Scorpaeniformes	Agonidae	44	●			●			●			
Scorpaeniformes	Anoplopomatidae	2	●				●		●			
Scorpaeniformes	Apistidae	3	●				?		●			
Scorpaeniformes	Aploactinidae	37	●				?		●			
Scorpaeniformes	Bathylutichthyidae	1	●				?		●			
Scorpaeniformes	Bembridae	8	●				?		●			
Scorpaeniformes	Caracanthidae	4	●				●				●	
Scorpaeniformes	Comephoridae	2		●					●			
Scorpaeniformes	Congiopodidae	9	●				●		●			
Scorpaeniformes	Cottidae	300	●		●				●			
Scorpaeniformes	Cottocomephoridae	24	●		●				●			
Scorpaeniformes	Cyclopteridae	28	●		●				●			
Scorpaeniformes	Dactylopteridae	7	●				●		●			
Scorpaeniformes	Ereuniidae	2	●				?		●			
Scorpaeniformes	Eschmeyeridae	1	●				?		●			
Scorpaeniformes	Gnathanacanthidae	1	●				?		●			
Scorpaeniformes	Hemitripteridae	8	●		●				●			
Scorpaeniformes	Hexagrammidae	12	●		●				●			
Scorpaeniformes	Hoplichthyidae	10	●				●		●			
Scorpaeniformes	Liparidae	195	●		●				●			
Scorpaeniformes	Neosebastidae	12	●				?		●			
Scorpaeniformes	Normanichthyidae	1	●				?		●			
Scorpaeniformes	Parabembridae	2	●				?		●			
Scorpaeniformes	Pataecidae	5	●				?		●			
Scorpaeniformes	Peristediidae	30	●				●		●			
Scorpaeniformes	Platycephalidae	60	●				●		●	●		
Scorpaeniformes	Plectrogenidae	2	●				?		●			
Scorpaeniformes	Psychrolutidae	29	●				?		●			
Scorpaeniformes	Rhamphocottidae	1	●				?		●			
Scorpaeniformes	Scorpaenidae	172	●				●		●			
Scorpaeniformes	Sebastidae	128		●					●			
Scorpaeniformes	Setarchidae	5	●				?		●			
Scorpaeniformes	Synanceiidae	31	●				?		●			
Scorpaeniformes	Tetrarogidae	35	●				?		●			
Scorpaeniformes	Triglidae	100	●				●		●			

(*Table 9.1 Contd.*)

(*Table 9.1 Contd.*)

Class *and* Order	Family	No. of species	Oviparity	Viviparity	Nest spawners	Demersal eggs	Pelagic eggs	Special breeders	Gonochorism	Sex change (m to f)	Sex change (f to m)	simult. hermaprodit
Scorpaeniformes	Zaniolepididae	1	●					?	●			
Siluriformes	Akysidae	13							●			
Siluriformes	Amblycipitidae	10							●			
Siluriformes	Amphiliidae	47	●		●				●			
Siluriformes	Ariidae	120	●		◗			◗	●			
Siluriformes	Aspredinidae	32	●		◗	◗		◗	●			
Siluriformes	Astroblepidae	40	●			●			●			
Siluriformes	Auchenipteridae	60	●			●			●			
Siluriformes	Bagridae	210	●		●				●			
Siluriformes	Callichthyidae	130	●		◗	◗			●			
Siluriformes	Cetopsidae	24	●		?	?			●			
Siluriformes	Chacidae	2	●		?	?			●			
Siluriformes	Clariidae	100	●		◗	◗			●			
Siluriformes	Cranoglanididae	2	●		●				●			
Siluriformes	Diplomystidae	5	●		?	?			●			
Siluriformes	Doradidae	90	●		●				●			
Siluriformes	Erethistidae	14	●		?	?			●			
Siluriformes	Helogeneidae	4	●		?	?			●			
Siluriformes	Heptapteridae	203	●			●			●			
Siluriformes	Heteropneustidae	2	●		●				◗			
Siluriformes	Hypophthalmidae	3	●		?	?			●			
Siluriformes	Ictaluridae	45	●		●				●			
Siluriformes	Loricariidae	550	●		◗	◗		◗	●			
Siluriformes	Malapteruridae	2	●		?	?			●			
Siluriformes	Mochokidae	150	●		?	?			●			
Siluriformes	Nematogenyidae	1	●		?	?			●			
Siluriformes	Olyridae	4	●		?	?			●			
Siluriformes	Pangasiidae	21	●			●			●			
Siluriformes	Parakysidae	2	●		?	?			●			
Siluriformes	Pimelodidae	300	●			●			●			
Siluriformes	Plotosidae	32	●		●				●			
Siluriformes	Pseudopimelodidae	26	●			●			●			
Siluriformes	Schilbeidae	45	●			●			●			
Siluriformes	Scoloplacidae	4	●		?	?			●			
Siluriformes	Siluridae	100	●		◗	◗			●			
Siluriformes	Sisoridae	85	●		?	?			●			

(*Table 9.1 Contd.*)

(*Table 9.1 Contd.*)

Class *and* Order	Family	No. of species	Oviparity	Viviparity	Nest spawners	Demersal eggs	Pelagic eggs	Special breeders	Gonochorism	Sex change (m to f)	Sex change (f to m)	simult. hermaprodit
Siluriformes	Trichomycteridae	155			?	?						
Stomiiformes	Gonostomatidae	32	●				●		●	▶		
Stomiiformes	Other families (3)	170	●				●		●			
Synbranchiformes	Chaudhuriidae	5	●		?	?	?		●			
Synbranchiformes	Mastacembelidae	67	●			●			●			
Synbranchiformes	Synbranchidae	15	●		●				▶			●
Syngnathiformes	Syngnathidae	215	●					●	●			
Syngnathiformes	Aulostomidae	3	●				●		●			
Syngnathiformes	Centriscidae	10	●				●		●			
Syngnathiformes	Fistulariidae	4	●				●		●			
Syngnathiformes	Solenostomidae	3	●					●	●			
Tetraodontiformes	Balistidae	40	●		●		▶		●			
Tetraodontiformes	Diodontidae	19	●			▶	●		●			
Tetraodontiformes	Molidae	4	●				●		●			
Tetraodontiformes	Monacanthidae	95	●			●			●			
Tetraodontiformes	Ostraciidae	33	●				●		●			
Tetraodontiformes	Tetraodontidae	121	●		▶	▶			●			
Tetraodontiformes	Triacanthidae	7	●		?	?	?		●			
Tetraodontiformes	Triacanthodidae	20	●		?	?	?		●			
Tetraodontiformes	Triodontidae	1	●		?	?	?		●			
Zeiformes	All families (6)	41	●				●		●			
Sarcopterygii												
Coelacantthiformes	Latimeriidae	2		●					●			
Lepidosireniformes	All Families (2)	5	●		●				●			

OVIPARITY

The eggs are released by the female and mostly fertilized after the release. Internal fertilization occurs only in non-viviparous sharks and rays, as well as a few teleosts. Balon (1981) calls them: (a) **obligate lecithotrophic live-bearers**. Here the eggs are fertilized internally, incubated in the reproductive system of females until the end of the embryonic phase or beyond; yolk is the source of nourishment; and (b) **facultative internal bearers**. In these the eggs are sometimes fertilized internally by accident

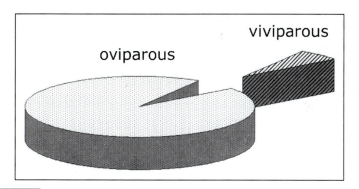

Fig. 9.1 Diagram showing different modes of offspring release, unfertilized eggs (oviparity) versus viviparity.

via close apposition of gonopores in normally oviparous fishes, and may be retained within the reproductive system to complete the early stages of embryonic development.

VIVIPARITY

Viviparity can be defined as a very intense form of parental care, in which the fertilized and developing eggs remain inside the body of the mother for a certain period of time and the offspring are released as free-swimming young. A prerequisite for this is internal fertilization by the male. This requires a copulatory organ which morphologically distinguishes the male from the female. Live-bearers produce only a small number of relatively large juveniles which have a good chance of survival to adulthood. Balon (1981) refers to this group as **viviparous trophoderms**.

Matrotrophy is the nourishment of viviparous embryos by resources provided by a source other than yolk between fertilization and parturition, while **lecithotrophy** describes embryo nourishment provided before fertilization. In fishes, embryo nourishment encompasses a continuum from solely lecithotrophic to primarily matrotrophic. Matrotrophy has evolved independently from lecithotrophic ancestors many times in many groups (Wourms and Lombardi, 1992; Texler and De Angelis, 2003). The benefits of matrotrophy are not Quite clear. It has been proposed that lecithotrophy is favored in fluctuating environment because all energies need for completion of abroad is prepackaged (Wourms and Lombardi, 1992).

Many shark species have evolved ways to increase the amount of food available to the unborn young, so that they are larger and better developed at birth, and thus presumably more likely to survive the first dangerous weeks of life in the ocean. In mackerel sharks, some species produce only a few fertile eggs—in some cases only one from each ovary—and many infertile ones which are used to feed the growing young inside the uterus, a process known as oophagy. In the Sandtiger Shark (*Carcharias taurus*, Chondrichthyes: Lamniformes: Odontaspididae), the developing embryos exhibit **intrauterine cannibalism**. They do not only feed on unfertilized eggs while still inside their mother, they also eat the other developing embryos in the ovary, until just one survives (Compagno *et al.*, 2005). In Balon (1981), this group is referred to as **matrotrophous oophages and adelphophages**. Another example is the Porbeagle Shark (*Lamna nasus*, Lamniformes: Lamnidae).

Oophagy has also been reported for Coelacanth (*Latimeria chalumnae*, Sarcopterygii: Coelacanthiformes) (Musick *et al.*, 1991). The females produce rather large eggs surrounded by a shell. After fertilization, not all embryos develop at the same rate. The first to leave the eggshell feed not only on their large yolk sac but also on the embryos which are still in an earlier phase of development. As a result, few offspring are produced, but the young are born well developed and, therefore, have an increased chance of becoming mature.

Greven (1995) gives an overview on the viviparity of several freshwater aquarium fishes. Members of the Poeciliidae, Anablepidae, Goodeidae (order Cyprinodontiformes) and Hemiramphidae (order Beloniformes) are live-bearing or viviparous. In the four teleost families mentioned above, the female always cares for the brood which develops in the ovary. In order to classify a taxon as viviparous, examination of trophic relations between offspring and mother may be useful. In lecithotrophic viviparity, the embryo depends exclusively on the yolk reserves laid down during oogenesis. In matrotrophic viviparity, however, the mother contributes at least in part to the nourishment of the young during gestation. The 'incubator' is always the ovary: oocytes develop in its follicles and are surrounded by a reduced envelope (zona radiata). After fertilization, the embryos develop either in the follicle (intrafollicular gestation in Poeciliidae, *Anableps* species and some Hemiramphidae) or else in the eggs—which in most cases are already fertilized—are discharged from the follicle and develop in the ovary (intraovarial gestation in Goodeidae, *Jenynsia* species and perhaps in some Hemiramphidae). A prerequisite for this reproductive strategy is internal fertilization, which is allowed for by the remodelling of the male analis into a copulatory organ. Spermatozoa of some species (excluding Goodeidae) are stored in folds of the ovarian epithelium. When embryonic tissue is closely attached to maternal tissue a placenta is formed. Depending on the tissue involved, a follicular placenta (Poeciliidae, *Anableps* species, and some Hemiramphidae), a trophotaenial placenta (Goodeidae) or a branchial placenta (*Jenynsia* species) can be distinguished. In the case of intrafollicular gestation and matrotrophy, e.g., in Dwarf Livebearer, *Heterandria formosa*, nutrients (among others macromolecules) from the maternal serum need to cross the capillary endothelium, the follicle epithelium and the reduced egg envelope. Thereafter, they are taken up by specialized cells of the embryonic surface. In the case of intraovarial gestation and matrotrophy (Goodeidae), macromolecules are transported from the maternal serum across the maternal capillaries and the ovarian epithelium into the ovarian lumen. Here they are endocytosed by the resorbing cells covering specialized nutrient cords (trophotaeniae), which are embryonic hindgut derivatives, and transported to embryonic blood vessels. Superfetation, demonstrated mainly for Poeciliidae, is related to matrotrophy in nearly all the described cases. Under certain conditions, viviparity is a successful reproductive strategy. Some disadvantages, such as the increased weight of the female are obvious, but are compensated for. Its advantages are evident, but difficult to prove experimentally. The same applies to studying the selection forces or conditions facilitating the evolution of viviparity, which are not necessarily the same as those for fostering (after establishing viviparity) an adaptive radiation of viviparous species. Furthermore, viviparity has evolved independently many times within related taxa, e.g., at least three times in Cyprinodontiformes (Greven, 1995).

SPAWNING TYPES

Thresher (1984) provides some general statements about demersal and pelagic eggs and spawners, respectively: (a) demersal eggs are not significantly larger than pelagic eggs; (b) demersal and pelagic spawners do not differ in fecundity; (c) demersal spawners are (in general) significantly smaller than pelagic ones; (d) the incubation period of eggs is much longer in demersal spawners; (e) newly hatched larvae of demersal spawners are larger than those of pelagic spawners; and (f) the degree of larval development is higher in demersal spawners. As shown in Fig. 9.2, around one-third of fish species are pelagic spawners and two-third demersal spawners. From the latter, about the half of species (33% of total) take care for their brood (parental care and most of special breeders).

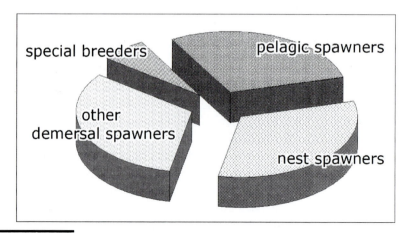

Fig. 9.2 Different types of spawning in fishes.

Four spawning strategies have been identified and described for teleost fish, depending on the frequency of spawning events (Fig. 9.3): (a) single spawners; (b) total spawners; (c) determinate and (d) indeterminate batch spawners. In both marine and limnic environments, most teleosteans seem to be batch spawners (George, 1995).

Nest Spawners/Parental Care

The reproductive behaviour of peacock blenny, *Salaria (=Blennius) pavo*, is similar to that of other blennioid fishes and is chosen here as an example of parental care. Spawning was studied in the northern part of the Adriatic Sea (Patzner *et al.*, 1986). Continuous observations lasted from dawn until the late afternoon and were terminated only at least two hours after the last spawning activity. Approximately 500 successful spawnings— related to more than 30 males of *S. pavo*—were observed during the course of the work. Male *S. pavo* live solitary in cavities formed by groups of stones or in clefts in artificial stone walls. They stay inside their refuges even when the entrance is exposed to air at extreme low tides. During the whole spawning season, the inner surface of each cavity is covered by a single layer of eggs at different stages of development. The quantity of eggs depends on the size of the cavity. In one cavity more than 10,000 eggs were counted. Generally, the male rests at the entrance of his cavity, showing the anterior part of his body with the pectoral fins inside or outside the cavity opening. The male has a characteristic helmet or crest (= adipose and connective tissue) on the head, which is bright yellow anteriorly and, therefore, visible over a relatively long distance. A dark

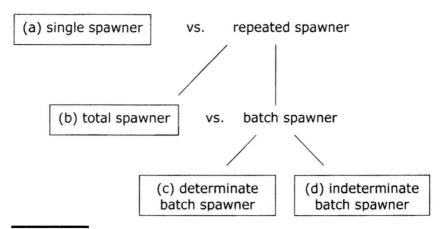

Fig. 9.3 Four strategies of spawning strategies of teleosts, depending on the frequency of spawning events (changed after George, 1995).

bar running parallel to the head profile through the eyes to the tip of the helmet emphasizes the bright colour which probably functions as a signal. Spawning behaviour is divided into different stages, almost none of them obligatory. Spawning starts when a female in nuptial coloration approaches the habitation of a male. In many cases, displaying females are chased away by the male for no obvious reason. However, when the female has been accepted, the male begins to display itself by rapidly changing his appearance. Many observations of two or more females displaying to one male showed that the males favour large females. The female approaches the cavity, still displaying (Fig. 9.4a), and enters tail first ('threading') (Fig. 9.4b). As soon as the female has completely entered the cavity, the male leaves to perform 'figure-8-swimming' in front of the entrance, either once or several times. Thereafter he enters the cavity headfirst, turns around and appears at the entrance. While the female remains inside the cavity, he fans with one or both pectoral fins and bends his body in wave-like motion. The male is usually more aggressive towards other females if there is already a female inside the cave. In 11.7% of all observations, however, a second female was allowed to enter. At the end of the spawning act, the female glides out of the hole and rests briefly nearby before she is chased away by a bite from the male. Thereafter, the male retreats into the cavity for some seconds before re-emerging. This is repeated several times until he returns to his normal behaviour. The actual fertilization of the eggs by the male has never been observed, neither in nature nor in the aquarium. Females of *S. pavo* do not inhabit

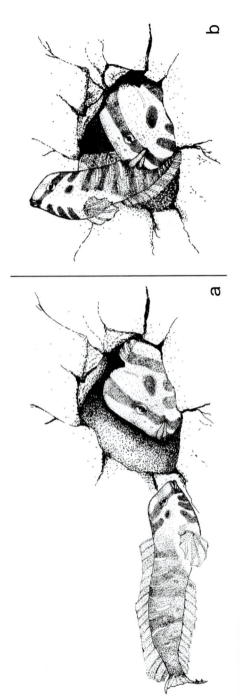

Fig. 9.4 Display behaviour of *Salaria pavo*. (a) Female approaching the cavity of the male. (b) Female 'threading' (from Patzner *et al.*, 1986).

cavities like the males but move around freely and only when resting sometimes visit recesses or holes. However, observations of marked animals show that they seem to be restricted to a certain area and that they are familiar with the locality. During egg deposition, the female is usually not visible to the observer. Studies in the aquarium revealed that her genital papilla makes contact with the inner surface of the cavity. When she has found a suitable place she fixes the eggs by pressing them onto the wall of the cavity, an activity which takes 5 to 7 seconds per egg. Continuous daily observations of spawning activity showed a clear activity peak between 06.00 and 07.30 h (European summer time) (Fig. 9.5) followed by a steady decrease until noon. After this only a single spawning occurred, and from 15.00 to 17.00 h no activity was observed. However, some spawning activity was observed towards the evening. Spawning frequency can vary considerably from male to male and from day to day. It has been observed that a male chased all females on one day and on the next day achieved seven successful spawnings. The maximum number of spawnings achieved by one male during one single day was 15 times. The duration of individual spawnings varied between one minute and 105 minutes with a mean of 19.6 minutes.

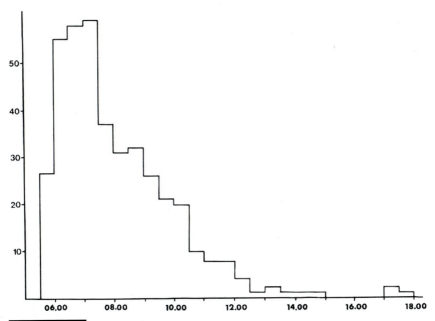

Fig. 9.5 Spawning activity of *Salaria pavo* in the course of one day. Ordinate shows number of observations (changed after Patzner *et al.*, 1986).

Observations on females showed that they spawn several times a day. Between two spawnings they sometimes rested only for a short while. It was never observed that one female spawned twice or more often with the same male on the same day. However, on consecutive days they visited the same cavities. There was no indication for a particular female to be allied with a particular male (Patzner et al., 1986).

In demersal eggs, the micropyle is usually either easy to reach by the spermatozoon after the attachment of the egg to the substratum or internal fertilization takes place (Wickler, 1957; Kinzer, 1960). It is still an unsolved problem as to how fertilization in blenniid and most gobiid fishes takes place, considering the fact that the micropyle is located in the middle of the adhesive disc. Authors describing the morphology of the egg came to the conclusion that fertilization has to take place prior to attachment (Eggert, 1931; Wickler, 1957; Patzner, 1984). On the other hand, authors studying the behaviour of blenniid and gobiid fish conclude that the eggs are fertilized after deposition (Casimir and Herkner, 1962; Abel, 1964; Cole, 1982). However, most authors were unable to observe the actual expulsion of sperm (Kinzer, 1960; Abel, 1964; Wirtz, 1978; Marraro and Nursall, 1983). If the egg is attached by the adhesive disc, it would be very difficult for the spermatozoa to reach the micropyle. In some gobies, the male has been found to put trails of spermatozoa on the substratum onto which the female deposits its eggs (Auty, 1987; Horsthemke, 1995; Mazzoldi et al., 1996; Ota et al., 1996). Whether this behaviour also occurs in other gobies and blennies so far has not been established.

Balon (1981) distinguishes between several groups of fish with parental care:

1. *Substrate choosers*: (a) **Pelagic spawners**: the positively buoyant eggs are guarded at the surface of hypoxic waters. (b) **Above water spawners**: the eggs are attached above the water line and are splashed periodically by the male. (c) **Rock spawners**: egg clusters are attached to a pole by fibres. (d) **Plant spawners**: the eggs are attached to aquatic plants, the embryos lack cement glands.
2. *Nest spawners*: (a) **Froth nesters**: the eggs are deposited in a cluster of mucous bubbles, the embryos have cement glands. (b) **Miscellaneous substrate and material nesters**: the eggs are attached singly or in clusters on any available substratum. (c) **Rock and gravel nesters**: the adhesive eggs are deposited in spherical or elliptical envelopes. (d) **Gluemaking nesters**: the eggs are deposited

in a nest held together by filamentous kidney secretions. (e) **Plant material nesters**: the eggs are attached to plants and embryos hang onto plants by secretions from their cement glands. (f) **Sand nesters**: the eggs are deposited in a nest and covered by sand. The sand is then gradually washed off. (g) **Hole nesters**: two modes prevail in this guild: cavity roof top nesters with moderately developed embryonic respiratory structures and bottom burrow nesters with strongly developed respiratory structures. (h) **Anemone nesters**: adhesive eggs are deposited in clusters at the base of a sea anemone.

The reader is also referred to the chapter 'parental care and predation' in this volume.

Other Demersal Spawners

Demersal spawning means laying eggs which hatch outside the female. Many marine and freshwater fishes have eggs with different surface structures for attaching them to the substrate (for review see Riehl, 1991). Several species of freshwater fishes live in running water and are adapted to strong currents. Under such conditions, it is expected that eggs should have a strong envelope and a sticky surface. On the other hand, it is well known that salmonids do not attach their eggs to the substrate, although they deposit them on the stream bottom. For fishes of standing waters, mainly cyprinids, it is important to have eggs with a sticky surface to allow them to be laid on macrophytes so as to avoid mud covering the eggs. In several cases, adhesive villi of varying lengths and densities were observed. In some Percidae, the outer egg surface showed a honeycomb-like structure. In all investigated species, the structures for attachment were found to be more or less regularly arranged all over the egg surface. This is in contrast to eggs of fishes with parental care where, in most cases, the adhesive filaments are only found at one pole of the egg (see above). In some representatives belonging to different families (Cyprinidae, Salmonidae, Esocidae and Gadidae), no villous structures were found (Patzner and Glechner, 1996).

Histochemical investigations revealed the presence of acid mucopolysaccharides in the outer egg envelope, the zona radiata externa, in several species. After contact with water the mucopolysaccharides swell, become adhesive and so help to attach the egg to the substrate (Arndt, 1960; Riehl, 1977; Guraya, 1978; Hagenmaier, 1985).

Figure 9.6 and Table 9.2 provide an overview of the egg surface structures of native European fishes. The data do not allow implications to be made about a connection between the ecology of a fish species and the surface structures of its eggs. Eggs of fishes living in running water show the same variety of attachment structures as are found in those living in standing water. Patzner *et al.* (1993) were also unable to find any connection between egg surface structure and taxonomy, at least within the family of Cyprinidae. In contrast, the thickness of the egg envelope is well correlated with the ecology of the fish taxa studied (Riehl, 1996).

Cyprinidae (Cypriniformes): The egg surface of several representatives of this family is sticky and has, in most cases, adhesive villi of different lengths and densities all around the egg (Table 9.2). In the dace, *Leuciscus leuciscus,* there are only a few, short villi (Fig. 9.6a). They are about 4 μm long, at a distance of about 6 μm from each other. In the roach, *Rutilus rutilus,* the villi stand very close together and have a length of about 11 μm (Fig. 9.6b). The eggs of the schneider, *Alburnoides bipunctatus,* have a surface covered with a large number of villi of different lengths, the longest reaching about 10 μm (Fig. 9.6c). The eggs of the Danube bleak, *Chalcalburnus chalcoides mento,* also possess a sticky surface but in contrast to the other cyprinids studied up to now no adhesive villi have been found (Fig. 9.6d).

Salmonidae (Salmoniformes): All the eggs studied up to now (brown trout, *Salmo trutta* f. *fario,* and huchen, *Hucho hucho*) are not sticky and have a smooth surface without any villi (Table 9.2).

Esocidae (Esociformes): The eggs of the pike, *Esox lucius,* the only representative of this family studied so far, are sticky. They have no adhesive villi on their surface (Table 9.2).

Gadidae (Gadiformes): The only freshwater representative of this family, the burbot, *Lota lota,* has eggs with a sticky surface. There are no adhesive villi (Fig. 9.6e, Table 9.2).

Percidae (Perciformes): In this family, eggs of the streber, *Zingel streber,* the zingel, *Zingel zingel,* and the ruffe, *Gymnocephalus cernuus,* were studied. Their surface shows a honeycomb-like structure all around the egg (Fig. 9.6f). The combs have a diameter of about 9 μm and a height of 13 μm (measured in *Z. streber*). The eggs also have a sticky surface (Table 9.2) (Patzner and Glechner, 1996).

Balon (1981) distinguishes between several groups of non-guarding, demersal spawners:

Table 9.2 Ecology, egg surface structures and stickiness of egg envelopes in different native fishes.

Order	Family	Genus	Ecology	Villi	Authors
Cypriniformes	Cyprinidae	*Alburnoides*	running	many, diff. length	Glechner et al. (1993)
Cypriniformes	Cyprinidae	*Aspius*	running	many	Riehl et al. (2002)
Cypriniformes	Cyprinidae	*Ballerus* (=*Abramis*)	running	no	Weidinger et al. (2005b)
Cypriniformes	Cyprinidae	*Chalcalburnus*	runn./stand.	no	Riehl et al. (1993a)
Cypriniformes	Cyprinidae	*Cyprinus*	standing	no	Riehl and Patzner (1994)
Cypriniformes	Cyprinidae	*Leuciscus*	running	few, many	Riehl et al. (1993b); Petz-Glechner et al. (1998)
Cypriniformes	Cyprinidae	*Rutilus*	standing	many, long	Patzner et al. (1996)
Cypriniformes	Cyprinidae	*Vimba*	running	many	Riehl et al. (1993c)
Esociformes	Esocidae	*Esox*	standing	no	Riehl and Patzner (1992)
Gadiformes	Gadidae	*Lota*	runn./stand.	no	Patzner and Riehl (1992)
Perciformes	Percidae	*Gymnocephalus*	runn./stand.	combs	Riehl and Meinel (1994)
Perciformes	Percidae	*Zingel*	running	combs	Patzner et al. (1994)
Salmoniformes	Salmonidae	*Hucho*	running	no	Weidinger et al. (2005a)
Salmoniformes	Salmonidae	*Salmo*	running	no	Riehl (1980)
Scorpaeniformes	Cottidae	*Cottus*	running	no	Patzner et al. (2001)

*parental care

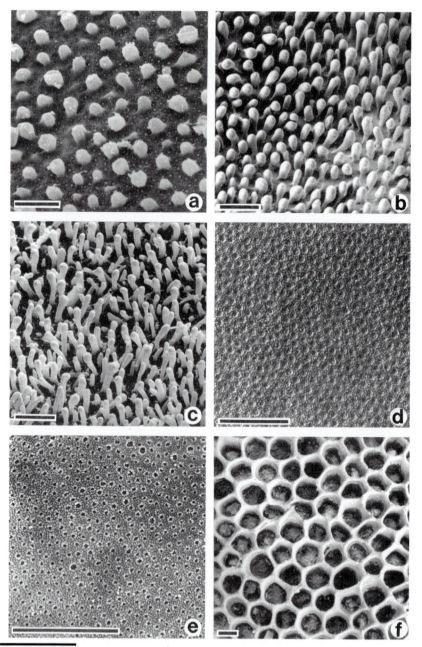

Fig. 9.6   Surface structures of the zona radiata externa of different eggs. The horizontal bars represent 10 μm, each. (a) dace, *Leuciscus leuciscus,* (b) roach, *Rutilus rutilus,* (c) schneider, *Alburnoides bipunctatus,* (d) Danube bleak, *Chalcalburnus chalcoides mento,* (e) burbot, *Lota lota,* (f) streber, *Zingel streber* (from Patzner and Glechner, 1996).

1. *Open substratum spawners*: (a) **Rock and gravel spawners with pelagic larvae**: the eggs are demersal at first, later buoyant; the larvae pelagic. (b) **Rock and gravel spawners with benthic larvae**: the embryos hide under stones. (c) **Non-obligatory plant spawners**: the eggs adhere to submerged items and free embryos have cement glands. (d) **Obligatory plant spawners**: the egg envelope sticks to submerged live or dead plants, the embryos also have cement glands. (e) **Sand spawners**: the eggs are deposited in running water on sand or fine roots over sand to which they adhere. The embryos do not have cement glands. (f) **Terrestrial spawners**: the adhesive eggs are scattered out of water over damp sod.
2. *Brood hiders*: (a) **Beach spawners**: spawning above the high tide waterline. (b) **Annual fishes**: with a facultative diapause and two further obligate resting intervals. Eggs and embryos can survive for many months in dry mud. (c) **Rock and gravel spawners**: the eggs are buried in gravel depressions or rock interstices, the early hatched free embryos are photophobic, the alevin large and emerging. (d) **Cave spawners**: A few large adhesive eggs are hidden in crevices, the emerging larvae are large.

In several sharks and some teleosts, the eggs are fertilized inside the body of the female and are kept in the ovary for many months before release. The advantage of this is that the time of exposure of the eggs to the environment and to predators is drastically reduced. In such cases, the final stages of development outside the body of the female may take only a short time.

Pelagic Spawners

In reef fishes, several behavioural strategies have been projected for pelagic spawning (reviewed by Shapiro *et al.*, 1988). The choice of spawning times and sites has been hypothesized to function primarily either to minimize egg predation by reef-associated fishes and invertebrates, to maximize dispersal, or to provide maximum opportunity for pelagic larvae to survive in waters with irregular distribution of food. The main reason is the advantage for pelagic eggs and larvae to be removed from shallow waters as quickly as possible (Table 9.3). However, the problem is that the larvae need to return to shallow water to colonize their final living habitats. Therefore, spawning often takes place when winds or currents are weak, in order to reduce the transport of larvae from where they originated (Johannes, 1978). Many larger species of reef fishes migrate to

Table 9.3 Reproductive strategies of tropical coastal marine fishes—timing and location of spawning. From Johannes (1978).

Reproductive event and location or time	Apparent selective advantage
Spawning near or over deep offshore waters	Reduces predation on eggs and larvae
Spawning off seaward promontories	Entrains larvae in near shore gyres, increasing chances of returning to coastal adult habitat
Spawning of pelagic eggs at night	Reduces predation on spawners and eggs
Hatching of demersal eggs at night	Reduces predation on larvae
Spawning during seasons when prevailing winds or currents are weakest	Minimizes washout of larvae from oceanic waters near shore thereby increasing their chances for recruitment
Release of pelagic eggs on outgoing spring tides	Reduces predation on eggs and larvae
Hatching of demersal eggs on outgoing spring tides	Reduces predation on larvae

spawn in deeper water from where their eggs and larvae commonly drift to the surface and are out of the range of benthic and demersal predators. Adults of small species are more prone to predation and normally remain near their normal life habitat to spawn. Many of them release their eggs near the water surface, out of the reach of benthic and demersal egg predators.

The larvae of reef fishes need to return to shallow waters to complete their life cycle. Therefore, several species place their eggs in the vicinity of circulating currents to return their larvae to shallow water. This increases the chances to recruit to their initial populations (Johannes, 1978).

SPECIAL BREEDERS

Mouthbrooders

Parents care for eggs and larvae by taking them into their mouth to protect them from predators and bacterial attack. Several forms of mouthbreeding are found in different fish species: (a) the male (father) keeps the eggs; (b) the female (mother) keeps the eggs—this is often the case in polygamous species; or (c) both parents care for the eggs—usually in species with strong pair bonding.

In most species, the male first prepares a nest for the offspring. The female then lays the eggs into the nest and the male fertilizes them. Thereafter, the male picks up the eggs with his mouth and oral incubation takes place. The eggs will never hatch if not incubated because they are easily succumb to bacterial attack.

The black-chinned tilapia (*Sarotherodon melanotheron*, Perciformes: Cichlidae) is a mouthbrooder in which the male keeps the eggs in his mouth for 14 to 18 days after fertilization (Specker and Kishida, 2000). The eggs remain inside his mouth until they depart as free-swimming larvae. Androgen and estradiol levels are found to be relatively high in prespawning males, to decrease significantly with the start of mouthbrooding, and to return to around prespawning levels by the end of mouthbrooding (Fig. 9.7). The black-chinned tilapia in captivity exhibits both paternal and maternal-parental behaviour. In both sexes, the initiation of mouthbrooding triggers a decrease in androgen and estradiol concentrations. Specker and Kishida (2000) conclude that the presence of eggs inhibits the pituitary-gonadal axis in both males and females and hypothesize that a chemical signal from the eggs is delaying the initiation of the next brood.

The mode of mouthbreeding has developed in several orders of Actinopterygii. In the following some examples are given (see also Table 9.1). In the order Gymnotiformes, banded knife fish, *Gymnotus carapo* (Gymnotidae), is a male mouthbreeder (Kirschbaum and Schugardt, 2002). In the order Perciformes, mouthbreeding has developed in several families,

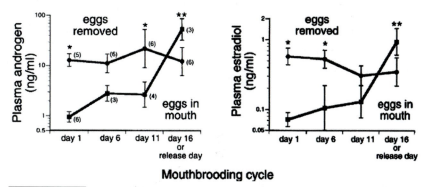

Fig. 9.7 Changes in plasma androgen concentrations (left) and in plasma estradiol concentrations (right) in male black-chinned tilapia that were allowed to keep their eggs (squares) versus males from which eggs were removed on the morning of Day 1 (circles). *N* is in parentheses. *Differs from males brooding eggs. **Differs from Day 1 (from Specker and Kishida, 2002).

e.g., the Channidae (*Channa bleheri*, Vierke, 1991), many Cichlidae, the Grammidae (*Gramma loretto*, Rosti, 1967), Osphronemidae (*Betta splendens*, Smith, 1993; *Luciocephalus* spp., Kokoscha, 1995), Plesiopidae (*Assessor macneilli*, Allen and Kuiter, 1975), as well as in some species of the Pseudaphritidae. In the Siluriformes, mouthbreeders are found in the Ariidae, Aspredinidae and Loricariidae.

Some North American cave fishes incubate the eggs in their gill cavities. This **gill-chamber breeding** probably derives from mouth breeding (Balon, 1975). The female takes the eggs into her mouth and subsequently transfers them into the gill cavity, where development proceeds until some time after hatching. In the end, the gill cover muscles usually slacken and the young fall out. About 70 eggs, each measuring 2.3 mm in diameter, are reported per female (Balon, 1975).

Breeding inside invertebrates

In the European bitterling, *Rhodeus sericeus*, the female deposits the eggs into a living mussel. Males establish territories in the vicinity of several species of living freshwater mussels (Unionidae). Females lay two to four eggs at a time into the gills of a mussel by inserting a long ovipositor into the exhalent siphon. They often spend several minutes inspecting a mussel before spawning, and are highly selective. Males fertilize the eggs by releasing sperm over the inhalant siphon both before and after egg laying (Wiepkema, 1961; Candolin and Reynolds, 2002a). Males release sperm one to four times per minute during courtship. As in many freshwater fishes, spermatozoa of bitterling are short lived. (Smith *et al.*, 2000; Mills and Reynolds, 2002a, b). Males become more aggressive after a female has spawned in a mussel (Candolin and Reynolds, 2002b). Male bitterling adjust their ejaculation rates to the density of competing males at the time before the first female starts to spawn. The rate of ejaculation by the dominant male reaches a peak when only one other male is present and then decreases with the number of competitors. The reduction in ejaculation rate when more than one competitor is present may be due to a decreased benefit per ejaculate (Candolin and Reynolds, 2002a).

In Japanese tubesnout (*Aulichthys japonicus*, Gasterosteiformes: Aulorhynchidae), one female deposits 40 to 70 eggs in the peribranchial cavity of the ascidian, *Halocynthia* (=*Cynthia*) *roretzi* (Uchida, 1934; Akagawa *et al.*, 2004). Gravid females visit male territories in groups from 5 to more than 20 individuals. There they gradually split up into small groups to examine each male's territory. Several females often approach the territory of one male together. Courtship has three stages: (a) The territorial male charges toward a female, stops and turns back to his territory. He repeats this until the female either leaves or enters his territory. (b) The male guides the female to a certain position inside his territory. (c) The male indicates an ascidian by approaching and stopping. After the female has examined the area she moves away, and both partners re-enter the tail beat-follow cycle. *A. japonicus* eggs are only found in the ascidian, *H. roretzi*. In the territory of *A. japonicus*, around 40% of ascidians contain eggs. One female deposits 40 to 70 eggs into an ascidian where they adhere to the peribranchial cavity through the atrial aperture. The eggs are demersal, slightly adhesive and nearly spherical with 2.2 to 2.6 mm in diameter. Inside each ascidian, the fertilized eggs are practically all at the same developmental stage. Males do not show any parental care behaviour (Akagawa *et al.*, 2004).

Pregnant males and skin brooding

Seahorses and also pipefishes (Syngnathiformes: Syngnathidae) have remarkable adaptations for parental care, investing much in the development of their young. Most species are monogamous, which means that they form pairs which are stable through multiple matings. In seahorses, the eggs mature in the spirally formed ovaries of the female. These ovaries contain relatively few oocytes for their size (Boisseau, 1967). The eggs are oval or pear-shaped, semi-transparent, and orange as a result of the carotenoids in their crustacean-dominated diet (Foster and Vincent, 2004). During mating, the female deposits her entire clutch of eggs in a special brood pouch on the male, the so-called 'marsupium'. The eggs are fertilized inside the pouch to ensure paternity. The embryos develop in the marsupium—which resembles a mammalian uterus. They are protected from predators and provided oxygen through a capillary network. Females do not play any role in parental care after mating. The duration of the male's pregnancy ranges from 9 to 45 days, depending on the species and water temperature (Foster and Vincent, 2004). At the end of the pregnancy, the male actively forces the brood out of his marsupium. The young resemble miniature adult seahorses, complete with hardened fin rays, trunk rings and pigmentation (Boisseau, 1967). After birth they receive no further parental care. All syngnathid males exhibit this form of parental care but the brooding structures vary, ranging from the simple ventral gluing areas of some pipefishes to the completely enclosed pouches found in seahorses (Wilson *et al.*, 2003).

Balon (1975) proposed **skin brooding** as one of the modes of reproduction in fish. This term means the attachment of eggs and developing embryos to the skin of one of the parents. Wetzel *et al.* (1997) described the morphological modifications of the integument for egg attachment in skin-brooding fishes such as the banded banjo (*Platystacus cotyleforus*, Siluriformes: Aspredinidae) and the ornate ghostpipefish (Solesnostomus paradoxus, Syngnathiformes: Solenostomidae).

In nurseryfish, *Kurtus gulliveri* (Perciformes: Kurtidae), the males carry eggs in clusters or balls attached to a hook on the head. This may be an adaptation to low oxygen and high turbidity environments (Berra, 2001).

GONOCHORISM AND SEX CHANGE

Gonochorism

Around 88% of all fish species are gonochorists (Fig. 9.8). The sexes are separate and genetically determined. No sex change takes place during the lifetime of a fish (Fig. 9.9). However, the final sexual state attained in mature fish may not reflect the initial gonadal developmental pathway taken (Devlin and Nagahama, 2002). Gonochorists may develop directly as males or females, or gonads may initially be hermaphroditic to later develop into functional ovaries or testes.

There are two types of gonochorists (Atz, 1964; Devlin and Nagahama, 2002): (a) in primary gonochorism early gonad development proceeds

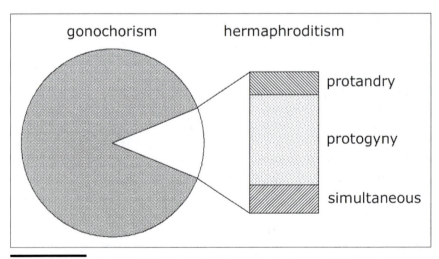

Fig. 9.8   Gonochorism and different modes of hermaphroditism in all fish species.

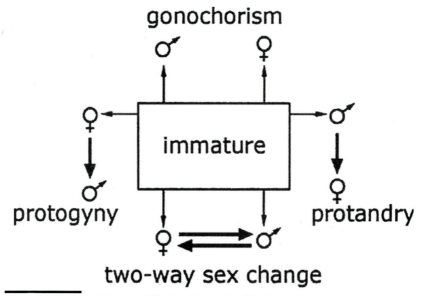

Fig. 9.9 Sexual pathways of life histories in fish. Thin lines indicate sexual maturation, thick lines functional sex change (changed after Kuwamura and Nakashima, 1998).

from an indifferent gonad directly to ovary or testis; or all the individuals initially develop an ovarian tissue which degenerates in about half the population and the gonad is invaded by additional somatic cells which develop into a normal testis. (b) In secondary gonochorists, all gonads are initially intersexual and later differentiate into either an ovary or a testis.

In a situation opposite to gonochorism, hermaphroditism, several types have been found in fish, including sequential hermaphroditism and simultaneous hermaphroditism.

Sex Change Male to Female: Protandry

Sexually mature males reduce their testis, develop an ovary and become sexually mature females. Protandric porgies (Sparidae) and flatheads (Platycephalidae) are peculiar insofar as they possess ovarian tissue from the very beginning of the functional male phase. But the oocytes in the ovarian lamellae do not develop beyond a certain growth stage (early perinucleolus stage), when their size corresponds roughly to the diameter of oocytes in functional ovaries between two successive spawning periods. Sooner or later, the testicular tissue becomes more and more reduced while the ovarian part proceeds to maturity. This basic process is apparently the same in all species (Reinboth, 1970).

Hesp *et al.* (2004) carried out macroscopic and histological investigations of the gonads of yellowfin seabream, *Acanthopagrus latus* (Perciformes: Sparidae), and demonstrated that this species is a protandrous hermaphrodite. In individuals smaller than 9 cm, the gonads are thin and cord like. First, gonial cells were detected in the gonads of fish 8 to 9 cm long. Gonads with testicular and ovarian tissue were found in fish *larger than* 11 cm (Fig. 9.10). At a total length between 11 and 19 cm, the testicular zone contains spermatogonia, spermatocytes and occasionally spermatids, while the ovarian zone contains immature oocytes.

The testicular zone contains either spermatogonia, spermatocytes or some spermatids in crypts, or substantial amounts of connective tissue and brown bodies and no crypts or evidence of spermatogenesis. In the time before spawning, the testicular zones of some ovotestes are far larger than their ovarian zones and contain spermatocytes, spermatids and spermatozoa. The authors assumed that fish with this type of gonad are destined to become functional males. During the spawning period, the crypts in the testicular zone of functional males break down and release their spermatozoa. Although a small amount of ovarian tissue is present in the gonads of functional males, the oocytes in this tissue never mature beyond the chromatin nucleolar stage.

After spawning, the testicular zone of such gonads decreases to a size similar to the ovarian zone. These gonads thus rever' to a form of ovotestes that is similar to that of juveniles of 11–19 cm, but of a larger size (Fig. 9.10). During the non-spawning period, the testicular zone contains substantial amounts of degenerating tissue and has become smaller than the ovarian zone (Fig. 9.10). Shortly before the spawning period, the ovarian zone contains several stages in oocyte development (Fig. 9.10) and the fish are destined to become functional females in spawning (Fig. 9.10). The testicular zone is very small and no longer contains detectable germ cells (Hesp *et al.*, 2004).

Members of the family Sparidae (Perciformes) are usually classified as protogynous hermaphrodites. A review showed, however, that sparids are not only protogynous but also protandrous: simultaneous and rudimentary hermaphroditism have been reported for this family (Buxton and Garratt, 1990). Histological investigations showed that even in those species reputed to have separate sexes, intersexuality is found in the juvenile condition. The authors suggested that two reproductive styles, sex change and 'late' gonochorism are present in the family. The bipotentiality of the sparid gonad is considered to be a pre-adaptation for

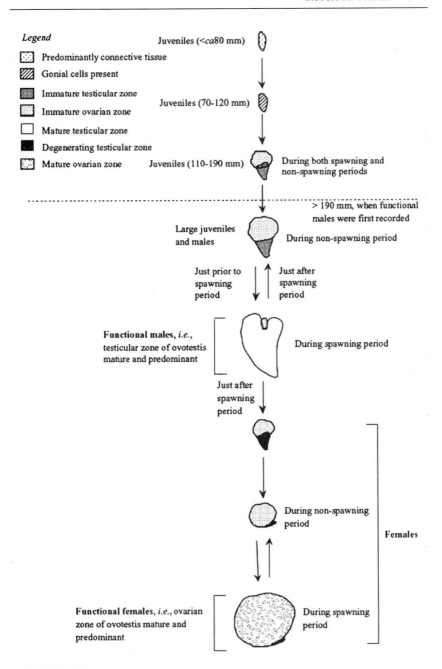

Legend

Juveniles (<*ca*80 mm)

Predominantly connective tissue

Gonial cells present

Immature testicular zone

Immature ovarian zone   Juveniles (70-120 mm)

Mature testicular zone

Degenerating testicular zone

Mature ovarian zone   Juveniles (110-190 mm)   During both spawning and
non-spawning periods

> 190 mm, when functional
males were first recorded

Large juveniles   During non-spawning period
and males

Just prior to   Just after
spawning   spawning
period   period

Functional males, *i.e.,*
testicular zone of ovotestis   During spawning period
mature and predominant

Just after
spawning
period

During non-spawning
period

Females

Functional females, *i.e.,* ovarian   During spawning
zone of ovotestis mature and   period
predominant

Fig. 9.10   Schematic representation of the sequence of changes that occur in the
ovotestes of yellowfin seabream, *Acanthopagrus latus*, during life (from Hesp *et al.*, 2004).

the development of sequential hermaphroditism in species in which reproductive success is size-related. In these species, sex change is an alternative reproductive style that enables individuals to maximize their lifetime reproductive success by functioning as one sex when small and the other sex when large (Buxton and Garratt, 1990).

Sex Change Female to Male: Protogyny

Sexually mature females reduce their ovary, develop a testis and become sexually mature males. Protogynous hermaphroditism has been demonstrated for several species of *Coryphopterus* (Perciformes: Gobiidae). This genus exhibits extensive geographical, morphological and ecological diversity. The occurrence of protogyny suggests that hermaphroditism is an ancestral rather than a recently derived condition (Cole and Shapiro, 1990). No testicular tissue was evident in the ovary, no ovarian tissue remained in the secondary testis once the sex change was complete, and testicular tissue usually first appeared at the periphery of the ovarian lumen. As with other protogynous gobiids, tissue masses were associated with the ovary and, subsequently, developed into apparently secreatory accessory gonadal structures associated with the secondary testis.

In some species, both gonochorism and protogyny may occur. For example, the Mediterranean rainbow wrasse, *Coris julis* (Perciformes: Labridae), develops male and female individuals with the same appearance. Those 'primary males' have a normally developed testis and produce normal spermatozoa. The females, however, after acting as such, degenerate their ovary and establish a functional testis. They grow to become larger than the primary males and obtain a different, more colourful appearance. They are called 'secondary males'. In the normal spawning act, only females and secondary males are involved. The primary males reproduce only as 'parasites' or 'satellites'. They try to eject their spermatozoa together with a normal spawning pair; a behaviour which is found in a variety of fishes and other vertebrates (Reinboth, 1957; Laurent and Lieune, 1988).

A further variant of hermaphroditism is the bi-directional or two-way sex change (Fig. 9.9). Munday *et al.* (1998) investigated into the social structure and pattern of sex change in the coral-dwelling fish, *Gobiodon histrio*. The social structure in a colony usually consists of a single juvenile or a heterosexual adult pair. *G. histrio* is primarily a protogynous hermaphrodite. All immature individuals are females and sex change from female to male occurs readily when two mature females are placed in a coral colony. In addition, males are able to change back to females when two mature males are placed in a coral. Sex changes from female to male, however, occur more than twice as often as sex changes from male to female.

Simultaneous Hermaphrodites

Each individual is simultaneously male and female and can release either eggs or sperm during spawning. This type of hermaphroditism is not as widespread as the two forms of sequential hermaphroditism. Up to now, it has been found only in a small number of fish families (Table 1).

Several Serranidae (Perciformes), especially representatives of the subfamily Serraninae, are simultaneous hermaphrodites. Thresher (1984) gives an overview over these fishes: In the gonads, ovarian and testicular tissues are separate, an arrangement referred to as 'territorial hermaphroditism'. Sperm and eggs are shed through different openings, which prevent their mixing and subsequent self-fertilization. Three openings are visible in front of the anal fin: the first is the ovarian pore, the central one is the opening of the sperm duct, and the hindmost is the anus. Smith (1965) suggests this type of gonad to be a primitive character and that sequential hermaphroditism has developed from it.

The factors which have lead to the evolution of simultaneous hermaphroditism in the serranines are not well understood (Thresher, 1984). Such bisexuality is normally considered stable only in low-density populations, where it is adaptive for each individual to be able to mate with any other individual it meets (Ghiselin, 1969). However, this argument clearly does not apply to the majority of serranines, which are often quite common.

Fischer (1981) examined the spawning system of the western Atlantic serranid black hamlet, *Hypoplectrus nigricans*. He suggested that each hermaphrodite protects its interests both as a male and as a female by engaging in a complex spawning behaviour. First, each individual engages in a courtship display before spawning several times as a female, releasing only a few eggs at a time. Thereafter, the spawning pair alternate roles: first, one individual releases eggs while the other releases sperm, and then the procedure is repeated, vice versa. The latter ensures that each individual will get at least some of its eggs fertilized. As a result, such hermaphrodites have a higher fecundity than a comparable non-hermaphrodite and will maintain themselves in the population even with high population densities (Thresher, 1984).

References

Abel, E.F. 1964. Freiwasserstudien zur Fortpflanzungsethologie zweier Mittelmeerfische, *Blennius canevae* Vinc. und *Blennius inaequalis*. C.V. *Zeitschrift für Tierpsychologie*. 21: 205-222.

Akagawa, I., T. Iwamoto, S. Watanabe and M. Okiyama. 2004. Reproductive behaviour of Japanese tubesnout, *Aulichthys japonicus* (Gasterosteiformes), in the natural habitat compared with relatives. *Environmental Biology of Fishes* 70: 353-361.

Allen, G.R. and R.H. Kuiter. 1976. A review of the plesiopid fish genus *Assessor*, with descriptions of two new species. *Records Western Australia Museum* 4: 201-215.

Arndt, E.A. 1960. Untersuchungen über die Eihüllen von Cypriniden. *Zeitschrift für Zellforschung*. 52: 315-327.

Atz, J.W. 1964. Intersexuality in fishes. In: *Intersexuality in Vertebrates Including Man*, C.N. Armstrong and A.J. Marshall (eds.). Academic Press, London, pp. 145-232.

Auty, E.H. 1978. Reproductive behaviour and early development of the empire fish *Hypselotris compressus* (Eleotridae). *Australian Journal of Marine and Fresh Water Research* 29: 585-597.

Balon, E.K. 1975. Reproductive guilds of fishes: A proposal and definition. *Journal of Fisheries Research Board of Canada* 32: 821-864.

Balon, E.K. 1981. Additions and amendments to the classification of reproductive styles in fishes. *Environmental Biology of Fishes* 6: 377-389.

Berra, T. 2001. *Freshwater Fish Distribution*. Academic Press, San Diego.

Boisseau, J. 1967. Les régulations hormonales de l'incubation chez un vertèbre male: recherches sur la reproduction de l'Hippocampe. Ph.D. Thesis, University Bordeaux, France.

Breder, C.M. Jr. and D.E. Rosen. 1966. *Modes of Reproduction in Fishes*. American Museum of Natural History, New York.

Buxton, C.D. and P.A. Garratt. 1990. Alternative reproductive styles in seabreams (Pisces: Sparidae). *Environmental Biology of Fishes* 28: 113-124.

Candolin, U. and J.D. Reynolds. 2002a. Adjustments of ejaculation rates in response to risk of sperm competition in a fish, the bitterling (*Rhodeus sericeus*). *Proceedings of the Royal Society London* B 269: 1549-1553.

Candolin, U. and J.D. Reynolds. 2002b. Why do males tolerate sneakers? Tests with the European bitterling, *Rhodeus sericeus*. *Behavioral Ecology and Sociobiology* 51: 146-152.

Casimir, M. and H. Herkner. 1962. Verhalten und Fortpflanzung von *Blennius pavo* Risso im Aquarium. *DATZ* 15: 141-144.

Cole, K.S. 1982. Male reproductive behaviour and spawning success in a temperate zone goby, *Coryphopterus nicholsi*. *Canadian Journal of Zoology* 60: 2309-2316.

Cole, K.S. and D.Y. Shapiro. 1990. Gonad structure and hermaphroditism in the gobiid genus *Coryphopterus* (Teleostei: Gobiidae). *Copeia* 1990: 996-1003.

Compagno, L., M. Dando and S. Fowler. 2005. *Sharks of the World*. Princeton University Press, Princeton.

Devlin, R.H. and Y. Nagahama. 2002. Sex determination and sex differentiation in fish: an overview of genetic, physiological, and environmental influences. *Aquaculture* 208: 191-364.

Eggert, B. 1931. Die Geschlechtsorgane der Gobiiformes und Blenniiformes. *Zeitschrift für wissenschaftliche Zoologie* 139: 249-559.

Fisher, E.A. 1981. Sexual allocation in a simultaneously hermaphroditic coral reef fish. *American Naturalist* 117: 64-82.

Foster, S.J. and A.C. J. Vincent. 2004. Life history and ecology of seahorses: implications for conservation and management. *Journal of Fish Biology* 65: 1-61.

Froese, R. and D. Pauly (eds.). 2004. FishBase. World Wide Web electronic publication. www.fishbase.org.

George, M. 1995. Intraovarielle Aspekte und Laichstrategien von Knochenfischen. In: *Fortpflanzungsbiologie der Aquarienfische*, H. Greven and R. Riehl (eds.). Birgit Schmettkamp Verlag, Bornheim, Germany, pp. 27-32.

Ghiselin, M.T. 1969. The evolution of hermaphroditism among animals. *Quarterly Review of Biology* 44: 189-208.

Glechner, R., R.A. Patzner and R. Riehl (1993): Die Eier heimischer Fische. 5. Schneider— *Alburnoides bipunctatus* (Bloch, 1782)—Cyprinidae. *Österreichs Fischerei* 46: 169-172.

Greven, H. 1995. Viviparie bei Aquarienfischen (Poeciliidae, Goodeidae, Anaplepidae, Hemiramphidae). In: *Fortpflanzungsbiologie der Aquarienfische*, H. Greven and R. Riehl (eds.). Birgit Schmettkamp Verlag, Bornheim, Germany, pp. 141-160.

Guraya, S.S. 1978. Maturation of the follicular wall of nonmammalian vertebrates. In: *The Vertebrate Ovary*, R.E. Jones (ed.). Plenum Press, New York, pp. 261-329.

Hagenmaier, H.E. 1985. Zum Schlüpfprozess bei Fischen. 8. Die chemische Zusammensetzung der Forelleneihülle (zona radiata) und ihre Veränderung durch die Einwirkung des Schlüpfenzyms. *Zoologisches Jahrbuch für Allgemeine Zoologie und Physiologie* 89: 509-520.

Hesp, S.A., I.C. Potter and N.G. Hall. 2004. Reproductive biology and protandrous hermaphroditism in *Acanthopagrus latus*. *Environmental Biology of Fishes* 70: 257-272.

Horsthemke, H. 1995. Fortpflanzungsbiologie von Grundeln. In: *Fortpflanzungsbiologie der Aquarienfische*, H. Greven and R. Riehl (eds.). Birgit Schmettkamp Verlag, Bornheim, Germany, pp. 115-128.

Howe, J.C. 1991. Egg surface morphology of *Dialommus fuscus* Gilbert (Pisces: Labrisomidae). *Journal of Fish Biology* 38: 149-152.

Johannes, R.E. 1978. Reproductive strategies of coastal marine fishes in the tropics. *Environmental Biology of Fishes* 3: 65-84.

Kinzer, J. 1960. Zur Ethologie und Biologie der Gobiiden, unter besonderer Berücksichtigung der Schwarzgrundel *Gobius jozo* L. *Zoologische Beiträge* 6: 207-290.

Kirschbaum, F. and C. Schugardt. 2002. Reproductive strategies and developmental aspects in mormyrid and gymnotiform fishes. *Journal of Physiology, Paris* 96: 557-566.

Kokoscha, M. 1995. Fortpflanzung bei der Gattung *Luciocephalus*. In: *Fortpflanzungsbiologie der Aquarienfische*, H. Greven and R. Riehl (eds.). Birgit Schmettkamp Verlag, Bornheim, Germany, pp. 33-36.

Kuwamura, T. and Y. Nakashima. 1998. New aspects of sex change among reef fishes: recent studies in Japan. *Environmental Biology of Fishes* 52: 125-135.

Laurent, L. and P. Lejeune. 1988. Coexistence en méditerranée de deux livrées terminales différentes chez la girelle *Coris julis* (Pisces, Labridae). *Cybium* 12: 91-95.

Marraro, C.H. and J.R. Nursall. 1983. The reproductive periodicity and behaviour of *Ophioblennius atlanticus* (Pisces: Blenniidae) at Barbados. *Canadian Journal of Zoology* 61: 317-325.

Mazzoldi, C., M. Scaggiante and M.B. Rasotto. 1996. Mode of fertilization and alternative male mating strategies in a Mediterranean goby. *ASIH, New Orleans*, p. 212.

Munday, P.L., J. Caley and G.P. Jones. 1998. Bi-directional sex change in a coral-dwelling goby. *Behavioral Ecology and Sociobiology* 43: 371-377.

Musick, J.A., M. Bruton, and E. Balon. 1991. The biology of *Latimeria chalumnae* and the evolution of coelacanths. *Environmental Biology of Fishes* 11: 1-446.

Ota, D., M. Marchesan and E.A. Ferrero. 1996. Sperm release behaviour and fertilization in the grass goby. *Journal of Fish Biology* 49: 246-256.

Patzner, R.A. 1984. The reproduction of *Blennius pavo* (Teleostei, Blenniidae). II. Surface structures of the ripe egg. *Zoologischer Anzeiger* 213: 44-50.

Patzner, R.A. and R. Glechner. 1996. Attaching structures in eggs of native fishes. *Limnologica* 26: 179-182.

Patzner, R.A. and R. Riehl. 1992. Die Eier heimischer Fische. 1. Rutte, *Lota lota* L. (1758), (Gadidae). *Österreichs Fischerei* 45: 235-238.

Patzner, R.A., R. Glechner and R. Riehl. 1994. Die Eier heimischer Fische-9. Streber, *Zingel streber* Siebold, 1863 (Percidae). *Österreichs Fischerei* 47: 122-125.

Patzner, R.A., M. Seiwald, M. Adlgasser and G. Kaurin. 1986. The reproduction of *Blennius pavo*. V. Reproductive behaviour in natural environment. *Zoologischer Anzeiger* 216: 338-350.

Patzner, R.A., R. Riehl and R. Glechner. 1996. Die Eier heimischer Fische. 11. Plötze—*Rutilus rutilus* (Linnaeus, 1758) und Perlfisch—*Rutilus frisii meidingeri* (Heckel, 1852) (Cyprinidae). *Fischökologie* 9: 15-26.

Patzner, R.A., S. Fischer and R. Riehl. 2001: Die Eier heimischer Fische. 13. Mühlkoppe-*Cottus gobio* (Linnaeus, 1758) (Cottidae). *Österreichs Fischerei* 54: 50-54.

Petz-Glechner, R., R.A. Patzner and R. Riehl. 1998. Die Eier heimischer Fische. 12. Hasel—*Leuciscus leuciscus* (L. 1758) und Strömer—*Leuciscus souffia agassizi* (Cuvier und Valenciennes, 1844) (Cyprinidae). *Österreichs Fischerei* 51: 83-90.

Reinboth, R. 1957. Sur la sexualité du teleostéen *Coris julis* (L.). *Comptes Rendus des Seances de la societe de Biologie, Paris* 245: 1662-1665.

Reinboth, R. 1970: Intersexuality in fishes. *Memoirs of the Society for Endocrinology* 18: 515-543.

Riehl, R. 1977. Histo- und ultrahistochemische Untersuchungen an den Oocyten von *Noemacheilus barbatulus* (L.) und *Gobio gobio* (L.) (Pisces, Teleostei). *Zoologischer Anzeiger* 198: 328-354.

Riehl, R. 1980. Micropyle of some salmonins and coregonins. *Environmental Biology of Fishes* 5: 59-66.

Riehl, R. 1991. Die Struktur der Oocyten und Eihüllen oviparer Knochenfische - eine Übersicht. *Acta Biologica Benrodis* 3: 27-65.

Riehl, R. 1996. The ecological significance of the egg envelope in teleosts with special reference to limnic species. *Limnologica* 26: 183-189.

Riehl, R. and W. Meinel. 1994. Die Eier heimischer Fische. 8. Kaulbarsch—*Gymnocephalus cernuus* (Linnaeus, 1758) mit Anmerkungen zum taxonomischen Status von *Gymnocephalus baloni* (Holcik and Hensel, 1974). *Fischökologie* 7: 25-33.

Riehl, R. and R.A. Patzner. 1992. Die Eier heimischer Fische. 3. Hecht *Esox lucius* L., 1758 (Esocidae). *Acta Biologica Benrodis* 3: 27-65.

Riehl, R. and R.A. Patzner. 1994. Die Eier heimischer Fische. 7. Karpfen *Cyprinus carpio* L., 1758 (Cyprinidae). *Acta Biologica Benrodis* 6: 1-7.

Riehl, R., R.A. Patzner and R. Glechner. 1993a. Die Eier heimischer Fische. 2. Seelaube, *Chalcalburnus chalcoides mento* (Agassiz, 1832)—Cyprinidae. *Österreichs Fischerei* 46: 138-140.

Riehl, R., R. Glechner and R.A. Patzner. 1993b. Die Eier heimischer Fische. 4. Döbel *Leuciscus cephalus* (L. 1758)—(Cyprinidae). *Zeitschrift für Fischkunde* 2: 45-55.

Riehl, R., R.A. Patzner and R. Glechner. 1993c. Die Eier heimischer Fische. 6. Zährte, *Vimba vimba elongata* (Valenciennes, 1844)—Cyprinidae. *Österreichs Fischerei* 46: 266-269.

Riehl, R., R.A. Patzner and K. Zanger. 2002. Die Eier heimischer Fische. 14. Rapfen oder Schied—*Aspius aspius* (Linnaeus, 1758) (Cyprinidae). *Österreichs Fischerei* 55: 275-281.

Rosti, P. 1967. Breeding of the royal gramma. *Salt Water Aquarium* 2: 106-108.

Shapiro, D.Y., D.A. Hensley and R.S. Appeldoorn. 1988. Pelagic spawning and egg transport in coral-reef fishes: A skeptical overview. *Environmental Biology of Fishes* 22: 3-14.

Smith, A. 1993. Broody betas. *Practical Fishkeeping* August: 12-13.

Smith, C.L. 1965. The patterns of sexuality and the classification of serranid fishes. *American Museum Novitates.* 2207: 1-20.

Specker, J.L. and M. Kishida. 2000. Mouthbrooding in the black-chinned tilapia, *Sarotherodon melanotheron* (Pisces: Cichlidae): The presence of eggs reduces androgen and estradiol levels during paternal and maternal parental behaviour. *Hormones and Behavior* 38: 44-51.

Thresher, R.E. 1984. *Reproduction in Reef Fishes.* T.F.H. Publications, Neptune City.

Trexler, J.C. and D.L. De Angelis. 2003. Resource allocation in offspring provisioning: an evaluation of conditions favoring the evolution of matrotrophy. *American Naturalist* 162: 574-585.

Uchida, K. 1934. Life history of *Aulichthys japonicus* Brevoort (Hemibranchii, Pisces). *Japanese Journal of Zoology* 5: 4-5.

Vierke, J. 1991. Der Regenbogen-Channa. Haltung und Zucht von *Channa bleheri*. *Aquarium (Bornheim)* 266: 15-19.

Wetzel, J., J.P. Wourms and J. Friel. 1997. Comparative morphology of cotylephores in *Platystacus* and *Solenostomus*: modifications of the integument for egg attachment in skin-brooding fishes. *Environmental Biology of Fishes* 50: 13-25.

Weidinger, C., R.A. Patzner and R. Riehl. 2005a. Die Eier heimischer Fische. 15. Huchen—*Hucho hucho* (Linnaeus, 1758) (Salmonidae). *Österreichs Fischerei* 58: 92-97.

Weidinger, C., R.A. Patzner and R. Riehl. 2005b. Die Eier heimischer Fische. 16. Zobel—*Ballerus* (=*Abramis*) *sapa* (Pallas, 1814) (Cyprinidae). *Österreichs Fischerei* 58: 202-206.

Wickler, W. 1957. Vergleichende Verhaltenmsstudien an Grundfischen. I. Beiträge zur Biologie, besonders zur Ethologie von *Blennius fluviatilis* Asso im Vergleich zu einigen anderen Bodenfischen. *Zeitschrift für Tierpsychologie* 14: 391-428.

Wiepkema, P.R. 1961. An ethological analysis of the reproductive behaviour of the bitterling (*Rhodeus amarus* Bloch). *Archives Néerlandaises de Zoologie* 14: 103-199.

Wilson, A.B., I. Ahnesjö, A.C.J. Vincent and A. Meyer. 2003. The dynamics of male brooding, mating patterns, and sex roles in pipefishes and seahorses (family Syngnathidae). *Evolution* 57: 1374-1386.

Wirtz, P. 1978. The behavior of the Mediterranean *Tripterygion* species (Pisces, Blennioidei). *Zeitschrift für Tierpsychologie* 48: 142-174.

Wourms, J.P. and J. Lombardi. 1992. Reflections on the evolution of piscine viviparity. *American Zoologist* 32: 276-293.

10

Energetic Trade-offs in Reproduction: Cost-Benefit Considerations and Plasticity

Ole Kristian Berg[1] and Anders Gravbrøt Finstad[2]

INTRODUCTION

Any higher organism must at some time allocate time and resources to reproduction in order to be represented by its genes in following generation. An individual's **fitness** is measured by the proportion of offspring in the next generation the individual contributes to. The optimal energy allocation to reproduction at any given time, is therefore a trade-off between the benefit in terms of number of offspring produced and the cost in terms of survival; to future breeding attempts (Williams, 1966; Stearns, 1992; Schaffer, 2004; Stevenson and Woods, 2006). Two basic questions face organism in regard to this trade-off (1) At **what age** shall the individual mature? and (2) If an individual decides to reproduce,

Authors' addresses: [1]Department of Biology, Norwegian University of Science and Technology, N-7491 Trondheim, Norway. E-mail: Ole.berg@bio.ntnu.no
[2]Norwegian Institute for Nature Research, Tungasletta 2, N-7485 Trondheim, Norway. E-mail: Anders.Finstad@nina.no

what proportion of available resources is allocated to each reproductive attempt and to each individual offspring. There is a multitude of factors, each with its set of trade-offs, that determines the individual choice in these questions.

The most taxing period of life for many organisms is the reproduction period. In the pre-reproductive period, resources are diverted from procuring food into the metabolism and structural tissue. In the reproductive period, these resources are procured into tasks necessary for successful breeding, including the allocation of energy to breeding behaviour, sexual characteristics and gonad tissue (Fig. 10.1). Energy used in current reproduction detracts energy that could be used to secure life of the individual, leading to a decrease in the probability of later reproduction. The use of resources into current reproduction is, therefore, negatively linked; with residual life span (Williams, 1966; Stearns, 1992; Schaffer, 2004). As each individual has only a limited amount of total energy available, there is **a trade-off between allocations of resources** into the different components associated with reproduction. The optimal solution to this trade-off depends on a number of biological characteristics of an individual such as sex, size, age, condition or energy stores of the individual, the reproduction system of the species as well as timing of reproduction and the physical characteristics of the reproduction site (e.g., Wootton, 1998; Schaffer, 2004).

Bioenergetics provides a functional framework with a standardized, common energetic currency, suited to link physiological processes with evolutionary and ecological theory. In this chapter, we will mainly focus on the patterns of energy allocation associated with reproduction in fishes. We will emphasize on teleost fishes, although the modelling tools are universal. Our approach is based on the framework of optimization models, such as life history theory, and how the optimal allocation pattern is influenced by various factors and states in both the individual, and population. We start with an outline of the basic principles of bioenergetics and how the energy budget of the reproductive expenditure may be modelled before we move on to discuss in detail what factors affect these trade-offs. It is a challenge to try to reveal some of the characteristic features of such a diverse and wide taxonomic array as fishes. As the published literature on the subject is overwhelming, we shall try to confine ourselves to general outlines and to illustrate the theory with a few examples rather than attempt a comprehensive review.

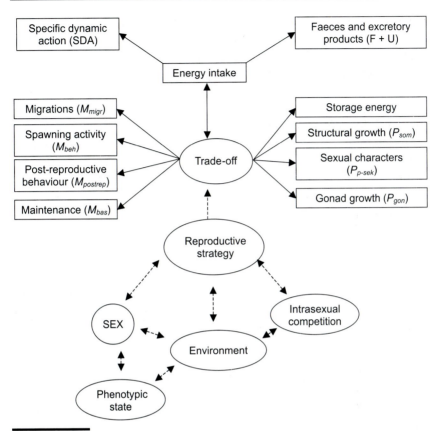

Fig. 10.1 The limited energy available to the individual through food intake has to be allocated into different tasks (solid lines) associated with both maintenance and growth. The trade-off between different allocation patterns is governed (stippled lines) by the reproductive strategy. The optimal trade-off between the different energy compartments depends on both the sex of the individual, various environmental characteristics and intrasexual competition for the access to mating partner(s) as well as interactions between these factors. The phenotypic state compartment covers individual state like energy stores and growth rate.

ALLOCATION OF ENERGY RESOURCES IN REPRODUCTION

The study of bioenergetics involves the examination of energy gains and losses and transfer of energy within the whole organism. Ingested energy is partitioned into the major physiological components leading to the universal energy budget equation:

$$C = F + P + M + U$$

Of the energy consumed as food (C) some is lost in faeces (F), some is invested into the growth of somatic tissue and gametes (P), some is expended in providing energy for a range of metabolic processes, both basic metabolic rates and metabolic costs associated with activity and digestion of food (M). Metabolism results in a production of partially oxidized products which is excreted as urea (U) (e.g., Kamler, 1992; Jobling, 1994; Wootton, 1998).

The different compartments of the original energy budget can be partitioned into different subunits. With regard to the reproductive period of the fish, energy invested into reproduction is divided into different components. Under circumstances with **no nutritional intake** and given no investment in structural somatic tissue, the total energy resources allocated to reproduction can be simplified, and total body energy change is diverted into the processes:

$$P_{gon} + P_{sek} + M_{bas} + M_{migr} + M_{beh} + M_{postrep}$$

The energy invested in growth of reproductive tissue is divided into gametes (P_{gon}) and growth of secondary sexual characteristics (P_{sek}). In addition to the basic metabolic demands during the reproductive period (M_{bas}), behavioural components associated with reproduction lead to increased metabolic costs either associated with migration to reproductive areas if such are a part of the fishes reproductive strategy (M_{migr}), or behaviour during the reproductive period such as defending of territories, courtship and mating (M_{beh}). Finally, there can also be increased metabolic costs associated with post-reproductive behaviour ($M_{postrep}$) such as parental care or nest defence.

With no or insignificant nutritional intake, the energy invested into reproduction is reallocated from storage energy into somatic tissue or gametes. The costs of reproduction equal the difference in body energy content prior to ($E_{tot\ pre}$) and after ($E_{tot\ post}$) spawning.

$$E_{tot\ pre} - E_{tot\ post} = P_{gon} + P_{sek} + M_{bas} + M_{migr} + M_{beh} + M_{postrep}$$

When **nutritional intake** has a significant role during the reproductive season, net energy intake is used in addition to stored somatic resources into reproduction

$$E_{net\ energy\ intake} + (E_{tot\ pre} - E_{tot\ post}) = P_{gon} + P_{s\text{-}sek} + M_{bas} + M_{migr} + M_{beh} + M_{postrep}$$

Since there are limited resources to be invested into the different purposes, a trade-off exists between the allocation of resources to the different parts of the energy budgets. For a number of species the energy use is known for most of these compartments. There is still, however, a marked lack of complete energy budgets of reproduction.

Energy budgeting during spawning has been especially popular with fish belonging to the family Salmonidae. Salmonids typically show profound niche-shifts, and the reproductive area may typically harbour numerous but small, foraging organisms. The large reproductive individuals have a limited or negative cost benefit relationship to feeding in the reproductive area. A number of species of salmonids therefore rely completely on stored energy during reproduction. Detailed energetic "tour de France"—have been given for a number of species (e.g., Gilhousen, 1980; Brett, 1995; Jonsson et al., 1997; Hendry and Berg, 1999; Kinnison et al., 2001; Finstad et al., 2002; Crossin et al., 2004; Hendry and Beall, 2004).

Trade-offs between the different energy allocation compartments is evident from the equations above. With limitations in the total energy level, diversion of energy to one purpose necessarily leaves less to other purposes. A huge amount of literature describes these trade-offs between initial and final energy level and investment into different compartments, the refinement of methods increases and, e.g., Crossin et al. (2004), determines most of the different compartments of energy in sockeye salmon populations differing in upriver migration distance and elevation.

MALE AND FEMALE ALLOCATION PATTERNS

Male and Female Sex Roles

Sexual reproduction is completely dominating in fishes, and as for all sexual reproducing organisms (except bacteria and virus) haploid, uninucleate gametes are produced that join in fertilization to form the diploid zygote (e.g., Otto and Nuismer, 2004). At some later stage in the life history of the organism, the chromosome number is again reduced by meiosis to form the next generation of gametes (e.g., Jobling, 1995). The gametes may be equal in size (isogamy), or one may be (slightly) larger than the other (anisogamy). In fishes, **the anisogamy** is the governing pattern in the form of a relative large egg and minute sperm which has a large impact on the differences in energetic trade-offs between sexes.

Fishes as a taxonomical group (e.g., Hickman et al., 2004) show a wide array of different sex-dependent reproductive allocations of energy and matter. Mating systems reflect the number of mates an individual acquires per breeding attempt and the **conflicting interests of the two sexes.** Mating systems, where an individual has several mates, are called polygamous in contrast to monogamous systems. Berglund (1997) reviews the allocation into reproduction in different mating systems from monogamy to polygamy (polygyny, polyandry and promiscuity).

Most fish species are gonochoristic (separate sexes) (Wootton, 1998). However, both sequential and simultaneous hermaphrodites are observed (Jobling, 1995; Wootton, 1998; Muñoz and Warner, 2004). Deep-sea fishes are examples of species where simultaneous hermaphrodites may occur (Jobling, 1995). The distance between the potential mates may be considerable, and when a potential mate is met a full and mutual exchange of both types of sexual products can be undertaken. Sequential hermaphrodites are found in species where reproductive success of one sex is highly energy or size dependent. The sex change is commonly determined by size-dependent differences in reproductive success, where typically the reproductive success in females are more closely linked to body size (number of eggs produced). The sex changing system may, accordingly, be energy or size determined, where the fish usually starts out as a male and change sex only after reaching a certain size. (e.g., Jobling, 1995; Allsop and West, 2004; Muñoz and Warner, 2004).

Early life is a period of high mortality for a wide variety of organisms, and many adaptations have evolved to minimize losses during this period (Roff, 1992). In addition to the investment into the pre-zygotic egg, **the behavioural investment in post-zygotic protection and parental care** (Clutton-Brock, 1991; Mousseau and Fox, 1998; Wilson et al., 2003) can represent increased energy investment into reproduction. Within the super class Gnathostomata, the class Chondrichthyes—the cartilaginous fishes (sharks, rays and chimeras) usually represent a reproductive system where the (female) parent allocate relatively large amounts of energy into each individual progeny. The teleost fishes usually have reproduction patterns with a high number of offspring, with relatively small amounts of energy allocated into each offspring, often as non-guarders of eggs and young. However, among a number of teleost (Osteichthyes) orders and families, the parent investment into each individual progeny is increased by the involvement of parental care as either guarders or bearers of eggs, externally or internally (Balon, 1975; Jobling, 1995; Wilson et al., 2003; Moyle and Cech, 2004; Mank and Avise, 2006). Parental care exists in 21% of families of bony fishes, with female only care being the rarest form (7% of the families with parental care) (Fleming and Reynolds, 2004). This increase in investment in the progeny, after fertilization, can further increase by the involvement by both sexes in the post-fertilization process of protecting and/or bearing the progeny. These post-fertilization processes may involve males or females or both and these processes may confound the typical male and female sex roles and reproductive budget associated with reproduction.

For the species who invest into **post-fertilization behaviour**, the number of progeny may be limited by the capacity of guarding and herding the progeny in larger numbers. The energy resources a female can invest into a given egg can be limited, but either sex can increase their investment into progeny with post-reproductive guarding or parental care of their offspring. Among the fishes there is, thus, an array of different reproductive allocation patterns in investment into each progeny in resources. The egg can vary in size and composition, or the egg can be protected after fertilization in organs corresponding to the mammalian womb or the egg or offspring can otherwise be protected and or fed with parental care from one or both sexes.

Gonadal Vs Secondary Sexual Characters Investment Vs Behavioural Components

The production of tissues in an organism is typically classified into **reversible somatic tissue (Ps) and released gonadal tissue (Pr)**. The universal energy budget illustrates the typical allocation conflict between these two sources

$$P_s + P_r = C - (F + U + M)$$

Provided that the other components of the energy budget remain unchanged during the reproduction period, P_s and P_r can be considered as directly competitive processes for resources.

The egg production of females represents a large energetic investment (anisogamy is treated above) compared to the male sperm. Females typically produce **energetically expensive eggs** (e.g., Kamler 1992), which are high-energy packages for the development of a zygote into a self-contained organism. The egg represents a major investment into the individual offspring. Since there is a direct correlation between the number of eggs and the potential number of offspring that a female may have, it must be expected that female reproductive output is largely determined by natural selection, giving priority to the number of viable offspring balanced by the effect investment into each egg will have on survival. The trade-offs related to investment into individual progeny (Smith and Fretwell, 1974; Einum et al., 2004), including the conflict between egg-size and number, will be treated in more detail later. The females shall, however, also be expected to invest into an array of morphological and behavioural adaptations in order to ensure mating success (e.g., Fleming and Reynolds, 2004). Because of the positive correlation between the gonad size/egg number and the number of offspring, there will therefore be a strong positive correlation between determinants of female size or quality and gonad size.

The male sperm is energetically a **cheap gamete** compared to the female eggs (e.g., Atlantic cod (*Gadus morhua*)) male/female mass specific gonad energy is 2.8 kJg^{-1} and 5.0 kJg^{-1} respectively (calculated from Lambert and Dutil, 1997). For salmonid fishes, the typical corresponding values are about 3.0 and 7.0 kJg^{-1} (e.g., Hendry and Berg, 1999; Berg et al., 2001). In addition comes the costs of constructing gonadal tissue (e.g., production costs of DNA), which may be higher for many small gametes than a few large. A review of energy content in eggs of 20 different fish species is given by Kamler (1992). A given male is likely to have enough energy to invest into gametes to be able to fertilize eggs from a number of females. This large male surplus capacity for fertilization gives rise to a commonly observed intense competition for access to females and eggs, where an increased volume of sperm is not likely to further increase the probability of fertilization success. A given amount may suffice for reproduction and there is probably limited increase in reproductive success with a further increase in male gonad size.

Because of the fundamental **difference between female and male sexual roles**, any analysis of energy use associated with reproduction must take sex into consideration. Budgets of energy for females and males together, e.g., in the form of GSI for both sexes will, therefore, be of limited value. This is illustrated by the relative weaker scaling of testis size with body size than the corresponding relationship between the female gonad size and body size (Fig. 10.2). As male testis has an energy density about half that of female gonads, the sex-dependent difference in gonad investment based on energy terms is, therefore, larger than on wet mass basis.

The comparably cheap gametes of the male commonly leads to the situation were males are the sex in surplus, since any given male may produce enough sperm to fertilize the offspring from a number of females (but see Forsgren et al., 2004 or Muñoz and Warner, 2004); for recent examples on sex role reversal). In order to increase the number of potential spawns and thus the reproductive success of any male, they regularly invest energy and matter into sexual characteristics—secondary sexual characters. The secondary sexual characters may be used as weapons against other individuals, as attractive ornaments or as indicators of handicap, usually indicating an individual's energy status or other indicators of the individual's possible fitness (e.g., Dugatkin and

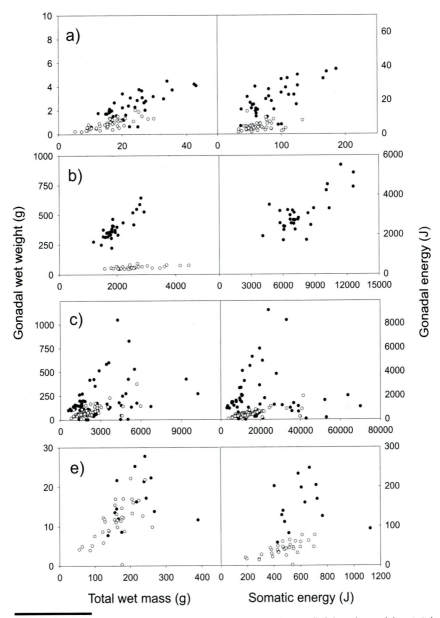

Fig. 10.2 Total wet mass plotted against gonadal wet mass (left hand panels) or total somatic energy plotted against gonadal energy (right hand panels) for: (a) Arctic char (*Salvelinus alpinus* (L.)); (b) Sockeye salmon (*Oncorhynchus nerka* (Walbaum)); (c) Atlantic salmon (*Salmo salar* L.); and (d) burbot (*Lota lota* (L.)). Open symbols represent males and closed symbols represent females.

FitzGerald, 1997; Kinnison *et al.*, 2003; Fleming and Reynolds, 2004; Svensson *et al.*, 2006). The evolution of these structures—which often may reduce the survival probability of the bearer—is a result of **sexual selection** often yielding pronounced sexual dimorphism. Sexual selection has been defined as selection for traits that enhance mating success (e.g., Fleming and Reynolds, 2004; Mank and Avise, 2006) and commonly acts more strongly on males (Darwin, 1871; Anderson, 1994; Fleming and Reynolds, 2004). This illustrates the possible difference between female and male energy budgets, as females usually invest considerable energy resources into gonads, whereas males invest more into secondary sexual characters and reproductive behaviour.

WHEN TO REPRODUCE?

Size and Age at First Maturity—Reproductive Strategies and Tactics

The onset of maturation represents a major transition in life. From a situation where energy, time and matter have been allocated to growth and survival, resources are (re-)allocated to reproduction, usually in direct opposition to growth and survival.

Life history theoretically provides a framework to **predict lifetime allocation patterns**, given defined environmental effects on age-specific mortality, growth and fecundity. It is assumed that selection processes favours genotypes that have age-specific energy allocation pattern that generates the highest per capita number of offspring, relative to other genotypes in the population. Mathematically, the instantaneous rate of natural increase of the population (r) is maximized (e.g., Stearns, 1992; Schaffer, 2004)

$$1 = \sum_{X=a}^{X=\infty} e^{(-rx)} l_x m_x$$

where x is age, r is as defined by the Euler-Lotka equation, l_x is the probability of surviving from birth to beginning of age class x and m_x is the expected number of offspring for a female in age class x. The per capita number of offspring will be maximized if the individual at each age maximizes the value of $m_i + (p_i \, v_{i+1} / v_0)$, where m_i is the fecundity of a fish aged i, p_i is the probability of a fish surviving from age i to $i+1$ and v_{i+1}/v_0 is the reproductive value of a fish aged $i+1$.

At each age (i), the per capita number of offspring is maximized if the fish maximizes the sum of its present fecundity (m_i) and its future expected fecundity ($p_i\, v_{i+1}\, /v_0$) (Stearns, 1992; Wootton, 1998; Schaffer, 2004). Reproductive values weighs the individual contribution of different aged individuals and its effect on population growth (Stearns, 1992; Wootton, 1998). The reproductive value of a female at age i can be expressed as the average number of young that a female aged i can expect to have over the remainder of her life, discounted back to present, and expressed relative to the reproductive value of a female at birth (Stearns, 1992; Wootton, 1998). The discounting is done to reflect the principle that early birth contributes more to future population growth than late births. This is comparable to the consequences of investing an amount of money in an interest bearing account in a bank now or 10 or 20 years from now.

Life history theory is based on Euler-Lotka stable-age-distribution equation, and assumes that fecundity, survival and growth rates are functions of age. However, the performance of an individual of a given age is likely to depend on other physiological characteristics of the individual (Wootton, 1998 and references therein). For example, in situations with lowered energy reserves, reproductive fish may respond by decreasing their allocation of energy into reproductive tissues (e.g., Berg et al., 1998; Lambert and Dutil, 2000). An alternative strategy is to maintain reproductive investment at the expense of somatic condition or energy and thereby increase the risk of mortality due to exhaustion of energy resources.

The age at first reproduction varies widely both between and within species of fish. This variation is largely shaped by the broad array of environments inhabited by fishes, affecting the trade-off between reproductive output and future survival. An extreme example is cyprinodonts, particularly those found in temporary ponds and pools in tropical and subtropical areas. These can reach maturity at the age of a few weeks (Miller, 1979; Simpson, 1979). In the other extreme we have most species of the Chondrichthyes or species of teleosts like flatfish of the genera *Hippoglossus* and *Hippoglossoides* where mature individuals are 15 years old or more. Even within one order, age of maturity may vary widely, for example within the flatfish order Pleuronectiformes, age at maturity may vary from one year (*Cynoglossus semifasciatus*) to 15 years or more (*Hippoglossoides platessoides*) (e.g., Wootton, 1998). A common response to environmental changes that increases growth of a fish is a decrease the age at maturity. This may be accompanied by a change in the size when the fish is maturing (Alm, 1959; Stearns and Koella, 1986; Wootton, 1998). With higher growth rate, the predicted current reproductive output in both energy terms and in number of gametes increases compared to slower growing individuals.

Breeding Costs and Survival

What amount of energy should an individual invest in each reproductive event? The **conflicts between current investment and future reproductive output** are illustrated by the bioenergetic equations, where increased allocation to one purpose reduces residual energy available for other purposes. This principle has also been illustrated in a large number of empirical studies. In particular, numerous studies have been based primarily on birds and mammals, most of which have shown a negative relationship between the investment of resources into current reproduction and parent survival until further reproduction or future fecundity (e.g., Stearns, 1992; Daan and Tinbergen, 1997; Berg *et al.*, 1998; Lambert and Dutil, 2000; Ruusila *et al.*, 2000).

High mortality on spawning grounds or low energy status may lead maturing fish to skip reproduction (Jorgensen *et al.*, 2006). A similar phenomenon, "resting year" or years when previous spawners do not mature has received little attention, probably due to methodological problems concerned with ascertaining of previous spawning history. For a number of important species, there are extensive evidence for "resting years" (Engelhard and Heino, 2006; Rideout and Rose, 2006). These phenomena are to be expected to occur based on the principles of maximized number of offspring (life history theory), as residual life-time reproductive output may increase through the skipping of spawning.

Iteroparous (Repeated Reproduction) vs Semelparous (One-time Reproduction)

Organisms may allocate such a relative high amount of resources and energy in current reproduction, that further life is corrupted; this is called semelparous reproduction. **Semelparous organisms reproduce by a single reproductive act**, represented by annual and biennial plants as are many insects and a few vertebrates, notably fishes in the families Salmonidae (salmonids) and Anguillidae (eels). The majority of fish species may reproduce repeatedly, iteroparous reproduction (from the mathematical process of iteration—repetition of a process) and the number of reproductive events is a more substantial component of fitness than a given breeding event (e.g., Schaffer, 2004). The positive correlation between increased energy use in reproduction and mortality, yields **an almost continuous scale for iteroparous spawners, where the semelparous situation represents one end of the iteroparous**

spectrum, where use of energy and matter in current reproduction is maximized.

Different species among the Salmonidae fishes represent cases along this reproductive allocation gradient. These species are often used as examples of life history traits and phylogenetic constraints in the use of energy in reproduction. A recent, extensive, review of the topic was given by Kinnison and Hendry (2004). Semelparity is generally found in Pacific salmonids (genus *Oncorhynchus* spp.), characterized by population parameters such as low adult survival and high juvenile survival (e.g., Groot and Margolis, 1991). This reduces the degree to which reproductive input must be changed to match the fitness achievable by the iteroparous alternative (Charnov and Schaffer, 1973; Schaffer, 2004), typically illustrated by Atlantic salmonids (genus *Salmo* spp.) (e.g., Fleming and Reynolds, 2004).

The semelparous Pacific salmonids have a **common phylogenetic origin,** where semelparity have been perfected to a degree where further life span is impossible due to the maximization of resources into current reproduction (Kinnison and Hendry, 2004). Semelparity can represent a discrete genetic innovation, but the difference between the semelparous Pacific salmon species and other iteroparous Salmonidae species appears to be a difference in degree of energy allocation and not a deeper fundamental difference (Kinnison and Hendry, 2004). Parallel evolution seems to have contributed strongly to the suites of traits found within the different species, as both *Oncorhynchus* spp. and *Salmo* spp. have populations that are predominantly non-anadromous and iteroparous, whereas the anadromous populations of the Pacific salmonids are semelparous. This indicates that parallel and convergent evolution have contributed to the current situation. Thus, the semelparous salmonids is not necessarily a matter of shared ancestral history as a diversity of convergent life histories arose in each genus (Kinnison and Hendry, 2004).

The anadromous Pacific salmonids are characterized by **long-distance migrations**, which will increase mortality between breeding episodes. These migrations may have been necessary for the origin of semelparity in Pacific salmon (Crespi and Teo, 2002). Similar arguments can easily be made for members of the eel family Anguillidae. Iteroparity may have a benefit **in unpredictable environments** where the reproduction may be risky, increasing the value of future reproductive efforts. In such environments, the conflict between current reproductive efforts versus survivorship of adults between successive breeding should be more evident than in less extreme environments (Sibly and Callow, 1983; Wootton, 1998; Schaffer, 2004).

The semelparous salmonids, relative to the iteroparous salmonids, appear to have increased investment with respect to secondary sexual characters, notably hump and increased jaw region. The most significant secondary trait development is among sockeye salmon (*Oncorhynchus nerka*) and pink salmon (*Oncorhynchus gorbuscha*). The semelparous strategy may give room for increased energy use into secondary sexual characters, however, these species are also experiencing the highest levels of breeding competition (Kinnison and Hendry, 2004). It is, therefore, difficult to conclude whether the potential for energy use or breeding competition or both are cause for these elaborate secondary sexual traits.

The difference in energy use between the Pacific and the Atlantic salmonids can be small. One reason may be bioenergetic constraints since only limited amounts of energy and matter can be invested into different reproductive tasks. Both sexes, therefore, has to conserve and maintain somatic tissue to defend the organism and to compete with conspecifics. Among the Pacific salmonids, this intraspecific competition is especially intense, and the Pacific salmon die with considerable amount of resourses left, especially protein (e.g., Hendry and Berg, 1999). Further allocation of resources into gonads will probably counterbalance the necessary conservation of the somatic tissue necessary to compete during reproduction. The digestion system is probably the only tissue which can be minimized in a semelparous fish—compared to an iteroparous one— without a major effect on spawning performance. Therefore, the potential difference in energy use between semelparous Pacific and iteroparous Atlantic salmonids can be highly limited.

REPRODUCTIVE ENERGY COMPARTMENTS

Egg

Each egg represents a **substantial energetic investment**, as the egg normally provides yolk needed for the development of the zygote until it can have an independent life-involving exogenous feeding.

Typical ovaries are either single or double structures, with or without symmetry. The anatomical features of **different taxonomic groups** have been reviewed by Bond (1996). In particular among Chondrichthyes, an ovary or a pouch functioning as an ovary occurs which secretes nutrition to the progeny (Balon, 1975). The composition of the egg may, therefore, be quite different from that of a typical teleost. The teleost egg normally consists of 20-40% dry matter (mostly protein and lipid) with corresponding energy content within the range 5-8 kJ g^{-1} (e.g., Jobling, 1994). There is little synthesis of yolk in the ovaries or oocytes, and the liver is the major site of the production of vitellogenin, which is transported in the bloodstream from the liver to the ovaries. In salmonids, for example, the diameter of the oocyte may increase a million folds in volume from 50 µm to approx 5 mm during the course of vitellogenesis (Jobling, 1994). Kamler (1992) reviews the inter- and intra-specific and intra-individual comparison of egg sizes and egg composition among a wide array of fishes. The between species variation in egg size is enormous, with the smallest eggs having a diameter below 0.3 mm (*Cymatogaster aggregata*) and the largest eggs being in the range of 85-90 mm diameter from the coelacanth (*Latimeria chalumnae*). The wet weight of the largest eggs is about 30 million times that of the smallest ones (Kamler, 1992).

Determining the **allocation into gonads** of energy and matter represents a considerable practical problem, especially for batch spawning species. Current methods used to measure batch fecundity may be inappropriate for measuring annual fecundity (Manning and Crim, 1998; Kjesbu et al., 2003). The measurements of reproductive investment based on mass can be complicated by the rapid hydration of the oocytes, contributing to considerable error when females of different degree of maturation are used in the analyses (Manning and Crim, 1998; Kjesbu et al., 2003 and references therein; Hesp et al., 2004).

Testes

The typical testes are either single or double structures, with or without symmetry. The anatomical features of different taxonomic groups have been reviewed by Bond (1996). Testes size may vary widely both absolutely and relative to body size. In contrast to females, where usually an increase in number of eggs is directly correlated with an increase in fitness, there is usually no such direct correlation between gonad size and fitness in males. Due to the large variation of reproductive systems, the investment into semen varies widely. Important factors in this connection include internal vs external fertilization, distance between male and female at spawning, and intra- and inter-sexual competition including sperm competition with other males and distance to other males.

The males typically also engage most actively in courtship, breeding-territory defence and cuffing with other males. These costs in reproductive behaviour are usually not calculated, but the sum of energy use for testes' growth and spawning activity for males may often exceed that of females, which usually use the bulk of their reproductive energy in gonad energy. The major reproductive costs of males are thus often invested into behaviour rather than testis, making the larger part of the investment in reproductive behaviour metabolism (M_{beh}), rather than production of gonads (P_{GON}).

Gonadosomatic Index—GSI

A number of different indices have been proposed to describe the reproductive effort (Bagenal, 1978; Kamler, 1992; Kjesbu *et al.*, 2003) and it is not our intention to review all of these. One of the most common indices of reproductive effort is the body-mass specific mass of gonads (gonadosomatic index—GSI). Other indices include the ratio of the energy (or matter) output in reproduction to input, given as symbols $P_{gon}*100/C$. The use of such indices should be restricted because C (consumption/food, see p. 354) is difficult to determine and P_{gon} will underestimate reproductive investments (Kamler, 1992).

The calculation of GSI is not standardized, as body-mass may be given with or without gonads

$$GSI_{gross} = 100 \text{ (mass of gonads)}/\text{(total body mass)}$$

or

$$GSI_{net} = 100 \text{ (mass of gonads)}/\text{(somatic body mass)}$$

where somatic body mass is the total mass of body excluding gonads. The two expressions of GSI tell slightly different stories and the GSI based on the total weight will, consequently, show a lower value than the GSI based on somatic body mass. The deviation between the two will be minor under circumstances where gonads are small relative to body mass. Since variation in wet weight often reflect variation in water content, GSI may also be given in terms of dry mass or energy directly, and thereby circumvent the pitfalls of using wet mass as a substitute for reproductive investment. Although the index provides a simple tool to describe the changes of gonads over time, it suffers from a range of disadvantages. The GSI based on somatic body mass will be a true mass specific predictor of gonad size for an individual fish at a given time. However, a change in GSI may reflect either changes in weight of either gonads or somatic body tissue. Between individual variations in GSI may also result from allometric relationships between somatic body mass and gonadal body mass. Whenever possible, GSI should, therefore, be replaced with appropriate regression analyses. This is particularly important when the relationship between somatic and gonad size or energy content are involved in statistical inference in order to test specific hypotheses.

The inter- and intra-specific ranges of a ripe female vary widely. Salmonids and cyprinids typically show values of GSI gross of 20-30% of their body mass (Wootton, 1998). In others, even when ripe, ovaries represent less than 5% of the body weight, whereas the ripe female cichlid *Oreochromis leucostictus* has a GSI of only 3% of the body mass (Welcomme, 1967). Even within a single suborder, the gobies (Gobioidei), GSI ranges between about 5% to 30% (Miller, 1979, 1984; Wootton, 1998).

Allocation of Resources to Individual Progeny

The amount of energy that the female invests into each single offspring may have pronounced effects on fitness (Crespi and Semeniuk, 2004). In oviparous fishes, a critical aspect of maternal provisioning is the size and energy content of eggs. Juveniles originating from larger eggs are larger, and typically have higher growth, survival and fitness (Hutchings, 1991; Roff, 1992; Heath and Blouw, 1998; Einum *et al.*, 2004). Thus, all else being equal, selection acting on offspring fitness should normally favour large, energy-rich eggs. Egg size cannot increase without bounds, of course, because females can only supply their eggs with finite resources, and because egg size is ultimately constrained by the size of the reproductive tract (Roff, 1992; Bernardo, 1996; Einum *et al.*, 2004; Kolm and Ahnesjo, 2005).

Mothers and offspring are often in conflict as to the optimal amount of allocation into individual eggs (Crespi and Semeniuk, 2004). From an evolutionary perspective, females should maximize their own fitness rather than that of their individual offspring. The distinction is important because maternal fitness is the product both of offspring fitness and the number of offspring produced (Smith and Fretwell, 1974; Einum *et al.*, 2004). If offspring were genetically identical, parents and offspring are not in conflict, and optimal egg size is the same from the perspective of maternal or offspring fitness. In sexually-reproducing, diploid organisms in outbreeding populations, however, siblings are at most an average of 50% related, and so an offspring should attempt to acquire extra resources at some expense to their siblings. A given unit of parental investment into egg production must be partitioned between individual eggs (e.g., size) and their number. From the perspective of maternal fitness, optimal egg size is that at which the increase in offspring fitness with a further increase in egg size no longer exceeds the increase in maternal fitness that would attend the same proportional investment into egg number (Smith and Fretwell, 1974; Trivers, 1974; Mock and Parker, 1997; Einum *et al.*, 2004).

Parent-offspring conflict has been proposed to play a fundamental and causal role in the evolution of viviparity and placentation in vertebrates (Crespi and Semeniuk, 2004; Perrand Roitberg, 2006). These reproductive traits have been suggested to result from offspring manipulation to increase the parental energetic investment into each offspring. In addition to the reproductive organ development, manipulation may include the release of hormones and other compounds into the maternal bloodstream (Crespi and Semeniuk, 2004 and references therein). In viviparous vertebrates, two main nutritional patterns are typical for the transfer of maternal energy into offspring: lecithotrophy (where energy is derived from yolk) and matrotrophy (energy directly supplied from the mother). The latter patterns involve the widest scope for mother-offspring conflict, because of the more intimate contact. Matrotrophy is categorized into three main forms: (1) oophagy, where the developing foetus feeds on "sibling" ova—seen in many Chondrichthyes (sharks and rays) and teleost fishes (Gilmore, 1993); (2) histophagy involves ingestion of maternal secretions in the embryos seen in Chondrichthyes; and (3) placental viviparity involves transfer of energy via an intimate fusion between fetal and maternal organs for physiological exchange of energy and matter as seen in both Chondrichthyes and teleost fishes (Crespi and Semeniuk, 2004). With increased contact between offspring and mother, the offspring ability to manipulate maternal input increases. Notably, foetus-produced hormones or other compounds may play an increasing effect in the regulation of transfer of energy and matter from parent into offspring (Crespi and Semeniuk, 2004). The role of parent-offspring conflict in the diversification of various reproductive systems is currently under debate, and future investigators have to integrate physiological, morphological and ecological data to understand the ultimate factors.

Nearly all studies addressing maternal provisioning and egg quality in fish consider only various measures of egg and alevin size (Brooks *et al.*, 1997; Heath and Blouw, 1998; Balon, 1999). This approach has been defended because: (1) many fish eggs are so small that individual energy content cannot be determined reliably; (2) determination of energy content kills the embryo, thereby rendering subsequent evaluation of performance impossible; and (3) it seems intuitively reasonable that larger eggs will contain greater energy stores. This approach can be questioned, however, because energy stores or other factors may vary among eggs of a given size (Brooks *et al.*, 1997; Balon, 1999; Berg *et al.*, 2001; Jaworski and Kamler, 2002). A few studies have shown that the total energy content of individual eggs is correlated with their size (e.g., Kristjánsson and Vøllestad, 1996), but only few studies have determined the fat, protein, and energy content of individual eggs (Einum and Fleming, 1999, 2000; Keckeis *et al.*, 2000; Berg *et al.*, 2001). In addition to the obvious environmental characters influencing the early growth of juveniles in three-spined stickleback (*Gasterosteus aculeatus*), a strong genetic component appears to be present (Wright *et al.*, 2004).

As a result of their relatively large egg size, Salmonidae fishes have been relatively well investigated as to egg size and offspring quality. Larger eggs give rise to larger juveniles (e.g., Thorpe *et al.*, 1984; Hutchings, 1991; Hayashizaki *et al.*, 1995), and larger juveniles enjoy increased survival, competitive ability, swimming performance, growth, and overall fitness (Bagenal, 1969; Ojanguren *et al.*, 1996; Cutts *et al.*, 1999; Einum and Fleming, 1999, 2000a; Einum *et al.*, 2004). Offspring fitness should, therefore, exert a strong influence on maternal provisioning, favouring large and energy-rich eggs. However, a strong trade-off between egg size and number within populations indicates that, for a given level of investment into egg production, females producing larger eggs must produce less of them (Thorpe *et al.*, 1984; Bromage *et al.*, 1992; Quinn *et al.*, 1995; Heath *et al.*, 1999; Jonsson and Jonsson, 1999). These conditions set the stage for a classic parent-offspring conflict, one which mothers should dominate by stopping investment at fertilization (before offspring can exert any influence). Recent evidence confirms that average egg size within Atlantic salmon populations does indeed maximize maternal fitness (Einum and Fleming, 2000b; Einum *et al.*, 2004).

Variation in egg size has been shown to have a genetic, maternal and environmental basis (Thorpe *et al.*, 1984; Jonsson *et al.*, 1996; Heath *et al.*, 1999; Jonsson and Jonsson, 1999). Among salmonid fishes, variation among populations can be large, and has been explained by reference to selection imposed by incubation temperature (Fleming and Gross, 1990; Jonsson and Jonsson, 1999), gravel size (Quinn *et al.*, 1995), and migration distance (Beacham and Murray, 1993). Variation between years within populations has been explained in some cases as a plastic response to growth conditions experienced by females (Lobon-Cervia *et al.* 1997). Variation among females within populations can be explained in part by body size (larger females produce larger eggs (Quinn *et al.*, 1995) and growth rate (Morita *et al.*, 1999), but much remains unexplained. Finally, variation within females may be small or large, and has not as yet been explained.

Conclusion

We regard reproduction as the single most important and interrelated process in living creatures, and most of the different processes that an organism has evolved have a direct or indirect involvement in reproduction. In this chapter, we have focused on factors governing energy transfer between different parts of the reproductive systems of a large and taxonomically diverse group, the fishes. As illustrated by bioenergetic principles, each individual must always make trade-offs between investing energy into the various tasks associated with reproduction or into non-reproductive tasks. A range of interacting factors, including individual characteristics such as sex and phenotypic state, intrasexual competition as well as environmental factors affecting growth and survival will determine the optimal energy allocation pattern. Fishes are the most diverse vertebrate group, inhabiting a wide range of environments. Energy allocation patterns associated with reproduction vary accordingly, and it is a vast challenge to try to review this variation. The vast majority of articles citing results concerning reproduction in fish have, therefore, not been cited.

Bioenergetics yields a general currency and a model toolkit where scientists with diverse backgrounds like physiologists, ecologists or behavioural ecologists can find a common platform to develop comprehension and predictability of the processes that steer life. We encourage further studies describing reproductive allocation patterns as well as studies trying to understand the ultimate causes of the large variation in energetic trade-offs, that underlies the variety of reproductive investment patterns found in living organisms, including fishes.

Acknowledgements

Chris Bingham is acknowledged for valuable comments on the manuscript.

References

Allsop, D.J. and S.A.West. 2004. Sex-ratio evolution in sex changing animals. *Evolution* 58: 1019-1027.

Alm, G. 1959. Connection between maturity, size and age in fishes. *Report of the Institute of Freshwater Research, Drottningholm* 40: 5-145.

Andersson, M.B. 1994. *Sexual Selection*. Princeton University Press, Princeton.

Bagenal, T.B. 1969. Relationship between egg size and fry survival in brown trout *Salmo trutta* L. *Journal of Fish Biology* 1: 349–353.

Bagenal, T.B. 1978. *Methods for Assessment of Fish Production in Fresh Waters*. Blackwell Scientific Publications, London.

Balon, E.K. 1975. Reproductive guilds of fishes: a proposal and definition. *Journal of the Fisheries Research Board of Canada* 32: 821-864.

Balon, E.K. 1999. Alternative ways to become a juvenile or a definitive phenotype and on some persisting linguistic offences. *Environmental Biology of Fishes* 56: 17–38.

Beacham, T.D. and C.B. Murray. 1993. Fecundity and egg size variation in North American Pacific salmon *Oncorhynchus*. *Journal of Fish Biology* 42: 485–508.

Berg, O.K., A.P. Hendry, B. Svendsen, C. Bech, J.V. Arnekleiv and A. Lohrmann. 2001. Maternal provisioning of offspring and the use of those resources during ontogeny: Variation within and between Atlantic salmon families. *Functional Ecology* 15: 13-23.

Berg, O.K., E. Thronæs and G. Bremset. 1998. Energetics and survival of virgin and repeat spawning brown trout *Salmo trutta*. *Canadian Journal of Fisheries and Aquatic Sciences* 55: 47-53.

Berglund, A. 1997. Mating systems and sex allocation. In: *Behavioural Ecology of Teleost Fishes*, J.-G.J. Godin (ed.). Oxford University Press, Oxford, pp. 237-265.

Bernardo, J. 1996. The particular maternal effect of propagule size, especially egg size, patterns, models, quality of evidence and interpretations. *American Zoologist* 36: 216–236.

Bond, C.E. 1996. *Biology of Fishes*. Saunders College Publishing, New York.

Brett, J.R. 1995. Energetics. In: *Physiological Ecology of Pacific Salmon*, C. Groot, L. Margolis and W.C. Clarke (eds.). University of British Columbia Press, Vancouver, pp. 1-68.

Bromage, N., J. Jones, C. Randall, M. Thrush, B. Davies, J. Springate, J. Duston and G. Barker. 1992. Broodstock management, fecundity, egg quality and the timing of egg production in the rainbow trout *Oncorhynchus mykiss*. *Aquaculture* 100: 141-166.

Brooks, S., C.R. Tyler and J.P. Sumpter. 1997. Egg quality in fish: what makes a good egg? *Reviews in Fish Biology and Fisheries* 7: 387-416.

Charnov, E.L. and W.W. Schaffer. 1973. Life history consequences of natural selection: Cole's result revisited. *American Naturalist* 106: 791-793.

Clutton-Brock, T.H. 1991. *The Evolution of Parental Care*. Princeton University Press, Princeton.

Crespi, B.J. and C. Semeniuk. 2004. Parent offspring conflict in the evolution of vertebrate reproductive mode. *American Naturalist* 163: 635-653.

Crespi, B.J. and R. Teo. 2002. Comparative phylogenetic analysis of the evolution of semelparity and life-history in salmonid fishes. *Evolution* 56: 1008-1020.

Crossin, G.T., S.G. Hinch, A.P. Farrell, D.A. Higgs, A.G. Lotto, J.D. Oakes and M.C. Healey. 2004. Energetics and morphology of sockeye salmon: Effects of upriver migratory distance and elevation. *Journal of Fish Biology* 65: 788-810.

Cutts, C.J., B. Brembs, N.B. Metcalfe and C. Taylor. 1999. Prior residence, territory quality and life history strategies in juvenile Atlantic salmon *Salmo salar* L. *Journal of Fish Biology* 55: 784-794.

Daan, S. and J.M. Tinbergen. 1997. Adaptation of life histories. In: *Behavioural Ecology: An Evolutionary Approach*, J.R. Krebs and N.B. Davies (eds.). Blackwell Scientific Publications, London, pp. 311-333.

Darwin, C. 1871. *The Descent of Man and Selection in Relation to Sex.* J. Murray, London.

Dugatkin, L.A. and G.J. FitsGerald. 1997. Sexual selection. In: *Behavioural Ecology of Teleost Fishes*, J.-G.J. Godin (ed.). Oxford University Press, Oxford, pp. 266-291.

Einum, S. and I.A. Fleming. 1999. Maternal effects of egg size in brown trout *Salmo trutta*: norms of reaction to environmental quality. *Proceedings of the Royal Society of London* B 266: 2095-2100.

Einum, S. and I.A. Fleming. 2000a. Selection against late emergence and small offspring in Atlantic salmon *Salmo salar. Evolution* 54: 628-639.

Einum, S. and I.A. Fleming. 2000b. Highly fecund mothers sacrifice offspring survival to maximize fitness. *Nature (London)* 405: 565-567.

Einum, S., M.T. Kinnison and A.P. Hendry. 2004. Evolution of egg size and number. In: *Evolution Illuminated—Salmon and Their Relatives*, A.P. Hendry and S.C. Stearns (eds.). Oxford University Press, Oxford, pp. 126-153.

Engelhard, G. and M. Heino. 2006. Climate change and condition of herring (*Clupea harengus*) explain long-term trends in extent of skipped reproduction. *Oecologia* 149: 593-603.

Finstad, A.G., O.K. Berg, A. Langeland and A. Lohrmann. 2002. Reproductive investment and energy allocation in an alpine Arctic charr, *Salvelinus alpinus*, population. *Environmental Biology of Fishes* 65: 63-70.

Fleming, I.A. and J.D. Reynolds. 2004. Salmonid breeding systems. In: *Evolution Illuminated— Salmon and Their Relatives*, A.P. Hendry and S.C. Stearns (eds.). Oxford University Press, Oxford, pp. 264-295.

Fleming, I.A. and M.R. Gross. 1990. Latitudinal clines: a trade-off between egg number and size in Pacific salmon. *Ecology* 71: 1-11.

Forsgren, E., T. Amundsen, Å. Borg and J. Bjelvenmark. 2004. Unusually dynamic sex roles in a fish. *Nature (London)* 429: 551-554.

Gilhousen, P. 1980. Energy sources and expenditures in Fraser River Sockeye salmon during their spawning migration. *International Pacific Salmon Fisheries Commission Bulletin* 22.

Gilmore, R.G. 1993. Reproductive biology of lamnoid sharks. *Environmental Biology of Fishes* 38: 95-114.

Groot, C. and L. Margolis. 1991. *Pacific Salmon Life Histories.* University of British Columbia Press, Vancouver.

Hayashizaki, K., M. Hirohashi and H. Ida. 1995. Effect of egg size on the characteristics of embryos and alevins of chum salmon. *Fisheries Science* 61: 177-180.

Heath, D.D. and M. Blouw. 1998. Are maternal effects in fish adaptive or merely physiological side effects. In: *Maternal Effects as Adaptations*, T.A. Mousseau and C.W. Fox (eds.). Oxford University Press, Oxford, pp. 178-201.

Heath, D.D., C.W. Fox and J.W. Heath. 1999. Maternal effects on offspring size: Variation through early development of chinook salmon. *Evolution* 53: 1605-1611.

Hendry, A.P. and O.K. Berg. 1999. Secondary sexual characters, energy use, senescence, and the cost of reproduction in sockeye salmon. *Canadian Journal of Zoology* 77: 1663-1675.

Hendry, A.P. and E. Beall. 2004. Energy use in spawning Atlantic salmon. *Ecology of Freshwater Fish* 13: 185-196.

Hesp, S.A., I.C. Potter and S.R.M. Schubert. 2004. Factors influencing the timing and frequency of spawning and fecundity of the goldlined seabream (*Rhabdosargus sarba*) (Sparidae) in the lower reaches of an estuary. *Fishery Bulletin (Washington)* 102: 648-660.

Hickman, C.P., L.S. Roberts, A. Larson and H. l'Anson. 2004. *Integrated Principles of Zoology*. McGraw-Hill, New York.

Hutchings, J.A. 1991. Fitness consequences of variation in egg size and food abundance in brook trout *Salvelinus fontinalis*. *Evolution* 45: 1162-1168.

Jaworski, A. and E. Kamler. 2002. Development of a bioenergetics model for fish embryos and larvae during the yolk feeding period. *Journal of Fish Biology* 60: 785-809.

Jobling, M. 1994. *Fish Bioenergetics*. Chapman and Hall, London.

Jobling, M. 1995. *Environmental Biology of Fishes*. Chapman and Hall, London.

Jonsson, N. and B. Jonsson. 1999. Trade-off between egg mass and egg number in brown trout. *Journal of Fish Biology* 55: 767-783.

Jonsson, N., B. Jonsson and I.A. Fleming. 1996. Does early growth cause a phenotypically plastic response in egg production of Atlantic Salmon? *Functional Ecology* 10: 89-96.

Jonsson, N., B. Jonsson and L.P. Hansen. 1997. Changes in proximate composition and estimates of energetic costs during upstream migration and spawning in Atlantic salmon *Salmo salar*. *Journal of Animal Ecology* 66: 425-436.

Jorgencen, C., B. Ernande, O. Fiksen and U. Dieckman. 2006. The logic of skipped spawning in fish. *Canadian Journal of Fisheries and Aquatic Sciences* 63: 200-211.

Kamler, E. 1992. *Early Life History of Fish: An Enegetics Approach*. Chapman and Hall, London.

Keckeis, H., E. Bauer-Nemeschkal, V.V. Menchutkin, H.L. Nemeschkal and E. Kamler. 2000. Effects of female attributes and egg properties on offspring viability in a rheophilic cyprinid, *Chondrostoma nasus*. *Canadian Journal of Fisheries and Aquatic Sciences* 57: 789-796.

Kinnison, M.T. and A.P. Hendry. 2004. From macro- to micro-evolution: Tempo and mode in salmonid evolution. In: *Evolution Illuminated—Salmostet and Their Relatives*, A.P. Hendry and S.C. Stearns (eds.). Oxford University Press, Oxford, pp. 208-231.

Kinnison, M.T., M.J. Unwin, A.P. Hendry and T.P. Quinn. 2001. Migratory costs and the evolution of egg size and number in introduced and indigenous salmon populations. *Evolution* 55: 1656-1667.

Kjesbu, O.S., J.R. Hunter and P.R. Witthames. 2003. Moden approaches to assess maturity and fecundity of warm and cold-water fish and squids. Bergen: *Institute of Marine Research, Fisken og havet* 12. (Available at: http://www.imr.no/dokumenter/fisken og havet).

Kolm, N. and I. Ahnesjö. 2005. Do egg size and parental care coevolve in fishes? *Journal of Fish Biology* 66: 1499-1515.

Kristjánsson, L.T. and L.A. Vøllestad. 1996. Individual variation in progeny size and quality in rainbow trout, *Oncorhynchus mykiss* Walbaum. *Aquaculture Research* 27: 335-343.

Lambert, Y. and J.-D. Dutil. 1997. Condition and energy reserves of Atlantic cod *Gadus morhua*. During the collapse of the northern Gulf of St. Lawrence stock. *Canadian Journal of Fisheries and Aquatic Sciences* 54: 2388-2400.

Lambert, Y. and J.-D. Dutil. 2000. Energetic consequences of reproduction in Atlantic cod *Gadus morhua*. In relation to spawning level of somatic energy reserves. *Canadian Journal of Fisheries and Aquatic Sciences* 57: 815-825.

Lobon-Cervia, J., C.G. Utrilla, P.A. Rincón and F. Amezcua. 1997. Environmentally induced spatio-temporal variations in the fecundity of brown trout *Salmo trutta* L. tradeoffs between egg size and number. *Freshwater Biology* 38: 277-288.

Mackay, I. and K.H. Mann. 1969. Fecundity of two cyprinid fishes in the River Thames, Reading, England. *Journal of the Fisheries Research Board of Canada* 26: 2795-2805.

Mank, J.E. and J.C. Avise. 2006. The evolution of reproductive and genomic diversity in ray-finned fishes: insights from phylogeny and comparative analysis. *Journal of fish Biology* 69: 1-27.

Manning, A.J. and L.W. Crisp. 1998. Maternal and interannual comparison of the ovulatory periodicity, egg production and egg quality of the batch-spawning yellowtail flounder. *Journal of Fish Biology* 53: 954-972.

Miller, P.J. 1979. Adaptiveness and implications of small size in teleosts. *Symposium of the Zoological Society of London No.* 44: 263-306.

Miller, P.J. 1984. The tokology of gobioid fishes. In: *Fish Reproduction: Strategies and Tactics*, G.W. Potts and R.J. Wootton (eds.). Academic Press, London, pp. 119-153.

Mock, D.W. and G.A. Parker. 1997. *The Evolution of Sibling Rivalry.* Oxford University Press, Oxford.

Morita, K., S. Yamamoto, Y. Takashima, T. Matsuishi, Y. Kanno and K. Nishimura. 1999. Effect of maternal growth history on egg number and size in wild whitespotted char *Salvelinus leucomaenis*. *Canadian Journal of Fisheries and Aquatic Sciences* 56: 1585-1589.

Mousseau T.A. and C. Fox. 1998. The adaptive significance of maternal effects. *Trends in Ecology and Evolution* 13: 403-407.

Moyle, P.B. and J.J. Cech Jr. 2004. *Fishes: An Introduction to Ichthyology.* 5th Edition. Prentice Hall, London.

Muoz, R.C. and R.R. Warner. 2004. Testing a new version of the size-advantage hypothesis for sex change: sperm competition and size-skew effects in the bucktooth parrotfish, *Sparisoma radians. Behavioural Ecology* 15: 129-135.

Ojanguren, A.F., F.G. Reyes-Gavilán and F. Braña. 1996. Effects of egg size on offspring development and fitness in brown trout, *Salmo trutta* L. *Aquaculture* 147: 9-20.

Otto, S.P. and S.L. Nuismer. 2004. Species interactions and the evolution of sex. *Science* 304: 1018-1020.

Perry, J.C. and B.D. Roitberg. 2006. Trophic egg laying: hypotheses and tests. *Oikos* 112:706-714.

Quinn, T.P., A.P. Hendry and L.A. Wetzel. 1995. The influence of life history trade-offs and the size of incubation gravels on egg size variation in sockeye salmon *Oncorhynchus nerka. Oikos* 74: 425-438.

Rideout, R.M. and G.A. Rose. 2006. Suppression of reproduction in Atlantic cod *Gadus morhua. Marine Ecology-Progress Series* 320: 267-277.

Roff, D.A. 1992. *The Evolution of Life Histories.* Chapman and Hall, New York.

Ruusila, V., A. Ermala and H. Hyvarinen. 2000. Costs of reproduction in introduced female Canadian beavers *Castor canadensis. Journal of Zoology (London)* 252: 79-82.

Schaffer, W.M. 2004. Life histories, evolution and salmonids. In: *Evolution Illuminated— Salmon and Their Relatives*, A.P. Hendry and S.C. Stearns (eds.). Oxford University Press, Oxford, pp. 20-51.

Sibly, R.M. and P. Callow. 1986. *Physiological Ecology of Animals: An Evolutionary Approach.* Blackwell Scientific Publications, Oxford.

Simpson, B.R.C. 1979. The phenology of annual killifishes. *Symposium of the Zoological Society of London* No. 44: 243-261.

Smith, C.C. and S.D. Fretwell. 1974. The optimal balance between size and number of offspring. *American Naturalist* 108: 499-506.

Stearns, S.C. 1992. *The Evolution of Life Histories*. Oxford University Press, Oxford.

Stearns, S.C. and J. Koella. 1986. The evolution of phenotypic plasticity in life-history traits: Predictions for norms of reaction for age- and size-at maturity. *Evolution*: 40: 893-913.

Stevenson, R.D. and W.A. Woods. 2006. Condition indices for conservation: new users for evolving tools. *Integrative and Comparative Biology* 46: 1169-1190.

Svensson, P.A., C. Pelabon, J.D. Blount, P.F. Surai and T. Amundsen. 2006. Does female nuptial coloration reflect egg carotenoids and clutch quality in the Two-Spotted Goby (Gobiusculus flavescens, Gobiidae)? *Functional Ecology* 20: 689-698.

Thorpe, J.E., M.S. Miles and D.S. Keay. 1984. Developmental rate, fecundity and egg size in Atlantic salmon, *Salmo salar* L. *Aquaculture* 43: 289-305.

Trivers, R.L. 1974. Parent–offspring conflict. *American Zoologist* 14: 249-264.

Welcomme, R.L. 1967. The relationship between fecundity and fertility in the mouthbreeding cichlid fish, *Tilapia leucosticte*. *Journal of Zoology (London)* 151: 453-468.

Williams, G.C. 1966. Natural selection, the cost of reproduction, and a refinement of Lack's principle. *American Naturalist* 100: 687-690.

Wilson, A.B., I. Ahnesjö, A.C.J. Vincent and A. Meyer. 2003. The dynamics of male brooding, mating patterns, and sex roles in pipefishes and seahorses Family Sygnathidae. *Evolution* 57: 1374-1386.

Wootton, R.J. 1998. *Ecology of Teleost Fishes*. Fish and Fisheries Series 24. Kluwer Academic Publishers, London.

Wright, H.A., R.J. Wootton and I. Barber. 2004. Interopopulation variation in early growth of threespine sticklebacks (*Gasterosteus aculeatus*) under laboratory conditions. *Canadian Journal of Fisheries and Aquatic Sciences* 61: 1832-1838.

The Influence of Reproductive Behaviour on the Social Organization of Shoaling Fish

**Darren P. Croft[1,*], Siân W. Griffiths[2],
Anne E. Magurran[3] and Jens Krause[1]**

OVERVIEW

Shoaling behaviour is extremely common amongst teleost fishes. Of the 27,000 species, approximately 50% live in shoals as juveniles and 25% maintain the behavioural trait throughout their lives (Shaw, 1978). The decision to join a shoal represents the outcome of a cost benefit trade-off, which may be influenced by a number of factors including predation risk, parasite load, body size, nutritional state, the availability and distribution of resources, and reproductive strategies (Krause and Ruxton, 2002). As a result, there is great variability in the degree of shoaling tendency both

Authors' addresses: [1]School of Biology, University of Leeds, Leeds LS2 9JT, U.K.
[2]Cardiff School of Biosciences, Main Building Cardiff University, PO Box 915, Cardiff, CF10 3TL, U.K.
[3]University of St. Andrews, Gatty Marine Laboratory, KY16 8BL, Scotland, U.K.
Corresponding author: Presently at School of Biological Sciences, University of Wales Bangor, Brambell Building, Deiniol Road, Bangor, Gwynedd, LL57 2UW, UK.
E-mail: darren_croft@hotmail.com

within and between species. For example, some species (particularly pelagic oceanic species) form polarized groups (schools) which may consist of many thousands of individuals, whilst other species form more dynamic non-polarized groups (shoals) which often comprise only a few individuals, particularly in freshwater species.

Whilst there has been a number of reviews of shoaling behaviour (e.g., Godin, 1986; Magurran, 1990; Hoare *et al.*, 2000; Krause *et al.*, 2000), the influence of reproductive strategies has received little attention (see Magurran and Macias-Garcia, 2000). As a taxon teleost fishes are unusual, in the sense that they exhibit a diverse array of mating systems. For example, some species use internal fertilization whilst others use external fertilization; parental care may or may not be provided by one (either male or females) or both sexes, sex changes are relatively common and males may even become 'pregnant' (Helfman *et al.*, 1997). The degree of sexual selection varies across different mating systems (Kodric-Brown, 1990). For example, in polygynous species, the potential reproductive rate of males is much higher than that of females, which may result in intense sexual selection leading to sexual dimorphism in body size, elaborate sexual ornaments and complex mating displays (Kodric-Brown, 1990). In contrast, when levels of sexual selection are low, for example in monogamous species, the sexes may be monomorphic. However, patterns of sexual dimorphism in fish are complex, and sexual dimorphism may be greatest in some monogamous species (e.g., deep-sea fishes of the genus *Photocorynus*—see Magurran and Macias-Garcia, 2000).

The nature of the mating system and the degree of sexual dimorphism may have both direct and indirect influences on patterns of social organization (Magurran and Macias-Garcia, 2000). For example, reproductive strategies may influence habitat use, and movement strategies, potentially limiting opportunities for social interactions to occur. In contrast, reproductive strategies may have a direct influence on social behaviour in the context of shoal choice decisions. For example, males may prefer to associate with shoals containing a greater proportion of females in an attempt to maximize reproductive success.

In this chapter, we will explore how selection of reproductive behaviour may influence social organization of shoaling fish. Initially, we will review the costs and benefits of group living, providing a foundation from which we examine the indirect and direct effects of reproductive behaviour on social organization. Finally, we will describe the ecological and evolutionary implications of such effects.

THE COSTS AND BENEFITS OF GROUP LIVING

Living in groups may confer a number of important anti-predator benefits (Magurran, 1990; Pitcher and Parrish, 1993; Fuiman and Magurran, 1994). For example, in a classic experiment, Neill and Cullen (1974) demonstrated that the success of various predators (including both cephalopod and fish predators) was reduced when attacking shoals of larger size (Fig. 11.1). Such anti-predator benefits are thought to result from a combination of factors. For example, as the size of a group increases, vigilance may also increase, resulting in earlier predator detection (Magurran *et al.*, 1985). After a predator has been detected, shoals may employ complex predator evasive tactics (e.g., flash expansion) reducing the capture success of the predator (see Pitcher and Parrish, 1993 for a review). Furthermore, when attacking shoals composed of individuals of a similar phenotype, the simultaneous movement of shoal members, may make it difficult for a predator to focus on a particular target, reducing the predator's attack to kill ratio and resulting in confusion (Neill and Cullen, 1974; Landeau and Terborgh, 1986). Finally, grouping individuals can also benefit from the encounter-dilution effect provided a predator can only eat a fraction of the overall group (see Turner and Pitcher, 1986 for details). In addition to the anti-predator benefits of group living, there may also be foraging benefits for shoaling fish. For example, fish in larger shoals may locate food with greater speed (Pitcher *et al.*, 1982). However, as the benefits of group living increase with increasing group

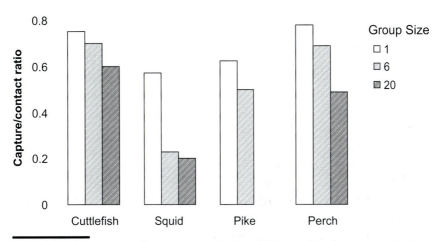

Fig. 11.1 The success of rate per attack of cuttlefish, squid, pike and perch when attacking prey fish either in groups or singly. (Adapted from Neill and Cullen, 1974).

size, so do the costs due to increased competition for scarce resources (Pitcher and Parrish, 1993). For example, fish in larger shoals may allocate more time to feeding to compensate for the increased competition for food (Grand and Dill, 1999; Johnsson, 2003).

The nature of the mating system may influence the costs and benefits of social behaviour. For example, in guppies (*Poecilia reticulata*), females give birth to live young and the reproductive success is dependent on investment in their offspring and their selection of high quality mates (Magurran and Seghers, 1994). In contrast, in males it is the total number of successful copulation that determines their reproductive output (Magurran and Seghers, 1994). As a result, female guppies spend more time schooling and invest highly in anti-predator behaviour (Magurran and Seghers, 1994). Males, in contrast, appear to trade-off the anti-predator benefits of schooling against increased mobility in search of mating opportunities frequently moving between shoals (Croft *et al.*, 2003b).

INDIRECT EFFECTS OF REPRODUCTION ON SOCIAL ORGANIZATION

Opportunities for social interactions to occur and persist may be influenced by a number of factors (Croft *et al.*, 2003c) which, in turn, may be shaped by reproductive behaviour. For example, reproductive behaviour may affect the distribution and movement of individuals in the habitat, influencing the frequency of encounters between individuals and the opportunity for social interactions to occur. Furthermore, locomotion speed and activity budgets may differ as a result of reproductive strategies, which may influence group fission events. In the following section, we shall examine the implications of such effects on the social organization of shoaling fishes.

Habitat Use

It has been well documented that habitat use is often non-random with respect to body length (Greenberg *et al.*, 1996; Lightfoot and Jones, 1996; Bremset and Berg, 1999; Croft *et al.*, 2003b). Size-specific habitat use may be an important mechanism contributing to the phenotypic assortment of social groups. For example, in an investigation on guppies, Croft *et al.* (2003b) found that larger fish occurred more frequently in deeper waters and further away from the river bank than smaller fish. Such a non-random distribution may limit opportunities for social interactions to occur between size classes, and contribute to the phenotypic assortment of social groups by body length (see Croft *et al.*, 2003b).

Habitat use results from a trade-off between foraging gains and predation risk (Gilliam and Fraser, 1987), and size-specific habitat segregation may be explained by body-length differences in competitive ability (Bremset and Berg, 1999), and predation risk (Post and Evans, 1989; Fuiman and Magurran, 1994). Reproductive strategies may influence predation risk, which in turn may influence the distribution of individuals in the habitat (Ruckstuhl and Neuhaus, 2002). For example, in the guppy, sex differences in body size (female guppies attain a larger maximum body length), coloration (females are cryptically coloured, whereas males are polymorphic for colour and brightly ornamented (Houde, 1997)), and behaviour (females invest more time in anti-predator behaviour (Magurran and Nowak, 1991)), result in males suffering higher mortality rates due to predation (Seghers, 1973; Rodd and Reznick, 1997). In a previous investigation on guppies, Croft *et al.* (2004b, 2006) observed sexual segregation in the habitat with males occupying shallow water habitats associated with reduced predation risk (Mattingly and Butler, 1994), whilst a larger proportion of females were observed in deep water (Fig. 11.2). Such a non-random distribution of individuals in the habitat will influence opportunities for encounters between groups which may

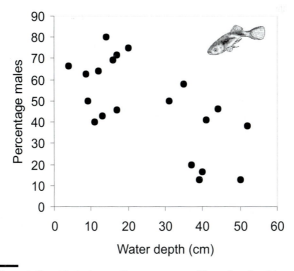

Fig. 11.2   The relationship between the sex composition of a shoal (expressed as the percentage of males) and water depth (N=20) in the Arima River, Northern Mountain Range of Trinidad (Pearson's Correlation: r_{19}=-0.66, P=0.001) (Redrawn from Croft *et al.*, 2004b).

contribute to social segregation in the species (a non-random distribution of males between groups of females) (see Croft *et al.*, 2004b).

Although predation risk may be important in driving habitat use, other mechanisms may also have a role to play. For instance, sex differences in reproductive strategies may result in sex differences in environmental requirements. Noltie and Johansen (1986) showed that male and female guppies in a laboratory experiment differ in their preferences for water depth, with female guppies showing a greater preference for shallow water. Johansen and Cross (1986) demonstrated—in a separate investigation— that male guppies prefer cooler temperatures in comparison to females. These observations may be potentially explained by the need of viviparous females to increase the metabolic rate of the embryos (Magurran and Macias-Garcia, 2000), and may contribute to differential habitat use by sexes in the wild, and ultimately social segregation.

Swimming Speeds

The phenotypic composition of a group will be dependent on assortment during fission and fusion events, which may be further influenced by locomotion speed. In fish, the optimal swimming speed is body length dependent (Beamish, 1978). Individuals in a group of others of dissimilar body length may be forced to travel and forage at sub-optimal speeds, which may incur an energetic cost and potentially contribute to group assortment by size. A positive relationship between body-length and speed of locomotion has been proposed as a mechanism for creating assortment by size in a number of other taxa (krill: Watkins *et al.*, 1992; African ungulates: Gueron *et al.*, 1996). In species that are sexually dimorphic in body size (as a result of sex differences in reproductive strategies), sex differences in locomotion speeds may result in sexual segregation (Conradt, 1998; Ruckstuhl, 1998). For example, if males and females differ substantially in their locomotion speeds (as a result of sexual dimorphism), mixed sexed groups may become energetically costly and result in fission into groups that are structured by sex and age (Ruckstuhl, 1999; Ruckstuhl and Kokko, 2002). Whilst work on ungulates suggests that differential activity budgets are the most likely mechanism explaining components of sexual segregation not accounted for by habitat segregation (Ruckstuhl and Neuhaus, 2002), there has been no experimental work on fishes, despite the potential of field and laboratory studies. Future work should concentrate on examining the link between reproductive strategies, body length and sex differences in activity budgets and locomotion speeds in fishes.

Movement Strategies

The social organization of free-ranging groups may be affected by the movement of individuals in the habitat (Krause *et al.*, 2000), with an increase in movement increasing the probability of encounters (and mixing) occurring between individuals (see Croft *et al.*, 2003c). Remaining in one location may confer direct advantages; for example, site fidelity may allow individuals to remain in environmentally favourable habitats (Winker *et al.*, 1995; Aparicio and De Sostoa, 1999) and acquire information on the location of food, refugia, and other resources (Wootton, 1998; Perrin and Goudet, 2001). In contrast, animals may move to reduce predation risk (Gilliam and Fraser, 2001; Schaefer, 2001), escape harsh environmental conditions (Railsback *et al.*, 1999; Labbe and Fausch, 2000), avoid inbreeding or competition with kin (Pusey and Wolf, 1996) and increase access to potential mates (Greenwood, 1980).

The advantages of movement may not be distributed evenly between the sexes, which has led to the evolution of sex-biased dispersal in many species. Although multiple factors (e.g., avoidance of inbreeding and kin competition) probably play a role, these biases are strongly correlated with mating systems in both birds and mammals (Greenwood, 1980). In contrast, there is no well-established relationship between dispersal and mating systems in fishes, which probably reflects the diversity of the 27,000 species of teleost fish and their mating systems. Indeed, many studies have found no relationship between sex and movement (e.g., Aparicio and De Sostoa, 1999; Schleusner and Maughan, 1999). However, there is some evidence for a correlation between mating systems and sex-biased dispersal in shoaling fishes. For example, there is some evidence for male-biased movement in live-bearing species (Brown, 1985; Chapman and Kramer, 1991; Croft *et al.*, 2003a), where male mating success is dependent on gaining access to females, whereas females are dependent on investing resources into the nourishment of young. As a result of male-biased movement, female-female social relationships may be more stable than interactions involving males in such species (see Croft *et al.*, 2004c). Future studies should address the influence of sex and mating systems on movement patterns in fish and the implications for social organization.

DIRECT EFFECTS OF REPRODUCTION ON SOCIAL ORGANIZATION

Reproductive strategies may have a direct effect on social organization by influencing shoal choice decisions. For example, reproductive strategies may influence association preferences in the context of body length, sex, familiarity and kin preferences. In the following section, we shall discuss the influence of reproductive strategies on shoal choice behaviour and their implications for the social organization of shoaling fish.

Body Length

Animal groups are frequently assorted by phenotype (Watkins *et al.*, 1992; Gueron *et al.*, 1996; Kearney *et al.*, 2001). Previous investigations have found the composition of free-ranging fish shoals to be non-random and typically size structured, in both fresh-water (e.g., guppies Croft *et al.*, 2003b; banded killifish *Fundulus diaphanus* Krause *et al.*, 1996; threespine stickleback *Gasterosteus aculeatus* Peuhkuri *et al.*, 1997) and marine species (mackerel *Scomber scambrus* Pitcher *et al.*, 1985; capelin *Mallotus villosus* and gilt sardines *Sardinella aurita* Fréon and Misund, 1999).

The phenotypic assortment of social groups has adaptive significance in terms of reducing predation risk by minimizing phenotypic oddity (see above, the cost and benefits of group living). Support for the oddity effect selecting for group assortment is provided by an investigation by Theodorakis (1989), who presented bass with shoals containing either 25 large fish and five small fish or 25 small fish and five large fish. The minority size was taken more often than would be predicted by chance. There may be other benefits to phenotypic assortment. Using a simulation model, Ranta *et al.* (1994) demonstrated that phenotypic assortment of shoals may be adaptive when differences in food finding and food sharing between individuals are body size dependent. Accordingly, laboratory investigations have found active partner preferences based on body length in a number of species including zebrafish, *Brachydanio rerio* (McCann *et al.*, 1971), threespine sticklebacks (Ranta and Lindström, 1990), banded killifish (Krause and Godin, 1994) and guppies (Lachlan *et al.*, 1998).

It has been suggested that selection against oddity is so strong that it has suppressed the evolution of sexual dimorphism in pelagic shoaling fish species in marine environments (Hobson, 1968). Interestingly, males of some of these species, such as the haddock *Melanogrammus aeglefinus*, use sound during courtship (Hawkins, 1986). However, in species with lower levels of shoaling tendency, reproductive strategies may disrupt body length assortment. For example, in an investigation on guppies where female fecundity increases with body size, Herdman *et al.* (2004) found that male guppies preferred to associate (mate) with the larger of two virgin females (thus potentially increasing reproductive success). However, the benefits of mating with larger females (that are more fecund) may be reduced if there is larger degree of sperm competition (due to multiple matings). Accordingly, Herdman *et al.* (2004) suggested that male preferences for larger females may only be expressed when males can accurately assess the mating status of females that differ in body size (also see Ojanguren and Magurran, 2004). In support of this prediction, Herdman *et al.* (2004) observed that males showed no preference for larger females when limited to visual stimuli only, and thus presumably were unable to assess the reproductive status of the female.

Sex Composition

Reproductive strategies may influence the shoaling decisions based on the sex composition of shoals. For example, Lindström and Ranta (1993) observed that the sex ratio of a shoal influenced association patterns of male guppies, with males preferring to associate with shoals having a larger proportion of females. The mating success of a male will, in part, be dependent on the shoal sex ratio, with males having a higher mating probability in a shoal with a more female-biased sex ratio. However, males may have to trade-off the possibility of increased mating opportunities with exposure to predators. Accordingly Lindström and Ranta (1993) observed that males preferred to spend time in larger over smaller groups overriding decisions based on shoal composition. Furthermore, the proximity of other males may affect the shoaling behaviour of males. For example, Dugatkin and Sargent (1994) found that male guppies preferred to associate with other males that were further away from potential mates than they were themselves, potentially increasing a male's chances of being the individual chosen by a female assessing nearby males.

In contrast, one sex may actually try to avoid the other. For example, in guppies, sexual harassment can be costly to females (Magurran and Seghers, 1994) by reducing time available for foraging (Griffiths, 1996). Furthermore, sneaky mating by males may overthrow a female's choice of a particular male (Luyten and Liley, 1991). As a result, females may actively avoid males to reduce levels of sexual harassment (see Croft *et al.*, 2006), or associate with other females to dilute it. Sexual harassment and agonistic behaviour by males has been proposed as a mechanisms driving sexual segregation in other taxa, including whales (Smultea, 1994) and primates (see Lyons *et al.*, 1992).

There may be other benefits for females avoiding interactions with males. For example, predation risk to female guppies by *Crenicichla* sp. has been shown to be greater when the female is in close proximity to a male (Pocklington and Dill, 1995). It is thought that sexual activity by the male advertises the presence of females to the predator, which may actively prefer to feed on females that are energetically more profitable prey (Pocklington and Dill, 1995; Johansson *et al.*, 2004).

Trade-offs between Sex and Shoaling

Shoaling provides both safety and a source of sexual partners and, as a result, leads to interesting trade-offs between predator avoidance and courtship. In species such as the guppy that are continually sexually active, these decisions are constantly appraised. Magurran and Nowak (1991) modelled the interactions between males and females and discovered that under moderate risk no stable outcome emerges. Instead, there is a cyclical game in which the behaviour of one sex is dependent upon what the other is doing. Seasonally breeding species, for instance, three-spined sticklebacks (*Gasterosteus aculeatus*) and European minnows (*Phoxinus phoxinus*) play out these trade-offs on a different time scale. Outside the breeding season, males and females are monomorphic and form cohesive shoals. With the onset of reproductive activity, males set up breeding territories and start to solicit matings from females. Early arrival is typically traded off against increased predation risk. Candolin and Voigt (2003) discovered that larger male sticklebacks postpone territory formation until the females are present and predation risk is reduced. However, since they are able to evict small males from their nests, the large males suffer no disadvantage by delay. Small males, in contrast, must pay the increased predation risk costs of early arrival to gain any reproductive opportunities at all.

Familiarity and Kinship

Associations between familiar individuals may have important adaptive benefits (Griffiths and Ward, 2006). For example, in an experiment on fathead minnows (*Pimephales promelas*), Chivers *et al.* (1995) found that in response to a predation threat from the northern pike (*Esox lucius*), minnows in groups that were familiar showed greater shoal cohesion in comparison to unfamiliar groups, which may result in increased anti-predator protection (also see Griffiths *et al.*, 2004). It may also be of benefit for individuals to have information on the competitive ability of others (Metcalfe and Thomson, 1995), which may result in reduced levels of aggression (Utne-Palm and Hart, 2000). And finally, associations between familiar individuals are also probably very important in the evolution of co-operative interactions (Chivers *et al.*, 1995; Dugatkin, 1997).

There may be a number of mechanisms underlying familiarity preferences in fish (Griffiths, 2003; Ward and Hart, 2003). Firstly, individuals may distinguish between potential shoal partners based on previous experience in association with specific cues (e.g., size and colour). Such recognition is known as condition-dependent recognition. A good example of condition-dependent recognition is provided by work on predator inspection (where fish leave a shoal and approach a predator to gain information on and the probability of attack (Pitcher *et al.*, 1986)). For example, Milinski *et al.* (1990) found that sticklebacks recognized and preferred to associate with co-operative individuals during inspection behaviour. Such preferences were established after only a few hours of association. In contrast, individuals may distinguish between potential shoal mates based on previous experience alone, in the absence of specific cues (condition-independent recognition). Condition-independent recognition (familiarity) may take substantially longer to develop. For example, Griffiths and Magurran (1997), while investigating guppies, observed that familiarity took approximately 12 days to develop, and may be based on both visual and olfactory cues (Griffiths and Magurran, 1999). In contrast, recent work by Ward *et al.* (2004) documented that three-spined sticklebacks developed preferences for others, which were 'familiar' after just 24 hours, and such preferences were based on olfactory habitat cues and not direct experience and individual recognition. The importance of such mechanisms in free-ranging shoals in the wild is an exciting area for future research.

The role of familiarity in structuring fish shoals may be influenced by reproductive strategies. For example, in guppies, intra-sexual preferences for familiar individuals are expressed by both males and females under laboratory conditions (Griffiths and Magurran, 1997; Croft *et al.*, 2004a). In contrast, inter-sexual interactions may show the opposite patterns; for example, male guppies prefer to associate and direct more courtship behaviour toward novel females (Kelley *et al.*, 1999), presumably to increase reproductive success by maximizing the number of females mated with. Furthermore, males and females may differ in the stability of social interactions (due to sex differences in reproductive strategies), which may limit the opportunities for familiarity to occur under natural conditions. For example, from information on pair-wise interactions (obtained by repeatedly sampling the combustion of shoals of marked individuals over a 7-day period), Croft *et al.* (2004c) observed that repeated female-females pair-wise interactions occurred more frequently than expected by random assortment. In contrast, male-male and inter-sexual repeated pair-wise interactions did not differ from random. Accordingly, Griffiths and Magurran (1998) observed that female guppies from wild caught shoals exhibited a preference to associate with their natural shoal mates over females from another shoal. Males, in contrast, showed no such preference.

Advantages of familiarity may be enhanced through associating with kin, which may lead to indirect fitness benefits when shoaling (Pitcher and Parrish, 1993). For example, associations between kin may reduce inter-group aggression, and maximize the benefits of participation in co-operative behaviours, through kin selection (see Ward and Hart, 2003 for a review). However, associations between kin may be influenced by reproductive strategies. For example, in an investigation on rainbowfish (*Melanotaenia eachamensis*), Arnold (2000) found that females preferred to shoal with female relatives but avoided male relatives, potentially reducing costs due to inbreeding. Indeed, the possibility that kin recognition allows individuals to maximize inclusive fitness benefits by choosing optimally related mates may explain a paradox. Studies of kin-biased behaviour among fish have produced diametrically opposing results. On the one hand, evidence that fish are able to recognize kin is widespread (reviewed by Griffiths, 2003; Ward and Hart, 2003). On the other hand, evidence that wild shoals are composed of related individuals is sparse, even in those species that have demonstrated preferences for close relatives in laboratory experiments (Van Havre and Fitzgerald, 1988; Fitzgerald and Morrissette, 1992; Peuhkuri and Seppa, 1998; Russell *et al.*, 2004). Perhaps the benefits of kin-recognition are not afforded by preferentially shoaling with relatives, but rather by discriminating among potential partners during mate-choice decisions.

Recent work, for example, shows that female sticklebacks use water-borne odour cues to choose males with suitable immune-genes (i.e., MHC (major histocompatibility complex) alleles). Females prefer males with genes which, when combined with her own genes in their offspring, optimize the immune system (Aeschlimann *et al.*, 2003). Offspring with optimal allelic diversity at the MHC are infected with a significantly lighter parasite burden than sticklebacks with sub-optimal MHC diversity (Kurtz *et al.*, 2004). It seems that judicious choice of mating partner can confer fitness benefits to parents and offspring. The increasing evidence for MHC-biased association preferences is suggestive of an underlying role for MHC in kin recognition. However, although kinship has been shown to influence shoal composition in a number of species under laboratory conditions, the importance of kinship in structuring free ranging wild shoals remains equivocal (see Ward and Hart, 2003 for a review), and warrants further investigation.

ECOLOGICAL AND EVOLUTIONARY IMPLICATIONS

Co-operative Behaviour

Cooperation has been documented in fish in the form of helpers at the nest, nest raiding, foraging and predator inspection (see Dugatkin, 1997 for a review). Early theoretical exploration identified that cooperation could evolve between related individuals, whereby individuals are altruistic towards kin, gaining a selective advantage via 'inclusive fitness' (Hamilton, 1964a, b). In addition, Trivers (1971) pointed out that cooperation could evolve between unrelated individuals via reciprocity, whereby individuals are altruistic to those who have been altruistic to themselves (termed 'reciprocal altruism').

The evolution of cooperative behaviour may depend on the rate at which social groups break up and new groups form, and on assortment by individuals during encounters between groups (Michod and Sanderson, 1985; Toro and Silio, 1986). In particular, assortative interactions based on co-operation may allow partnerships to develop between individuals (Peck, 1993) and minimize costs due to defection (Wilson and Dugatkin, 1997). Reproductive strategies may influence the structure of assortative interactions (see above) and thus the evolution of cooperation. For example, in the guppy males exchange groups frequently in search of females (see above). In contrast, females appear to form stable social units (Griffiths and Magurran, 1998; Croft *et al.*, 2004c), fulfilling an underlying assumption for the evolution of co-operation (see Dugatkin, 1997 for a review). Thus, in wild guppy populations, cooperation may be more likely to evolve between females than males due to sex differences in reproductive strategies. Furthermore, in species such as the rainbowfish where individuals prefer to associate with non-kin during intra-sexual relationships and kin during inter-sexual relationships, cooperative behaviour may be more likely to emerge via kin selection within rather than between the sexes (Arnold, 2000). Future work should examine assortative interactions in free-ranging populations and examine the degree of co-operation between pairs of individuals that frequently associate.

Social Learning and Information Exchange

Social groups offer opportunities for individuals to learn about the environment as a result of observing or interacting with others. Social learning has been demonstrated in shoaling fish in a number of areas, including foraging locations, foraging routes, predator recognition, mate choice and migration routes (see Brown *et al.*, 2003 for a comprehensive review). Early theoretical models of social learning and information exchange generally assumed that individuals in a population interacted at random (Boyd and Richerson, 1985), and the larger picture of 'who learnt from whom' was largely neglected. However, in wild animal populations, social interactions rarely occur at random (Whitehead, 1999), and may be influenced by a number of factors including reproductive strategies (as discussed above). Such non-random social interactions may have implications for the transmission of learned information through populations (Coussi-Korbel and Fragaszy, 1995). For example, in the guppy sex differences in habitat use (Fig. 11.1) can restrict social interactions occurring between individuals. As a result, social information may be more likely to be transmitted within rather than between sexes. Furthermore, associations between familiar individuals may facilitate the transmission of information. For example, in an investigation on guppies, Swaney *et al.* (2001) demonstrated that social leaning occurred more rapidly between familiar individuals. In species such as the guppy where familiarity appears to be more important in structuring female-female interactions (see above), social information may spread more quickly between females than males.

Disease Transmission

The influence of reproductive strategies on the social organization of shoaling fishes may have implications for the transmission of disease. For example, the dynamics of the spread of pathogens through a population may be influenced by the spatial structure of a population and the duration, frequency and outcome of interactions (Loehle, 1995; Mollison and Levin, 1995), all of which may be influenced by reproductive strategies. Reproductive behaviour may select for the exchange of individuals between groups (see above) which may facilitate the spread of a pathogen through a population (Loehle, 1995). In contrast, differential habitat use by males and females resulting in sexual segregation may restrict the spread of a pathogen limiting social interactions between the sexes.

Acknowledgements

DPC would like to acknowledge funding NERC and the FSBI, and JK from the Leverhulme Trust.

References

Aeschlimann, P.B., M.A. Haberli, T.B.H. Reusch, T. Boehm and M. Milinski. 2003. Female sticklebacks *Gasterosteus aculeatus* use self-reference to optimize MHC allele number during mate selection. *Behavioral Ecology and Sociobiology* 54: 119-126.

Aparicio, E. and A. De Sostoa. 1999. Pattern of movements of adult *Barbus haasi* in a small Mediterranean stream. *Journal of Fish Biology* 55: 1086-1095.

Arnold, K.E. 2000. Kin recognition in rainbowfish (*Melanotaenia eachamensis*): sex, sibs and shoaling. *Behavioral Ecology and Sociobiology* 48: 385-391.

Beamish, F.W.H. 1978. Swimming capacity. In: *Fish Physiology*, W.S. Hoar and D.J. Randall (eds.). Academic Press, New York, Vol. 7, pp. 101-187.

Boyd, R. and P.J. Richerson. 1985. *Culture and the Evolutionary Process.* Chicago University Press, Chicago.

Bremset, G. and O.K. Berg. 1999. Three-dimensional microhabitat use by young pool-dwelling Atlantic salmon and brown trout. *Animal Behaviour* 58: 1047-1059.

Brown, C., K.N. Laland and J. Krause (eds.). 2003. *Learning in Fishes: Why they are smarter than you think?* Special edition of Fish and Fisheries. Blackwell Science, Oxford.

Brown, K.L. 1985. Demographic and genetic characteristics of dispersal in the mosquitofish, *Gambusia affinis* (Pisces, Poeciliidae). *Copeia* 1985: 597-612.

Candolin, U. and H.R. Voigt. 2003. Size-dependent selection on arrival times in sticklebacks: why small males arrive first. *Evolution* 57: 862-871.

Chapman, L.J. and D.L. Kramer. 1991. The consequences of flooding for the dispersal and fate of Poeciliid fish in an intermittent tropical stream. *Oecologia* 87: 299-306.

Chivers, D.P., G.E. Brown and R.J.F. Smith. 1995. Familiarity and shoal cohesion in fathead minnows (*Pimephales promelas*)—Implications for antipredator behaviour. *Canadian Journal of Zoology-Revue Canadienne De Zoologie* 73: 955-960.

Conradt, L. 1998. Could asynchrony in activity between the sexes cause intersexual social segregation in ruminants? *Proceedings of the Royal Society of London Series* B 265: 1359-1363.

Coussi-Korbel, S. and D.M. Fragaszy. 1995. On the relation between social dynamics and social learning. *Animal Behaviour* 50: 1441-1453.

Croft, D.P., B. Albanese, B.J. Arrowsmith, M. Botham, M. Webster and J. Krause. 2003a. Sex biased movement in the guppy (*Poecilia reticulata*). *Oecologia* 137: 62-68.

Croft, D.P., B.J. Arrowsmith, J. Bielby, K. Skinner, E. White, I.D. Couzin, A.E. Magurran, I. Ramnarine and J. Krause. 2003b. Mechanisms underlying shoal composition in the Trinidadian guppy (*Poecilia reticulata*). *Oikos* 100: 429-438.

Croft, D.P., J. Krause, I.D. Couzin and T.J. Pitcher. 2003c. When fish shoals meet: outcomes for evolution and fisheries. *Fish and Fisheries* 4: 138-146.

Croft, D.P., B.J. Arrowsmith, M. Webster and J. Krause. 2004a. Intrasexual preferences for familiar individuals in male guppies (*Poecilia reticulata*). *Journal of Fish Biology* 64: 279-283.

Croft, D.P., M.S. Botham and J. Krause. 2004b. Is sexual segregation in the guppy, *Poecilia reticulata*, consistent with the predation risk hypothesis? *Environmental Biology of Fishes* 71: 127-133.

Croft, D.P., J. Krause and R. James. 2004c. Social networks in the guppy (*Poecilia reticulata*). *Proceedings of the Royal Society of London Biology Letters* 271: 516-519.

Croft, D.P., *et al.* 2006. Predation risk as a driving force for sexual segregation: A cross-population comparison. *American Naturalist* 167(6): 867-878.

Dugatkin, L.A. 1997. *Cooperation Among Animals: An Evolutionary Perspective.* Oxford University Press, Oxford.

Dugatkin, L.A. and R.C. Sargent. 1994. Male-male association patterns and female proximity in the guppy, *Poecilia reticulata*. *Behavioural Ecology and Sociobiology* 35: 141-145.

Fitzgerald, G.J. and J. Morrissette. 1992. Kin recognition and choice of shoal mates by threespine sticklebacks. *Ethology, Ecology and Evolution* 4: 273-283.

Fréon, P. and O.A. Misund. 1999. Schooling behaviour. In: *Dynamics of Pelagic Fish Distribution and Behaviour: Effects on Fisheries and Stock Assessment.* Fishing News Books, Blackwell Science, Oxford, pp. 56-101.

Fuiman, L.A. and A.E. Magurran. 1994. Development of predator defences in fishes. *Reviews in Fish Biology and Fisheries* 4: 145-183.

Gilliam, J.F. and D.F. Fraser. 1987. Habitat selection under predation hazard: Test of a model with foraging minnows. *Ecology* 68: 1856-1862.

Godin, J.G.J. 1986. Antipredator functions of shoaling in teleost fishes: a selective review. *Le Naturaliste Canadien* 113: 241-250.

Grand, T.C. and L.M. Dill. 1999. The effect of group size on the foraging behaviour of juvenile coho salmon: reduction of predation risk or increased competition? *Animal Behaviour* 58: 443-451.

Greenberg, L., P. Svendsen and A. Harby. 1996. Availability of microhabitats and their use by brown trout (*Salmo trutta*) and grayling (*Thymallus thymallus*) in the River Vojman, Sweden. *Regulated Rivers—Research and Management* 12: 287-303.

Greenwood, P.J. 1980. Mating systems, philopatry and dispersal in birds and mammals. *Animal Behaviour* 28: 1140-1162.

Griffiths, S.W. 1996. Sex differences in the trade-off between feeding and mating in the guppy. *Journal of Fish Biology* 48: 891-898.

Griffiths, S.W. 2003. Learned recognition of conspecifics by fishes. In: *Fish are smarter than you think: Learning in fishes*, C. Brown, K.N. Laland and J. Krause (eds.). *Fish and Fisheries*, Special Ed. 4: 256-268.

Griffiths, S.W., S. Brockmark, J. Hojesjo and J.I. Johnsson. 2004. Coping with divided attention: the advantage of familiarity. *Proceedings of the Royal Society of London B* 271: 695-699.

Griffiths, S.W. and A.E. Magurran. 1997. Familiarity in schooling fish: How long does it take to acquire? *Animal Behaviour* 53: 945-949.

Griffiths, S.W. and A.E. Magurran. 1998. Sex and schooling behaviour in the Trinidadian guppy. *Animal Behaviour* 56: 689-693.

Griffiths, S.W. and A.E. Magurran . 1999. Schooling decisions in guppies (*Poecilia reticulata*) are based on familiarity rather than kin recognition by phenotype matching. *Behavioral Ecology and Sociobiology* 45: 437-443.

Griffiths, S.W. and A.J.W. Ward. (2006). Learned recognition of conspecifics. In: *Fish Cognition and Behavior*, C. Brown, K.N. Laland and J. Krause (eds.). Blackwell Publishing, pp. 139-165.

Gueron, S., S.A. Levin and D.I. Rubenstein. 1996. The dynamics of herds: From individuals to aggregations. *Journal of Theoretical Biology* 182: 85-98.

Hamilton, W.D. 1964a. Genetical evolution of social behaviour 2. *Journal of Theoretical Biology* (1): 17-52.

Hamilton, W.D. 1964b. Genetical evolution of social behaviour I. *Journal of Theoretical Biology* (1): 1-16.

Hawkins, A.D. 1986. Underwater sound and fish behaviour. In: *The Behaviour of Teleost Fishes*, T.J. Pitcher (ed.). Croom Helm, London, pp. 114-151.

Helfman, G.S., B.B. Collette and D.E. Facey. 1997. *The Diversity of Fishes*. Blackwell Science, Malden, MA.

Herdman, E.J.E., C.D. Kelly and J.G.J. Godin. 2004. Male mate choice in the guppy (*Poecilia reticulata*): Do males prefer larger females as mates? *Ethology* 110: 97-111.

Hoare, D.J., J. Krause, N. Peuhkuri and J.G.J. Godin. 2000. Body size and shoaling in fish. *Journal of Fish Biology* 57: 1351-1366.

Hobson, E.S. 1968. Predatory behaviour of some shore fishes in the Gulf of California. *United States Bureau of Sport Fisheries and Wildlife Research Report* 73: 1-91.

Houde, A.E. 1997. *Sex, Color, and Mate Choice in Guppies*. Princeton University Press, Princeton.

Johansen, P.H. and J.A. Cross. 1986. Effects of sexual maturation and sex steroid hormone treatment on the temperature preferences of the guppy (*Poecilia reticulata* Peters). *Canadian Journal of Zoology* 58: 586-588.

Johansson, J., H. Turesson and A. Persson. 2004. Active selection for large guppies, *Poecilia reticulata*, by the pike cichlid, *Crenicichla saxatilis*. *Oikos* 105: 595-605.

Johnsson, J.I. 2003. Group size influences foraging effort independent of predation risk: an experimental study on rainbow trout. *Journal of Fish Biology* 63: 863-870.

Kearney, M., R. Shine, S. Comber and D. Pearson. 2001. Why do geckos group? An analysis of 'social' aggregations in two species of Australian lizards. *Herpetologica* 57: 411-422.

Kelley, J.L., J.A. Graves and A.E. Magurran. 1999. Familiarity breeds contempt in guppies. *Nature* (*London*) 401: 661-662.

Kodric-Brown, A. 1990. Mechanisms of sexual selection—Insights from fishes. *Annales Zoologici Fennici* 27: 87-100.

Krause, J. and J.G.J. Godin. 1994. Shoal choice in the banded killifish (*Fundulus diaphanus*, Teleostei, Cyprinodontidae)—Effects of predation risk, fish size, species composition and size of shoals. *Ethology* 98: 128-136.

Krause, J. and G.D. Ruxton. 2002. *Living in Groups*. Oxford University Press, Oxford.

Krause, J., J.G.J. Godin and D. Brown. 1996. Size-assortativeness in multi-species fish shoals. *Journal of Fish Biology* 49: 221-225.

Krause, J., R.K. Butlin, N. Peuhkuri and V.L. Pritchard. 2000. The social organization of fish shoals: a test of the predictive power of laboratory experiments for the field. *Biological Reviews* 75: 477-501.

Kurtz, J., M. Kalbe, P.B. Aeschlimann, M.A. Haberli, K.M. Wegner, T.B.H. Reusch and M. Milinski. 2004. Major histocompatibility complex diversity influences parasite resistance and innate immunity in sticklebacks. *Proceedings of the Royal Society of London* B 271: 197-204.

Lachlan, R.F., L. Crooks and K.N. Laland. 1998. Who follows whom? Shoaling preferences and social learning of foraging information in guppies. *Animal Behaviour* 56: 181-190.

Landeau, L. and J. Terborgh. 1986. Oddity and the confusion effect in predation. *Animal Behaviour* 34: 1372-1380.

Lightfoot, W. and V. Jones. 1996. The relationship between the size of 0+roach, *Rutilus rutilus*, their swimming capabilities, and distribution in an English river. *Folia Zoologica* 45: 355-360.

Lindström, K. and E. Ranta. 1993. Social preferences by male guppies, *Poecilia reticulata*, based on shoal size and sex. *Animal Behaviour* 46: 1029-1031.

Loehle, C. 1995. Social barriers to pathogen transmission in wild animal populations. *Ecology* 76: 326-335.

Luyten, P.H. and N.R. Liley. 1991. Sexual selection and competitive mating success of males guppies (*Poecilia reticulata*) from 4 Trinidad populations. *Behavioral Ecology and Sociobiology* 28: 329-336.

Lyons, D.M., S.P. Mendoza and W.A. Mason. 1992. Sexual segregation in squirrel-monkeys (*Saimiri sciureus*)—A transactional analysis of adult social dynamics. *Journal of Comparative Psychology* 106: 323-330.

Magurran, A.E. 1990. The adaptive significance of schooling as an antipredator defense in fish. *Annales Zoologici Fennici* 27: 51-66.

Magurran, A.E. and C. Macias-Garcia. 2000. Sex differences in behaviour as an indirect consequence of mating system. *Journal of Fish Biology* 57: 839-857.

Magurran, A.E. and M.A. Nowak. 1991. Another battle of the sexes—The consequences of sexual asymmetry in mating costs and predation risk in the guppy, *Poecilia reticulata*. *Proceedings of the Royal Society of London* B 246: 31-38.

Magurran, A.E. and B.H. Seghers. 1994. Sexual conflict as a consequence of ecology—Evidence from guppy, *Poecilia reticulata*, populations in Trinidad. *Proceedings of the Royal Society of London* B 255: 31-36.

Magurran, A.E., W.J. Oulton and T.J. Pitcher. 1985. Vigilant behaviour and shoal size in minnows. *Zeitschrift fur Tierpsychologie* 67: 167-178.

Mattingly, H.T. and M.J. Butler. 1994. Laboratory predation on the Trinidadian guppy—Implications for the size-selective predation hypothesis and guppy life-history evolution. *Oikos* 69: 54-64.

McCann, L.I., D.J. Koehn and N.J. Kline. 1971. The effects of body size and body markings on nonpolarized schooling behaviour of zebra fish (*Brachydanio rerio*). *The Journal of Psychology* 79: 71-75.

Metcalfe, N.B. and B.C. Thomson. 1995. Fish recognise and prefer to shoal with poor competitors. *Proceedings of the Royal Society of London* B 259: 207-210.

Michod, R.E. and M.J. Sanderson. 1985. Behavioural structure and the evolution of co-operation. In: *Evolution—Essays in honour of John Maynard Smith*, J. Greenwood and M. Saltkin, Cambridge University Press, Cambridge, pp. 95-104.

Milinski, M., D. Pfluger, D. Kulling and R. Kettler. 1990. Do sticklebacks cooperate repeatedly in reciprocal pairs? *Behavioral Ecology and Sociobiology* 27: 17-21.

Mollison, D. and S.A. Levin. 1995. Spatial dynamics of parasitism. In: *Ecology of Infectious Diseases in Natural Populations*, B.T. Grenfell and A.P. Dobson (eds.). Cambridge University Press, Cambridge, pp. 384-398.

Neill, S.R., St.J. and J.M. Cullen. 1974. Experiments on whether schooling by their prey affects the hunting behaviour of cephalopods and fish predators. *Journal of Zoology* 172: 549-569.

Noltie, D.B. and P.H. Johansen. 1986. Laboratory studies of microhabitat selection by the guppy, *Poecilia reticulata* (Peters). *Journal of Freshwater Ecology* 3: 299-307.

Ojanguren, A.F. and A.E. Magurran. 2004. Uncoupling the links between male mating tactics and female attractiveness. *Proceedings of the Royal Society of London Biology Letters.* (In Press).

Peck, J.R. 1993. Friendship and the evolution of co-operation. *Journal of Theoretical Biology* 162: 195-228.

Peuhkuri, N., E. Ranta and P. Seppa. 1997. Size-assortative schooling in free-ranging sticklebacks. *Ethology* 103: 318-324.

Peuhkuri, N. and P. Seppa. 1998. Do three-spined sticklebacks group with kin? *Annales Zoologici Fennici* 35: 21-27.

Pitcher, T.J. and J.K. Parrish. 1993. Functions of shoaling behaviour in teleosts. In: *Behaviour of Teleost Fishes*, T.J. Pitcher (ed.). Chapman and Hall, London, pp. 363-439.

Pitcher, T.J., A.E. Magurran and I. Winfield. 1982. Fish in larger shoals find food faster. *Behavioural Ecology and Sociobiology* 10: 149-151.

Pitcher, T.J., A.E. Magurran and J.I. Edwards. 1985. Schooling mackerel and herring choose neighbors of similar size. *Marine Biology* 86: 319-322.

Pitcher, T.J., D.A. Green and A.E. Magurran . 1986. Dicing with death—predator inspection behaviour in minnow shoals. *Journal of Fish Biology* 28: 439-448.

Pocklington, R. and L.M. Dill. 1995. Predation on females or males—who pays for bright male traits. *Animal Behaviour* 49(4): 1122-1124.

Post, J.R. and D.O. Evans. 1989. Experimental-evidence of size-dependent predation mortality in juvenile Yellow Perch. *Canadian Journal of Zoology* 67: 521-523.

Pusey, A. and M. Wolf. 1996. Inbreeding avoidance in animals. *Trends in Ecology and Evolution* 11: 201-206.

Ranta, E. and K. Lindström. 1990. Assortative schooling in 3-spined sticklebacks. *Annales Zoologici Fennici* 27: 67-75.

Ranta, E., N. Peuhkuri and A. Laurila. 1994. A theoretical exploration of antipredatory and foraging factors promoting phenotype-assorted fish schools. *Ecoscience* 1: 99-106.

Rodd, F.H. and D.N. Reznick. 1997. Variation in the demography of guppy populations: The importance of predation and life histories. *Ecology* 78: 405-418.

Ruckstuhl, K.E. 1998. Foraging behaviour and sexual segregation in bighorn sheep. *Animal Behaviour* 56: 99-106.

Ruckstuhl, K.E. 1999. To synchronise or not to synchronise: a dilemma for young bighorn males? *Behaviour* 136: 805-818.

Ruckstuhl, K.E. and H. Kokko. 2002. Modelling sexual segregation in ungulates: effects of group size, activity budgets and synchrony. *Animal Behaviour* 64: 909-914.

Ruckstuhl, K.E. and P. Neuhaus. 2002. Sexual segregation in ungulates: a comparative test of three hypotheses. *Biological Reviews* 77: 77-96.

Russell, S.T., J.L. Kelley, J.A. Graves and A.E. Magurran. 2004. Kin structure and shoal composition dynamics in the guppy, *Poecilia reticulata*. *Oikos* 106: 520-526.

Schleusner, C.J. and O.E. Maughan. 1999. Mobility of largemouth bass in a desert lake in Arizona. *Fisheries Research* 44: 175-178.

Seghers, B.H. 1973. An analysis of geographic variation in the anti-predator adaptations of the Guppy, *Poecilia reticulata*. Ph. D. Thesis Zoology, The University of British Columbia, Vancouver, Canada.

Shaw, E. 1978. Schooling fishes. *American Scientist* 66: 166-175.

Smultea, M.A. 1994. Segregation by humpback whale (*Megaptera novaeangliae*) cows with a calf in coastal habitat near the Island of Hawaii. *Canadian Journal of Zoology* 72: 805-811.

Swaney, W., J. Kendal, H. Capon, C. Brown and K.N. Laland. 2001. Familiarity facilitates social learning of foraging behaviour in the guppy. *Animal Behaviour* 62: 591-598.

Theodorakis, C.W. 1989. Size segregation and the effects of oddity on predation risk in minnow schools. *Animal Behaviour* 38: 496-502.

Toro, M. and L. Silio. 1986. Assortment of encounters in the two-strategy game. *Journal of Theoretical Biology* 123: 193-204.

Trivers, R.L. 1971. Evolution of reciprocal altruism. *Quarterly Review of Biology* 46: 35-57.

Turner, G.F. and T.J. Pitcher. 1986. Attack abatement—A model for group protection by combined avoidance and dilution. *American Naturalist* 128: 228-240.

Utne-Palm, A.C. and P.J.B. Hart. 2000. The effects of familiarity on competitive interactions between threespined sticklebacks. *Oikos* 91: 225-232.

Van Havre, N. and G.J. Fitzgerald. 1988. Shoaling and kin recognition in the threespine stickleback (*Gasterosteus aculeatus* L.). *Biology of Behaviour* 13: 190-201.

Ward, A.J.W. and P.J.B. Hart. 2003. The effects of kin and familiarity on interactions between fish. *Fish and Fisheries* 4: 348-358.

Ward, A.J.W., P.J.B. Hart and J. Krause. 2004. The effects of habitat- and diet-based cues on association preferences in three-spined sticklebacks. *Behavioral Ecology* 15: 925-929.

Watkins, J.L., F. Buchholz, J. Priddle, D.J. Morris and C. Ricketts. 1992. Variation in reproductive status of Antarctic Krill swarms evidence for a size-related sorting mechanism? *Marine Ecology Progress Series* 82: 163-174.

Whitehead, H. 1999. Testing Association Patterns of Social Animals. *Animal Behaviour* 57: F26-F29.

Wilson, D.S. and L.A. Dugatkin. 1997. Group selection and assortative interactions. *American Naturalist* 149: 336-351.

Parental Care and Predation Risk in Fish—Defend, Desert or Devour

Carin Magnhagen

INTRODUCTION

In fish, the majority of species do not care for their offspring after spawning, but disperse their gametes freely or, at the most, hide the eggs in the substrate (Balon, 1975). Nevertheless, 22% of all teleost fish families include species which display some kind of parental care (Blumer, 1979; Sargent and Gross, 1993). Among fish families with parental care, paternal care is found in half, maternal care in about a third, and biparental care in a fifth of these families (Blumer, 1979; Sargent and Gross, 1993). Parental care can consist of guarding the eggs from predators, fanning the eggs to provide oxygen, constructing nests to protect the offspring, or carrying the eggs, internally or externally (Balon, 1975). Parental care usually ends at the hatching of the brood, although there are species that also guard free-swimming larvae or fry, for example, some cichlids

Author's address: Department of Aquaculture, Swedish University of Agricultural Sciences, SE - 901 83 Umeå, Sweden. E-mail: Carin.Magnhagen@vabr.slu.se

(Fryer and Iles, 1972), centrarchids (bass and sunfish) (Gross, 1982; Brown, 1984) and sticklebacks (Wootton, 1976).

Parental investment can be defined as any investment by the parent in the offspring that will increase the offspring's chance of survival, at the cost of the parent's ability to invest in other offspring (Trivers, 1972). This cost can result from decreased energy reserves and even survival, or due to time constraints (Smith and Wootton, 1995a). For parental care to evolve, lifetime fitness for caring parents has to be higher than for non-caring parents, and thus the increased survival of the offspring more than compensate for the decrease in future reproductive success (Clutton-Brock, 1991).

In general, predation on young fish is very high (Sogard, 1997), and only a small fraction of all fish survive their first months of life. Fish species with parental care generally produce fewer but larger eggs than those without care, and the size of eggs increases with the degree of parental care (Sargent *et al.*, 1987). Parental care is expected to substantially decrease the risk of the offspring to be eaten, especially during the early stages (Perrone and Zaret, 1979). Since predation on young fish usually is size dependent (Sogard, 1997), survival rates in fishes with parental care may be higher than in other species even after independence, due to the production of larger progeny. At the same time, parental care may increase the predation risk of the parent (Magnhagen, 1992; Smith and Wootton, 1995a), and there would then be a trade-off between predator avoidance and brood defence (Sargent and Gross, 1986; Clark and Ydenberg, 1990).

Wisenden (1999) suggested that parental care might have evolved mainly in species with a low general level of predation risk on adults. This is probably true for larger species like cichlids and centrarchids, such as the colonially nesting bluegill sunfish *Lepomis macrochirus*, in which predation on parental males is low (Gross and McMillan, 1981). Small-sized fish species that live in shallow areas often have parental care, and many of these are also subjected to substantial predation by both fish and birds. Yet, in some species, for example, gobies and blennies, which spawn in mussel shells, burrows or crevices, both eggs and the parent fish may be protected at the nesting site (Potts, 1984; but see examples below). In contrast, other fishes have more conspicuous nest sites, like in the fifteen-spined stickleback *Spinachia spinachia* where females prefer males with nests built high up in the vegetation, probably because the eggs are more protected from predation by crabs (Östlund-Nilsson, 2000). Also, in a marine population of three-spined stickleback

Gasterosteus aculeatus, males marked their nests' entrance with a deviant colour of algae (Östlund-Nilsson and Holmlund, 2003). In aquaria, males decorated their nests with shiny and colourful artificial objects, thereby apparently reducing the nest camouflage. Females preferred to mate with males with decorated nests compared to those with plain nests (Östlund-Nilsson and Holmlund, 2003). Hence, it is likely that while in some species, parental care only affects predation on the offspring, in other species an increased predation risk for the parent is influencing decisions of parental investments.

In this chapter I will look at different aspects of predation risk in connection with parental care in fish. Topics included are: (1) predation on fish performing care and adaptations to decrease this risk; (2) the parent's risk-taking and defence of the offspring against predators in relation to the reproductive value of the current brood; (3) the occurrence of, and explanation for, filial cannibalism (eating of own offspring); and (4) cases where the risk of parental care may not be worth it.

PREDATION RISK OF PARENT

During the parental phase the caring parent may face an increased risk of predation due to one or several effects of its parental status (Magnhagen, 1991, 1992). For example, defending the nest against intruders and tending to the eggs might decrease vigilance, while fanning behaviour would make a nest-guarding fish more conspicuous. Further, egg carrying could increase visibility and decrease swimming performance both in males brooding eggs, externally or internally, and in live-bearing females. An indirect effect of parental care on predation risk may be a decrease in condition, making the ability to escape an attack low, especially in the end or just after a reproductive cycle.

Few studies have actually shown a direct effect of parental care on predation risk, but in the pipefish *Nerophis ophidion*, egg-carrying males were preyed upon to a higher extent than were females or non-caring males (Svensson, 1988). This was explained by the higher visibility of males, with bright yellow eggs attached ventrally to their skin, compared to the usual cryptic appearance in pipefish. Similarly, nest guarding sand goby *Pomatoschistus minutus* males were preyed upon by piscivorous birds (Lindström and Ranta, 1992). The nests seemed to be a cue for the hunting birds and non-cryptic nests were attacked more often than nests covered with sand (Lindström and Ranta, 1992).

The proportion of males in populations of the two-spotted goby *Gobiusculus flavescens* decreased dramatically over the breeding season (Forsgren *et al.*, 2004; Fig. 12.1), and male-biased predation was one suggestion for this decrease. In other fish species, the higher visibility of males caused by breeding coloration and ornaments have been suggested to lead to a higher mortality due to predation of males compared to the more cryptic females (Garcia *et al.*, 1998). However, in the two-spotted goby, also the females have a sexually selected bright breeding coloration (Amundsen and Forsgren, 2001). Hence, conspicuousness should not differ between the sexes because of the nuptial colours. Instead, differences in behaviour may affect predation risk. The males guard and fan the eggs at their nest site, while females and non-breeding males are shoaling. The solitary life-style compared to group living may lead to differential vulnerability to predators (Pitcher and Parrish, 1993). Another possible explanation for the higher mortality in male gobies as compared to females is that the condition of males decreases over the season (Forsgren *et al.*, 2004). The males may, therefore, risk starving to death (Marconato *et al.*, 1993), but then the poor condition would also make them easy targets for predators. An experimental study on this species showed that a decrease in the condition of males over a reproductive

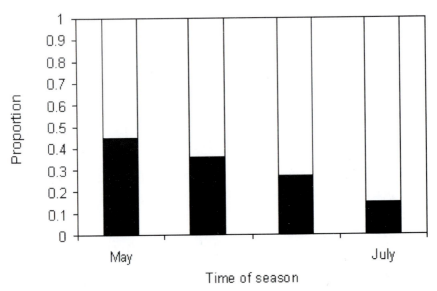

Fig. 12.1 Proportion of males (black bars) and females (white bars) in populations of two-spotted goby *Gobiusculus flavescens* on the Swedish west coast, at different times of the breeding season. Data from Forsgren *et al.* (2004).

cycle was caused by a gradual increase in fanning rate and lower feeding rates with increasing age of the brood (Skolbekken and Utne-Palm, 2001). Energetic costs of parental care, shown by a decline in body condition, have been found also in several other fish species (e.g., Hinch and Collins, 1991; Lindström and Hellström, 1993; Marconato et al., 1993).

In the three-spined stickleback Gasterosteus aculeatus, males were bolder than females, showed a more exploratory behaviour and recovered faster after being exposed to a predator (Giles, 1984). This difference in behaviour was explained by the parental role of the males, which would force them to take risks in order not to lose the brood they were guarding. The higher boldness, in connection with their bright red nuptial colour, suggests that predation on nest guarding males would be higher than that of females. There are some indications of this being the case (Ulrika Candolin, pers. comm.) but no published articles have so far confirmed it. However, in some populations of stickleback, a black form has evolved, with only a low proportion of the males developing the usual red throat (Moodie, 1972). Experiments showed that cutthroat trout Salmo clarki attacked red-throated males to a higher degree than duller males (Moodie, 1972), suggesting that nest-guarding males in full breeding coloration would be vulnerable to predation.

In females of live-bearing species, pregnancy can impair mobility, which can further lead to an increased risk of predation. In the guppy Poecilia reticulata, females from high-predation fish communities allocate more to reproduction, i.e., the mass of the brood in relation to female size is larger than in females that experience lower predation risk (Reznick and Endler, 1982; Reznick et al., 1990, 1996). When alarmed, fast-start performance differed between females from different populations, with fish from high-predation sites having a faster acceleration, reaching higher velocities, and travelling greater distances as compared to females from low-predation sites (Ghalambor et al., 2004). Velocity and distance travelled decreased with the progress of pregnancy, but more so in females from high-predation populations. The larger increase in body mass, surface area and body stiffness due to a higher allocation to reproduction could explain the differences between the populations. Thus, at high predation risk, there seems to be a trade-off between investing in reproduction and avoiding predation in female guppies (Ghalambor et al., 2004). The decrease in swimming performance over the pregnancy cycle most likely affects the risk of being caught by an attacking predator. At the same time, life-history evolution in guppies is driven by age-specific mortality with early maturation and the production of more offspring at high predation risk (Reznick and Endler, 1982; Reznick *et al.*, 1990, 1996). Whether female guppies compensate for the increased vulnerability by altering their behaviour is not known, but could be expected. Accordingly, a change of habitat use and escape behaviour was found in gravid three-spined sticklebacks, seemingly to decrease the vulnerability to predators (Rodewald and Foster, 1998).

In the mosquitofish *Gambusia affinis*, a decrease in swimming performance in pregnant females was also found (Plaut, 2002) with the critical swimming speed (U_{crit}), that measures the aerobic swimming ability, decreasing over time of pregnancy. Whether this reduction in swimming ability increases predation risk, however, was not investigated.

ADAPTATIONS IN PARENT BEHAVIOUR TO AVOID BEING EATEN

The reason that not very many studies find an effect of parental care on predation of parents may be due to the difficulties of quantifying this effect, but also, because of ethical considerations, this kind of experiments is generally not performed anymore. However, there would be a strong selection for behaviour reducing the potential predation risk. Being eaten while caring for the offspring would, of course, totally counteract the effort put into the current reproductive event (and put a stop for future ones). Thus, adaptation to a potentially higher predation risk during the parental phase should be commonly found among fish caring for their offspring. Accordingly, the three-spined stickleback *Gasterosteus aculeatus* chose safer breeding habitats over those that would be more favourable regarding reproductive rate in the beginning—but not in the end—of the breeding season (Candolin and Voigt, 2003). Further, Kraak *et al.* (2000) found that male three-spined sticklebacks preferred to build nests in areas concealed by macrophytes. In shallow areas, where predation by herons was high, males that occupied concealed sites were larger and redder than in open sites. This may reflect the higher competitive ability, but also the greater conspicuousness of more colourful males (Kraak *et al.*, 2000). Blackspotted stickleback *Gasterosteus wheatlandi* males adjusted boldness to nest location (FitzGerald and Lachance, 1993). Males with nests in deep water, who were less vulnerable to predation from birds, returned sooner after a simulated predator attack compared to those with a shallow water nest. Also, male sand gobies *Pomatoschistus minutus* preferred nest sites in deeper areas that were protected from bird predation (Lindström and Ranta, 1992). Males nesting in shallow areas were smaller than the ones in deeper areas, indicating a male-male competition for safe nest sites. Male fathead minnows *Pimephales promelas* nested more frequently in areas with low densities of pike *Esox lucius* than in those where pike were abundant (Jones and Paszkowski, 1997). Also, nest guarding males that had encountered a pike had a lower egg rubbing frequency compared to control males, seemingly to reduce the risk of being detected by the predator (Jones and Paszkowski, 1997).

CHANGES IN RISK-TAKING WITH REPRODUCTIVE VALUE OF THE BROOD

Risks taken by a parental individual are expected to vary with factors influencing the value of the current brood, such as number and age of offspring and the parent's probability of future reproduction (Sargent and Gross, 1993). This has been confirmed in many studies on fish with parental care. Three-spined stickleback *Gasterosteus aculeatus* males in the presence of a predator defended the nest more vigorously against intruders if the nest contained eggs compared to an empty nest (Ukegbu and Huntingford, 1988). The boldness of sticklebacks towards a pike also increased from the non-reproductive stage to egg-guarding and further to guarding newly hatched young (Huntingford, 1976). Furthermore, risk-taking by nest-guarding stickleback males in the presence of a dummy predator increased with the number and age of eggs in his nest (Pressley, 1981).

Field experiments on the common goby *Pomatoschistus microps* showed that nest-guarding males that were chased away from their nests in the presence of a predator returned sooner with increasing age and size of the brood in the nest (Magnhagen and Vestergaard, 1991, 1993; Fig. 12.2). Similar results were found in the closely related sand goby *P. minutus* (Lindström and Wennström, 1994). In addition, males of both species that were approached by the researcher stayed longer in the nest when they guarded larger and/or older egg masses (Magnhagen and Vestergaard, 1991; Lindström and Wennström, 1994). There was also an effect of time of season in the risk-taking of male common gobies (Magnhagen and Vestergaard, 1991). These gobies are short-lived and reproduce repeatedly during one season only. At the end of the season, with low probability of future reproduction, risk-taking was higher than in the beginning of the season, and was then also independent of the age of the offspring (Magnhagen and Vestergaard, 1991). The three-spined stickleback is also a short-lived species, often with only one reproductive season. Even in this species, risk-taking increased over a period of time.

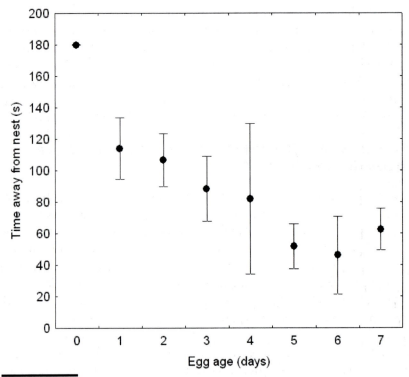

Fig. 12.2 Time (s) (mean ± SE) before returning to the nest in male common gobies *Pomatoschistus microps*, after being chased away in the presence of a predator, versus egg age (days). The graph is modified from Magnhagen and Vestergaard (1993).

Both in field studies and in aquaria, nests were located in more exposed areas in the end as compared to the beginning of the breeding season (Candolin and Voigt, 2003). In the field, shallow and structurally less complex habitats would be costly due to an increased risk of predation by terns. However, due to higher temperatures compared to deeper areas, the reproductive rate is higher (Candolin and Voigt, 2003).

PREDATION ON OFFSPRING

One of the most important effects of parental care in fish is a decrease in egg predation and, thus, an inverse increase in offspring survival. Parental fish are commonly seen attacking predators on eggs and fry (e.g., Gross and MacMillan, 1981; Karino, 1997; Östlund-Nilsson, 2002). Also, if the parent abandons the nest or if the guarding parent is removed, the offspring is usually taken by predators (Bain, 1983; Jan, 1991; Warner *et al.*, 1994;

Neff, 2003), even though exceptions can be found (Almada *et al.*, 1992). The cost of defending the offspring can be exemplified by a study of the convict cichlid *Cichlasoma nigrofasciatum*, where interspawning intervals increased in parental pairs exposed to potential brood predators compared to pairs without predators present (Smith and Wootton, 1995b). This spawning delay was explained by an increase in energy expenditure by attacking the predators, leading to lower growth rates in females and slower oocyte development compared to females not exposed to predators.

Naturally, the brood defence is not always successful. Females in several species of fish have been found to choose mates who will be efficient in defending their progeny against predation. Survival of progeny from eggs until swim-up was positively related to the site tenacity of males in smallmouth bass, *Micropterus dolomieui* (Wiegmann et al., 1992). Females preferred large males that were more site-tenacious than small ones, and thus choose males that were good parents (Wiegmann and Baylis, 1995). In the sand goby *Pomatoschistus minutus*, females chose partners that had a higher hatching success in the presence of nest predators than had non-preferred males (Forsgren, 1997; Fig. 12.3). Similarly, in the freshwater goby *Padogobius martensi*, females showed a marked preference for spawning in large nest sites, which were always occupied by the larger males (Bisazza et al., 1989). Smaller males had a lower reproductive success than larger, due to courtship interference, nest take-over and egg cannibalism. In the bicolour damselfish, *Stegastes partitus* where males guard the eggs, females broke their usual nest fidelity if earlier broods had been lost to predation (Knapp, 1993). The females used visual cues to assess brood loss, but also reacted with nest avoidance even by the odour of egg-feeding brittlestars.

Nest site concealment was the most important determinant of female nest-choice and male reproductive success in experiments on the three-spined stickleback *Gasterosteus aculeatus* (Sargent and Gebler, 1980). One reason for this was that intra-specific egg predation was lower in concealed nests, which was also found by Kynard (1978). However, this is in contrast with the findings in a marine population, that female sticklebacks chose males with nests decorated with shiny objects (Östlund-Nilsson and Holmlund, 2003). Differences in the environment, for example, regarding predation risk, might explain these differences in female choice of nest characteristics. This would then be comparable to the variation in female preference of colourful male guppies across populations with different predation risk (Houde and Endler, 1990).

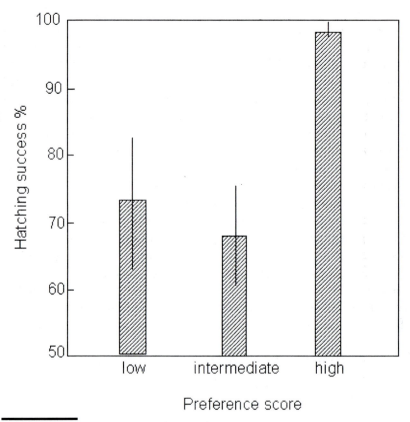

Fig. 12.3   Hatching success (%) (mean ± SE) of eggs spawned by female sand gobies *Pomatoschistus minutus* with different preference scores for the parental male. The graph is modified from Forsgren (1997).

Females of many fish species choose to spawn in nests that already contain eggs, for example, in three-spined stickleback *Gasterosteus aculeatus* (Ridley and Rechten, 1981; but see also Jamieson and Colgan, 1989), river bullhead *Cottus gobio* (Marconato and Bisazza, 1986), fathead minnow, *Pimephales promelas* (Unger and Sargent, 1988), fantail darter, *Ethiostoma flabellare* (Knapp and Sargent, 1989), sphinx blenny *Aidablennius sphynx* (Kraak and Groothuis, 1994), and sand goby *Pomatoschistus minutus* (Forsgren *et al.*, 1996). This behaviour is believed to have evolved because it reduces the predation risk of own offspring. Several hypotheses have been put forward to explain the mechanisms behind the risk reduction. Predation of one's own progeny may be reduced with the presence of other potential prey in the nest, due to a dilution effect (Ridley and Rechten, 1981; Unger and Sargent, 1988). Furthermore, the guarding activity of the male should increase with an increase in brood size (Coleman et al., 1985; Sargent, 1988 and below), and the risk of the male consuming the entire clutch may be lower in a large brood (Forsgren et al., 1996; Svensson et al., 1998 and below). Also, for the nest guarding male, there might be a benefit if the brood was diluted with unrelated eggs, as a way of decreasing the predation of his own progeny (Wisenden, 1999). Accordingly, egg adoption is found in several fish species with paternal care, for example, the fathead minnow *Pimephales promelas* (Unger and Sargent, 1988) and three-spined stickleback *Gasterosteus aculeatus* (Ridley and Rechten, 1981). This phenomenon may also be explained by female preference for nests with eggs. However, the adaptive value of egg adoption is still unclear in some species (Jamieson and Colgan, 1992; Östlund-Nilsson, 2002). In the fifteen-spined stickleback *Spinachia spinachia*, where females do not prefer to mate in nests with eggs, adopted eggs have instead been suggested to be used as a food reserve for the nest guarding male (Östlund-Nilsson, 2002). Further benefits of alloparental care in fish have been discussed in Wisenden (1999).

FACTORS AFFECTING THE INTENSITY OF DEFENCE

In the same way as parents adjust their own risk-taking, the caring fish should also adjust their efforts to defend the offspring according to the reproductive value of the brood, and, furthermore, to its vulnerability. Females of the mouth-brooding cichlid *Oreochromis mossambicus* increased aggression directed towards other females over the brood cycle (Oliveira and Almada, 1998). Agonistic interactions peaked when the fry started to feed outside the mother's mouth and became vulnerable to predation by non-incubating females. The aim of the higher aggression was probably to defend a space where the fry could forage more safely (Oliveira and Almada, 1998). In the smallmouth bass *Micropterus dolomieui*, males increased the frequency of attacks towards a potential predator from egg to wriggler stages, but decreased defence with further development, even though conspicuousness would increase when fry started to be more active (Ridgway, 1988). This change in behaviour may reflect an interaction between the increased value of the brood with age and the increased difficulty to successfully defend a brood as the fry become more dispersed.

Male desertion of the mate and the offspring has been observed in the biparental convict cichlid *Cichlasoma nigrofasciatum* (Wisenden, 1994).

This was most frequently done when predation risk of the fry was low, and when they were close to independence. When the risk of brood predation is high, biparental care would be more important to secure the survival of the offspring, and desertion may therefore be inhibited, as suggested for *Cichlasoma panamense* (Townsend and Wootton, 1985).

The number of offspring will also reflect the value of the brood for the parent, and defence might be adjusted also to brood size. In the cichlid *Neolamprologus moorii* attack rate towards fry-eating fishes was in both parents positively correlated with brood size (Karino, 1997). Male sand gobies *Pomatoschistus minutus* lost their brood to an egg-eating crab more often if they had a small brood rather than a large one in an aquarium study (Lindström, 1998). Correspondingly, a field study of nest guarding common gobies *Pomatoschistus microps* showed that males attacking a nest disturber (the researcher's finger) had larger broods and older eggs in his nest than had males that did not attack (Magnhagen and Vestergaard, 1993; Fig. 12.4).

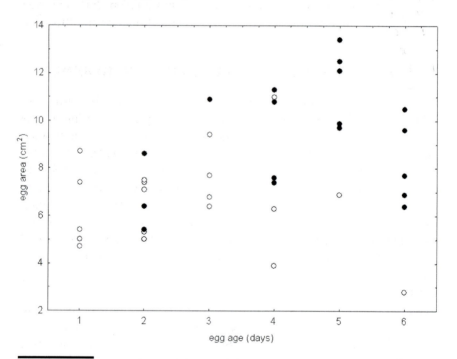

Fig. 12.4 Area covered with eggs versus egg age in nests of common goby *Pomatoschistus microps* males showing no aggression (o), or attacking (•) the researcher's finger after disturbance of nest. Data from Magnhagen and Vestergaard (1993).

FILIAL CANNIBALISM

Rowher (1978) suggested, in a theoretical article, that filial cannibalism (eating one's own progeny) could be expected to occur in fishes with paternal care. Since then, filial cannibalism have been found to be quite prevalent in fish (FitzGerald, 1992), and many recent studies have investigated this phenomenon (reviewed in Manica, 2002a). Filial cannibalism is suggested to increase lifetime reproductive success of fish performing parental care, and thus to be an adaptive behaviour (Manica, 2002a). This behaviour should be found more often in males than in females. Females invest more in the eggs before spawning and can, therefore, not obtain extra energy by eating them, while males actually gain energy by eating instead of caring for the offspring (FitzGerald, 1992). However, in the maternal mouthbrooding cichlid *Pseudocrenilabrus multicolor*, the mothers sometimes consume small clutches of their own young (Mrowka, 1987). In species with biparental care, the male is more likely to cannibalize the clutch than is the female (Schwanck, 1986; Lavery and Keenleyside, 1990). The examples on filial cannibalism mentioned below mainly deal with male parental fish.

Body condition commonly decreases during nest guarding, and if the availability of other food items is low, eating part of the brood may be a strategy to survive until the hatching of the eggs, as suggested for the river bullhead, *Cottus gobio* (Marconato *et al.*, 1993). Male river blennies *Salaria fluviatilis* increased their body condition during the breeding season, even in the face of decreased prey density (Vinyoles *et al.*, 1999), and filial cannibalism by the parental males seemed to explain this pattern. Nest-guarding common goby *Pomatoschistus microps* males in aquarium experiments ate more of their eggs when given a low compared to a high food ration (Kvarnemo *et al.*, 1998). Similar results were found while feeding nest guarding sphinx blenny *Aidablennius sphynx*, but only males with small broods decreased the number of eggs eaten when fed (Kraak, 1996). Males with larger broods were believed to eat only those eggs that were already dead, since numbers of eggs cannibalized in large broods in the field corresponded to egg mortality rate (due to other causes than predation) (Kraak, 1996). Contrary to expectations, in the fantail darter *Etheostoma flabellare*, fed parental males did not eat less of their brood than did unfed ones, and initial body condition did not correlate with number of eggs eaten (Lindström and Sargent, 1997). However, parental bluegill sunfish *Lepomis machrochirus* were less likely to partially cannibalize their brood if they were in good condition (Neff, 2003). Also, the parental energy reserve was an important factor affecting filial cannibalism in the goby *Rhinogobius* sp. (Okuda *et al.*, 2004). Most of these studies thus indicate that energy requirements, to a high extent, can explain the occurrence of filial cannibalism. By eating some of the eggs in a brood (partial brood cannibalism), the survival probability of the parent and hence of the rest of the brood can be increased.

Under some circumstances, the parental fish not only eats part of the brood but also consumes all of the eggs, i.e., performs total brood cannibalism. While partial cannibalism can improve both current and future reproductive output, total cannibalism can be seen as an investment only into future reproductive events (Manica, 2002a). By eating the entire brood, the male improves his body condition, which may lead to a higher reproductive success during the next breeding cycle. Furthermore, he should be able to remate earlier than if he had stayed with the current brood. Reproductive rate would, of course, also depend on the availability of mates, and it has been predicted that mate availability would influence filial cannibalism in paternal fish (Kondoh and Okuda, 2002). This was verified in the freshwater goby, *Rhinogobius* sp., where total filial cannibalism increased as the sex ratio became more female-biased (Okuda *et al.*, 2004). Also, in the cardinal fish *Apogon doederleini*, a connection between total filial cannibalism and density of potential mates was indicated (Okuda and Yanagisawa, 1996). In contrast, in an experiment with the two-spotted goby *Gobiusculus flavescens*, mate availability did not affect the incidence of total clutch cannibalism (Bjelvenmark and Forsgren, 2003).

Total filial cannibalism should be performed when the effort to care for the offspring is not matched by the expected gain. Accordingly, while large broods were only partially cannibalized in the fantail darter *Etheostoma flabellare*, small broods were totally devoured (Lindström and Sargent, 1997). Also, in other fish species, small broods are more often completely cannibalized as compared to larger broods, for example, in the sand goby *Pomatoschistus minutus* (Forsgren *et al.*, 1996; Lissåker *et al.*, 2003), the common goby *Pomatoschistus microps* (Kvarnemo *et al.*, 1998), and the scissors-tail sergeant *Abudefduf sexfasciatus* (Manica, 2002b; Fig. 12.5).

IS CARING ALWAYS WORTH IT?

For parental care to exist, lifetime fitness for caring parents should be higher than for non-caring parents. To cannibalize on your own offspring

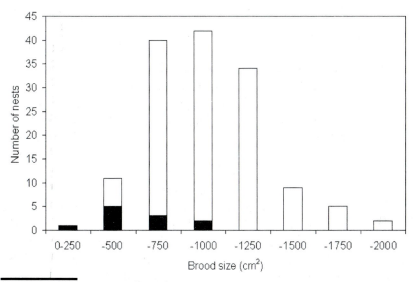

Fig. 12.5 Size distribution (area covered with eggs) of scissors-tail sergeant *Abudefduf sexfasciatus* nests at the beginning of the parental phase, on a reef in the Seychelles. Black bars indicate cases of total filial cannibalism. Data from Manica (2002b).

if the brood is too small to be worth the effort—as discussed above and in Manica (2002a)—is obviously profitable for future reproductive opportunities. Furthermore, the benefit of parental care to the care giver would depend on the relatedness between the care giver and the offspring (Owens, 1993). To care for young that are not genetically related to yourself would seem maladaptive. However, this is what happens in species with alloparental care, i.e., fish that steal and adopt eggs from other fish's nests. For such a behaviour to occur, it should either have no extra costs involved, or have some benefits, such as the decrease of predation of your own progeny due to the dilution with other eggs (see above and Wisenden, 1999). Accordingly, in experimentally augmented broods of convict cichlids (*Cichlasoma nigrofasciatum*), parents that adopted smaller foreign fry had an increased survival of their own fry, probably by the combined effects of brood dilution and differential predation on small fry (Wisenden and Keenleyside, 1994). In Lake Malawi, two species of cichlids, *Ctenopharynx pictus* and *Copadichromis pleurostigmoides*, placed their young in the nest of a biparental catfish, *Bagrus meridionalis* (McKaye et al., 1992). Both the cichlid parents and the catfish defended the mixed school of young, but the catfish also forced the young cichlids out into the periphery of the school. This led to a lower predator attack rate on catfish fry, compared to schools of only catfish fry. The catfish parents thus benefited from the adoption of young of the other species by the dilution effect.

The fathead minnow *Pimephales promelas* showed differential treatment of egg clutches of different parentage in an aquarium study (Sargent, 1989). Males with both sired and adopted clutches cared less for the adopted clutch in terms of egg rubbing, and defended this clutch less against an egg predator compared to the sired clutch. The male also cannibalized more on the adopted clutch and, consequently, survival rate was lower than in the sired clutch.

The common occurrence of alternative mating strategies in fish, i.e., males fertilizing eggs in nests of other males (reviewed in Taborsky, 1998) also raise the same question. Why would nest-guarding males take risks and spend energy to increase the survival of non-related young? Only a few studies have found an effect of decreased paternity due to sneak fertilisations on parental care in fish. Parental male scissortail sergeant, *Abudefduf sexfasciatus* ate more of their brood at a perceived decrease of paternity compared to in the absence of potential sneakers, but did not change the time spent guarding the brood (Manica, 2004). In the bluegill sunfish *Lepomis macrochirus*, the willingness to defend eggs and fry against a predator decreased with the increasing proportion sired by a cuckolder (Neff and Gross, 2001). Also, males that deserted their broods (which were then cannibalized) had been more heavily cuckolded by sneakers than those that did not desert (Neff, 2003). Paternity tests of the broods revealed that number of pecks at the nest (correlated with number of offspring eaten) by the parental male was positively correlated with proportion of sneaker and satellite paternity in the brood (Neff, 2003). This was found at the fry phase of care, but not at the egg stage. Bluegill sunfish parental males were able to use olfactory cues to recognize their own progeny only at the fry stage (Neff and Sherman, 2003). This study indicated that the males are able to selectively prey on non-related fry. The ability to recognize your own offspring should be crucial for an increase in partial filial cannibalism with decreased paternity to be adaptive. However, scissortail sergeant males increased cannibalism when sneaking was only simulated, and thus recognition of own and foreign offspring, respectively, did not seem to occur in this species (Manica, 2004). Maybe these males instead used the presence of potential cuckolders as an indication of low paternity. The lack of recognition was suggested to have been a reason that neither the readiness to defend the brood against egg predators, nor filial cannibalism was connected with the presence or absence of sneakers in the common goby *Pomatoschistus microps* (Svensson et al., 1998). Also in the fifteen-spined stickleback *Spinachia spinachia*, there was a lack of correlation between nest defence and level of sneak fertilizations (Östlund-Nilsson, 2002). In nest-guarding fish, care is considered to be sharable (Clutton-Brock, 1991). Thus, adding strange offspring into the brood may not lead to extra costs for the parent (Taborsky, 1994). The selection pressure for adjusting parental care with paternity level may thus be small if there is either: (a) a low cost per brood to care for more offspring; (b) a lack of information on paternity or (c) paternity is constant or varies in an unpredictable way (further discussed by Owens, 1993 and Reynolds, 1996).

In the peacock wrasse *Symphodus tinca*, care is facultative; females either place their eggs in a nest cared for by a male or they disperse their eggs widely with no protection (Warner *et al.*, 1995). This variation is probably unique among fish and is suggested to depend on the change in benefit of care over the breeding season. The hatching success within nests is always higher than outside nests but the difference varies over the season. In the beginning and the end of the season, most females spawned outside nests, since the cost to search for and find an acceptable nest was high (Fig. 12.6). In mid-season, a high predation rate on eggs not cared for lead to a five times higher hatching-success in nests compared

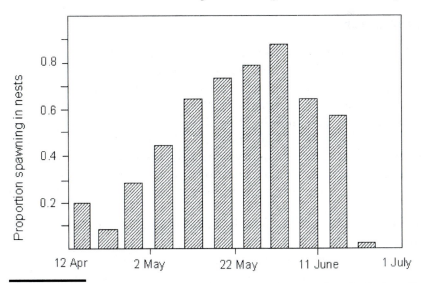

Fig. 12.6 Proportion of observed mating female peacock wrasse *Symphodus tinca*, that spawned in nests at different times during a breeding season. The graph is modified from Warner *et al.* (1995).

to eggs placed outside of nests. This survival difference in combination with a higher density of nests resulted in a high proportion of females that spawned in nests during mid-season (Warner *et al.*, 1995; Fig. 12.6). This is an excellent example of the assumption that the benefit of parental care must be worth the effort.

If survival of the offspring is equally high without care as with care, the benefit of caring obviously disappears. In Canada, a new form of stickleback within the *Gasterosteus* complex, 'the white stickleback', was recently discovered (Blouw and Hagen, 1991). In this stickleback, males build nests and court females, but when a clutch is fertilized the eggs are removed from the nest and dispersed among filamentous algae (Jamieson *et al.*, 1992) or on bare rock (MacDonald *et al.*, 1995). The survival of the unguarded embryos was high in laboratory and field experiments (MacDonald *et al.*, 1995; Blouw, 1996), but mortality due to predation was not accounted for here. However, the filamentous algae were suggested to be able to deter potential predators, due to an increase of the crypticity of the developing embryos and the physical refuge in the algae that are difficult to penetrate (Blouw, 1996). The reversed evolution from paternal care to desertion thus seems to be adaptive, since parental care may be a constraint on sexual activity and mating success (Jamieson *et al.*, 1992).

Conclusions

In fish, as in many other taxa, there is often a trade-off between parental care behaviour and other behaviours important for lifetime fitness of the parent, such as foraging and hiding from predators (Clutton-Brock, 1991). I have here shown that fish can adjust their parental behaviour in response to threats in the environment, directed to themselves or to their offspring. The risk of being eaten while guarding a brood could be decreased by the right choice of habitat and by decreasing certain activities that would be conspicuous to a potential predator. If guarding prevents the parent from foraging, eating part of the brood can be a way to avoid starvation and thereby increase survival probabilities for both the parent and the rest of the brood. Risk-taking by the parent is adjusted to the reproductive value of the current brood, indicated by brood size, age, and probability of future reproduction. Defence of the offspring against predators is also adjusted to its reproductive value. If the brood is small or if the perceived paternity is low due to the presence of sneakers, a nest-guarding male can benefit by deserting the brood or even devour it, so as to be able to remate earlier than would otherwise have been the case. When the brood is big enough to be worth defending, the risk of predation on own progeny will be diluted if egg clutches of different parentage is added to the defended nest site. This is also beneficial to females who often choose to spawn in nests already containing eggs. Furthermore, females have been found to prefer to spawn with paternal males that are more successful at defending the offspring than are other males. For an evolution of parental care to occur, the benefits of caring should outweigh the costs. However, if survival of offspring not cared for approach the survival rates of defended broods, evolution can be reversed, from parental care to brood desertion.

Acknowledgements

I thank Elisabet Forsgren, Andrea Manica and Bob Warner for permission to use their data for graphs in this chapter. Ulrica Candolin provided information on yet unpublished results. I am also grateful for useful comments on the manuscript provided by Elisabet Forsgren and Lotta Kvarnemo.

References

Almada, V.C., E.J. Conçalves, R.F. Oliveira and E.N. Barata. 1992. Some features of the territories in the breeding males of the intertidal benny *Lipophrys pholis* (Pisces: Blenniidae). *Journal of the Marine Biological Association of the United Kingdom, Plymouth* 72: 187-197.

Amundsen, T. and E. Forsgren. 2001. Male mate choice selects for female coloration in a fish. *Proceedings of the National Academy of Sciences of the United States of America* 98: 13155-13160.

Bain, M.B. and L.A. Helfrich. 1983. Role of male parental care in survival of larval bluegills. *Transactions of the American Fisheries Society* 112: 47-52.

Balon, E.K. 1975. Reproductive guilds of fishes: a proposal and definition. *Journal of Fisheries Research Board Canada* 32: 821-864.

Bisazza, A., A. Marconato and G. Marin. 1989. Male competition and female choice in *Padogobius martensi* (Pisces, Gobiidae). *Animal Behaviour* 38: 406-413.

Bjelvenmark, J. and E. Forsgren. 2003. Effects of mate attraction and male-male competition on paternal care in a goby. *Behaviour* 140: 55-69.

Blouw, D.M. 1996. Evolution of offspring desertion in a stickleback fish. *Ecoscience* 3: 18-24.

Blouw, D.M. and D.W. Hagen. 1991. Breeding ecology and evidence of reproductive isolation of a widespread stickleback fish (Gasterosteidae) in Nova Scotia, Canada. *Biology Journal of the Linnean Society* 39: 195-217.

Blumer, L.S. 1979. Male parental care in the bony fishes. *Quarterly Review of Biology* 54: 149-161.

Brown, J.A. 1984. Parental care and the ontogeny of predator-avoidance in two species of centrarchid fish. *Animal Behaviour* 32: 113-119.

Candolin, U. and H.R. Voigt. 2003. Do changes in risk-taking affect habitat shifts of sticklebacks? *Behavioral Ecology and Sociobiology* 55: 42-49.

Clark, C.W. and R.C. Ydenberg. 1990. The risks of parent-hood. I. General theory and applications. *Evolutionary Ecology* 4: 21-34.

Clutton-Brock, T.H. 1991. *The Evolution of Parental Care*. Princeton University Press, Princeton.

Coleman, R.M., M.R. Gross and R.C. Sargent. 1985. Parental investment decision rules: a test in bluegill sunfish. *Behavioral Ecology and Sociobiology* 18: 59-66.

FitzGerald, G.J. 1992. Filial cannibalism in fishes: why do parents eat their offspring? *Trends in Ecology and Evolution* 7: 7-10.

FitzGerald, G.J. and S. Lachance. 1993. Paternal investment in the blackspotted stickleback *Gasterosteus wheathlandi. Acta Oecologia—International Journal of Ecology* 14: 17-22.

Forsgren, E. 1997. Female sand gobies prefer good fathers over dominant males. *Proceedings of the Royal Society of London* B 264: 1283-1286.

Forsgren, E., A. Karlsson and C. Kvarnemo. 1996. Female sand gobies gain direct benefits by choosing males with eggs in their nests. *Behavioral Ecology and Sociobiology* 39: 91-96.

Forsgren, E., T. Amundsen, Å.A. Borg and J. Bjelvenmark. 2004. Unusually dynamic sex roles in a fish. *Nature (London)* 429: 551-554.

Fryer, G. and T.D. Iles. 1972. *The Cichlid Fishes of the Great Lakes of Africa.* Oliver and Boyd, Edinburgh.

Garcia, C.M., E. Saborío and C. Berea. 1998. Does male-biased predation lead to male scarcity in viviparous fish? *Journal of Fish Biology* 53 (Supplement A): 104-117.

Ghalambor, C.K., D.N. Reznick and J.A. Walker. 2004. Constraints on adaptive evolution: the functional trade-off between reproduction and fast-start swimming performance in the Trinidadian guppy (*Poecilia reticulata*). *American Naturalist* 164: 38-50.

Giles, N. 1984. Implications of parental care of offspring for the anti-predator behaviour of adult male and female three-spined sticklebacks, *Gasterosteus aculeatus* L. In: *Fish Reproduction: Strategies and Tactics,* G.W. Potts and R.J. Wootton (eds.). Academic Press, London. pp. 275-290.

Gross, M.R. and A.M. McMillan. 1981. Predation and the evolution of colonial nesting in bluegill sunfish (*Lepomis macrochirus*). *Behavioral Ecology and Sociobiology* 8: 163-174.

Hinch, S.G. and N.C. Collins. 1991. Importance of diurnal and nocturnal nest defense in the energy budget of male smallmouth bass: Insights from direct video observations. *Transactions of the American Fisheries Society* 120: 657-663.

Houde, A.E. and J.A. Endler. 1990. Correlated evolution of female mating preferences and male color patterns in the guppy *Poecilia reticulata. Science* 248: 1405-1408.

Huntingford, F.A. 1976. A comparison of the responses of sticklebacks in different reproductive conditions towards conspecifics and predators. *Animal Behaviour* 24: 694-697.

Jamieson, I.G. and P.W. Colgan. 1989. Eggs in the nests of males and their effect on mate choice in the three-spined stickleback. *Animal Behaviour* 38: 859-865.

Jamieson, I.G. and P.W. Colgan. 1992. Sneak spawning and egg stealing by male threespined sticklebacks. *Canadian Journal of Zoology* 70: 963-967.

Jamieson, I.G., D.M. Blouw and P.W. Colgan. 1992. Parental care as a constraint on male mating success in fishes—A comparative study of threespined and white sticklebacks. *Canadian Journal of Zoology* 70: 956-962.

Jan, R.Q. 1991. Malicious neighbors in leks of sergeant major damselfish, *Abudefduf vaigiensis*. *Bulletin of the Institute of Zoology Academia Sinica* 30: 49-53.

Jones, H.M. and C.A. Paszkowski. 1997. Effects of northern pike on patterns of nest use and reproductive behavior of male fathead minnows in a boreal lake. *Behavioral Ecology* 8: 655-662.

Karino, K. 1997. Influence of brood size and offspring size on parental investment in a biparental cichlid fish, *Neolamprologus moorii*. *Journal of Ethology* 15: 39-43.

Knapp, R.A. 1993. The influence of egg survivorship on the subsequent nest fidelity of female bicolor damselfish, *Stegastes partitus*. *Animal Behaviour* 46: 111-121.

Knapp, R.A. and R.C. Sargent. 1989. Egg-mimicry as a mating strategy in the fantail darter, *Etheostoma flabellare*—Females prefer males with eggs. *Behavioral Ecology and Sociobiology* 25: 321-326.

Kondoh, M. and N. Okuda. 2002. Mate availability influences filial cannibalism in fish with paternal care. *Animal Behaviour* 63: 227-233.

Kraak, S.B.M. 1996. Female preference and filial cannibalism in *Aidablennius sphynx* (Teleostei, Blenniidae): A combined field and laboratory study. *Behavioural Processes* 36: 85-97.

Kraak, S.B.M. and T.G.G. Groothuis. 1994. Female preference for nests with eggs is based on the presence of the eggs themselves. *Behaviour* 131: 189-206.

Kraak, S.B.M., T.C.M. Bakker and S. Hocevar. 2000. Stickleback males, especially large and red ones, are more likely to nest concealed in macrophytes. *Behaviour* 137: 907-919.

Kvarnemo, C., O. Svensson and E. Forsgren. 1998. Parental behaviour in relation to food availability in the common goby. *Animal Behaviour* 56: 1285-1290.

Kynard, B.E. 1978. Breeding behavior of a lacustrine population of threespine sticklebacks (*Gasterosteus aculeatus* L.). *Behaviour* 67: 178-207.

Lavery, R.J. and M.H.A. Keenleyside. 1990. Filial cannibalism in the biparental fish *Cichlasoma nigrofasciatum* (Pisces, Cichlidae) in response to early brood reductions. *Ethology* 86: 326-338.

Lindström, K. 1998. Effects of costs and benefits of brood care on filial cannibalism in the sand goby. *Behavioral Ecology and Sociobiology* 42: 101-106.

Lindström, K. and E. Ranta. 1992. Predation by birds affects population structure in breeding sand goby, *Pomatoschistus minutus*, males. *Oikos* 64: 527-532.

Lindström, K. and M. Hellström. 1993. Male size and parental care in the sand goby, *Pomatoschistus minutus*. *Ethology, Ecology and Evolution* 5: 97-106.

Lindström, K. and R.C. Sargent. 1997. Food access, brood size and filial cannibalism in the fantail darter, *Etheostoma flabellare*. *Behavioral Ecology and Sociobiology* 40: 107-110.

Lindström, K. and C. Wennström. 1994. Expected future reproductive success and paternal behavior in the sand goby, *Pomatoschistus minutus* (Pisces, Gobiidae). *Journal of Fish Biology* 44: 469-477.

Lissåker, M., C. Kvarnemo and O. Svensson. 2003. Effects of a low oxygen environment on parental effort and filial cannibalism in the male sand goby, *Pomatoschistus minutus*. *Behavioral Ecology* 14: 374-381.

MacDonald, J.F., S.M. MacIsaac, H. Bekkers and D.M. Blouw. 1995. Experiments on embryo survivorship, habitat selection and competitive ability of a stickleback fish (*Gasterosteus*) which nests in the rocky intertidal zone. *Behaviour* 132: 1207-1221.

Magnhagen, C. 1991. Predation risk as a cost of reproduction. *Trends in Ecology and Evolution* 6: 183-186.

Magnhagen, C. 1992. Parental care and predation risk in fish. *Annales Zoologici Fennici* 29: 227-232.

Magnhagen, C. and K. Vestergaard. 1991. Risk taking in relation to reproductive investments and future reproductive opportunities; field experiments on nest guarding common gobies, *Pomatoschistus microps*. *Behavioral Ecology* 2: 351-359.

Magnhagen, C. and K. Vestergaard. 1993. Brood size and offspring age affect risk-taking and aggression in nest-guarding common gobies. *Behaviour* 125: 233-243.

Manica, A. 2002a. Filial cannibalism in teleost fish. *Biological Reviews* 77: 261-277.

Manica, A. 2002b. Alternative strategies for a father with a small brood: mate, cannibalise or care. *Behavioral Ecology and Sociobiology* 51: 319-323.

Manica, A. 2004. Parental fish change their cannibalistic behaviour in response to the cost-to-benefit ratio of parental care. *Animal Behaviour* 67: 1015-1021.

Marconato, A. and A. Bisazza. 1986. Males whose nests contain eggs are preferred by female *Cottus gobio* L. (Pisces, Cottidae). *Animal Behaviour* 34: 1580-1582.

Marconato, A., A. Bisazza and M. Fabris. 1993. The cost of parental care and egg cannibalism in the river bullhead, *Cottus gobio* L. (Pisces, Cottidae). *Behavioral Ecology and Sociobiology* 32: 229-237.

McKaye, K.R., D.E. Mughogho and T.J. Lovullo. 1992. Formation of the selfish school. *Environmental Biology of Fishes* 35: 213-218.

Moodie, G.E.E. 1972. Predation, natural selection and adaptation in an unusual threespined stickleback. *Heredity* 28: 155-167.

Mrowka, W. 1987. Filial cannibalism and reproductive success in the maternal mouthbrooding cichlid fish *Pseudocrenilabrus multicolor*. *Behavioral Ecology and Sociobiology* 21: 257-265.

Neff, B.D. 2003. Paternity and condition affect cannibalistic behavior in nest-tending bluegill sunfish. *Behavioral Ecology and Sociobiology* 54: 377-384.

Neff, B.D. and M.R. Gross. 2001. Dynamic adjustment of parental care in response to perceived paternity. *Proceedings of the Royal Society of London* B 268: 1559-1565.

Neff, B.D. and P.W. Sherman. 2003. Nestling recognition via direct cues by parental male bluegill sunfish (*Lepomis macrochirus*). *Animal Cognition* 6: 87-92.

Okuda, N. and Y. Yanagisawa. 1996. Filial cannibalism in a paternal mouthbrooding fish in relation to mate availability. *Animal Behaviour* 52: 307-314.

Okuda, N., S. Ito and H. Iwao. 2004. Mate availability and somatic condition affect filial cannibalism in a paternal brooding goby. *Behaviour* 141: 279-296.

Oliveira, R.F. and V.C. Almada. 1998. Maternal aggression during the mouthbrooding cycle in the cichlid fish, *Oreochromis mossambicus*. *Aggressive Behavior* 24: 187-196.

Östlund-Nilsson, S. 2000. Are nest characters of importance when choosing a male in the fifteen-spined stickleback (*Spinachia spinachia*)? *Behavioral Ecology and Sociobiology* 48: 229-235.

Östlund-Nilsson, S. 2002. Does paternity or paternal investment determine the level of paternal care and does female choice explain egg stealing in the fifteen-spined stickleback? *Behavioral Ecology* 13: 188-192.

Östlund-Nilsson, S. and M. Holmlund. 2003. The artistic three-spined stickleback (*Gasterosteous aculeatus*). *Behavioral Ecology and Sociobiology* 53: 214-220.

Owens, I.P.F. 1993. When kids just aren't worth it: Cuckoldry and parental care. *Trends in Ecology and Evolution* 8: 269-271.

Perrone, M. and T.M. Zaret. 1979. Parental care patterns of fishes. *American Naturalist* 113: 351-361.

Pitcher, T.J. and J.K. Parrish. 1993. Functions of shoaling behaviour in teleosts. In: *Behaviour of Teleost Fishes*, T.J. Pitcher (ed.). Chapman and Hall, London, pp. 363-439.

Plaut, I. 2002. Does pregnancy affect swimming performance of female mosquitofish, *Gambusia affinis? Functional Ecology* 16: 290-295.

Potts, G.W. 1984. Parental behaviour in temperate marine teleosts with special reference to the development of nest structures. In: *Fish Reproduction: Strategies and Tactics*, G.W. Potts and R.J. Wootton (eds.). Academic Press, London, pp. 223-244.

Pressley, P.H. 1981. Parental effort and the evolution of nest-guarding tactics in the threespine stickleback, *Gasterosteus aculeatus*. *Evolution* 35: 282-295.

Reynolds, J.D. 1996. Animal breeding systems. *Trends in Ecology and Evolution* 11: 68-72.

Reznick, D.N. and J.A. Endler. 1982. The impact of predation on life-history evolution in Trinidadian guppies (*Poecilia reticulata*). *Evolution* 36: 160-177.

Reznick, D.N., H. Bryga and J.A. Endler. 1990. Experimentally induced life history evolution in a natural population. *Nature* (*London*) 346: 357-359.

Reznick, D.N., M.J. Butler, F.H. Rodd and P. Ross. 1996. Life history evolution in guppies (*Poecilia reticulata*). 6. Differential mortality as a mechanism for natural selection. *Evolution* 50: 1651-1660.

Ridgway, M. 1988. Developmental stage of offspring and brood defense in smallmouth bass (*Micropterus dolomieui*). *Canadian Journal of Zoology* 66: 1722-1728.

Ridley, M. and C. Rechten. 1981. Female sticklebacks prefer to spawn with males whose nests contain eggs. *Behaviour* 93: 82-100.

Rodewald, A.D. and S.A. Foster. 1998. Effects of gravidity on habitat use and antipredator behaviour in three-spined sticklebacks. *Journal of Fish Biology* 52: 973-984.

Rowher, S. 1978. Parent cannibalism of offspring and egg raiding as a courtship strategy. *American Naturalist* 112: 429-440.

Sargent, R.C. 1988. Paternal care and egg survival both increase with clutch size in the fathead minnow, *Pimephales promelas*. *Behavioral Ecology and Sociobiology* 23: 33-37.

Sargent, R.C. 1989. Allopaternal care in the fathead minnow, *Pimephales promelas*—stepfathers discriminate against their adopted eggs. *Behavioral Ecology and Sociobiology* 25: 379-385.

Sargent, R.C. and J.B. Gebler. 1980. Effects of nest site concealment on hatching success, reproductive success, and paternal behaviour of the threespine stickleback, *Gasterosteus aculeatus*. *Behavioral Ecology and Sociobiology* 7: 137-142.

Sargent, R.C. and M.R. Gross. 1993. William's principle: an explanation of parental care in teleost fishes. In: *Behaviour of Teleost Fishes*, T.J. Pitcher (ed.). Chapman and Hall, London, pp. 333-361.

Sargent, R.C., P.D. Taylor and M.R. Gross. 1987. Parental care and the evolution of egg size in fishes. *American Naturalist* 129: 32-46.

Schwanck, E. 1986. Filial cannibalism in *Tilapia mariae*. *Journal of Applied Ichthyology* 2: 65-74.

Skolbekken, R. and A.C. Utne-Palm. 2001. Parental investment of male two-spotted goby, *Gobiusculus flavescens* (Fabricius). *Journal of Experimental Marine Biology and Ecology* 261: 137-157.

Smith, C. and R.J. Wootton. 1995a. The costs of parental care in teleost fishes. *Reviews in Fish Biology and Fisheries* 5: 7-22.

Smith, C. and R.J. Wootton. 1995b. Experimental analysis of some factors affecting parental expenditure and investment in *Cichlasoma nigrofasciatum* (Cichlidae). *Environmental Biology of Fishes* 42: 289-302.

Sogard, S.M. 1997. Size-selective mortality in the juvenile stage of teleost fishes: A review. *Bulletin of Marine Science* 60: 1129-1157.

Svensson, I. 1988. Reproductive costs in two sex-role reversed pipefish species (Syngnathidae). *Journal of Animal Ecology* 57: 929-942.

Svensson, O., C. Magnhagen, E. Forsgren and C. Kvarnemo. 1998. Parental behaviour in relation to confidence of paternity in the common goby. *Animal Behaviour* 56: 175-179.

Taborsky, M. 1994. Sneakers, satellites and helpers: parasitic and cooperative behavior in fish reproduction. *Advances in the Study of Behavior* 23: 1-100.

Taborsky, M. 1998. Sperm competition in fish: 'Bourgeois' males and parasitic spawning. *Trends in Ecology and Evolution* 13: 222-227.

Townsend, T.J. and R.J. Wootton. 1985. Variation in the mating system of a biparenatal cichlid fish, *Cichlasoma panamense*. *Behaviour* 95: 181-197.

Trivers, R.L. 1972. Parental investment and sexual selection. In: *Sexual Selection and the Descent of Man*, B. Campbell (ed.). Aldine, Chicago. pp. 13-179.

Ukegbu, A.A. and F.A. Huntingford. 1988. Brood value and life expectancy as determinants of parental investment in male three-spined sticklebacks, *Gasterosteus aculeatus*. *Ethology* 78: 72-82.

Unger, L.M. and R.C. Sargent. 1988. Allopaternal care in the fathead minnow, *Pimephales promelas*—Females prefer males with eggs. *Behavioral Ecology and Sociobiology* 23: 27-32.

Warner, R.R., F. Wernerus, P. Lejeune and E. van den Berghe. 1995. Dynamics of female choice for parental care in a fish species where care is facultative. *Behavioral Ecology* 6: 73-81.

Wiegmann, D.D. and J.R. Baylis. 1995. Male body size and paternal behaviour in smallmouth bass, *Micropterus dolomieui* (Pisces: Centrarchidae). *Animal Behaviour* 50: 1543-1555.

Wiegmann, D.D., J.R. Baylis and M.H. Hoff. 1992. Sexual selection and fitness variation in a population of smallmouth bass, *Micropterus dolomieui* (Pisces, Centrarchidae). *Evolution* 46: 1740-1753.

Vinyoles, D., I.M. Cote and A. de Sostoa. 1999. Egg cannibalism in river blennies: The role of natural prey availability. *Journal of Fish Biology* 55: 1223-1232.

Wisenden, B.D. 1994. Factors affecting mate desertion by males in free-ranging convict cichlids (*Cichlasoma nigrofasciatum*). *Behavioral Ecology* 5: 439-447.

Wisenden, B.D. 1999. Alloparental care in fishes. *Reviews in Fish Biology and Fisheries* 9: 45-70.

Wisenden, B.D. and M.H.A. Keenleyside. 1994. The dilution effect and differential predation following brood adoption in free-ranging convict cichlids (*Cichlasoma nigrofasciatum*). *Ethology* 96: 203-212.

Wootton, R.J. 1976. *The Biology of the Sticklebacks*. Academic Press, London.

13

Involvement of Endocrine and Environmental Factors in Gonadal Sex Differentiation in Gonochoristic Fish

Takeshi Kitano[1],* and Shin-ichi Abe[2]

INTRODUCTION

In vertebrates, including fish, sex is ordinarily determined by the genotype under normal circumstances. However, various environmental factors such as temperature have also been shown to influence sex determination greatly in some non-mammalian species such as reptiles, amphibians and fishes (Adkins-Regan, 1987). Pioneering experiments by Yamamoto (1969) using medaka (*Oryzias latipes*) demonstrated that androgens and estrogens induce complete masculinization and feminization, respectively, leading to a hypothesis that steroids are endogenous inducers, namely estrogens and androgens are the female and male inducers, respectively. Since

Authors' addresses: [1]Department of Materials and Life Science, Graduate School of Science and Technology, Kumamoto University, 2-39-1 Kurokami, Kumamoto 860-8555, Japan.
[2]Department of Materials and Life Science, Graduate School of Science and Technology, Kumamoto University, Kumamoto 860-8555, Japan.
*Corresponding author: E-mail: tkitano@kumamoto-u.ac.jp

then, many studies have been performed in some gonochoristic fishes on the effect of steroids on gonadal sex differentiation.

Here, we shall mainly focus on five excellent fish models (medaka, Nile tilapia, rainbow trout, common carp, and Japanese flounder) to study the mechanism of sex determination and differentiation in gonochoristic fish. In the medaka, which has a XX (female)/XY (male) sex determination system (Aida, 1921), DMY (DM-related gene on the Y chromosome), also known as DMRT1Y (doublesex and mab-3-related transcription factor 1 on the Y chromosome), has been identified as a strong candidate for the sex-determining gene localized on the Y chromosome (Matsuda et al., 2002; Nanda et al., 2002). This species has several advantages, including small body size, short generation time, small genome size, and also several useful strains (Ishikawa, 2000). Therefore, medaka is an excellent vertebrate model in molecular genetics for analyzing various biological phenomena including embryonic development and sex determination. Nile tilapia (Oreochromis niloticus), rainbow trout (Oncorhynchus mykiss), common carp (Cyprinus carpio), and Japanese flounder (Paralichthys olivaceus) are also useful model fishes for studying sex differentiation, because the genetically all-male (XY) and all-female (XX) populations can be produced by fertilizing eggs of the genetic females (XX) with sperm from homogametic males (YY) and sex-reversed males (XX), respectively (Tabata, 1991; Gimeno et al., 1996; Guiguen et al., 1999). Moreover, in the Japanese flounder, whose sex is genetically determined by male heterogametic (XX/XY) system (Tabata, 1991), all phenotypic female and male populations can be produced by rearing the genetically-female (XX) broods at 18°C and 27°C, respectively (Kitano et al., 1999), suggesting that this species provides an excellent model to investigate temperature-dependent sex determination mechanism in fish.

In this chapter, we will review recent findings on the basic mechanisms of sex determination and differentiation, and then discuss about effects of sex steroid hormones, water temperature, and endocrine disrupters on gonadal sex differentiation in gonochoristic fish including five fish models.

SEX DETERMINATION IN GONOCHORISTIC FISH

Sex-determining gene Sry has been identified in mammals (Sinclair et al., 1990). However, no comparable genes have been found in any non-mammalian vertebrates. Recently, in the medaka (Oryzias latipes), a strong candidate for the sex-determining gene on the Y chromosome, DMY, was isolated using positional cloning and shotgun sequencing

(Matsuda *et al.*, 2002). The full-length cDNA sequence of DMY encodes a putative protein of 267 amino acids, including the highly conserved DM domain. The DM domain was originally described as a DNA-binding motif shared between *doublesex* (*dsx*) in *Drosophila melanogaster* and *mab-3* in *Canorhabditis elegans*. Interestingly, DMY appears to be derived from the duplication of DMRT1, which is another DM domain gene involved in male development in other vertebrates (Nanda *et al.*, 2002). DMY mRNA is expressed specifically in the somatic cells of the XY gonads around the time of sex differentiation, but not in the XX gonads (Matsuda *et al.*, 2002), and detected even in the ovary of the XY females sex-reversed by 17β-estradiol treatment (Nanda *et al.*, 2002). To elucidate the role of DMY in sexual differentiation, wild medaka populations with naturally occurring DMY mutants were screened and two XY females with distinct mutants in DMY were obtained from different populations (Matsuda *et al.*, 2002). The first heritable mutant (single insertion in exon 3 and the subsequent truncation of DMY) resulted in all XY female offspring. Similarly, the second XY mutant female had a reduced DMY expression with a high proportion of XY female offspring. These findings strongly suggest that the sex-specific DMY is required for testicular development and is also a prime candidate for the medaka sex-determining gene. DMY gene is also expressed in *Oryzias curvinotus* (Matsuda *et al.*, 2003). However, it is absent from the other fish species, *Oryzias celebensis*, *O. mekongensis*, guppy, tilapia, zebrafish, and fugu (Kondo *et al.*, 2003). Therefore, DMY appears to be a recently evolved gene specific to some species of the genus *Oryzias*.

GONADAL SEX DIFFERENTIATION IN GONOCHORISTIC FISH

Ovarian Differentiation

In Nile tilapia, ovarian differentiation is initially marked by the appearance of a narrow space in the stromal tissue representing the formation of the ovarian cavity (Nakamura and Nagahama, 1985). On the other hand, in medaka and rainbow trout, the criterion of the initial ovarian differentiation is the occurrence of meiotic germ cells for oogenesis subsequently to the increase of germ cells before the formation of the ovarian cavity (Yamamoto, 1958; Lebrun *et al.*, 1982). In tilapia, positive immunoreaction to the polyclonal antibodies against four steroidogenic enzymes, cytochrome P450 cholesterol side chain cleavage (P450scc),

3β-hydroxysteroid dehydrogenase (3β-HSD), cytochrome P450 17α-hydroxylase/17,20 lyase (P450c17), and cytochrome P450 aromatase (P450arom), the last of which catalyzes the conversion of androgens to estrogens, are observed in the gonads during the ovarian differentiation (Nakamura *et al.*, 1998), whereas those are not detected in the testis during the sex differentiation. Thereafter weakly positive reactions to P450scc, 3β-HSD and P450c17 antibodies appear first in the testis at 30 days, while P450arom immunoreactivity has never been found. These findings strongly suggest that endogenous estrogens act as the natural inducers of ovarian differentiation in tilapia. Moreover, Guiguen *et al.* (1999) reported that treatment of broods with an aromatase inhibitor (1,4,6-androstatriene-3,17-dione) results in a high percentage of masculinization of all-female (XX) population in rainbow trout and tilapia. Studies on *P450arom* and *estrogen receptor* (ER) gene expression in gonads sampled before, during and after sex-differentiation of rainbow trout showed that in spite of a complete absence of differences in the ER gene expression between sexes, *P450arom* gene is expressed at least a few hundred times more in female gonads than in male ones, suggesting an important role of P450arom in the ovarian differentiation process of the rainbow trout (Guiguen *et al.*, 1999). In chinook salmon (*Oncorhynchus tshawytscha*), brief treatment with an aromatase inhibitor (fadrozole) during the sex differentiation causes genetic females to develop into normal phenotypic males (Piferrer *et al.*, 1994). These results indicate that endogenous estrogens are indispensable for the sex differentiation into females in some teleost fishes.

Testicular Differentiation

In many species of fish, behavior of the somatic cells in the gonad during the course of morphological sex differentiation is often sex-specific, although germ cells in the gonads destined to testes remain quiescent for a long time (Nakamura *et al.*, 1998). In tilapia, a slit-like space extending from the proximal region of the gonad into the central stromal tissue is observed, when the germ cells remain at the gonial stage (Nakamura and Nagahama, 1989). Subsequently, the space is recognized as the efferent duct during testicular development process and spermatogenesis begins at 50-70 days after hatching. Therefore, the criterion of the initial testicular differentiation is the formation of the efferent duct in tilapia. Judging from the fact that no positive immunostaining for P450scc, 3β-HSD, P450c17 and P450arom antibodies can be detected during the testicular differentiation (Nakamura *et al.*, 1998), neither endogenous androgens nor estrogens appear to be involved in the testicular differentiation in tilapia. However, it may be necessary to study steroidogenesis during testicular differentiation in the other fish species because only a few studies have been performed on the steroidogenic potentialities of the differentiating gonads during sex differentiation.

Recently, it was reported in tilapia that *DMRT1* mRNA is expressed specifically in the testis (Guan *et al.*, 2000). In rainbow trout, semi-quantitative RT-PCR and virtual Northern analysis showed that *DMRT1* mRNA is highly expressed in the differentiating testis when compared to the differentiating ovary (Marchand *et al.*, 2000). Therefore, DMRT1 may play important roles in testicular differentiation in teleost fishes.

EFFECT OF TEMPERATURE ON GONADAL SEX DIFFERENTIATION

Sex determination in some reptiles, amphibians, and fishes is known to be influenced markedly by environmental factors (Adkins-Regan, 1987). Japanese flounder, a teleost fish that has a XX (female) / XY (male) sex determination system (Tabata, 1991), exhibits temperature-dependent sex determination (TSD). The details of temperature sensitivity and gonadal differentiation have been documented in this species, making it useful for studies investigating the physiology and molecular biology of TSD (Yamamoto, 1995, 1999; Kitano *et al.*, 1999). The percentage of females in the normal (XX+XY) and genetically female (all XX) larvae reared at various water temperatures from days 30 to 100 after hatching (the critical period of gonadal sex differentiation in the flounder) is shown in Table 13.1. The percentages of females in normal and genetically female broods reared at 18°C are 53% and 100%, respectively. However, they decrease rapidly in both broods when reared at higher (25°C) or

Table 13.1   Effect of water temperature treatment during sex differentiation (30-100 days after hatching) in Japanese flounder (*Paralichthys olivaceus*).

Water temperature (°C)	Percentage of females in XX+XY group (%)	Percentage of females in all XX group (%)
15	28.1	78.9
18	53.0	100.0
25	4.5	21.8
27	—	0.0

lower (15°C) water temperature. Particularly, the genetically female broods reared at 27°C masculinize completely. Thus, the temperature conditions are determined for inducing genetically female broods to all phenotypic females or males, namely rearing at 18°C or 27°C, respectively, from 30 to 100 days after hatching (Kitano *et al.*, 1999). A similar thermal influence on fishes has previously been demonstrated by studies on the atherinids (Conover and Kynard, 1981; Strüssmann *et al.*, 1996, 1997), cichlids (Baroiller *et al.*, 1995; Romer and Beisensherz, 1996; Desprez and Mélard, 1998), and channel catfish *Ictalurus punctatus* (Patiño *et al.*, 1996). In atherinids, low temperature treatment during the period of sex determination results in a high percentage of females, and high temperature brings about a high percentage of males. In a cichlid *Oreochromis niloticus*, the percentage of males in all-genetically female progenies is increased by rearing at high temperatures (32-37°C) during the thermosensitive period (Baroiller *et al.*, 1995). On the other hand, the sex ratio in channel catfish is skewed toward females at 34°C but no effects at 20°C or 27°C (Patiño *et al.*, 1996). These results suggest that both genotypic and temperature-dependent sex determination (TSD) mechanisms are functional in some fishes and that this phenomenon in fishes is more widespread than previously believed.

According to current evolutionary theories, this mechanism is adaptive to environments with moderate to large fluctuations in environmental conditions (Bulmer, 1987). It is suggested that environmental sex determination is favorable when environmental effects have different consequences for the two sexes. Conover and Heins (1987) showed that the natural populations at different latitudes of *Menidia menidia*, whose sex is determined by the interaction of genotype and temperature during the specific period of larval development (Conover and Kynard, 1981), compensate for differences in thermal environment and seasonality by adjusting the response of sex ratio to temperature. Therefore, TSD must be one of the strategies for the constant maintenance of the natural populations in some species.

In Japanese flounder, mRNA expression patterns of *P450arom* and *Müllerian inhibiting substance* (MIS) (also known as anti-Müllerian hormone), the latter of which is responsible for regression of Müllerian ducts in mammals, were analyzed during gonadal sex differentiation (Kitano *et al.*, 1999; Yoshinaga *et al.*, 2004). Phenotypic females and males were produced by rearing the genetically female larvae (XX) at normal (18°C) and higher (27°C) water temperatures, respectively, during the sex differentiation. On day 50 after hatching when the gonad is sexually indifferent (Yamamoto, 1995; Kitano *et al.*, 1999), *P450arom* and *MIS* mRNAs are detected in both the female and male gonads. However, after initiation of the sex differentiation (day 60), *P450arom* mRNA is expressed strongly in female gonads reared at 18°C, whereas it is hardly detected in male gonads reared at 27°C. In contrast, *MIS* mRNA is expressed strongly in male gonads, but not in female gonads. These results suggest that masculinization of the genetically female flounder by 27°C treatment is due to the suppression of *P450arom* mRNA expression and induction of *MIS* mRNA expression.

In order to address the question whether inhibition of the P450arom induces masculinization of the genetically female flounder, we investigated the effect of P450arom inhibitor (fadrozole) during the gonadal sex differentiation. When the genetic female larvae are provided with control diet at 18°C, all of them develop into normal females. In contrast, treatment with fadrozole dose-dependently induces masculinization of the genetically female flounder (Kitano *et al.*, 2000). The effect of fadrozole is counteracted by co-administration of 17β-estradiol. Therefore, inhibition of P450arom and the resultant decrease in the amount of estrogen may trigger masculinization of the genetic females by high water temperature treatment in Japanese flounder.

EFFECTS OF SEX STEROID HORMONES ON GONADAL SEX DIFFERENTIATION

Effects of Estrogens

Exogenous estrogens are known to induce feminization of the genetic males in many species (reviewed in Hunter and Donaldson, 1987). In medaka, it is reported that estrogens such as estrone (E_1), 17β-estradiol (E_2), and ethynylestradiol (EE_2) induce complete sex-reversal of the genetic males (Yamamoto, 1969). In the common carp (*Cyprinus carpio*), exposure of E_2 to the all-male population for 90 days results in regular phenotypic female gonads containing oviducts and oocytes at various stages of development, but no testicular tissues (Gimeno *et al.*, 1996). In all-male populations of rainbow trout obtained by fertilizing normal eggs (XX) with sperm of phenotypic males (YY), *DMRT1* mRNA expression in gonads is decreased by a short term E_2 treatment during the gonadal sex differentiation, suggesting that feminizing treatment with E_2 down-regulates the expression of *DMRT1* in rainbow trout (Marchand *et al.*, 2000). In tilapia, EE_2 causes complete feminization of the genetic males as well as up-regulation of *P450arom* mRNA expression and down-regulation of *DMRT1* mRNA expression (Kobayashi *et al.*, 2003). Therefore, estrogens are likely to induce sex-reversal of the genetic males by regulating the expression of sex-specific genes, such as *P450arom* and *DMRT1*.

We investigated the effects of E_2 on TSD in Japanese flounder. When the genetic female larvae are provided with control diet and reared at a masculinizing temperature (27°C), all of them develop into normal males with gonads differentiating into testes. In contrast, treatment with E_2 (1 ug/g diet) induces complete feminization of the larvae reared at a masculinizing temperature (Kitano *et al.*, 2007). Moreover, anti-estrogen (tamoxifen) treatment leads to masculinization of the genetic females, suppresses *P450arom* mRNA expression and induces MIS mRNA expression (Kitano *et al.*, 2007). These results strongly suggest that estrogens counteract masculinization of the genetically female flounder at a high water temperature by regulating the expression of *P450arom* and MIS through ERs.

Effects of Androgens

Exogenous androgens are known to induce masculinization of the genetic females in many species (reviewed in Hunter and Donaldson, 1987). In medaka, it is reported that androgens such as testosterone (T), 17α-methyltestosterone (MT), 17α-ethynyltestosterone and androstenedione induce complete sex-reversal of the genetic females (Yamamoto, 1969). In a genotypic all-female population of chinook salmon, the relative potency of several androgens to induce the phenotype male was compared in two separate experiments (Piferrer *et al.*, 1993). Aromatizable androgen T, nonaromatizable androgen 11-ketotestosterone (11KT), synthetic aromatizable androgen MT, and nonaromatizable androgen 17α-methyldihydrotestosterone (MDHT), were administered to newly hatched alevins. Essentially 100% masculinization occurs in a group exposed to MDHT. Treatment with MT results in a dose-dependent masculinization, but 11KT and T are less potent than the synthetic androgens. Though the number of males produced after treatment with 11KT follows a dose-dependent pattern, T shows virtually no masculinizing effect in inducing male phenotype in these studies. These results indicate that MDHT, MT and 11KT have sex-reversal action during sex differentiation in chinook salmon. In tilapia, MT treatment causes masculinization of the genetic females and increases *DMRT1* mRNA expression in gonads (Guan *et al.*, 2000). In Japanese flounder, MT and 11KT induce complete masculinization of the genetic females as well as up-regulation of *MIS* mRNA expression and down-regulation of *P450arom* mRNA expression (Kitano *et al.*, 2000; Yoshinaga *et al.*, 2004). Hence, there may be some differences in effective androgens among the species, and the androgenic action appears to regulate the expression of sex-specific genes, such as *DMRT1*, *MIS* and *P450arom*.

EFFECTS OF ENDOCRINE DISRUPTERS ON GONADAL SEX DIFFERENTIATION

Endocrine disrupters damage wildlife by disrupting the endocrine function. Effects of estrogenic compounds, which are one class of these compounds, are mediated mainly through two ER subtypes, ERα and ERβ, which are members of nuclear receptor family, and the estrogenic effects are produced *in vivo*. In the common carp, the effect of 4-tert-pentylphenol (TPP) exposed to the genetic males during sex differentiation was investigated (Gimeno *et al.*, 1996). The range of TPP concentrations tested corresponds to the range of concentrations of some alkylphenolic compounds found in sewage-treatment works. TPP treatment dose-dependently induces in the genetic males' development of oviducts, which are female phenotypic characteristics in the gonads. Moreover, exposure to the highest TPP concentration forms ovo-testis which possesses several oocytes within the testicular structure and in which spermatogenesis is severely inhibited. These results demonstrate that TPP has a potent estrogenic effect during sex differentiation in the carp. In medaka, exposure to 4-nonylphenol (4NP) skewed the sex ratio toward females when estimated by the appearance of their secondary sex characteristics (Yokota *et al.*, 2001). Histological observation shows that some percentage of the gonads have ovo-testes. Therefore, 4NP is likely to be estrogenic compounds which have the potential to alter sexual development in medaka. Recently, we investigated effects of *p*-nonylphenol (NP), bisphenol A (BPA), genistein, and *p,p'*-DDE on gonadal sex differentiation in Japanese flounder. We had already confirmed that their compounds have the estrogenic activity in the flounder estrogen receptor-mediated reporter gene assay (Kitano *et al.*, 2006). When the genetic female larvae were provided with control diet and reared at a masculinizing temperature (27°C), most of them developed into normal males with gonads differentiating into testes. In contrast, treatment with NP, BPA, genistein, or *p,p'*-DDE induces feminization of the larvae reared at a masculinizing temperature (Table 13.2). No ovo-testis is observed in any samples examined. The highest percentage (96.7%) of feminization is observed when the larvae are treated with 100 ug/g genistein. These results indicate that NP, BPA, genistein, and p,p'-DDE have estrogenic action on gonadal sex differentiation in Japanese flounder. Moreover, RT-PCR analysis shows that *P450arom* mRNA expression is induced in the gonads of some NP, BPA, p,p'-DDE, or genistein-treated fishes, while *MIS* mRNA expression is suppressed in the gonads (data not shown). Therefore, these compounds seem to induce feminization of the genetically female flounder reared at a masculinizing temperature as well as upregulation of *P450arom* mRNA expression and down-regulation of *MIS* mRNA expression. Thus, it is suggested that TSD in the Japanese flounder provides an excellent model to investigate the effect of endocrine disrupters in fishes.

We also examined the effect of tributyltin oxide (TBTO) on the sex-differentiation process in Japanese flounder (Shimasaki *et al.*, 2002). TBTO induces sex reversal of genetically female flounder into phenotypic males. The proportion of sex-reversed males treated with TBTO significantly increases to 25.7% in the 0.1 ug/g diet treatment group and to 31.1% in the 1.0 ug/g diet treatment group, when compared to the control (2.2%). Judging from morphological and histological examination of the gonads in the TBTO-treated fishes, sex-reversed males have typical testes. Moreover, *P450arom* mRNA expression is suppressed in the gonads of TBTO-treated fishes. Therefore, TBTO appears to induce sex-reversal of the genetic females by inhibiting *P450arom* mRNA expression in the gonads. For a detailed overview of endocrine disruption, see Chapters 15 and 16.

Table 13.2 Effects of estrogenic compounds during sex differentiation at a masculinizing temperature (27°C) in Japanese flounder (*Paralichthys olivaceus*).

Chemicals (ug/g diet)	n	Percentage of females in all XX group (%)
Control	40	0.0
p-Nonylphenol (100)	20	30.0
Bisphenol A (100)	21	57.0
Control	30	16.7
Genistein (10)	30	46.7
Genistein (100)	30	96.7
Control	30	16.7
p,p'-DDE (10)	40	75.0
p,p'-DDE (100)	42	83.3

CONCLUSIONS

The results by various investigators shown in this chapter demonstrate that endogenous estrogens are indispensable for the sex differentiation into females in some gonochoristic fishes, such as tilapia, rainbow trout, and Japanese flounder. However, the physiological roles of endogenous androgens during gonadal sex differentiation remain unclear. On the other hand, since masculinization of the genetic females by high water temperature treatment in Japanese flounder is shown to be due to inhibition of *P450arom* mRNA expression and the resultant decrease in the amount of estrogen, we strongly suggest that endogenous estrogens play key roles in the regulation of gonadal sex differentiation in gonochoristic fish including fishes exhibiting TSD. Future studies should be focused on molecular mechanism of testicular differentiation including roles of androgens, DMRT1, and MIS during sex differentiation using some fish models.

SUMMARY

In vertebrates, including fish, sex is ordinarily determined by the genotype under normal circumstances. In the medaka (*Oryzias latipes*), a gonochoristic fish that has a XX (female)/XY (male) sex determination system, *DMY* (DM-related gene on the Y chromosome), also known as *DMRT1Y* (DM-related transcription factor 1 on the Y chromosome), has been identified as a strong candidate for the sex-determining gene localized on the Y chromosome. However, the sex-determining gene in the other non-mammalian vertebrates remains unclarified. On the other hand, sex determination in some reptiles, amphibians and fishes is known to be greatly influenced by environmental factors. For example, although the genetic sex determination system in the Japanese flounder (*Paralichthys olivaceus*) is basically XX (female)/XY (male) type, the genetic females can be sex-reversed to phenotypic males by rearing the larvae at high or low water temperature during the gonadal sex differentiation. In addition, the phenotypic sex of many teleost fishes including the flounder can be experimentally altered by treatment with sex steroid hormones, suggesting an important role of sex steroid hormones in gonadal sex differentiation in fish.

In this chapter, we shall review recent findings on the basic mechanisms of sex determination and gonadal sex differentiation, and then discuss the effects of sex steroid hormones, water temperature, and endocrine disrupters on gonadal sex differentiation in gonochoristic fish.

References

Aida, T. 1921. On the inheritance of colour in a freshwater fish, *Aplocheilus latipes* Temminck and Schlegel, with special reference to sex-linked inheritance. *Genetics* 6: 554–573.

Adkins-Regan, E. 1987. Hormones and sexual differentiation. In: *Hormones and Reproduction in Fishes, Amphibians and Reptiles*, D.O. Norris and R.E. Jones (eds.). Plenum Press, New York, pp. 1-29.

Baroiller, J.F., D. Chourrout, A. Fostier and B. Jalabert. 1995. Temperature and sex chromosomes govern sex ratios of mouthbrooding cichlid fish *Oreochromis niloticus*. *Journal of Experimental Zoology* 273: 216-223.

Bulmer, M. 1987. Sex determination in fish. *Nature (London)* 326: 440-441.

Conover, D.O. and B.E. Kynard. 1981. Environmental sex determination: Interaction between temperature and genotype in a fish. *Science* 213: 577-579.

Conover, D.O. and S.W. Heins. 1987. Adaptive variation in environmental and genetic sex determination in a fish. *Nature (London)* 326: 496-498.

Desprez, D. and C. Mélard. 1998. Effect of ambient water temperature on sex determination in the blue tilapia *Oreochromis aureus*. *Aquaculture* 162: 79-84.

Gimeno, S., A. Gerritsen, T. Bowmer and H. Komen. 1996. Feminization of male carp. *Nature (London)* 384: 221-222.

Guan, G., T. Kobayashi and Y. Nagahama. 2000. Sexually dimorphic expression of two types of DM (*Doublesex/Mab-3*)-domain genes in a teleost fish, the tilapia (*Oreochromis niloticus*). *Biochemical Biophysical Research Communications* 272: 662-666.

Guiguen, Y., J.F. Baroiller, M.J. Ricordel, K. Iseki, O.M. McMeel, S.A.M. Martin and A. Fostier. 1999. Involvement of estrogens in the process of sex differentiation in two fish species: The rainbow trout (*Oncorhynchus mykiss*) and a tilapia (*Oreochromis niloticus*). *Molecular Reproduction and Development* 54: 154-162.

Hunter, G.A. and E.M. Donaldson. 1987. Hormonal sex control and its application to fish culture. In: *Fish Physiology*, W.S. Hoar, D.J. Randall and E.M. Donaldson (eds.). Academic Press, New York, Vol. 6, pp. 223-291.

Ishikawa, Y. 2000. Medaka fish as a model system for vertebrate developmental genetics. *BioEssays* 22: 487-495.

Kitano, T., K. Takamune, T. Kobayashi, Y. Nagahama and S-I. Abe. 1999. Suppression of P450 aromatase gene expression in sex-reversed males produced by rearing genetically female larvae at a high water temperature during a period of sex differentiation in the Japanese flounder (*Paralichthys olivaceus*). *Journal of Molecular Endocrinology* 23: 167-176.

Kitano, T., K. Takamune, Y. Nagahama and S-I. Abe. 2000. Aromatase inhibitor and 17α-methyltestosterone cause sex-reversal from genetical females to phenotypic males and suppression of P450 aromatase gene expression in Japanese flounder (*Paralichthys olivaceus*). *Molecular Reproduction and Development* 56: 1-5.

Kitano, T., T. Koyanagi, R. Adachi, N. Sakimura, K. Takamune and S-I. Abe. 2006. Assessment of estrogenic chemicals using an estrogen receptor α (ERα)- and ERβ-mediated reporter gene assay in fish. *Marine Biology* 149: 49-55.

Kitano, T., N. Yoshinaga, E. Shiraishi, T. Koyanagi and S-I. Abe. 2007. Tamoxifen induces masculinization of genetic females and regulates P450 aromatase and Mullerian inhibiting substance mRNA expression in Japanese flounder (*Paralichthys olivaceus*). *Molecular Reproduction and Development*. 74: 1171-1177.

Kobayashi, T., H. Kajiura-Kobayashi and Y. Nagahama. 2003. Induction of XY sex reversal by estrogen involves altered gene expression in a teleost, tilapia. *Cytogenetic and Genome Research* 101: 289-294.

Kondo, M., I. Nanda, U. Hornung, S. Asakawa, N. Shimizu, H. Mitani, M. Schmid, A. Shima and M. Schartl. 2003. Absence of the candidate male sex-determining gene dmrt1b(Y) of medaka from other fish species. *Current Biology* 13: 416-420.

Lebrun, C., R. Billard and B. Jalabert. 1982. Changes in the number of germ cells in the gonads of the rainbow trout (*Salmo gairdneri*) during the first 10 post-hatching weeks. *Reproduction Nutrition Development* 22: 405-412.

Marchand, O., M. Govoroun, H. D'Cotta, O. McMeel, J.-J. Lareyre, A. Bernot, V. Laudet and Y. Guiguen. 2000. DMRT1 expression during gonadal differentiation and spermatogenesis in the rainbow trout, *Oncorhynchus mykiss*. *Biochimica et biophysica Acta* 1493: 180-187.

Matsuda, M., Y. Nagahama, A. Shinomiya, T. Sato, C. Matsuda, T. Kobayashi, C.E. Morrey, N. Shibata, S. Asakawa, N. Shimizu, H. Hori, S. Hamaguchi and M. Sakaizumi. 2002. DMY is a Y-specific DM-domain gene required for male development in the medaka fish. *Nature* (London) 417: 559-563.

Matsuda, M., T. Sato, Y. Toyazaki, Y. Nagahama, S. Hamaguchi and M. Sakaizumi. 2003. *Oryzias curvinotus* has DMY, a gene that is required for male development in the medaka, *O. latipes*. *Zoological Science* 20: 159-161.

Nakamura, M. and Y. Nagahama. 1985. Steroid producing cells during ovarian differentiation of the tilapia *Sarotherodon niloticus*. *Development Growth and Differentiation* 27: 701-708.

Nakamura, M. and Y. Nagahama. 1989. Differentiation and development of Leydig cells, and change of testosterone levels during testicular differentiation of tilapia *Oreochromis niloticus*. *Fish Physiology and Biochemistry* 7: 211-219.

Nakamura, M., T. Kobayashi, X.T. Chang and Y. Nagahama. 1998. Gonadal sex differentiation in teleost fish. *Journal of Experimental Zoology* 281: 362-372.

Nanda, I., M. Kondo, U. Hornung, S. Asakawa, C. Winkler, A. Shimizu, Z. Shan, T. Haaf, N. Shimizu, A. Shima, M. Schmid and M. Schartl. 2002. A duplicated copy of DMRT1 in the sex-determining region of the Y chromosome of the medaka, *Oryzias latipes*. *Proceedings of the National Academy of Sciences of the United States of America* 99: 11778-11783.

Piferrer, F., I.J. Baker and E.M. Donaldson. 1993. Effects of natural, synthetic, aromatizable, and nonaromatizable androgens in inducing male sex differentiation in genotypic female chinook salmon (*Oncorhynchus tshawytscha*). *General and Comparative Endocrinology* 91: 59-65.

Piferrer, F., S. Zanuy, M. Carrillo, I.I. Solar, R.H. Devlin and E.M. Donaldson. 1994. Brief treatment with an aromatase inhibitor during sex differentiation causes chromosomally female salmon to develop as normal, functional males. *Journal of Experimental Zoology* 270: 255-262.

Patiño, R., K.B. Davis, J.E. Schoore, C. Uguz, C.A. Strüssmann, N.C. Parker, B.A. Simco and C.A. Goudie. 1996. Sex differentiation of channel catfish gonads: normal development and effects of temperature. *Journal of Experimental Zoology* 276: 209-218.

Romer, U. and W. Beisensherz. 1996. Environmental determination of sex in *Apistogramma* (Cichlidae) and two other fresh water fishes (Teleostei). *Journal of Fish Biology* 48: 714-725.

Shimasaki, Y., T. Kitano, Y. Oshima, S. Inoue, N. Imada and T. Honjo. 2002. Tributyltin causes masculinization in fish. *Environmental Toxicology and Chemistry* 22: 141-144.

Sinclair, A.H., P. Berta, M.S. Palmer, J.R. Hawkins, B.L. Griffiths, M.J. Smith, J.W. Foster, A.M. Frischauf, R. Lovell-Badge and P.N. Goodfellow. 1990. A gene from the human sex-determining region encodes a protein with homology to a conserved DNA-binding motif. *Nature* (*London*) 346: 240-244.

Strüssmann, C.A., S. Moriyama, E.F. Hanke, J.C.C. Cota and F. Takashima. 1996. Evidence of thermolabile sex determination in pejerrey. *Journal of Fish Biology* 48: 643-651.

Strüssmann, C.A., T. Saito, M. Usui, H. Yamada and F. Takashima. 1997. Thermal thresholds and critical period of thermolabile sex determination in two atherinid fishes, *Odontesthes bonariensis* and *Patagonina hatcheri*. *Journal of Experimental Zoology* 278: 167-177.

Tabata, K. 1991. Induction of gynogenetic diploid males and presumption of sex determination mechanisms in the hirame *Paralichthys olivaceus*. *Bulletin of the Japanese Society of Scientific Fisheries* 57: 845-850.

Yamamoto, T. 1958. Artificial induction of functional sex-reversal in genotypic females of the medaka (*Oryzias latipes*). *Journal of Experimental Zoology* 137: 227-260.

Yamamoto, T. 1969. Sex differentiation. In: *Fish Physiology*, W.S. Hoar and D.J. Randall (eds.). Academic Press, New York, Vol. 3, pp. 117-175.

Yamamoto, E. 1995. Studies on sex-manipulation and production of cloned populations in hirame flounder, *Paralichthys olivaceus* (Temminek et Schlegel). *Bulletin of the Tottori Prefectural Fisheries Experimental Station* 34: 1-145.

Yamamoto, E. 1999. Studies on sex-manipulation and production of cloned populations in hirame, *Paralichthys olivaceus* (Temminck et Schlegel). *Aquaculture* 173: 235-246.

Yokota, H., M. Seki, M. Maeda, Y. Oshima, H. Tadokoro, T. Honjo and K. Kobayashi. 2001. Life-cycle toxicity of 4-nonylphenol to medaka (*Oryzias latipes*). *Environmental Toxicology and Chemistry* 20: 2552-2560.

Yoshinaga, N., E. Shiraishi, T. Yamamoto, T. Iguchi, S-I. Abe and T. Kitano. 2004. Sexually dimorphic expression of a teleost homologue of Müllerian inhibiting substance during gonadal sex differentiation in Japanese flounder, *Paralichthys olivaceus*. *Biochemical Biophysical Research Communications* 322: 508-513.

A Review of Shark Reproductive Ecology: Life History and Evolutionary Implications

Glenn R. Parsons[1,*], Eric R. Hoffmayer[2],
Jill M. Hendon[1] and Will V. Bet-Sayad[1]

INTRODUCTION

The accumulation of shark reproductive data has been a slow process and, for the most part, it has been left up to a few ardent biologists and naturalists. Although over 350 species of sharks are known worldwide, we have relatively complete, reproductive life history data for no more than about 10 species. For the most part, the life histories of the vast majority of sharks are either completely unknown or have to be assembled piece-meal from different populations, regions, and even ocean basins. Indeed, when it comes to shark biology in general, we continue to labor in the dark. This dearth of knowledge, sadly contrasts with the fact that many shark species are believed to be on the decline due to environmental

Authors' addresses: [1] Biology Department, The University of Mississippi, University, MS 38677, U.S.A.
[2] Center for Fisheries Research and Development, Gulf Coast Research Laboratory, Ocean Springs, MS 39564, U.S.A.
*Corresponding author: E-mail: bygrp@olemiss.edu

degradation, habitat loss, and over-fishing (Manire and Gruber, 1993; Camhi, 1998; Baum and Myers, 2004), although the question of how shark populations change temporally and how they are influenced by the environment remains largely unanswered. With practically no knowledge of natural population fluctuations, our ability to recognize anthropogenic changes is poor at best (Parsons and Hoffmayer, 2005a).

If one accepts the assertion that natural selection recognizes a single valid currency, successful offspring, then an argument could be made that ultimately every aspect of an organism's biology is a reproductive life history parameter. However, in this chapter, we will address primarily those life history characteristics that have classically been defined as reproductive, such as birth size, size at maturation, maximum size of females, and number of offspring. We will use these metrics to make comparisons across species and habitats, and to also consider aspects such as reproductive strategies, sexual segregation, sexual dimorphism, mating behavior, multiple paternity, sperm storage, nursery grounds, and latitudinal variation in reproductive parameters.

SHARK REPRODUCTIVE STRATEGIES

All elasmobranchs use internal fertilization and, their reproductive strategies run the gamut from oviparity to placental viviparity (Table 14.1). Although sharks are often described incorrectly as primitive, viviparity in some shark species is as sophisticated as that found in mammals. Viviparity is observed in about 60% of elasmobranch species and is the predominant reproductive strategy in a number of shark families. Based on the mode of maternal input, viviparous sharks can be divided into four groups: yolk-dependent, oophagy, aldelphophagy, and placental nutrition (Table 14.1) (Dulvy and Reynolds, 1997).

Table 14.1   Reproductive modes of sharks.

Parity mode	Maternal input
Single oviparity	Yolk dependency
Multiple oviparity	Yolk dependency
Viviparity	Yolk dependency
Viviparity	Oophagy
Viviparity	Adelphophagy
Viviparity	Placental

Based on Wourms (1981), Compagno (1988), and Dulvy and Reynolds (1997).

Oviparity Oviparous or egg-laying sharks enclose their fertilized eggs in protective egg cases and deposit them on or attached to substrate on the ocean bottom. Embryos are nourished solely from the yolk sac and no further parental care is afforded to the young. The tough protective egg case provides the only protection for the developing embryo from predators, and mortality can be extremely high at this vulnerable life stage (Wourms, 1977). Nakaya (1975) distinguished between two types of oviparity in sharks: single oviparity and multiple oviparity (Table 14.1). Single oviparity occurs when a single egg case is formed in each oviduct and deposited before the embryo develops, resulting in an extended hatching time ranging from several months to a year. This occurs in only two families, the Heterodontidae and Scyliorhinidae (Table 14.2), representing approximately 109 shark species (Compagno, 2001). Multiple oviparity occurs when several egg cases are formed in the oviduct over a period of time and the embryos develop to varying degrees before deposition, resulting in a relatively short hatching time (Compagno, 1988). Multiple oviparity occurs in three families, the Parascyllidae, Hemiscyllidae, and Stegostomidae (Table 14.2), representing approximately 20 species of sharks (Compagno, 2001).

Table 14.2 Reproductive mode for 8 orders and 34 families of sharks. Orders are arranged phylogenetically. Based on Wourms (1981), Compagno (1988), and Dulvy and Reynolds (1997).

	Parity mode	*Maternal input*
Order Hexanchiformes		
Family Chlamydoselachidae	Viviparity	Yolk dependency
Family Hexanchidae	Viviparity	Yolk dependency
Order Squaliformes		
Family Echinorhinidae	Viviparity	Yolk dependency
Family Squalidae	Viviparity	Yolk dependency
Family Centrophoridae	Viviparity	Yolk dependency
Family Etmopteridae	Viviparity	Yolk dependency
Family Somniosidae	Viviparity	Yolk dependency
Family Oxynotidae	Viviparity	Yolk dependency
Family Dalatiidae	Viviparity	Yolk dependency
Order Pristiophoriformes		
Family Pristiophoridae	Viviparity	Yolk dependency
Order Squatiniformes		
Family Squatinidae	Viviparity	Yolk dependency

(Table 14.2 Contd.)

(Table 14.2 Contd.)

	Parity mode	Maternal input
Order Heterodontiformes		
Family Heterodontidae	Single oviparity	Yolk dependency
Order Lamniformes		
Family Mitsukurinidae	Viviparity	Oophagy
Family Odontaspididae	Viviparity	Adelphophagy
Family Pseudocarchariidae	Viviparity	Oophagy
Family Megachasmidae	Viviparity	Oophagy
Family Alopiidae	Viviparity	Oophagy
Family Cetorhinidae	Viviparity	Oophagy
Family Lamnidae	Viviparity	Oophagy
Order Orectolobiformes		
Family Parascyllidae	Multiple oviparity	Yolk dependency
Family Brachaeluridae	Viviparity	Yolk dependency
Family Orectolobidae	Viviparity	Yolk dependency
Family Hemiscyllidae	Multiple oviparity	Yolk dependency
Family Ginglymostomatidae	Viviparity	Yolk dependency
Family Stegostomatidae	Multiple oviparity	Yolk dependency
Family Rhincodontidae	Viviparity	Yolk dependency
Order Carcharhiniformes		
Family Scyliorhinidae	Single oviparity*	Yolk dependency
Family Proscyllidae	Viviparity*	Yolk dependency
Family Pseudotriakidae	Viviparity	Yolk dependency
Family Leptochariidae	Viviparity	Yolk dependency
Family Hemigaleidae	Viviparity	Yolk dependency
Family Triakidae	Viviparity	Placental/yolk dependency
Family Carcharhinidae	Viviparity*	Placental/yolk dependency
Family Sphyrnidae	Viviparity	Placental

*indicates that sharks in this family utilize more than one reproductive strategy.

Several interesting commonalities were identified among the oviparous shark species. The majority are benthic or demersal in habit (Tortonese, 1950). Many of these species live in complex habitats (i.e., coral reefs, hard bottom, etc.), which offers refuge for their small offspring and reduces predation risks. Additionally, the majority of oviparous species are relatively small in size (Parascyllidae 31 to 49 cm, total length, TL; Hemiscyllidae 65 to 105 cm, TL; Scyliorhinidae 26 to 115 cm, TL; Heterodontidae 59 to 135 cm, TL; Compagno, 1984; Cortés, 2000). Offspring size at hatching is also small (7 to 22 cm, TL), which is not surprising because of the relatively small size of the adults and the limited supply of nutrients provided by the yolk sac. Likewise, this is supported by the fact that there is no significant relationship between maximum maternal size and maximum offspring size in either the single oviparous or multiple oviparous groups (Fig. 14.1).

Small offspring size may make these species vulnerable to predation which suggests that an increase in fecundity may offset mortality. For most oviparous species, detailed information about annual fecundity is lacking, however, estimates suggest that females lay large numbers of eggs per year (Compagno, 1984). In benthic environments characterized by high mortality, more fecund oviparous species may be at a competitive advantage over viviparous species.

Yolk Dependent Viviparity Yolk-dependent viviparity, the least specialized form of live birth, appears to have evolved from oviparity (Breder and Rosen, 1967; Springer and D'Aubrey, 1972; Dulvy and Reynolds, 1997). Yolk-dependent viviparity is the most common reproductive strategy utilized by sharks and is represented by six orders (Hexanchiformes, Squaliformes, Pristiophoriformes, Squatiniformes, Orectolobiformes, and Carcharhiniformes), 19 families, and approximately 166 species (Wourms,

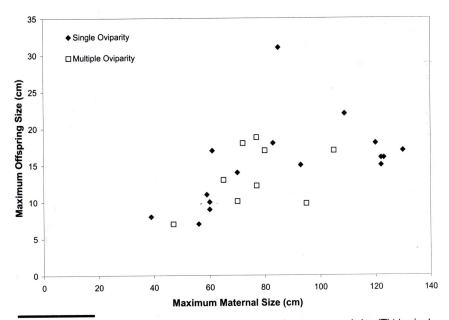

Fig. 14.1 Maximum offspring size (TL) in relation to maximum maternal size (TL) in single oviparous (F = 4.54, R^2 = 0.25, p = 0.051, n = 16) and multiple oviparous (F = 1.76, R^2 = 0.18, p = 0.223, n = 9) shark species. Data taken from Compagno (1984) and Cortés (2000).

1977; Compagno, 2001). Yolk-dependent viviparous sharks rely solely on the yolk sac for nourishment and receive no additional nutrients from the mother during gestation. The embryos are retained within the uteri of the female until hatching. This allows the developing embryos to be protected during this highly vulnerable stage and increases their chances of survival.

The high diversity of sharks that utilize this strategy make it difficult to find similar attributes among these species, although some interesting trends are evident. Similar to oviparity, the majority of the species that utilize this reproductive strategy are small and demersal to benthic, except for the large, pelagic whale shark, *Rhincodon typus*. Since many of these species are constantly on the move for gill ventilation, live bearing is a benefit because they do not have to stop swimming for egg deposition. Offspring size is also relatively small and consistent across taxa (70% of known species are 20 to 40 cm, TL; Fig. 14.2) with few exceptions (i.e., six gill shark, *Hexanchus griseus* 60 to 70 cm, TL; tiger shark, *Galeocerdo cuvier* 70 to 85 cm, TL) (Compagno, 1984; Cortés, 2000). In addition, a

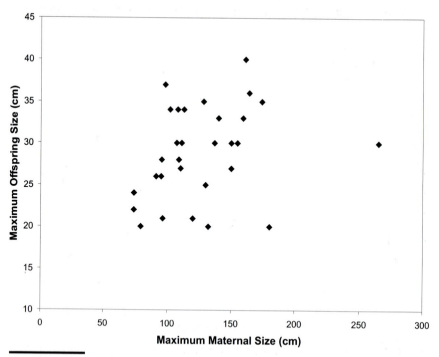

Fig. 14.2 Maximum offspring size (20-40 cm, TL) plotted against maximum maternal size (TL) in yolk dependent species (n = 31). Data taken from Cortés (2000).

strong positive relationship is evident between maximum maternal size and maximum fecundity (# of offspring) for that species (Fig. 14.3). Maximum fecundity appears to be influenced by the amount of space available for growth within the uteri of the female. Small shark species (i.e., the dogfish, *Squalus megalops* 76 cm, TL) tend to have a lower fecundity (n = 3) and larger species (i.e., *R. typus* 1900 cm, TL) tend to have a higher fecundity (n = 300) (Joung *et al.*, 1996; Cortés, 2000). Small yolk-dependent species may be both nutrient and space limited, while large species may be only nutrient limited. Internal development is an advantage in this group, although long gestation periods and small uterine space limit the number of young that can be produced.

Oophagous Viviparity Oophagy, or egg eating, may have evolved as a solution to the problem of limited nourishment from the yolk sac. In this strategy, additional nutrients are provided to the developing embryo in the form of unfertilized eggs. Only a few fertilized eggs are deposited in each uterine horn at the beginning of the gestation period. The embryos utilize the yolk sac for a short period (a few weeks), at which time the female provides the young with unfertilized eggs as nourishment. The

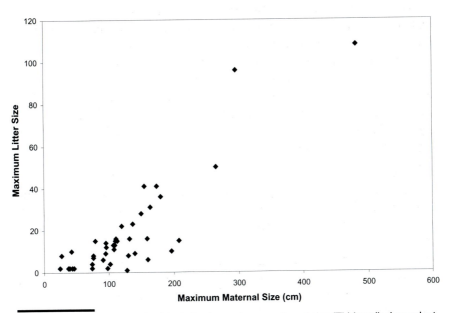

Fig. 14.3 Maximum litter size in relation to maximum maternal size (TL) in yolk-dependent viviparous species (F = 128.6, R^2 = 0.76, p < 0.0001, n = 45). Data taken from Cortés (2000).

embryos develop a cardiac stomach, which is a large distended stomach full of yolk from the unfertilized eggs, and can resemble a yolk sac. Because of the large demand for nutrients, females typically have an extremely large ovary that can weigh over 5 kg (Gilmore, 1993). At about 5 cm (TL), the embryo develops temporary teeth that allow them to rupture the unfertilized eggs and ingest the contents. The limited competition between the small numbers of developing young allow them to attain a relatively large birth size, which lowers predation risk. Oophagy is common among the Lamniformes and is used by the Pseudocarchariidae, Alopiidae, Cetorhinidae, Laminidae, and Odontaspidae. Oophagy has not been confirmed for the Mitsukurinidae or the Megachasmidae, although both are suspected to utilize this strategy (Compagno, 2001).

Only a small number (n = 15) of species utilize this reproductive strategy. All oophagous species produce relatively large offspring (70 to 190 cm, TL), have a relatively long gestation period (typically 9 to 18 months), are considered coastal-pelagic to pelagic in habit, and are obligate ram ventilators. Sharks in the families Pseudocarchariidae, Alopiidae, and Megachasmidae are predominately pelagic, whereas sharks in the families Cetorhinidae, Laminidae, and Odontaspidae are considered coastal-pelagic. In general, coastal-pelagic species have a higher fecundity (6-18 young/female) with relatively smaller offspring (30% of maternal size at maturity) compared to the pelagic species, which appear to have a lower fecundity (2-4 young/female) with relatively larger offspring (48% of maternal size at maturity). There appears to be a trade off between the number of offspring produced and the size of the offspring, which is not observed with the oviparous and yolk-dependent viviparous species. Sharks that utilize this strategy may have solved the problem of limited nutrients but may be space limited.

Adelphophagy Adelphophagy or sibling eating has only been observed in one family, the Odontaspidae, with particular attention paid to the sandtiger shark, *Eugomphodus* (=*Carcharias) taurus* (Gilmore *et al.*, 1983), where the largest fetus ingests its siblings and any unfertilized eggs. In this species, as many as 12 fertilized eggs may be found in each uterine horn at the beginning of the gestation period. Once these fertilized eggs are deposited, the female switches to producing egg cases of unfertilized eggs, which are used to feed the developing embryos. Once the first embryo attains a size of 10 to 12 cm, it seeks out and kills the remaining embryos by biting them. The surviving embryo then ingests the dead siblings (intrauterine cannibalism), along with any unfertilized egg cases, and develops an engorged yolk stomach full of nutrients. The *E. taurus* embryos will stop feeding one to two months prior to birth and will reabsorb the nutrients stored within their yolk stomachs. Much like other Lamniformes, Odontaspids produce large young that are likely to suffer lower mortality compared to other shark groups.

Placental Viviparity Placental viviparous sharks provide nutrients during the majority of gestation via a yolk-sac placenta, a direct connection between the mother and embryo. Initially, the embryos are yolk-sac dependent. However, after the first few weeks, the yolk-sac elongates and attaches to the uterine wall of the mother, forming a yolk-sac placenta. Once this connection is made, the needs of the developing embryo are supplied via maternal input. Because of the supply and demand nature of this relationship, placental viviparity is considered the most efficient reproductive strategy among sharks. This strategy is restricted to the order Carcharhiniformes and is represented by five families (Leptochariidae, Triakidae, Hemigalidae, Carcharhinidae, and Sphyrnidae) and 73 species (Compagno, 1988).

A significant, positive relationship was detected between maximum maternal size and maximum offspring size among placental viviparous sharks, suggesting that large sharks have large offspring (Fig. 14.4). Additionally, maximum fecundity increased with maximum maternal size (Fig. 14.5) with the largest species producing 30 to 40 offspring. Relative offspring size (maximum offspring size/maternal size at maturation), however, decreased with maximum maternal size (Fig. 14.6). The offspring of small species were 40 to 60% of maternal size, whereas the offspring of large species were 20 to 40%.

The majority of sharks that utilize placental viviparity are obligate ram ventilating species and are coastal, coastal-pelagic, or pelagic in nature. Similar to oophagous species, it appears that the environment in which these sharks occur may influence offspring size. To investigate the effect of environment on offspring size, only sharks from the family Carcharhinidae were examined. When environment was included in the analysis, it was observed that most of the coastal sharks were small (< 160 cm, TL) and produced smaller offspring (absolute size: 44.9 ± 3.3 cm, TL) and most of the coastal-pelagic to pelagic sharks were large (> 160 cm, TL) and produced larger offspring (absolute size: 68.4 ± 3.0 cm, TL) (Fig. 14.4). Patchy food distribution and predation pressure may have selected for large offspring size.

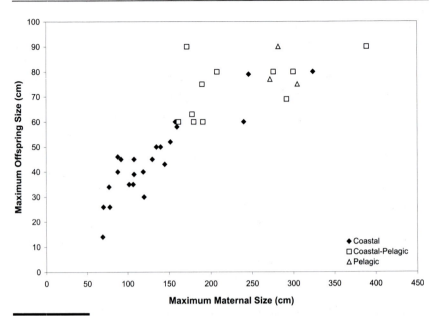

Fig. 14.4 Maximum offspring size (TL) in relation to maximum maternal size (TL) in placental viviparous species ($F = 102.5$, $R^2 = 0.73$, $p < 0.0001$, $n = 60$) from different environments (coastal, coastal-pelagic, and pelagic). Data taken from Compagno (1984) and Cortés (2000).

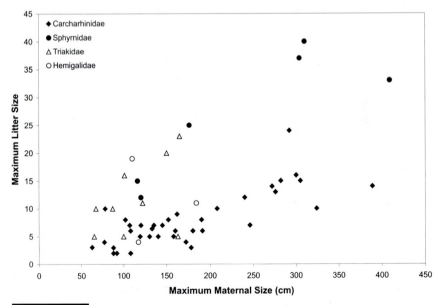

Fig. 14.5 Maximum litter size in relation to maximum maternal size (TL) in placental viviparous species ($F = 28.4$, $R^2 = 0.35$, $p < 0.0001$, $n = 59$). Data taken from Cortés (2000).

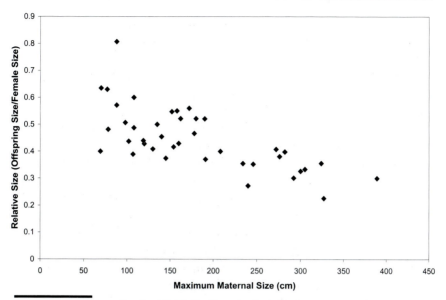

Fig. 14.6   Relative offspring size (offspring size/female size at maturity) in relation to maximum maternal size (TL) in placental viviparous species ($F = 29.3$, $R^2 = 0.48$, $p < 0.0001$, $n = 56$). Data taken from Compagno (1984) and Cortés (2000).

There may be several advantages to this reproductive strategy. First, since the young develop inside the female, they are protected during this crucial life stage. Second, because of the placental connection, they are provided with an increased supply of nutrients relative to other reproductive strategies. If placental viviparous sharks are indeed able to supply increased nutrients to their developing young, we reasoned that they should have larger birth size when compared to yolk-dependent species. A significant ($F_{3,93} = 10.6$, $p < 0.0001$) effect of reproductive strategy on relative birth size was observed (Fig. 14.7). Additionally, in pairwise comparisons, placental shark species have significantly larger birth size when compared to yolk-dependent species. Interestingly, when birth size for oviparous and yolk-dependent-viviparous species was compared, there was no significant difference.

The Evolution of Viviparity   Viviparity is estimated to have evolved independently at least 100 times and occurs in mammals, reptiles, amphibians and fishes (Blackburn, 1995). Oviparity is considered the plesiomorphic state in sharks from which other reproductive strategies evolved. However, in at least one case, the zebra shark, *Stegostoma fasciata*, oviparity apparently arose from a viviparous ancestor (Dulvy and Reynolds, 1997). Among elasmobranchs, viviparity is estimated to have evolved independently nine to 10 times (Dulvy and Reynolds, 1997) and unlike other groups of fishes, it is a common reproductive strategy. Placental viviparity is considered to be the most recently evolved reproductive strategy (Wourms, 1977, 1981; Dulvy and Reynolds, 1997).

It is difficult to identify the selective forces that have resulted in the development of life history traits. However, some hypotheses to explain the evolution of viviparity have been proposed. The *cold climate hypothesis* for the evolution of viviparity in reptiles suggests that in colder environments, species that retain the eggs *in-utero* may use behavioral mechanisms of thermoregulation to affect faster development of young (Shine, 1983). The thermal stability of marine habitats relative to the terrestrial environment, and the scant evidence for behavioral thermoregulation in elasmobranchs (Carey and Scharold, 1990; Morrissey and Gruber, 1993), suggest that this hypothesis may not apply to sharks.

It is possible that viviparity evolved in *response to environmental forces* that favored production of larger offspring at the expense of increased fecundity. In support of this hypothesis is our observation that a shark's reproductive strategy is a significant ($F_{3,93}$ = 10.6, p < 0.0001) predictor of relative offspring size (Fig. 14.7). Sharks that use placental viviparity

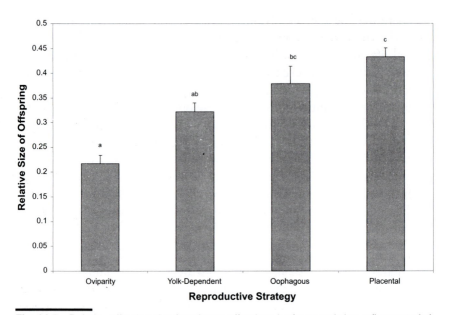

Fig. 14.7   Relative offspring size (maximum offspring size/maternal size at first maturity) by reproductive strategy. A significant ($F_{3,93}$ = 10.6, p < 0.0001) difference in relative offspring size was evident. Means and standard errors are plotted. Letters not in common indicate significant differences.

have significantly greater relative birth size than either oviparous or yolk-dependent sharks. This lends support to the hypothesis that placental viviparity has evolved for increased offspring size. However, our results show that as maximum maternal size increases, relative offspring size actually decreases while, at the same time, fecundity increases. This suggests a trade-off between offspring size and fecundity. Rather than producing proportionally larger offspring as female size increases, it may be more advantageous to invest in producing larger litters, i.e., small sharks produce few, relatively large offspring and large sharks produce more, relatively small offspring (Figs. 14.5 and 14.6).

More recently, Crespi and Semeniuk (2004) proposed that *parent-offspring conflict* plays an important role in the evolution of viviparity and has been responsible for the diversity observed in animal reproductive mode. Parent-offspring conflict (Trivers, 1974) refers to the supposition that offspring, even *in-utero*, may compete with the parent over the level of parental investment. In viviparous species, offspring may be selected to obtain greater investment from the mother than she is selected to provide. Put another way, 'the viviparity-conflict hypothesis implies antagonistic co-evolution between parents and offspring in traits that influence pre-natal investment' (Crespi and Semeniuk, 2004). This reproductive 'arms race' has led to great diversity in placentation and the forms of viviparity.

We propose a final hypothesis to explain the appearance of viviparity in sharks. Many shark species are pelagic, obligate ram ventilators that can never stop swimming. It is possible that viviparity co-evolved with obligate ram ventilation during the invasion of the pelagic environment. The necessity to swim continuously may have precluded oviparity as a viable reproductive strategy. Ram-ventilating sharks must give birth 'on the go. Parsons (1991) observed bonnethead sharks, *Sphyrna tiburo*, using their forward movement in a "cloaca dragging" and "quick turn" behavior to facilitate birth. It is, likewise, interesting that among the batoids (skates and rays) those families that are the largest and the most pelagic (the stingrays, Dasyatidae, eagle rays, Myliobatidae, cownose rays, Rhinopteridae, and manta/devil rays, Mobulidae) are also viviparous. In contrast, the more benthic skates (Rajidae) are generally smaller and oviparous. In the pelagic environment, it may be particularly advantageous to produce large, more competitively fit offspring that are less likely to be preyed upon. Larger offspring may be more effective hunters, since swimming speed is a function of absolute size (Marshall, 1971). Food tends to be patchy in distribution in the pelagic environment and large body size and increased swimming speed may be an adaptation to increase the efficiency and likelihood of patch location. Perhaps the abundance of prey in the coastal environment has resulted in selection for larger numbers of smaller offspring. Large offspring in pelagic environments are also observed in placental viviparous species (Fig. 14.3).

SHARK MATING BEHAVIOR

The ocean is a concealing environment that makes *in situ* observation of the mating behavior of these highly active organisms exceedingly difficult. For this reason, much of our knowledge of mating behavior comes from captive animals. Further, limiting the general applicability of these observations is the fact that many of the species that are most amenable to captivity are benthic. Additionally, the possibility exists that the behavior of species mating in aquaria may have been altered by the captive environment. For example, Klimley (1980) found that captive nurse sharks, *Ginglymostoma cirratum*, rolled for copulation access immediately after the male grasped her pectoral, and once copulation began the male was able to release the pectoral hold without losing contact with his mate. This is in contrast to the behavior observed in wild G. *cirratum* where the male maintained grasp of the female's pectoral to keep the female in a copulation-conducive position and continued the grasp until the copulatory act had been completed (Carrier *et al.*, 1994).

Although scientists have had the opportunity to observe very few shark species while mating, it has become apparent that pre-copulatory and copulatory behaviors differ between species (Table 14.3). This is understandable given the diversity of lifestyles and anatomies among shark species. For example, many of the requiem (Carcharhinidae) and mackerel (Laminidae) sharks are obliged to swim continuously and have a relatively inflexible, muscular body. The requirement for continuous swimming in these species has no doubt shaped many aspects of their reproductive life histories. On the other hand, the catsharks (Scyliorhinidae), carpet sharks (Orectolobidae), and horn sharks (Heterodontidae), are not continuously active and have a more flexible body that can aid in reproductive activities. This flexibility allows for greater pre-copulatory and copulatory posturing.

Male Pre-copulatory Behavior   A suite of male pre-copulatory behaviors has been observed and defined for various species (Table 14.3). These behaviors include: (1) *following*, (2) *hierarchy*, (3) *nosing*, and (4) *lacerations*.

As defined by Carrier *et al.* (1994), *following* is when a male *G. cirratum* swims alongside a female in a slightly lagging position. The male's head is generally positioned just behind the female's pectoral fin. A similar *hierarchy* behavior has been observed between captive Sandtiger sharks *Eugomphodus* (= *Carcharias*) *taurus*. A dominant male will swim closely behind a competing male, inhibiting movement of the fore individual's caudal fin. At the same time, the male will display aggressive acts—such as biting in a quick, shallow manner—toward other species in the area (Gordon, 1993). *Nosing* behavior, where a male swims closely behind a female with his snout in close proximity to the female's cloaca was observed in the blacktip reef shark, *Carcharhinus melanopterus, E. taurus,* and the basking shark, *Cetorhinus maximus* (Johnson and Nelson, 1978; Gordon, 1993; Harvey-Clark *et al.*, 1999). This behavior is generally coupled with a female swimming with her caudal fin slightly elevated. Bite scars as a result of *lacerations* inflicted during mating have been noticed in many shark species and are often the only information that we have to infer reproductive behavior for a given species. Fresh bite wounds or bite scars are caused by males biting and sometimes orally holding females along the dorsum, flank, and pectoral or pelvic fins (Dempster and Herald, 1961; Stevens, 1974; Pratt, 1979; Klimley, 1980; Gilmore *et al.*, 1983; Tricas and LeFeurve, 1985; Castro *et al.*, 1988; West and Carter, 1990; Castro, 1993, 1996, 2000; Gordon, 1993; Carrier *et al.*, 1994; Hazin *et al.*, 1994, 2000; Ebert, 1996; Feldheim *et al.*, 2002). Stevens (1974) divided the types of observed bites into three categories: (1) full, semicircular jaw impressions (generally upper jaw), (2) tooth slashes, (3) tooth nicks. Initial bites by the male have been theorized to be used as a pre-copulatory releaser for females (Springer, 1960; Stevens, 1974). Thickening of the skin in the female blue shark, *Prionace glauca* is undoubtedly a response to lacerations during mating (Pratt, 1979).

In more flexible species, the male will bite the female (typically on the latter portion of the pectoral fin and part of the flank) and will hold on to aid in positioning the female for copulation. Mating scars have been noticed on wild, female night sharks, *Carcharhinus signatus,* but these scars were not prominent. This implies that either this species may not use biting during regular mating behavior, or that the wounds may have healed before capture (Hazin *et al.*, 2000). No mating scars have been observed on wild, female blacknose sharks, *Carcharhinus acronotus* (Hazin *et al.*, 2002). Bites in wild, female blacktip sharks *Carcharhinus limbatus* were typically 80 to 150 mm long and 15 to 20 mm deep, although they never punctured the peritoneum (Castro, 1996).

Table 14.3 Summary of reproductive behaviors in shark species divided by order. (W = observed in wild animals, C = observed in captive animals, ? = not observed in the species to date, NR = no reproductive information available).

Order species	♀ Elevated Tail ♂ Nosing	♀ Submission	♀ Avoidance	♀ Biting	♂ Lacerations	♂ Hierarchy	♂ Cooperation
Carcharhiniformes							
C. isodon	?	?	?	?	W[2]	?	?
C. limbatus	?	?	?	?	W[3]	?	?
C. melanopterus	W[4]	W[4]	?	?	?	?	?
C. plumbeus	?	?	?	?	W[5]	?	?
C. signatus	?	?	?	?	W[6]	?	?
N. brevirostris	?	?	?	?	W[7]	?	?
P. glauca	?	?	?	?	W[8,9,10]	?	?
R. terraenovae	?	?	?	?	W[11]	?	?
S. canicula	?	C[12]	?	?	?	?	?
S. retifer	?	C[13]	C[13]	?	C[13]	?	?
S. torazame	?	?	?	?	C[14]/W[15]	?	?
T. obesus	W[4]	?	?	?	C[14]	?	?
Lamniformes							
C. carcharias	?	?	?	?	W[16]	?	?
C. maximus	W[17]	?	?	?	W[18]	?	?
E. taurus	C[19]	C[19]	C[19]	C[19]	W[20]/C[19]	C[19]	?
Orectolobiformes							
G. cirratum	?	W[21]	W[21,22,23]/C[24]	?	W[21,22,23]/C[24]	?	W[21,22]
H. ocellatum	?	?	C[24]	C[24]	C[24]	?	?

(Table 14.3 Contd.)

(Table 14.3 Contd.)

Order species	♀ Elevated Tail ♂ Nosing	♀ Submission	♀ Avoidance	♀ Biting	♂ Lacerations	♂ Hierarchy	♂ Cooperation
Heterodontiformes							
H. francisci	?	C[26]	?	?	C[26]	?	?
Hexanchiformes							
N. cepedianus	?	?	?	?	W[27]	?	?
Squatiniformes							
	NR	NR	NR	NR	NR	NR	NR
Pristiophoriformes							
	NR	NR	NR	NR	NR	NR	NR
Squaliformes							
	NR	NR	NR	NR	NR	NR	NR

1—Hazin *et al.* (2002), 2—Castro (1993), 3—Castro (1996), 4—Johnson and Nelson (1978), 5—Springer (1960), 6—Hazin *et al.* (2000), 7—Feldheim *et al.* (2002), 8—Hazin *et al.* (1994), 9—Pratt (1979), 10—Stevens (1974), 11—Personal Observation, 12—Bolau (1881), 13—Castro *et al.* (1988), 14—Uchida (1990), 15—Tricas and Le Feuvre (1985), 16—Francis (1996), 17—Matthews (1950), 18—Harvey-Clark *et al.* (1999), 19—Gordon (1993), 20—Gilmore *et al.* (1983), 21—Pratt and Carrier (2001), 22—Carrier *et al.* (1994), 23—Castro (2000), 24—Klimley (1980), 25—West and Carter (1990), 26—Dempster and Herald (1961), 27—Ebert (1996)

Female Pre-copulatory Behavior   Several types of *avoidance* behaviors have been observed in female sharks. Female *G. cirratum* have been observed swimming into shallow water to avoid male pursuits (Carrier *et al.*, 1994). Unreceptive female *G. cirratum* may also anchor themselves by digging their pectoral fins into the bottom (Castro, 2000). Captive, female epaulette sharks, *Hemiscyllium ocellatum*, have been observed attempting to keep their ventral surfaces appressed to the substrate, thereby protecting their cloaca (West and Carter, 1990). Upon pectoral grabbing by males, some females will initiate a roll as an attempt to release the grasp (Pratt and Carrier, 2001). If the grasp cannot be released and copulatory positioning occurs, captive, female *H. ocellatum* will arch their backs in an attempt to prevent the male's claspers from reaching their cloaca (West and Carter, 1990; Pratt and Carrier, 2001).

Captive female *H. ocellatum* have also been observed *biting* the male's pectoral fin. This behavior was initiated prior to any reproductive behavior by the males and may be a mating initiative by the female (West and Carter, 1990). Captive female *E. taurus* were also observed biting sexually advancing males as retaliation (Gordon, 1993). In the *elevated tail* behavior, a female swims with her caudal fin slightly elevated (Johnson and Nelson, 1978; Gordon, 1993; Harvey-Clark *et al.*, 1999). It has been proposed that the female may actually be releasing a pheromone(s) during this act in order to attract or alert mature males (Johnson and Nelson, 1978). This behavior is often coupled with male *nosing*.

Copulation   Very few shark species have been observed during copulation but some behaviors have been noted. In some species, *male positioning* occurs where the male attempts to place the female in a favorable position for copulation. For carpet sharks (Orectolobidae), catsharks (Scyliorhinidae), and horn sharks (Heterodontidae), which have more flexible torsos, the female and male bodies wrap around one another. For example, males will wrap their tails around the dorsal surface of the female (in front of the second dorsal fin) to anchor his body, and then insert one clasper into her cloaca (Bolau, 1881; Dempster and Herald, 1961; Johnson and Nelson, 1978; Castro *et al.*, 1988; West and Carter, 1990). Many mating observations have also been recorded in *G. cirratum*. This species also exhibits wrapping behavior with either the male on top of the female (ventral sides together), or both the male and female lying next to each other ventral side up with the male's posterior angling to the female's cloaca for clasper insertion (Klimley, 1980).

In the lemon shark, *Negaprion brevirostris,* sevengill shark, *Notorynchus cepedianus,* the reef white-tip shark, *Triaenodon obesus,* and sand tiger shark, *E. taurus,* copulation takes place with the male and female swimming slow and synchronously side by side with the male's caudal fin angled toward the female's to allow for clasper insertion (Clark, 1963; Tricas and LeFeurve, 1985; Gordon, 1993; Ebert, 1996). In all the species, once a clasper has been inserted, it will flare to aid in maintaining penetration and sperm release.

Acceptance by a female can be observed through either complete submission, where the female appears to simply submit to copulation, or it may be associated with acceptance signals. Some females will appear to become limp or paralyzed and will not resist the male's advances (Bolau, 1881; Dempster and Herald, 1961; Johnson and Nelson, 1978; Castro *et al.,* 1988; Gordon, 1993). In *G. cirratum* and *E. taurus,* the pelvics fins of females will create a "cup," which is a curling of the pelvics downward, creating a bowl shape around the cloaca. This is often noticed during *nosing* by the male and may be associated with release of a pheromone(s) (Johnson and Nelson, 1978). The female may also "flare" her pelvics fins just before penetration of the male clasper. The flaring behavior splays the pelvics directly out to the sides of the flank, fully exposing the cloacal opening (Gordon, 1993; Pratt and Carrier, 2001). Female *E. taurus* have been described as "*stalling,*" or completely stopping swimming to cue to the male that she is ready for copulation (Gordon, 1993). During the copulatory act, males have been described as *thrusting* or undulating their bodies. In *G. cirratum,* a male was reported to "spasmodically thrust" against the female's body (Tricas and LeFeuvre, 1985; Carrier *et al.,* 1994), an act which may be necessary to promote sperm release from the seminal ducts.

In some species, *cooperation* by males during mating activities has been observed (Carrier *et al.,* 1994). In *G. cirratum,* a female was tightly surrounded by six males, all facing the same direction, with a seventh male laying perpendicular to the others, effectively blocking the female in. As copulatory biting and rolling was initiated, the blocking shark rose and sank in synch with the mating couple to inhibit forward movement. There have also been reports of several males grasping onto a single female's pectorals.

The duration of pre-copulatory behavior varies, depending upon species. In *E. taurus,* pre-copulatory behavior lasted three weeks to one month, while actual copulation only lasted 1-2 minutes (Gordon, 1993).

This contrasts with the horn shark, *Heterodontus francisci*, which remains in actual copulation for 30 minutes (Dempster and Herald, 1961).

Coloration changes have been noticed in several shark species during reproductive events. Gordon (1993) reports that captive male *E. taurus* became chromatically lighter during pre-copulatory behavior. The color change was most noticeable in the alpha male. Color differences were also noted in *G. cirratum* mating activities where the males were significantly darker than the females (Klimley, 1980). The senior author has observed male bonnethead sharks, *Sphyrna tiburo*, in Florida Bay, Florida, rapidly darken when approaching conspecifics. However, it was not clear whether this was pre-copulatory behavior or agonistic behavior toward other males.

The Evolution of Shark Mating Behaviors   The evolution of mating behaviors in general has been extensively studied using many animal models. For the most part, mating behaviors are typically discussed within the context of sexual selection theory. Darwin presented the theory of sexual selection which explains the physical and behavioral traits that an organism will express in an attempt to increase its odds of reproducing. Secondary sex characteristics—those used for display and mate attraction—are not naturally selected and generally cause a reduction in overall fitness for the displayer (Andersson, 1994). The possession of costly sexual characteristics may impart an advantage to the males that possess them over other males in the competition for mates.

Sexual selection, as it applies to shark mating behavior, presents some challenging problems. First, sharks can be poor models for behavioral studies, particularly the continuously active species; thus methods for testing sexual selection theories simply do not exist. Secondly, sharks are practically devoid of obvious secondary sex characteristics, i.e., those characteristics that are generally cited as the raw material for sexual selection. Certainly, female sharks could select mates based on the characteristics of the male's claspers, their most obvious secondary sex characteristic, but we are aware of no evidence for this behavior. One exception is the color change that occurs in males of some species prior to and during mating. This darkening of the skin may make the male more attractive to the female and may increase his likelihood of being selected by the female. Undoubtedly, other, more subtle characteristics have yet to be discovered. A particularly fruitful area for study may be the mechanisms for mate location in sharks.

The size and concealing nature of the marine environment creates interesting challenges for shark mate location and we suggest that pheromonal attraction may be an important behavioral mechanism. The aforementioned *nosing* behavior in some male sharks suggests that pheromones are involved. Additionally, the exceedingly sensitive olfactory ability of sharks coupled with their highly developed locomotory ability suggests they have the potential to detect and then locate females from great distances. Parsons and Hoffmayer (2005a) observed a sudden exodus of adult male sharpnose sharks, *Rhizoprionodon terraenovae*, from inshore waters and suggested that this migration may be reproductively motivated. Perhaps male *R. terraenovae* were stimulated to leave the Mississippi Sound in response to pheromonal attraction from females segregated offshore. It is possible that *R. terraenovae* and other shark species engage in *scrambles*, a behavior defined as early search and swift location of mates (Thornhill and Alcock, 1983). Species that engage in this behavior are characterized by well developed sensory and locomotory organs, a description that applies easily to sharks.

Gender differences in the energetic investment in reproduction might also be important in explaining shark-mating behaviors. Sharks exhibit extreme *anisogamy*, size differences between gametes. The eggs produced by females are large, contain yolk that is high in nutrients, and is more energetically expensive to produce than the millions of small sperm produced by males (Andersson, 1994). This, in turn, may have given rise to the more promiscuous behaviors of males, the eventual reserved behaviors found in females, and the competition that occurs between and within sexes. Females have more to lose if they mate with a less fit male. This concept may reveal itself in the aggressiveness that some males express toward each other and the development of male hierarchies based on dominance (*cooperation* may be a result of these hierarchies). The aggressiveness that males show toward females is also understandable under this theory as the male is simply trying to be the first to copulate and produce offspring.

POLYGYNY, POLANDRY, MULTIPLE PATERNITY, SPERM STORAGE AND SPERM COMPETITION

Multiple matings by males (polygyny) is a widespread phenomenon among organisms. Since male reproductive effort is generally considered to be less than females (see Adams and Parsons, 1998 for an exception to this) in an energetic sense, males may be better able to engage in multiple matings but may suffer increased mortality due to the risk associated with mating. This may be particularly true in some sharks when mating is violent and where males are significantly smaller than females (see section

Sexual Dimorphism). An attempt to mate with an unreceptive female could be disastrous for the male. However, the benefit of increasing the probability of siring offspring obviously outweighs the risk. Males may exhibit a genetic maximizing strategy by mating with as many partners as possible (Becher and Magurran, 2004).

Multiple mating by females (polyandry) is also widespread among taxa and multiple paternity is virtually the rule among sexual organisms (Crespi and Semeniuk, 2004). However, the evolution of such reproductive behaviors is not clearly understood (Clapham and Palsboll, 1997; Kichler et al., 1999; Haynie et al., 2003; Eakley and Houde, 2004; Sorin, 2004; Thom et al., 2004). The general theory states that mating with multiple individuals increases genetic variability of the offspring (Baer and Schmid-Hempel, 2001; Brown and Schmid-Hempel, 2003), although multiple mating is likely to increase mortality because of the risk involved in mate location and mating (Fjerdingstad and Keller, 2004). However, polyandrous females of some organisms demonstrate increased longevity and fecundity when compared to singly mated females (Wagner et al., 2001), whereas multiple matings in others may lead to diminished reproductive returns (Lewis, 2004). Female promiscuity serves to avoid inbreeding (Lehman and Perrin, 2003; Mateo, 2003; Stockley, 2003; Hanson et al., 2004). A female would find it advantageous to mate with many males to increase the genetic diversity of her litter (Jennisons and Petrie, 2000; Tregenza and Wedell, 2000; Zeh and Zeh, 2001). Another hypothesis to explain multiple mating is that some organisms may use polyandry as a means to manipulate sex ratios (Brown et al., 2004). In polyandrous females, sperm competition may also be an important evolutionary force in shaping the reproductive strategies of sexually reproducing organisms (Pilastro et al., 2002).

Only three shark species have been examined for evidence of polyandry. Feldheim et al. (2001a, b) first demonstrated polygamous mating and multiple paternity in the lemon shark, Negaprion brevirostris. In one case, a female N. brevirostris had a litter that was sired by at least three different males. Feldheim et al. (2002) found that 86% of N. brevirostris litters had multiple sires raising the possibility that polyandry may have evolved in viviparous sharks to reduce genetic incompatibilities between the mother and embryos. Both polygyny and polyandry are common in the nurse shark, Ginglymostoma cirratum (Pratt and Carrier, 2001). It is possible that mate encounter is reduced in species that exhibit strong site attachment, and polyandry may serve to decrease inbreeding. Both G. cirratum and N. brevirostris exhibit strong philopatry, returning to their respective nursery areas for mating (Pratt and Carrier, 2001; Feldheim *et al.*, 2002). Polyandry has also been observed in the bonnethead shark, *Sphyrna tiburo* (Chapman et al., 2004) although 18 of 22 females examined carried litters sired by a single male. Direct evidence of polyandry was observed by Chapman et al. (2003) in the southern stingray, *Dasyatis americana*, in which a mating event involved a single female that copulated with two males in rapid succession.

Sperm storage in the oviducal gland has been demonstrated in a number of shark species (Table 14.4). Metten (1941) first demonstrated sperm storage in a shark species, the dogfish, *Scyliorhinus canicula*. Hamlett *et al.* (2002) observed sperm storage in the Oman shark, *Iago omanensis*. Pratt (1993) divided sperm storage in western North Atlantic sharks into three categories: (1) non-storage/immediate insemination, (2) short-term storage/delayed insemination, and (3) long-term storage/repeated insemination. Pratt (1993) further suggested that the plesiomorphic Lamniform sharks do not exhibit sperm storage. However, the senior author has collected spermatophores from the seminal vesicles of the white shark, *Carcharodon carcharias* (Adams *et al.*, 1994). Sperm packaged into

Table 14.4 Sharks examined for evidence of sperm storage.

Species	Sperm storage			Citation
	Non-storage (Days)	Short-term (Wks to months)	Long-term (Months to yrs)	
Scyliorhinus caniculua	?			Metten (1941)
Iago omanensis			?	Hamlett *et al.* (2002)
Alopias vulpinus	x			Pratt (1993)
Alopias superciliosus	x			" "
Lamna nasus	x			" "
Isurus oxyrhynchus	x			" "
Carcharodon carcharias		?	?	Adams *et al.* (1994)
Carcharhinus obscurus			x	" "
Carcharhinus plumbeus			x	" "
Galeocerdo cuvieri			x	" "
Prionace glauca		x		Pratt (1979)
Rhizoprionodon terraenovae		x		Pratt (1993)
Sphyrna tiburo		x		Parsons (1993)
Sphyrna lewini			x	Pratt (1993)
Cetorhinus maximus	x			Matthews (1950)
Furgaleus macki			x	Simpfendorfer and Unsworth (1998)

spermatophores is a mechanism for sperm storage in females of some organisms and suggests sperm storage in *C. carcharias*. The many eggs produced by the Lamnid sharks to nourish the embryos (oophagy) may flush stored sperm from the oviducal gland (Pratt, 1993). However, spermatophores could perhaps prevent the removal of sperm from the gland. Parsons (1993a) used temporal differences between the mating season and conception, and the very narrow confidence intervals around the lengths of developing embryos to suggest sperm storage in *S. tiburo*.

Sperm competition is defined as the probability that the sperm of two or more males will compete for fertilization of ova. In sharks, both polyandry and sperm storage exist together in the same species, creating the potential for sperm competition. However, to our knowledge, sperm competition has never been documented in any elasmobranch species.

SEXUAL SEGREGATION AND SEXUAL DIMORPHISM IN SHARK POPULATIONS

Sexual segregation has been extensively studied, particularly among ungulates, and is known to exist among a wide variety of animal groups, but its causes are poorly understood (Ruckstuhl and Neuhaus, 2000). The hypotheses that could perhaps have been important in the evolution of sexual segregation in shark populations (discussed below) are those that relate sexual segregation and sexual dimorphism in body size. Unlike ungulates, wherein males are typically the largest sex, female sharks are typically larger. Using the maximum size information reported by Cortes (2000) for 169 species/populations within seven families of sharks, we observed that sexual dimorphism is common in most shark species. However, the degree and direction of dimorphism varied between families of sharks. For most families (e.g., Triakidae, Carcharhinidae, Sphyrnidae, Squalidae, Lamnidae, and Alopiidae) 92 to 100% of species/populations show larger females. On the other hand, in the catshark family, Scyliorhinidae, 58% of species/populations show larger females, whereas 42% show larger males. We examined the average dimorphism observed in only those species that show larger females (Figure 14.8). Females of the families Scyliorhinidae, Carcharhinidae, Alopiidae, Triakidae, Sphyrnidae, Lamnidae, and Squalidae are, on average, 1.0, 7.0, 9.8, 12.3, 12.7, 17.7, and 18.2% larger respectively, than males. It may be noteworthy that the oviparous family (Scyliorhinidae) shows the least sexual dimorphism. We suggest that, unlike viviparity, the oviparous reproductive strategy may not place as great an energetic demand upon the female and thus body size is similar between the sexes.

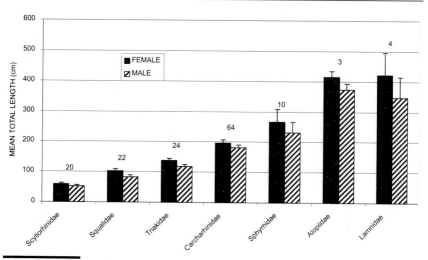

Fig. 14.8 Sexual dimorphism in seven families of sharks. Only species/populations that have females larger than males are plotted. Means and standard errors are shown.

A number of studies have identified sexual segregation by depth, particularly in deep-water shark species. Wetherbee (1996) observed apparent sex and size segregation in southern lantern sharks, *Etmopterus granulosus*, from New Zealand. Size and sex segregation was recorded for black dogfish, *Centroscyllium fabricii*, with females more numerous at deeper depths (Yano 1995). Girard and DuBuit (1999) observed differences in the distribution with depth of the Portuguese shark, *Centroscymnus coelolepis*, from the British Isles. Klimley (1985, 1987) found sexual segregation in groups of the scalloped hammerhead shark, *Sphyrna lewini*. Adult female *S. lewini* were observed circling a seamount in the Gulf of California and actively chased smaller females from the inner circle. Males were able to find and mate with reproductively receptive females by dashing into the inner circle. Female grey reef sharks, *Carcharhinus amblyrhynchos*, were observed to form aggregations during daylight hours only, around Johnson Atoll in the Central Pacific Ocean (Economakis and Lobel, 1998). Goldman (2002) observed sexual segregation in the North Pacific salmon shark, *Lamna ditropis*, with males dominating the western portion of the North Pacific and females the eastern. Sexual and size segregation in the Atlantic sharpnose shark, *Rhizoprionodon terraenovae*, has been known for some time (Springer, 1967). However, Parsons and Hoffmayer (2005a) observed extreme sexual segregation between male and female *R. terraenovae* in the north central Gulf of Mexico. Adult male *R. terraenovae* were collected in large numbers during much of the year from the Mississippi Sound but adult females were rarely collected. After maturation, female *R. terraenovae* moved from the Sound and rarely, if ever, returned, apparently spending their entire adult life offshore. The implications of these observations are that during the summer months, adult males must migrate offshore for mating. Springer (1967) noted that sharks often migrate to specific locations for reproductive reasons. While the mating grounds for some benthic shark species are known (Carrier *et al.*, 1994; Pratt and Carrier, 2001), we are aware of no pelagic shark species wherein this information is available. Additionally, because gravid female *R. terraenovae* were never collected in the Mississippi Sound (although newborn sharks with open umbilical scars were consistently taken during early summer), this likewise suggests that parturition occurs outside the Sound. To our knowledge, disjunct pupping and nursery grounds have not been reported for a shark species. The inshore migration of newborn sharks may be analogous to the scramble of newly hatched turtles across open beach (Parsons and Hoffmayer, 2005a).

There are currently at least five hypotheses proposed to explain the evolution of sexual segregation, but we suggest that only three, all relying on sexual dimorphism in body size, could be used to explain sexual segregation in shark populations. (1) The *predation risk hypothesis* states that the larger sex will seek habitats that have higher food availability irrespective of predation risk, whereas the smaller sex will choose habitats that are first safe from predators, with food abundance/quality being of secondary importance. (2) The *forage selection hypothesis* predicts that the sexes segregate because differences in body size lead to different energy requirements and hence food selection. (3) The *activity budget hypothesis* states that sexual differences in activity budgets and movement rates are the key factors of sexual segregation. Although sexual segregation has been frequently reported for shark species, our understanding of this phenomenon in sharks is too rudimentary to evaluate any of the above hypotheses.

SHARK NURSERY GROUNDS

A variety of organisms use nursery areas to maximize the chances that their young survive this critical life history stage. A discrete nursery implies the existence of a suite of environmental or biological characteristics conducive to growth and survival. Selection of a nursery area is probably dependent upon: (1) an appropriate food supply (quality and quantity), (2) physical conditions (temperature, salinity, etc.) conducive to growth and development, and (3) reduced biological interactions (predation, competition, etc.).

Nursery grounds have been suspected or identified for several shark species (Meek, 1916; Springer, 1967; Parsons 1983, 1993a; Castro, 1993; Pratt and Merson, 1996). Extreme philopatry in *N. brevirostris* was detected using microsatellite genotyping of both females and offspring. Female *N. brevirostris* consistently returned to the same area to give birth and young sharks remained in the area for several years (Gruber *et al.*, 1988; Feldheim *et al.*, 2002). Unfortunately, this species is benthic in nature and may not reflect the situation for more active shark species. Carlson (1999) observed neonate and juvenile sandbar sharks, *Carcharhinus plumbeus*, in the northeastern Gulf of Mexico which suggested the area was a nursery ground for that species. Parsons and Hoffmeyer (2005a) conducted an extensive survey in the north central Gulf of Mexico and identified important nursery areas for sharpnose sharks, *Rhizoprionodon terraenovae*, blacktip sharks, *Carcharhinus limbatus*, finetooth sharks, *C. isodon*, bull sharks, *C. leucas* and perhaps scalloped hammerhead, *Sphyrna lewini* and spinner sharks, *C. brevipinna*. The physical characteristics (dissolved oxygen, salinity, etc.) of sites where sharks were collected were typically not different from areas where sharks were not collected. The commonly measured environmental variables were poor predictors of shark relative abundance.

We suggest that the classic definition of a nursery ground—wherein a helpless and/or vulnerable life history stage must be afforded narrowly defined conditions for survival—may not apply to many shark populations. The extremely dynamic nature of the estuarine environment where newborn and juvenile sharks are often found, make it difficult or impossible to identify the physical parameters that could be used to predict their presence/abundance. Additionally, many sharks are pelagic, highly active, and efficient predators upon birth (some even before birth, see Adelphophagy section) and range over a wide area. Given the above, we suggest that large geographic areas are necessary to support newborn sharks and to maintain healthy shark populations. If the classic concept of a nursery area is applied to sharks, the best place to search for nursery grounds for many pelagic species may be *in-utero*.

LATITUDINAL VARIATION IN REPRODUCTIVE LIFE HISTORY PARAMETERS

The variation in life history parameters with latitude is a well documented phenomenon and has even been expressed as general ecological principles (Bergman's, Allen's, and Jordan's Rules). This variation is evident in bony fish populations (Leggett and Carscadden, 1978; L'Abee-Lund *et al.*, 1989) but the effect of latitude on elasmobranch life histories is not well studied. Parsons (1993a, b) compared two populations of *S. tiburo*, one from Tampa Bay and another from Florida Bay, and found differences in size and age at maturation, time of fertilization, rate of embryonic development, birth size, the energetic investment in producing offspring, gestation period, and the incidence of infertility. In North Pacific waters, Taniuchi *et al.* (1993) observed larger size at maturity at more northern latitudes in male and female shortspine spurdog, *Squalus mitsukurii*, and Yamaguchi *et al.* (2000) found a similar trend in female starspotted dogfish, *Mustelus manazo*. However, these differences were not statistically significant. Carlson and Parsons (1997) and Lombardi-Carlson *et al.* (2003) found clinal variation in *S. tiburo* reproductive parameters between three populations in the Gulf of Mexico and likewise found that growth rates in *S. tiburo* were fastest at more northern latitudes.

While it is clear that some shark populations demonstrate latitudinal variation, it is not always clear whether these differences are the result of ultimate or proximal factors. Management of shark fisheries requires the best available life history information. Recognizing that shark populations demonstrate clinal variation in life history traits means that shark management policies must be drafted that take into account this variation.

Acknowledgments

We are most grateful to Dr. Dennis Goulet and Dr. Trey Driggers for reviewing a draft of the manuscript.

References

Adams, D.H., M.E. Mitchell and G.R. Parsons. 1994. Seasonal occurrence of the white shark, *Carcharodon carcharias*, in waters off the Florida West Coast, with notes on its life history. *Marine Fisheries Review* 56: 24-28.

Adams, S.R. and G.R. Parsons. 1998. Laboratory based measurements of swimming performance and related metabolic rates of field sampled smallmouth buffalo (*Ictiobus bubalus*): A study of seasonal changes. *Physiological Zoology* 71: 350-358.

Andersson, M. 1994. The theory of sexual selection. In: *Sexual Selection*, J. Krebs and T.H. Clutton-Brock (eds.). Princeton University Press, Princeton, pp. 3-31.

Andres, M., M. Solignac and M. Perret. 2003. Mating system in mouse lemurs: Theories and facts, using analysis of paternity. *Folia Primatologica* 74: 355-366.

Baer, B. and P. Schmid-Hempel. 2001. Unexpected consequences of polyandry for parasitism and fitness in the bumblebee, *Bombus terrestris*. *Evolution* 2001: 1639-1643.

Baum, J.K. and R.A. Myers. 2004. Shifting baselines and the decline of pelagic sharks in the Gulf of Mexico. *Ecology Letters* 7: 135-145.

Becher, S.A. and A.E. Magurran. 2004. Multiple mating and reproductive skew in Trinidadian guppies. *Proceedings of the Royal Society of London* B 271: 1009-1114.

Blackburn, D.G. 1995. Saltationist and punctuated equilibrium models for the evolution of viviparity and placentation. *Journal of Theoretical Biology* 174: 199-216.

Breder, C.M. and D.E. Rosen. 1966. *Modes of Reproduction in Fishes*. American Museum of Natural History, New York.

Bretman, A., N. Wedell and T. Tregenza. 2004. Molecular evidence of post-copulatory inbreeding avoidance in the field cricket *Gryllus bimaculatus*. *Proceedings of the Royal Society of London* B 271: 159-164.

Brown, M. and P. Schmid-Hempel. 2003. The evolution of female multiple mating in social insects. *Evolution* 57: 2067-2081.

Brown, W.D., A. Bjork, K. Schneider and S. Pitnick. 2004. No evidence that polyandry benefits females in *Drosophila melanogaster*. *Evolution* 58: 1242-1250.

Bolau, H. 1881. Uber die Paarung und Fortpflanzung der Scyllium-Arten. *Zeitschrift fur Wissenschaftliche Zoologie* 35: 321-325.

Camhi, M.1998. *Sharks on the Line: A State-by-State Analysis of Sharks and Their Fisheries*. National Audobon Society, Islip, New York.

Carey, F.G. and J.V. Scharold. 1990. Movements of blue sharks (*Prionace glauca*) in depth and course. *Marine Biology* 106: 329-342.

Carlson, J.K. 1999. Occurrence of neonate and juvenile sandbar sharks, *Carcharhinus plumbeus* in the northeastern Gulf of Mexico. *Fishishery Bulletin* 97: 387-391.

Carlson, J.K. and G.R. Parsons. 1997. Age and growth of the bonnethead shark, *Sphyrna tiburo*, form northwest Florida, with comments on clinal variation. *Environmental Biology of Fishes* 50: 331-341.

Carrier, J.C., H.L. Pratt and L.K. Martin. 1994. Group reproductive behaviors in free-living nurse sharks, *Ginglymostoma cirratum*. *Copeia* 1994: 646-656.

Carrier, J.C., H.L. Pratt and J.I. Castro. 2004. Reproductive biology of elasmobranchs. In: *Biology of Sharks and Their Relatives*, J.C. Carrier, J.A. Musick and M.R. Heithaus (eds.). CRC Press, London, pp. 269-286.

Castro, J.I. 1993. The biology of the finetooth shark, *Carcharhinus isodon*. *Environmental Biology of Fishes* 36: 219-232.

Castro, J.I. 1996. Biology of the blacktip shark, *Carcharhinus limbatus*, off the southeastern United States. *Bulletin of Marine Science* 59: 508-522.

Castro, J.I. 2000. The biology of the nurse shark, *Ginglymostoma cirratum*, off the Florida east coast and the Bahama Islands. *Environmental Biology of Fishes* 58: 1-22.

Castro, J.I., P.M. Bubucis and N.A. Overstrom. 1988. The reproductive biology of the chain dogfish, *Scyliorhinus retifer*. *Copeia* 1988: 740-746.

Chapman, D.D., M.J. Corcoran, G.M. Harvey, S. Malan and M.S. Shivji. 2003. Mating behavior of southern stingrays, *Dasyatis americana* (Dasyatidae). *Environmental Biology of Fishes* 68: 241-245.

Chapman, D.D., P.A. Prodohl, J. Gelsleichter, C.A. Manire and M.S. Shivji. 2004. Predominance of genetic monogamy by females in a hammerhead shark, *Sphyrna tiburo*: Implications for shark conservation. *Molecular Ecology* 13: 1965-1974.

Clapham, P.J. and P.J. Palsboll. 1997. Molecular analysis of paternity shows promiscuous mating for female humpback whales (*Megaptera novaengliae*, Bovowski). *Proceedings of the Royal Society of London* B 264: 95-98.

Clark, E. 1963. The maintenance of sharks in captivity, with a report on their instrumental conditioning. In: *Sharks and Survival*, P.W. Gilbert (ed.). D.C. Heath, Boston, pp. 115-149.

Compagno, L.J.V. 1984. Sharks of the world. An annotated and illustrated catalogue of shark species known to date. *FAO Species Catalogue*. Vol. 4, Parts 1 and 2. FAO Fish. Synopsis 125. FAO, Rome, Italy.

Compagno, L.J.V. 1988. *Sharks of the Order Carcharhiniformes*. Princeton University Press, Princeton.

Compagno, L.J.V. 2001. Sharks of the World. An annotated and illustrated catalogue of shark species known to date. *FAO Species Catalogue*. Vol. 2. *FAO Species Catalogue for Fishery Purposes*. FAO, Rome, Italy.

Conrath, C.L. and J.A. Musick. 2002. Reproductive biology of the smooth dogfish, *Mustelus canis*, in the northwest Atlantic Ocean. *Environmental Biology of Fishes* 64: 367-377.

Cortés, E. 2000. Life history patterns and correlations in sharks. *Reviews in Fisheries Science* 8: 299-344.

Crespi, B. and C. Semeniuk. 2004. Parent-offspring conflict in the evolution of vertebrate reproductive mode. *The American Naturalist* 163: 635-653.

Dempster, R.P. and E.S. Herald. 1961. Notes on the hornshark, *Heterodontus francisci*, with observations on mating activities. *Occasional Papers of the California Academy of Science* 33: 1-7.

Dulvy, N.K. and J.D. Reynolds. 1997. Evolutionary transitions among egg-laying, live-bearing and maternal input in sharks and rays. *Proceedings of the Royal Society of London* B 264: 1309-1315.

Eakley, A.L. and A.E. Houde. 2004. Possible role of female discrimination against 'redundant' males in the evolution of coulour pattern polymorphism in guppies. *Proceedings of the Royal Society of London* B 271: 299-301.

Ebert, D.A. 1996. Biology of the sevengill shark, *Notorynchus cepedianus* (Peron, 1987) in the temperate coastal waters of Southern Africa. *South African Journal of Marine Science* 17: 93-103.

Emlen, S.T. and P.H. Wrege. 2004. Size dimorphism, intrasexual competition, and sexual selection in wattled jacana (*Jacana jacana*), a sex-role-reversed shorebird in Panama. *The Auk* 121: 391-403.

Feldheim, K.A., S.H. Gruber and M.V. Ashley. 2001a. Population genetic structure of the lemon shark (*Negaprion brevirostris*) in the western Atlantic: DNA microsatellite variation. *Molecular Ecology* 10: 295-303.

Feldheim, K.A., S.H. Gruber and M.V. Ashley. 2001b. Multiple paternity of a lemon shark litter (Chondrichthyes: Carcharhinidae). *Copeia* 2001: 781-786.

Feldheim, K.A., S.H. Gruber and M.V. Ashley. 2002. The breeding biology of lemon sharks at a tropical nursery lagoon. *Proceedings of the Royal Society of London* B 269: 1655-1661.

Feldheim, K.A., S.H. Gruber and M.V. Ashley. 2003. Reconstruction of parental microsatellite genotypes reveals female polyandry and philopatry in the lemon shark, *Negaprion brevirostris*. *Evolution* 58: 2332-2342.

Fjerdingstad, E.J. and L. Keller. 2004. Relationships between phenotype, mating behavior, and fitness of queens in the ant *Lasius niger*. *Evolution* 58: 1056-1063.

Francis, M.P. 1996. Observations on a pregnant white shark with a review of reproductive biology. In: *Great White Sharks: The Biology of Carcharodon Carcharias*, A.P. Klimley and D.G. Ainley (eds.). Academic Press, San Diego, pp. 152-172.

Gilmore, R.G. 1993. Reproductive biology of lamnoid sharks. *Environmental Biology of Fishes* 38: 95-114.

Gilmore, R.G., J.W. Dodrill and P.A. Linley. 1983. Reproduction and embryonic development of the sand tiger shark, *E. taurus* (Rafinesque). *Fishery Bulletin* 81: 201-225.

Girard, M. and M.-H. Du Buit. 1999. Reproductive biology of two deep-water sharks from the British Isles, *Centroscymnus coelolepis* and *Centrophorus squamosus*. *Journal of the Marine Biological Association of the United Kingdom, Plymouth* 79: 923-931.

Goldman, K.J. 2002. Aspects of age, growth, demographics and thermal biology of two Lamniform shark species. Ph.D. dissertation. College of William and Mary, School of Marine Science. Virginia Institute of Marine Science, U.S.A.

Gordon, I. 1993. Pre-copulatory behaviour of captive sandtiger sharks, *Carcharias taurus*. *Environmental Biology of Fishes* 38: 159-164.

Gruber, S.H., D. Nelson and J. Morrissey. 1988. Patterns of activity and space utilization of lemon sharks, *Negaprion brevirostris* in a shallow Bahamian lagoon. *Bulletin of Marine Science* 43: 61-76.

Hamlett, W.C., L. Fishelson, A. Baranes, C.K. Hysell and D.M. Sever. 2002. Ultrastructural analysis of sperm storage and morphology of the oviducal gland in the Oman shark, *Iago omanensis*. *Marine and Freshwater Research* 53: 601-613.

Hanson, B., D. Hasselquist and S. Bensch. 2004. Do female great reed warblers seek extra-pair fertilizations to avoid inbreeding? *Proceedings of the Royal Society of London* B 271: S290-S292.

Haynie, M.L., R.A. Van Den Bussche, J.L. Hoogland and D.A. Gilbert. 2002. Parentage, multiple paternity, and breeding success in Gunnison's and Utah prairie dogs. *Journal of Mammology* 84: 1244-1253.

Hazin, F.H.V., K. Kihara, K. Otsuka, C.E. Boeckman and E.C. Leal. 1994. Reproduction of the blue shark *Prionace glauca* in the south-western equatorial Atlantic Ocean. *Fisheries Science* 60: 487-491.

Hazin, F.H.V., F.M. Lucena, T.S.A.L. Souza, C.E. Boeckman, M.K. Broadhurst and R.C. Menni. 2000. Maturation of the night shark, *Carcharhinus signatus*, in the southwestern equatorial Atlantic Ocean. *Bulletin Marine Science* 66: 173-185.

Hazin, F.H.V., P.G. Oliveira and M.K. Broadhurst. 2002. Reproduction of the blacknose shark (*Carcharhinus acronotus*) in coastal waters off northeastern Brazil. *Fishery Bulletin* 100: 143-148.

Jennions, M.D. and M. Petrie. 2000. Why do females mate multiply? A review of the genetic benefits. *Biological Reviews* 75: 21-64.

Johnson, R.H. and D.R. Nelson. 1978. Copulation and possible olfaction-mediated pair formation in two species of carcharhinid sharks. *Copeia* 1978: 539-542.

Joung, S.J., C.T. Chen, E. Clark, S. Uchida and W.Y.P Huang. 1996. The whale shark *Rhincodon typus*, is a livebearer: 300 embryos found in one 'megamomma' supreme. *Environmental Biology of Fishes* 46: 219-223.

Kichler, K., M.T. Holder, S.K. Davis, R. Marquez and D.W. Owens. 1999. Detection of multiple paternity in the Kemp's Ridley sea turtle with limited sampling. *Molecular Ecology* 8: 819-830.

Klimley, A.P. 1980. Observations of courtship and copulation in the nurse shark, *Ginglymostoma cirratum*. *Copeia* 1980: 878-882.

Klimley, A.P. 1985. The areal distribution and autecology of the white shark, *Carcharodon carcharias*, off the west coast of North America. *Memoirs Southern California Academy of Science* 9: 15-40.

Klimley, A.P. 1987. The determinants of sexual segregation in the scalloped hammerhead shark, *Sphyrna lewini*. *Environmental Biology of Fishes* 18: 27-40.

Leggett, W.C. and J.E. Carscadden. 1978. Latitudinal variation in reproductive characteristics of American shad (*Alosa sapidissima*): evidence for population specific life-history strategies in fish. *Journal of Fisheries Research Board of Canada* 35: 1469-1478.

L'Abee-Lund, J.H., B. Jonson, A.J. Jensen, L.M. Saetten, T.G. Heggberget, B.O. Johnsen and T.F. Naesje. 1989. Latitudinal variation in life-history characteristics of sea-run migrant brown trout *Salmo trutta*. *Journal of Animal Ecology* 58: 525-542.

Lehman, L. and N. Perrin. 2003. Inbreeding avoidance through kin recognition: Choosy females boost male dispersal. *American Naturalist* 162: 638-652.

Lewis, S.M. 2004. Multiple mating and repeated copulations: Effects on male reproductive success in red flour beetles. *Animal Behavior* 67: 799-804.

Liu, H.C. and C.W. Li. 2000. Reproduction in the fresh-water crab *Candidiopotamon rathbunae* (Brachyura: Potamidae) in Taiwan. *Journal of Crustacean Biology* 20: 89-99.

Lombardi-Carlson, L.A., E. Cortés, G.R. Parsons and C.A. Manire. 2003. Latitudinal variation in life-history traits of bonnethead sharks, *Sphyrna tiburo* (Carcharhiniformes:Sphyrnidae) from the eastern Gulf of Mexico. *Marine and Freshwater Research* 54: 875-883.

Macias-Garcia, C. and E. Saborio. 2004. Sperm competition in a viviparous fish. *Environmental Biology of Fishes* 70: 211-217.

Mateo, J.M. 2003. Kin recognition in ground squirrels and other rodents. *Journal of Mammology* 84: 1163-1181.

Manire, C.A. and S.H. Gruber. 1993. Many sharks may be headed for extinction. *Conservation Biology* 4: 10-11.

Marshall, N.B. 1971. *Explorations in the Life of Fishes*. Harvard University Press, Cambridge.

Matthews, L.H. 1950. Reproduction in the basking shark, *Cetorhinus maximus* (Gunner). *Philosophical Transactions of the Royal Society of London* B 612: 247-316.

Meek, A. 1916. *The Migrations of Fishes*. Edward Arnold Publisher, London.

Metten, H. 1941. Studies on the reproduction of the dogfish. *Philosophical Transactions of the Royal Society of London* B 1: 217-238.

Morrissey, J.F. and S.H. Gruber. 1993. Habitat selection by juvenile lemon sharks, *Negaprion brevirostris*. *Environmental Biology of Fishes* 38: 311-319.

Nakaya, N. 1975. Taxonomy, comparative anatomy and phylogeny of Japanese catsharks, Scyliorhinidae. *Memoirs of the Faculty of Fisheries of Hokkaido University* 23: 1-94.

Parsons, G.R. 1983. Reproductive biology of the Atlantic sharpnose shark, *Rhizoprionodon terraenovae*, Richardson. *Fishery Bulletin* 81: 61-73.

Parsons, G.R. 1991. Notes on the behavior of the bonnethead shark, *Sphyrna tiburo* during birth. *Journal of Aquariculture and Aquatic Sciences* 6: 6-8.

Parsons, G.R. 1993a. Geographic variation in reproduction between two populations of the bonnethead shark, *Sphyrna tiburo*. *Environmental Biology of Fishes* 38: 25-35.

Parsons, G.R. 1993b. Age determination and growth of the bonnethead shark *Sphyrna tiburo*: A comparison of two populations. *Marine Biology* 117: 23-31.

Parsons, G.R. and E.R. Hoffmayer. 2005a. Seasonal changes in the distribution and availability of the Atlantic sharpnose shark *Rhizoprionodon terraenovae* in the north central Gulf of Mexico. *Copeia* 2005: 913-919.

Parsons, G.R. and E.R. Hoffmayer. 2005b. Identification and characterization of shark nursery grounds along the Mississippi and Alabama Gulf coasts. In: *Shark Nursery Grounds of the U.S. Atlantic and Gulf of Mexico*, C. McCandless (ed.). American Fisheries Society Publication.

Pilastro, A., M. Scaggiante and M. Rassotto. 2002. Individual adjustment of sperm expenditure accords with sperm competition theory. *Proceedings of the National Academy of Sciences of the United States of America* 99: 9913-9915.

Pratt, H.L. 1979. Reproduction in the blue shark, *Prionace glauca*. *Fishery Bulletin* 77: 445-470.

Pratt, H.L.J. 1993. The storage of spermatozoa in the oviducal glands of western North Atlantic sharks. *Environmental Biology of Fishes* 38: 139-149.

Pratt, H.L. and J.C. Carrier. 2001. A review of elasmobranch reproductive behavior with a case study of the nurse shark, *Ginglymostoma cirratum*. *Environmental Biology of Fishes* 60: 157-188.

Pratt, H.L., Jr. and R.R. Merson. 1996. Pilot Study: Pupping and nursery grounds of the sandbar shark in Delaware Bay. *Report and video to Highly Migratory Species Management Division, Office of Sustainable Fisheries, National Marine Fisheries Service, Silver Spring, MD*.

Ruckstuhl, K.E. and P. Neuhaus. 2000. Sexual segregation in ungulates: a new approach. *Behaviour* 137: 361-377.

Shine, R. 1983. Reptilian viviparity in cold climates: testing the assumptions of an evolutionary hypothesis. *Oceologia* 57: 397-405.

Simpfendorfer, C.A. and P. Unsworth. 1998. Reproductive biology of the whiskery shark, *Furgaleus macki*, off south-western Australia. *Marine and Freshwater Research* 49: 687-693

Sorin, A.B. 2004. Paternity assignment for white-tailed deer (*Odocoileus virginianus*): Mating across age classes and multiple paternity. *Journal of Mammology* 85: 356-362.

Springer, S. 1960. Natural history of the sandbar shark, *Eulamia milberti*. *Fishery Bulletin* 61: 1-38.

Springer, S. 1967. Social organization of shark populations. In: *Sharks, Skates and Rays*, P.W. Gilbert, R.F. Matheson and D.P. Rall (eds.). Johns Hopkins Press, Baltimore, Maryland, pp. 149-174.

Springer, S. and J.D. D'Aubrey. 1972. Two new scyliorhinid sharks from the east coast of Africa, with notes on related species. *South African Association of Marine Biological Research, Oceanogr. Res. Inst., Invest. Rep. #29*, 19 pages.

Stevens, J.D. 1974. The occurrence and significance of tooth cuts on the blue shark (*Prionace glauca* L.). from British waters. *Journal of the Marine Biological Association of the United Kingdom, Plymouth* 54: 373-378.

Stockley, P. 2003. Female multiple mating behaviour, early reproductive failure and litter size variation in mammals. *Proceedings of the Royal Society of London* B 270: 271-278.

Szekely, T., J.D. Reynolds and J. Figuerola. 2000. Sexual size dimorphism in shorebirds, gulls, and alcids: The influence of sexual and natural selection. *Evolution* 54: 1404-1413.

Taniuchi, T., H. Tachikawa and M. Shimizu. 1993. Geographical variation in reproductive parameters of shortspine spurdog in the North Pacific. *Nippon Suisan Gakkaishi* 59: 45-51.

Thom, M.D., D.W. Macdonald, G.J. Mason, V. Pedersen and P.J. Johnson. 2004. Female American mink, *Mustela vison*, mate multiply in a free-choice environment. *Animal Behavior* 67: 975-984.

Thornhill, R. and J. Alcock. 1983. *The Evolution of Insect Mating Systems*. Harvard University Press, Cambridge.

Tortonese, E. 1950. Studi sui Plagiostomi. III. La viviparita: Un fondamentale carattere biologico degli Squali. *Archivio Zoologico of Italiano* 35: 101-155.

Tregenza, T. and N. Wedell. 2000. Genetic incompatibility, mate choice and patterns of parentage: Invited review. *Molecular Evolution* 9: 1013-1027.

Tricas, T.C. and E.M. LeFeuvre. 1985. Mating in the white tip reef shark, *Triaenodon obesus*. *Marine Biology* 84: 233-237.

Trivers, R.L. 1974. Parent-offspring conflict. *American Zoologist* 14: 249-264.

Uchida, S., M. Toda and Y. Kamei. 1990. Reproduction of elasmobranches in captivity. In: *Elasmobranchs as Living Resources: Advances in the Biology, Ecology, Systematics and Status of the Fisheries*, H.L. Pratt, S.H. Gruber and T. Taniuchi (eds.). NOAA Technical Report NMFS 90: 211-237.

Vieites, D.R., S. Nieto-Roman, M. Barluenga, A. Palanca, M. Vences and A. Meyer. 2004. Post-mating clutch piracy in an amphibian. *Nature (London)* 431: 305-308.

Wagner, W.E., J.K. Robert, R.T. Kayleen and C.J. Harper. 2001. Females receive a life-span benefit from male ejaculates in a field cricket. *Evolution* 55: 994-1001.

West, J.G. and S. Carter. 1990. Observations on the development and growth of the epaulette shark, *Hemiscyllium ocellatum* (Bonnaterre) in captivity. *Journal of Aquaculture and Aquatic Sciences* 5: 111-117.

Wetherbee, B.M. 1996. Distribution and reproduction of the southern lantern shark from New Zealand. *Journal of Fish Biology* 49: 1186-1196.

Wourms, J.P. 1977. Reproduction and development in chondrichyan fishes. *American Zoologist* 17: 473-515.

Wourms, J.P. 1981. Viviparity: The maternal-fetal relationship in fishes. *American Zoologist* 21: 473-515.

Yamaguchi, A., T. Taniuchi and M. Shimizu. 2000. Geographic variation in reproductive parameters of the starspotted dogfish *Mustelus manazo* from five locations in Japan and in Taiwan. *Environmental Biology of Fishes* 57: 221-233.

Yano, K. 1995. Reproductive biology of the black dogfish, *Centroscyllium fabricii*, collected from waters off western Greenland. *Journal of the Marine Biological Association of the United Kingdom, Plymouth* 75: 285-310.

Zeh, J.A. and D.W. Zeh. 1996. The evolution of polyandry. I. Intragenomic conflict and genetic incompatibility. *Proceedings of the Royal Society of London* B 264: 69-75.

Zeh, J.A. and D.W. Zeh. 2001. Reproductive mode and the genetic benefits of polyandry. *Animal Behavior* 61: 1051-1063.

Fish Estrogenic Pathways: Chemical Disruption and Related Biomarkers

Augustine Arukwe

INTRODUCTION

The terms environmental estrogens, endocrine disruptors, endocrine modulators, eco-estrogens, environmental hormones, xenoestrogens, hormone-related toxicants, and phytoestrogens all have one thing in common. They describe synthetic chemicals and natural plant or animal compounds that may affect the endocrine system (the biochemical messengers or communication systems of glands, hormones and cellular receptors that control the body's internal functions) of various organisms. Many of the effects caused by these substances have been associated with developmental, reproductive and other health problems such as cancer, in wildlife and laboratory animals (for reviews: see Colborn et al., 1993; Gray et al., 1998). There is also growing concern that these compounds may be affecting humans in similar ways (Toppari et al., 1996; Gray, 1998).

Author's address: Department of Biology, Norwegian University of Science and Technology (NTNU), Høgskoleringen 5, 7491 Trondheim, Norway. E-mail: arukwe@bio.ntnu.no

Fish reproductive endocrinology is governed by a complex set of intricate pathways that require communication and interaction along system axes. Therefore, it is not surprising that hormones and other physiological systems not typically associated with reproduction may play a role in ensuring that reproduction occurs at a time of the year optimal for survival of the offspring (Duston and Bromage, 1987; Davies et al., 1999). In the temperate regions, seasonal cyclicity in reproduction is important to ensure that the timing of spawning coincides with compatible environmental condition and nutritional supplies necessary for fry survival and growth. In salmonids, for example, reproduction is a seasonal event (Ueda et al., 1984; Cyr et al., 1988). As a consequence, maturation only occurs if a reaction norm is intersected prior to a critical time of the year to ensure an optimal timing of spawning. Otherwise, maturation is delayed by one year, thus implying age-specific size threshold rather than continuous reaction norm (Stearns, 1983; Stearns and Koella, 1986). During sexual maturation, the internal responses that are synchronized by external signals depend upon some genetically determined performance threshold, and maturation processes will continue if this performance exceeds a set point at this critical time in winter or spring (Thorpe, 1994), thereby rendering the maturation process and the associated energetic demand extremely susceptible to pollutant exposure with potentially deleterious ecological consequences.

ESTROGEN PATHWAYS

Along with the nervous system, the endocrine systems, which are composed of glands that secrete chemical messengers, are also composed of communication systems that regulate all responses and functions of the body. Unlike the nervous system, which sends rapid-fire signals via electrochemical transmission along neuron conduits to the brain where they are deciphered and relayed to appropriate parts of the body to elicit instantaneous responses, the endocrine system provides cellular instructions more subtly and slowly via chemical messengers. Hormones that are produced in the endocrine glands in one part of the body travel through the bloodstream until they encounter special receptors with which they interact to initiate essential biological responses in specific target tissues (Gustafsson, 2000; Moggs and Orphanides, 2001; Nilsson and Gustafsson, 2002; Oettel, 2002).

Reproductive development is a continuous process throughout ontogeny and, therefore, may be susceptible to chemical insult. Many environmental chemicals (see Fig. 15.1) exhibit estrogenic activity (Arukwe, 2001). Some of these chemicals occur naturally in plants and fungi; others are manmade by-products that are present in agricultural and industrial chemicals. Endocrine-modulating compounds, both environmental and endogenous, may interact with steroid hormones and their receptors, or other hormones and transcription factors in the biochemical pathway of hormonal activity. The detailed mechanisms by which xenoestrogenic compounds affect fish estrogenic pathways are not fully understood, but it is known that they can bind with high affinity to the ER (as agonists) and initiate cell synthetic processes typical of natural estrogens (e.g., induction of oogenic proteins). Some compounds also have the ability to bind to the receptor, but not eliciting estrogenic activities (as antiestrogens or antagonists), thereby blocking the binding site of natural estrogens (Spink *et al.*, 1994; Ahlborg *et al.*, 1995; Safe *et al.*, 2000; Wormke *et al.*, 2000; Safe and Wormke, 2003).

ESTROGEN RECEPTORS

In organisms, steroid hormone actions are mediated by specific receptors whose physiological activities control the transcription of target genes and elicit a cascade of cellular events. The estrogen receptors (ERα and ERβ) are members of the nuclear receptor (NR) super family of ligand-activated transcription factors that includes steroid, thyroid, and retinoid receptors (along with other orphan receptors with undefined ligands). Structurally, the ERs are composed of six domains (A-F; with various degrees of sequence conservation). The C-domain (i.e., DNA-binding domain; DBD) is responsible for specific interaction with the estrogen response element (ERE), while the E-domain (i.e., ligand-binding domain; LBD) interacts specifically with agonists and antagonists (hormones and non-hormones). The amino (NH_2) terminal (A/B), carboxy (COOH) terminal (F), and the hinge (D) regions, as well as the transcription activation functions AF-1 (ligand-independent transcription activation domain) and AF-2 (ligand-inducible activation domain) are all structural aspects of the ERs. Between the ERα and ERβ gene variants, there is a modest overall sequence identity (47%). Specifically, the A/B domain is poorly conserved between the two ER isoforms (17% identity). No clear secondary structure can be identified in these regions and no structural data have until now been obtained. In contrast, there is functional and structural data available for C and E regions (Ruff *et al.*, 2000; Gangloff *et al.*, 2001). The C and E regions harboring the DBD and LBD, respectively, are highly conserved between the ER isoforms (97 and 59%, respectively, Gustafsson, 1999; Nilsson and Gustafsson, 2002). The two remaining regions, D and F, are again of variable size and are not conserved. The D region can be considered as a linker peptide between the DBD and the LBD, whereas F is a C-terminal extension region of the LBD.

The ERs bind to estrogen response elements (EREs) and activate transcription in an estrogen concentration dependent manner (Katzenellenbogen, 1996). This transcriptional activation requires the recruitment of co-activator complexes (Tremblay et al., 1997; Tremblay and Giguere, 2002). Recently, ER complementary DNA (cDNA) have been cloned and characterized in several vertebrate species (Pakdel et al., 1990; Kuiper et al., 1996; Mosselman et al., 1996; Tremblay et al., 1997; Tchoudakova et al., 1999; Socorro et al., 2000). Ligand–activated receptors bind to promoter regions of specific genes, recruiting co-activators or co-repressors and other transcription machinery that brings about gene expression (Nilsson and Gustafsson, 2002). After the alteration of specific mRNA transcripts, nuclear mRNA is processed and transported to the cytoplasm, where protein translation occurs. Thus, the ER-ERE interaction is an important link in the event cascade that leads to changes in cellular functions (Nilsson and Gustafsson, 2002).

STEROIDOGENIC ENZYMES

Tissue-specific enzymes belonging to the cytochrome P450 super family mediate steroid hormone synthesis and metabolism. Recent studies have shown that environmental pollutants known to mimic the action of endogenous estrogens (e.g., estradiol-17β) also cause variations in isoforms of hepatic cytochrome P450 dependent steroid- and xenobiotic-metabolizing enzymes (Arukwe et al., 1997) and variations in aromatase (CYP19) isotypes gene expression and activity patterns in the brain and liver (Ankley et al., 2002). The CYP19 gene that belongs to the CYP gene super family encodes aromatase. Aromatase catalyzes the conversion of C19 androgens to estrogens, a reaction that essentially involves the removal of the C19 carbon and aromatization of the A rings of the steroid skeleton. In addition to steroidogenic tissues, this reaction occurs at a number of non-steroidogenic sites and serves various physiological roles through the production of estrogens. The regulatory pathways involving gonadotropins, steroid hormones, and growth factors control the expression of aromatase (Orlando et al., 2002; Kagawa et al., 2003; Min et al., 2003).

Similar to other CYP enzymes, aromatase expression is also responsive to environmental factors, including nutritional elements and chemical exposures (Liu et al., 1996; Om and Chung, 1996). A number of synthetic compounds have been shown to affect aromatase expression, both in vivo (Rocha Monteiro et al., 2000; McAllister and Kime, 2003) and in vitro (Pelissero et al., 1996). Compared to humans with a single CYP19 gene in the haploid genome that shows multiple tissue-specific promoters and first exons (Conley and Hinshelwood, 2001), teleost fish have at least two separate and distinct CYP19 loci with subdivided expression domains (Callard et al., 2001; Tchoudakova et al., 2001). For example, in both goldfish (Tchoudakova et al., 2001) and zebrafish (Kishida and Callard, 2001), CYP19b encodes the P450 aromB isoform in the brain, retina and pituitary, tissues that express exceptionally high enzyme and mRNA levels, whereas low levels of enzyme activity in ovary are associated with low levels of a second mRNA isoform (P450 aromA) encoded by one or more CYP19a genes.

In fish, several constitutive cytochrome P450 forms are developmentally regulated and display sex-specific expression (Andersson and Förlin, 1992; Winzer et al., 2002; Sepulveda et al., 2004). In fish species that exhibit a strong sex difference the similarities between juvenile and reproductively active males and females have led to suggestions that the sex differences are due to a suppression of cytochrome P450 isoenzyme expression in reproductively active females (Stegeman and Hahn, 1994). There is a good number of reports describing higher levels of cytochrome P450 and higher activities of several monoxygenase reactions in males of different fish species (Hansson et al., 1980; Stegeman and Chevion, 1980; Förlin and Hansson, 1982; Larsen et al., 1992; Arukwe and Goksøyr, 1997) supporting the concept that a sexual differentiation of the hepatic microsomal cytochrome P450 monooxygenase activities occur in late stages of sexual maturation in fish. The sex differences in P450 expressions indicate that estradiol-17β (E_2) or other sex steroids may affect the catalytic activity of P450 isoenzymes directly. Recent studies have shown that estrogen mimics also affect P450 gene, enzyme and protein expressions in similar ways. It is most likely that such in situ regulation of enzyme activity may be of critical importance in the signal transduction processes that regulate maturation and spawning in fish.

XENOBIOTIC EFFECTS ON ESTROGENIC PATHWAYS

It has been known since the 1930s (Dodds and Lawson, 1936; Bradbury and White, 1954; Tullner, 1961; Bitman et al., 1968; Bitman and Cecil,

1970) that common contaminants in the environment act in a similar way as hormones, with a resulting ability of influencing endocrine regulations. For example, apart from plant constituents (phytoestrogens), synthetic compounds such as methoxychlorine, DDT (1,1,1-trichloro-2,2-bis(chlorophenyl)ethane), and polychlorinated biphenyls (PCBs) may produce estrogenic activities in laboratory animals (Fig. 15.1). Recently, the increased interest in the problem of environmental estrogens can be explained by: (1) recent findings on sex differentiation and reproductive biology in wildlife populations including, e.g., fish, reptiles and birds (for review see Colborn and Clement, 1992; Arukwe, 2001; Arukwe and Goksøyr, 2003); (2) results of experiments using fish as bioindicators in draining sections of British sewage treatment plants (Purdom *et al.*, 1994); and (3) recent epidemiological studies on the development and function of human genitals, especially of males (Kavlock *et al.*, 1996; Toppari *et al.*, 1996).

Many estrogenic contaminants are produced for specific purposes and are used in pesticides, plastics, electrical transformers and other products (for review see Caldwell, 1985; Ahlborg *et al.*, 1995). Other substances

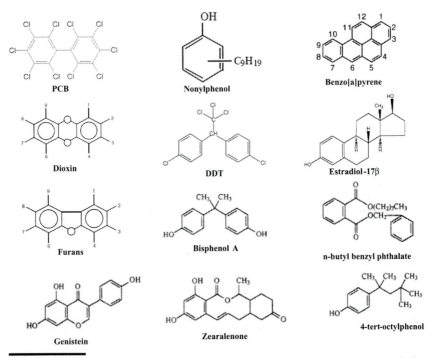

Fig. 15.1 Structures of selected naturally occurring and industrial estrogenic chemicals.

are generated as by-products during manufacturing or are breakdown products of some other chemical, and some, like 17β-ethinylestradiol and diethylstilbestrol (DES), are drugs. Natural compounds capable of disrupting estrogenic pathways, such as the phytoestrogens and mycoestrogens which occur in a variety of plants and fungi, have also been studied in fish (Pelissero *et al.*, 1991, 1993; Arukwe *et al.*, 1999a; Celius *et al.*, 1999). Regardless of the source or original intended use, substantial amounts of these chemicals end up in the aquatic environment due to physico-chemical, hydrologic and atmospheric processes (Guardans and Gimeno, 1994; Bjerregaard *et al.*, 1998).

A large number of reports in chemical modulation of estrogen pathways have used mammalian and non-mammalian estrogen receptor based *in vitro* assays such as purified ER assays; breast cancer cell and hepatic cell-lines; yeast assays with transfected human ER genes. In addition, some reports have used *in vitro* assays for detecting Vtg and Zr-protein induction in fish hepatocytes (Celius *et al.*, 1999; Smeets *et al.*, 1999; Kordes *et al.*, 2002; Rankouhi *et al.*, 2002). Metabolic activation might results in the formation of more or less potent estrogenic compounds. Given that *in vitro* systems lack the essential metabolic competence that is characteristic of *in vivo* system and, therefore, not be able to incorporate the endocrine effects caused by metabolites or endocrine modulation caused by interference with steroid metabolism, some caution should be used when extrapolating to fish *in vivo* cases. Furthermore, there may still be important differences between species in the steroid hormone structure and function, despite being highly conserved during the evolution of the vertebrates. Finally, the various endocrine pathways in intact animals are extremely complex and interdependent (Cyr and Eales, 1996); so, effects in an *in vitro* assay will not necessarily be replicated *in vivo*. Nevertheless, *in vitro* systems provide good models for mechanistic evaluation of the disruption of estrogenic pathways by chemical contaminants in the environment. Laboratory studies have been conducted to evaluate the impact of fish exposure to toxicants on ovarian development. Several effects have been observed and these include inhibition of oocyte development and maturation, increased follicular atresia of both yolked and previtellogenic oocytes, abnormal yolk deposition and formation within oocytes, and abnormal egg maturation and production (for reviews: see Lam, 1983; Kime, 1995; Arukwe and Goksøyr, 2003).

Chemical disruption of estrogen pathways may either be of activational or organizational nature (Guillette Jr *et al.*, 1995). Unfortunately, most studies of estrogen disruptors have focused on their activational modifications on adult model systems. Usually, activational effects occur as transitory actions during adulthood (Phoenix *et al.*, 1959), while organizational effect occurs early in an individual lifetime and induces permanent effects. Both activation and organization effects have been useful concepts in explaining the role of hormones in the differentiation of vertebrate sexual dimorphism with morphological, physiological or behavioral origin (Arnold and Breedlove, 1985; Guillette Jr *et al.*, 1995). Given the pivotal role played by hormones in sexual development and reproduction, it is obvious that the organizational and activational concepts are central to the role of environmental estrogen disrupting contaminants. For a complete assessment of the environmental impact of EDCs, their organizational and re-organizational effects of embryos and/ or adult individuals should be a major concern. In this respect, it is important to note that many organizational modifications do not become apparent until later in life.

BIOMARKERS AND BIOASSAYS

Biological indicators (biomarkers) for monitoring environmental contamination have developed to an informative tool for detecting exposure and effects of chemical pollutants (McCarthy and Shugart, 1990). Biomarkers have been defined as biological responses that can be related to an exposure to, or toxic effects of, an environmental chemical(s) (Peakall, 1994). Early detection of environmental pollution is important because once the impact has advanced to a stage where populations or even the entire ecosystem changes, irreparable damage may have occurred. Almost without exception, the effects of pollutants have molecular basis prior to clinical manifestation. Therefore, changes in cell homeostasis, and cytological defence mechanisms will respond through compensatory processes for the toxic effect of the pollutants (Huggett *et al.*, 1992; Peakall, 1992). This compensatory mechanism (i.e., adaptation) will progress until the cell is no longer able to cope with the stress, and the effect will be transmitted to the next level of biological organization (i.e., tissue or organ, see Fig. 15.2). Dysfunctional tissue or organ may affect the fitness and survival of fish (or other organisms) and effects of pollutants might thus be transmitted to the higher levels of biological organization (i.e., population, community and ecosystem level). At the population, community and ecosystem levels of biological organizations, pollutants effects have already reached an irreversible stage.

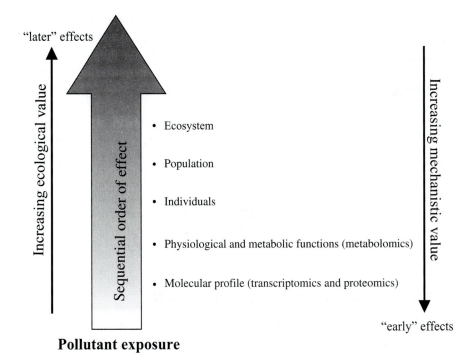

"later" effects

Increasing ecological value

Sequential order of effect

Increasing mechanistic value

- Ecosystem

- Population

- Individuals

- Physiological and metabolic functions (metabolomics)

- Molecular profile (transcriptomics and proteomics)

"early" effects

Pollutant exposure

Fig. 15.2   Visualization of the sequential order of effects to pollutant exposure within a biological system. Effects at higher levels are normally preceded by changes in the "earlier" biological processes, allowing the development of biomarker for early warning signals of effects at "later" response level. Effects at higher organization levels has increased ecological significance, while effects at lower levels has increased mechanistic value.

The use of biomarkers has several advantages compared to traditional chemical analyses, because it integrates the cumulative effects of multiple pollutants and is inexpensive to perform (Peakall, 1992). In evaluating the feasibility of using an induced response as a biomarker for chemical contamination, the following major factors should be considered: (a) the environmental relevance of the response, (b) its relationship to biological effects at higher level of organization, and (c) its utility in the field (Stegeman *et al.*, 1992). To be useful as a diagnostic screening tool, a biomarker response/assay must possess the following criteria: (1) it should have an early warning capacity, i.e., the response should be obtained before the effects occur at higher levels of biological organization, (2) it should be sensitive, reliable, and easy to measure, (3) it should respond to pollutants in a dose-response manner over environmentally relevant concentration ranges of a given pollutant, (4) it must be carefully evaluated with respect to endogenous and exogenous factors that may affect the response measured, (5) it should be fairly specific in order to identify the active chemical or group of chemicals, and (6) it must be reproducible and quantifiable in order to be related specifically to a certain degree of physiological disorder such as impaired growth or reproduction, or energy utilization that may directly affect survival of an organism. Basically, increases in the levels of a biomarker should be correlated with "health" and "fitness" of an organism (Depledge, 1989, 1994).

A vast majority of biomarker analyses implies the sacrifice of organism to be studied. The use of non-destructive biomarkers has become important in order to minimize the animals' sacrifice, notably in performing experiments with endangered, protected or threatened species, or when measurement of a biomarker response are to be obtained from a given individual subjected to constant and variable chemical insult (Fossi et al., 1992, 1994; Fossi, 1994).

EGG YOLK AND EGGSHELL PROTEINS

In oviparous animals, accumulation of yolk materials into oocytes during oogenesis and their mobilization during embryogenesis are key processes for successful reproduction. Most oocyte yolk proteins and lipids are derived from the enzymatic cleavage of complex precursors, predominantly vitellogenin (Vtg) and very low-density lipoprotein (Weigand, 1982; Tyler and Sumpter, 1996). Vitellogenesis is defined as E_2-induced hepatic synthesis of egg yolk protein precursor, Vtg, its secretion and transport in blood to the ovary and its uptake into maturing oocytes (Norberg and Haux, 1985; Mommsen and Walsh, 1988). For a detailed overview of vitellogenesis see **chapter Four** of this book. Vtg is an important source of nutrients for egg and larvae, making the vitellogenesis an important developmental process. In addition, teleost eggs contain maternal sex steroids (Feist et al., 1990), cortisol, and other lipophilic hormones like thyroxin that may enter the egg through Vtg (Mommsen et al., 1999). The biological role(s) of hormones in eggs is not well understood. However, it has been hypothesized that they may act as metabolites or as synergists with other substances during early development (Matsubara et al., 1999; Mommsen et al., 1999; Sorbera et al., 2001; Wong and Zohar, 2004).

The envelope surrounding the animal egg plays significant roles in the reproductive and developmental processes; firstly as an interface between the egg and sperm, and secondly as an interface between the embryo and its environment (Grierson and Neville, 1981). The egg envelope is a major structural determinant of the eggshell in fish, and is often referred to as *zona radiata* because of its striated appearance under the light microscope (Ginsburg, 1963). In mammals, these proteins function as sperm receptors and undergo a hardening process (also in fish) after fertilization. This process is important for the prevention of polyspermy, because the fish eggshell contains only one narrow canal or micropyle through which sperm gain access to the egg. In fish, the egg envelope is much thicker than in mammals, providing physical protection from the environment and playing a role in diffusive exchange of gases (Grierson and Neville, 1981). The micropyle is closed within minutes after the eggs are activated by exposure to fresh water, which initiates a cortical reaction necessary for development of fertilized eggs (Ginsburg, 1963). Ionic concentration of the medium lower than 0.1 M is needed for complete activation (Yamamoto, 1951). After activation, the *zona radiata* takes up water, gains resistance to breakage and can support up to 100 times more weight than oviductal eggs (Manery *et al.*, 1947; Eddy, 1974). Zonagenesis is the E_2-induced hepatic synthesis of eggshell proteins, *zona radiata* proteins (Zr-*proteins*), their secretion and transport in blood to the ovary and uptake into maturing oocytes.

MOLECULAR MECHANISMS FOR VITELLOGENESIS AND ZONAGENESIS

Vitellogenesis and zonagenesis are crucial for the reproduction of oviparous animals. The cellular and molecular events that occur in tissues producing these proteins and in the ovary provide ideal systems for the study of several fundamental biological processes (Wahli, 1988). For example, the abundantly transcribed Vtg genes are being used to analyze stage-, sex-, tissue- and hormone-specific gene expression. One research area that has received a lot of attention in recent time is xenobiotic modulation of gene expression in organisms. Thus, selective gene expression is considered to be central to our understanding of cellular differentiation and the regulation of developmental processes (Hough-Evans *et al.*, 1979). The term 'gene expression' is not always well defined, but it is most often used to indicate a change in the nature of, or rate at which, different genes are transcribed (Tata and Smith, 1979; Tata *et al.*, 1983). Recent advances in studies of the organization of eukaryotic genomes have also focused attention on the importance of structural features of expressed and unexpressed genes and on the post-transcriptional mechanisms that would determine the processing of primary transcripts into the correct messenger sequences (Clemens *et al.*, 1975; Tata, 1976; Dimitriadis and Tata, 1982).

A simplified general model of the molecular mechanisms that leads to the production of Vtg and Zr-proteins in the hepatocyte is shown in Figure 15.3. E_2 produced by the ovarian follicular cells in response to GtH I (Fig. 15.4) enters the cell by diffusion. There, the E_2 is retained in target cells by high affinity binding to a specific steroid-receptor protein (such as the estrogen receptor, ER). The hormone-receptor dimerizes

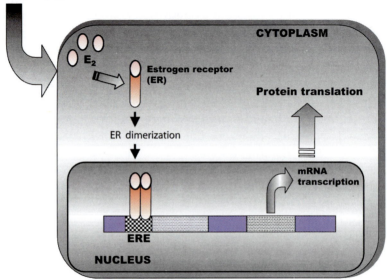

Fig. 15.3   Molecular mechanism of estradiol-17β (E_2) or estrogen mimics stimulated hepatic protein synthesis, their secretion and transport in blood to the ovary and their uptake into maturing oocytes) in female teleosts. E_2 produced by the ovarian follicular cells in response to GtH I enters the cell by diffusion and retained in target cells by high affinity binding to a specific steroid-receptor protein (such as the E_2-receptor, ER). There, the hormone-receptor complex binds tightly (after dimerization); in the nucleus at estrogen responsive elements (ERE) located upstream of, or within the estrogen-responsive genes in DNA and results in the activation or enhanced transcription of genes (e.g., vitellogenin and *zona radiata* protein genes) and subsequent increase and stabilization of messenger RNA (mRNA). At present, ERE for *zona radiata* protein genes have not been identified in fish. Given the speculation that different EREs on the DNA may be temporarily masked by associated proteins, thus resulting in sequential or partial induction of various estrogenic responses (Ruh *et al.*, 1988), it is possible that there may be subtle differences in the responsive elements for vitellogenin and *zona radiata* protein.

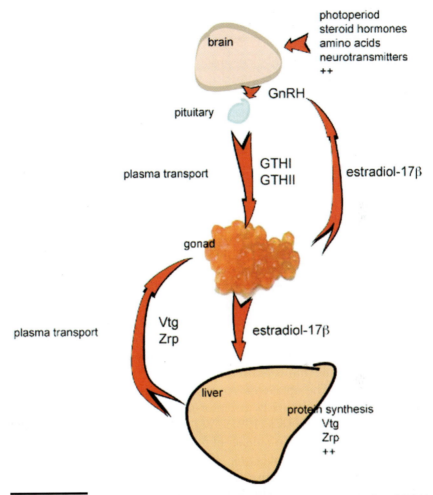

Fig. 15.4 Schematic representation of the hypothalamus-pituitary-gonadal-liver (HPGL) axis during reproductive protein synthesis in female teleosts. The HPGL is regulated through the negative feedback mechanism by estradiol-17β. The hypothalamus, pituitary, gonad and liver are all potential targets for endocrine disruptors, as discussed in the text. GtH = gonadotropin II & I. Reproduced with permission from Arukwe and Goksøyr (2003).

and binds tightly in the nucleus at estrogen responsive elements (ERE) located upstream of, or within the estrogen-responsive genes in DNA. This results in the activation or enhanced transcription of Vtg genes and subsequent increase and stabilization of Vtg messenger RNA (mRNA). Recent reports have shown that the ERs are actually located in the cell nucleus. At present, ERE for Zr-proteins genes have not been identified in fish. Given the speculation that different EREs on the DNA may be temporarily masked by associated proteins, thus resulting in sequential or partial induction of various estrogenic responses (Spelsberg *et al.*, 1988), it is possible that there may be subtle differences in the responsive elements for Zr-proteins and Vtg. Zr-proteins and Vtg precursors are modified extensively in the rough endoplasmic reticulum (RER) and modified Zr-protein and Vtg are secreted into the serum for transport to the ovary.

The post-translational modifications occurring to the Zr-proteins prior to secretion into the systemic tracks are not well understood. However, more is known about Vtg post-translational modifications in teleost fish. Prior to secretion into the blood stream, the biochemical information concerning Vtg clearly indicates that substantial post-translational modification must occur in the liver cell in order to reach the end product seen in the serum. Several changes in hepatic morphology such as proliferation of RER and Golgi apparatus also accompany estrogen stimulation. Firstly, the protein backbone of the Vtg is synthesized on membrane-bound ribosomes. Vtg shares this feature with other proteins destined for secretion from the hepatocytes (Lewis *et al.*, 1976). Thereafter, the Vtg molecule is lipidated, glycosylated and phosphorylated. Although some information exists concerning the nature and extent of modifications of the Vtg molecule, rather limited information is available for fish with respect to the mechanisms, sequential events or location of these transformations. The uptake of Vtg by growing oocytes is rapid, specific and saturable, and occurs by receptor-mediated endocytosis (Opresko and Wiley, 1987; Opresko, 1991). Vtg receptors (VTGRs) have been identified in the ovary of a number of fish species (Chan *et al.*, 1991; Tao *et al.*, 1996; Tyler and Lubberink, 1996), and was recently cloned and sequenced in rainbow trout and winter flounder (Prat *et al.*, 1998; Hiramatsu *et al.*, 2002). The fish VTGRs are 70-80% similar to the chicken very low-density lipoprotein receptor VLDLR (ibid.).

CHEMICAL MODULATION OF ESTROGEN-INDUCED GENE AND PROTEIN EXPRESSION

The detailed mechanisms by which xenoestrogenic compounds mediate their induction of protein expressions is not fully understood, but it is known that they can bind with high affinity to the ER (as agonists) and initiate cell synthetic processes typical of natural estrogens (Routledge *et al.*, 2000; Thomas, 2000; Arukwe and Goksøyr, 2003). Some compounds also have the ability to bind to the receptor, but not eliciting estrogenic activities (as antiestrogens or antagonists), thereby blocking the binding site of natural estrogens (Safe and Krishnan, 1995; Ramamoorthy *et al.*, 1999). During ovarian recrudescence, incorporation of reproductive proteins accounts for the major growth of the developing oocytes. Therefore, a probable indirect measure of altered hepatic reproductive protein synthesis in fish exposed to xenobiotics is reduced or increased gonadosomatic index (GSI). A more direct quantification of these alterations can be obtained from plasma, hepatic and ovarian oogenic (Vtg and Zr-proteins) protein concentrations (Kime, 1995). Modern and advanced molecular biology techniques are revolutionizing the process of oogenic protein quantitation in oviparous species (Snape *et al.*, 2004).

Laboratory studies have been conducted to evaluate the impact of fish exposure to toxicants on ovarian development. Several effects have been observed and these include inhibition of oocyte development and maturation, increased follicular atresia of both yolked and previtellogenic oocytes, abnormal yolk deposition and formation within oocytes, and abnormal egg maturation and production (Lam, 1983; Susani, 1986; Kime, 1995; Arukwe and Goksøyr, 1998, 2003). Wester and Canton (1986) observed the development of testis-ova in males and induced vitellogenesis in either sex of medaka (*Oryzias latipes*) exposed to β-HCH, demonstrating estrogenic effects of this compound. Similar responses have been observed when medaka was exposed to 4-nonylphenol (NP) and to bisphenol in more recent studies (Gimeno *et al.*, 1996; Gray and Metcalfe, 1997; Kang *et al.*, 2002).

ASSAYS FOR CHEMICAL DISRUPTION OF ESTROGEN PATHWAYS

Prior to clinical manifestation, the first interaction between toxic compounds and organisms occurs at the molecular and cellular levels. Therefore, changes in gene expression as a result of environmental stressors and the subsequent molecular processes that lead to adverse health effects may be used as quantitative and qualitative marker for cellular, organismal and population effects. Piferrer and Donaldson (1989) showed that exposure of male fish larvae to estrogenic hormones during the sensitive part of gonadal development can completely feminize the phenotype. Male genotypes become apparently normal female phenotypes as a result of this feminization effect. Recently, it was reported by Jobling *et al.* (1996) that exposure of rainbow trout to four different alkylphenolic chemicals that induced Vtg synthesis, also inhibited testicular growth.

Exposure of juvenile fish to sewage or oil refinery treatment plant effluents in the laboratory resulted in the induction of Zr-protein and Vtg synthesis (Jobling *et al.*, 1996; Arukwe *et al.*, 1997b).

In designing a bioassay for chemical disruption of fish estrogen pathways, toxicologists have used the induction of Vtg and Zr-protein in male and juvenile oviparous vertebrates as an effective and sensitive biomarker for xenoestrogens (Heppell *et al.*, 1995; Folmar *et al.*, 1996; Arukwe *et al.*, 1997b, 2001; Celius and Walther, 1998; Palmer *et al.*, 1998). Both genomic and proteomic assay methods can be used in evaluating Vtg and Zr-protein inductions in oviparous species (Fig. 15.5). Depending on the target organ or tissue, a wide variety of assays have been developed to measure oogenic protein expression in fish. These include radioimmunoassays, enzyme-linked immunosorbent assays (ELISAs) and immunohistochemistry using monoclonal and polyclonal antibodies (Abs), RNA protection assay and transcript analysis by Northern blotting or various variants of polymerase chain reaction (PCR). Recently, the use of real-time (quantitative) PCR is increasingly becoming a valuable tool in assessing chemical disruption of estrogenic pathways. In plasma samples, these assays vary in their sensitivity, but some have the ability to detect very low levels of protein expression, i.e., 1 ng/ml or

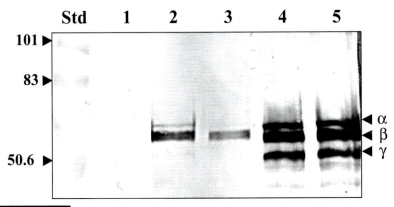

Fig. 15.5   Immunoblot analysis of *zona radiata* proteins *(Zr*-proteins), in plasma of control (1), 4-nonylphenol (NP; single i.p. injection at 125 mg/kg body weight; **Lanes 2 and 3**) and estradiol-17β (E_2 : 5 mg/kg body weight **Lanes 4 and 5**) treated juvenile salmon. *Zr*-proteins were probed with homologous primary polyclonal rabbit anti-salmon *Zr*-proteins antiserum; Goat anti-rabbit Horseradish peroxidase (GAR-HRP) was used as secondary antibody. **Std:** Broad Range Prestained Molecular weight standard (Bio-Rad). 50 nL (control and NP treated) and 10 nL (E_2 treated) plasma was applied per well. Molecular weight of reacting proteins are: VTG = 180 kDa; *Zr*-p: α = 60 kDa; β = 55 kDa; γ = 50 kDa.

less (Brion *et al.*, 2002; Tyler *et al.*, 2002; Nilsen *et al.*, 2004). Vtg assays based on polyclonal antibodies are generally restricted for use with the homologous species, but some antibodies do cross-react with Vtg in other species (Benfey *et al.*, 1989; Tyler *et al.*, 1996; Nilsen *et al.*, 1998; Fig. 15.6).

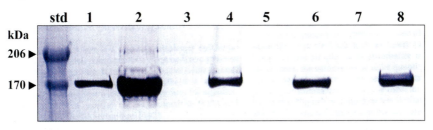

Fig. 15.6 Cross-reactivity of a monoclonal zebrafish (*Danio rerio*) vitellogenin antibody to different cyprinid fish species. Monoclonal mouse anti-zebrafish vitellogenin IgG JE-10D4 (Biosense Laboratories AS, Bergen, Norway) was used to probe a Western blot with samples of: std, Pre-stained molecular weight standard (Bio-Rad), (1) purified zebrafish Vtg, (2) whole-body homogenate sample of estradiol-17β (E$_2$) treated zebrafish, (3) whole-body homogenate sample of control zebrafish, (4) plasma sample of E$_2$ treated carp (*Cyprinus carpio*), (5) plasma sample of control carp, (6) plasma sample of E$_2$ treated fathead minnow (*Pimephales promelas*), (7) plasma sample of control fathead minnow, (8) Plasma sample of E$_2$ treated roach (*Rutilus rutilus*). Reproduced with permission from Arukwe and Goksøyr (2003).

The basic principle of a radioimmunoassay (RIA) is the use of radio labeled Abs or antigens (Ags) to detect Ag:Ab reactions. The Abs or Ags are labeled with the ^{125}I (iodine125) isotope, and the presence of Ag:Ab reactions is detected using a gamma counter. RIA techniques are well developed for egg yolk (Vtg) analysis (Norberg and Haux, 1985; Pavlidis *et al.*, 2000), but have not been developed for the *zona radiata* proteins. Because this technique requires the use of radioactive substances, RIAs are more and more being replaced by other immunologic assays such as ELISAs, that over the last decade have reached similar levels of sensitivity. The ELISA technique is a sensitive laboratory technique widely used to detect and quantitate Ags or Abs in a variety of biological samples. It can be quantitative (with a standard curve) or semi-quantitative (without a standard curve). The two most widely used principles for quantitative detection of proteins are the competitive ELISA and the sandwich ELISA techniques. In addition to the general issues of antibody specificity and sensitivity, there are some specific challenges related to the development of quantitative immunoassays for Vtg and Zr-proteins. For Vtg, although it is relatively easily purified from plasma of estrogenized fish (where it can reach levels of 50–150 mg/ml), it is an inherently unstable protein. The instability of Vtg is due to its role as a precursor for shorter peptide fragments, and it is very sensitive to proteolytic breakdown into these fragments. Care must, therefore, be taken during sampling to avoid proteolytic breakdown by adding suitable protease inhibitors (Hiramatsu et al., 2002).

This instability leads to some problems with immunization, since breakdown products may be more immunogenic than Vtg itself. In addition, it creates an important problem for the use of Vtg as a standard in quantitative assays, since users must ensure that each batch of standard is stored under conditions that prevent breakdown, and is quantitated in a consistent manner (see below). In our own laboratory, we have had success in finding conditions for stabilizing Vtg by lyophilization, although this has not been a straightforward task, and different species behave differently in this process (A. Goksøyr, pers. comm.). The dynamic range of Vtg concentrations found in fish plasma creates another problem. Plasma Vtg can vary maybe 100 million-fold, from a few ng/ml in unexposed male fish, to the 50–150 or above mg/ml found in estrogenized salmonids (Tyler et al., 2002). To be able to quantify this enormous range in blind samples, the working range of the assay should preferentially be as wide as possible. Nevertheless, even with an assay covering several hundred-fold variation, all samples need to be serially diluted at least 3–4 times to ensure that at least one dilution falls within the working range of the assay. Many of the recent assays published obtain this range (Nilsen et al., 2004). The assay also needs to be robust and reproducible, and current experience in our laboratory demonstrates that the sandwich type ELISA is more robust and reproducible over the working range of the assay compared to the competitive format.

The method used to quantify the standard must be consistent and reliable. For Vtg, many different methods are presented. In some cases, Vtg is weighed after a lengthy purification procedure. Others have used different protein quantification methods such as Lowry et al. (1951), Bradford (1976), or the simple A280 absorbency measure. In all these cases, the sample needs to be quantitated towards a known sample. When bovine serum albumin (BSA), ovalbumin, or Immunoglobulin G is used, an assumption is made that Vtg behaves more or less similar to the chosen standard. Generally, this is not the case, and some laboratories develop their own "gold standard" of Vtg, which is used as the standard in quantitation. Again, this gold standard needs to be verified, and this can be done by quantitative amino acid analysis. In this case, one may want to take into account the non-proteinaceous parts of the Vtg, i.e., the lipid, phosphate, and carbohydrate parts. The lipid and phosphate parts have been reported for some species to represent 15–20% and 0.6–0.8%, respectively (e.g., Silversand and Haux, 1995), whereas the carbohydrate portion is not well studied. In general, however, the protein part of the molecule is calculated to represent around 65–75% of the weight of the whole molecule, depending on species. The most important aspect of a protein to be used as a standard in an immunoassay is of course that the epitope(s) involved in the immunoassay maintain their stability. This can only be checked by a quality control using the immunoassay itself, so the question becomes a "hen or egg" issue. One way to manufacture a Vtg standard that maintains both proteolytic and epitope stability is to produce a synthetic peptide fragment that contains the epitope(s) of interest.

For Zr-proteins, the challenges are somewhat different. Zr-proteins are found in lower concentrations in plasma compared to Vtg, but recent analyses show that they may reach levels of 1–10 mg/ml in estrogenized rainbow trout (Berg et al., 2002). The protein is much more stable than Vtg, probably due to the different natures of their fate in the oocyte. Whereas Vtg needs to be broken down to fulfill its role as nutrient for the embryo, the Zr-proteins needs to be incorporated into the eggshell intact. In the eggshell, the Zr-proteins will cross-link by a transglutaminase reaction to form the robust zona radiata structure upon fertilization and hardening (Oppen-Berntsen et al., 1990). The solubilization of Zr-proteins from eggshells requires harsh conditions (ibid.), whereas it is more easily obtained from plasma. Although polyclonal antibodies for Zr-proteins have been developed and used for some time (Arukwe et al., 1997b; Celius and Walther, 1998), monoclonal antibodies (MAbs) to Zr-proteins have only recently become available (Berg et al., 2002).

Disruption estrogenic pathways can be assayed by mRNA analysis using reverse transcriptase polymerase chain reaction (RT-PCR, Islinger et al., 2002), or quantitative PCR techniques (qPCR; Fig. 15.7). Quantitative (real-time)-PCR is a rather new method for the quantification of target mRNA sequences. Unlike conventional PCR, qPCR systems are probe-based PCR product detection. During amplification, annealing of the probe to its target sequence generates a substrate that is cleaved by the 5' nuclease activity of Taq DNA polymerase when the enzyme extends from an upstream primer into the region of the probe. This dependence on polymerization ensures that cleavage of the probe occurs only if the target sequence is being amplified. The

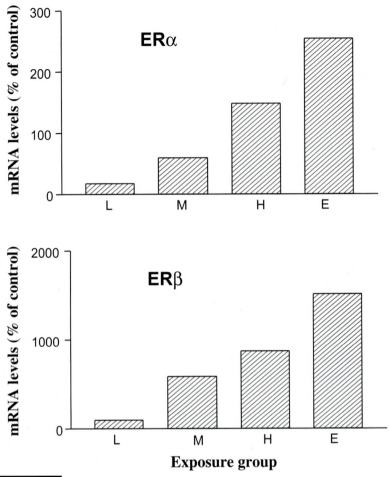

Fig. 15.7   Gene expression patterns of estrogen receptor isoforms (ERα and ERβ) in Atlantic cod (*Gadus morhua*) larvae exposed to produced water discharge (PWD) from an oil platform at Low (L: 1:10000 dilution); Medium (M: 1:1000 dilution); High (H: 1:100 dilution); and estradiol-17b (E: 10 ppb). Larvae were exposed from egg and for a period of 90 days. Total RNA were isolated from whole larvae homogenate and quantified for gene expression using real-time polymerase chain reaction (real-time PCR) with gene specific primers. Each data point represents mRNA level for the respective ER isoform of a pool of 5 individual cDNA samples. Note the higher ERβ transcript levels compared to ERα. From Arukwe *et al.* (unpublished).

development of fluorogenic probes made it possible to eliminate post-PCR processing for the analysis of probe degradation. The probe is an oligonucleotide with both a reporter fluorescent dye and a quencher dye attached. While the probe is intact, the proximity of the quencher greatly reduces the fluorescence emitted by the reporter dye by fluorescence resonance energy transfer (FRET) through space.

Probe design and synthesis has been simplified by the finding that adequate quenching is observed for probes with the reporter at the 5' end and the quencher at the 3' end. The qPCR has several advantages compared to other hybridization techniques. This includes fluorogenic probes over DNA binding dyes require specific hybridization between probe and target to generate fluorescent signal. Thus, with fluorogenic probes, non-specific amplification due to mis-priming or primer-dimer artifact does not generate signal. Another advantage of fluorogenic probes is that they can be labeled with different, distinguishable reporter dyes. By using probes labeled with different reporters, amplification of two distinct sequences can be detected in a single PCR reaction. The disadvantage of fluorogenic probes is that different probes must be synthesized to detect different sequences. Other mRNA targeting assays for oogenic proteins, such as the RNA protection assay (Islinger *et al.*, 1999), have also been developed.

Cellular localization of hepatic oogenic protein synthesis has also been demonstrated using immunohistochemical analysis of exposed fish with specific antibodies (Arukwe *et al.*, 1999b; Bieberstein and Braunbeck, 1999). Immunohistochemistry is a valuable tool in the studies of estrogen and estrogen mimicking compound induced hepatic synthesis of Vtg and *Zr-proteins* in oviparous vertebrates, especially in situations where blood samples are difficult to collect, e.g., when studying small-sized species. Although this technique is time-consuming, localization of Vtg in liver sections may provide insight into responses of different cell types that are important for understanding the role and mechanisms of the estrogens and estrogen mimicking compounds.

EFFECTS AND ASSAYS FOR HORMONE SYNTHESIS AND METABOLISM

Hormone production is regulated by a complicated negative-feedback pathway that is turned on and off in response to fluctuating hormone levels. In contrast to eggshell and egg yolk protein inductions, alterations in plasma sex steroid hormone concentrations can result from several different mechanisms of action, including direct effects on steroidogenic enzymes or indirect modifications associated with altered feedback loops. For example, when hormone production peaks, the hormone acts as an inhibitor and causes the pathway producing the substance to shut down (see Fig. 15.4).

Endocrine modulation may occur through several pathways, not only by compounds mimicking natural hormones. Endocrine response pathways are very complex, and involve synthesis, release and transport of signaling molecules to target cells and interactions with cellular membrane or intracellular receptors. In addition, the involvement of co-activators and repressors (chaperones) in mediating estrogenic actions further reflects the added complex nature of the endocrine pathways. The formation of hormone receptor complexes results in the induction of a signaling cascade or directly interacts with specific genomic sequences and modulates gene transcription (or repression) and translation, leading to an altered biochemical or physiological response. Homeostasis in cells is maintained through these complex signaling pathways, and unscheduled modulation of any of these processes can lead to the development of endocrine modulation or toxicity at various levels. Furthermore, steroid hormone synthesis and metabolism is catalyzed by a series of enzymatic steps involving a number of different steroid hydroxylases, reductases and other enzymes that modify the parent cholesterol molecule. These steps take place in several different tissues and cell types, that all are possible target cells for endocrine modulators.

It is well known that a number of cytochrome P450 (CYP or P450) subfamilies participate in steroid metabolism in fish (e.g., CYP2K and CYP3A; Stegeman and Hahn, 1994). *In vitro* and *in vivo* studies with mammals have also demonstrated a role of AhR-regulated inducible CYP enzymes in E_2 metabolism. For example, it was reported recently that both E_2 and NP inhibited BNF-induced microsomal EROD activity in rat and fish (Lee *et al.*, 1996; Arukwe *et al.*, 1997a). Elsewhere, TCDD, a potent CYP1A-inducer, has been shown to increase the hydroxylation of E_2 at 2- and 16α-positions in human breast cancer cells (Gierthy *et al.*, 1988). Hydroxylation of E_2 was also reported using purified rat CYP1A2 enzymes (Spink *et al.*, 1990).

During sexual maturation and reproduction, P450 activities (measured as 7-ethoxyresorufin O-deethylase [EROD] and aryl hydrocarbon hydroxylase [AHH] activities) are negatively correlated with increasing cellular levels of 17β-estradiol and gonadosomatic index (GSI; Fig. 15.8) in female fish (e.g., Moore *et al.*, 1997). This implies a crosstalk mechanism between the metabolic and synthetic pathways. For example, several studies (Hansson and Gustafsson, 1981; Pajor *et al.*, 1990; Gray

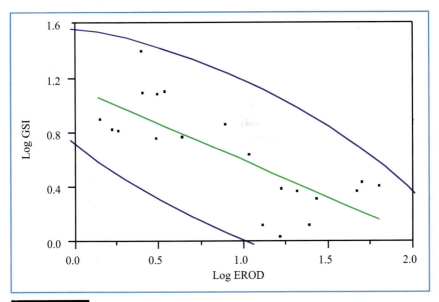

Fig. 15.8 Regression analysis on a scatter plot of log-transformed female gonad somatic index (GSI) and microsomal cytochrome P4501A1 mediated enzyme activity measured as 7-ethoxyresorufin O-deethylase (EROD) in turbot (*Scophthalmus maximus*) at the sexual maturation (April–July). Correlation coefficient = −0.7 and p-value = 0.001. Data from Arukwe (1994).

et al., 1991; Larsen *et al.*, 1992; Arukwe and Goksøyr, 1997) have reported increased levels of liver microsomal steroid and xenobiotic metabolizing enzyme levels in males, compared to females, over a wide range of fish species. The differences in metabolizing enzyme levels between male and female fish are known to reverse soon after reproduction, when estradiol levels are low. Paradoxically, the coplanar halogenated CYP1A-inducer, 3,3′4,4′-tetrachlorobiphenyl (PCB-77), with documented antiestrogenic effects in mammals was shown under the same conditions to potentiate NP-induced zonagenesis and vitellogenesis in juvenile salmon (Arukwe *et al.*, 2001). Whether the above-mentioned disturbances may translate into reproductive toxicity, developmental toxicity or other types of toxicity will depend upon a number of factors, among others the dose of contaminant, co-exposure with other interacting compounds, the timing in relation to gonad development and maturation, temperature, and season. Very little is known about how such modulations may transfer into higher level effects, although some studies do suggest their role. These findings, reviewed below, must therefore be considered in the light that the mechanistic relationships remain speculative.

There are a number of studies that have linked fish exposure to toxic contaminants and development of neoplasia or other adverse health effects (Myers *et al.*, 1991; Moore and Myers, 1994). Adequate answers must consider the full range of possible toxic mechanisms involving the P450 function or AhR pathways. However, depending on the target cells for CYP1A induction, intricate balances of hormone or signal production or immune system functioning may be disabled, thus resulting in growth, health and behavioural changes (see below).

Since the P450 system metabolizes both endogenous and exogenous substances, interactions between foreign chemicals and physiological processes are possible. In this respect, the relationships between induction of biotransformation enzymes in fish liver and altered steroid metabolism *in vitro* and *in vivo* deserve more attention. In several investigations, a relationship between elevated P450 activities and disturbed physiological endocrine functions essential for successful reproduction has been found (Andersson *et al.*, 1988; Thomas, 1990). Although no direct links between the induction of P450 and impaired reproductive functions have yet been established, it is nevertheless important that the mechanism by which potential P450-inducers may affect sexual development and fertility is elucidated.

EFFECTS AND ASSAYS FOR POPULATION AND COMMUNITY LEVEL EFFECTS

The potential transfer of chemically-induced reproductive disturbances from the lower to the higher order of biological organization in fish is shown in Fig. 15.2. Fish exposure to estrogens (and probably also to androgens) during the sensitive part of development means that genotypic males may be completely feminized, developing apparently normal ovaries and oviducts and reproducing normally as adult. The documented feminization caused by estrogen exposure of larvae is impossible to detect without sophisticated genetic analysis, but it implies that the sex ratios of adult populations can become skewed (at least in the short term) towards females (Larsson *et al.*, 1999). The implications of skewed sex ratios for the reproductive capabilities of a fish population might be profound. However, such effects have rarely been observed in wild fish populations, indeed, the best example involves a sex ratio bias towards males caused by plant sterols (Larsson *et al.*, 1999) and are very difficult to distinguish from effects caused by a range of natural stressors, including climate and fishing pressure (Lang *et al.*, 1995). Whether effects of chemical disruption of estrogen pathways on fish sex ratios are more widespread than currently suspected remains to be seen.

Intersex is a much more common condition caused by early exposure of fish larvae to estrogen pathway disrupting chemicals. The intersex condition in males usually takes the form of ovotestis. The subject has been reviewed and reported, particularly with respect to the masculinization of females caused by pulpmill effluents (Howell *et al.*, 1980; Bortone *et al.*, 1989; Drysdale and Bortone, 1989; Davis and Bortone, 1992). A partial feminization in which oocytes may appear in otherwise normal testes is termed ovotestis. Little is known about the implications of this condition for reproductive functionality, as species with ovotestis does not show severe reproductive abnormality above this condition. Ovotestis can be induced in the laboratory by exposing fish larvae to weak estrogens like NP (Gimeno *et al.*, 1996; Gray and Metcalfe, 1997), and have also been observed at high prevalence in wild fish populations in United Kingdom freshwaters (van Aerle *et al.*, 2001).

Despite the lack of conclusive evidence of causal links between toxicological effects on individuals and the response of populations, there is ample evidence of reduced abundance in wild populations caused by pollution-related reductions in recruitment. Such effects were observed in, for example, brown trout (*Salmo trutta*, Kubecka and Matena, 1991), stoneloach, bullhead (*Cottus cottus*) and minnow (*Phoxinus phoxinus*, Bagge and Hakkari, 1992), salmon (Hesthagen *et al.*, 1995), striped bass (Weisberg *et al.*, 1996), and whole fish communities have suffered from the effects of pollution (Lyons *et al.*, 1998). Reduction in the abundance of more sensitive species has been used in species diversity indices to estimate the effect of pollutants on fish communities (Paller *et al.*, 1996). It is, thus, neither the effect of pollution on individual fish, nor their consequences for entire populations or communities that need investigation, but the links between the two, possibly leading to biomarkers for imminent population collapse.

The effects of pollution on the population level is comparable in magnitude to the effects of fishing pressure (Landahl *et al.*, 1997), and considerable research has been done in attempting to apply well developed fishery models to predict the population effects of pollution (Griswold, 1997). However, most models are extremely sensitive to changes in survival estimates of eggs and larvae, which are often difficult to estimate due to complex spatial and temporal dynamics of ichthyoplankton and to sampling problems (Horst, 1997), but may be more affected by pollution than adult fish. Furthermore, while most models are capable of incorporating density dependent factors, in practice they have not (Boreman, 1997), and may thus be of limited value to predict long-term effects of reduced recruitment to wild populations. They are, however, useful in identifying important factors and the kind of data needed for improved analysis.

THE "OMICS" APPROACHES—SHIFTING THE PARADIGM

Because molecular and cellular events are causal and occur in a defined temporal sequence, and because these molecular events branch out into different directions, up- and down-regulation of genes and proteins (i.e., molecular profiles of gene and protein expression patterns) are molecular approaches with potential to serve as predictors of toxicological (i.e., endocrine) and provide crucial and reliable information for specific mechanisms of hormonal action, including xenoestrogens on organismal health and reproduction. This implies that environmental contaminants may induce genomic responses in an organism. The combination of *toxicology*; the study of the nature and effects of poisons, with *genomics*; the investigation of the way organismal genetic make-up, the genome, translates into biological functions (*toxicogenomics*; most often associated with mammalian toxicology) has become a scientific sub-discipline.

Depending upon the severity and duration of the toxicant exposure, genomic measures may be short-term toxicological responses leading to impacts on survival and reproduction (i.e., parental and offspring fitness). Chapman (2001) introduced the term *ecogenomics* to describe the application of genomics to ecology. The proposed application of genomics to organisms outside the laboratory aims to provide insight into their physiological status and to translate this into an understanding of their responses to each other and to the environment. Recently, another term *ecotoxicogenomics* was introduced to describe the integration of genomic-based science into ecotoxicology (Snape *et al.*, 2004). Given the fact that these terms are most often used synonymously, no distinction between these words will be made here and *toxicogenomics* is used as a unifying term. *Toxicogenomics* deal primarily with the effects of toxic compounds on gene and protein expression patterns in target cells or tissues and may reveal genetic signatures in organisms that can be used to predict toxicity of these compounds. An understanding of the molecular regulation of these processes require the identification of relevant subsets of differentially expressed genes and proteins of interest to be cloned, characterized, and studied in details. Efforts on genomic sequencing cover a wide range of the animal phyla and include populations of microbes, plants, insects, nematodes, amphibians, fish and mammals. These genomic informations are spread over and between several discrete databases. Endocrine toxicology has taken advantage of these genomic informations.

Toxicogenomics has emerged as a key and mainstream approach in screening new and emerging endocrine disruptors because it may reveal genetic signatures in organisms that can be used to predict their effects on the endocrine systems. Therefore, the development of powerful molecular tools presents unprecedented opportunities to elucidate biological responses to environmental toxicants and stressors at the genomic level. In the environment, chemical interactions may have profound consequences since organisms are exposed to complex mixtures of environmental pollutants. However, these complex interactions have only recently become the focus of systematic investigations (Arukwe *et al.*, 2001). There is no doubt that genome- and proteome-based biomarkers (of exposure to environmental hazards, of effects to environmentally-induced cellular and molecular changes and of genetic susceptibility) are revolutionizing the science of risk assessment. As a paradigm shift in investigative toxicology, toxicogenomic approaches is replacing the current *hypothesis-driven* toxicology research based on "is my gene or protein affected by a given exposure?" with a *discovery-driven* research based on "what genes or proteins are affected by a given exposure?" (Fig. 15.9).

Conventional approach in toxicology is based on the concept that a certain chemical exposure will affect the expression of particular gene(s) and/or protein(s) in target tissues. Therefore, the hypothesis is to expose individuals or tissue/cell culture to a particular chemical or mixture of chemicals and evaluate the expression of individual genes or proteins in target tissues. Various methods for profiling of gene transcripts (mRNA) and their protein complements have been described previously, with the most well-established being the Northern and Western blots for mRNA and protein analysis, respectively. These techniques are now being replaced with newer techniques such gene chips (microarray). In hypothesis-driven toxicology, mechanisms are sometimes well investigated and understood at one level but not known at all at another level. There are examples where the initial interaction of a toxic compound with a target macromolecule in the cell is well understood at the molecular level, yet the downstream molecular processes that lead to toxicity are still puzzling. Similarly, initial cascades of pathophysiological events that

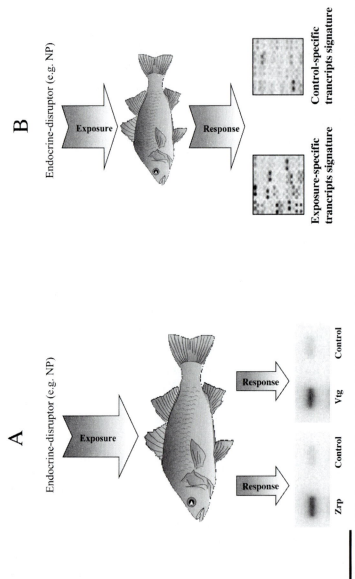

Fig. 15.9 Schematic representation of hypothesis-driven and discovery-driven toxicology research. In example A, fish exposure to an endocrine-disruptor such as nonylphenol (NP) will cause changes in vitellogenin (Vtg) and eggshell *zona radiata* protein (*Zrp*) transcript patterns compared to control fish, while in example B, exposure to the endocrine-disruptor, NP results in a unique transcripts pattern for the given exposure condition, compared to control (unexposed fish). In some sense, this is an extension of the hypothesis-driven research, with intent to discovering new up-regulated and down-regulated transcripts.

lead to toxic effect may be well characterized, while the types of molecular interactions of a toxic compound with the initial target are completely unknown. Thus, hypothesis-driven toxicology investigates effector mechanisms of biological responses (i.e., mechanism of defense, tolerance, repair or activation) singly or individually in target tissue or organ. Since the mechanisms of toxicity can be studied at different molecular levels, following the sequence of events that occur after exposure to potentially toxic compound, hypothesis-driven toxicology is becoming less favored and as such making way for discovery-driven toxicology research.

Because the first interaction between toxic compounds and organisms occurs at the molecular and cellular levels, because these molecular and cellular events are causal and occur in a defined temporal sequence, and because these molecular events branch out into different directions, discovery-driven toxicology research is becoming the mainstream of investigative toxicology. The techniques of hypothesis-driven toxicology research has been superseded by technologies such as differential display (DD; Dakis and Kouretas, 2002), suppression subtractive hybridization (SSH; Diatchenko et al., 1996), serial analysis of gene expression (SAGE; Larose et al., 2004), expressed sequence tags (ESTs) for gene expression (Douglas et al., 1999) and matrix-assisted laser-desorption/ionization time-of-flight mass spectrometry (MALDI-TOF MS) for proteome/genome analysis (Jurinke et al., 2004). These techniques allow the simultaneous analysis of multiple genes and proteins in samples. For these purposes, technological platforms have been developed and are currently in use. These include cDNA microarrays and oligonucleotide arrays and protein arrays. cDNA microarrays and oligonucleotide arrays consist of large number of genes and gene fragments deposited on a glass slide used for multiplex reaction. Nucleic acid (usually DNA) is spotted, in a grid arrangement, onto a solid support such as glass or filter membranes. These arrays serve as hybridization targets for cDNA made from tissue or cell lysates.

Compared to hypothesis-driven research, discovery driven research has an added significance because differentially expressed genes and proteins as molecular tools are used to assess toxicological properties of known and emerging environmental contaminants, and complex mixtures that are the reality of exposure in all taxa, species and at all levels of biological organization. Thus, *toxicogenomics*-based approach to toxicology and ecotoxicology provides critical "snapshot" information on how the regulation and expression of genes control the physiological responses induced by exposure to a toxic compound and complex mixtures at a particular time, and within a particular tissue and species. Stress from toxicant exposure at various levels of biological organization is generally a complex process, which can involve numerous related and unrelated biochemical processes. An integrated genomics-proteomics screening approach accelerates the assessment of affected targets (e.g., receptors, proteins, etc.). Therefore, the linkage between DNA response, mRNA transcription and protein expression resulting from toxicant stressors is essential in understanding mechanism of action, and hence effective mitigation. The long-term goals of toxicogenomic research programs are to develop comprehensive models (using chemicals and global genome-wide endpoints) of toxicant effects on gene and protein expression patterns, endocrine control of individual reproduction, individual health and fitness and to relate these molecular/physiological indicators of stress to population level effects.

Nevertheless, *toxicogenomics* will probably dominate the mainstream of analytical techniques used in toxicological and biomedical sciences as it offers big opportunities in elucidating the potential impacts and limitations of emerging technologies on risk assessment and environmental decision-making, on the impact of new technologies on toxicology research and knowledge management. Above all, toxicogenomic tools may provide us with a better mechanistic understanding of ecotoxicology. However, it is not expected (at least not in the nearest future) that *toxicogenomics* will replace the conventional methods of toxicology, but it will rather be a very important tool in discovery-oriented toxicology particularly for chemical that affect the organismal endocrine pathways (endocrine disruptors), for finding new chemical targets and defining potential consequences. In order to meet these expectations and fulfill its potentials, collaborative and intra- and interdisciplinary efforts are necessary through the parallel use of model microorganisms together with aquatic and terrestrial plants, animals and microorganisms.

THE WAY FORWARD

Despite the large volume of research articles on the modulation of endocrine pathways by chemical pollutants, very few studies have conclusively established the cause and effect connections (Arukwe and Goksøyr, 2003). An interdisciplinary approach is a prerequisite in achieving an overall view on the problems of xenobiotic-induced reproductive disturbances in fish. At present, the greatest challenge to this problem is the translation of subtle functional deficits within individuals into population-level effects and this will require better field observations and laboratory studies to simulate field exposures. Furthermore, a complete risk assessment of the effects of xenobiotics or xenoestrogens on fish reproduction will require good reproductive toxicity testing, defined as the occurrence of adverse effects on the reproductive system that may result from exposure to agents from exogenous sources. The toxicity may be expressed as alterations to the reproductive organs or the related endocrine system or to progenies (Fig. 15.5).

While all the life stages of endocrine disrupting effects have been examined in an individual fashion in various studies, there have been no reports for comparative life-stage evaluation of developmental and reproductive outcomes in a single study (Gierthy, 2002). Some immediate challenges in this regard are studies on the possible effects of xenobiotics on population heterogeneity through selection pressure. Because of the potential long-term impacts on both individuals and populations, better definition of normal variability in reproductive parameters and more comprehensive temporal data are needed, so that potential trends can be identified more readily and reliably, and hypotheses tested regarding their causation (for review, see Arukwe and Goksøyr, 1998; Gillesby and Zacharewski, 1998; Hutchinson *et al.*, 2000). A comprehensive method of data collection should involve extensive studies of highly exposed fish populations, in addition to more information on normal population variation, as well as regional and seasonal variations. The persistent and bioaccumulative properties of the lower potency endocrine modulators, coupled with their widespread use, certainly warrant continued investigation. Furthermore, greater attention should be directed towards the examination of potent natural and synthetic estrogens in the aquatic systems, especially those originating from human female contraceptives such as ethinylestradiol (Larsson *et al.*, 1999).

CONCLUSION

Defined as an exogenous substance that causes adverse health effects in an intact organism, or its progeny, subsequent to changes in endocrine function, endocrine disrupters have raised wide scientific and societal concerns within the past ten years. Various definite or possible reproductive abnormalities in both male and female fish caused by endocrine disrupters have been identified, but in a majority of the reported cases, it is not known whether adverse effects have occurred in the population level of biological organization. Disruption of the estrogen functions in fish may have effects on a number of events, including sexual maturation, gamete production and transport, sexual behaviour, fertility, gestation, lactation or modifications in other functions that are dependent on the integrity of the reproductive system. Although several reproductive effects have been reported, but the degree of causality established between the abnormalities observed and exposure to particular chemicals is variable, and understanding of the mechanism(s) is limited. Disruption of androgenic pathways is treated elsewhere in this book and this chapter have dealt with the effects that are related to the estrogenic pathways and biological indicators used in establishing these effects in both male and female fish.

References

Ahlborg, U.G., L. Lipworth, L. Titus-Ernstoff, C.-C. Hsieh, A. Hanberg, T. Baron, D. Trichopoulos and H.-O. Adami. 1995. Organochlorine compounds in relation to breast cancer, endometrial cancer, and endometriosis: An assessment of the biological and epidemiological evidence. *Critical Reviews in Toxicology* 25: 463-531.

Andersson, T. and L. Förlin. 1992. Regulation of the cytochrome P450 enzyme system in fish. *Aquatic Toxicology* 24: 1-20.

Andersson, T., L. Förlin, J. Härdig and Å. Larsson. 1988. Physiological disturbances in fish living in coastal water polluted with bleached kraft pulp mill effluents. *Canadian Journal of Fisheries and Aquatic Science* 45: 1525-1536.

Ankley, G.T., M.D. Kahl, K.M. Jensen, M.W. Hornung, J.J. Korte, E.A. Makynen and R.L. Leino. 2002. Evaluation of the aromatase inhibitor fadrozole in a short-term reproduction assay with the fathead minnow (*Pimephales promelas*). *Toxicological Sciences* 67: 121-130.

Arnold, A.P. and S.M. Breedlove. 1985. Organizational and activational effects of sex steroids on brain and behavior: A reanalysis. *Hormone and Behaviour* 10: 469-498.

Arukwe, A. 1994. Biochemical changes of the hepatic cytochrome P 450 biotransformation enzyme system in turbot (*scophthalmus maximus* L.) during a reproductive cycle. characterization of sex differences and induction response. Thesis for the degree of candidatus scientarium, Department of Fisheries and Marine biology, University of Bergen, Norway. Pp. 110.

Arukwe, A. 2001. Cellular and molecular responses to endocrine-modulators and the impact on fish reproduction. *Marine Pollution Bulletin* 42: 643-655.

Arukwe, A. and A. Goksøyr. 1997. Changes in three hepatic cytochrome P450 subfamilies during a reproductive cycle in Turbot (*Scophthalmus maximus* L.). *Journal of Experimental Zoology* 277: 313-325.

Arukwe, A. and A. Goksøyr. 1998. Xenobiotics, xenoestrogens and reproduction disturbances in fish. *Sarsia* 83: 225-241.

Arukwe, A. and A. Goksøyr. 2003. Eggshell and egg yolk proteins in fish: hepatic proteins for the next generation: Oogenetic, population, and evolutionary implications of endocrine disruption. *Comparative Hepatology* 2: 4.

Arukwe, A., L. Förlin and A. Goksøyr. 1997a. Xenobiotic and steroid biotransformation enzymes in Atlantic salmon (*Salmo salar*) liver treated with an estrogenic compound, 4-nonylphenol. *Environmental Toxicology and Chemistry* 16: 2576-2583.

Arukwe, A., F.R. Knudsen and A. Goksøyr. 1997b. Fish zona radiata (eggshell) protein: A sensitive biomarker for environmental estrogens. *Environmental Health Perspectives* 105: 418-422.

Arukwe, A., T. Grotmol, T.B. Haugen, F.R. Knudsen and A. Goksøyr. 1999a. Fish model for assessing the in vivo estrogenic potency of the mycotoxin zearalenone and its metabolites. *The Science of Total Environment* 236: 153-161.

Arukwe, A., B.M. Nilsen, K. Berg and A. Goksøyr. 1999b. Immunohistochemical analysis of the vitellogenin response in the liver of Atlantic salmon exposed to environmental oestrogens. *Biomarkers* 4: 373-380.

Arukwe, A., S.W. Kullman and D.E. Hinton. 2001a. Differential biomarker gene and protein expressions in nonylphenol and estradiol-17beta treated juvenile rainbow trout (*Oncorhynchus mykiss*). *Comparative Biochemistry and Physiology* C 129: 1-10.

Arukwe, A., F. Yadetie, R. Male and A. Goksøyr. 2001b. *In vivo* modulation of nonylphenol-induced zonagenesis and vitellogenesis by the antiestrogen, 3,3'4,4'-tetrachlorobiphenyl (PCB-77) in juvenile fish. *Environmental Toxicology and Pharmacology* 10: 5-15.

Bagge, P. and L. Hakkari. 1992. Effects of paper mill effluents on the fish fauna of stony shores of Lake Paijanne. *Hydrobiology* 243: 413-420.

Benfey, T.J., E.M. Donaldson and T.G. Owen. 1989. An homologous radioimmunoassay for coho salmon (*Oncorhynchus kisutch*) vitellogenin, with general applicability to other Pacific salmonids. *General and Comparative Endocrinology* 75: 78-82.

Berg, K., K. Bringsvor, M.V. Nilsen, B.T. Walther, A. Goksøyr and B.M. Nilsen. 2002. Monoclonal antibodies against zona radiata proteins for purification of individual Zrp-monomers and development of a quantitative Zrp-ELISA. *Marine Environmental Research* 54: 745.

Bieberstein, U. and T. Braunbeck. 1999. Immunohistochemical localization of vitellogenin in rainbow trout (*Oncorhynchus mykiss*) hepatocytes using immunofluorescence. *The Science of Total Environment* 233: 67-75.

Bitman, J. and H.C. Cecil. 1970. Estrogenic activity of DDT analogs and polychlorinated biphenyls. *Journal of Agriculture and Food Chemistry* 18: 1108-1112.

Bitman, J., H.C. Cecil, S.J. Harris and G.F. Fries. 1968. Estrogenic activity of o,p'-DDT in the mammalian uterus and avian oviduct. *Science* 162: 371-372.

Bjerregaard, P., B. Korsgaard, L.B. Christiansen, K.L. Pedersen, L.J. Christensen, S.N. Pedersen and P. Horn. 1998. Monitoring and risk assessment for endocrine disruptors in the aquatic environment: a biomarker approach. *Archives of Toxicology* 20 (Supplement): 97-107.

Boreman, J. 1997. Methods for comparing the impact of pollution and fishing on fish populations. *Transactions of American Fisheries Society* 126: 506-513.

Bortone, S.A., W.P. Davis and C.M. Bundrick. 1989. Morphological and behavioral characters in mosquitofish as potential bioindication of exposure to kraft mill effluent. *Bulletin of Environmental Contamination and Toxicology* 43: 370-377.

Bradbury, R.B. and D.E. White. 1954. Estrogens and related substances in plants. *Vitamins and Hormones* 12: 207-233.

Bradford, M.M. 1976. A rapid and sensitive method for the quantitation of microgram quantities of protein utilizing the principle of protein-dye binding. *Analytical Biochemistry* 72: 248-254.

Brion, F., B.M. Nilsen, J.K. Eidem, A. Goksøyr and J.M. Porcher. 2002. Development and validation of an enzyme-linked immunosorbent assay to measure vitellogenin in the zebrafish (*Danio rerio*). *Environmental Toxicology and Chemistry* 21: 1699-1708.

Caldwell, J. 1985. Novel xenobiotic-lipid conjugates. *Biochemical Society Transactions* 13: 852-854.

Callard, G.V., A.V. Tchoudakova, M. Kishida and E. Wood. 2001. Differential tissue distribution, developmental programming, estrogen regulation and promoter characteristics of cyp19 genes in teleost fish. *Journal of Steroid Biochemistry and Molecular Biology* 79: 305-314.

Celius, T. and B.T. Walther. 1998. Oogenesis in Atlantic salmon (*Salmo salar*) occurs by zonagenesis preceding vitellogenesis *in vivo* and *in vitro*. *Journal of Endocrinology* 158: 259-266.

Celius, T., T.B. Haugen, T. Grotmol and B.T. Walther. 1999. A sensitive zonagenetic assay for rapid in vitro assessment of estrogenic potency of xenobiotics and mycotoxins. *Environmental Health Perspectives* 107: 63-68.

Chan, S.L., C.H. Tan, M.K. Pang and T.J. Lam. 1991. Vitellogenin purification and development of assay for vitellogenin receptor in oocyte membranes of the Tilapia (*Oreochromis niloticus*, Linnaeus 1766). *Journal of Experimental Zoology* 257: 96-109.

Chapman, R.W. 2001. EcoGenomics—A consilience for comparative immunology? *Development and Comparative Immunology* 25: 549-551.

Clemens, M.J., R. Lofthouse and J.R. Tata. 1975. Sequential changes in the protein synthetic activity of male *Xenopus laevis* liver following induction of egg-yolk proteins by Estradiol-17 beta. *Journal of Biological Chemistry* 250: 2213-2218.

Colborn, T. and C. Clement. 1992. *Chemically-Induced Alterations in Sexual and Functional Development: The Wildlife/Human Connection*. Princeton Scientific Publishing, New Jersey.

Colborn, T., F.S. vom Saal and A.M. Soto. 1993. Developmental effects of endocrine-disrupting chemicals in wildlife and humans. *Environmental Health Perspectives* 101: 378-384.

Conley, A. and M. Hinshelwood. 2001. Mammalian aromatases. *Reproduction* 121: 685-695.

Cyr, D.G. and J.G. Eales. 1996. Interrelationship between thyroidal and reproductive systems in fish. *Review in Fish Biology and Fisheries* 6: 165–200.

Cyr, D.G., N.R. Bromage, J. Duston and J.G. Eales. 1988. Seasonal patterns in serum levels of thyroid hormones and sex steroids in relation to photoperiod-induced changes in spawning time in rainbow trout, *Salmo gairdneri*. *General and Comparative Endocrinology* 69: 217-225.

Dakis, D. and D. Kouretas. 2002. Differential display of m-RNAs in fish gonads by modified DD-PCR. *In Vivo* 16: 103-105.

Davies, B., N. Bromage and P. Swanson. 1999. The brain-pituitary-gonadal axis of female rainbow trout *Oncorhynchus mykiss*: Effects of photoperiod manipulation1. *General and Comparative Endocrinology* 115: 155-166.

Davis, W.P. and S.A. Bortone. 1992. Effects of kraft mill effluents on the sexuality of fishes: An environmental early warning. In: *Chemically-Induced Alterations in Sexual and Functional*

Development: The Wildlife/Human Connection, T. Colborn and C. Clement (eds.). Princeton Scientific Publishing, New Jersey, pp. 113-127.

Depledge, M. 1989. The rational basis for detection of the early effects of marine pollutants using physiological indicators. *Ambio* 18: 301-302.

Depledge, M. 1994. The rationale basis for the use of biomarkers as ecotoxicological tools. In: *Nondestructive Biomarkers in Vertebrates*, M.C. Fossi and C. Leonzio (eds.). CRC Press, Boca Raton, pp. 271-295.

Diatchenko, L., Y.F. Lau, A.P. Campbell, A. Chenchik, F. Moqadam, B. Huang, S. Lukyanov, K. Lukyanov, N. Gurskaya, E.D. Sverdlov and P.D. Siebert. 1996. Suppression subtractive hybridization: A method for generating differentially regulated or tissue-specific cDNA probes and libraries. *Proceedings of National Academy of Sciences of the United States of America* 93: 6025-6030.

Dimitriadis, G.J. and J.R. Tata. 1982. Differential sensitization to deoxyribonuclease I of *Xenopus* vitellogenin and albumin genes during primary and secondary induction of vitellogenesis by oestradiol. *Biochemical Journal* 202: 491-497.

Dodds, E.C. and W. Lawson. 1936. Synthetic estrogenic agents without the phenanthrene nucleus. *Nature (London)* 137: 996.

Douglas, S.E., J.W. Gallant, C.E. Bullerwell, C. Wolff, J. Munholland and M.E. Reith. 1999. Winter flounder expressed sequence tags: establishment of an EST database and Identification of novel fish genes. *Marine Biotechnology* 1: 458-464.

Drysdale, D.T. and S.A. Bortone. 1989. Laboratory induction of intersexuallty in the mosquitofish, *Gambusia affinis*, using paper mill effluent. *Bulletin of Environmental Contamination and Toxicology* 43: 611-617.

Duston, J. and N. Bromage. 1987. Constant photoperiod regimes and the entrainment of the annual cycle of reproduction in the female rainbow trout (*Salmo gairdneri*). *General and Comparative Endocrinology* 65: 373-384.

Eddy, F.B.. 1974. Osmotic properties of the perivitelline fluid and some properties of the chorion of Atlantic salmon eggs (*Salmo salar*). *Journal of Zoology* 174: 237-243.

Feist, G., C.B. Schreck, M.S. Fitzpatrick and J.M. Redding. 1990. Sex steroid profiles of coho salmon (*Oncorhynchus kisutch*) during early development and sexual differentiation. *General and Comparative Endocrinology* 80: 299-313.

Folmar, C.L., N.D. Denslow, V. Rao, M. Chow, D.A. Crain, J. Enblom, J. Marcino and J.J. Guillette Jr. 1996. Vitellogenin induction and reduced serum testosterone concentrations in feral male carp (*Cyprinus carpio*) captured near a metropolitan sewage treatment plant. *Environmental Health Perspectives* 104: 1096-1101.

Fossi, M.C. 1994. Nondestructive biomarkers in ecotoxicology. *Environmental Health Perspectives* 102(Supplement 12): 49-54.

Fossi, M.C., L. Marsili, C. Leonzio and S. Focardi. 1992. The hazard assessment of cetaceans by the use of a non-destructive biomarker in skin biopsy. SC-CAMLR/WG-CEMP 92/47.

Fossi, M.C., C. Leonzio and D.P. Peakall. 1994. The use of nondestructive biomarkers in the hazard assessments of vertebrate populations. In: *Nondestructive Biomarkers in Vertebrates*, M.C. Fossi and C. Leonzio (eds.). CRC Press, Boca Raton, pp. 3-34.

Förlin, L. and T. Hansson. 1982. Effects of oestradiol-17ß and hypophysectomy on hepatic mixed function oxidases in rainbow trout. *Journal of Endocrinology* 95: 245-252.

Gangloff, M., M. Ruff, S. Eiler, S. Duclaud, J.M. Wurtz and D. Moras. 2001. Crystal structure of a mutant hERalpha ligand-binding domain reveals key structural features for the mechanism of partial agonism. *Journal of Biological Chemistry* 276: 15059-15065.

Gierthy, J.F. 2002. Testing for endocrine disruption: How much is enough? *Toxicological Science* 68: 1-3.

Gierthy, J.F., D.W. Lincoln, S.J. Kampcik, H.W. Dickerman, H.L. Bradlow, T. Niwa and G.E. Swaneck. 1988. Enhancement of 2- and 16 alpha-estradiol hydroxylation in MCF-7 human breast cancer cells by 2,3,7,8-tetrachlorodibenzo-P-dioxin. *Biochemical and Biophysical Research Communication* 157: 515-520.

Gillesby, B.E. and T.R. Zacharewski. 1998. Exoestrogens: Mechanisms of action and strategies for identification and assessment. *Environmental Toxicology and Chemistry* 17: 3-14.

Gimeno, S., A. Gerritsen, T. Bowmer and H. Komen. 1996. Feminization of male carp. *Nature* (London) 384: 221-222.

Ginsburg, A.S. 1963. Sperm-egg association and its relationship to the activation of the egg in salmonid fishes. *Journal of Embryology and Experimental Morphology* 31:13-33.

Gray, E.S., B.R. Woodin and J.J. Stegeman. 1991. Sex differences in hepatic monooxygenases in winter flounder (*Pseudopleuronectes americanus*) and scup (*Stenotomus chrysops*) and regulation of P450 forms by estradiol. *Journal of Experimental Zoology* 259: 330-342.

Gray, L.E. 1998. Xenoendocrine disrupters: Laboratory studies on male reproductive effects. *Toxicology Letters* 102: 331-335.

Gray, L.E., J. Ostby, C. Wolf, C. Lambright and W. Kelce. 1998. The value of mechanistic studies in laboratory animals for the prediction of reproductive effects in wildlife: Endocrine effects on mammalian sexual differentiation. *Environmental Toxicology and Chemistry* 17: 109-118.

Gray, M.A. and C.D. Metcalfe. 1997. Induction of testis-ova in japanese medaka (*Oryzias latipes*) exposed to p-nonylphenol. *Environmental Toxicology and Chemistry* 16: 1082-1086.

Grierson, J.P. and A.C. Neville. 1981. Helicoidal architecture of fish eggshell. *Tissue and Cell* 13: 819-830.

Griswold, B. 1997. Fisheries and pollution. *Transaction of American Fisheries Society* 126: 504-505.

Guardans, R. and B.S. Gimeno. 1994. Long distance transportation of atmospheric pollutants and its effects on ecosystems. *Microbiology* 10: 145-158.

Guillette Jr, J.L., D.A. Crain, A.A. Rooney and D.B. Pickford. 1995. Organization versus activation: The role of endocrine-disrupting contaminants (EDCs) during embryonic development in wildlife. *Environmental Health Perspectives* 103(Supplement 7): 157-164.

Gustafsson, J.Å. 1999. Estrogen receptor beta—A new dimension in estrogen mechanism of action. *Journal of Endocrinology* 163: 379-383.

Gustafsson, J.Å. 2000. An update on estrogen receptors. *Seminars in Perinatology* 24: 66-69.

Hansson, T. and J.Å. Gustafsson. 1981. In vitro metabolism of 4-androstene-3,17-dione by hepatic microsomes from the rainbow trout (*Salmo gairdnerii*): Effects of hypophysectomy and oestradiol-17ß. *Journal of Endocrinology* 90: 103-112.

Hansson, T., J. Rafter and J.Å. Gustafsson. 1980. Effects of estradiol-17ß on the cytochrome P-450-dependent steroid hydroxylases in fish liver. In: *Biochemistry, Biophysics and Regulation of Cytochrome P-450*, J.Å. Gustafsson (ed.). Elsevier, Amsterdam, pp. 191-194.

Heppell, S.A., N.D. Denslow, L.C. Folmar and C.V. Sullivan. 1995. Universal assay of vitellogenin as a biomarker for environmental estrogens. *Environmental Health Perspectives* 103(Supplement 7): 9-15.

Hesthagen, T., H. Berger, B.M. Larsen and R. Saksgard. 1995. Monitoring fish stocks in relation to acidification in Norwegian watersheds. *Water and Soil Pollution* 85: 641-646.

Hiramatsu, N., K. Hiramatsu, K. Hirano and A. Hara. 2002. Vitellogenin-derived yolk proteins in a hybrid sturgeon, bester (*Huso huso × Acipencer ruthenus*): identification, characterization and course of proteolysis during embryogenesis. *Comparative Biochemistry and Physiology A* 131: 429-441.

Hiramatsu, N., N. Ichikawa, H. Fukada, T. Fujita, C.V. Sullivan and A. Hara. 2002. Identification and characterization of proteases involved in specific proteolysis of vitellogenin and yolk proteins in salmonids. *Journal of Experimental Zoology* 292: 11-25.

Horst, T. 1997. Use of the Leslie matrix for assessing environmental impact with an example for a fish population. *Transaction of American Fisheries Society* 106: 253-257.

Hough-Evans, B.R., S.G. Ernst, R.J. Britten and E.H. Davidson. 1979. RNA complexity in developing sea urchin oocytes. *Developmental Biology* 69: 258-269.

Howell, W.M., D.A. Black and S.A. Bortone. 1980. Abnormal expressions of secondary sex characters in a population of mosquito fish, *Gambusia affinis* holbrookii: Evidence for environmentally-induced masculinization. *Copeia* 1980: 676-681.

Huggett, R.J., R.A. Kimerle, P.M. Mehrle Jr. and H.L. Bergman (eds.). 1992. *Biomarkers: Biochemical, Physiological, and Histological Markers of Antropogenic Stress.* Lewis Publishers, Chelsea, MI, USA.

Hutchinson, T.H., R. Brown, K.E. Brugger, P.M. Campbell, M. Holt, R. Lange, *et al.* 2000. Ecological risk assessment of endocrine disruptors. *Environmental Health Perspectives* 108: 1007-1014.

Islinger, M., S. Pawlowski, H. Hollert, A. Volkl and T. Braunbeck. 1999. Measurement of vitellogenin-mRNA expression in primary cultures of rainbow trout hepatocytes in a non-radioactive dot blot/RNAse protection-assay. *The Science of Total Environment* 233: 109-122.

Islinger, M., H. Yuan, A. Voelkl and T. Braunbeck. 2002. Measurement of vitellogenin gene expression by RT-PCR as a tool to identify endocrine disruption in Japanese medaka (*Oryzias latipes*). *Biomarkers* 7: 80-93.

Jobling, S., D. Sheahan, J.A. Osborne, P. Matthiessen and J.P. Sumpter. 1996. Inhibition of testicular growth in rainbow trout (*Oncorhynchus mykiss*) exposed to estrogenic alkylphenolic chemicals. *Environmental Toxicology and Chemistry* 15: 194-202.

Jurinke, C., P. Oeth and D. van den Boom. 2004. MALDI-TOF mass spectrometry: a versatile tool for high-performance DNA analysis. *Molecular Biotechnology* 26: 147-164.

Kagawa, H., K. Gen, K. Okuzawa and H. Tanaka. 2003. Effects of luteinizing hormone and follicle-stimulating hormone and insulin-like growth factor-I on aromatase activity and P450 aromatase gene expression in the ovarian follicles of red seabream, *Pagrus major*. *Biology of Reproduction* 68: 1562-1568.

Kang, I.J., H. Yokota, Y. Oshima, Y. Tsuruda, T. Yamaguchi, M. Maeda, *et al.* 2002. Effect of 17beta-estradiol on the reproduction of Japanese medaka (*Oryzias latipes*). *Chemosphere* 47: 71-80.

Katzenellenbogen, B.S. 1996. Estrogen receptors: bioactivities and interactions with cell signaling pathways. *Biology of Reproduction* 54: 287-293.

Kavlock, R.J., G.P. Daston, C. DeRosa, P. Fenner-Crisp, L.E. Gray, S. Kaattari, *et al.* 1996. Research needs for the risk assessment of health and environmental effects of endocrine disruptors: A report of the U.S. EPA-sponsored workshop. *Environmental Health Perspectives* 104 (Supplement 4): 715-740.

Kime, D.E. 1995. The effects of pollution on reproduction in fish. *Reviews of Fish Biology and Fisheries* 5: 52-96.

Kishida, M. and G.V. Callard. 2001. Distinct cytochrome P450 aromatase isoforms in zebrafish (*Danio rerio*) brain and ovary are differentially programmed and estrogen regulated during early development. *Endocrinology* 142: 740-750.

Kordes, C., E.P. Rieber and H.O. Gutzeit. 2002. An in vitro vitellogenin bioassay for oestrogenic substances in the medaka (*Oryzias latipes*). *Aquatic Toxicology* 58: 151-164.

Kubecka, J. and J. Matena. 1991. Downstream regeneration of the fish populations of 3 polluted trout streams in southern Bohemia. *Ekologia* CSFR 10: 389-404.

Kuiper, G.G., E. Enmark, M. Pelto-Huikko, S. Nilsson and J.Å. Gustafsson. 1996. Cloning of a novel receptor expressed in rat prostate and ovary. *Proceedings of National Academy of Sciences of the United States of America* 93: 5925-5930.

Lam, T.J. 1983. Environmental influences on gonadal activity in fish. In: *Fish Physiology*, W.S. Hoar, D.J. Randall and E.M. Donaldson (eds.). Academic Press, New York, Vol. 9B, pp. 65-116.

Landahl, J.T., L. Johnson, J.E. Stein, T.K. Collier and U. Varanasi. 1997. Marine pollution and fish population parameters: English sole (*Pleuronectes vetulus*) in Puget Sound, WA. *Transaction of American Fisheries Society* 126: 519-535.

Lang, T., U. Damm and V. Dethlefsen. 1995. Changes in the sex ratio of North Sea dab (*Limanda limanda*) in the period 1981-1995. In: *International Council for the Exploration of the Sea* (ICES), Copenhagen, Denmark ICES CM 1995/G :25 Ref E, 11.

Larose, M., J. St-Amand, M. Yoshioka, P. Belleau, J. Morissette, C. Labrie, *et al.* 2004. Transcriptome of mouse uterus by serial analysis of gene expression (SAGE): Comparison with skeletal muscle. *Molecular Reproduction and Development* 68: 142-148.

Larsen, H.E., M. Celander and A. Goksøyr. 1992. The cytochrome P450 system of Atlantic salmon (*Salmo salar*): II. Variations in hepatic catalytic activities and isozyme patterns during an annual reproductive cycle. *Fish Physiology and Biochemistry* 10: 291-301.

Larsson, D.G.J., M. Adolfsson-Erici, J. Parkkonen, M. Pettersson, A.H. Berg, P.E. Olsson and L. Förlin. 1999. Ethinyloestradiol—An undesired fish contraceptive. *Aquatic Toxicology* 45: 91-97.

Lee, P.C., S.C. Patra, C.T. Stello, W. Lee and M. Struve. 1996. Interaction of nonylphenol and hepatic CYP1A in rats. *Biochemical Pharmacology* 52: 885-889.

Lewis, J.A., M.J. Clemens and T.R. Tata. 1976. Morphological and biochemical changes in the hepatic endoplasmic reticulum and golgi apparatus of male *Xenopus laevis* after induction of egg-yolk protein synthesis by oestradiol-17 beta. *Molecular and Cellular Endocrinology* 4: 311-329.

Liu, R.C., M.E. Hurtt, J.C. Cook and L.B. Biegel. 1996. Effect of the peroxisome proliferator, ammonium perfluorooctanoate (C8), on hepatic aromatase activity in adult male Crl:CD BR (CD) rats. *Fundamental and Applied Toxicology* 30: 220-228.

Lowry, O.H., N.J. Rosebrough, A.L. Farr and R.J. Randall. 1951. Protein measurement with the folin phenol reagent. *Journal of Biological Chemistry* 193: 265-275.

Lyons, J., G. Gonzalez-Hernandez, E. Soto-Galera and M. Guzman-Arroyo. 1998. Decline of freshwater fishes and fisheries in selected drainages of west-central Mexico. *Fisheries* 23: 10-18.

Manery, J.F., K.C. Fisher and E. Moore. 1947. Water intake and membrane hardening of fish eggs. *Federation Proceedings* 65.

Matsubara, T., N. Ohkubo, T. Andoh, C.V. Sullivan and A. Hara. 1999. Two forms of vitellogenin, yielding two distinct lipovitellins, play different roles during oocyte maturation and early development of barfin flounder, *Verasper moseri*, a marine teleost that spawns pelagic eggs. *Developmental Biology* 213: 18-32.

McAllister, B.G. and D.E. Kime. 2003. Early life exposure to environmental levels of the aromatase inhibitor tributyltin causes masculinisation and irreversible sperm damage in zebrafish (*Danio rerio*). *Aquatic Toxicology* 65: 309-316.

McCarthy, J.F. and L.R. Shugart (eds.). 1990. *Biomarkers of Environmental Contamination*. Lewis Publishers Inc. Boca Raton.

Min, J., S.K. Lee and M.B. Gu. 2003. Effects of endocrine disrupting chemicals on distinct expression patterns of estrogen receptor, cytochrome P450 aromatase and p53 genes in *Oryzias latipes* liver. *Journal of Biochemical and Molecular Toxicology* 17: 272-277.

Moggs, J.G. and G. Orphanides. 2001. Estrogen receptors: Orchestrators of pleiotropic cellular responses. *EMBO Reproduction* 2: 775-781.

Mommsen, P.T. and P.J. Walsh. 1988. Vitellogenesis and oocyte assembly. In: *Fish Physiology*, W.S. Hoar, D.J. Randall and E.M. Donaldson (eds.). Academic Press, New York, Vol. 11A, pp. 347-406.

Mommsen, T.P., M.M. Vijayan and T.W. Moon. 1999. Cortisol in teleosts: dynamics, mechanism of action and metabolic regulation. *Reviews of Fish Biology and Fisheries* 9: 211-268.

Moore, M., M. Mustain, K. Daniel, I. Chen, S. Safe, T. Zacharewski, B. Gillesby, A. Joyeux and P. Balaguer. 1997. Antiestrogenic activity of hydroxylated polychlorinated biphenyl congeners identified in human serum. *Toxicology and Applied Pharmacology* 142: 160-168.

Moore, M.J. and M.S. Myers. 1994. Pathobiology of chemical-associated neoplasia in fish. In: *Aquatic Toxicology: Molecular, Biochemical, and Cellular Perspectives*, D.C. Malins and G.K. Ostrander (eds.). Lewis Publishers Inc., Boca Raton, pp. 327-386.

Mosselman, S., J. Polman and R. Dijkema. 1996. ER beta: Identification and characterization of a novel human estrogen receptor. *FEBS Letters* 392: 49-53.

Myers, M.S., J.T. Landahl, M.M. Krahn and B.B. McCain. 1991. Relationships between hepatic neoplasms and related lesions and exposure to toxic chemicals in marine fish from the U.S. West Coast. *Environmental Health Perspectives* 90: 7-15.

Nilsen, B.M., K. Berg, A. Arukwe and A. Goksøyr. 1998. Monoclonal and polyclonal Antibodies against fish vitellogenin for use in pollution monitoring. *Marine Environmental Research* 46: 153-157.

Nilsen, B.M., K. Berg, J.K. Eidem, S.I. Kristiansen, F. Brion, J-M. Porcher and A. Goksøyr. 2004. Development of quantitative vitellogenin-ELISAs for fish test species used in endocrine disruptor screening. *Analytical and Bioanalytical Chemistry* 378: 621-633.

Nilsson, S. and J.Å. Gustafsson. 2002a. Biological role of estrogen and estrogen receptors. *Critical Reviews in Biochemistry and Molecular Biology* 37: 1-28.

Nilsson, S. and J.Å. Gustafsson. 2002b. Estrogen receptor action. *Critical Reviews in Eukaryotic Gene Expression* 12: 237-257.

Norberg, B. and C. Haux. 1985. Induction, isolation and a characterization of the lipid content of plasma vitellogenin from two Salmo species: Rainbow trout (*Salmo gairdneri*) and sea trout (*Salmo trutta*). *Comparative Biochemistry and Physiology* 81: 869-876.

Oettel, M. 2002. Is there a role for estrogens in the maintenance of men's health? *Aging Male* 5: 248-257.

Om, A.S. and K.W. Chung. 1996. Dietary zinc deficiency alters 5 alpha-reduction and aromatization of testosterone and androgen and estrogen receptors in rat liver. *Journal of Nutrition* 126: 842-848.

Oppen-Berntsen, D.O., J.V. Helvik and B.T. Walther. 1990. The major structural proteins of cod (*Gadus morhua*) eggshells and protein crosslinking during teleost egg hardening. *Developmental Biology* 137: 258-265.

Opresko, L.K. 1991. Vitellogenin uptake and in vitro culture of oocytes. *Methods in Cell Biology* 36: 117-132.

Opresko, L.K. and H.S. Wiley. 1987. Receptor-mediated endocytosis in *Xenopus* oocytes. II. Evidence for two novel mechanisms of hormonal regulation. *Journal of Biological Chemistry* 262: 4116-4123.

Orlando, E.F., W.P. Davis and L.J. Guillette Jr. 2002. Aromatase activity in the ovary and brain of the eastern mosquitofish (*Gambusia holbrooki*) exposed to paper mill effluent. *Environmental Health Perspectives* 110 (Supplement 3): 429-433.

Pajor, A.M., J.J. Stegeman, P. Thomas and B.R. Woodin. 1990. Feminization of the hepatic microsomal cytochrome P-450 system in brook trout by estradiol, testosterone, and pituitary factors. *Journal of Experimental Zoology* 253: 51-60.

Pakdel, F., F. Le Gac, P. Le Goff and Y. Valotaire. 1990. Full-length sequence and in vitro expression of rainbow trout estrogen receptor cDNA. *Molecular and Cellular Endocrinology* 71: 195-204.

Paller, M., M. Reichert and J.M. Dean. 1996. Use of fish communities to assess environmental impacts in South Carolina coastal plain streams. *Transaction of American Fisheries Society* 125: 633-644.

Palmer, B.D., L.K. Huth, D.L. Pieto and K.W. Selcer. 1998. Vitellogenin as a biomarker for xenobiotic estrogens in an amphibian model system. *Environmental Toxicology and Chemistry* 17: 30-36.

Pavlidis, M., L. Greenwood, B. Mourot, C. Kokkari, F. Le Menn, P. Divanach and A.P. Scott. 2000. Seasonal variations and maturity stages in relation to differences in serum levels of gonadal steroids, vitellogenin, and thyroid hormones in the common dentex (*Dentex dentex*). *General and Comparative Endocrinology* 118: 14-25.

Peakall, D. 1992. *Animal Biomarkers as Pollution Indicators*. Chapman and Hall, London.

Peakall, D.B. 1994. Biomarkers: The way forward in environmental assessment. *Toxicology and Ecotoxicology News* 1: 55-60.

Pelissero, C., B. Bennetau, P. Babin, F. Le Menn and J. Dunogues. 1991. The estrogenic activity of certain phytoestrogens in the siberian sturgeon *Acipenser baeri*. *Journal of Steroid Biochemistry and Molecular Biology* 38: 293-299.

Pelissero, C., G. Flouriot, J.L. Foucher, B. Bennetau, J. Dunogues, F. Le Gac and J.P. Sumpter. 1993. Vitellogenin synthesis in cultured hepatocytes; an in vitro test for the estrogenic potency of chemicals. *Journal of Steroid Biochemistry and Molecular Biology* 44: 263-272.

Pelissero, C., M.J. Lenczowski, D. Chinzi, B. Davail-Cuisset, J.P. Sumpter and A. Fostier. 1996. Effects of flavonoids on aromatase activity, an *in vitro* study. *Journal of Steroid Biochemistry and Molecular Biology* 57: 215-223.

Phoenix, C.H., R.W. Goy, A.A. Gerall and W.C. Young. 1959. Organizing action of prenatally administered testosterone propionate on the tissues mediating mating behavior in the female guinea pig. *Endocrinology* 65: 369-382.

Prat, F., K. Coward, J.P. Sumpter and C.R. Tyler. 1998. Molecular characterization and expression of two ovarian lipoprotein receptors in the rainbow trout, *Oncorhynchus mykiss*. *Biology of Reproduction* 58: 1146-1153.

Purdom, C.E., P.A. Hardiman, V.J. Bye, N.C. Eno, C.R. Tyler and J.P. Sumpter. 1994. Estrogenic effects of effluents from sewage treatment works. *Chemical Ecology* 8: 275-285.

Ramamoorthy, K., M.S. Gupta, G.L. Sun, A. McDougal and S.H. Safe. 1999. 3,3',4,4'-tetrachlorobiphenyl exhibits antiestrogenic and antitumorigenic activity in the rodent uterus and mammary cells and in human breast cancer cells. *Carcinogenesis* 20: 115-123.

Rankouhi, T.R., I. van Holsteijn, R. Letcher, J.P. Giesy and M. van Den Berg. 2002. Effects of primary exposure to environmental and natural estrogens on vitellogenin production in carp (*Cyprinus carpio*) hepatocytes. *Toxicological Sciences* 67: 75-80.

Rocha Monteiro, P.R., M.A. Reis-Henriques and J. Coimbra. 2000. Polycyclic aromatic hydrocarbons inhibit in vitro ovarian steroidogenesis in the flounder (*Platichthys flesus* L.). *Aquatic Toxicology* 48: 549-559.

Routledge, E.J., R. White, M.G. Parker and J.P. Sumpter. 2000. Differential effects of xenoestrogens on coactivator recruitment by estrogen receptor (ER) alpha and ERbeta. *Journal of Biological Chemistry* 275: 35986-35993.

Ruff, M., M. Gangloff, J.M. Wurtz and D. Moras. 2000. Estrogen receptor transcription and transactivation: Structure-function relationship in DNA- and ligand-binding domains of estrogen receptors. *Breast Cancer Research* 2: 353-359.

Safe, S. and V. Krishnan. 1995. Cellular and molecular biology of aryl hydrocarbon (Ah) receptor-mediated gene expression. *Archives of Toxicology* 17 (Supplement): 99-115.

Safe, S. and M. Wormke. 2003. Inhibitory aryl hydrocarbon receptor-estrogen receptor alpha cross-talk and mechanisms of action. *Chemical Research and Toxicology* 16: 807-816.

Safe, S., M. Wormke and I. Samudio. 2000. Mechanisms of inhibitory aryl hydrocarbon receptor-estrogen receptor crosstalk in human breast cancer cells. *Journal of Mammary Gland Biology and Neoplasia* 5: 295-306.

Schneider, W.J. 1996. Vitellogenin receptors: Oocyte-specific members of the low-density lipoprotein receptor supergene family. *International Review of Cytology* 166: 103-135.

Sepulveda, M.S., E.P. Gallagher, C.M. Wieser and T.S. Gross. 2004. Reproductive and biochemical biomarkers in largemouth bass sampled downstream of a pulp and paper mill in Florida. *Ecotoxicology and Environmental Safety* 57: 431-440.

Silversand, C. and C. Haux. 1995. Fatty acid composition of vitellogenin from four teleost species. *Journal of Comparative Physiology* 164: 593-599.

Smeets, J.M., T.R. Rankouhi, K.M. Nichols, H. Komen, N.E. Kaminski, J.P. Giesy and M. van den Berg. 1999. In vitro vitellogenin production by carp (*Cyprinus carpio*) hepatocytes as a screening method for determining (anti)estrogenic activity of xenobiotics. *Toxicology and Applied Pharmacology* 157: 68-76.

Snape, J.R., S.J. Maund, D.B. Pickford and T.H. Hutchinson. 2004. Ecotoxicogenomics: the challenge of integrating genomics into aquatic and terrestrial ecotoxicology. *Aquatic Toxicology* 67: 143-154.

Socorro, S., D.M. Power, P.E. Olsson and A.V.M Canario. 2000. Two estrogen receptors expressed in the teleost fish, *Sparus aurata*: cDNA cloning, characterization and tissue distribution. *Journal of Endocrinology* 166: 293-306.

Sorbera, L.A., J.F. Asturiano, M. Carrillo and S. Zanuy. 2001. Effects of polyunsaturated fatty acids and prostaglandins on oocyte maturation in a marine teleost, the European sea bass (*Dicentrarchus labrax*). *Biology of Reproduction* 64: 382-389.

Spelsberg, T.C., T. Ruh, M. Ruh, A. Goldberger, M. Horton, J. Hora and R. Singh. 1988. Nuclear acceptor sites for steroid hormone receptors: comparisons of steroids and antisteroids. *Journal of Steroid Biochemistry* 31: 579-592.

Spink, D.C., D.W. Lincoln, H.W. Dickerman and J.F. Gierthy. 1990. 2,3,7,8-Tetrachlorodibenzo-p-dioxin causes an extensive alteration of 17-beta-estradiol metabolism in MCF-7 breast tumor cells. *Proceedings of National Academy of Sciences of the United States of America* 87: 6917-6921.

Spink, D.C., C.L. Hayes, N.R. Young, M. Christou, T.R. Sutter, C.R. Jefcoate and John F. Gierthy. 1994. The effects of 2,3,7,8-tetrachlorodibenzo-p-dioxin on estrogen metabolism in MCF-7 breast cancer cells: evidence for induction of a novel 17 beta-estradiol 4-hydroxylase. *Journal of Steroid Biochemistry and Molecular Biology* 51: 251-258.

Stearns, S.C. 1983. The evolution of life-history traits in mosquitofish since their introduction to the Hawaii in 1905: Rates of evolution, heritabilities, and developmental plasticity. *American Zoologist* 23: 65-75.

Stearns, S.C. and J. Koella. 1986. The evolution of phenotypic plasticity in life-history traits: Predictions of reaction norms for age and size at maturity. *Evolution* 40: 893-913.

Stegeman, J.J. and M. Chevion. 1980. Sex differences in cytochrome P-450 and mixed-function oxygenase activity in gonadally mature trout. *Biochemical Pharmacology* 29: 553-558.

Stegeman, J.J. and M.E. Hahn. 1994. Biochemistry and molecular biology of monooxygenases: Current perspectives on forms, functions and regulation of Cytochrome P450 in aquatic species. In: *Aquatic Toxicology: Molecular, Biochemical, and Cellular Perspectives*, D.C. Malins and G.K. Ostrander (eds.). Lewis Publishers Inc., Chelsea, pp. 87-204.

Stegeman, J.J., M. Brouwer, T.D.G. Richard, L. Förlin, B.A. Fowler, B.M. Sanders and P.A. van Veld. 1992. Molecular responses to environmental contamination: Enzyme and protein systems as indicators of chemical exposure and effect. In: *Biomarkers: Biochemical, Physiological and Histological Markers of Anthropogenic Stress*, R.J. Huggett, R.A. Kimerle, P.M. Mehrle Jr. and H.L. Bergman (eds.). Lewis Publishers Inc., Chelsea, pp. 235-335.

Susani, L. 1986. Effects of contaminants on teleost reproduction: past and ongoing studies. NOAA Technical Memorandum NOS OMA 29: 18p.

Tao, Y., D.L. Berlinsky and C.V. Sullivan. 1996. Characterization of a vitellogenin receptor in white perch (*Morone americana*). *Biology of Reproduction* 55: 646-656.

Tata, J.R. 1976. The expression of the vitellogenin gene. *Cell* 9: 1-14.

Tata, J.R., T.C. James, C.S. Watson, J.L. Williams and A.P. Wolffe. 1983. Hormonal regulation and expression of vitellogenin multigene family. *Ciba Foundation Symposium* 98: 96-110.

Tata, J.R. and D.F. Smith. 1979. Vitellogenesis: A versatile model hormonal regulation of gene expression. *Recent Progress in Hormone Research* 35: 47-90.

Tchoudakova, A., S. Pathak and G.V. Callard. 1999. Molecular cloning of an estrogen receptor beta subtype from the goldfish, *Carassius auratus*. *General and Comparative Endocrinology* 113: 388-400.

Tchoudakova, A., M. Kishida, E. Wood and G.V. Callard. 2001. Promoter characteristics of two cyp19 genes differentially expressed in the brain and ovary of teleost fish. *Journal of Steroid Biochemistry and Molecular Biology* 78: 427-439.

Thomas, P. 1990. Molecular and biochemical responses of fish to stressors and their potential use in environmental monitoring. *American Fishery Society Symposium* 8: 9-28.

Thomas, P. 2000. Chemical interference with genomic and nongenomic actions of steroids in fishes: Role of receptor binding. *Marine Environmental Research* 50: 127-134.

Thorpe, J.E. 1994. Reproductive strategies in Atlantic salmon, *Salmo salar* L. *Aquaculture and Fisheries Management* 25: 77-87.

Toppari, J., J.C. Larsen, P. Christiansen, A. Giwercman, P. Grandjean, L.J. Guillette Jr., B. Jegou, T.K. Jensen, P. Jouannet, N. Keiding, H. Leffers, J.A.. Mclachlan, O. Meyer, J. Muller, E. Rajpert-de Meyts, T. Scheike, R. Sharpe, J. Sumpter and N.E. Skakkebaek. 1996. Male reproductive health and environmental xenoestrogens. *Environmental Health Perspectives* 104 (Supplement 4): 741-803.

Tremblay, G.B., and V. Giguere. 2002. Coregulators of estrogen receptor action. *Crit. Rev. Eukaryot. Gene Expression* 12: 1-22.

Tremblay, G.B., A. Tremblay, N.G. Copeland, D.J. Gilbert, N.A. Jenkins, F. Labrie and V. Giguere. 1997. Cloning, chromosomal localization, and functional analysis of the murine estrogen receptor beta. *Molecular Endocrinology* 11: 353-365.

Tullner, W.W. 1961. Uterotrophic action of the insecticide methoxychlor. *Science* 133: 647-648.

Tyler, C.R. and K. Lubberink. 1996. Identification of four ovarian receptor proteins that bind vitellogenin but not other homologous plasma lipoproteins in the rainbow trout, *Oncorhynchus mykiss*. *Journal of Comparative Physiology* 166: 11-20.

Tyler, C.R. and J.P. Sumpter. 1996. Oocyte growth and development in teleost. *Reviews of Fish Biology and Fisheries* 6: 287-318.

Tyler, C.R., B. van der Eerden, J.P. Sumpter, S. Jobling and G. Panter. 1996. Measurement of vitellogenin a biomarker for exposure to oestrogen in a wide variety of cyprinids. *Journal of Comparative Physiology* 166: 418-426.

Tyler, C.R., R. van Aerle, M.V. Nilsen, R. Blackwell, S. Maddix, B.M. Nilsen and A. Goksøyr. 2002. Monoclonal antibody enzyme-linked immunosorbent assay to quantify vitellogenin for studies on environmental estrogens in the rainbow trout (*Oncorhynchus mykiss*). *Environmental Toxicology and Chemistry* 21: 47-54.

Ueda, H., O. Hiroi, A. Hara, K. Yamauchi and Y. Nagahama. 1984. Changes in serum concentrations of steroid hormones, thyroxine, and vitellogenin during spawning migration the chum salmon, *Oncorhynchus keta*. *General and Comparative Endocrinology* 53: 203-211.

van Aerle, R., T.M. Nolan, S. Jobling, L.B. Christiansen, J.P. Sumpter and C.R. Tyler. 2001. Sexual disruption in a second species of wild cyprinid fish (the gudgeon, *Gobio gobio*) in United Kingdom freshwaters. *Environmental Toxicology and Chemistry* 20: 2841-2847.

Wahli, W. 1988. Evolution and expression of vitellogenin genes. *TIG* 4: 227-232.

Weigand, M.D. 1982. Vitellogenesis in fishes. In: *Reproductive Physiology of Fish*, C.J.J. Richter and H.J.T. Goos (eds.). Pudoc. Wagenningen, Netherlands, pp. 136-146.

Weisberg, S., P. Himchak, T. Baum, H.T. Wilson and R.E. Allen. 1996. Temporal trends in abundance of fish in the tidal Delaware River. *Estuaries* 19: 723-729.

Wester, P.W. and J.H. Canton. 1986. Histopathological study of *Oryzias latipes* (Medaka) after long-term β-hexachlorocyclohexane exposure. *Aquatic Toxicology* 9: 21-45.

Winzer, K., C.J. Van Noorden and A. Kohler. 2002. Sex-specific biotransformation and detoxification after xenobiotic exposure of primary cultured hepatocytes of European flounder (*Platichthys flesus* L.). *Aquatic Toxicology* 59: 17-33.

Wong, T.T. and Y. Zohar. 2004. Novel expression of gonadotropin subunit genes in oocytes of the gilthead seabream (*Sparus aurata*). *Endocrinology* 145: 5210-5220.

Wormke, M., M. Stoner, B. Saville and S. Safe. 2000. Crosstalk between estrogen receptor alpha and the aryl hydrocarbon receptor in breast cancer cells involves unidirectional activation of proteasomes. *FEBS Letters* 478: 109-112.

Yamamoto, K. 1951. Activation of the egg of the dog-salmon by water and the associated phenomena. *Journal of Faculty of Science*, Hokkaido University Series VI, Zoology 10: 303-319.

Molecular Markers of Androgen Disruption

Per-Erik Olsson[1,*], Johnny Karlsson[1], Anders Larsson[1], Carina Modig[1] and Ian Mayer[2]

INTRODUCTION

There is increasing awareness over the adverse effects of environmental contaminants on the health and well being of both humans and wildlife (Olsson et al., 1998; Tyler et al., 1998). While evidence for humans is still somewhat circumstantial, the occurrence of a variety of developmental and reproductive disorders in wildlife species have clearly been linked to the exposure to environmental contaminants. Due, in part, to the fact that most man-made contaminants will eventually end up in the aquatic environment, most of the more pronounced developmental and reproductive disorders have been reported in aquatic species. For example, probably the most clearly identified case of endocrine disruption to date is that of imposex (development of a penis and vas deferens in females) in marine gastropods, resulting from the exposure to tributyltin

Authors' addresses: [1]Department of Natural Sciences, Life Science Center, Örebro University, SE-701 82 Örebro, Sweden.
[2]Department of Biology, University of Bergen, N-5020 Bergen, Norway.
*Corresponding author: E-mail: per-erik.olsson@nat.oru.se

(Gibbs *et al.*, 1990). Another widely documented example of endocrine disruption in aquatic species is that of feminization of male freshwater fish (Jobling *et al.*, 1998). In this case, as a result of exposure to primarily domestic effluents, male fish show elevated plasma levels of the female-specific yolk protein precursor, vitellogenin (VTG), as well as a high incidence of intersex (ovo-testis).

As most of these developmental and reproductive disorders have been attributed to environmental contaminants that interfere with the endocrine system, these compounds are commonly referred to as endocrine disrupting substances (EDS). An increasing number of substances, with possible endocrine disrupting potential, are being released into the environment and many of these substances may disturb vertebrate reproduction. While most work has been aimed at identifying oestrogenic effects, many other endocrine systems, including the androgenic system, may be affected by chemicals released into the environment. This review is focused on androgenic effects of EDS on fish.

Effects caused by EDS have rendered much attention due to their profound impact on human, domestic and wildlife welfare. However, it is important to understand that hormonal effects are systemic and that it is, therefore, difficult to identify EDS without performing *in vivo* experiments. This was emphasized at the European workshop held in Weybridge, U.K., in 1996, were it was agreed that "An endocrine disrupter is an exogenous substance that causes adverse health effects in an intact organism, or its progeny, consequent to changes in endocrine function". It was also agreed that "a **potential** endocrine disrupter is a substance that possesses properties that might be expected to lead to endocrine disruption in an intact organism" (EUR 17549, 1996). As these definitions suggest, it is important to develop *in vivo* systems for determination of EDS while *in vitro* systems may be helpful in the selection of candidate substances. In 1998, the US Environmental Protection Agency defined EDSs as being "exogenous agents that interfere with the synthesis, secretion, transport, binding, action, or elimination of natural hormones in the body that are responsible for the maintenance of homeostasis, reproduction, development, and/or behaviour" (EPA, 1998). At present, a large number of substances that either directly or indirectly affect many endogenous hormone pathways have been identified. Especially regulatory pathways based around steroid hormones and sex hormones in particular, have proved to be sensitive and easily modified by the interference of exogenous endocrine disruptors. One of the reasons for this is that these regulatory pathways are based around lipid soluble ligands that penetrate the cells' lipid membrane to bind and activate their target steroid receptors therein. Exogenous substances with chemical properties similar to steroid hormones are, thereby, not only constantly threatening to compete with these endogenous hormones, but also likely to remain solved in fatty tissues of any higher organism where they can bioaccumulate until they reach a hazardous concentration.

There are four general mechanisms by which endocrine disrupting substances exert their effects: (1) by mimicking endogenous hormones and causing activation of regular downstream pathways, (2) by antagonizing endogenous hormones and blocking regular downstream pathways, (3) by interfering with synthesis and metabolism of endogenous hormones, and (4) by interfering with regulation of hormone receptor levels.

ANDROGENS

Androgens,essential for male sexual differentiation and development, are needed for the development of the male secondary sex characteristics. Testosterone (T) is synthesized from cholesterol in the Leydig cells of the testis. All natural androgens are C19-steroids. The naturally occurring androgens are T and 5α-dihydrotestosterone (DHT) in mammals. In teleosts 11-ketotestosterone (KT) is also present and has been shown to stimulate development of male characteristics. Androgens increase protein synthesis and stimulate growth, leading to increased growth rate. There are some conflicting data on the effect of androgens on androgen receptor (AR) mRNA levels (Jänne and Shan, 1989; Abdelgadir *et al.*, 1993; Gonzales-Cavadid *et al.*, 1993). *In vivo*, it appears that both T and DHT down-regulate the AR mRNA levels. However, *in vitro*, it appears that T down-regulates the androgen receptor while DHT up-regulates it (Gonzales-Cavadid *et al.*, 1993; Lin *et al.*, 1993). The exact mechanisms by which androgens may regulate AR remain to be elucidated.

The hormone systems vary markedly between different teleost species. Testosterone is produced in the gonads of both sexes and displays increased plasma levels in pre-spawning and spawning fish (Fostier *et al.*, 1983; Borg, 1994). In most of the fishes studied, including salmonids, the T levels are higher in females than in males. However, the 11-oxygenated androgens, mainly KT occur at higher levels in males than in females. They are effective in stimulating male secondary sexual characteristics and at least in some fish, spermatogenesis and male reproductive behaviour.

ANDROGEN RECEPTORS

Androgens mediate their physiological function through interactions with receptors belonging to the steroid hormone receptor super-family. The receptors activate genes through binding to a variety of different gene response elements. Nuclear steroid hormone receptors, including the oestrogen, androgen and progesterone receptors, bind to DNA as homodimers. The genes for the receptors and the binding proteins are highly conserved between species and show specific expressions in tissues and during development. The nuclear hormone receptors share a common structure and have a common mechanism of action.

The receptors can be divided into six regions (A-F) with regard to amino acid similarities in different parts of the protein (Krust *et al.*, 1986). The most conserved region is the C region that is responsible for the DNA-binding specificity. The ligand-binding domain is located in region E and is also conserved. A schematic representation of androgen receptors is given in Figure 16.1.

The main transcription-activating domain of these receptors is located in the ligand-binding domain (E-region) in the C-terminus (Webster *et al.*, 1988). The receptors contain two independent acidic transcriptional activation function (AF) domains, AF-1 and AF-2 (Lees 1989; Tora *et al.*, 1989). The receptors are inducible transcription factors that modulate specific gene expression by binding to short DNA sequences located on hormone-regulated genes. These short DNA sequences are—in the case of androgens—termed the androgen response elements (ARE),

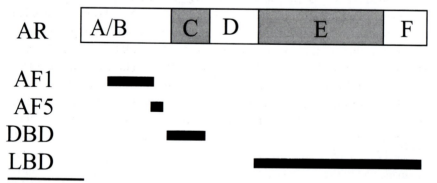

Fig. 16.1 Schematic representation of AR and the localization of specific domains. The AF1 (activator function 1) domain corresponds to amino acids (aa) 102-370 in the human androgen receptor (AR). The AF5 domain corresponds to aa 360-385 in the human AR. The DNA binding domain (DBD) corresponds to aa 550-635 in the human AR. The ligand-binding domain (LBD) corresponds to aa 672-919 in the human AR.

and consist of inverted repeat sequences with the core sequence TGTTCT (Ham *et al.*, 1988). While the half-sites are shared between different hormone receptors, the spacing in-between the two half-sites may vary. The palindromic nature of the hormone response elements indicates that the receptors bind their targets as dimers (Kumar and Chambon, 1988). A region within the hormone-binding domain has been identified to be required for both hormone receptor dimerization and high affinity DNA-binding (Fawell *et al.*, 1990).

Androgen receptors have been isolated and characterized from all vertebrate phyla and show high similarities between the different groups (Fig. 16.2). Most, but not all, vertebrates have two individual AR with high affinities for either T or DHT, the major androgens in mammals. Unlike mammals, the most important androgen in fish is generally KT (Borg, 1994).

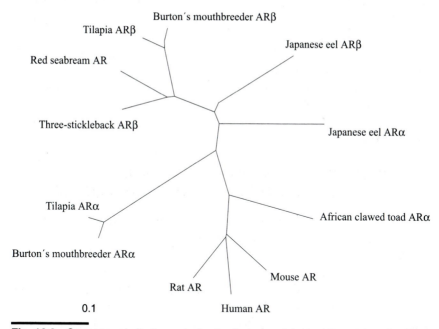

Fig. 16.2 Sequence similarity analysis of selected vertebrate AR proteins. GenBank Accession Nos.: Human AR (M34233); Mouse AR (X53779); Rat AR (M20133); African clawed toad AR (U67129); Tilapia ARa (AB045211); Tilapia ARb (AB045212); Burton's mouthbreeder AR (AF121257); Burton's mouthbreeder ARb (AY082342); Japanese eel ARa (AB023960); Japanese eel ARb (AB023961); Red seabream (AB017158); Three-spined stickleback (AY24707); The protein sequences were aligned by the Clustal W algorithm (Version 1.7) and the tree was subsequently constructed using Tree View (Version 1.6.2).

AR have been cloned and characterized from several fish species, including Japanese eel (*Anguilla japonica*), rainbow trout (*Oncorhynchus mykiss*) and red seabream (*Pagrus major*) (Ikeuchi *et al.*, 1999; Takeo and Yamashita, 1999; Todo *et al.*, 1999; Touhata *et al.*, 1999). Both rainbow trout and Japanese eel have two AR isoforms and while only one is active in rainbow trout (ARα), both Japanese eel AR are active. Transfection of the human embryonic kidney 293 cell line with AR expression vectors showed that KT and DHT were the best activators of both AR isoforms while T was significantly less potent in activating an MMTV-LTR driven luciferase reporter vector (Ikeuchi *et al.*, 1999; Todo *et al.*, 1999). Cloned red seabream AR was transfected into COS-7 cells and both T and KT were equally potent at activating the MMTV promoter system (Touhata *et al.*, 1999). The rainbow trout ARα was equally activated by DHT, KT and T in the carp EPC cell line, while determination of binding affinity showed that DHT bound stronger than T that in turn bound stronger than KT (Takeo *et al.*, 1999). Recently, the three-spined stickleback (*Gasterosteus aculeatus*) AR was cloned and characterized (Olsson *et al.*, 2005). Interestingly, while the stickleback AR was found to bind DHT stronger than KT and T, transactivation assays showed that KT was better at activating the AR than DHT or T. The identification of a KT-regulated AR in a fish species opens up the possibility that species differences should be expected and that EDSs may act differently on fish and mammals.

ANDROGENIC AND ANTI-ANDROGENIC SUBSTANCES OBSERVED IN THE ENVIRONMENT

An increasing number of man-made chemicals, having a very diverse array of chemical structures, have been identified as endocrine disrupters. To date, most attention has been focused on those chemicals that interact with the oestrogen receptor, the so-called xenoestrogens. Indeed, most reported reproductive disturbances reported in aquatic wildlife have been attributed to exposure to chemicals having oestrogenic properties. For this reason, a number of both *in vitro* and *in vivo* assays have been developed to screen chemicals for oestrogenic properties. For example, one of the most commonly used *in vivo* assays used for the identification of oestrogenic effects is the fish VTG assay (Denslow *et al.*, 1999). However, an increasing number of man-made chemicals are now being found to have either androgenic or more commonly anti-androgenic properties (Kelce *et al.*, 1998; Sohoni *et al.*, 2001), or in some cases can interact with more than one steroid receptor. For example, o,p'-DDT, the principal isomer of DDT, has both oestrogenic and anti-androgenic properties (Kelce *et al.*, 1995).

The androgen-regulated pathways are much less investigated than the corresponding oestrogenic systems. Increased androgenic effects can occur not only through AR-ligand interactions but also through substances that interfere with endogenous aromatase activity and the conversion of androgens into 17β-estradiol. Exposure to aromatase inhibitors result in increasing levels of circulating androgens and, thereby, an indirect route to activation of androgenic downstream pathways in females. Similarly, in males, substances evoking anti-androgenic effects may do so by specifically blocking steroidogenic enzyme activity and thereby inhibiting the conversion of cholesterol into any of the endogenous androgenic hormones.

There is limited information regarding identified EDSs with documented agonistic androgenic effects in the environment (Parks *et al.*, 2001). However, androgenic effects have been observed on mosquitofish (*Gambusia affinis*) in a river system receiving pulp mill effluents (Jenkins *et al.*, 2001). Using liquid chromatography-mass spectrometry, androstenedione was identified in the water. The authors hypothesized that the androgens were a result of bacteria transforming plant sterols into androgenic compounds (Jenkins *et al.*, 2004). They further showed that this conversion could be mimicked in the lab by using *Mycobacterium smegmatis*. Further on, a brominated flame retardant, 1,2-dibromo-4-(1,2-dibromoethyl) cyclohexane (BCH), was identified as a potent agonist to the human AR. By using a combination of computational modelling and a luciferase in vitro reporter system, the BCH was found to interact with, and activate, the receptor (Larsson *et al.*, 2006). These studies clearly show that androgenic compounds are present in the environment and that they can be produced by the bacterial population in areas receiving pulp mill effluents.

While few androgenic compounds are being detected in nature, the list of compounds displaying anti-androgenic effects is slowly growing more extensive. Among these anti-androgenic substances are several chlorinated pesticides, such as the fungicide vinclozolin and procymidone, and the insecticides DDT (2,2-bis(*p*-chlorophenyl)-1,1,1-trichloroethane) and methoxychlor (2,2-bis(*p*-methoxyphenyl)-1,1,1-trichloroethane). One common feature of many of these substances is that they are not particularly hazardous by themselves, until they are metabolized by the liver and converted to their more active metabolites. Vinclozolin, which is predominantly used for treatment of grains, grapes and turf grass, is metabolized into active 2-{((3,5-dichlorophenyl)-car-bamoyl)oxy}-2-methyl-3-butenoic acid (M1) and 3',5'-dichloro-2-hydroxy-2-methylbut-3-enanilide (M2) (Szeto et al., 1989). DDT is metabolized into active p-p-DDE (p,p'-dichlorodiphenyldichloroethylene), and methoxychlor, the currently predominant substitute for DDT, is converted into the active 2,2-bis(p-methoxyphenyl)-111-trichloroethane (HPTE) (Bulger et al., 1978).

Both vinclozolin metabolites M1 and M2, as well as procymidone, p-p'-DDE and methoxychlor, are proposed to down regulate androgen-dependent gene expression and to disrupt normal sexual differentiation in male rats (Kelce et al., 1997; Gray et al., 1999; Ostby et al., 1999). However, the anti-androgenic effect of methoxychlor has been questioned and recent reports instead indicated that this substance may act by interfering with endogenous steroid hormone synthesis (Murono and Derk, 2004). Vinclozolin and p-p-DDE has also been tested in different fish species and have been shown to exert anti-androgenic effects in guppy (*Poecilia reticulata*) with significant reduction in testes size, ejaculated sperm cells and disrupted coloration and male courtship behaviour (Baatrup and Junge, 2001), while no effects were observed in fathead minnow (*Pimephales promelas*) (Makynen et al., 2000). In addition, methoxychlor has further been reported to display mild oestrogenic effects and ability to induce VTG mRNA synthesis and serum VTG accumulation in fathead minnow (Hemmer et al., 2001).

While mammalian systems are commonly appreciated when investigating potential anti-androgenic effects of a chemical, they are more limited in regard to compounds displaying an androgenic effect. This is partially due to the fact that sex hormones in mammals are kept under strict control of *overlaying* genetic control mechanisms. This is not the case in fish however, and it is quite common, especially among sexually dimorphic fish species, to find males and females displaying pure androgen- or oestrogen-dependent secondary sex characteristics. Androgens are also proposed to relate to social behaviours in fish such as mating and aggression (Olivera et al., 2002). Since androgens in particular often induce changes in both morphology and behaviour in many fish species, especially during reproductive phases, any of these features can be utilized to monitor the influence of potential environmental androgenic disrupters. Several field observations in both the U.S.A. and Europe have already reported female fish showing proposed evidence of masculinization by displaying typical male morphological changes such as gonopodium development (Angus et al., 2001; Parks et al., 2001).

While most of the identified EDS display weak hormonal activity as an unintended side effect, other substances have been developed synthetically with the intent to alter the endocrine system. These are chemicals designed for medical aspects, as growth promoters in farmed animals, or to increase performance among dubious professional athletics. One of the most commonly used synthetic anti-androgens is the therapeutic drug hydroxyflutamide (OHF). This compound binds with high affinity to ARs and effectively blocks any downstream activity in both mammals and fish (Kemppainen *et al.*, 1992; Ankley *et al.*, 2004). This feature has been utilized not only in the treatment of androgen-dependent prostate cancer, but has also made OHF a much appreciated tool as an effective competitor in AR binding assays.

The anabolic steroid trenbolone acetate (TBA) is widely used on farmed cattle in the U.S.A. It is converted into 17β-trenbolone (17β-hydroxy-estra-4,9,11-trien-3-one; TB) in cattle. TB has not only been shown having high affinity for the human AR, but also for the bovine progestin receptor, as well as displaying potent anti-glucocorticoid properties (Danhaive and Rousseau, 1988; Bauer *et al.*, 2000). Studies made in cattle feedlot facilities in the U.S.A. have revealed that TB, and other related metabolites such a TB and triendione, are released into the environment. Once in the environment, TB is a comparatively stable compound and has a half-life of approximately 260 days in liquid manure (Schiffer *et al.*, 2001). Effects on fish have been observed, with incidences of disturbed fecundity and reproductive performance in fathead minnow populations living down stream of cattle feedlot facilities (Ankley *et al.*, 2003; Miller and Ankley, 2004).

There is an increasing consumption of androgenic anabolic steroids (AASs), and there is a concern that these substances may find their way into the environment through sewage treatment plants. This is a well-documented route of environmental exposure of xenobiotic oestrogenic substances originating from an extensive use of oral contraceptives (Larsson *et al.*, 1999). Many of these AASs such as nandrolone decanoate (ND) are highly potent androgenic agonists used both therapeutically as also to increase performance among professional as well as amateur athletics. Studies on acute health effects in humans due to abusive AAS consumption have shown enhanced impulsiveness, irritability, aggression and hostility (Choi *et al.*, 1990; Thiblin and Parlklo, 2002; Perry *et al.*, 2003). These results have also been verified in rat (Lindqvist and Fahlke, 2005) but so far, no studies have shown any effect on fish. Endogenous and exogenous substances with androgenic and anti-androgenic characteristics are shown in Figure 16.3.

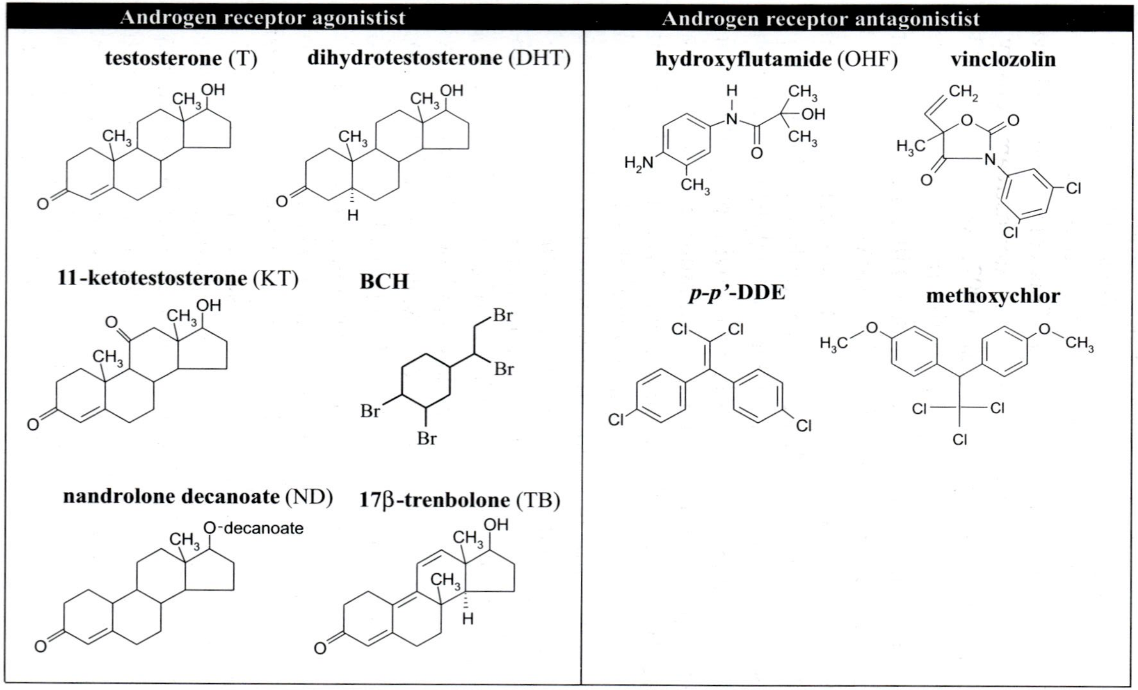

Fig. 16.3   Chemical structure and properties of endogenous and exogenous substances known to bind and interfere with androgen receptor activated pathways. A graphical presentation of the chemical structure of a number of substances with known AR agonistic and antagonistic effects. Among the AR agonists are the endogenous androgens T, DHT and KT, while ND and TB are anabolic androgenic substances (AASs). The synthetic therapeutic drug, OHF, has strong anti-androgenic properties, while vinclozolin, p-p'-DDE and methoxychlor are pesticides with reported antagonistic effects in both mammals and teleosts.

THE STICKLEBACK AS AN INDICATOR SPECIES

It has been known for many years that exposure of fish to pulp mill effluents, which by their nature consist of a complex mixture of chemicals, can result in masculinization. For example, even before the concept of endocrine disruption became topical, it was reported that exposure to pulp mill effluents induced certain masculine traits in female mosquitofish, notably the induction of male sexual behaviour and the development of the gonopodium (Howell *et al.*, 1980; Bortone *et al.*, 1989). The observation that viviparous eelpout (*Zoarces viviparus*) caught near a Swedish pulp mill showed a significant male embryo bias (Larsson *et al.*, 2000), and that exposure to pulp mill effluent in laboratory studies induced spiggin synthesis in female sticklebacks (Katsiadaki *et al.*, 2002a) further supported the view that pulp mill effluents exhibited androgenic properties. Even with these reported cases of putative androgenic effects, there has until recently been a lack of sensitive *in vivo* assays for the identification of chemicals having androgenic or anti-androgenic properties. This situation has now been addressed by the development of a robust and sensitive *in vivo* assay based on the induction of the male specific kidney protein, spiggin, produced by the three-spined stickleback.

The three-spined stickleback is a small teleost fish that occurs widely throughout the northern temperate region. While the stickleback has traditionally been the subject of many studies on behaviour, ecology and evolution (Bell and Foster, 1994), in recent years it has attracted increasing attention as an indicator species for the identification of endocrine disrupters. The interest in the stickleback centres on the fact that the male displays a number of pronounced androgen-dependent male secondary sexual characters, including breeding colour and kidney hypertrophy. During the breeding season, the kidney of male sticklebacks undergoes pronounced hypertrophy, and its function changes from its normal excretory role to one of producing glue, which the male uses to build its nest. This glue consists primarily of a single 203 kDa cysteine-rich glycoproprotein called 'spiggin' (Jakobsson *et al.*, 1999), and its synthesis and secretion is controlled by androgens, notably KT (Borg *et al.*, 1993), the major androgen found in male teleost fishes (Borg *et al.*, 1994). However, following the molecular cloning and characterization of spiggin, it now appears that spiggin is not a single protein but is a protein complex consisting of three distinct subunits synthesized in the kidney, and later assembled in the urinary bladder where it is stored (Jones *et al.*, 2001).

The ability of the male stickleback kidney to produce spiggin in response to androgens opened up the unique possibility of developing the stickleback as a test-species for EDS having androgenic properties. This potential was realized by producing a polyclonal antiserum to spiggin, and the subsequent development and validation of a sensitive ELISA for the measurement of spiggin in kidney tissue (Katsiadaki et al., 2002b). The feasibility of employing the spiggin ELISA as a viable in vivo assay for the identification of androgenic chemicals was heightened by two points. Firstly, although the kidneys of female sticklebacks do not normally produce spiggin, it was found that when exposed to androgens via the water, females produced spiggin in much the same way as males (Katsiadaki et al., 2002b). Female sticklebacks exposed to four concentrations of the synthetic androgen, 17α-methyltestosterone (MT) ranging from 10 ngL^{-1} to 10 μgL^{-1} showed a clear dose response after a 21-day exposure, with a 100,000-fold difference between the lowest and highest ELISA values (spiggin units/g body weight). In many ways, the spiggin response shown by female sticklebacks in response to androgen exposure is analogous to the VTG response shown by male fish, in the sense that male fish that do not normally synthesize VTG will show high plasma levels of this lipoprotein if exposed to oestrogenic chemicals in the water. While the kidneys of male sticklebacks also produce more spiggin in response to waterborne androgens, the response in females is a more sensitive biomarker, as male sticklebacks will often have higher background levels of spiggin.

The potential of the spiggin assay was greatly increased after discovering that it could be used as a sensitive in vivo assay for the identification of anti-androgens as well as androgens. This is of importance in view of the fact that environmental contaminants are generally more prone to act as anti-androgens rather than androgens. The simultaneous exposure of female sticklebacks with the anti-androgen flutamide (500 μgL^{-1}) and either of two androgens DHT or MT for 21 days totally or significantly reduced spiggin production by the kidneys, as compared to kidneys of females exposed to the androgens alone (Katsiadaki et al., 2002a).

While the stickleback has many characters that make it an ideal sentinel species for the identification of (anti-) androgenic chemicals, two of the most important characters are that the species is indigenous to both Europe and North America, making it an ideal species for environmental biomonitoring, and it is presently the only species with a quantifiable androgen (and anti-androgen) endpoint. With the development and validation of a stickleback VTG assay, the stickleback is now being promoted as simultaneous biomarker for both environmental androgens (spiggin induction in females) and oestrogens (VTG induction in males) (Katsiadaki *et al.*, 2002a; Hahlbeck *et al.*, 2004b). This will potentially result in reduced testing costs, and at the same time will result in a dramatic reduction in the number of fish used for testing— this is important from an ethical and welfare point of view, an increasingly important consideration.

It is increasingly being recognized that one of the possible consequences of exposure to EDS in aquatic wildlife is disrupted sexual differentiation. Sexual differentiation in fishes is a very labile process, and depending on the species, it occurs at a critical period some time during early embryo or larval development (reviewed by Devlin and Nagahama, 2002). The balance between endogenous androgens and estrogens during this short window of sensitivity determines the future gonadal sex of the progeny. It is now known that exposure to exogenous steroids during this sensitive period can profoundly influence the resultant sexual differentiation. For example, exposure coho salmon (*Oncorhynchus kisutch*) to either androgen or oestrogen during their critical period, corresponding to the first week post-hatch resulted in complete masculinization or feminization, respectively (Piferrer and Donaldson, 1989). Although not as effective as natural steroids, increasing evidence indicates that exposure to endocrine disrupters can also influence sexual differentiation in teleost fishes (Scholz and Gutzeit, 2000; van Aerle *et al.*, 2001). Rather than complete sex reversal, exposure to such EDS more generally results in the condition termed intersex, in which the progeny are characterized by the presence of both ovarian and testicular tissue in the gonad.

To date, the occurrence of intersex in teleost fish has been almost exclusively attributed to exposure to environmental oestrogens, with intersex individuals (presence of ovo-testis) being assumed to be feminized genetic males. For example, the high incidence of intersex in roach (*Rutilus rutilus*) and other freshwater species in U.K. rivers has generally been attributed to exposure to xenoestrogens, primarily those originating from sewage treatment plants (Jobling *et al.*, 1998; van Aerle *et al.*, 2001). However, in all likelihood, many cases of reported intersex could alternatively be due to exposure to androgenic contaminants, that is the masculinization of genetic females. A clear indication of masculinization, is that of the skewed sex ratios in favour of males found in the embryos of gravid viviparous eelpout females sampled close to a Swedish pulp mill (Larsson *et al.*, 2000). However, in order to confirm that masculinization had occurred, the genetic sex of the embryos should have been determined. This clearly indicates how valuable the availability of genetic markers for the sex chromosomes would be in these studies. Chromosome-linked sex markers have now been developed for a number of species including the Japanese medaka, *Oryzias latipes* (Scholz *et al.*, 2003) the chinook salmon, *Oncorhynchus tshawytscha* (Afonso *et al.*, 2002) and the three-spined stickleback (Griffiths *et al.*, 2000). Recent evidence suggests that the stickleback possesses a simple chromosomal mechanism for sex determination, with male or female sexual development being controlled by a single major chromosomal region (Peichel *et al.*, 2004). The availability of DNA sex marker for the stickleback represents a very powerful tool for investigating the effect of endocrine disrupters on sex differentiation, not just in laboratory studies but also in field surveys. By applying this DNA sex marker, the effect of androgens and oestrogens (both natural and synthetic) on sexual differentiation has been investigated in juvenile sticklebacks (Hahlbeck *et al.*, 2004a). These studies have demonstrated that steroid exposure can readily induce intersex in both genetic males and females, as well as in some cases complete sex reversal (genetic males changing to gonadal females and *vice versa*).

In conclusion, the three-spined stickleback is an ideal sentinel species for the study of endocrine disruption in fish, both in laboratory and field studies. The stickleback has many advantages, not the least that it is presently the only species with a quantifiable (anti-) androgenic endpoint. In view of the current lack of valid *in vivo* assays for the identification of contaminants having androgenic or anti-androgenic properties, this makes the stickleback a unique species. In addition, the availability of stickleback DNA-linked sex markers now offers the possibility of determining the genetic sex of stickleback progeny. This is an invaluable tool in determining the possible impact of environmental contaminants on sexual differentiation in stickleback populations, both in the laboratory and in the field.

Acknowledgements

Research finances by grants from CFN, Sweden (to PEO) and KK Stiftelsen, Sweden (PEO and CM).

References

Abdelgadir, S.E., P.B. Connolly and J.A. Resko. 1993. Androgen regulation of androgen receptor messenger ribonucleic acid differs in rat prostate and selected brain areas. *Molecular and Cellular Neuroscience* 4: 532-537.

van Aerle, R., T.M. Nolan, S. Jobling, L.B. Christiansen, J.P. Sumpter and C.R. Tyler. 2001. Sexual disruption in a second species of wild cyprinid fish (the gudgeon, *Gobio gobio*) in United Kingdom freshwaters. *Environmental Toxicology and Chemistry* 20: 2841-2847.

Afonso, L.O.B., J.L. Smith, M.G. Ikonomu and R.T. Giulio. 2002. Y-chromosome DNA markers for discrimination of chemical substance and effluent effects on sexual differentiation in salmon. *Environmental Health Perspectives* 110: 881-887.

Angus, R.A., H.B. McNatt, W.M. Howell and S.D. Peoples. 2001. Gonopodium development in normal male and 11-ketotestosterone-treated female mosquitofish (*Gambusia affinis*): a quantitative study using computer image analysis. *General and Comparative Endocrinology* 123: 222-234.

Ankley, G.T., K.M. Jensen, E.A. Makynen, M.D. Kahl, J.J. Korte, M.W. Hornung, T.R. Henry, J.S. Denny, R.L. Leino, V.S. Wilson, M.C. Cardon, P.C. Hartig and L.E. Gray. 2003. Effects of the androgenic growth promoter 17-beta-trenbolone on fecundity and reproductive endocrinology of the fathead minnow. *Environmental Toxicology and Chemistry* 22: 1350-1360.

Ankley, G.T., D.L. Defoe, M.D. Kahl, K.M. Jensen, E.A. Makynen, P. Hartig, L.E. Gray, M. Cardon and V. Wilson. 2004. Evaluation of the model anti-androgen flutamide for assessin mechanistic basis of responses to an androgen in the fathead minnow (*Pimephales promelas*). *Environmental Science and Technology* 38: 6322-6327.

Baatrup, E. and M. Junge. 2001. Antiandrogenic pesticides disrupt sexual characteristics in the adult male guppy (*Poecilia reticulata*). *Environmental Health Perspectives* 109: 1063-1070.

Bauer, E.R., A. Daxenberger, T. Petri, H. Sauerwein and H.H. Meyer. 2000. Characterisation of the affinity of different anabolics and synthetic hormones to the human androgen receptor, human sex hormone binding globulin and to the bovine progestin receptor. *APMIS* 108: 838-846.

Bell, M.A. and S.A. Foster (eds.) 1994. *The Evolutionary Biology of the Three-spined Stickleback.* Oxford University Press, Oxford.

Borg, B. 1994. Androgens in teleost fishes. *Comparative Biochemistry and Physiology* C 109: 219-245.

Borg, B., E. Antonopoulou, E. Andersson, T. Carlberg and I. Mayer. 1993. Effectiveness of several androgens in stimulating kidney hypertrophy, a secondary sexual character, in castrated male three-spined sticklebacks, *Gasterosteus aculeatus. Canadian Journal of Zoology* 71: 2327-2329.

Bortone, S.A., W.P. Davis and C.M. Brundrick. 1989. Morphological and behavioral characters in mosquitofish as potential bioindication of exposure to kraft mill effluent. *Bulletin of Environmental Contamination and Toxicology* 43: 370-377.

Bulger, W.H., R.M. Muccitelli and D. Kupfer. 1978. Studies on the *in vivo* and *in vitro* estrogenic activities of methoxychlor and its metabolites. Role of hepatic mono-oxygenase in methoxychlor activation. *Biochemical Pharmacology* 27: 2417-2423.

Choi, P.Y.L., A.C. Parrot and D. Cowan. 1990. High-dose anabolic steroid in strenght athletes: effect upon hostility and aggression. *Human Psychopharmacology* 5: 349-356.

Danhaive, P.A. and G.G. Rousseau. 1988. Evidence for sex-dependent anabolic response to androgenic steroids mediated by muscle glucocorticoid receptors in the rat. *Journal of Steroid Biochemistry* 29: 575-581.

Denslow, N.D., M.C. Chow, K.J. Kroll and L. Green. 1999. Vitellogenin as a biomarker of exposure for oestrogen or oestrogen mimics. *Ecotoxicology* 8: 385-398.

Devlin, R.H. and Y. Nagahama. 2002. Sex determination and sex differentiation in fish: an overview of genetic, physiological and environmental influences. *Aquaculture* 208: 191-364.

EUR 17549. 1996. European workshop on the impact of endocrine disrupters on human health and wildlife. *Report of proceedings*, pp. 125.

EPA. United States Environmental Protection Agency. 1998. *ORD Research plan for endocrine disruptors*. EPA/600/R-98/087.

Fawell, S.E., J.A. Lees, R. White and M.G. Parker. 1990. Characterization and colocalization of steroid binding and dimerization activities in the mouse oestrogen receptor. *Cell* 60: 953-962.

Fostier, A., B. Jalabert, B. Billard and Y. Zohar. 1983. The gonadal steroids. In: *Fish Physiology*. W.S. Hoar, D.J. Randall and E.M. Donaldson (eds.). Academic Press, New York, Vol. 9, pp. 277-372.

Gibbs, P.E., G.W. Bryan, P.L. Pascoe and G.R. Burt. 1990. Reproductive abnormalities in female *Ocenebra erinacea* (Gastropoda) resulting from tributyltin-induced imposex. *Biological Journal of Marine Biological Association of the United Kingdom, Plymouth* 70: 639-656.

Gonzales-Cavadid, N.F., D. Vernet, A. Fuentes Navarro, J.A. Rodriguez, R.S. Swerdloff and J. Rafjer. 1993. Up-regulation of the levels of androgen receptor and its mRNA by androgens in smooth-muscle cells from rat penis. *Molecular and Cellular Endocrinology* 90: 219-229.

Gray, L.E., J. Ostby, R.L. Cooperand and W.R. Kelce. 1999. The estrogenic and antiandrogenic pesticide methoxychlor alters the reproductive tract and behaviour without affecting pituitary size and prolactin secretion in male rats. *Toxicology and Industrial Health* 15: 37-47.

Griffiths, R., K.J. Orr, A. Adam and I. Barber. 2000. DNA sex identification in the three-spined stickleback. *Journal of Fish Biology* 57: 1331-1334.

Hahlbeck, E., R. Griffiths and B.-E. Bengtsson. 2004a. The juvenile three-spined stickleback (*Gasterosteus aculeatus* L.) as a model organism for endocrine disruption I—Sexual differentiation. *Aquatic Toxicology* 70: 287-310.

Hahlbeck, E., I. Katsiadaki, I. Mayer, M. Adolfsson-Erici, J. James and B.-E. Bengtsson. 2004b. The juvenile three-spined stickleback (*Gasterosteus aculeatus* L.) as a model organism for endocrine disruption II—kidney hypertrophy, vitellogenin and spiggin induction. *Aquatic Toxicology* 70: 311-326.

Ham, J., A. Thomson, M. Needham, P. Webb and M. Parker. 1988. Characterization of response elements for androgens, glucocorticoids and progestins in mouse mammary tumour virus. *Nucleic Acids Research* 16: 5263-5276.

Hemmer, M.J., B.L. Hemmer, C.J. Bowman, K.J. Kroll, L.C. Folmar, D. Marcovich, M.D. Hoglund and N.D. Denslow. 2001. Effects of p-nonylphenol, methoxychlor, and endosulfan on vitellogenin induction and expression in sheepshead minnow (*Cyprinodon variegatus*). *Environmental Toxicology and Chemistry* 202: 336-343.

Howell, W.M., D.A. Black and S.A. Bortone. 1980. Abnormal expression of secondary sexual characters in a population of mosquitofish, *Gambusia affinis holbrooki*: evidence for environmentally-induced masculinization. *Copeia* 1980: 676-681.

Ikeuchi, T., T. Todo, T. Kobayashi and Y. Nagahama. 1999. cDNA cloning of a novel androgen receptor subtype. *Journal of Biological Chemistry* 274: 25205-25209.

Jakobsson, S., B. Borg, C. Haux and S.J. Hyllner. 1999. An 11-ketotestosterone induced kidney-secreted protein: The nest building glue from the male three-spined stickleback, *Gasterosteus aculeatus*. *Fish Physiology and Biochemistry* 20: 79-85.

Jänne, O.A. and L. Shan. 1989. Structure and function of the androgen receptor. *Annals of the New York Academy of Sciences* 626: 81-91.

Jenkins, R., R.A. Angus, H. McNatt, W.M. Howell, J.A. Kemppainen, M. Kirk and E.M. Wilson. 2001. Identification of androstenedione in a river containing paper mill effluent. *Environmental Toxicology and Chemistry* 20: 1325-1331.

Jenkins, R.L., E.M. Wilson, R.A. Angus, W.M. Howell, M. Kirk, R. Moore, M. Nance and A. Brown. 2004 Production of androgens by microbial transformation of progesterone in vitro: a model for androgen production in rivers. *Environmental Health Perspectives* 112: 1508-1511.

Jobling, S., M. Nolan, C.R. Tyler, G. Brighty and J.P. Sumpter. 1998. Widespread sexual disruption in wild fish. *Environmental Science and Technology* 32: 2498-2506.

Jones, I., C. Lindberg, S. Jakobsson, U. Hellman, A. Hellqvist, B. Borg and P.-E. Olsson. 2001. Molecular cloning and characterization of spiggin: an androgen regulated extraorganismal adhesive with structural similarities to von Willebrand Factor-related proteins. *Journal of Biological Chemistry* 276: 17857-17863.

Katsiadaki, I., A.P. Scott and I. Mayer. 2002a. The potential of the three-spined stickleback (*Gasterosteus aculeatus* L.) as a combined biomarker for oestrogens and androgens in European waters. *Marine Environmental Research* 54: 725-728.

Katsiadaki, I., A.P Scott, M.R. Hurst, P. Matthiessen and I. Mayer. 2002b. Detection of environmental androgens: A novel method based on Enzyme-Linked Immunosorbent Assay of spiggin, the stickleback (*Gasterosteus aculeatus*) glue protein. *Environmental Toxicology and Chemistry* 21: 1946-1954.

Kelce, W.R., C.R. Stone, S.C. Laws, L.E. Gray, J.A. Kemppainen and E.M. Wilson. 1995. Persistent DDT metabolite *p,p'*-DDE is a potent androgen receptor antagonist. *Nature (London)* 375: 581-585.

Kelce, W.R., C.R. Lambright, L.E. Jr. Gray and K.P Roberts. 1997. Vinclozolin and p,p'-DDE alter androgen-dependent gene expression: in vivo confirmation of an androgen receptor-mediated mechanism. *Toxicology and Applied Pharmacology* 142: 192-200.

Kelce, W.R., L.E. Gray and E.M. Wilson. 1998. Antiandrogens as environmental endocrine disruptors. *Reproduction Fertility and Development* 10: 105-111.

Kemppainen, J.A., M.V. Lane, M. Sar and E.M. Wilson. 1992. Androgen receptor phosphorylation, turnover, nuclear transport, and transcriptional activation. Specificity for steroids and antihormones. *Journal of Biological Chemistry* 267: 968-974.

Krust, A., S. Green, P. Argos, V. Kumar, P. Walter, J.M. Bornert and P. Chambon. 1986. The chicken oestrogen receptor sequence: homology with v-erbA and the human oestrogen and glucocorticoid receptors. *EMBO Journal* 5: 891-897.

Kumar, V. and P. Chambon. 1988. The oestrogen receptor binds tightly to its responsive element as a ligand-induced homodimer. *Cell* 55: 145-156.

Larsson, A., L. A. Eriksson, P.L. Andersson, P. Ivarsson and P.-E. Olsson. 2006. Identification of the brominated flame retardant 1,2-dibromo-4-(1,2-dibromoethyl)cyclohexane as an androgen agonist. *Journal of Medicinal Chemistry* 49: 7366-7372.

Larsson, D.G.J., M. Adolfsson-Erici, J. Parkkonen, P.-E. Olsson, A.H. Berg and L. Förlin. 1999. Ethinylestradiol—An undesired fish contraceptive. *Aquatic Toxicology* 45: 91-97.

Larsson, D.G.J., H. Hällman and L. Förlin. 2000. More male fish embryos near a pulp mill. *Environmental Toxicology and Chemistry* 19: 2911-2917.

Lees, J.A., S.E. Fawell and M.G. Parker. 1989. Identification of constitutive and steroid-dependent transactivation domains in the mouse oestrogen receptor. *Journal of Steroid Biochemistry and Molecular Biology* 34: 33-39.

Lin, M.C., J. Rajfer, R.S. Swerdloff and N.F. Gonzales-Cavadid. 1993. Testosterone down-regulates the levels of androgen receptor mRNA in smooth muscle cells from rat corpora cavernosa via aromatization to estrogens. *Journal of Steroid Biochemistry and Molecular Biology* 45: 333-343.

Lindqvist, A.-S. and C. Fahlke. 2005. Nandrolone decanoate has long-term effects on dominance in a competitive situation in male rats. *Physiology and Behavior* 84: 45-51.

Makynen, E.A, M.D. Kahl, K.M. Jensen, J.E. Tietge, K.L. Wells, G.Van Der Kraak and G.T. Ankley. 2000. Effects of the mammalian antiandrogen vinclozolin on development and reproduction of the fathead minnow (*Pimephales promelas*). *Aquatic Toxicology* 48: 461-475.

Miller, D.H. and G.T. Ankley. 2004. Modeling impacts on populations: fathead minnow (*Pimephales promelas*) exposure to the endocrine disruptor 17β-trenbolone as a case study. *Ecotoxicology and Environmental Safety* 59: 1-9.

Murono, E.P. and R.C. Derk. 2004. The effects of the reported active metabolite of methoxychlor, 2,2-bis(p-hydroxyphenyl)-1,1,1-trichloroethane, on testosterone formation by cultured Leydig cells from young adult rats. *Reproductive Toxicology* 19: 135-146.

Olivera, R.F., K. Hirschenhauser, L.A. Carnerio and V.M. Canario. 2002. Social modulation of androgen levels in male teleost fish. *Comparative Biochemistry and Physiology* B 132: 203-215.

Olsson, P.E., B. Borg, B. Brunström, H. Håkansson and E. Klasson-Wehler. 1998. Endocrine disrupting substances—Impairment of reproduction and development. *Swedish Environmental Protection Agency Report* 4859. pp. 150.

Olsson, P.-E., A.H. Berg, J. von Hofsten, B. Grahn, A. Hellqvist, A. Larsson, J. Karlsson, C. Modig, B. Borg and P. Thomas 2005. Molecular cloning and characterization of a nuclear androgen receptor activated by 11-ketotestosterone. *Reproductive Biology and Endocrionology.* (In Press).

Ostby, J., W.R. Kelce, C. Lambright, C.J. Wolf, P. Mann and L.E. Jr. Gray. 1999. The fungicide procymidone alters sexual differentiation in the male rat by acting as an androgen-receptor antagonist in vivo and in vitro. *Toxicology and Industrial Health* 15: 80-93.

Parks, L.G., C.S. Lambright, E.F. Orlando, L.J. Jr. Guillette, G.T. Ankley and L.E. Jr. Gray. 2001. Masculinization of female mosquitofish in Kraft mill effluent-contaminated Fenholloway River water is associated with androgen receptor agonist activity. *Toxicological Sciences* 62: 257-267.

Peichel, C.L., J.A. Ross, C.K. Matson, M. Dickson, J. Grimwood, J. Schmutz, R.M. Myers, S. Mori, D. Schlute and D.M. Kingsley. 2004. The master sex-determination locus in threespine sticklebacks is on a nascent Y chromosome. *Current Biology* 14: 1416-1424.

Perry, P.J., E.C. Kutscher, B.C. Lund, W.R. Yates, T.L. Holman and L. Demers. 2003. Measures of aggression and mood changes in male weightlifters with and without androgenic anabolic steroid use. *Journal of Forensic Sciences* 48: 646-651.

Piferrer, F. and E.M. Donaldson. 1989. Gonadal differentiation in coho salmon, *Oncorhynchus kisutch*, after a single treatment with androgen or oestrogen at different stages during ontogenesis. *Aquaculture* 77: 251-262.

Schiffer, B., A. Daxenberger, K. Meyer and H.H. Meyer. 2001. The fate of trenbolone acetate and melengestrol acetate after application as growth promoters in cattle: Environmental studies. *Environmental Health Perspectives* 109: 1145-1151.

Scholz, S. and H.O. Gutzeit. 2000. 17-α-ethinylestradiol affects reproduction, sexual differentiation and aromatase gene expression of the medaka (*Oryzias latipes*). *Aquatic Toxicology* 50: 363-373.

Scholz, S., S. Rösler, M. Schäffer, U. Hornung, M. Schartl and H.O. Gutzeit. 2003. Hormonal induction and stability of monosex populations in the Medaka (*Oryzias latipes*): Expression of sex-specific marker genes. *Biology of Reproduction* 69: 673-678.

Sohoni, P., P.A. Lefevre, J. Ashby and J.P. Sumpter. 2001. Possible androgenic/anti-androgenic activity of the insecticide fenitrothion. *Journal of Applied Toxicology* 21: 173-178.

Szeto, S.Y., N.E. Burlinson, J.E. Rettig and J. Trotter. 1989. Identification of hydrolysis products of vinclozolin by spectroscopic and X-ray crystallographic methods. *Journal of Agricultural and Food Chemistry* 37: 1103-1108.

Takeo, J. and S. Yamashita. 1999. Two distinct isoforms of cDNA encoding rainbow trout androgen receptors. *Journal of Biological Chemistry* 274: 5674-5680.

Thiblin, I. and T. Parlklo. 2002. Anabolic androgenic steroids and violence. *Acta Psychiatrica Scandinavica* 412 (Supplement): 125-128.

Todo, T., T. Ikeuchi, T. Kobayashi and Y. Nagahama. 1999. Fish androgen receptor: cDNA cloning, steroid activation of transcription in transfected mammalian cells, and tissue mRNA levels. *Biochemical and Biophysical Research Communications* 254: 378-383.

Tora, L., J. White, C. Brou, D. Tasset, N. Webster, E. Scheer and P. Chambon. 1989. The human oestrogen receptor has two indepenent nonacidic transcriptional activation functions. *Cell* 59: 477-487.

Touhata, K., M. Kinoshita, Y. Tokuda, H. Toyohara, M. Sakaguchi, Y. Yokoyama and S. Yamashita. 1999. Sequence and expression of a cDNA encoding the red seabream androgen receptor. *Biochimimica Biophysica Acta* 1450: 481-485.

Tyler, C.R., S. Jobling and J.P. Sumpter. 1998. Endocrine disruption in wildlife: a critical review of the evidence. *Critical Reviews of Toxicology* 28: 319-361.

Webster, N.J.G., S. Green, R. Jin-Jia and P. Chambon. 1988. The hormone-binding domains of the oestrogen and glucocorticoid receptors contain an inducible transcription activation function. *Cell* 54: 199-207.

17

Fish Reproduction in Relation to Aquaculture

Sena S. De Silva[1,2,*], Thuy T.T. Nguyen[2] and Brett A. Ingram[3]

INTRODUCTION

Aquaculture, an ancient practice, is thought to have originated in China, over 3000 years ago, and Fan Li is credited for the first written account on aquaculture in 460 BC (Chinese Aquaculture Society, 1986). Aquaculture, for all intents and purposes, remained an 'art' until the second half of the last century, when the world began to realize that the increasing demand for aquatic food products could be met from the traditional supplies, in particular the marine capture fisheries. It is in this context that the science of aquaculture began to develop, and over the last 30 years this sector has recorded the highest rate of growth amongst all primary production sectors in the world (De Silva, 2001).

Authors' addresses: [1]School of Ecology & Environment, Deakin University, P.O. Box 423, Warrnambool, Victoria, Australia 3280.
[2]Network of Aquaculture Centres in Asia-Pacific, P.O. Box 1040, Kasetsart Post Office, Bangkok 10903, Thailand.
[3]PIRVic, Department of Primary Industries, Snobs Creek, Private Bag 20, Alexandra, Victoria, Australia 3714.
Corresponding author: E-mail: sena.desilva@deakin.edu.au

Currently, the world aquaculture production is estimated at 51.4 million tons, valued at US$ 59.9 billion, and accounts for approximately 33% of all aquatic food products that are consumed (FAO, 2002). Over the last 20 years, aquaculture production has recorded a mean annual growth rate of approximately 10%, and all projections indicate that the sector is still growing, although the rate of growth has decreased, being only 6.5% between 2000 and 2001 (De Silva, 2001). Cultured finfish production accounts for over 70% of the world aquaculture production, and this is a trend that has been witnessed throughout its growth history, and it is expected to remain so to the foreseeable future. Also, the aquaculture industry is dominated by the developing world, with PR China being the main producer (FAO, 2002). Consequent to this is that the bulk of cultured finfish production is non-carnivorous species, feeding low in the food chain, and is dominated by species such as the silver carp, *Hypophthalamichthys molitrix*, bighead carp, *Aristichthys nobilis*, grass carp, *Ctenopharyngodon idellus*, common carp, *Cyprinus carpio*, rohu, *Labeo rohita*, catla, *Catla catla* and mrigal, *Cirrhinus mrigla*, all belonging to the Family Cyprinidae and the tilapias, notably Nile tilapia, *Oreochromis niloticus*, of the Family Cichlidae. These species are cultured in both tropical and sub-temperate climates. Amongst the cold-water species, Salmonid species account for the highest production, notably, Atlantic salmon *Salmo salar* and rainbow trout *Oncorhynchus mykiss*.

At present, the relevance of aquaculture as a main source of supply for aquatic food is becoming increasingly important for a number of reasons. Firstly, it is now generally accepted that the marine capture fisheries have reached a static state, and that its contribution to the world fish production is unlikely to exceed 100 million tons, at best, as nearly 50% of the commercially important fish stocks have been already overfished (Botsford *et al.*, 1997). Secondly, and on the other hand, the aquatic fish food consumption in the world is growing (Ye, 1999; Delgado *et al.*, 2003) and, coupled with human population growth the gap between supply and demand for these food resources is widening. It is in this regard that aquaculture will have an increasing role in narrowing this widening gap. The increase in consumption patterns for aquatic foods have been further exacerbated by relatively recent findings that finfish are a very good source of highly unsaturated fatty acids of the omega three and six series, commonly referred to as n-3 and n-6 fatty acids (HUFA), such as eicosapentaenoic (EPA, 20:5n-3), docosahexaenoic acid (DHA, 22:6n-3) and arachidonic acid (AA, 20:4n-6), amongst others. The fatty acids EPA and DHA in particular are known to have positive effects on human health and well being, the foremost being their influence on coronary thrombosis (Stansby, 1990; Ulbricht and Southgate, 1991; De Deckere *et al.*, 1998). The other functions thus far attribute to HUFA fatty acids include:

- as structural elements of cell membranes,
- act as precursors to eicosanoids, a heterogenous group of highly active 'local hormones',
- contribute to osmo-regulation,
- influences reproduction and egg quality,
- important for brain development, and
- important for development of vision.

Furthermore, the development of the human brain—and hence what we are today—and in particular from an evolutionary viewpoint, has also been linked to food sources rich in n-3 (EPA and DHA) and n-6 (AA) PUFAs. Indeed, a large quantum of evidence has been brought forward to show that *Homo sapiens* evolved not in a savannah habitat but in a habitat that was rich in fish and shellfish resources (Crawford *et al.*, 1999). More and more medical studies are emerging on the positive aspects of fish in the diet on human health, growth and general well-being. However, no attempt will be made here to review these in detail, but suffice is to say that in all probability *H. sapiens* is what it is today because of its early dependence on fish.

From the above discussion it is increasingly evident that aquatic food supplies are becoming an increasingly important component in the human diet and consequently, the importance and reliance of aquaculture to fulfil some of these needs is also growing. On the other hand, the aquaculture sector cannot and will not be able to grow unabated, particularly so with limitations on land and water resources and environmental concerns. It is, therefore, important that all developments in aquaculture take into account the importance of artificial propagation and improvements to brood stocks and hence the special relevance here of this chapter.

RELEVANCE OF FISH REPRODUCTION TO AQUACULTURE DEVELOPMENT

The success of large-scale aquaculture production has been underpinned by access to large numbers of seed stock or juveniles for grow-out. Historically, few farmed fish species were capable of breeding naturally in captivity (see Jhingran and Gopalakrishnan, 1974). Some species such as tilapias (*Tilapia* spp.), common carp, channel catfish (*Ictalurus punctatus*) and Asian catfish (*Clarias batrachus*) readily spawn in ponds in captivity. But often the ability to control spawning in these species was limited as the timing of spawning was difficult to predict and survival of eggs and larvae was highly variable (Harvey and Hoar, 1979).

In many traditional aquaculture industries, early pioneers relied on procuring stock directly from the wild with eggs, larvae, fry or juveniles being harvested. The only source of Chinese carps (grass carp, silver carp, bighead carp, black carp—*Mylopharyngodon piceus* and mud carp—*Cirrhina molitorella*) and major or Indian carps (catla, rohu, calbasu—*L. calbasu* and mrigal) for culture was from eggs, larvae or fry collected from streams or rivers during the spawning seasons (Pillay, 1990; Landau, 1992). Mullet (*Mugil* spp.) and milkfish (*Chanos chanos*) culture, which has occurred over centuries, was initially based on wild harvest of fry from coastal and estuarine areas (Nash and Shehadeh, 1980; Pillay, 1990; Su et al., 2002). Eel (*Anguilla* spp.) farming is a major industry in several Asian and European countries, yet even today farms rely solely on the capture of glass eels or elvers from the wild (Brusle, 1990; Heinsbroek, 1991; Gousset, 1992).

For other species, mature, running ripe fish were captured from the wild during the spawning seasons, and gametes were stripped and fertilized artificially. In the case of trouts (e.g., rainbow trout and brown trout—*Salmo trutta*) and salmons (Atlantic salmon and Pacific salmon—*Oncorhynchus* spp.), spawning adults were trapped and stripped during upstream spawning migrations, a technique that is still used today for some species. Adult striped bass (*Morone saxatilis*) were captured as they congregated during the spawning season (Landau, 1992). Until recently, seed stock of *Tor* spp., the mahseers, a group that is highly regarded as sport fish and considered to have a high aquaculture potential in Asia, could be obtained from hormone-induced spawning of mature females and males caught from the wild, during the reproductive season (Joshi, 1988; Ogale, 2002).

The inability to breed fish in captivity has hampered the progress of aquaculture of several important species. Relying on wild caught fish, either juveniles or spawning adults, as the sole source of seed stock for grow-out was problematic for many species and industries. Availability of spawning adults or seed stock from year to year was highly inconsistent and variable. Predicting when fish were spawning and when juveniles could be caught was often difficult, and for many species this occurred either seasonally or for a short period only. The quality of seed stock collected was also often highly variable. Frequently demand outstripped supply, and impeded industry development. Further, dwindling stocks in the wild, associated with over-exploitation, habitat destruction, environmental degradation, etc., meant that such harvesting was unsustainable in the long term.

Undoubtedly finfish aquaculture production the world over has made great strides in the last three to four decades, and has become a stable source of finfish supplies to the global market place. It is relevant and appropriate to question how this sector managed to achieve such high production levels in a relatively short time period. Obviously, a number of factors, technical, managerial, financial, socio-economic and policy-related (Subasinghe *et al.*, 2001), acting singly or in combination, have been responsible, and the impact of these factors being variable from nation to nation, and region to region, for the developments in the sector. However, amongst the technological factors, one of the foremost developments was the development of artificial propagation techniques, through hypophysation, which enabled the culture of finfish to be independent of the vagaries of natural seed supplies, and over the years also enabled selective breeding programs to be developed in order to improve the quality of seed stock.

Overall, presently, globally, about 50 finfish species are cultured commercially. Of these, perhaps with the exception of anguillid eels, almost all other finfish cultures are based on artificially propagated seed stocks. Therefore, artificial propagation and, hence, reproduction of all these finfish species is crucial to aquaculture.

CONTROL OF REPRODUCTION

In general, in cultured species a commonly used phrase is, 'closing the life cycle'. This phrase is used to denote those species in which the whole life cycle can be completed in a culture environment, and preferably when such a closing can be applied for purposes of commercial production. Closing of the life cycle enables culturists to be independent of wild stocks, except when they require genetic improvements, i.e., increase the genetic diversity, of the existing stock(s). The majority of cultured finfish species of commercial importance in the world fall into this category.

Development of fish reproduction techniques has largely been driven by aquaculture needs. Large-scale rearing is a requirement in the domestication of animals and is needed to take advantage of genetic selection, as has been the case in agricultural industries. Achieving control of reproduction of captive fish of aquaculture importance is a major step in industry development. This eliminates the need to rely on collection of running ripe adults or seed stock from wild, provides for more predictable and reliable production, and is a substantial step forward in meeting seed stock demands of expanding industries.

However, this has not always been the case. In Australian native species, for example, development of captive breeding techniques for several species, were initially developed to produce juveniles for release into the wild to replenish stocks of species that had undergone dramatic declines in numbers and distribution in the wild and to enhance fisheries and stocking farm dams. Several of these species particularly silver perch (*Bidyanus bidyanus*) and Murray cod (*Maccullochella peelii peelii*), are now being farmed commercially for human consumption in Australian and other countries (Rowland and Bryant, 1995; Ingram and De Silva, 2004).

Several major reviews in reproduction cycles and control of reproduction for fish have been published (e.g., Harvey and Hoar, 1979; Lam, 1982; Shelton, 1989; Zohar, 1989, 1996; Mylonas and Zohar, 2001). The reproduction cycle of many fish species are regulated by environmental stimuli such as light, temperature, rainfall and habitat (Fig. 17.1). When the brain, through sensory receptors, receives appropriate environmental information, the hypothalamus is stimulated to produce releasing hormones (RH) which, in turn, stimulate the pituitary gland to release gonadotrophic hormones, which act on the gonads. In turn, the gonads are stimulated to produce sex steroids that are responsible for gamete development (gametogenesis) as well as secondary sexual characteristics such as nuptial colouration changes and breeding behaviour. In females, gametogenesis (oogenesis) involves multiplication and transformation of oogonia into oocytes, which then grow in size through accumulation of yolk (vitellogenesis). Following vitellogenesis, oocytes reach the final oocyte maturation (FOM) stage, and then undergo ovulation. In males, gametogenesis (spermatogenesis) comprises a phase of formation and multiplication of spermatogonia followed by accumulation of spermatozoa and their subsequent emission (spermiation). Understanding these mechanisms provides the basis for controlling and inducing reproduction, particularly through provision of appropriate environmental stimuli and administration of hormone treatments to facilitate gametogenesis, ovulation and spermiation.

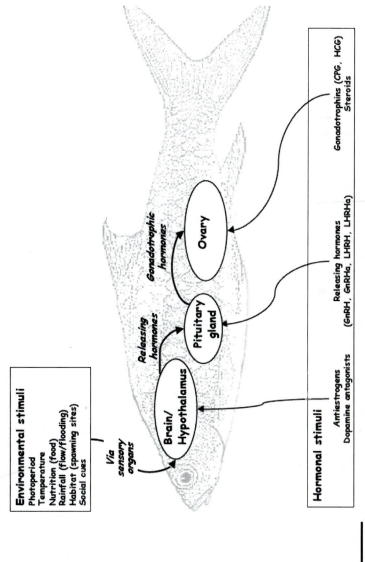

Fig. 17.1 Schematic relationship of environmental and hormonal stimuli controlling reproduction in fish.

There are three general patterns of ovarian development in fish (Tyler and Sumpter, 1996). Synchronous fish reproduce only once in their lifetime (e.g., Pacific salmons and eels). Group-synchronous fish have two or more distinct populations of oocytes present and ovulate once in a season (e.g., rainbow trout, silver perch and Murray cod) or undergo multiple ovulations over a few days or weeks within the spawning season (sea bass— *Dicentrarchus labrax* and certain tilapias). Asynchronous fish, which have a population of primary oocytes and a heterogenous population of vitellogenic oocytes, are capable of ovulating on a regular or semi-regular basis.

In a culture environment, often in a confined space, the natural cues that trigger the reproductive process are often absent. The operative natural cues are mostly environmental, and these may be changes in temperature and day length, water quality, rainfall and associated flooding, and so on. These cues act either singly and/ or in combination, and different fish species react to different cues. For example, most temperate fish species will respond to changes in day length and associated change in temperature, whereas tropical fish species may respond to flooding and associated changes in water quality. In the latter case, it is often difficult, if not impossible, to discern whether the fish respond to the flooding *per se* or the associated changes in water quality, such as increased turbidity.

In the absence of natural cues, the fish species can be induced to reproduce only through intervention of the hormonal cycle, which was initially achieved through hypophysation.

Early aquaculture industries relied on species that readily spawned naturally in captivity such as in ponds. Carp, were spawned and reared in China some 2,500 years ago (Landau, 1992). During the latter half of the nineteenth and early twentieth centuries, fish farms and hatcheries were established on several continents; and artificial fertilization techniques (e.g., hand stripping and mixing of gametes) were developed and used on several species such as salmonids.

Probably the most common type of reproduction dysfunction in captive fish is the failure of females to undergo final oocyte maturation (FOM) and thus fail to ovulate and spawn. Instead, oocyte development is arrested at the stage of vitellogenesis and oocytes became atretic and are resorbed. This may be due to an absence of appropriate environmental stimuli to stimulate release of hormones by the pituitary gland to, in turn, effect FOM and trigger ovulation. Another type of reproduction dysfunction occurs when FOM and ovulation occurs but ovulated eggs are not released to the water. These fish must be manually stripped before the eggs become 'overripe' and artificially fertilized. However, the most severe type of reproduction dysfunction in captivity is exemplified by eels, which fail to undergo vitellogenesis (and spermatogenesis). In males, poor quality and/ or small amounts of milt may be produced (Zohar and Mylonas, 2001).

Environmental Manipulation

Several species of farmed fish readily spawn in captivity, providing the appropriate environmental cues that trigger the reproductive cycle, are present. For example, some tropical grouper species, *Epinephalus* spp., tend to spawn on their own accord in culture ponds, provided that the brood stock is in good condition and the environmental conditions in the ponds, such as salinity and temperature are conducive. Other examples include tilapias, common carp and channel catfish and Murray cod. Environmental factors such as water temperature, photoperiod, light intensity, changes to volume and velocity of water, flooding, and access to spawning habitats influence the reproduction cycle of fish (Munro *et al.*, 1990), and manipulation of environmental conditions has also been successfully used to control spawning in captive fish. Heating culture water by covering brood ponds with a hothouse structure may advance spawning events. A range of habitats and structures have been used provided to induce spawning. These include gravel for nest construction (e.g., bass —*Micropterus* spp. and some catfish species), and plants, spawning mats (e.g., strips of raffia of plastic) and spawning boxes for egg deposition or attachment (e.g., common carp, channel catfish and Murray cod). Increasing water level in ponds has been used to induce spawning in golden perch (*Macquaria ambigua*) (Lake, 1967). Environmental manipulation of temperature and photoperiod have been applied to induce spawning in fish on demand and out-of-season so as to provide a constant and reliable supply of seed stock year-round for aquaculture (Bromage *et al.*, 1993; Carrillo *et al.*, 1993; Bromage, 2001).

Hormonal Manipulation

Where environmental manipulation has had limited success or failed to induce spawning, or it has been impractical to simulate complex environmental conditions required for spawning in the hatchery, efforts have instead shifted to the application of hormones to control the reproduction cycle. The focus of induced breeding involves the hormone induction of final oocyte maturation (FOM) and/or ovulation in female fish, induction or enhancement of spermiation in males, followed by either natural spawning or artificial fertilization. In some species, hormonal treatments have been found to be the only means of inducing maturation and ovulation in captivity. Hormone dosage regimes (number of injections, dose rates and timing of injection(s)) vary quite considerably between species. Some types of hormones and treatments may be more effective on some species than others.

Consequently, a considerable body of work has been undertaken on the major aquaculture species to identify appropriate hormone treatments and dosage regimes to optimize production of gametes.

Hypophysation

Hypophysation refers to induction of ovulation by injection of ground fish pituitary glands or crude extracts of pituitary glands. Houssay (1931) first demonstrated the effectiveness of this technic to induce spawning in fish, with later successes being reported by von Ihering (1937). Injection of pituitary gland extracts simulates the natural gonadotropin surge, which by-passes to some extent the effects of environmental stimuli, such as temperature or photoperiod, on gonadotropin production.

Initial problems in hypophysation related to supply and dosage, mainly due to the crudity of collection, refinement, preservation, storage and administration techniques. However, these techniques have since been developed and refined and today, carp pituitary glands (CPG), either fresh or refined and preserved, are widely used for many species, particularly for carps. Methods for use of CPG in carps have more or less been standardized and typically involve an initial injection of a small priming dose (10-20% of total dose) followed by a larger resolving dose 12-24 hours later (Lam, 1982). CPG with known purity of active ingredients and quality are now readily available from a number of suppliers. There are, however, several drawbacks to the use of pituitary glands. The potency (hormone content) varies considerably due to the weight, sex and age of the donor fish, time of year and time between collection and preservation. Continued handling of fish to administer several does of CPG may have a negative effect on spawning due to stress, and there is a risk of disease transmission between donor fish and recipient fish.

HCG

Mammalian gonadotropin and human chorionic gonadotropin (HCG), typically administered in a single dose (100-4,000 μ/kg), have been found to be effective in inducing maturation and ovulation (and spermiation) in numerous fish species (Lam, 1982; Zohar and Mylonas, 2001). Mammalian gonadotropins act directly at the level of the gonad and do not require the existence of either stores of releasing hormones or activation of the pituitary gonadotropins. Highly purified mammalian gonadotropins of standardized biological activity are widely used in domestic animal industries (pets and livestock) and as such are readily available through commercial suppliers. Yet in some species such as carps, mammalian gonadotropins are ineffective and fish pituitary extracts are more potent in inducing ovulation. Further, recent research indicates that some fish which have been regularly treated with HCG develop antibodies against HCG, which may reduce the effectiveness of this hormone in subsequent applications (Zohar and Mylonas, 2001).

GnRHa

Research in the early 1970s showed that release of gonadotropins from the pituitary gland was controlled by a hypothalamic hormone, gonadotropin-releasing hormone (GnRH) (Schally, 1978). Since then, numerous studies have shown that GnRH agonists (GnRHa), synthetic decapeptides that are 30-100 times more potent in inducing gonadotropin release rather than native GnRH, are highly effective in inducing FOM and ovulation of many species. Consequently, the application of various GnRHa peptides in aquaculture has had a significant impact on controlling and improving the reliability of reproduction. In particular, the use of GnRHa has been especially effective in spawning of cyprinids.

GnRHa is most often administered by injection with either a saline solution or in the form of a sustained-release delivery system. Dopamine antagonists such as domperidone and pimozide that inhibit negative steroidal feedback on gonadotropin secretion and enhances the stimulatory effect of subsequent GnRH doses have often been used in combination with GnRHa (Zohar, 1989). A number of GnRHa peptides are now commercially available for use in finfish, such as Ovaprim (Syndel Laboratories Ltd., Canada), a synthetic salmon GnRHa combined with domperidone suspended in propylene glycol, Gonazon (Azagyl-nafarelin) ((Intevet International, The Netherlands) and Ovapel (Interfish, Hungary), a synthetic GnRHa combined with metoclopramide.

With many species multiple injections of hormones (CPG, HCG or GnRHa) may be required to induce FOM, ovulation and spermiation (Zohar and Mylonas, 2001). However, the repeated handling associated these treatments can be stressful to broodstock and may adversely affect progression of FOM and spermiation, and even result in mortalities of valuable stock. Repeated hormone treatment is also laborious and time consuming. These problems may be alleviated by administration of hormones in a sustained manner, which will improve the efficacy of treatment and reduce the amount of bloodstock handling.

A range of sustained-release delivery systems have been developed for GnRHa to control FOM, ovulation and spermiation (Mylonas and Zohar, 2001). Various preparations of cholesterol and cellulose have been used to produce implantable pellets that are 'slow' (weeks) or 'fast' (days) releasing, depending on composition (Crim, 1985; Lee et al., 1986; Sherwood et al., 1988). Cholesterol implants for use in fish are commercially available (e.g., Ovaplant (Syndel Canada), which contain salmon GnRHa in a cholesterol matrix). Other hormone delivery systems include microspheres (95-500 μm diameter) of co-polymers of lactic acid and glycolic acid, wherein hormone release is immediate and can last for months depending on the ratio of lactic acid:glycolic acid and the length of the polymer, solid monolithic implants using non-degradable co-polymer of ethylene and vinyl acetate and emulsions of lipophilized gelatin with various lipid anhydrites (Mylonas and Zohar, 2001).

Although GnRHa delivery systems have been used to induce FOM, ovulation and spermiation in fish, they have also been effective in advancing and synchronising ovulation, especially in fish with multiple-batch group-synchronous or asynchronous ovarian development, and in some instances have also enhanced vitellogenesis and spermatogenesis (Crim and Glebe, 1984; Mylonas and Zohar, 2001).

While application of appropriate environmental stimuli and/or administration of an appropriate hormone treatment(s) have been shown to be important factors in inducing FOM, ovulation and spermiation in fish, the condition of broodstock also plays a major role in the reproductive cycle. Disease, poor nutrition and other stressors associated with captivity may impede FOM, ovulation and spermiation and even affect the quantity and quality of eggs and fry.

Despite the significant developments in fish reproduction, there are still major aquaculture species for which captive techniques are yet to be developed. Notably, while progress has been made in maturing and spawning eels under laboratory conditions, large-scale seedstock production has not been achieved and production of A. *anguilla*, A. *japonica*, A. *rostrata* and A. *australis* is still based entirely on wild catches of glass eels and elvers, during their migration into freshwaters (Tachiki and Nakagawa, 1993; Ohta *et al.*, 1997; Tanaka *et al.*, 2001; Pedersen, 2003; Tanaka *et al.*, 2003). Nevertheless, the advent of induced spawning techniques has led to greater control over production through reliability of supply and quality of seedstock, and has also made it possible to apply a range of other reproduction techniques aimed to improve and enhance production. These include out-of-season spawning to provide for year-round supply of juveniles, genetic improvement by selective breeding, hybridization between closely related species, and other chromosome manipulation techniques such as ploidy induction by heat or pressure shock shortly after fertilization.

Egg Incubation

Egg incubation techniques are many and varied and depend mainly on the type of egg and number of eggs being incubated. Buoyant or pelagic eggs are more common to marine fish, such as sea bass and bream and porgy, though some freshwater species also produce pelagic eggs (golden perch and silver perch). Non-buoyant or demersal eggs are more common in freshwater stream-spawning fish such as salmonids carps and catfish. Most pelagic eggs are non-sticky, but demersal eggs have varying degrees of stickiness, which allows them to adhere to various surfaces. Fecundity varies according to many factors including age, size and species. Relatively low fecund fish such as rainbow trout (2,200 eggs/kg) and channel catfish (7,000 eggs/kg) produce just a few thousand eggs per/kg, whereas as high fecund fish can spawn in excess of 100,000 eggs/kg (e.g., silver perch — 250,000 eggs/kg, common carp — 150,000 eggs/kg). Fecundity is broadly related to the level of care accorded to the eggs by the parent. High-fecund fish may simply broadcast their eggs into the environment and provide no care whatsoever. Low-fecund fish may provide varying levels of care and protection to eggs, such as laying eggs in a protected 'nest' and guarding the eggs from predation. Species that produce fewer and larger eggs tend to have larger and, therefore, hardier larvae than species that produce many and smaller eggs.

Pelagic eggs are typically incubated in large tanks that are hydraulically designed to ensure that the eggs remain in suspension during incubation. This may involve the use of tanks with a conical or hyperbolic base and sufficient water flow and/or aeration to prevent the eggs from settling out of suspension (Fig. 17.2a). Incubation techniques for demersal eggs depend on the degree of stickiness. Non-sticky eggs may be incubated in a mono-layer in flat trays or baskets suspended in troughs that are provided with a constant flow of well-aerated water (Fig. 17.2b). Alternatively, demersal and semi-demersal eggs may be incubated in specially designed hatching jars such as MacDonald or Zuger jars (Fig. 17.2c). In upwelling jars such as used for trout and salmon, constantly flowing water enters the base of the jar and passed up through the eggs without disturbing them (Fig. 17.2d). It should be noted that at certain times of embryonic development, some eggs may be very susceptible to disturbance. In MacDonald jars water is injected onto the conical surface at the bottom of the jar, which causes the eggs to continually roll or tumble. This movement helps to keep the surface of the eggs clean and prevents fungal infection. For some species, when hatching is imminent or commences, the eggs are transferred from the jars to trays, baskets for hatching. Trout and salmon eggs are moved to trays when pigmented eyes become visible through the chorion (eyed stage).

Due to their nature, sticky eggs tend to be more difficult to incubate. In species where the broodstock are allowed to spawn naturally in earthen ponds, such as common carp and channel catfish, the eggs may simply be left alone to incubate and larvae or fry are harvested from the ponds following hatching. For some pond-spawning species, artificial spawning structures, such as spawning boxes or spawning mats (e.g., raffia rafts) are provided in the ponds, as for example in the case of the Australian native species the Murray cod (Fig. 17.3a). In some cases, once spawning has occurred and eggs detected, these structures may be removed from the ponds and the eggs incubated in hatchery facilities under more controlled conditions (Fig. 17.3b). In fish that are induced to spawn and hand-stripped in hatchery facilities, following artificial fertilization, the eggs may be simply transferred to flat trays or baskets, where they adhere to the sides or bottom. However, for some species, techniques have been developed to remove the adhesive layer so that eggs can be incubated as for non-stocky demersal eggs (Kowtal et al., 1986; Krise, 1988; Michaels, 1988).

In some cases with extended incubation periods, eggs may need to be treated with a fungicide to prevent fungal growth, which can greatly reduce hatch rates.

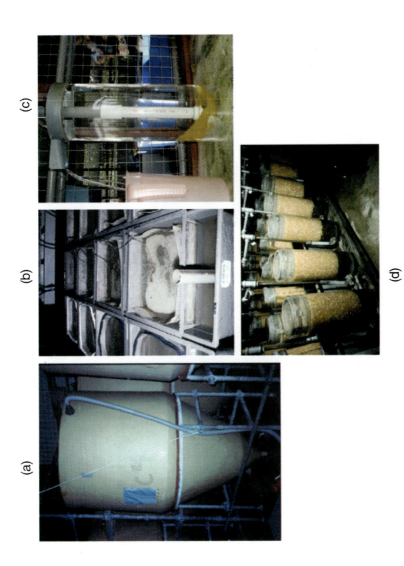

Fig. 17.2   Egg incubators: (a) 10,000 L halibut egg incubator; (b) Ewos incubation trough; (c) MacDonald jar; and (d) trout upwelling jars.

Fig. 17.3 (a) Artificial spawning structure for Murray cod. (b) Murray cod eggs adhered to mesh being placed into an incubator.

BROODSTOCK MANAGEMENT

Broodstock is the commonly used term for a group of individual fish that are used for producing the seed stock for grow-out purposes. As such, the health and well being of the broodstock is crucial for the production of viable, good quality seed stock, and especially so as aquaculture is becoming more and more independent of natural seed stock supplies, except in a few cases such as for anguillid species. As such, the management of broodstock in aquaculture is crucial to the success and well being of the sector.

Very often in aquaculture—as in the case of most terrestrial animal husbandry practices—broodstock rearing and larval and fry/ fingerling production are often conducted separately from grow-out facilities (Fig. 17.4). This is to be expected for a number of reasons. Firstly, the capital layout will be higher if an operation is to have all stages of production in one entity, and secondly and more importantly, the expertise needed for maintaining and managing broodstock, and associated production of the young are widely different to those needed for grow-out operations.

Broodstock management entails a number of facets, each distinct from the other. Therefore, it is appropriate that these are dealt with separately.

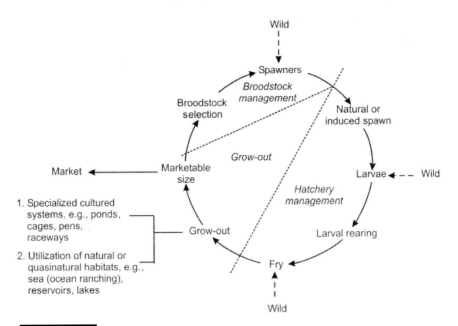

Fig. 17.4 Stages in aquaculture operations (modified after De Silva and Anderson, 1995).

Broodstock Procurement and Rearing

In general, most cultured finfish species are relatively highly fecund, apart from a few exceptions such as catfishes and the tilapias. Also, broodstock of most cultured species are relatively large. For example, the average size of broodstocks of the Chinese major carps could range from 6 to about 20 kg and salmon in excess of 6 to 8 kg. Grouper broodstock can be as high as 25 kg. On the other hand, that of Nile tilapia may be in the region of 0.75 to 1.5 kg. In general, broodstock are maintained in large, indoor or outdoor facilities, depending on the species, and are rarely kept in high densities.

More often than not, the original broodstock of a species, for a facility, are obtained from the wild or from an ongoing aquaculture operation. In most operations dealing with species for which the life cycle has been closed, the original broodstock are replenished from progeny from the operation itself. However, continued reliance for replacement of broodstock from the same original stock, over a few generations, could bring about loss in genetic quality of the progeny.

Broodstock Nutrition and Egg Quality

Nutrition of the broodstock is the singlemost non-genetic factor that would potentially determine the viability and quality of the eggs and pre-feeding larvae (yolk-sac larvae). In fact, broodstock nutrition is a subject by itself, and has been dealt with comprehensively in the past (see for, e.g., Bromage and Roberts, 1995; Sargent, 1995). However, it is important to consider, in brief, some of the more important aspects pertaining to this area as the whole purpose of the reproductive process is lost if unviable or poor quality seed are the final result of spawning.

In finfish, the fertilized egg will have to carry all the nutrients required for its development, and the subsequent larval stages until the larvae change over from an endogenous food supply, the yolk, to an exogenous food supply. During this phase of development, crucial morphological changes occur and vital organ development, such as the brain, eyes and the skeletal system begin to take form. Also, during these stages of development, the metabolic rate is relatively high (per unit weight) and consequently for development to proceed the embryo will need to have adequate energy supplies. All of the above means that the spawned egg needs to have all the energy and other nutrient requirements needed for growth and development; in essence the mother will have to be properly nourished to enable her to deposit all the developing embryo requirements prior to ovulation, and hence the importance of broodstock nutrition in aquaculture.

In general, most good grow-out diets used in aquaculture practices fulfil the broodstock requirements. Indeed, Sargent (1995), in his synthesis on origins and functions of egg lipids and their nutritional implications, stated very candidly, 'Fortunately, the theoretical problems of fish lipid nutrition become largely academic in practice, especially in broodstock nutrition, where, in the light of our present understanding, currently available fish oils closely approach what may be regarded as very satisfactory balance of the four fatty acid families...'. However, this may be an oversimplification, especially with regard to some diets available in the developing world, which may not be effective. A case in point was the inability to artificially propagate two mahseer species (Cyprinidae), *Tor tambroides* and *T. douronensis* in Sarawak, Malaysia, using long-term pond reared broodstock for a number of years (due to non-ripening ova), and the final success was attributed, amongst other factors, to a change in the diet, which had a suitable fatty acid profile (Ingram *et al.*, 2005).

The influence of dietary fatty acids in broodstock nutrition is the one area that has received most attention. The earliest studies, in this regard were conducted by Watanabe and his co-workers on the red seabream (*Pagrus major*), and were then reviewed by Watanabe and Kiron (1995). These studies showed the importance of fatty acid fractions in different ingredients, such as krill oil, in broodstock diets, and also the importance of Astaxanthin (3,3'-dihydroxy-β, β-carotene-4,4'-dione) on egg quality. Other studies, for example, have demonstrated the influence of the level of n-3 HUFA in broodstock diets on egg quality in the Japanese flounder, *Paralichthys olivaceus* (Furuita *et al.*, 2000); and European sea bass (Bell *et al.*, 1997); effects of 20: 4n-6 (arachidonic acid) on growth, survival, resistance to handling in gilthead seabream, *Sparus aurata* (Koven *et al.*, 2001); and the reproductive performance in male European sea bass (Asturiano *et al.*, 2001); and Bruce *et al.* (1999) emphasized the importance of n-3 and n-6 HUFA in broodstock diet development in for this species. Czesny and Dabrowski (1998) demonstrated that in walleye (*Stizostedion vitreum*) deficiency of n-3 fatty acids was associated with impaired development, and thus poor larval and juvenile viability. In contrast to fatty acid influences on spawning performance, egg quality and larval viability, little attention has been paid to influences of amino acid influences on these aspects. It may be that in most instances a good quality grow-out diet more often than not satisfies the amino acid requirements needed. In this regard, Gunasekera *et al.* (1998) suggested that the 'swollen yolk sac syndrome (SYSS)', in Murray cod was related to broodstock nutrition, resulting from a cumulative dietary deficiency of some of the essential amino acids.

In general, it can be concluded that a good, effective grow-out diet, with the appropriate amino acid and fatty acid balance can be used for broodstock. However, there could be exceptions and regular monitoring will ensure that ovarian development progresses as expected.

Egg quality determinants could be physical (diameter, weight, shape, specific gravity and colour), structural (membrane stability, quantity of yolk, location of the yolk sac) and chemical. For all intents and purposes, physical and structural determinants could be considered to be genetically determined to a very great extent. For example, the egg diameter of cultured finfish species could range from as much as small as a 0.2 to 0.5 mm to a few mm, as in the case of most cyprinids and salmonids, respectively. Within each species, the egg diameter will have a narrow range, genetically determined. The specific gravity of the eggs would determine whether they are free floating and or demersal, each of these also being a species characteristic. Most cultured cyprinids, except common carp, lay free-floating eggs, or mildly adhesive eggs, often less dense than water, whereas salmonids lay demersal eggs, which are denser than water.

The most important aspect of the egg quality, particularly from a broodstock nutritional viewpoint, is the chemical composition. The gross chemical composition often does not vary to a great extent amongst species; in all species the eggs tend to have a protein content (by dry weight) ranging from about 60 to 70%, and the biggest difference occur in the amount of water in the egg. Numerous studies on the fatty acid composition have been conducted on eggs of different species of fish, and all such studies indicate the relatively high proportion of *n*-3 and *n*-6 HUFA in the lipid component, indirectly indicating their importance to egg development and larval well-being. Obviously, and as expected, there are species differences the most notable being the *n*-3 to *n*-6 ratio in fresh water and marine species, it being higher in marine species and vice versa, reflecting the basic nutritional requirement differences in the two groups of fish.

Genetic Aspects

Selection of broodstock

Selection of broodstock is the initial requirement for any breeding program and often depends on the objectives of the programs. For example, if hatchery production is for the purpose of restocking or stock enhancement then it is often encouraged that broodstock should be acquired from the local populations in the environment where the juveniles are being released. The main reason for this is that introduction of exotic populations into a new environment may have adverse effects on the native/local gene pool. Also, it may capitalize on any adaptation to local conditions (Shaklee, 1983).

If the objective of the breeding program is for food fish production, many factors should be taken into account in broodstock selection. Firstly, the species of interest must be fully domesticated, e.g., full life-cycle is controlled in captivity. Secondly, selective breeding program should be undertaken only when aquaculture of species under consideration is sustainable. Thirdly, there should be genetic variations associated with traits of commercial interest. With the advanced developments in the field of molecular genetics, it is encouraged to measure levels of genetic variation within broodstock, if possible, at the initial stage of domestication of a new species for aquaculture, as it would help to develop an appropriate management strategy and to monitor genetic changes between generations. It is important to establish a base-line broodstock population with a significantly wide level of genetic variation. However, if there is no or little genetic variation in the initial founder broodstock, one possible solution is to introduce new genetic material by bringing in individuals from alternative populations or broodstocks. In such instances, care needs to be taken to ensure this does not risk the inadvertent introduction of alien individuals into the wild.

Broodstock management

From a genetic perspective, broodstock management is mainly aimed to avoid problems associate with inbreeding and random genetic drift. Inbreeding is defined as the mating of closely related individuals and random genetic drift is random changes in gene frequency that occur as a result of sampling error (e.g., broodstock selection). It is common that deleterious recessive alleles are hidden in heterozygous individuals, and inbreeding will provide an opportunity for these alleles to combine together in offspring, where potentially lethal phenotypes will begin to be expressed.

Negative impacts on productivity, such as reduced fecundity, reduced disease resistance, reduced survival of seed stock and increased incidence of abnormalities, can also be brought about by limited genetic variability resulting from inbreeding and genetic drift.

Avoiding inbreeding and random genetic drift is critical for the maintenance of genetic variance in cultured stocks. It is a problem that aquaculture of highly fecund species, such as Indian and Chinese carps, as well as marine species such as grouper are likely to encounter. Because of the high fecundity of these species, generally, there is a tendency to use a fewer number of broodstock to meet production targets. Furthermore, as considerable volumes of fry and fingerlings are produced in backyard hatcheries—as often practiced in developing countries—there is more likelihood for the broodstock numbers maintained and used in such practices to be less than desirable, an almost unavoidable consequence of the practices. Consequently, genetic problems associated with small gene pools, such as inbreeding, have a greater probability to occur.

Inbreeding is measured by the 'inbreeding coefficient', F, and the objective is to prevent F from reaching 0.25—the level where inbreeding depression is likely to occur in fish (Dunham, 2004). The simplest method to calculate the accumulation of inbreeding per generation with random mating is as follows:

$$F = \frac{1}{8N_{em}} + \frac{1}{8N_{ef}}$$

where N_{em} and N_{ef} are numbers of males and females that successfully breed, respectively.

Avoidance of inbreeding is often primarily resolved around population size. Maintaining effective population size (N_e) together with avoiding mating among closely related individuals of a hatchery stock are important measures that are generally recommended for controlling genetic erosion in hatchery produced seed. Genetic variability decreases rapidly if the effective population size of the broodstock is small.

In a random mating population, effective population size is calculated as follows:

$$N_e = \frac{4N_{em} \, N_{ef}}{N_{em} + N_{ef}}$$

As such, the effective population size can be increased in one of two ways: (1) increase the number of breeding individuals; and (2) bring the breeding population close to 1:1 sex ratio.

Effective population size is an important concept in broodstock management, as it is inversely related to both inbreeding and genetic drift. When N_e decreases, inbreeding and variance in changes of allele frequencies resulting from genetic drift increase. The relationship between inbreeding coefficient F and effective population size N_e is described below:

$$F = \frac{1}{2N_e}$$

Genetic diversity of bottlenecked broodstock can be increased without bringing in new brooders, as described by Doyle *et al.* (2001). The mean relatedness of each potential breeder to the whole population is estimated using microsatellites, by the formula proposed by Ritland (2000). A subset of breeders is then selected to maximize the number of founder lineages, in order to carry the fewest redundant copies of ancestral genes. This approach is particularly effective when the available number of captive broodstock is small (e.g., endangered species).

To estimate relatedness between pairs of individuals, an indicator variable 'δ_s' ('Kronecker operator') is used. At each diploid locus, two paired individuals have four alleles, denoted by A_i and A_j for the first individual and A_k and A_l for the second individual. If allele A_i and A_j are the same then $\delta_{ij}=1$, otherwise $\delta_{ij}=0$. There are six δ_s among the four sampled alleles, one for each comparison between two alleles, both within and between individuals. Pairwise relatedness is estimated as:

$$\hat{r} = \frac{[(\delta_{ik} + \delta_{il})/p_i] + [(\delta_{jk} + \delta_{jl})/p_i] - 1}{4(n-1)}$$

The mean kinship of the ith individual, mk_i, is the average kinship values for that individual with every individual in the population, including it. A low mean kinship value indicates that an individual has few relatives in the population, and thus is valuable in maintaining genetic diversity.

Apart from genetic monitoring, some other strategies can be applied to maintain genetic diversity. For example, fertilization of a batch of eggs with sperms from several males can help to maximise N_e. The result of mixing of sperms from several males to fertilize eggs may not be desirable as sperms from one male may be more competitive and, thus, dominate the fertilization process. As such, it might be more practical to divide eggs from one female into sub-samples then fertilize each sample with sperms from different males (Tave, 1993). Recently, cryopreservation of sperm has become routine for some species, which enables the hatcheries to use stored sperm from a large number of males.

GENETIC MANIPULATIONS IN AQUACULTURE

Selective Breeding

It is an accepted fact that selective breeding and consequent improvements have enabled the current production levels from husbanded terrestrial animals such as poultry, cattle (meat and milk), pigs, etc., to be achieved, the genetic improvements often contributing more than 25% to the overall increases in productivity. On the other hand, in respect of cultured aquatic organisms, significant and notable improvements in production through genetic improvements have been achieved in the case of Atlantic salmon and trout in Norway, and is reputed to be similar to those with livestock and poultry (Gjoen and Bentsen, 1997). Similarly, a multi-nation effort on selective breeding on Nile tilapiain Asia have resulted in the production of the GIFT Strain (Genetically Improved Farmed Tilapias), that is known to perform considerably better than those used previously in most Asian countries (Gupta and Acosta, 2004). Gupta and Acosta (2004) reviewed the development of the GIFT strain in Asia, using fresh germ plasm from Africa, and the socio-economic impacts and the future of the tilapia culture industry based on these improved strain. In general, however, genetic improvement of cultured finfish species have lagged far behind that of husbanded terrestrial animals. Perhaps, one of the primary reasons is that the number of commercially important cultured species in the world exceeds 50, and to develop selective breeding programs for each of these species will take a considerable amount of time and effort.

Most of the economically important traits in relation to aquaculture are those described as being quantitative. These characteristics can be measured (e.g., body length, body weight, growth rate, etc.), in contrast to qualitative characteristics, which are descriptive (e.g., albino, black, spotted, unspotted, etc.). Common quantitative characteristics include body weight, length, fecundity, proximate composition, dressing percentage, etc. The nature of quantitative measurements means that such traits are continuously distributed throughout the population, or in other words, distribution of the frequency of these traits follows a normal distribution (Fig. 17.5). Quantitative traits show continuous distributions because of two reasons: (a) the large number of genes and alleles involved

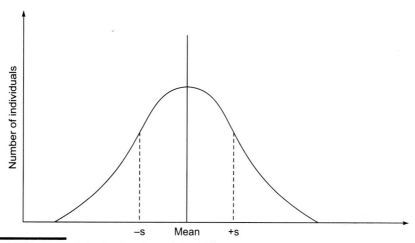

Fig. 17.5 Normal distribution of a trait in a population (s is standard deviation and the variance is equal to s²).

and the various interactions they have with each other; and (b) the environment.

The population variation for a character is expressed as V_P, and this is a sum of the genetic variance (V_G), the environmental variance (V_E), and the variation contributed by the interaction between the environment and the genes (V_{G-E}). As such:

$$V_P = V_G + V_E + V_{G-E}$$

The genetic variance, V_G, is the one of most interest. V_G or genetic variance can be further partitioned into three components: these are V_A, V_D and V_I, which represent additive, dominance, and epistatic genetic variance, respectively.

$$V_G = V_A + V_D + V_I$$

Of these three components of V_G, V_I is difficult to exploit; therefore the two important sources are V_D and V_A. They are very different and are exploited in different ways: V_D by hybridization, and V_A by selection.

Selection is the process of selecting individuals or families, which have desirable characteristics, then breed from them to try and change the population mean for one or a number of quantitative traits in the next generation. Fish that do not meet the minimum selection criteria will not be crossed and their alleles will not contribute to future generations. One important parameter in this process is the heritability, which is an indicator of the contribution to the total population variance (V_P), which is made by V_A. Heritability is designated by h^2:

$$h^2 = \frac{V_A}{V_P}$$

Values of h^2 range from 0 (where phenotypic variance is entirely the result of environmental effects) to 1.0 (where phenotypic variance is the result of genetic effects).

Heritability can be estimated by various methods. The first method is parent-offspring regression, which involves a series of matings and all families being reared under the same environmental condition. A scatter plot is carried out, each point presents one family (mean of parent mean versus mean of offspring of a particular trait), and the resulting slope of the regression gives a measure of the heritability of the trait. The second method involves the design of fill-sib and half-sib mating experiments, the full details of which can be obtained from Falconer (1981) and Kearsey and Pooni (1996). However, the only method available to estimate the true heritability is to carry out trial experiments and measure the actual response to selection is schematically represented in Fig. 17.6.

Several selection strategies are available for genetically improving performance of fish. These include:

• Mass selection, or individual selection: This is a strategy where the selection of breeding individuals is based on their phenotype, involving the selecting the best-performing individuals for the particular trait of interest from a population, and the next generation will be produced from these selected individuals. It is the simplest type and the most frequently applied selection strategy for fish. However, it can only be used to improve traits that are recorded on the breeding candidates while they are still alive (e.g., growth rate, body shape, colour), and it is not efficient to improve discrete traits with low heritability such as survival rate. Also, mass selection may encounter the problem of inbreeding, which can be prevented by using only a restricted number of individuals from each full and half-sib family to be tested as breeding candidates when applying mass selection.

• Family selection: In family selection, the breeding individuals are ranked on the basis of records taken on their full and half-sibs. This selection strategy can be applied to improve traits that cannot be recorded in the breeding individuals while they are alive (e.g., carcass quality traits),

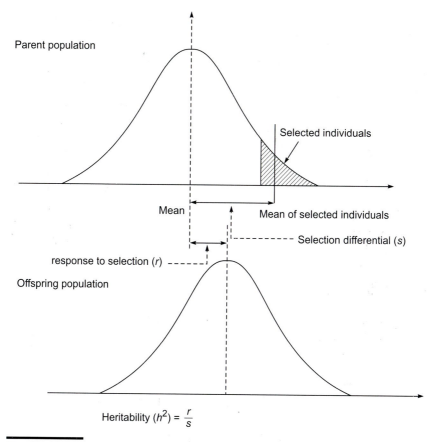

Fig. 17.6 Response to selection (modified from Beaumont and Hoare, 2003).

traits that can only be recorded for groups of fish (e.g., feed utilization), and discrete traits of high or low incidences (e.g., survival rate, age at sexual maturity). Furthermore, family selection is more effective than mass selection when the heritability of a trait is low (<0.30). However, family selection requires that each family be produced and kept separately and that all fish are tagged before testing in common test environment. This may not be a major issue with the advanced developments in molecular genetic techniques, from which molecular markers can be generated and then used as tags for these purposes.

Usually, genetic breeding programmes are designed to obtain desired traits such as fast growth, delay of early sexual maturity, disease resistance and carcass quality, etc., and is mostly accomplished through traditional genetic improvement methods (Tave, 1993). The more recent availability of molecular genetic techniques offers several ways of improving the efficiency of these breeding programmes and is increasingly being incorporated into many of them. Phenotypic characteristics of fish species likely to be most important to genetic improvement programs in the aquaculture industry will almost always be inherited quantitatively (Falconer, 1981), making it difficult to determine the genetic basis of a desirable trait (Ferguson, 1995). The new DNA-based technologies are very powerful in identifying marker loci associated with nuclear loci that control commercially important traits (i.e., quantitative trait loci, or QTLs). Once such markers have been identified, they can be used in selection programs (marker-assisted selection, or MAS). Lande and Thompson (1990) used theoretical analyses and suggested that molecular genetic polymorphisms could be incorporated into traditional methods of artificial selection to achieve substantial increases in the efficiency of selective breeding.

Hybridization

In aquaculture (and in stock enhancement) inter-specific hybridization is utilized for many purposes, either singly and or in combination. These purposes include:

- Increase growth rate,
- Transfer and or combine desirable traits between species into a single group,
- Reduce unwanted reproduction through production of sterile or mono-sex progeny,
- Increase harvestability/production, and
- Increase environmental tolerance and general hardiness in culture environments (Bartley et al., 2001).

Bartley et al. (2001) listed 35 commonly used hybridizations in aquaculture and stocking programs, and included cyprinids (4), salmonids (5), tilapia (5), miscellaneous freshwater fish (6), miscellaneous marine and diadromous fish (10) and catfish (5). We will not attempt to expand on these hybridizations except to focus on selected examples. An instance of using hybridization to obtain a better growth/production and overall performance that exceeds that of either of the parent species, as well as better flesh quality is the use of hybrids of the Thai walking catfish *Clarias macrocephalus*, and the African catfish, *Clarias gariepinus*. This hybrid, the culture of which has became very popular with in a decade or so, is almost exclusively used in Thailand catfish culture operations, and the hybrid is reputed to grow faster, to a larger size and be of superior flesh quality (Na-Nakorn, 2004). On the other hand, the continued use of hybrid catfish in Thailand has had negative consequences on the native catfish populations, in particular through its impact on the native gene pools (Na-Nakorn *et al.*, 2004; Senanan *et al.*, 2004).

The use of inter-specific hybridization between closely-related species carried out to produce progeny of a single sex, with better growth performance and also to avoid prolific, unwanted breeding resulting in overcrowding of the culture systems, and finally leading to stunting of the reared stock(s), is best exemplified in tilapia culture. Indeed, hybridization amongst tilapia species was one of the first instances when hybridization was utilized in aquaculture (Hickling, 1960, 1963). Most species of tilapias are prolific breeders, and also in general in most tilapia species males grow faster and perform better than the females. In order to avoid overcrowding and consequent stunting, as well as to utilize the faster growth trait of males, hybridization between tilapia species that results in preponderance of males was utilized, commencing in the early 1950s. For example, female *O. mossambicus* to male *O. hornorum* (Hickling, 1960, 1963) cross generated nearly all-male hybrids, and was followed by other crosses, some of which resulted in all male and/or nearly all male hybrids. Numerous crosses, mostly those that produced a preponderance of males, have been utilized in tilapia culture and are summarized in Table 17.1.

Table 17.1 Summary of different hybrid combinations that have known to produce monosex male progeny (from Mair, 2001).

Female parent	Male parent	Note
O. niloticus	*O. aureus*	Applied commercially but results inconsistent
O. niloticus	*O. macrochir*	—
O. niloticus	*O. urolepis hornorum*	Majority of broods are all male; some commercial application
O. niloticus	*O. variables*	All progenies monosex
O. mossambicus	*O. aureus*	—
O. mossambicus	*O. urolepis hornorum*	All progenies monosex
O. spilurus niger	*O. macrochir*	—
O. spilurus niger	*O. urolepis hornorum*	All progenies monosex
O. aureus	*O. urolepis hornorum*	—
T. zillii	*O. andersonii*	All progenies monosex

Overall, although a large number of inter-specific hybrids are used in aquaculture, those currently making an impact on production amounts to about 35, the most significant being Thai catfish culture and characid hybrids in Venezuela and tilapia in Israel, which in order contribute about 80, 29 and 100% to the production of these groups in each of these countries. Bartley *et al.* (2001) evaluated the 'status of hybridization' from a genetic viewpoint and concluded that hybridization does not fall within the realm of 'genetically modified organisms'. However, these authors stress the need for proper broodstock management, maintenance of correct parental lines, avoidance of inbreeding and inadvertent hybridization and or backcrossing to fully utilize this method of genetic improvement in aquaculture.

Sex-reversal Using Hormones

Initially, sex-reversal was conducted on progeny, through hormonal treatment (e.g., bath, feed or implant depending on developmental and culture characteristics of species) to utilize the faster growth rate of one of the sexes, e.g., males in tilapias, and also to avoid undesirable breeding amongst grow-out stocks, resulting in lower production and stunting of the stocks. Some androgenic hormones have been used to produce monosex male populations, and the most efficacious and widely used hormone is 17α-methyltestosterone (Yamazaki, 1983; Dunham, 1990). Several oestrogenic compounds are used to produce female monosex populations. An example of this is the use of hormone 3-oestradiol in producing all-females populations in salmonids species (Ashby, 1957; Donaldson and Hunter, 1982). The use of hormones on sex-reversal of progeny is beyond the scope of this chapter and the readers are referred to recent reviews on the subject (Pandian and Sheela, 1995; Penman and McAndrew, 2000).

Chromosome Manipulation

Polyploidy

Polyploidy has been well researched in fish. Individuals with extra sets of chromosome are called polyploid. The common individuals with two sets of chromosomes are diploid, while triploid refers to individuals with three sets of chromosomes and tetraploidy are those with four sets. Of these, triploidy is the major focus of aquaculture.

From the aquaculture point of view, culture of triploid fish is advantageous as triploid production has great potential to enhance performance in fish. Potential benefits of triploidy include improved feed conversion ratio, higher survival rate and greater growth rate. Triploid individuals are sterile; as such reduced gonadal development may allow energy that are normally used in reproductive process to be used for growth of somatic tissues (e.g., the case of tilapia). Sterile characteristic of triploid also helps to minimize the potential risks to endemic wild populations of fish in areas where exotic species, hybrids or transgenic stocks are being cultured.

Triploids can be produced in two different ways. The most common method used in fish to produce 'meiotic' triploids involves applying either thermal, pressure or chemical shock to newly-fertilized eggs with the resultant disruption of the mechanisms that would otherwise force the second polar body out of the egg. Alternately, 'interploid' triploids can also be produced by crossing tetraploid (4N) individuals with normal diploids. Tetraploid fish produce diploid (2N) gametes, when fertilized with normal haploid gametes (1N) will form triploid (3N) zygotes.

Many attempts have been made to produce triploids of many fish species that are of importance in aquaculture. It is reported that the heat-shock induced triploid *Oreochromis niloticus* showed similar growth rates to the diploid counterpart up to the age of maturation, however, at the end of the experiment, triploid males and females exceed their diploid counterparts by an average of 95% and 66% on body weight, respectively (Brämick *et al.*, 1995).

However, it is also noted that the results are not always desirable nor consistent. For example, in a study of meiotic triploid common carp, Cherfas *et al.* (1994) reported that overall survival of heat-shock treated triploids was only approximately 70% of that observed in diploid controls by one year of age. In addition, although triploids appeared to be functionally sterile, their mean body weight was approximately 85% of that of diploid controls.

Gynogenesis and androgenesis

With increasing opposition to use of androgens and oestrogens in animal husbandry, there has been an attempt to develop fish broodstock that would continue to produce monosex progeny, through the use of phenotypic males or females. As a sequel to the hormonal sex-reversal of tilapias, and the relative inconsistency in the production of single sex progeny through hybridization, and with a greater degree of understanding of the sex-determination systems in a number of the key species (Mair and Little, 1991; Mair *et al.*, 1991), it became possible to develop techniques for the production of genetically male tilapia (GMT), initially of *O. niloticus* (Scott *et al.*, 1989; Mair *et al.*, 1991). A schematic representation of the steps involved in the production of genetically all male tilapia is given in Fig. 17.7. This technique is known to produce all males, and at worst only a small percentage of females. Genetically male *O. mossambicus* have also been produced, but in this instance through gynogenesis of XY neofemales (Varadaraj and Pandian, 1989). The technique of producing GMT, particularly in the case of *O. niloticus*, is well established and is now commercially adopted, and has been achieved through several generations of breeding (Mair and Little, 1991).

Similarly, using a combination of sex-reversal and gynogenesis can also produce all female progeny. If the female is homogametic sex, the gynogenetic progeny are all XX. If the fry are sex-reversed to phenotypic males and if these are fertile, they can be mated with normal XX females to produce all XX female offspring. The technique has been applied to produce all female progeny in grass carp (Boney *et al.*, 1984; Shelton, 1986), and silver barb, *Puntius gonnionotus* (Pongthana *et al.*, 1999).

Gene Manipulation

One of the most recent advances in genetic improvement of aquatic organism is the application of gene transfer technologies, or often referred

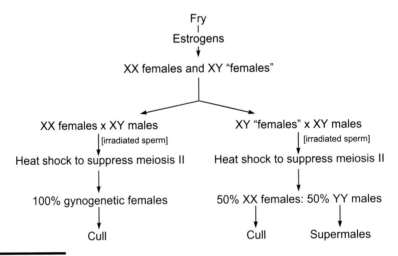

Fig. 17.7   Schematic representation of the steps involved in the production of supermale tilapia.

to as transgenesis. The technique involves the incorporation of genes from one individual into the nuclear DNA of another, which could be of the same or different species. Transgenesis was first successfully demonstrated in 1982 when a rat growth hormone gene was incorporated and expressed in mice (Palmiter *et al.*, 1982). Not long after, production of transgenic fish were reported, e.g., rainbow trout (Maclean and Talwar, 1984) and goldfish, *Carassius auratus* (Zhu *et al.*, 1985). To date, transgenic experiments have been performed on approximately 20 fish species worldwide.

Simply, transgenesis involves the introduction of foreign DNA into the nucleus of a host cell, which can also be a newly fertilized egg. The foreign DNA is often known as a DNA construct, which typically includes: (1) a promoter which acts to switch on and off the transcription and translation of the target gene, (2) a structural gene or target gene, (3) a reporter gene to enable to ascertain whether the target gene has combined successfully with the construct, and (4) a termination sequence. If the promoter does not precede the structural gene, there is possibility that the target trait may not be expressed in the transgenic individual.

Various techniques have been used to produce transgenic fish, and procedures are often straightforward. Details on these techniques are summarized as follows:

- Microinjection is the most widely used method in producing transgenic fish. A large number of transgenic fish from a wide range of species (e.g., Atlantic salmon, coho salmon, trout, tilapia, carp, catfish) have been produced using this method. The technique involves the use of a very fine glass tube to introduce a number of copies of constructs directly into newly fertilized eggs; the constructs are then incorporated into the chromosome before the first cell division.

- Electroporation involves the suspension of cells in a solution containing high concentration of DNA constructs, then high voltage pulses of electricity is applied to open pores in the cell membranes, through which foreign DNA can pass.

- Biolistic or biological ballistics, involves coating microscopic particles, usually of gold or tungsten, with DNA constructs and explosively firing these particles directly into the cell through the cell membrane.

- Lipofection or liposome-mediated transfection, involves the encapsulation of DNA constructs in synthetic lipid vesicles and subsequently bringing the vesicles into contact with the target cell.

- Sperm-mediated transfer involves the binding of the DNA construct to the outer coat of spermatozoa before it fertilizes the egg. This technique has the potential to be one of the most favourable gene transfer methods because of the nature of the procedure, and counters the problem of limited number of DNA constructs that entering the egg as there is only one sperm can enter the egg at a time.

A number of genes with useful aquaculture traits have been successfully transferred into genomes of various aquaculture species. The gene that is of most interest in aquaculture is that of the growth hormone (GH). In fish, most work done to date has involved injecting a mammalian GH gene and mammalian metallothionein promoter into an embryonic fish. The overall conclusion is that growth rates of GH transgenic fish are significantly higher than that of non-transgenic controls. For example, GH transgenic salmonids were reported to grow 3-5 times faster than non-transgenic fish (Devlin et al., 1994). In addition, considerable effort has been expended to introduce the antifreeze protein genes (from winter flounder, *Pleuronectes americanus*) into other species, such as Atlantic salmon (Hew et al., 1999), goldfish (Wang et al., 1995), Nile tilapia, and milkfish (Wu et al., 1998). A number of other target traits such as salinity tolerance, sterility, disease resistance (Mialhe et al., 1995), and greater production of n-3 HUFA (Donaldson, 1997) also offer considerable potential for transgenic work. Although all of these still need a great effort, they have the potential to bring great benefits to the aquaculture industry.

One of the major issues relating to the expression of the transgene is its integration into a position on a chromosome. The factors determining sites of integration are still poorly understood. Transgene expression in fish is highly variable, even among fish independently transformed with the same construct. There is also no guarantee that primary transgenic fish showing strong transgene expression will give rise to progeny with the same characteristics.

Criticisms over the use of transgenesis, however, have been raised with concerns regarding the risks brought about to human health, biodiversity, animal welfare and poor communities (Dunham, 1999; Maclean and Laight, 2000). Transgenesis being a process that falls within the realm of genetically modified organisms (GMO) it has to encounter the ethical issues that confront all GMOs which are increasingly subjected to public scrutiny and controversy. The final decisions on the wider application of transgenesis in aquaculture will finally be a political decision, and in this regard its applications in aquaculture will not be different to other food organisms.

In conclusion, it is evident that aquaculture, an important global food industry, has been able to sustain a very high growth rate, indeed the highest amongst all food-producing sectors (De Silva, 2001), primarily as a result of the developments of artificial propagation techniques for most of the important cultured species, thereby making it independent of wild seed stock. In the foreseeable future the developments in this regard are likely to mostly focus on increasing the efficacy of hormones and or their analogues used in hypophysation, and in larval rearing techniques, particularly of some marine fin fish species, in which the larval survival is very low. However, the most important developments will perhaps be related to the use of modern genetic techniques in improving broodstock management, which will contribute significantly to underpinning the overall sustainability of the aquaculture industry in the long term.

References

Ashby, K.R. 1957. The effect of steroid hormones on the brown trout *Salmo trutta* L. during the period of gonadal differentiation. *Journal of Embryology and Experimental Morphology* 5: 249-255.

Asturiano, J.F., L.A. Sorbera, M. Carrillo, S. Zanuy, J. Ramos, J.C. Navarro and N. Bromage. 2001. Reproductive performance in male European sea bass (*Dicentrarchus labrax* L.) fed two PUFA-enriched experimental diets: A comparison with males fed a wet diet. *Aquaculture* 194: 173-190.

Bartley, D.M., K. Rana and A.J. Immink. 2001. The use of inter-specific hybrids in aquaculture and fisheries. *Reviews in Fish Biology and Fisheries* 10: 325-337.

Beaumont, A.R. and K. Hoare. 2003. *Biotechnology and Genetics in Fisheries and Aquaculture.* Blackwell Scientific Publications, London.

Bell, J.G., B.M. Farndale, M.P. Bruce, J.M. Navas and M. Carillo. 1997. Effect of broodstock dietary lipid on fatty acid compositions of eggs from sea bass (*Dicentrarchus labrax*). *Aquaculture* 149: 107-119.

Boney, S.W., W.L. Shelton, S.L. Yang and L.O. Wilken. 1984. Sex reversal and breeding of grass carp. *Transactions of the American Fisheries Society* 113: 348-353.

Botsford, L.W., C.J. Castilla and C.H. Peterson. 1997. The management of fisheries and marine ecosystems. *Science* 277: 509-514.

Brämick, U., B. Puckhaber, H.-J. Langholz and G. Hörstgen-Schwark. 1995. Testing of triploid tilapia (*Oreochromis niloticus*) under tropical pond conditions. *Aquaculture* 137: 343-353.

Bromage, N. 2001. Recent Developments in the Control of Reproduction of Farmed Fish, *Modern Aquaculture in the Coastal Zone—Lessons and Opportunities*. NATO Science Series: Series A: Life Sciences. Vol. 314. IOS Press, Amsterdam, pp. 242-260.

Bromage, N. and R.J. Roberts (eds.). 1995. *Broodstock Management and Egg and Larval Quality*. Blackwell Scientific Publications, Oxford.

Bromage, N., C. Randall, J. Duston, M. Thrush and J. Jones. 1993. Environmental control of reproduction in salmonids. In: *Recent Advances in Aquaculture*, J.F. Muir and R.J. Roberts (eds.) Blackwell Scientific Publications, London, Vol. 4, pp. 55-66.

Bruce, M., F. Oyen, G. Bell, J.F. Asturiano, B. Farndale, M. Carrillo, S. Zanuy, J. Ramos and N. Bromage. 1999. Development of broodstock diets for the European sea bass (*Dicentrarchus labrax*) with special emphasis on the importance of n-3 and n-6 highly unsaturated fatty acid to reproductive performance. *Aquaculture* 177: 85-97.

Brusle, J. 1990. Eels and eel farming. In: *Aquaculture*, G. Barnabé (ed.). Ellis Horwood Ltd, West Sussex, Vol. 2, pp. 756-785.

Carrillo, M., S. Zanuy, F. Prat, R. Serrano and N. Bromage. 1993. Environmental and hormonal control of reproduction in sea bass. In: *Recent Advances in Aquaculture*, J.F. Muir and R.J. Roberts (eds.). Blackwell Scientific Publications, London, Vol. 4, pp. 43-54.

Cherfas, N.B., B. Gomelsky, N. Ben-Dom, Y. Peretz and G. Hulata. 1994. Assessment of triploid common carp (*Cyprinus carpio* L.) for culture. *Aquaculture* 127: 11-18.

Chinese Aquaculture Society. 1986. Fan Li on Pisciculture. In: Chen, L.-X. (ed.). Chinese Agriculture Publishers, Beijing.

Crawford, M.A., M. Bloom, C.L. Broadhurst, W.F. Schmidt, S.C. Cunnane, C. Galli, K. Gehbremeskel, F. Linseisen, J. Lloyd-Smith and J. Parkington. 1999. Evidence for the unique function of docosahexaenoic acid during the evolution of the modern hominid brain. *Lipids* 34: S39- S47.

Crim, L.W. 1985. Methods for acute and chronic hormone administration in fish. In: *Reproduction and Culture of Milkfish*, C.S. Lee and I.C. Liao (eds.). Oceanic Institute and Tungkang Marine Laboratory, pp. 1-13.

Crim, L.W. and B.D. Glebe. 1984. Advancement and synchrony of ovulation in Atlantic salmon with pelleted LHRH analog. *Aquaculture* 43: 48-56.

Czesny, S. and K. Dabrowski. 1998. The effect of egg fatty acid concentrations on embryo viability in wild and domesticated walleye (*Stizostedion vitreum*). *Aquatic Living Resources* 11: 371-378.

De Deckere, E.A.M., O. Korver, P.M. Verschuren and M.B. Katan. 1998. Health aspects of fish and n-3 polyunsaturated fatty acids from plant and marine origin. *European Journal of Clinical Chemistry* 52: 749- 753.

De Silva, S.S. 2001. A global perspective of aquaculture in the new millennium. In: *Aquaculture in the Third Millennium*, R.P. Subasinghe, P. Bueno, M.J. Phillips, C. Hough, S.E. McGladdery and J.R. Arthur (eds.), *Technical Proceedings of the Conference on Aquaculture in the Third Millennium, Bangkok, Thailand*, NACA, Bangkok and FAO, Rome, pp. 431-459.

De Silva, S.S. and T.A. Anderson. 1995. *Fish Nutrition in Aquaculture*. Chapman and Hall, London.

Delgado, C.L., N. Wada, M.W. Rosegrant, S. Meijer and M. Ahmed. 2003. Fish to 2020. Supply and demand in changing global market. International Food Policy Research Institute, Washington, D.C., 226 pp.

Devlin, R.H., T.Y. Yesaki, C.A. Biagi, E.M. Donaldson, P. Swanson and K.C. Whoon. 1994. Growth enhancement of salmonids through transgenesis using an "all-salmon" gene construct.

In: *High performance Fish: Proceedings of an International Fish Physiology Symposium*, D.D. MacKinlay (ed.), The University of British Columbia, Vancouver, Canada, pp. 343-345.

Donaldson, E.M. 1997. The role of biotechnology in sustainable aquaculture. In: *Sustainable Aquaculture*, J.E. Basdad (ed.). John Wiley & Sons, New York, pp. 101-126.

Donaldson, E.M. and G.A. Hunter. 1982. Sex control in fish with particular reference to salmonids. *Canadian Journal of Fisheries and Aquatic Sciences* 39: 99-110.

Doyle, R.W., R. Perez-Enriquez, M. Takagi and N. Taniguchi. 2001. Selective recovery of founder genetic diversity in aquacultural broodstocks and captive endangered fish populations. *Genetica* 111: 291-304.

Dunham, R.A. 1990. Production and use of monosex or sterile fishes in aquaculture. *Reviews in Aquatic Sciences* 2: 401-406.

Dunham, R.A. 1999. Utilisation of transgenic fish in developing countries: Potential benefits and risks. *Journal of the World Aquaculture Society* 30:1.

Dunham, R.A. 2004. *Aquaculture and Fisheries Biotechnology*. CABI Publishing, 372 pp.

Falconer, D.S. 1981. *Introduction to Quantitative Genetics*. Longman, London.

FAO, 2002. Fishstat Plus: Universal software for fishery statistical times series. Version 2.3, Fisheries Department. Fishery Information. Data and Statistical Unit.

Ferguson, M.M. 1995. The role of molecular genetic markers in the management of cultured fishes. In: *Molecular Genetics in Fisheries*, G.R. Carvalho and T.J. Pitcher (eds.). Chapman and Hall, London, pp. 81-104.

Furuita, H., H. Tanaka, T. Yamamoto, M. Shiraishi and T. Takeuchi. 2000. Effect of n-3 HUFA levels in broodstock diet on the reproductive performance and egg and larval quality of the Japanese flounder, *Paralichthys olivaceus*. *Aquaculture* 187: 387-398.

Gjoen, H.M. and H.B. Bentsen. 1997. Past, present and future of genetic improvement in salmon aquaculture. *ICES Journal of Marine Science* 54: 1009-1014.

Gousset, B. 1992. Eel Culture in Japan. *Bulletin de l'Institut océanographique*, Monaco, no spécial 10, Monaco, pp. 128.

Gunasekera, R.M.G., G.J. Gooley and S.S. De Silva. 1998. Characterisation of 'swollen yolk sac syndrome' in the Australian freshwater fish Murray cod, *Maccullochella peelii peelii*, and associated nutritional implications for large scale aquaculture. *Aquaculture* 169: 69-85.

Gupta, M.V. and B.O. Acosta. 2004. A review of global tilapia farming practices. *Aquaculture Asia* 9: 7-12.

Harvey, B.J. and W.S. Hoar. 1979. *The Theory and Practice of Induced Breeding in Fish*. International Development Research Centre, Ottawa, 48 pp.

Heinsbroek, L.T.N. 1991. A review of eel culture in Japan and Europe. *Aquaculture and Fisheries Management* 22: 57-72.

Hew, C.L., R. Poon, F. Xiong, S. Gauthier, M. Shears, M. King, P. Davies and G. Fletcher. 1999. Liver-specific and seasonal expression of transgenic Atlantic salmon harbouring the winter flounder antifreeze protein gene. *Transgenic Research* 8: 405.

Hickling, C.F. 1960. The Malaca tilapia hybrid. *Journal of Genetics* 57: 1-10.

Hickling, C.F. 1963. The culture of tilapias. *Scientific American* 208: 143-151.

Houssay, B.A. 1931. Action sexuelle de l'hypophyse sur les poissons et les reptiles. *Comptes Rendus des Seances de la Societe de Biology*, Paris 106: 377-378.

Ingram, B.A. and S.S. De Silva (eds.) 2004. *Development of Intensive Commercial Aquaculture. Production Technology for Murray Cod.* Final Report to the Fisheries Research and Development Corporation (Project No. 1999/328). Primary Industries Research, Victoria, DPI, Alexandra, Victoria, Australia, 202 pp.

Ingram, B.A., S. Sungan, G. Gooley, Y.S. Sim, D. Tinggi and S.S. De Silva. 2005. Induced spawning, larval development and rearing of two indigenous Malaysian mahseer, *Tor tambroides* and *T. douronensis. Aquaculture Research* 36: 1001-1014.

Jhingran, V.G. and V. Gopalakrishnan. 1974. *A Catalogue of Cultivated Aquatic Organisms.* FAO, Rome.

Joshi, C.B. 1988. Induced breeding of mahseer, *Tor putitora* (Hamilton). *Journal of Inland Fisheries Society, India* 20: 66-67.

Kearsey, M.J. and H.S. Pooni. 1996. *The Genetical Analysis of Quantitative Traits.* Chapman and Hall, London.

Koven, W., Y. Barr, S. Lutzky, I. Ben-Atia, R. Weiss, M. Harel, P. Behrens and A. Tandler. 2001. The effect of dietary arachidonic acid (20: 4n-6) on growth, survival and resistance to handling stress in gilthead seabream (*Sparus aurata*) larvae. *Aquaculture* 193: 107-122.

Kowtal, G.V., J. W.H. Clark and G.N. Cherr. 1986. Elimination of adhesiveness in eggs from the white sturgeon, *Acipenser transmontanus:* Chemical treatment of fertilized eggs. *Aquaculture* 55: 139-143.

Krise, W.F. 1988. Optimum protease exposure time for removing adhesiveness of walleye eggs. *Progressive Fish-Culturist* 50: 126-127.

Lake, J.S. 1967. Rearing experiments with five species of Australian freshwater fishes. I. Inducement to spawning. *Australian Journal of Marine and Freshwater Research* 18: 137-153.

Lam, T.J. 1982. Applications of endocrinology to fish culture. *Canadian Journal of Fisheries and Aquatic Sciences* 39: 111-137.

Landau, M. 1992. *Introduction to Aquaculture.* John Wiley & Sons, Brisbane.

Lande, R. and R. Thompson. 1990. Efficiency of marker-assisted selection in the improvement of quantitative traits. *Genetics* 124: 743-756.

Lee, C.-S., C.S. Tamaru and C.D. Kelley. 1986. Technique for making chronic-release LHRH-α and 17α-Methyltestosterone pellets for intramuscular implantation in fishes. *Aquaculture* 59: 161-168.

Maclean, N. and R.J. Laight. 2000. Transgenic fish: An evaluation of benefits and risks. *Fish and Fisheries* 1: 146-172.

Maclean, N. and S. Talwar. 1984. Injection of cloned genes with rainbow trout eggs. *Journal of Embryology and Experimental Morphology* 82: 187.

Mair, G.C. 2001. Genetics in Tilapia Culture. In: *Tilapia: production, marketing and technological development,* S. Subasinghe and T. Sing, (eds.). *Proceedings of the Tilapia 2001 International Technical and Trade Conference,* 28-30 May 2001, Kuala Lumpur, Malaysia, INFOFISH, 136-148.

Mair, G.C. and D.C. Little. 1991. Population control in farmed tilapias. *NAGA the ICLARM Quarterly* 14: 8-13.

Mair, G.C., A.G. Scott, D.J. Penman, J.A. Beardmore and D.O.F. Skibinski. 1991. Sex determination in the genus *Oreochromis.* 1. Sex reversal, gynogenesis and triploidy in *O. niloticus. Theoretical and Applied Genetics* 82: 144- 152.

Mialhe, E., E. Bachere, V. Boulo, J.P. Cadoret, C. Rousseau, V. Cedeño, E. Sorairo, L. Carrera, J. Calderon and R.R. Colwell. 1995. Future of biotechnology-based control of disease in marine invertebrates. *Molecular Biology and Biotechnology* 4: 275-283.

Michaels, V.K. 1988. *Carp Farming*. Fishing News Books Ltd., Surrey, London.

Munro, A.D., A.P. Scott and T.J. Lam. 1990. *Reproductive Seasonality in Teleosts: Environmental Influences*. CRC Press, Boca Raton.

Mylonas, C.C. and Y. Zohar. 2001. Use of GnRHa-delivery systems for the control of reproduction in fish. *Reviews in Fish Biology and Fisheries* 10: 463-491.

Na-Nakorn, U. 2004. A perspective on breeding and genetics of walking catfish in Thailand. *Aquaculture* Asia 9: 10-12.

Na-Nakorn, U., W. Kamonrat and T. Ngamsiri. 2004. Genetic diversity of walking catfish, *Clarias macrocephalus*, in Thailand and evidence of genetic introgression from induced farmed C. *gariepinus*. *Aquaculture* 237: 73-88.

Nash, C.E. and Z.H. Shehadeh (eds.). 1980. *Review of Breeding and Propagation Techniques for Grey Mullet, Mugil cephalus L. ICLARM Studies and Reviews 3*. International Center for Living Aquatic Resources Management, Manila, Philippines, 87 pp.

Ogale, S.N. 2002. Mahseer breeding and conservation and possibilities of commercial culture: the Indian experience. In: *Cold Water Fisheries in the Trans-Himalayan Countries*, T. Petr and S.B. Swar (eds.). FAO Fisheries Technical paper No. 431. FAO of the UN, Rome, pp. 193-212.

Ohta, K., H. Kagawa, H. Tanaka, K. Okuzawa, N. Iinuma and K. Hirose. 1997. Artificial induction of maturation and fertilization in the Japanese eel, *Anguilla japonica*. *Fish Physiology and Biochemistry* 17: 163-169.

Palmiter, R.D., R.L. Brinster, R.E. Hammer, M.E. Trumbauer, M.G. Rosenfeld, N.C. Birnberg and R.M. Evans. 1982. Dramatic growth of mice that develop from eggs microinjected with metallothionein-growth hormone fusion genes. *Nature (London)* 300: 611-615.

Pandian, T.J. and S.G. Sheela. 1995. Hormonal induction of sex reversal in fish. *Aquaculture* 138: 1- 22.

Pedersen, B.H. 2003. Induced sexual maturation of the European eel *Anguilla anguilla* and fertilisation of the eggs. *Aquaculture* 224: 323-338.

Penman, D.J. and B.J. McAndrew. 2000. Genetics for the management and improvement of cultured tilapias. In: *Tilapias: Biology and Exploitation*, M.C.M. Beveridge and B.J. McAndrew (eds.). Kluwer Academic Press, London, pp. 227- 266.

Pillay, T.V.R. 1990. *Aquaculture Principles and Practices*. Fishing News Books, Oxford.

Pongthana, N., D.J. Penman, P. Baoprasertkul, M.G. Hussain, M.S. Islam, S.F. Powell and B.J. McAndrew. 1999. Monosex female production in the silver barb (*Puntius gonionotus* Bleeker). *Aquaculture* 173: 247-256.

Ritland, K. 2000. Marker-inferred relatedness as tool for detecting heritability in nature. *Molecular Ecology* 9: 1195-1204.

Rowland, S.J. and C. Bryant. 1995. Silver Perch Culture. *Proceedings of Silver Perch Aquaculture Workshops, Grafton and Narrandera, April 1994*. Austasia Aquaculture, Sandy Bay, Tasmania. 125 pp.

Sargent, J.R. 1995. Origins and functions of egg lipids: nutritional implications. In: *Broodstock Management and Egg and Larval Quality*, N. Bromage and R.J. Roberts (eds.). Blackwell Scientific Publishers, London. pp. 353-372.

Schally, A.V. 1978. Aspects of hypothalamic regulation of the pituitary gland. *Science* 202: 18-28.

Scott, A.G., D.J. Penman, J.A.B. Beardmore and D.O.F. Skibinski. 1989. The 'YY' supermale in *Oreochromis niloticus* (L.) and its potential in aquaculture. *Aquaculture* 78: 237-251.

Senanan, W., A.R. Kapuscinski, U. Na-Nakorn and L.M. Miller. 2004. Genetic impacts of hybrid catfish farming (*Clarias macrocephalus* × *C. gariepinus*) on native catfish populations in central Thailand. *Aquaculture* 235: 167-184.

Shaklee, J.B. 1983. The utilisation of allozymes as gene markers in fisheries management and conservation. *Current Topics in Biological and Medical Research* 11: 213-247.

Shelton, W.L. 1986. Broodstock development for monosex production of grass carp. *Aquaculture* 57: 311-319.

Shelton, W.L. 1989. Management of finfish reproduction for aquaculture. *Reviews in Aquatic Sciences* 1: 497-535.

Sherwood, N.M., L.W. Crim, J. Carolsfeld and S.M. Walters. 1988. Sustained hormone release. I. Characteristics of in vitro release of gonadotropin-releasing hormone analogue (GnRH-A) from pellets. *Aquaculture* 74: 75-86.

Stansby, M.E. 1990. Nutritional properties of fish oil for human consumption- aspects. In: *Fish Oils in Nutrition*, M.E. Stansby (ed.). Van Nostrand Reinhold, New York, pp. 289-308.

Su, M.S., C.S. Lee and I.C. Liao. 2002. Technical responses to challenges in milkfish aquaculture. *Reviews in Fisheries Science* 10: 451-464.

Subasinghe, R.P., P. Bueno, M. Phillips, C. Hough and J.R. Arthur McGladdery (eds.). 2001. *Aquaculture in the Third Millennium. Technical proceedings of the Conference on Aquaculture in the Third Millennium.* NACA/FAO, Bangkok.

Tachiki, K. and T. Nakagawa. 1993. Induction of spawning in female cultured eel *Anguilla japonica. Bulletin Aichi Fisheries Research Institute* 1: 79-83.

Tanaka, H., H. Kagawa and H. Ohta. 2001. Production of leptocephali of Japanese eel (*Anguilla japonica*) in captivity. *Aquaculture* 201: 51-60.

Tanaka, H., H. Kagawa, H. Ohta, T. Unuma and K. Nomura. 2003. The first production of glass eel in captivity: fish reproductive physiology facilitates great progress in aquaculture. *Fish Physiology and Biochemistry* 28: 493-497.

Tave, D. 1993. *Genetics for Fish Hatchery Managers.* Second Edition. Chapman and Hall, New York.

Tyler, J.R. and J.P. Sumpter. 1996. Oocyte growth and development in teleosts. *Reviews of Fish Biology and Fisheries* 6: 287-318.

Ulbricht, T.L.V. and D.A.T. Southgate. 1991. Coronary heart disease: Seven dietary factors. *The Lancet* 338: 985-992.

Varadaraj, K. and T.J. Pandian. 1989. First report on production of supermale tilapia by integrating endocrine sex reversal with gynogenetic technique. *Current Science* 58: 434-441.

von Ihering, R. 1937. A method for inducing fish to spawn. *Progressive Fish Culturist* 34: 15-16.

Wang, R., P. Zhang, Z. Gong and C.L. Hew. 1995. Expression of the antifreeze gene in transgenic goldfish (*Carassius auratus*) and its implication in cold adaptation. *Molecular Marine Biology and Biotechnology* 4: 20-26.

Watanabe, T. and V. Kiron. 1995. Red sea bream (*Pagrus major*). In: *Broodstock Management and Egg and Larval Quality*, N. Bromage and R.J. Roberts (eds.). Blackwell Scientific Publications, London, pp. 398-413.

Wu, S.M., P.P. Hwang, C.L. Hew and J.L. Wu. 1998. Effects of antifreeze protein on cold tolerance in juvenile tilapia (*Oreochromis mossambicus* Peters) and milkfish (*Chanos chanos* Forskål). *Zoological Science* 37: 39-44.

Yamazaki, F. 1983. Sex control and manipulation in fish. *Aquaculture* 33: 329-354.

Ye, Y. 1999. Historical consumption and future demands for fish and fishery products: exploratory calculations for the years 2015/2030. FAO Fisheries Circular 946. FAO, Rome, Italy.

Zhu, Z.Y., G. Li, L. He and S. Chen. 1985. Novel gene transfer into the fertilized eggs of the goldfish *Carassius auratus* L. 1758. *Journal of Applied Ichthyology* 1: 31-34.

Zohar, Y. 1989. Fish reproduction: its physiology and artificial manipulation. In: *Fish Culture Warm Water Systems: Problems and Trends*, M. Shilo and S. Sarig (eds.). CRC Press, Boca Raton, pp. 65-119.

Zohar, Y. 1996. New approaches for the manipulation of ovulation and spawning in farmed fish. *Bulletin of the National Research Institute of Aquaculture* Supplement 2: 43-48.

Zohar, Y. and C.C. Mylonas. 2001. Endocrine manipulations of spawning in cultured fish: From hormones to genes. *Aquaculture* 197: 99-136.

Cultivation of
Coldwater Marine Species

Birgitta Norberg[1,*], Inger-Britt Falk-Petersen[2], Torstein Harboe[1], Terje van der Meeren[1] and Anders Mangor-Jensen[1]

Atlantic Cod

Anders Mangor-Jensen, Terje van der Meeren and Birgitta Norberg

The Atlantic cod (*Gadus morhua* L.) is considered a promising species for cold-water marine aquaculture. In addition, cod is one of the most economically important species for the fisheries and an average of more than 100,000 metric tons are harvested each year. In spite of the large catches of wild cod, the market has responded very positively to the farmed alternative, presumably due to high quality and freshness, and year-round availability. However, there are several important criteria that have to be fulfilled to approve a species candidature for aquaculture. The most important among these is a stable, seasonally independent production and the basis are methods for effective juvenile production. Cod farming is far from new. Towards the last part of the nineteenth century, trials with hatching of cod larvae in small incubators for releasing

Authors' addresses: [1]Institute of Marine Research, N-5392 Storebø, Norway.
[2]Norwegian College of Fishery Science, University of Tromsø, N-9037 Tromsø, Norway.
Corresponding author: E-mail: birgittan@imr.no

yolk sac larvae into the Norwegian fjords were commenced at Flødevigen Biological Station. This program was sustained for nearly a decade, although the documented impact on the local stocks was questionable. To overcome criticism about survival potential of the released yolk sac larvae in the sea, in 1886 the Norwegian captain G.M. Dannevig performed a startfeeding trial with cod in a 2500 m^3 large concrete enclosure on land (Rognerud, 1887). The intention was not aquaculture, but he managed to produce several thousands of cod fry in this system. Inspired by the results of Dannevig and other Norwegian pond experiments with plaice (*Pleuronectes platessa*) and flounder (*Platichtys flesus*) in the 1930s, the idea of producing marine fish for aquaculture purpose in enclosure systems was developed in the 1970s (Øiestad et al., 1976). The method was further escalated to large-scale production of cod fry in lagoon systems at the aquaculture station of the Institute of Marine Research at Austevoll, Norway, with a breakthrough in 1983 when 60,000 cod fry were harvested from the 60,000 m^3 enclosed lagoon "Hyltropollen" (Kvenseth and Øiestad, 1984). The lagoon method diversified into the semi-intensive method where cod larvae were start-fed in floating plastic bag enclosures or large tanks on land, with the lagoon as a plankton production unit where copepods were fractionated and concentrated by means of wheel filters (van der Meeren and Naas, 1997). Towards the millennium, Canada succeeded with large-scale intensive production methods of cod juveniles based on production of warm water marine fish. Based on both the Norwegian experience and the Canadian results, Norway produced around 8 million cod fry in 2004, among this one million in lagoon systems.

Broodstock Criteria and Selection

Broodstock individuals may be selected from catches, or preferably from genetically selected strains. So far, little effort has been directed to selective breeding programs, and the most active broodstocks are little more than individuals selected on size. There are several local strains of Atlantic cod, and a number of trials have been performed to find out which is most suited for domestication and aquaculture. The results from these trials have in most cases revealed little difference between strains, and other factors may be determinant for rearing success. Eggs from captive broodstocks, or from ripe individuals in grow-out facilities will occasionally be released into the local waters. Questions have, therefore, been raised whether only local stock should be allowed used as parent fish in order to avoid genetic mixing with unknown consequences. Before

Fig. 18.1 Atlantic cod broodstock in spawning tank with egg collector mounted on the side.

these questions have found their answers, the aquaculture of cod may face serious challenges with respect to egg- and juvenile trades between regions.

The Atlantic cod is a determinate, group-synchronous spawner, and captive females release up to 20 egg batches at controlled intervals. The intervals vary with temperature, but each ovulatory cycle typically takes 48-60 hours (Kjesbu, 1989; Kjesbu *et al.*, 1990). Egg batches contain a relatively lower number of eggs in the beginning and end of the spawning period, but overall, cod have a very high fecundity, averaging ca 5×10^5 eggs $\times$ kg^{-1}. The eggs are small, 1.2–1.5 mm in diameter, with decreasing egg size throughout the spawning period (Mangor-Jensen *et al.*, 1994; Kjesbu *et al.*, 1996). First-time spawners tend to have lower fecundity as well as smaller eggs, but there is a large variation among different families (Kjesbu *et al.*, 1996).

Cod spawn willingly in captivity even under primitive conditions. When stocked in tanks or submerged containers, the fish will produce a vast amount of fertilized eggs that can be collected at the surface and transferred to suitable egg-incubators (Fig. 18.1). Hand stripping of ripe individuals have occasionally been conducted, but mostly with poor results, as the cod female contracts the wall muscles in the oviduct on handling. In most cases when single batches are required, the fish has to be anaesthetized or sacrificed.

Broodstock Nutrition

A well-composed broodstock diet is of utmost importance for at least two reasons:

1. An optimal nutrient composition of the eggs is vital for larval viability, and is dependent on a balanced diet, as diet composition will influence the biochemical composition of the eggs.
2. Reproductive exhaustion of broodstock is a growing concern among cod farmers and may occur if the fish are given a diet, which has only a slight nutrient deficiency or imbalance. As the fish are kept over several consecutive years, such a diet can eventually lead to serious health problems for the fish.

Nutrient requirements for cod broodstock have been described with particular attention paid to dietary carbohydrate:protein ratio (Hemre et al., 1995), content of vitamin C (Mangor-Jensen et al., 1994) and the effect of dietary fatty acids on ovarian fatty acid composition (Lie et al., 1991). Cod do not easily digest carbohydrate, and broodstock feeds need to contain a relatively high proportion of protein. Addition of varying levels of vitamin C to the broodstock diet had no effects on egg quality, hatch rate or larval quality. In apparent contrast, there were large differences in the amount of vitamin C sequestered by the ovaries in the different experimental groups.

The fatty acid composition of the ovaries was affected by the fatty acid composition of the diet (Lie et al., 1991), especially for arachidonic acid (ARA, 20:4n-6) that was greatly enriched in the PI fraction. On the other hand, increase in dietary DHA (22:5n-3) and EPA (20:5n-3) resulted in increases in DHA and EPA in neutral lipids, but less so in the phospholipid fraction. There is, however a lack of information on the effects of increased ARA in relation to DHE and EPA on egg and larval quality.

An important consideration is the amount of feed given to the broodstock. Cod deposit a very large amount of energy into egg production, and the energy deposited increases with the amount of energy available. Overfeeding will result in an extremely high egg production, but the associated energetic cost will lead to a big weight loss during spawning, which will result in poor condition and even increased post-spawning mortality among the broodstock (Kjesbu *et al.*, 1991).

Year-round Spawning

In temperate teleost species, seasonal reproduction is an important feature to obtain optimal survival conditions for the offspring. Sexual maturation is controlled by endogenous rhythms that are entrained by external factors, *zeitgebers*, such as photoperiod and water temperature.

To ensure year-round availability of viable eggs and larvae, it is necessary to control the timing of spawning. In cod, as in many other species including Atlantic halibut, photoperiod is considered the most important *zeitgeber* for reproduction, and is used to manipulate sexual maturation and spawning to occur outside of the normal reproductive season (for review see Bromage *et al.*, 2000). The most widely used methods of photoperiod manipulation are either based on compression or extension of the annual photoperiod cycle, followed by a phase-shifted 12-month cycle (Norberg *et al.*, 2004a; Fig. 18.2) or exposure to continuous light followed by a phase-shifted cycle (Hansen *et al.*, 2001; Fig. 1b). Atlantic cod broodstock readily respond to shifts in photoperiod; however, spawning time is usually not completely shifted to the first reproductive season after manipulation. The endogenous rhythm controlling reproduction will be adjusted, but not completely so in the first year. In addition, broodstock groups that are kept for extended time periods on shifted photoperiods in covered tanks may slowly become desynchronized, especially between males and females, and there may be problems with irregular spawning in females, which fail to ovulate and/or release the eggs. The reasons for these problems are unclear, but the artificial light and daylength experienced by a fish in a tank will differ somewhat from the natural, where the horizon is influenced by the depth where the fish are found. There is also very little information on the effect of spectral composition and light intensity on experienced photoperiod.

Water temperature influences reproductive development, and is a critical factor for successful spawning in cod. Optimum spawning temperature is around 8°C, while the upper tolerance limit for spawning is as low as 10°C, i.e., the temperature interval where spawning may occur successfully is narrow (van der Meeren and Ivannikov, 2006). High water temperatures will make the ovulatory cycles shorter and less regular, and the occurrence of "over-ripening", i.e., cell death in ovulated eggs,

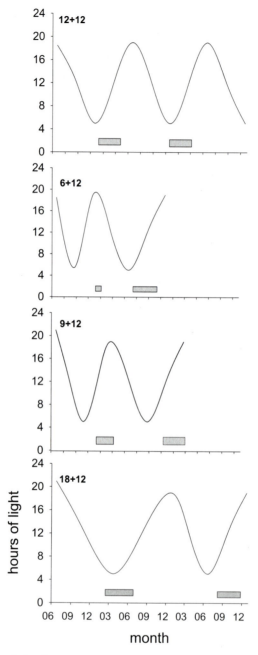

Fig. 18.2   Photoperiod regimes used to manipulate spawning time in cod. Grey bars denote spawning activity (males and females). Drawn after Norberg *et al.* (2004).

will increase. A large proportion of dead eggs is commonly observed as a result of holding spawning cod at temperatures above 10°C, and the total egg yield is lower from such fish. The mechanisms behind the deleterious effects of high temperature on cod reproduction are not known. There is, however, documentation in other species, such as red seabream (*Pagrus major*; Lim *et al.*, 2003) and Atlantic salmon (*Salmo salar*; Andersson *et al.*, 2003; Taranger *et al.*, 2003), that temperature has direct effects on the expression of key components in the brain-pituitary-gonad axis.

Eggs, Fertilization and Incubation

Eggs are collected from the surface of the spawning tanks by a surface skimmer, and conducted into a collecting tank. Eggs that fail to keep buoyant in normal strength seawater are either dead or of low quality, and are removed at the bottom by flushing the tank. After collection from the spawning tank the egg quality is evaluated. Important quality criteria are fertilization rate, mitotic cell symmetry of the blastodisc,

Fig. 18.3 Cod embryo prior to hatching.

clearness, buoyancy and non-fused cortical alveoli (Kjørsvik *et al.*, 1990) These can all be evaluated by a skilled technician, and thus make a good basis for the selection of batches for further incubation.

At 7°C cod eggs require approximately 13 days from fertilization to hatching (Fig. 18.3). In this period, the eggs are incubated in flow-trough incubators where temperature and oxygen are well regulated. Dead eggs are removed daily, and embryonic development is inspected at least 4-5 times during incubation. Cod eggs are normally stocked at densities of 5-10000 individuals/litre and the normal hatching rate is >80%. Newly hatched larvae depend entirely on their yolk sac as a source for nutrients. During the period between hatching and first feeding, cod larvae are kept in the egg incubator.

First Feeding

Three days after hatching, at 10-12°C, cod larvae may ingest small live prey like small copepod nauplii or rotifers (Fig. 18.4). In the early days of cod aquaculture, large numbers of newly hatched larvae were released into closed seawater lagoons were they foraged on zooplankton naturally produced in the system. This technique is called "extensive" and differs from the modern "intensive" technique where environmental factors are

Fig. 18.4 First-feeding cod larva.

controlled and the feed organisms (rotifers) are cultivated. Although more than 90% of today's juveniles are produced intensively, the fish raised in ponds on natural zooplankton exhibit a generally higher quality with respect to exterior development and general fitness. During the first days of first-feeding, the cod larvae need small food, and rotifers are perfect in this respect (150-300 µm; Fig. 18.5). However, the nutritional value of such prey do not meet the requirements of the cod larvae, and enrichment with several chemical components are necessary to secure the quality demands. Compared to copepods, rotifers contain less phospholipids, highly unsaturated fatty acids (particularly EPA and DHA), pigments (astaxanthin), and free amino acids. High-density cultures of rotifers are routinely performed at all cod hatcheries but are laborious and, therefore, expensive. One of the main concerns both in industry and research is, therefore, to find a method for rotifer production that will stabilize quality and reduce production cost.

The cod larvae are fed live rotifers for a period of approximately 30-40 days, often with a short switch to *Artemia* the last 10 days. During this period, the cod larvae increase their length from around 4 mm at hatching to ca 12 mm at onset of metamorphosis. The early juvenile period starts

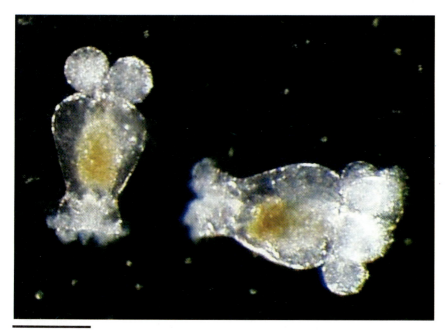

Fig. 18.5   Rotifers used as first feed for cod larvae.

at 18-20 mm length, but, according to Pedersen and Falk-Petersen (1992), the digestion system is not fully metamorphosed until the fish grow to 45 mm length. The period with live feed is called the "first feeding period", and is followed by a "weaning" period where live prey is substituted with a formulated diet. The question of early weaning has been addressed in several studies, but so far a period of live prey seems to be required to achieve a reasonable survival rate. Weaning was earlier considered the most critical event in juvenile production. However, improved feeding techniques combined with new diets have made this stage almost without complication. During the first feeding period, cod larvae are stocked at high densities, which may exceed 100 larvae per litre. Survival up to 50-60% during the larval stages has been reported.

In commercial, intensively operated hatcheries first feeding tanks are normally 2.5-3 m in diameter. Distribution of larvae and prey is maintained by aeration of the tank. To remove the surface film that inevitably arises from residue lipids of the live feed enrichment, surface skimmers (blowers) are used. Photoperiod is recommended to be 24-hour light. Green water, where algae are added, is commonly used for the first 20-30 days of larval rearing, corresponding to the period when rotifers are used as food. Green water is mostly obtained by use of algal paste, but also live algae may be used. Algal species like *Nannochloropsis* sp. or *Isochrysis* sp. are commonly grown to be used for green water. Green water has been demonstrated to have a profound positive effect on larval feeding probably by enhancing the feeding environment (turbidity, light scattering, and prey contrast). The presence of algae may also maintain nutritional quality of the rotifers in the rearing tank.

Larval Growth and Feed Requirements

In intensive systems, survival and growth have been found to be dependent on prey density, (Brown *et al.*, 2003). In lagoon systems, larval-specific growth has been demonstrated to be higher than in intensive systems. This may partly be caused by the opportunity to forage on a variety of sizes and types of zooplankton rather than the uniform diet in intensive systems.

When the larvae approach metamorphosis, cannibalistic behaviour may occasionally be a problem. However, this seems to be controlled by ensuring that sufficient food particles are present in the rearing tank. Up to 50% of the live feed added to the tanks may be lost in the drain and by mortality. In the high larval and prey densities used in intensive cod farming, accumulation of faeces and food particles on the bottom requires thorough cleaning. Most cod hatcheries, therefore, use an automatic wiper arm system for removal of wastes from the tank bottom (van der Meeren *et al.*, 1998). The introduction of automatic cleaning has improved water quality at the hatcheries, particularly during weaning to formulated diets.

Water Quality

Different species of fish have different requirements with regard to water quality. Water quality is closely linked to exchange rate and eventual recirculation treatment. Water exchange rate in the first-feeding tanks may vary between hatcheries, but is usually around 1 tank volume day^{-1} at the start of exogenous feeding, increasing to between 5 and 10 tank volumes day^{-1} at metamorphosis. Low water exchange rates during early juvenile stages, accompanied by increased ammonium (NH_4^+), reduced oxygen saturation, and reduced pH (indicating increased carbon dioxide), have significant negative effects on growth rate and survival of Atlantic cod. Thus, without oxygenation a water exchange rate of more than 4 litre g^{-1} fish day^{-1} is recommended.

Fig 18.6 Newly metamorphosed cod fry.

A particular problem that may occur occasionally is over-inflation of the swimbladder, causing the fish to float ventral side up at the surface. It is not known what causes this problem, but some attention has been drawn to a possible effect of gas super saturation, which is commonly seen at hatcheries. Another problem is the occurrence of skeletal deformities in juvenile cod. The main deformities are lordosis and neck bends (the so-called "stargazers"). The deformity problem is of prime interest to the research currently performed in optimizing cod farming methods.

In recirculated systems, accumulation of metabolites like ammonia (NH_3) and carbon dioxide (CO_2) may rapidly become a problem if proper precautions are not taken. Short-term exposure (48 hours) of cod larvae to ammonia levels up to 0.1 mg NH_3 l^{-1} appears not to be lethal. However, larval sizes between 1 and 2mg seem to be more sensitive than earlier or later stages. The effects and levels of CO_2 in cod farming are virtually unknown, but based on the experience from smolt production of Atlantic salmon (*Salmo salar*) this needs attention, particularly during the juvenile stages.

Ongrowth

When the cod juveniles have reached a size of about 0.5 g, they can be stocked in dense juvenile cultures (Figs. 18.6 and 18.7) and grown onto a size of about 50 g when they are vaccinated against vibriosis. At this size, the cod juveniles are normally transferred to sea cages where they are offered a formulated dry food. Juvenile cod has been demonstrated to have a higher requirement for dietary protein than salmonids, and also a requirement for lower fat content. Based on such knowledge new diets have been formulated that greatly has improved the quality of the market fish. Given good conditions the cod juveniles display good growth with SGRs ranging from 0.7 to 0.9 in the stages between 200 and 800 grams (Rosenlund et al., 2004). As for other species farmed cod tend to mature at young stages. A varying proportion (15-80%) of males mature at one year, while there is close to 100% incidence of maturation in both sexes at two years post-hatch. Moreover, maturation before two years, at weights of <500 g, is an increasing problem among offspring of both sexes from photoperiod-manipulated broodstock. For both sexes this implies reduced or negative growth for a prolonged period, and it is considered the largest bottleneck in commercial cod production. Energy deposition is switched from somatic growth to gonad growth months in advance of spawning,

Fig. 18.7   Cod juveniles, ca 6 months post-hatch.

and during spawning the fish display reduced or absent appetite, and weight loss is common. A large research effort is currently directed towards understanding how puberty is regulated in cod (cf. Hansen *et al.*, 2001; Norberg *et al.*, 2004a,b; Davie, 2005; Karlsen *et al.*, 2005; Skjæraasen *et al.*, 2005). Whereas maturation may be blocked by exposure to continuous light in light-proof tanks (Hansen *et al.*, 2001), these results have not been possible to transfer to cod kept in sea cages, where only a 6-month delay in spawning has been possible (Karlsen *et al.*, 2000). Apparently, the natural light cycle is able to override artificial light, even at very high intensities (10^4 lux). Attempts to control puberty by reduced growth during the first year of life have yielded mixed results. Even though incidence of maturation was affected, particularly in females, the fish were not able to compensate completely for the growth reduction (Norberg *et al.*, 2004b). Early maturation thus remains a major problem for cod farmers, and future research efforts should be directed to the production of sterile fish in addition to development of rearing methods that block or delay puberty.

Early quality problems with farmed cod were related to oversized livers. This was mostly due to a high dietary fat content (Rosenlund *et al.*,

2004), and has through the development of new feeds now become a lesser problem.

Cod thrive well in traditional net pens, and salmon farm technology may therefore to a large extent be applied. Market size is approximately 3 kg, which is reached within 36 months provided that maturation is avoided. Except for the susceptibility for vibrio infections in young stages, infective diseases have not been a problem. On the other hand, escapes are common, and this has to be addressed in the development of cod rearing technology.

Atlantic halibut

Torstein Harboe, Anders Mangor-Jensen and Birgitta Norberg

The Atlantic halibut is an attractive candidate species for aquaculture, due to its high prize and large size. Largest of the flatfishes (*Pleuronectiformes*), the halibut may grow to 3-4 m length and 400 kg (Haug, 1990) (Fig. 18.8).

The first attempts to grow halibut larvae beyond metamorphosis were successfully carried out at Flødevigen Biological Station in the beginning

Fig. 18.8 Wild caught halibut female, weight ca 280 kg; length ca 250 cm.

of the 1970s. On this occasion two individuals survived, an event that enjoyed great attention in media as well as among scientists. However, more than ten years elapsed before the first commercially produced halibut juveniles were seen. In this period, international scientific cooperation brought about trial production and early life stage research in several countries, thus making the halibut a promising species for marine cold-water aquaculture. At present, only Iceland and Norway are continuing industrial production of halibut fry. Problems in juvenile production are the main causes for this lapse. Lack of knowledge on water quality requirements, infective diseases and insufficient availability of viable eggs are the key reasons. However, the industry now seems to have overcome the birth problems and halibut farmers are producing a small but stable number of juveniles each year.

Halibut eggs are heavier that most other pelagic fish eggs and are, therefore, often referred to as mesopelagic. They undergo embryonic development at 50-250 m depth, in the zone between Atlantic water of high salinity and the less saline surface water. At hatch, at approximately 84-day degrees, the poorly developed halibut larvae differ from other marine fish by a long yolk-sac stage and relatively large size. Forty days after hatching, the larva has resorbed most of its yolk reserve and depends thereafter on exogenous food for further growth and development. In aquaculture, live prey is used for first feeding, and most common is a compound diet of rotifers and artemia fed in traditional green-water cultures. Sixty days after first feeding, the larvae are weaned onto a dry diet. Similar to other flatfish, the halibut larvae undergo a distinct metamorphosis where the exterior changes from a larval look to the adult shape, and the juvenile goes from a pelagic to a demersal habitat.

Broodstock Selection and Reproductive Biology

The Atlantic halibut is found on both sides of the North Atlantic, outside the Faroe Islands, Iceland and south of Greenland (Haug, 1990). Males mature at a smaller size and younger age than females: in wild halibut outside the Faroe Islands, age and weight at first maturity in males average 4.5 years and 1.7 kg, respectively, while females mature at 7 years of age and an average weight of 18 kg (Jakupsstovu and Haug, 1988). As growth is severely impaired in mature fish compared to immature (Norberg *et al.*, 2001), two of the most important criteria for selection of halibut broodstock will thus be late sexual maturation as well as fast growth.

In their natural habitat, halibut spawn at specific locations, spawning grounds, and at 300-1000 m depth (Devold, 1938). The spawning season varies both between and within different geographical locations, but generally takes place between December and May (Kjørsvik et al., 1987; Jakupstovu and Haug, 1988). At those depths, the water temperature normally varies between 5 and 7°C, with a salinity of 3.45-3.5%. Little is known about the natural spawning behaviour of halibut.

Female Atlantic halibut are determinate spawners with group synchronous ovarian development. In the spawning season, eggs mature and are released in batches. Captive halibut release 5-15 batches of eggs, at 65-80 hour intervals. The eggs are comparatively large, 3.0-3.3 mm, with decreasing diameter through the spawning period (Riple, 2000). Depending on the size of the female, egg batch volumes may vary between 0.5 and >10 litres, containing 20 000->400 000 eggs. The total mass of spawned eggs per individual may be up to 40% of the female's body weight (Haug and Gulliksen, 1988a; Norberg et al., 1991).

At present, as a result of failed spontaneous spawning in captivity, culture techniques include hand stripping of eggs from ripe females. The broodfish are usually not handled and, thus, not disturbed before eggs are observed floating in the tank. These eggs originate from spontaneous releases, and are usually not fertilized. However, all females may not release eggs spontaneously in the tank. To avoid extensive handling, such females may be checked by application of a gentle pressure on the abdomen while they are still at the tank bottom. If the female is ripe, eggs will come out of the ovipore. Eggs are collected by gently moving a ripe female onto a padded table, which is lifted above the surface level where eggs are stripped into a bucket and fertilized with milt obtained in a similar manner. Different individuals respond differently to handling, and highly stressed females should not be used as breeders. Halibut eggs will become overripe and die within a few hours if the female is not stripped at the right time (Norberg et al., 1991). To optimize the output of eggs from a broodstock it is, therefore, important to pinpoint the ovulatory rhythm of each individual female used for egg production.

Some females that do not release their eggs spontaneously may have a closed or blocked ovipore, and have irregular ovulatory cycles. The mechanisms underlying these reproductive disturbances are not known but may be stress related. In apparent conflict, irregular spawners appear to be a more common observation among cultured fish, and fish that have been kept in captivity from juvenile stages than among wild-caught, adult females. This suggests that stress may not be the only factor causing irregular spawning patterns, as cultured fish are more adapted to captive conditions than wild-caught fish. An alternative explanation may be that wild-caught adult females are established spawners and that the regular spawning pattern is maintained once it is established in an individual female.

Broodstock Nutrition

The Atlantic halibut is an iteroparous spawner, and since broodstock are highly valuable, they are normally kept for several consecutive spawning seasons. A good diet is, therefore, important to keep the fish in good health while also ensuring optimal egg quality and yield. The energy demand during ovarian growth is very high in halibut, and nutrients are mobilized from liver and muscle depots, as well as from the feed (Haug and Gulliksen, 1988a, b). The nutrient requirements of halibut broodstock are not well documented, but some knowledge can be transferred from other species.

Initially, halibut broodstock were fed whole herring or mackerel, so as to present a diet that resembled the natural diet as closely as possible. The drawbacks with giving whole fish as feed are, however, considerable. Nutrient composition will be unstable, and vary with season. Herring contains large amounts of thiaminase, and the amounts increase with storage. If the fish are fed herring over some time, this may lead to reduced levels of thiamine (vit B1) in the eggs, which will impair larval viability. In Baltic salmon, decreased ovarian thiamine levels as a consequence of an increase in the proportion of herring in the diet have been suggested as a cause for high mortalities at the yolk-sac stage (M74 syndrome; Amcoff, 2000). A further concern with using whole fish as broodstock feed is the hygiene, and possibility of introducing pathogens.

Challenges with formulating halibut broodstock feed include size of feed particles, as well as taste and nutrient composition. Halibut will only accept large feed particles, with a size of at least 30-60 mm. Commercial halibut farms often use a vitaminized fish meal mixture to produce a "sausage" that is prepared immediately before the fish are fed. Adult halibut do not feed more than 1-2 days every week and do not need to be fed daily. However, to ensure that all fish in a broodstock group have an adequate feed intake, it is recommended to feed the fish 2-3 days a week.

The impact of broodstock diet on gamete quality has only been partly described in halibut. The biochemical composition of the eggs varies greatly, both between females and within different batches from the same female. Parameters such as timing of stripping and handling stress are likely to influence the egg composition, and make it difficult to draw reliable conclusions on possible dietary influences. However, research attention has been paid to the polyunsaturated fatty acids, docosahexanoic acid (DHA), eicosapentaeinoic acid (EPA) and arachidonic acid (ARA), and how the ratio between (ARA) and DHA/EPA influences egg and larval survival (Mazorra et al., 2003). An increased proportion of ARA in the diet resulted in higher fecundity as well as better egg quality. ARA is a precursor for prostaglandins, which act at different stages of the reproductive cycle. Further, vitamins present in the yolk sac, such as vitamin C, B and E, are utilized during embryogenesis (Rønnestad et al., 1997, 1999). The egg content of vitamin B9 (pholate) varies between egg batches in halibut (Mæland et al., 2002). Pholate is necessary for normal embryonic development and a deficiency may be one of the underlying mechanisms for deformities and/or mortalities that occur during larval stages. It is, thus, important to ensure that the broodstock feed contains adequate amounts of vitamins.

Environmental Control of Sexual Maturation and Spawning

As in Atlantic cod, photoperiod control is used to manipulate the timing of spawning in halibut broodstock (Næss et al., 1996; Björnsson et al., 1998). Although halibut respond to shifted photoperiod, it will normally take at least one year before a new annual rhythm is established. A compressed photoperiod will advance spawning correspondingly, but a proportion of the females may skip spawning entirely the year after the artificial photoperiod was applied (Björnsson et al., 1998). This is in line with results obtained in other species such as rainbow trout (Duston and Bromage, 1988), and suggests that the compressed photoperiod may not give the females enough time to complete a full maturation cycle.

As the halibut is a deep-sea spawner, it is adapted to a spawning environment characterized by stable, low temperatures around 5-6°C, comparatively high salinity and low levels of light. If water temperature increases above 8°C, ovulatory rhythms become irregular and egg viability is impaired (see Brown et al., 1995). These problems increase with increased temperature and in addition broodstock mortalities become more frequent both during and post-spawning at temperatures >11-12°C. Establishment of broodstock groups that spawn out of the normal reproductive season thus requires control of temperature as well as of photoperiod. Effects of varying light intensities have not been investigated, but it is generally recommended that the intensity be kept as low as possible in order to remain close to natural spawning environmental conditions.

Induced Spawning

Towards the end of the spawning season, spawning may become desynchronized between male and female halibut. This is more pronounced in photo-manipulated broodstocks. Specifically, milt will become viscous, "spaghetti-like", with high spermatocrite but decreased sperm motility and fertilization capacity. Treatment with gonadotropin-releasing hormone analogue (GnRHa) implants efficiently reversed this condition: milt volume and sperm motility increased at the same time as spermatocrite decreased and motility increased (Vermeirssen *et al.*, 2004). Plasma levels of the gonadal steroids 5β-pregnane-3β,17,20β-triol 20-sulphate (3β,17,20β-P-5β-S), a major metabolite of 17,20β-dihydroxy-4-pregnen-3-one (17,20β-P), as well as 17,20α-dihydroxy-4-pregnen-3-one (17,20α-P) increased after GnRHa treatment, while plasma testosterone decreased. It was thus suggested that 17,20β-P is involved in milt hydration. Further, the study demonstrated that sperm quality and milt volume could be improved in male halibut approaching the end of the spawning season by treatment with GnRHa implants.

Egg Fertilization and Incubation

Fertilized eggs are incubated in up-welling units in darkness (Mangor-Jensen *et al.*, 1998a). Water flow is adjusted so that the eggs are uniformly distributed in the incubator. Some farmers use gentle aeration to secure even dispersal and to keep the eggs off the outlet sieve. Dead eggs are separated form live eggs by temporarily introducing a layer of high salinity water in the bottom of the incubator (Fig. 18.9). The amount of removed eggs normally peaks at approximately 24-day degrees after fertilization, when the fraction of non-fertilized eggs lose buoyancy. At approximately 70-day degrees, the eggs are collected by a net, disinfected (Harboe *et al.*, 1994a) in a mild solution of seawater and glutaric aldehyde (1:3000), and transferred to yolk sac units. However, some farmers let the eggs hatch in the egg incubators and transfer the larvae immediately after hatch.

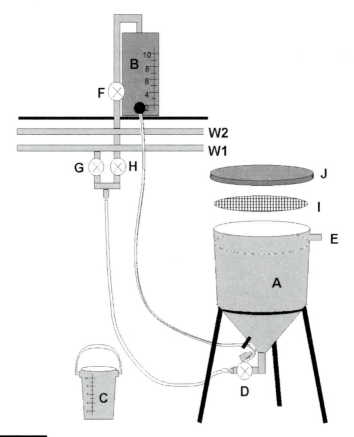

Fig. 18.9 Egg incubator (A). Seawater inlet is at the bottom of the tank (D) and outlet above the sieve (E). High saline water (B) is also introduced in the bottom, and equal volume is taken out together with dead eggs (C).

Halibut Yolk Sac Stage

Large plastic bags floating in lagoons were initially used for incubation of larvae during the yolk sac period, but have later been replaced by 5-15 m^3 indoor silos (Harboe et al., 1994b). These tanks have a conical bottom with a central water inlet, and with an outlet sieve near the top (Fig. 18.10). After introduction of the eggs, hatching is arrested by exposure to light (Helvik et al., 1992) until the eggs are beyond normal hatching age. By turning off the light, the hatching will synchronously be completed within 90 minutes (Harboe, unpublished results). Unhatched eggs and eggshells are easily removed from the bottom of the conical bottom. Newly hatched halibut larvae are very fragile, and contact

A   B

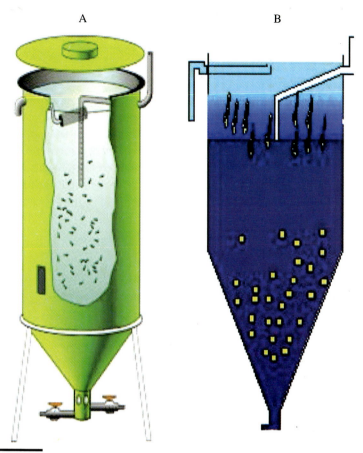

Fig. 18.10   Yolk sac incubator (A). Seawater inlet at the bottom and outlet near the surface. Low saline water is introduced through the horizontal tube in the centre of the incubator to create a density gradient. Dead eggs and larvae are removed from the bottom. Figure (B) shows the situation before and after hatching.

with the outlet sieve has been a major cause of mortality. A continuous salinity gradient preventing the larvae to reach the sieve is made by introducing low-salinity water in a vertical tube in the upper part of the silo. The outlet salinity is adjusted to a level lower than the neutral buoyancy of the larvae thus preventing them to reach the surface of the sieve. The main difficulty during the yolk sac stage is to obtain even larval dispersion in the silo. Immediately after hatching, the larvae will have an adequate distribution as a consequence of the salinity gradient. Later in the period, the larvae become positively rheotactic and will tend to congregate in the bottom part of the silo at the water inlet. To avoid this behaviour, oxygen is gently added as small bobbles at the bottom to create non-directive water currents. This is an effective method for redistribution of the larvae in the water column.

First Feeding

At an age of approximately 265-day degrees (Harboe *et al.*, 1998a), the larvae are transferred to first feeding tanks. At this stage they are positively phototactic (Mangor-Jensen and Naas, 1993) and are, therefore, easily attracted to the surface by use of a dim light. The larvae are gently collected along with water in buckets and released into the first feeding tanks. Halibut larvae are first fed at 12°C in tanks supplied with a central air bubbling to create slow water currents that prevents patching of the prey and also disperse the larvae in the tank, thus leading the prey to the larvae rather to make larvae search for prey (Harboe *et al.*, 1998b) (Fig. 18.11). Algae or inorganic particles are added to the water to achieve

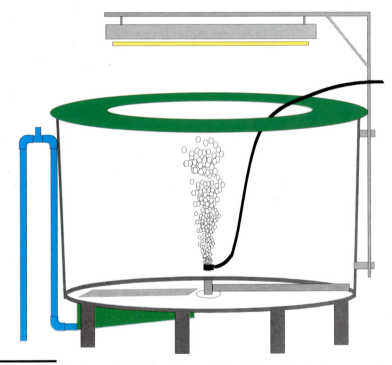

Fig. 18.11 First feeding tank. The cleaning arm in the bottom of the tank removes dead material. The central aeration prevents the prey from schooling and also disperses the larvae in the water column.

a turbidity of 1-2 NTU (units of turbidity — 1 NTU algae gives the water a transparent soft green colour), which has been demonstrated to have a profound positive effect on foraging (Naas *et al.*, 1992). Experiments have showed that turbidity made by a range of different micro algae species have more or less the same positive effect on feed intake. Algae pastes have also been used successfully, although both micro algae and algal pastes has been substituted with clay in a recently developed first feeding

Fig. 18.12 Halibut larvae from hatching to partly metamorphosed fry. The two upper larvae are in the yolk sac stage. The next three are in the first feeding stage and the sixth larva is close to be weaned to dry feed.

protocol. This method is more cost efficient compared to application of algae, and also affect water quality positively in that clay aggregates and precipitates organic matter. An automatic tank cleaning system was first made for intensive halibut rearing, and this system has later been widely used also for other species.

Weaning

After the initial period of first feeding with artemia, the halibut larvae are weaned onto a formulated food (dry pellets) with pellet size about 1-2 mm. At this stage (Fig. 18.12) the halibut larvae has attained the juvenile appearance with the lateral flattened body with both eyes and spotted pigmentation on the dorsal side (Fig. 18.12). Weaning may be carried out in shallow raceways or directly in the first feeding tank. During the first period on formulated diet, food has to be administrated in great excess. Unattended accumulation of debris and feces at the tank bottom will readily take place, and create unacceptable conditions. The introduction of automatic bottom cleaners have eliminated this problem and improved production stability. Weaning in raceways is more hygienic due to self-cleaning.

Several studies have been carried out concerning the weaning of halibut fry. Using live prey involves much labour, and one of the main goals of these experiments has been to shorten, or preferably eliminate the live prey period. Several commercial feeds have been tested in different rearing systems and at different larval sizes, separately or in conjunction with *Artemia*. The different commercial formulated diets intended for halibut larvae are continuously being improved. However, while formulated diets are developed for juveniles (e.g., Hamre et al., 2005) all trials aiming at using a formulated diet as first food have failed. This is also consistent with that a prolonged period with *Artemia* has been demonstrated to reduce mortality.

Ongrowth

Ongrowth systems for halibut need to be designed with special consideration both to the size and shape of the fish, as well as their environmental requirements. Technical solutions for ongrowth of halibut have been reviewed by other authors (e.g., Brown, 2002), and will not be addressed here.

Growth in juvenile halibut is temperature-dependent, with optimum temperatures that are inversely related to fish size (Björnsson and

Trygvadottir, 1995; Hallaråker *et al.*, 1995; Jonassen *et al.*, 1999). Halibut fry (5-10 g) have an optimum growth temperature of ca 15°C, which decreases to ca 13°C for 70 g juveniles. In larger fish, optimum growth temperature decreases further, to ca 10-12°C. At the same time, temperature tolerance increases as the fish becomes larger, so that the impact of temperature on growth becomes less important. However, temperatures >15-16°C are deleterious to Atlantic halibut, and high mortalities occur at temperatures >18°C (pers. obs. 1997, 2002).

A positive effect of continuous light on growth has been reported in juvenile halibut (Simensen *et al.*, 2000; Norberg *et al.*, 2001). This effect appeared to be due to increased feed conversion, and the relative impact of continuous photoperiod was larger and lasts longer at low (6°C) than at high (12°C) temperature (Jonassen *et al.*, 2000a). However, growth rates as well as environmental effects on growth may also vary between populations. Thus, when populations from Norway, Iceland and Canada were compared, Northern populations appeared to have the lowest optimum temperature for growth, as well as highest growth rate independent of temperature (Jonassen *et al.*, 2000b).

The very large sex dimorphism in growth is a special challenge in halibut production. This is further accentuated by the large difference in age at puberty: while males in captivity undergo their first sexual maturation at 2-3 years of age, and a weight around 1-2 kg, females do not mature until 5-6 years, at weights >10 kg. The energy expenditure associated with gonadal growth and maturation is considerable in both sexes, and mature males have significantly reduced, or negative, growth rates during their spawning season from December to May (Haug and Gulliksen, 1988a, b; Weltzien *et al.*, 2004). Attempts to use photoperiod manipulation to block or delay maturation in halibut have not been entirely successful, and it was suggested that the previously reported stimulation of growth associated with holding the fish on continuous light also triggered early male maturation (Norberg *et al.*, 2001). Male maturation is the most important factor that impairs growth. There is however, a small but significant sex difference in growth also in immature fish, and females grow faster than males (Norberg *et al.*, 2002). An alternative strategy that may be more useful in a long-term perspective would, therefore, be to avoid production of male halibut altogether. Studies with the aim to provide a basis for production of all-female halibut populations have been carried out in Canada (Hendry *et al.*, 2002, 2003). Gonadal sex differentiation appeared to occur after metamorphosis, at

36-38 mm length, and the preceding weeks could be defined as a window where phenotypic sex was labile and could be altered by treatment with methyltestosterone. Consequently, genetic females could be transformed into phenotypic males (neomales). Sperm from neomales can be used to produce all-female stocks. In most cultivated species, production of all-female stocks need to be accompanied by sterilisation of the fish, e.g., by induced triploidy. Production of sterile fish appears unnecessary in halibut, as females reach a market size of 5-10 kg several years before onset of puberty.

Spotted wolffish

Inger-Britt Falk Petersen

Spotted wolffish (*Anarhichas minor* Olafsen) is a promising candidate for cold-water aquaculture in both Norway, Iceland and Canada (Falk-Petersen *et al.*, 1999; Foss *et al.*, 2005). Large fillet yield and a high quality, firm, semi-fat meat make the wolffish popular at restaurants as well as in private households. The fish are caught in restricted amounts in the north Atlantic, either during seasonal line-fisheries for the species or as by-catches in other fisheries. In Norwegian waters, mean catches of 5,700 tons of wolffishes (common and spotted) have been reported; around Iceland 1,000-2,000 tons of spotted wolffish are taken, and in Canadian waters combined catches have been reduced from more than 10,000 tons to less than 2,000 tons the last 20 years (Foss et al., 2004).

Culture techniques were modified from those used in connection with pilot cultivation studies of the common wolffish (*Anarhichas lupus* L.) (Pavlov and Novikov, 1987; Moksness, 1994). The first artificial fertilization of spotted wolffish eggs was carried out 11 years ago at the Aquaculture Research Station of the University of Tromsø and some of these offspring were recruited to the broodstock of the first wolffish production site in northern Norway, Troms Steinbit A/S (Falk-Petersen *et al.*, 1999). Since then, research has focused on optimizing the production line of the species and the basic aspects of reproduction, larval development and early growth are fairly well known.

Due to the low temperature optima of the species, the land-based facilities established in Norway till now have used deep-water with stable salinity and low temperatures (Foss *et al.*, 2004). The species has high specific growth rates in captivity, is non-aggressive and apparently not particularly vulnerable to diseases (Moksness, 1994; Lundamo, 1999; Falk-Petersen *et al.*, 2003; Foss *et al.*, 2004).

Broodstock and Spawning

Broodstock individuals have been caught from the sea as also recruited from cultivated fish for several generations. Wolffish broodstock appear to thrive at high stocking densities and formulated marine fish feeds fulfil the nutritional requirements of the species (Falk-Petersen *et al.*, 1999; Foss *et al.*, 2004). It is recommended to keep this cold-water species at relatively low temperatures, around 4-6°C during the pre-spawning and spawning season, in order to produce high quality eggs. Exposure of sexually mature common wolffish (*A. lupus*) to higher temperature (8 and 12°C) during vitellogenesis delayed ovulation compared to fish kept at 4°C, and egg cleavage and survival to the eye-stage were also influenced by the parental temperature regimes. Abnormalities and mortalities were significantly higher among the high-temperature (12°C) group (Tveiten *et al.*, 2001).

In nature, wolffishes are reported to be territorial, at least for periods of the year. Sexually mature common wolffish may pair long before the spawning season (Keats *et al.*, 1985) and fertilization is reported to be internal (Johannesen *et al.*, 1993). After spawning, the male guards the spawned clutch of eggs during the long incubation period (800-1,000-day degrees, d°). Mating and copulation of spotted wolffish has not been documented in cultivation (Foss *et al.*, 2004), but the fact that the spermatozoa of wolffishes are deactivated when transferred to seawater supports the internal fertilization theory or at least that fertilization of the eggs takes place in ovarian fluid while they are being spawned by the female (Kime and Tveiten, 2002). Analysis have shown that a high Na^+/Mg^{2+} ionic ratio in the composition of the seminal fluid is related to high activity of the sperm cells and vice versa, such analysis could serve as quality determination of wolffish sperm (Kime and Tveiten, 2002).

In culture, wolffishes are stripped, as natural spawning does not seem to be a practical solution. Broodstock females, easily recognized by their bulging abdomens, are carefully observed when spawning approaches during autumn (under "natural" ambient light and temperature regimes). Increasing abdominal width, followed by obvious expansion of the ventral region (ovulation) as well as behavioural characteristics and the first appearance of ovarian mucous and eggs in the widened genital opening marks the reproductive event. Females between c. 4-13 kg produce between c. 8,000-30,000 eggs per season; the fecundity of the females generally increases with increasing size of the fish. Male wolffishes produce very restricted volumes of sperm, 0.5-6 ml in each portion, but good producers can often be stripped several times during the reproductive season (Falk-Petersen and Hansen, 2003) (Fig. 18.13). Broodstock individuals are anaesthetized before gentle stripping, the eggs fertilized by undiluted sperm from two to three males, carefully stirred and left for 2-3 hours before being transferred to seawater. They have to be gently stirred during the first hours in seawater to prevent them from sticking together.

Fig. 18.13 Male spotted wolffish ready for stripping.

Broodstock of spotted wolffish can be light-manipulated to spawn at other seasons of the year, preferably at times when the seawater temperatures are optimal (low) for high quality egg production (Foss *et al.*, 2004).

Incubation of Eggs

Fertilized, separated eggs (Fig. 18.14) are transferred to upstream incubators or rectangular raceways in single layers with water flow 2-2.5 l/min. Temperatures between 4-8°C as well as ambient temperatures (from 8 or 6°C decreasing gradually to 3-4°C) have produced viable larvae; incubation at c. 6°C has generally resulted in the highest survival until hatch though (Falk-Petersen *et al.*, 1999; Hansen and Falk-Petersen, 2001). Larvae from eggs incubated at 6°C also showed higher survival during startfeeding and subsequent growth. Eggs incubated at 8°C hatch at a more premature stage and with a larger yolk rest but at higher number of day degrees (>940) compared to eggs incubated at 6°C (c. 900-day degrees) and 4°C (c. 800-day degrees) (Hansen and Falk-Petersen, 2001;

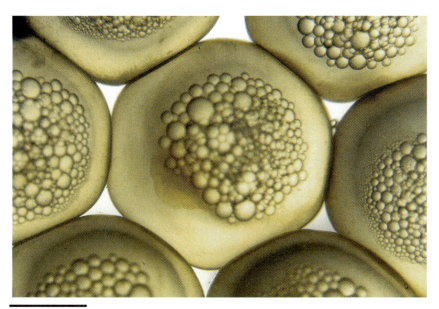

Fig. 18.14 Eggs from spotted wolffish (8-cell stage), 5.5 mm in diameter.

Sund and Falk-Petersen, 2005). The latter are apparently most advanced and have less yolk left. The embryonic development of spotted wolffish is illustrated in Falk-Petersen and Hansen (2003).

The incubation period of the large wolffish eggs is long (800-1000-day degrees). Mortality is generally highest during the first few weeks and relatively low after the eyed-stage (c. 240-day degrees) (Falk-Petersen *et al.*, 1999; Hansen and Falk-Petersen, 2001; Sund and Falk-Petersen, 2005). Egg batches with low fertilization rates and high initial mortality should be discarded, as removal of dead eggs in the hatcheries are labour intensive. Sudden environmental as well as mechanical disturbances during the egg phase may result in elevated mortality (Falk-Petersen, 2002). Microorganism infections may represent a problem during the long incubation period and until now disinfection with glutardialdehyde (150 ppm for five minutes at biweekly or monthly intervals the first 2/3 of the incubation period) has been used. New and less toxic disinfectants are tested and it is probable that better water treatment techniques in hatcheries can reduce or eliminate the use of chemical disinfectants.

Hatching may represent a critical phase during early life history in wolffishes (Fig. 18.15). Observed mortalities during this period may be caused by either environmental events during the incubation period, frequent disinfectant treatments as well as genetic differences between offspring affecting overall egg quality. Embryos may hatch prematurely, or in other cases have difficulties with hatching at all and die within the eggshell or hatch in a reduced physical state. Factors controlling hatching are probably not fully understood (Sund and Falk-Peteren, 2005).

The newly hatched "larvae" are developmentally advanced and relatively active soon after emerging from the egg shell. They have few larval characters, but still a significant amount of yolk left and a partly transparent, but pigment-banded skin as well as a cartilaginous skeleton (Falk-Petersen and Hansen, 2003). A stomach is present in the digestive channel and digestive enzyme activity is registered from hatch (Lamarre et al., 2003). Larval lengths vary between 20-25 mm, wet weights between 80-110 mg (Falk-Petersen et al., 1999; Falk-Petersen and Hansen, 2003). Sizes and lengths at hatch are related to initial egg size as well as egg incubation temperature (Hansen and Falk-Petersen 2001, 2003; Sund and Falk-Petersen 2005).

Survival until hatch varies between batches, but is reported to be c. 50% in hatcheries.

First Feeding

Soon after hatch, the larvae are transferred to low water-level raceways and feeding is initiated within the first few days. Various commercially formulated feed diets used for marine larvae or salmonids have been used as start-feeds (Hansen and Falk-Petersen, 2002; Andreassen et al., 2005). Larvae which do not feed or do not digest the feed particles die during the first three to four weeks. Mean survival during this period is between 30-50%, but in individual batches survival may be as high as 90% (Andreassen et al., 2005). In shallow raceways, it is important that the feed particles float for a certain period to secure an even distribution of nutrients to all larvae in the system. The larvae are positively phototactic immediately after hatch and are found in the upper 50 m of the water column in nature for several weeks. In culture, they rest on the bottom between feeding intervals and apparently thrive at high stocking densities in the raceways.

The larvae and early juveniles grow relatively fast. Larval weights generally double during the first two weeks (at 8°C) and the juveniles are between 3 and 5 g after 2-3 months. Increased temperatures (up to c. 14°C) results in increased growth rates, but higher mortalities (Hansen and Falk-Petersen, 2002).

Fig. 18.15   Newly hatched wolffish larvae in 20 L-upstream incubator.

Juveniles and Ongrowth

Growth of juvenile spotted wolffish is relatively high (Fig. 18.16). Apparently, the optimum temperature for growth decreases with increasing fish size (Foss *et al.*, 2004). Experiments have shown that fish from 4-35 g grew best at 8°C (pers. obs.) and fish of c. 700 g grew better at 6°C than at 8°C or ambient temperatures from 3.5-6.5°C (Lundamo, 1999).

Fig. 18.16   Spotted wolffish yearlings.

Nutritional requirements of spotted wolffish have not been investigated in detail. Generally, protein-rich feeds have been used. A negative correlation was found between particularly high fat levels and growth in feeding trials (Jonassen, 2002), and Foss *et al.* (2005) speculate on the possible non-ideal mineral composition of the feeds currently used in wolffish culture.

Experiments by Foss *et al.* (2001) showed that juvenile spotted wolffish could be reared at a wide range of salinity levels (12-34‰) and the results indicated in fact better growth and food conversion efficiency at intermediate salinities. Similar results have been reported from experiments with common wolffish (Le Francois *et al.*, 2004).

Both in Norway and Iceland, early spotted wolffish juveniles are reared in shallow raceways with high densities of fish, up to 40 kg m^{-2}, corresponding to 400 kg m^{-3} (Foss *et al.*, 2005). Spotted wolfish apparently prefer densely packed conditions; the fish gather in clusters when densities are low (Øiestad, 1999). It has, however, been shown in other cultivation systems (tanks) that fish grew better at lower compared to higher densities. Specific growth rates of larger fish (1.5-3 kg) were reduced when stocking densities exceeded 70 kg m^{-2} (Foss *et al.*, 2005), but biomass increase at both 110 and 150 kg m^{-2} was higher. Light regimes can also promote growth in marine fish and Lundamo (1999) found that 110 g juvenile spotted wolffish grew better when transferred to short day-lengths compared to continuous light or long days.

Wolffishes are bottom-dwelling fish and relatively area-demanding in cultivation. Land-based cultivation sites with supply of stable-temperature deep-water are efficient but expensive and pilot experiments with spotted wolffish cultivation in sea cages (flat-bottom net cages) have been initiated (Foss *et al.*, 2004; Andreassen *et al.*, 2005). Area-expanding constructions within the cages like cubes or shelves have been occupied by the fish after a while (Andreassen *et al.*, 2005). Growth rates and mortality rates have not differed from those observed in land-based facilities. January-March is the season with the lowest growth, probably due to a combination of low temperatures and teeth-renewal. Growth is best during autumn (Andreassen *et al.*, 2005). High temperatures in surface waters during summer (14°C) resulted in reduced appetite, but no disease outbreaks. It is important to protect the fish from the sun during spring and summer and the cages should preferably be placed at relatively sheltered locations to decrease wind and wave movements.

After 3-3.5 years, the fish is 4-4.5 kg and ready for the market. Optimalization of feeds and breeding conditions as well as selection programmes will probably reduce the production time. Early sexual maturation is not a problem in spotted wolffish cultivation.

Diseases and Welfare

Spotted wolffish appear to be very robust and few potential disease problems have been detected till now (Espelid, 2002). Immune responses are present early after hatching (Grøntvedt, 2003). Atypical furunculosis is the only bacterial disease registered in spotted wolffish and might represent a potential problem in wolffish farming, but experimental effective vaccines have been developed (Lund *et al.*, 2002, 2003). Susceptibility studies with viruses (nodavirus NV) have shown that wolffishes developed the disease, but no outbreaks have been observed at culture sites. Parasites may represent a problem in wolffish production. Ectoparasites like Costia (*Ichtyoboda necator*) and *Trichodina* sp., which attach to the gills and skin, are most common (Grøntvedt, 2003). Intracellular *Pleistophora* parasites that infect muscle cells can be a problem in natural populations, reducing the quality of the fish.

Wolffishes apparently like high stocking densities and have also been shown to tolerate and grow well under persistent environmental changes in water oxygen content (Foss *et al.*, 2002, 2003).

SUMMARY

With decreasing spawning stocks, along with improved methods for intensive juvenile production, of some of the commercially most important cold water marine species such as Atlantic cod (*Gadus morhua* L.), Atlantic halibut (*Hippoglossus hippoglossus* L.) and spotted wolffish (*Anarhichas minor*), the interest in cultivating these species has increased. A stable, seasonally independent supply of juveniles is a prerequisite for a commercially viable production. As temperate species in general reproduce seasonally, it is of great importance to develop methods in order to control and manipulate the timing of spawning in captive broodstocks. This requires in-depth knowledge of the biological mechanisms that regulate the timing of puberty and final sexual maturation, and how these mechanisms are affected by external factors such as photoperiod, water temperature and feed availability. In addition, the possibility to keep breeders over several consecutive seasons necessitates a carefully balanced broodstock diet, in order to maintain fish health as well as an optimal nutrient composition of the eggs. Intensive methods for larval rearing of cold water species have the potential to give high numbers of juveniles independent of season. Juvenile quality and production stability depend on a thorough knowledge within fields such as embryology, developmental biology, larval physiology and marine ecology, in order to ensure optimal rearing environment and nutrition.

The present chapter describes the biology of, and production methods for three cold water marine fish species of the Northern Hemisphere: Atlantic cod, Atlantic halibut and spotted wolffish. The species were chosen for their economic significance but also because they are biologically diverse and represent three different production strategies.

Conclusions and Research Priorities

Atlantic cod juvenile production increased exponentially between 2000 and 2005 and the cod now dominates cold water marine aquaculture. However, intensive production methods have resulted not only in a stable supply of juveniles, but production-related problems such as deformities and other health concerns, as well as precocious maturation and escapes have to be solved before cod cultivation can have a major impact on aquaculture production. Production of sterile cod may overcome both the problems with early maturation and also eliminate the possibility of release of fertilized eggs into the marine environment. Vaccine development against the most common diseases is also a highly prioritized task.

After some years of low and unpredictable juvenile survival, Atlantic halibut fry production methods have improved and high quality juveniles are now available. Ongoing research is mainly focussed on larval and juvenile nutritional requirements. However, there is need for an increased effort to improve appetite and growth in ongrowing systems, and early maturation of male halibut is a major problem.

No biological or technological bottlenecks associated with spotted wolffish culture really exist and the species is definitely ready for industrial upscaling (Foss *et al.*, 2005). Farming is apparently restricted to areas where water of relatively stable low temperatures can be found throughout the year. There is always a need for streamlining the cultivation protocol of a "new" species in aquaculture, though, and probably more species specific market research. The species is not well known among the general public outside some few markets in northern Europe and Canada. Spotted wolffish is a robust species, well suited for cultivation, grows fast at low temperatures, has a very high fillet-yield and is not fished in large quantities commercially.

References

Amcoff, P. 2000. The role of thiamine in Baltic salmon developing the "M74-syndrome". Doctoral thesis, Swedish University of Agricultural Sciences, Uppsala, Sweden.

Andreassen, I., I.B. Falk-Petersen, A. Pedersen and H. Tveiten. 2005. Flekksteinbit i oppdrett. *Kyst Og Havbruk* pp. 146-149.

Andersson, E., S.R. Kumar, J. Ackers, G.L. Taranger, S.O. Stefansson and J.M. Trant. 2003. Photoperiod and temperature affects seasonal ovarian gene expression of P450 aromatase and gonadotropin receptors in Atlantic salmon (*Salmo salar* L.) broodstock. *Fish Physiology and Biochemistry* 28: 411.

Björnsson, B. Th. and S.V. Trygvadottir. 1995. Effect of size on optimal temperature for growth and growth efficiency of immature Atlantic halibut (*Hippoglossus hippoglossus* L.). *Aquaculture* 142: 33-42.

Björnsson, B. Th. O. Halldorsson, C. Haux, B. Norberg and C. L. Brown. 1998. Photoperiod control of sexual maturation of the Atlantic halibut (*Hippoglossus hippoglossus*): Plasma thyroid hormone and calcium levels. *Aquaculture* 166: 117-140.

Brown, J.A., G. Minkoff and V. Puvanendran. 2003. Larviculture of Atlantic cod (*Gadus morhua*): Progress, protocols and problems. *Aquaculture* 227: 357-372.

Brown, N.P., N.R. Bromage and R.J. Shields. 1995. The effect of spawning temperature on egg viability in the Atlantic halibut (*Hippoglossus hippoglossus*). In: *Proceedings of the Fifth International Symposium on the Reproductive Physiology of Fish*, F. W. Goetz and P. Thomas (eds.). Fish Symposium 95, Austin, p. 181.

Davie, A. 2005. Effects of photoperiod manipulation on growth and reproduction in Atlantic cod (*Gadus morhua*). Ph.D. Thesis, University of Stirling, U.K.

Devold, F. 1938. The North Atlantic halibut and net fishing. *Fiskeridirektoratets Schrrifter Serie Havendursøketser* 5: 1-47.

Duston, J. and N. Bromage. 1988. The entrainment and gating of the endogenous circannual rhythm of reproduction in the female rainbow trout (*Salmo gairdneri*). *Journal of Comparative Physiology* A 164: 259-268.

Espelid, S. 2002. The susceptibility of spotted wolffish to infectious diseases and use of immune prophylaxis. *Bulletin of the Aquaculture Association of Canada* 102: 38-42.

Falk-Petersen, I.B. 2002. Factors affecting survival and growth of early life stages of spotted wolffish (*Anarhichas minor* Olafsen). *Bulletin of the Aquaculture Association of Canada* 102: 29-32.

Falk-Petersen, I.B. and T.K. Hansen. 2003. Early ontogeny of the spotted wolffish (*Anarhichas minor* Olafsen). *Aquaculture Research* 34: 1059-1067.

Falk-Petersen, I.B., T.K. Hansen, R. Fieler and L.M. Sunde. 1999. Cultivation of the spotted wolffish (*Anarhichas minor* Olafsen)—A new candidate for coldwater farming. *Aquaculture Research* 30: 711-718.

Foss, A., T.H. Evensen, A.K. Imsland and V. Øiestad. 2001. Effects of reduced salinities on growth, food conversion efficiency and osmoregulatory status in the spotted wolffish. *Journal of Fish Biology* 59: 416-426.

Foss, A., T.H. Evensen and V. Øiestad. 2002. Effects of hypoxia and hyperoxia on growth and food conversion efficiency in the spotted wolffish, *Anarhichas minor* (Olafsen). *Aquaculture Research* 33: 437-444.

Foss, A., T. Vollen and V. Øiestad. 2003. Growth and oxygen consumption at normal and O_2 supersaturated water, and interactive effects of O_2 saturation and ammonia on growth in the spotted wolffish *Anarhichas minor* (Olafsen). *Aquaculture* 224: 105-116.

Foss, A., A. Imsland, I.B. Falk-Petersen and V. Øiestad. 2004. A review of the culture potential of spotted wolffish *Anarhichas minor* Olafsen. *Reviews in Fish Biology and Fisheries* 14: 277-294.

Grøntvedt, R.N. 2003. Immune competence and protective responses in vaccinated spotted wolffish (*Anarhichas minor* O.) juveniles. Dr. scient. thesis, University of Tromsø, Norway.

Hallaråker, H., A. Folkvord, K. Pittman and S.O. Stefansson. 1995. Growth of juvenile halibut (*Hippoglossus hippoglossus*) related to temperature, daylength and feeding regime. *ICES Marine Science Symposia 201: Mass Rearing of Juvenile Fish Bergen, Norway, 1993.*

Hamre, T. K., G. Bæverfjord and T. Harboe. 2005. Macronutrient composition of formulated diets for Atlantic halibut (*Hippoglossus hippoglossus*) juveniles, II: Protein / lipid levels at low carbohydrate. *Aquaculture* 244: 283-291.

Hansen, T.K. and I.B. Falk-Petersen. 2001. The influence of rearing temperature on early development and growth of spotted wolffish *Anarhichas minor* (Olafsen). *Aquaculture Research* 32: 369-378.

Hansen, T.K. and I.B. Falk-Petersen. 2002. Growth and survival of spotted wolffish larvae (*Anarhichas minor* Olafsen) at various temperature regimes. *Aquaculture Research* 33: 119-127.

Hansen, K., O. Karlsen, G.L. Taranger and G.I. Hemre. 2001. Growth, gonadal development and spawning time of Atlantic cod (*Gadus morhua*) reared under different photoperiods. *Aquaculture* 203: 51-67.

Harboe, T. and A. Mangor-Jensen. 1998. Time of first feeding of atlantic halibut larvae. *Aquaculture Research* 29: 913-919.

Harboe, T., I. Huse and G. Øie. 1994. Effects of egg disinfecton on yolk sac and first feeding larvae of halibut (*Hippoglossus hippoglossus*) larvae. *Aquaculture* 119: 57-165.

Harboe, T., S. Tuene, A. Mangor-Jensen, H. Rabben and I. Huse. 1994. Design and operation of an incubator for yolk-sac larvae of atlantic halibut (*Hippoglossus hippoglossus*). *The Progressive Fish-Culturist* 56: 188-193.

Harboe, T., A. Mangor-Jensen, K. E. Naas and T. Næss. 1998. A tank design for first feeding of halibut, *Hippoglossus hippoglossus* L., larvae. *Aquaculture Research* 29: 919-925.

Haug, T. 1990. Biology af the Atlantic halibut (*Hippoglossus hippoglossus* L.1758). *Advances in Marine Biology* 26: 1-70.

Haug, T. and B. Gulliksen. 1988a. Fecundity and oocyte sizes in ovaries of females Atlantic halibut *Hippoglossus hippoglossus* (L.). *Sarsia* 73: 259-261.

Haug, T. and B. Gulliksen. 1988b. Variations in liver and body condition during gonad development of Atlantic halibut *Hippoglossus hippoglossus* (L.). *Fiskeridirektoratets Skrifter Serie Havendursøkelser* 18: 351-363.

Helvik, J.V. and B.T. Walther. 1992. Photo-regulation of the hatching process of halibut (*Hippoglossus hippoglossus*) eggs. *Journal of Experimental Zoology* 263: 204-209.

Hemre, G.-I., A. Mangor-Jensen, G. Rosenlund, R. Waagbø and Ø. Lie. 1995. Effect of dietary carbohydrate on gonadal development in broodstock cod, *Gadus morhua* L. *Aquaculture Research* 26: 399-408.

Hendry, C.I., D.J. Martin-Robichaud and T.J. Benfey. 2002. Gonadal sex differentiation in Atlantic halibut. *Journal of Fish Biology* 60: 1431-1442.

Hendry, C.I., D.J. Martin-Robichaud and T.J. Benfey. 2003. Hormonal sex reversal of Atlantic halibut (*Hippoglossus hippoglossus* L.). *Aquaculture* 219: 769-781.

Jakupsstovu, S.H.I. and T. Haug. 1988. Growth, sexual maturation and spawning season of Atlantic halibut, *Hippoglossus hippoglossus*, in Faroesea waters. *Fisheries Research* 6: 201-205.

Jonassen, T.M., A.K. Imsland and S.O. Stefansson. 1999. The interaction of temperature and fish size on growth of juvenile halibut (*Hippoglossus hippoglossus*). *Journal of Fish Biology* 54: 556-572.

Jonassen, T.M., A.K. Imsland, S. Kadowaki and S.O. Stefansson. 2000a. Interaction of temperature and photoperiod on growth of Atlantic halibut *Hippoglossus hippoglossus* L. *Aquaculture Research* 31: 219-227.

Jonassen, T.M., A.K. Imsland, R. Fitzgerald, S.W. Bonga, E.V. Ham, G. Nævdal, M.Ö. Stefansson and S.O. Stefansson. 2000b. Geographic variation in growth and food conversion efficiency of juvenile Atlantic halibut related to latitude. *Journal of Fish Biology* 56: 279-294.

Johannesen, T., J. Gjøsæter and E. Moksness. 1993. Reproduction, spawning behaviour and captive breeding of the common wolffish *Anarhichas lupus* L. *Aquaculture* 115: 41-51.

Jonassen, T.M. 2002. Effects of photoperiod, stocking density amd diet on growth of young spotted wolffish (*Anarhichas minor* Olafsen). *Aquaculture International* 10: 411-420.

Karlsen, Ø., B. Norberg, O.S. Kjesbu and G.L. Taranger. 2006. Effects of photoperiod and excercise on age at puberty in farmed Atlantic cod. *ICES Journal of Marine Science* 163(2): 355-364.

Keats, D.W., G.R. South and D.H. Steele. 1985. Reproduction and egg guarding by Atlantic wolffish (*Anarhichas lupus* L.: Anarhichidae) and ocean pout (*Macrozoarces americanus*: Zoarcidae) in Newfoundland waters. *Canadian Journal of Zoology* 63: 2565-2568.

Kime, D.E. and H. Tveiten. 2002. Unusual motility characteristics of sperm of spotted wolffish. *Journal of Fish Biology* 61: 1549-1559.

Kjesbu, O. S. 1989. The spawning activity of cod, *Gadus morhua* L. *Journal of Fish Biology* 34: 195-206.

Kjesbu, O.S., J. Klungsøyr, P.R. Witthames, H. Kryvi and M. Greer Walker. 1991. Fecundity, atresia, and egg size of captive Atlantic cod (*Gadus morhua*) in relation to proximate body composition. *Canadian Journal of Fisheries and Aquative Sciences* 48: 2333-2343.

Kjesbu, O.S., P. Solemdal, P. Bratland and M. Fonn. 1996. Variation in annual egg production in individual captive Atlantic cod (*Gadus morhua*). *Canadian Journal of Fisheries and Aquatic Sciences* 53: 610-620.

Kjørsvik, E., A. Mangor-Jensen and I. Holmefjord. 1990. Egg quality in fishes. *Advances in Marine Biology* 26: 71-113.

Kjørsvik, E., T. Haug and J. Tjemsland. 1987. Spawning season of the Atlantic halibut (*Hippoglossus hippoglossus*) in northern Norway. *Journal du consoil de l Exploration de la Mer* 43: 285-293.

Kvenseth, P.G. and V. Øiestad. 1984. Large-scale rearing of cod fry on the natural food production in an enclosed pond. In: *The Propagation of cod Gadus morhua* L., E. Dahl, D.S. Danielssen and E. Moksness (eds.). *Flødevigen Rapportser.* 1: 645-656.

Lamarre, S.G., N. R. Le Francois, I.B. Falk-Petersen and P.U. Blier. 2003. Can digestive and metabolic enzyme activity levels predict growth rate and survival of newly hatched Atlantic wolffish (*Anarhichas lupus* Olafsen)?. *Aquaculture Research* 35: 608-613.

Le Francois, N.R., S.G. Lamarre and P.U. Blier. 2004. Tolerance, growth and haloplasticity of the Atlantic wolffish (*Anarhichas lupus*) exposed to various salinities. *Aquaculture* 236: 659-675.

Lie, Ø. 1991. The fatty acid composition of phospholipids of roe and milt in cod—Influence of the feed. In: *16th Scandinavian Symposium on Lipids*, pp. 185-192. Lipidforum, Hardanger, Norway.

Lim, B., H. Kagawa, K. Gen and K. Okuzawa. 2003. Effects of water temperature on the gonadal development and expression of steroidogenic enzymes in the gonad of juvenile red seabream, *Pagrus major*. *Fish Physiology and Biochemistry* 28: 161-162.

Lund, V., J.A. Arnesen and G. Eggset. 2002. Vaccine development for atypical furunculosis in spotted wolfish *Anarhichas minor* O.: Comparison of efficacy of vaccines containing different strains of atypical *Aeromonas salmonicida*. *Aquaculture* 204: 33-44.

Lund, V., S. Espelid and H. Mikkelsen. 2003. Vaccine efficacy in spotted wolffish *Anarhichas minor*: Relationship to molecular variation in A-layer protein of atypical *Aeromonas salmonicida*. *Diseases of Aquatic Organisms* 56: 34-42.

Lundamo, I. 1999. Growth and Survival in Spotted Wolffish (*Anarhichas minor*). Effects of Temperature and Photoperiod Cand. Scient. Thesis, Norwegian College of Fishery Science, University of Tromsø, Tromso, Sweden.

Mæland, A., I. Rønnestad and R. Waagbø. 2002. Folate in eggs and developing larvae of Atlantic halibut (*Hippoglossus hippoglossus*). *Aquaculture Nutrition* 9: 185-188.

Mangor-Jensen, A. and I. Huse. 1991. On the changes in buoyancy of halibut, *Hippoglossus hippoglossus* (L.), larvae caused by hatching—A theoretical view. *Journal of Fish Biology* 39: 133-135.

Mangor-Jensen, A. and K.E. Naas. 1993. Phototaxis of halibut larvae (*Hippoglossus hippoglossus* L.). In: *Physiological and Biochemical Aspects of Fish Larval Development*, B.T. Walther and H.J. Fyhn (eds.). University of Bergen, Bergen, Norway, pp. 132-138.

Mangor-Jensen, A., J.C.H. Holm, G. Rosenlund, Ø. Lie and K. Sandnes. 1994. Effects of dietary vitamin C on maturation and egg quality of cod *Gadus morhua* L. *Journal of the World Aquaculture Society* 25: 30-40.

Mangor-Jensen, A., T. Harboe, J.S. Hennø and R. Troland. 1998. Design and operation of halibut egg incubator. *Aquaculture Research* 29: 887-892.

Mangor-Jensen, A., T. Harboe, R. Shields, B. Gara and K.E. Naas. 1998. Review of halibut cultivation literature, 1998. *Aquaculture Research* 29: 857-886.

Mazorra, C., M. Bruce, J.G. Bell, A. Davie, N. Alorend, N. Jordan, J. Rees, N. Papanikos, M. Porter and N. Bromage. 2003. Dietary lipid enhancement of broodstock reproductive performance and egg and larval quality in Atlantic halibut (*Hippoglossus hippoglossus*). *Aquaculture* 227: 21-33.

Moksness, E. 1994. Growth rates of the common wolffish *Anarhichas lupus* L. and spotted wolffish *A. minor* Olafsen, in captivity. *Aquaculture and Fishery Management* 25: 363-371.

Naas, K.E., T. Næss and T. Harboe. 1992. Enhanced first feeding of halibut larvae (*Hippoglossus hippoglossus* L.) in green water. *Aquaculture* 105: 143-156.

Næass, T., T. Harboe, A. Mangor-Jensen, K.E. Naas and B. Norberg. 1996. Successful first feeding of Atlantic halibut (*Hippoglossus hippoglossus*) larvae from a photoperiod adjusted broodstock. *The Progressive Fish Culthrist* 58: 212-214.

Norberg, B., V. Valkner, J. Huse, I. Karlsen and G. Lerøy Grung. 1991. Ovulatory rhythms and egg viability in Atlantic halibut (*Hippoglossus hippoglossus*). *Aquaculture* 97: 365-371.

Norberg, B., F-A. Weltzien, Ø. Karlsen and J.C. Holm. 2001. Effects of photoperiod on sexual maturation and somatic growth in Atlantic halibut (*Hippoglossus hippoglossus* L.). *Comparative Biochemistry and Physiology* B129: 357-366.

Norberg, B., C.L. Brown, O. Halldorsson, K. Stensland and B.T. Björnsson. 2004. Photoperiod regulates the timing of sexual maturation, spawning, sex steroid and thyroid hormone profiles in the Atlantic cod (*Gadus morhua*). *Aquaculture* 229: 451-467.

Pavlov, D.A. and G.G. Novikov. 1986. On the development of biotechnology for rearing White Sea wolffish *Anarhichas lupus marisalbi*. 1. Experiments on obtaining mature sex products, incubation of eggs and rearing of the young fish. *Journal of Ichthyology* 26: 95-106.

Pedersen, T. and I.-B. Falk-Petersen. 1992. Morphological changes during metamorphosis in cod (*Gadus morhua* L.), with particular reference to the development of the stomach and pyloric caeca. *Journal of Fish Biology* 41: 449-461.

Riple, G.L. 2000. Ovarian development and egg viability aspects in turbot (*Scophthalmus maximus*) and Atlantic halibut (*Hippoglossus hippoglossus*). Dr Scient. thesis, Department of Fisheries and Marine Biology, University of Bergen, Norway.

Rognerud, C. 1887. Hatching cod in Norway. *Bulletin of the United States Fish Commission* 7 (8): 113-119. (Translation from original text in Norwegian: Dannevig, G.M. 1886. Aarsberetning for Arendal og Omegns Filial. *Årsberetning Selskab for Norske Fisheriers Fremmende*: 37-44.)

Rosenlund, G., Ø. Karlsen, K. Tveit, A. Mangor-Jensen and G.-I. Hemre. 2004. Effect of feed composition and feeding frequency on growth, feed utilisation and nutrient retention in juvenile Atlantic cod, *Gadus morhua* L. *Aquaculture Nutrition* 10: 371-378.

Rønnestad, I., Ø. Lie and R. Waagbø. 1997. Vitamin B6 in Atlantic halibut, *Hippoglossus hippoglossus*-endogenous utilisation and retention in larave fed natural zooplankton. *Aquaculture* 157: 337-345.

Rønnestad, I., K. Hamre, Ø. Lie and R. Waagbø. 1999. Ascorbic acid and α-tocopherol in larvae of Atlantic halibut before and after exogenous feeding. *Journal of Fish Biology* 55: 720-731.

Simensen, L., T.M. Jonassen, A.K. Imsland and S.O. Stefansson. 2000. Growth of juvenile halibut (*Hippoglossus hippoglossus*) under different photoperiods. *Aquaculture* 190: 119-128.

Skjæraasen, J. E., A.G.V. Salvanes, Ø. Karlsen, R. Dahle and B. Norberg. 2004. The effect of photoperiod on sexual maturation, appetite and growth in wild Atlantic cod (*Gadus morhua*) held at two light regimes. *Fish Physiology and Biochemistry* 30: 163-174.

Sund, T. and I.B. Falk-Petersen. 2005. The effects of incubation temperature on development and yolk sac conversion efficiencies of spotted wolfish (*Anarhichas minor* Olafsen) embryos until hatch. *Aquaculture Research* 36(11): 1133-1143.

Taranger, G.L., E. Vikingstad, U. Klenke, I. Mayer, S.O. Stefansson, B. Norberg, T. Hansen, Y. Zohar and E. Andersson. 2004. Effects of photoperiod, temperature and GnRHa treatment on the reproductive physiology of Atlantic salmon (*Salmo salar* L.) broodstock. *Fish Physiology and Biochemistry* 28: 403-406.

Tveiten, H., S.E. Solevåg and H.K. Johnsen. 2001. Holding temperature during the breeding season influences final maturation and egg quality in common wolffish. *Journal of Fish Biology* 58: 374-385.

van der Meeren, T. and K.E. Naas. 1997. Development of rearing techniques using large enclosed ecosystems in the mass production of marine fish fry. *Reviews in Fisheries Science* 5: 367-390.

van der Meeren, T., T. Harboe, J.C. Holm and R. Solbakken. 1998. A new cleaning system for rearing tanks in larval fish culture. *ICES C.M. L:13*: 11.

van der Meeren, T. and V. Ivannikov. 2006. Seasonal shift in spawning of Atlantic cod (*Gadus morhua* L.) by photoperiod manipulation. Egg quality in relation to temperature, and intensive larval rearing. *Aquaculture Research* 37: 898-913.

Vermeirssen, E.L.M., C. Mazorra de Quer°, R.J. Shields, B. Norberg, D.E. Kime and A.P. Scott. 2004. Fertility and motility of sperm from Atlantic halibut (*Hippoglossus hippoglossus*) in relation to dose and timing of of gonadotrophin-releasing hormone agonist implant. *Aquaculture* 230: 547-567.

Weltzien, F-A., Ø. Karlsen and B. Norberg. 2003. Growth patterns and plasma levels of 11-ketotestosterone, testosterone and IGF-1 in male Atlantic halibut (*Hippoglossus hippoglossus*). *Fish Physiology and Biochemistry* 28: 227-228.

Øiestad, V., B. Ellertsen, P. Solemdal and S. Tilseth. 1976. Rearing of different species of marine fish fry in a constructed basin. In: *Proceedings of the 10th European Symposium on Marine Biology, Ostend, Belgia Sept. 17-23, 1975*, G. Persoone and E. Jaspers (eds.). Universal Press, Wetteren, Belgium, Vol. 1, pp. 303-329.

Index